Marine Sediment Transport
and Environmental Management

MARINE SEDIMENT TRANSPORT AND ENVIRONMENTAL MANAGEMENT

Edited by

DANIEL JEAN STANLEY

Smithsonian Institution
Washington, D.C.

DONALD J. P. SWIFT

Atlantic Oceanographic and
Meteorological Laboratories
Miami, Florida

**Under the Auspices of the
American Geological Institute**

A Wiley-Interscience Publication

JOHN WILEY & SONS
New York • London • Sydney • Toronto

Library of Congress Cataloging in Publication Data:

Marine sediment transport and environmental management.

   "A Wiley-Interscience publication."
   "Outgrowth of lectures presented at a short course,
'The New concepts of continental margin sedimentation,
II', sponsored by the American Geological Institute . . .
convened at Key Biscayne, Florida, on November 15 to 17,
1974."
   Includes index.
   1. Marine sediments. 2. Sediment transport.
3. Environmental engineering. I. Stanley, Daniel J.
II. Swift, Donald J. P. III. American Geological
Institute. IV. Title: The New concepts of continental
margin sedimentation, II.

GC308.M34   551.3'6   75-41396
ISBN 0-471-02540-2

Printed in the United States of America

10 9 8 7 6 5 4 3 2 1

Contributors

David E. Drake
Marine Geology Branch, U. S. Geological Survey
Menlo Park, California 94025

David B. Duane
National Sea Grant Program
National Oceanic and
Atmospheric Administration
Rockville, Maryland 20852

Kenneth O. Emery
Woods Hole Oceanographic Institution
Woods Hole, Massachusetts 02543

Monty A. Hampton
Marine Geology Branch
U. S. Geological Survey
Menlo Park, California 94025

Carl G. Hard
Project Management Branch
New England Division, Corps of Engineers
424 Trapelo Road
Waltham, Massachusetts 02154

Patrick G. Hatcher
Atlantic Oceanographic and
Meteorological Laboratories
15 Rickenbacker Causeway, Miami, Florida 33149

Gilbert Kelling
Department of Geology and Oceanography
University of Wales at Swansea
Swansea, Great Britain SA2 8PP

Paul D. Komar
School of Oceanography
Oregon State Univeristy
Corvallis, Oregon 97331

Ants Leetmaa
Atlantic Oceanographic and
Meteorological Laboratories
15 Rickenbacker Causeway, Miami, Florida 33149

John C. Ludwick
Institute of Oceanography
Old Dominion University
Norfolk, Virginia 23508

Ole S. Madsen
Ralph M. Parsons Laboratory
Massachusetts Institute of Technology
Cambridge, Massachusetts 02134

Gerard V. Middleton
Department of Geology
McMaster University
Hamilton, Ontario, Canada

Harold O. Mofjeld
Atlantic Oceanographic and
Meteorological Laboratories
15 Rickenbacker Causeway, Miami, Florida 33149

Christopher N. K. Mooers
Rosenstiel School of Marine and Atmospheric Science
10 Rickenbacker Causeway, Miami, Florida 33149

Harold D. Palmer
Westinghouse Ocean Research Laboratory
Annapolis, Maryland 21404

Jack W. Pierce
Division of Sedimentology
Smithsonian Institution
Washington, D.C. 20560

Saul B. Saila
Graduate School of Oceanography
University of Rhode Island
Kingston, Rhode Island 02981

Douglas A. Segar
Atlantic Oceanographic and
Meteorological Laboratories
15 Rickenbacker Causeway, Miami, Florida 33149

Francis P. Shepard
Geological Research Division
Scripps Institute of Oceanography
La Jolla, California 92093

John B. Southard
Department of Earth and Planetary Sciences
Massachusetts Institute of Technology
Cambridge, Massachusetts 02139

Daniel Jean Stanley
Division of Sedimentology
Smithsonian Institution
Washington, D.C. 20560

Donald J. P. Swift
Atlantic Oceanographic and
Meteorological Laboratories
15 Rickenbacker Causeway, Miami, Florida 33149

Harold R. Wanless
Rosenstiel School of Marine and Atmospheric Science
10 Rickenbacker Causeway, Miami, Florida 33149

Foreword

More than 50 years ago I was taught by my venerable professors at Harvard and the University of Chicago that the undertow from the waves breaking along the open coasts carried sediment seaward, depositing the coarse particles near shore and transporting the fine sediments to greater depths, developing a seaward grading of sizes. In 1923, I had the opportunity to collect bottom samples in Massachusetts Bay. I was amazed to find that seaward decrease in sediment sizes had little validity, at least in that area. To check this finding elsewhere, I obtained charts from various parts of the world and, using the bottom notations on the continental shelves, made rough sediment maps of the bottom types. In general, this study confirmed my discovery in Massachusetts Bay. Clearly, the old concepts needed investigation.

Now, students of marine sedimentation would not be at all disturbed by collecting coarse sediment on the outer shelf seaward of mud bottom areas. Meanwhile, we have learned that currents, partly tidal, are capable of maintaining relict sand or even of producing coarse sediment by eroding the bottom locally, regardless of depth or distance from the shore. Even undertow is largely discredited. In the long intervening period, marine geology has come into its own. The realization at the beginning of World War II that the U. S. Navy had a desperate need for oceanographic information, including the character of the seafloor, gave a great impetus to underwater geological studies with emphasis on continental shelf sediments. Fortunately, the program did not end but even expanded in postwar years. The Office of Naval Research and other Navy departments have continued subsidizing marine institutes of the principal universities in the United States. Other bureaus of the government, notably the National Oceanic and Atmospheric Administration (formerly the Coast and Geodetic Survey), have also become research-minded and have made great strides in advancing marine geology. The National Science Foundation has provided increasing support to marine geology studies, particularly in the generous provisions for deep-sea drilling. For many years, the U. S. Geological Survey virtually confined its research efforts to the lands. They even left untouched large collections of bottom samples from the Coast and Geodetic Survey in storerooms of the National Museum. A real change in policy took place in 1962, when the Geological Survey with the collaboration of Woods Hole Oceanographic Institution started a large project under K. O. Emery to investigate the continental margins off the East and Gulf Coasts. Somewhat later, extensive marine geology operations were undertaken from the Menlo Park, California branch of the Survey and, on a smaller scale, elsewhere.

Since World War II, most marine geology has been focused on the deep sea, but coastal studies and continental shelf sediment investigations have been very extensive, especially along the East and Gulf coasts. These studies are helping to alleviate the critical petroleum shortages of the United States and are therefore of great economic importance. The petroleum companies have contributed generously in the form of grants by the American Petroleum Institute, and have made gifts to various oceanographic departments. Much additional information is unfortunately maintained in the confidential files of the large companies.

Most of the Chapters included in the present volume are written by scientists from government and university marine laboratories. The papers serve a particularly important function in providing geologists and engineers with an insight into the nature of the marine processes that transport sediments to the various environments on the continental margins. These processes have been poorly understood in the past. This new information, presented by many of the most competent investigators in the field, is most timely.

Since World War II, the number of marine geologists with good financial backing and adequate instrumentation has greatly increased. However, there are still some serious gaps in our basic information. As far back as the late 1930s, Henry Stetson of Woods Hole Institution and Roger Revelle and I at Scripps Institution began a study of bottom currents, particularly in the shallow parts of submarine canyons. Unfortunately, this work came to a halt after a few records had been obtained, and it is only in recent years that our program at Scripps Institution has begun to make up for lost time. We have found so much to do in studying submarine canyon currents that we have scarcely any records of shelf or slope currents. A few exceptions are discussed in the present volume. Theoretically, the continental shelves should have relatively strong currents, especially near their outer margins. The presence of these currents is clearly indicated by bottom photographs and deep-diving vehicle operations, but we need many current-meter records to determine their nature and cause.

Finally, we have virtually no records of turbidity currents. Indirect evidence suggests that they reach high velocities. Most evidence is from cable breaks but this is open to various interpretations. We need some way to obtain more exact information. Available current meters can only measure a weak variety of turbidity current, because strong currents carry away the devices or stop their operation by entangling them in debris. Thus, there is much to be done in this field, and one can hope that for a future edition of *New Concepts of Continental Margin Sedimentation* more documentation on these unusual currents will have become available.

FRANCIS P. SHEPARD
*Scripps Institution
of Oceanography*

*La Jolla, California
January 1976*

Preface

This volume has been prepared for oceanographers, earth scientists, and environmental scientists and managers concerned with problems of sediment transport in the marine environment. It is an outgrowth of lectures presented at a short course, "The *New* Concepts of Continental Margin Sedimentation, II," sponsored by the American Geological Institute which we convened at Key Biscayne, Florida on November 15 to 17, 1974.

The subject of sediment transport is considered in three ways. The first pertains to understanding continental margin sediment transport which requires not only an appreciation of entrainment and transport mechanisms but also a thorough grasp of large-scale flow patterns as interpreted by physical oceanographers. Attention is then directed toward regional patterns of sediment transport utilizing the insights of classical sedimentology. In the final section, principles of fluid dynamics and sedimentation are applied to the problems of environmental management, a rapidly expanding field that has not been heretofore systematically described in the literature. This present work is a departure from our 1969 short course volume which presented a broad spectrum of problems pertaining to sediments on continental terraces and had as its major theme the application of concepts to the stratigraphic record.

Preparation of the volume, including synthesizing and ordering of new information, has been a valuable experience for us. To our parent organizations, the Smithsonian Institution and the Atlantic Oceanographic and Meteorological Laboratories NOAA, goes the credit of bearing the secretarial burden and support for the one thousand and one needs that face an editor. We thank Donna Copeland and Claire Ulanoff for their secretarial help, and L. Isham, A. Ramsey and H. Sheng for assistance with preparation of the illustrations. Appreciation is also expressed to Steve Waitzman, Smithsonian Institution, for extensive help with editing and preparation of the index.

Many scientists accepted the role of outside critic and supplemented the large pool of authors in reviewing the manuscripts. We are indebted to the following who interrupted their own work to provide us with the necessary criticism:

J. R. L. Allen, R. J. Byrne, D. A. Cacchione, D. Charnell, J. S. Creager, G. Csanady, C. Everts, D. Hamilton, D. Hansen, J. F. Kennedy, J. C. Kraft, J. W. Lavelle, R. L. Miller, J. D. Milliman, A. W. Niedoroda, J. Pearce, O. H. Pilkey, P. E. Potter, A. F. Richards, S. R. Riggs, D. R. Schink, J. S. Schlee, J. R. Schubel, C. Sonu, K. K. Turekian, and G. L. Weatherly.

The map used as one of the end-papers depicts areas inshore of the 200 m isobath (the Mercator projection distorts the polar regions) and was provided by D. A. Ross, Woods Hole Oceanographic Institution, from his article in *Oceanus* (1973). The front end-paper photograph, furnished by H. D. Palmer, Westinghouse Ocean Research Laboratory, shows core recovery operations at depths of about 20 m on the sandy sea floor near Santa Catalina Island off southern California (photo courtesy Dames & Moore). The color photograph of the ebb-flood channel system on the Great Bahama Banks used on the jacket was provided by Charles True, National Marine Fisheries Service Laboratory, Miami, Florida.

We also take this opportunity to thank the production side of our venture: T. F. Rafter and W. H. Matthews, III, of the American Geological Institute and R. Woolson and his associates at Wiley-Interscience.

DANIEL JEAN STANLEY
DONALD J. P. SWIFT

Washington, D.C.
Miami, Florida
January 1976

Contents

Marine Sediment Transport
and Environmental Management

CHAPTER 1

Introduction

DONALD J. P. SWIFT

Atlantic Oceanographic and Meteorological Laboratories, Miami, Florida

DANIEL JEAN STANLEY

Division of Sedimentology, Smithsonian Institution, Washington, D.C.

The six years since the publication of our first short course volume (Stanley, 1969) have seen major advances in the field of continental margin sedimentation. The surveys of surficial sediment character and distribution undertaken in earlier years are now more meaningful because we are beginning to understand the mechanisms of continental margin sediment transport.

The past few years have also seen some major changes in the nature of data consumption. In the past, we talked to each other primarily as scientists in order to advance our general understanding of our field. It has been tacitly understood that our ultimate data consumers were in large part petroleum geologists who wished to make uniformitarian interpretations of the rock record, in the search for petroleum deposits. This need is still with us, and has in fact increased drastically as a result of the ongoing energy crisis. However, it is now only a portion, albeit a major portion, of a complex array of information required by our society in its attempt to manipulate its environment and maintain and improve its quality of living. The need for data on the continental margin surface can be summarized under the title of environmental management. The continental margin must be managed in a thoughtful and rational way because it is being subjected to rapidly increasing usage for a variety of conflicting purposes.

The oldest of these, and still a major concern, is the use of the continental margin surface for food resources, primarily finfish and shellfish. A second major usage, which tends to conflict with the first, is for mineral resources. The continental margin has long been a source of placer minerals. Since population pressures have so greatly increased the value of land deposits, the continental margin is now a major source of sand and gravel as well. The coming exploitation of the continental margins for oil will of course overshadow all of these mineral uses. The continental margin is becoming real estate. We plan to export to it some of our problems of land management in the form of offshore nuclear generating stations, offshore deep-water ports, and possibly floating airports. The first offshore nuclear generating station is scheduled to begin construction in 1985 off the coast of New Jersey. The continental margin is currently a major site for waste disposal. Coastal cities dispose of sewage on the inner shelf by outfalls or by barging. Acid and industrial wastes and dredge spoil are also barged out on the inner shelf. Toxins and explosives are often dumped close to or over the shelf edge. Finally, the continental margin is used for recreation and navigation. Beaches and harbors must be maintained, and sports fisheries are important aspects of coastal economies.

Three parties are concerned with this complicated roster of usages. Environmental engineers must design the particular mode of continental margin exploitation. The public, represented by the environmental managers, must reconcile the competing demands made on the shelf. Both turn to the environmental scientist whose role is to develop a systematic body of basic information on the continental margin ecosystem, and the physical systems that support it.

Major studies of the continental margin environment are now in progress in the United States. The National Oceanic and Atmospheric Administration has formed an agency-wide task force, MESA (Marine Ecosystems Analysis), to investigate environmental impacts on the continental margin. Its prototype study is the New York Bight, where massive ocean dumping schedules are creating twenty-first century managerial problems. The Environmental Protection Agency is attempting to formulate systematic guidelines for ocean dumping. Both the U.S. Geological Survey and the Bureau of Land Management are concerned with the impact of the proposed large-scale exploitation of the nation's continental margins for oil. The Energy Research and Development Administration is concerned with extensive usage of the inner continental shelf for offshore power plants.

A new scheme of large-scale interdisciplinary study is developing whereby governmental and academic scientists in a variety of marine disciplines unite to study threatened areas. Information flow in such studies tends to follow the course depicted in Fig. 1. Physical oceanography is the core discipline. Geologists need to know how fluid flow drives the particulate system; chemists need to resolve both fluid transport and substrate storage of chemical components; and biologists must determine how all physical systems support the ecosystem. There is, of course, feedback through the various disciplines.

The physical disciplines are to a certain extent internal service groups in that they support biology; the public's concern with the environment is largely focused on aspects of food resources and public health. However,

each discipline may also make a direct contribution to the needs of environmental managers. Physical oceanographers may describe the effect of flows on dumped material. Geologists may describe whether dumping is occurring on an eroding or an aggrading surface, and hence whether the waste will be contained or dispersed. Chemists describe directly the quality of water.

The role of the geological oceanographer in this scheme is twofold. He must first inventory the substrate, to determine its bathymetry, its shallow stratigraphy, and the character and distribution of surficial sediments. Second, he must attempt to determine quantitatively the flux of particulate matter across the substrate, since this flux carries nutrients and contaminants, affects the composition of the benthic community, and directly affects the suitability of the surface for dumping and building.

We have planned our volume as an aid to this kind of investigation. The first two sections of the text deal with the large-scale flows of the continental margin, and the mechanisms by which they entrain, transport, and deposit sediment. The central portion deals with regional aspects of water-sediment interaction. The last section discusses the application of this material to environmental management.

This sequence of topics is broader than the classical curriculum of shelf geology. The field has been traditionally concerned with petrography and stratigraphy. In the last decade, we were exposed as well to the views of hydraulicists and engineers, as described in the Inman chapter of Shepard's (1963) text and in Middleton's (1965) volume on primary sedimentary structures. We are still consolidating the gains accruing from this merging of disciplines.

However, the limits of the petrographic-hydraulic approach to continental margin sedimentation are in sight. Analysis of water-grain interaction, no matter how detailed, will never lead us to the prediction of regional patterns of sediment transport. For this we must seek the insights of the physical oceanographers; regional analysis of sediment transport requires detailed knowledge of the temporal and spatial characteristics of the shelf flow field. It is with only moderate exaggeration that we may describe a continental margin sedimentologist as an undereducated physical oceanographer, specializing in sea-substrate interaction, just as some other physical oceanographers specialize in sea-air interaction.

Kuhn (1970) has argued that scientific disciplines do not evolve in linear fashion, but rather by steps. A scientific breakthrough does not necessarily require new data, but it does require a new approach to existing data. Such a breakthrough is followed by a period of consolida-

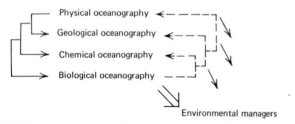

FIGURE 1. *Information flow in an interdisciplinary study of the continental margin.*

tion, until the new concepts are so throroughly explored that they too appear deficient. Kuhn refers to such sets of concepts as *paradigms*. Paradigms may include specific conceptual models, but are more general and basic than this; they are coherent and systematic ways of viewing the universe, or that portion of it considered by one's discipline. Geologists should be sympathetic to Kuhn's argument, for we have all just witnessed such a "scientific revolution" or change of paradigms: from classical tectonics to global, plate tectonics.

It may be that such a transformation is beginning within the narrower field of continental margin sedimentation. This change in outlook, perhaps prematurely identified, involves a more holistic approach to the velocity structure and circulation patterns of the shelf water column on one hand, and to the response of the shelf surface on the other. Just as the petrologist-stratigrapher stresses depositional environments and the deposits therein, so does the dynamical sedimentologist stress the regional pattern of flow that links these environments and the associated sediment budget.

We suggest that this transformation of viewpoint is contained within the spectrum of opinions expressed by the authors of our volume. The interested reader is invited to test our hypothesis by tracing the varied usage of the adjectives "relict" and "modern" as applied to the genesis of surficial sediment, from the Foreword to the concluding chapter.

Because we feel that principles of physical oceanography are becoming increasingly important in the analysis of shelf sediment transport, we have included a much more detailed and rigorous treatment of this topic than is customary in texts on sedimentation. We expect that our readership will include persons in all oceanographic disciplines. The level of presentation will cause no problem for readers with backgrounds in physical oceanography, who will be interested in the new information and new approaches to shelf dynamics described in Part I of the volume. Readers with conventional backgrounds in geology, such as ourselves, are urged to read these chapters also. All subsequent chapters draw on the basic material presented in these five chapters. What cannot be absorbed in the first reading may be profitably reread when geological aspects of hydrodynamical problems are encountered in the following chapters.

We feel that by inviting the participation of a variety of persons engaged in basic and applied research relating to continental margin sedimentation, we have been able to assemble a broad-gaged and timely statement. Organization has inevitably suffered in that the same topic is in some cases dealt with by more than one author. We as editors have attempted to minimize such overlap, but have allowed some redundancy to occur, on the grounds that some diveristy of viewpoint and approach is beneficial.

REFERENCES

Kuhn, T. S. (1970). *The Structure of Scientific Revolutions*. Chicago: University of Chicago Press, 210 pp.

Middleton, G. V., ed. (1965). *Primary Sedimentary Structures and Their Hydrodynamic Interpretation*. Tulsa: Soc. Econ. Paleontologists and Mineralogists Spec. Publ. 12, 265 pp.

Shepard, F. P. (1963). *Submarine Geology*, 2nd ed. New York: Harper & Row, 557 pp.

Stanley, D. J., ed. (1969). *The New Concepts of Continental Margin Sedimentation*. Washington, D.C.: Am. Geol. Inst., 400 pp.

Continental Margin Circulation

Part I describes the fluid motions that determine the patterns of sediment transport on the continental margin. While some important studies of continental margin circulation occurred at the onset of systematic physical oceanography, the present intensive approach to the problem is a very recent phenomenon. In the introductory chapter, C. N. K. Mooers describes the kind of investigations that are currently in progress on the continental margin, and explains their relevance to problems of sedimentation. He also develops the basic equations of fluid motion that are used in the ensuing chapters of Part I.

In Chapter 3, A. Leetmaa describes some simple steady state models for estuarine and shelf circulation, in which the principal forcing mechanisms are thermohaline density gradients and wind stress on the sea surface. The pattern of flow is seen to vary markedly as a function of the viscosity of the flow, as determined by depth and the value of the eddy coefficient. Such steady state modeling of fluid motion determines the patterns of suspended fine sediment dispersal that are described in Chapter 9, Part II, by D. E. Drake and Chapter 18, Part III, by J. W. Pierce.

The remaining chapters examine transient components of fluid motion on the continental margin. In Chapter 4, Mooers discusses long-period, time-dependent responses to wind stress. These are the peak flow events that mobilize the sandy floor of the shelf, as described in Chapters 10, 14, and 15 by D. J. P. Swift. In Chapter 5, H. O. Mofjeld describes the propagation of the tidal wave onto the continental margin and the resulting currents. Currents generated by tidal flows dominate sediment transport in estuaries and estuary mouths and on the floor of some shelves as described in Chapters 9, 10, 12, 14, and 15.

In Chapter 6, O. S. Madsen describes the regime of surface wind waves on the continental margin. He presents a concise summary of wave mechanics, and proceeds to discuss the frictional modification of wave motion by the bottom. This topic is a prerequisite for the analysis by P. D. Komar of the entrainment and transport of sand in the nearshore zone and shelf floor (Chapters 8 and 13).

CHAPTER 2

Introduction to the Physical Oceanography and Fluid Dynamics of Continental Margins

CHRISTOPHER N. K. MOOERS

Rosenstiel School of Marine and Atmospheric Science, University of Miami, Miami, Florida

The purpose of this chapter is to provide some unifying structure for the four chapters that follow, each of which considers one or more specific aspects of the physical oceanography and fluid dynamics of continental margins. The goals are threefold: to make several general remarks on the physical oceanography and fluid dynamics of continental margins, to provide an overview of the coverage in the four chapters, and to develop and discuss a general system of equations. The specialized systems of equations employed in each of the succeeding chapters are then derived from this general system. For brevity, the phrase "physical dynamics of continental margins" is occasionally used in lieu of "physical oceanography and fluid dynamics of continental margins."

GENERAL REMARKS ON THE PHYSICAL OCEANOGRAPHY AND FLUID DYNAMICS OF THE CONTINENTAL MARGINS

The physical dynamics of the continental margins is a vast, new field of physical oceanography and geophysical fluid dynamics. It is a vast field because most of the processes that occur in the ocean also occur some-

where on the continental margins, and because there may be additional processes unique to continental margins. It is a new field because, until very recent times, only limited studies had been devoted to the physical dynamics of continental margins. The reasons for this are probably the following: (1) extraordinary variability was observed and recognized as being beyond the sampling capability of earlier technologies; (2) the combined effect of density stratification and strong bottom topographic variations was without a firm fluid dynamical foundation; (3) the lure of deep-sea exploratory physical oceanography was irresistible; (4) the central problem in physical oceanography is the general oceanic circulation, which has required observational and theoretical investigation; and (5) the hope has been that basic process problems could be more readily resolved in the deep sea where circulation patterns have been presumed simple. With rapidly increasing pressure for utilization of the continental margins, an impetus has been provided for the investigation of this rather "messy" portion of physical oceanography and fluid dynamics.

At the present time, a predictive knowledge of water motions on the continental margins is still a distant goal. On the other hand, with improved observations and theoretical concepts, significant physical processes are being recognized, fostering a glimmer of hope that, at some level, predictability will eventually be achieved.

Contribution from the Rosenstiel School of Marine and Atmospheric Science, University of Miami, Miami, Florida 33149.

It is, then, with due caution and humility that the subject must be approached. Part of the task involves clarifying where certain knowledge exists, where the collective ignorance is most severe, and where general patterns of understanding are emerging. In subsequent sections, an attempt is made to engender an ordered awareness of the ensemble of considerations a physical oceanographer/geophysical fluid dynamicist must encompass when addressing the physical dynamics of the continental margins.

PRESENT LEVEL OF KNOWLEDGE

The present level of knowledge is rather primitive compared to what it could be with modern technological and intellectual resources. An excellent review of progress in shelf circulation during the past five years has been given by Niiler (1975). The dawning age of numerical modeling holds much promise once a better comprehension of the predominant processes is acquired. For this, more theoretical, observational, and experimental work is required. In recent years, rather exciting sets of observations have been acquired from several shelf regions; they are helping first to identify and then understand the major processes. The most significant observational advances are based on the advent of recording current meters which can be placed on moorings for a month or longer. (Nearshore current measurement remains a serious sampling problem because of contamination of current-meter records by vigorous shallow water, surface gravity wave particle velocities.) Earlier studies based primarily on modest amounts of hydrographic and drift data yielded an approximate but not satisfying "picture" of the circulation and hydrographic variability. Ironically, the direct current measurements at discrete points have produced a demand for vastly more hydrographic information in order to document the baroclinic (i.e., those caused by density stratification) structures and events over the full water column of a particular experimental area and period. Some of the best examples using earlier methods are found in the articles by Bumpus (1973), Stefansson et al. (1971), and Wyatt et al. (1972). Good examples using modern methods are found in the works of Huyer et al. (1975), Huyer and Smith (1974), Schmitz (1974), and Zimmerman (1971). Slowly the kinematical aspects, i.e., the space-time variation of temperature, salinity, and at least the horizontal velocity, are beginning to emerge. In the next stages, the dynamical aspects, i.e., the term balances in the equations of motion, need to be determined. For this, the advent of bottom-mounted pressure gages will probably be invaluable. Alongshore and transshore pressure gradients can be important in the transient response of waters over the continental margin and in the geostrophic balance of steady flows over the margins. One of the few examples of estimates of (steady) alongshore pressure gradients is provided by Sturges (1974). Also, it will be necessary to obtain a precise accounting of generation and dissipation processes. In this endeavor, a strong interaction between theoretical modeling and experimental design is required. Considering how strong the eddylike (turbulent) motions may be, the energetic aspects, i.e., the term balances in the energy equations, of the time-varying circulation problem may have to be examined experimentally and formulated theoretically.

A major objective of contemporary observational studies has been, and continues to be, the determination of the space, time, and amplitude scales of the predominant motions. In some cases it has also been possible to identify the processes associated with the predominant motions. Fortunately, the spectrum of motions is neither white, i.e., uniform as in a purely random noise process, nor red, i.e., energy levels rising indefinitely with lower frequencies. In contrast, there generally appears to be an ordered sequence of broadband energy peaks. Ordering the spectrum of motions by increasing frequency, a gross generalization of the spectrum of horizontal velocity, is estimated in Table 1. These conceptions are useful for several reasons: They provide guidelines for planning process-oriented experiments; they suggest a hierarchy of process-oriented models; and they yield an early (hopefully not premature) estimate of the most energetic scales.

For example, these values can be used to define minimal time-space record durations and maximal time-space sampling rates to avoid seriously biased and aliased estimates. If the motions are wavelike, they obey well-defined dispersion relations. Hence, space sampling requirements can be traded for time sampling. Otherwise, the motion is eddylike (turbulent) and the coherence scales dictate spatial sampling requirements; the coherence scales must be determined empirically or from first principles. The sorting out of the wavelike versus eddylike viewpoints, or possibilities, is a central, contemporary task. It is the author's view that the phenomena are correctly viewed as crossbreeds, i.e., eddywaves.

On the theoretical side, the existence of a banded spectrum, with spectral gaps implied, is an exciting prospect. This means that models can be created for processes governing different parts of the spectrum. These models can be interactive as necessary. There is an economy to be achieved in not having to solve (i.e., resolve in a model) for all processes simultaneously.

TABLE 1. Physical Processes and Their Scales

Process	Time Scale	Horizontal Velocity Scale (cm/sec)	Alongshore* Scale (km)	Transshore Scale (km)	Vertical Structure†
Long-term mean (climatic variability)	Decade	~1	~10^3-10^4	~Margin width	Barotropic and perhaps one baroclinic mode
Seasonal (year-to-year variability)	Annual and its harmonics	~10	~10^3	~Shelf width	Barotropic and at least one baroclinic mode
Wind event (season-to-season and year-to-year variability)	Several days to several weeks	~100	~10^2-10^3	A few horizontal modes, ~shelf width	Mainly barotropic and perhaps one or two baroclinic modes
Tidal (seasonal variability)	Diurnal and semidiurnal	~10	~10^3-10^4	Entire margin ~1-10^2	Barotropic and a few baroclinic modes
Inertial (wind event-to-wind event variability)	Depends on latitude >diurnal, equatorward of 30° latitude; <lunar semidiurnal poleward of 72° latitude	~10	(?) 10	~(?) 10	~10 baroclinic modes
Edge wave (day-to-day variability)	Minutes	~10	~1	~10^{-1}-10	Uniform (barotropic)
Surface gravity wave (sea and swell) (day-to-day variability)	Seconds	~10	~10^{-1}	10^{-3} — 10^{-1}	Uniform (barotropic)

* Alongshore scale is an ambiguous quantity; here it has been defined to mean the scale of organized motions, i.e., the scale of coherent motion.

† The comments on vertical structure are appropriate to a density-stratified continental margin; for unstratified margins, the vertical structure is dominated by effects of surface and bottom (frictional) boundary layers.

9

On the applications side, the transfers that occur only during extreme velocity conditions can be identified as responses to wind events associated with various classes of storms. Of course, the most extreme velocities occur through the constructive interference of different processes.

FORCES

The circulation on the continental margins is driven by several categories of external forces of meteorological, tidal, oceanic, and continental origin. In addition, there may be significant internal forces produced by various flow instabilities. [For theoretical examples of turbulent boundary layer and large-scale boundary current instabilities, see Faller and Kaylor (1966), Lilly (1966), or Kaylor and Faller (1972), and Orlanski (1969) or Orlanski and Cox (1973), respectively.] Looking seaward from the shoreline, a hierarchy of regions under varying degrees of influence from the various forces can be imagined. For the sake of argument, the hierarchy is considered to be composed of the following regimes: nearshore, inner shelf, outer shelf, and continental slope. The definitions of these regimes are evolved below.

The nearshore regime is defined to extend from the shoreline to the seaward edge of the shoreface, a depth of perhaps 20 m. It encompasses the region of strong surf zone influence, and it is frequently vertically homogeneous in properties. Near the shoreline, the generally persistent breaking of incident surface gravity waves provides a momentum transfer mechanism that can result in alongshore currents and transshore circulation cells. River runoff provides a thermohaline forcing through horizontal gradients of heat and salt, which is generally distributed at "points" along the coast. These heat and salt gradients often produce gradients of buoyancy, and thus pressure too, and hence a dynamic force. Both the surface gravity wave and river runoff forcing can be expected to exhibit strong seasonal and geographical dependence.

The inner shelf regime is defined as the regime between the nearshore and the outer shelf; the latter may be delimited on its shoreward side by the 50 to 200 m isobaths. This is the "true shelf regime"; i.e., it is not overwhelmingly dominated by nearshore or offshore influences. It can be expected to be vertically stratified in the summer and horizontally stratified in the winter. On the inner shelf, the influence of the wind-driven circulation may be expected to be most prominent. Here, the circulation response to wind events ("storms") is observed to be dramatically vigorous. The storms are also the most significant mechanism for producing mixed

layers, both at the surface and the bottom, except in areas of high tidal range. The seasonal time scale of the meteorological forcing, which includes the heat and moisture fluxes as well as momentum (wind stress) flux across the air-sea interface, is often, if not usually, the predominant influence in the net drift or circulation.

The outer shelf regime is defined as the transition zone between shelf and oceanic waters. It may be bounded by the 50 to 200 m isobaths on its shoreward side and the 100 to 500 m isobaths on its seaward side. It is generally a region of strong density stratification, which is often tilted from the horizontal, sometimes sufficiently to form surface or bottom fronts. On the outer shelf and over the continental slope, the influence of the general oceanic circulation can be prominent. In these regions, there is often a major oceanic boundary current. The mass and momentum fields of such currents must continuously "connect" to the slope and outer shelf regimes through frictional boundary layer processes. Hence, the major boundary currents can be expected to induce a steady circulation in the slope and outer shelf regimes. Of course, the major boundary currents fluctuate, and thus can be expected to induce transient circulations on the shelf, too. In such boundary layers, large eddies are thought to occur which can provide exchanges of mass, momentum, vorticity, etc. To the (unknown) extent that these eddies are spontaneously developed because of flow instabilities, they are examples of phenomena produced by internal forces.

Seaward of the outer shelf, above 200 to 500 m deep, the offshore (oceanic) region is encountered. Below that depth and seaward to the edge of the continental rise, the continental slope regime per se exists. The nearly omnipresent winds produce either wind-induced coastal upwelling or downwelling, which also produces exchange between the oceanic and shelf regions.

Tidal influences are felt throughout the continental margin. The current and sea surface fluctuations of the barotropic (surface) tide generally have only relatively minor significance in the outer reaches of the continental margin. But, because of shoaling, they become quite influential nearshore, as for example, in the flushing of estuaries and lagoons. On the other hand, the interaction of the surface tides with the bottom topography and density stratification over the continental slope and shelf can lead to the generation of baroclinic (internal) tides that propagate both shoreward and seaward and have significant horizontal and vertical velocities, with large shears that can be important for mixing.

In actuality, all the forces may have seasonal modulations and be experienced throughout the continental margin. Whether or not there is any separation into the distinct domains indicated above depends on the width of the margin and the absolute and relative vigor of the applied forces.

CONSTRAINTS

There are certain constraints placed on the circulation of the continental margin. They can be categorized as geometric, hydrographic, and dynamic.

The geometric constraints are largely provided by the topology of the continental margin. The very existence of the shoreline provides a profound barrier to flow. The bottom slope of the continental margin provides yet another barrier. In the continuously density-stratified ocean, the locus of intersection of each constant density (isopycnic) surface with the bottom can be considered to be the "shoreline" for that isopycnic surface. One effect of the shoreline barrier is to tend to organize the motion in the along-isobath direction.

All anomalies in bottom and coastline topography have an influence on the circulation. Some form barriers to flow, e.g., submarine banks, islands, or coastal capes. Others form channels for flow, e.g., submarine canyons, depressions, or coastal embayments. Large depressions in the shelf (basins) and large bays may have their own circulation systems, which are only partially coupled to the general circulation of the continental margin. The topographic anomalies may be very influential in producing dissipation, including mixing of water masses. On a finer scale, the various bed forms provide roughness elements for frictional dissipation, through production of disorder (turbulence) in the flow. Especially in the nearshore zone, where the bed forms seem so dynamic, there is probably a strong interaction between the circulation and sedimentation.

The water mass structure of offshore oceanic waters is cited as an example of a hydrographic constraint. While substantial water mass transformation can occur over a shelf, through air-sea transfers of heat and moisture, freshwater runoff, and mixing, the ultimate sources and sinks for the bulk of the shelf waters are found in the deep ocean, not necessarily immediately offshore of a shelf region of interest. The deep ocean cannot be considered a magic reservoir that will accept or yield on an isopycnic surface any temperature, salinity, dissolved O_2, or nutrient structure that is believed to be necessary for shelf waters and their circulation. Thus, there are compatibility or consistency requirements between shelf and oceanic waters that must be taken into account when constructing models or interpreting observations.

As an example of a dynamic constraint, the combined effect of variations in bottom topography and the vertical component of the Coriolis parameter is considered. In a frictionless flow, the law of conservation of potential vorticity holds; i.e., the quantity $(f + \zeta)\, \partial_z \rho$ is constant along a streamline on an isopycnic surface, where f is the planetary vorticity or the vertical component of the Coriolis parameter, ζ is the horizontal component of relative vorticity on an isopycnic surface $\zeta = \partial_x v - \partial_y u$, with ρ (the density) held constant, and $\partial_z \rho$ is the vertical gradient of density. Since f increases with latitude, either ζ or $\partial_z \rho$ (essentially the inverse thickness of an isopycnic layer) must decrease along a streamline for poleward motion. In the simple case of uniform density and negligible relative vorticity ($\zeta \ll f$), the law becomes f/h is constant, where h is the water column depth. Contours of f/h are called geostrophic contours; they all emanate from the coast at the equator and diverge poleward. When they reach the base of the continental slope and enter a (idealized) flat-bottomed, deep ocean, they acquire a zonal (constant latitude) direction; the flow could be in either direction until other factors are considered. At midlatitudes, over short distances (the order of 100 km), the geostrophic and bottom contours appear essentially identical. But, since f is proportional to the sine of the latitude, in moving from 30 to 45° latitude, a water column originally on the 30 m isobath would tend to reach the 70 m isobath. Little concrete evidence exists for this mechanism, probably because of baroclinic (i.e., those caused by density stratification) and transient effects; yet, the law is so fundamental that it must tend to govern at least the mean motion.

For transient motions, the law can provide a restoring mechanism. If a water parcel is displaced from its equilibrium contour, it will tend to return to its original contour. In the northern hemisphere, if a fluid column is displaced into shallower water, it will develop negative relative vorticity or anticyclonic (clockwise) motion. If displaced into deeper water, it will develop positive relative vorticity or cyclonic (anticlockwise) motion. In both hemispheres, the net effect is to produce a wave motion with poleward phase progression along the west coast of continents or equatorward phase progression along the east coast. These vorticity waves are often called continental shelf waves because the sloping bottom is essential for their existence. More will be said about such waves in Chapter 4, but a useful reference is Tareyev (1971). The governing horizontal scale for these motions is the width of the continental margin.

Another example of a dynamic constraint is provided by the length scale associated with steady geostrophic and potential vorticity-conserving motions in a density-stratified fluid. There is a countable infinity of such length scales (L_n, $n = 0, 1, 2, \ldots$) in a continuously stratified fluid. They are determined by the formula $L_n = c_n/f$, where f is again the Coriolis parameter and c_n is the nth vertical mode gravity wave speed. For the barotropic mode, $c_0 = (gh)^{1/2}$, where g is the gravitational acceleration and h is the water depth. For the baroclinic modes ($n \geq 1$), $c_n = c_1/n$, where $c_1 = \tilde{N}h$ and $\tilde{N}$ is the depth-averaged Väisälä-Brunt frequency (N is the frequency of oscillation of a displaced water parcel within a density-stratified water column). The formula for N is

$$ N = \left(-\frac{g}{\rho_0} \frac{d\rho}{dz} \right)^{1/2} $$

where $d\rho/dz$ is the in situ minus the adiabatic density gradient and ρ_0 is the mean density. For an offshore water depth of 1 km, a typical value of $\tilde{N}$ is $3 \times 10^{-3}\,\text{sec}^{-1}$ and a midlatitude value of $f = 10^{-4}\,\text{sec}^{-1}$, $c_0 = 100$ m/sec, $c_1 = 3$ m/sec, $c_2 = 1.5$ m/sec, etc. Thus, $L_0 = 10^3$ km, $L_1 = 30$ km, $L_2 = 15$ km, etc. Commonly, L_0 is called the barotropic (Rossby) radius of deformation, and L_1 is called the baroclinic (Rossby) radius of deformation. The response of the various vertical modes of motion to external forcing will to a large extent be determined by how well their horizontal scales and those of the forcing are matched. It is useful to note that the barotropic length scale is much larger than a typical margin width, whereas the baroclinic length scales are less than the margin width, except for very narrow shelves. Thus, the barotropic processes trapped on a shelf can be expected to be controlled by the bottom slope and to have a length scale of the order of the continental margin width. In contrast, the baroclinic motions can be trapped on a shelf at scales considerably smaller than the continental margin width. Noteworthy, too, is the decrease of the L_n's with increasing distance from the equator.

On a large scale, there are constraints provided by the combination of all three factors. These factors conspire to produce regional regimes of continental margin circulation. The concept of a regional regime implies some continuity of flow within the regional regime and only partial coupling to contiguous regional regimes. A regional regime may be delimited by a topographic barrier to flow, e.g., a prominent cape or embayment. Or it may be delimited by a topographic separation or attachment point of a major boundary current. On the other hand, a regional regime may be delimited by a transition in the offshore water mass. Often the boundary current and water mass transitions are virtually coincident. Precise knowledge for a rational definition of regional regimes does not exist at present, but the author provides a speculative grouping for the con-

tiguous United States as a useful "strawman" in Table 2. Generally, there also appears to be evidence from the geological and biological disciplines for a regime classification system along these lines. The major implications of the regional regime concept are that (1) an ecosystem may exist for each regime, (2) the general shelf and slope circulation need be determined on the regional scale for ecosystem analysis purposes, and (3) a regional regime can be treated as a semi-enclosed circulation system with leaky, yet restricted, alongshore end conditions and with a wide-open aperture to the deep ocean. The regional regime concept does not preclude the existence of subregimes nor the validity of studying component dynamic processes on a scale smaller than a regional regime. Interestingly, many of the proposed regional regimes coincide with different meteorological regimes. Considering that the general circulation of no single regional regime is adequately documented, let alone understood, it seems staggering that 10 such regimes are identified for the contiguous United States alone. If the forces are adequately quantified, and if the predominant processes are adequately understood and parametrized, there is the possibility that unifying principles will be found that will obviate the necessity for an intensive and extensive observational program to be mounted in every regime. In other words, a dynamic classification scheme for shelf and slope environments based on a few non-dimensional parameters is conceivable on one hand, and corresponding, broadly applicable numerical or other types of models are conceivable on the other hand.

PHENOMENA

Several phenomena have been touched upon in the preceding sections. In this section, several additional phenomena are brought to the reader's attention.

Surface and subsurface plumes of natural and man-made substances are prevalent on the continental margins. Some are associated with river and sewage discharges; others are wind or eddy induced. How these property anomalies disperse is not fully known. Certainly one mechanism involves mixing and sinking at frontal discontinuities along plume edges. Another dispersal agent may be the many classes and scales of irregular wave-type motions that are known to exist.

Reversal of the alongshore circulation on a seasonal time scale is a common occurrence on the inner shelf. The outer shelf and slope regimes are generally density stratified, both vertically and horizontally. There, under-currents flowing counter to the near-surface currents are prevalent. Frequently, the horizontal gradients in

TABLE 2.   Contiguous U.S. Regional Circulation Regimes

Region	Regime	Boundaries
East Coast	Gulf of Maine	Cape Sable (Canada) to Cape Cod
	Middle Atlantic Bight	Cape Cod to Cape Hatteras
	South Atlantic Bight	Cape Hatteras to Cape Canaveral
	East Florida shelf	Cape Canaveral to Dry Tortugas Island
Gulf Coast	West Florida shelf	Dry Tortugas Island to Cape San Blas
	MAFLA (Mississippi–Alabama–Florida Panhandle) shelf	Cape San Blas to Mississippi Delta
	Texas shelf	Mississippi Delta to Rio Grande River
(non-U.S.)	Mexican shelf	Rio Grande River to Yucatan Straits
West Coast	Southern California Bight	Punta Eugenia (Mexico) to Point Conception
	Northern California shelf	Point Conception to Cape Mendocino
	Northwest shelf	Cape Mendocino to Vancouver Island

properties are sufficiently intense to classify the situation as frontal. Such fronts are loci of intense variability and presumably exchange.

Because of the ubiquitous nature of the winds, either wind-induced coastal upwelling or downwelling is generally occurring at any given moment and location. These processes can occur in either the transient mode, i.e., on the synoptic (several day to several week) time scale, or the quasi-steady mode, i.e., on the seasonal (several month) time scale. Relatively little is known about downwelling compared to upwelling.

Winter storms in particular can produce cooling and subsequent sinking over a shelf, with the possibility of producing a water mass that subsequently cascades off the edge of the shelf and into the deep sea.

Especially in summer, the diurnal sea breeze regime generally penetrates 10 to 100 km seaward of the coastline. The associated winds can be quite vigorous and, hence, influential in amplifying the wind-stirring and exciting oscillatory motions.

The surface and bottom boundary layers are generally thought to be turbulent an appreciable fraction of the time. On the continental margins, the surface and bottom mixed layers, when they exist, are generally much thinner than in the open ocean, on the order of 10 or 20 m as opposed to 100 m or so. Whether these boundary layers are best thought of as turbulent Ekman layers has yet to be determined. Under conditions of light airs, shoreward-progressing patterns of surface slicks are often seen. To date these have been thought of as surface manifestations of internal gravity waves or as a convective phenomenon such as Langmuir cells or Ekman layer instabilities. Examples of recent studies of

the bottom boundary layer are found in the papers by Weatherly (1972, 1975).

An area in which geological oceanographers have provided many of the physical oceanographic observations is that of submarine canyons (e.g., Shepard et al., 1974). The evidence suggests that canyons focus tidal energy, resulting in vigorous, baroclinic tidal currents. They also funnel strong transient responses to meteorological storms (Cannon, 1972, Shepard and Marshall, 1973).

IMPLICATIONS FOR AND OF SEDIMENT TRANSPORT

Logically there are several facets of continental margin circulation which have implications for sediment transport. For example, those concerned with sediment transport need to know answers to the following questions: (1) Under what environmental conditions are sediments put into or taken out of suspension? (2) Once suspended, where and how fast do suspended sediments move horizontally? (3) How are sediments redistributed in the vertical? (4) Under what environmental conditions does the bed load move or become vertically redistributed? (5) What environmental conditions cause the various bed forms? The answers to most of these questions are embedded in the general problem of the continental margin circulation. Issues of suspension, bed forms, and vertical redistributions are probably linked to extreme storm events. Some of the horizontal distribution issue is linked to extreme events, too, but much of this issue for suspended sediments is connected with the vertical density stratification and the general background circulation associated with seasonal, tidal, and moderate wind-induced transient circulations. The generally scattered reports of continental margin circulation abound with accounts of surface jets, undercurrents, and countercurrents moving alongshore. Thus, there is no shortage of circulations to move sediments along isobaths. More subtle, but equally important, are the transisobath circulations providing conduits to and from the deep sea. Likely candidates are wind-induced upwellings and downwellings, tidal surges up and down canyons and troughs, wintertime convective sinking and subsequent offshore flow, and stirring of bottom layers which may then lead to density currents downslope.

For the physical oceanographer there may be several rewards for contributing to the understanding of sediment transport processes. Foremost, the sediments provide the roughness elements for bottom friction. Precious little attention has been paid to the distribution of grain sizes and bed form types by physical oceanographers when parametrizing bottom friction. The flow–bed form interaction problem appears to have been untouched by fluid dynamicists. Second, suspended sediments may make the effective density of seawater in certain areas of the continental margin appreciably different from that calculated from temperature, salinity, and pressure alone. Third, suspended sediments may serve as a tracer in circulation studies. Finally, mean flow patterns may be inferred from sediment distributions and bed forms. The shallow seafloor is a faithful recorder of the time-averaged, near-bottom circulation pattern, gated for the threshold velocity of the prevailing sediment grain size.

In the earlier discussion of the nearshore regime, the existence of river runoff influences was briefly mentioned. Actually, the coastline is as much a leaky sieve as it is a barrier to flow. In addition to river runoff per se, there are problems of tidal jets running in and out of inlets to estuaries and lagoons; the ebb, flood, and groundwater flow in saltwater marshes; the dissipation of wave energy in nearshore seaweed beds; and the flushing of fjords which are only partially coupled to the shelf because of shoal sill depths. The interaction of these flows associated with the discontinuities or imperfections in the coastline barrier and the general circulation of the continental margin is a virgin field of some practical importance to fisheries, pollution, and ecological, as well as sedimentation, interests.

Surface gravity wave and shelf circulation studies are seldom performed by the same investigators or even in a coordinated fashion. Surface gravity wave investigators point to the probable influence of extreme, storm-generated waves on sediment transport *across* the entire

shelf. Circulation investigators point to the influence of strong, storm-generated transient (several-day time scale) circulations and their probable influence on sediment transport *along* the entire shelf. These extreme views may be mutually consistent; their interrelationships or interactions have not been investigated.

Similarly, plume studies are usually conducted without a thorough consideration of the ambient shelf circulation. These studies provide fundamental information on plume and suspended sediment dispersion, yet for more than initial or local effects, it would seem that the interaction of plumes with transient circulations must be important.

Biological and chemical oceanographers have contributed information useful to physical oceanographers during the geographical and kinematical phases of the physical dynamics of continental margins. The geological and physical oceanographers probably have more interests in common during the dynamic phases, especially in determining the controlling processes in the bottom boundary layer. As the physical oceanographers become more meteorologically aware in their investigations of the surface boundary layer and transient circulations on the continental margin in general, it is safe to forecast that the geological oceanographers will develop a similar awareness for the role of the many aspects of atmospheric forcing in sediment transport.

The prospects for significant progress are probably greater with process and situational studies than geographical studies. For example, it is probably neither necessary nor sufficient to make a cursory study of circulation in all submarine canyons, over all submarine banks, around all coastal capes, or in the proximity of all tidal inlets. It will probably be adequate to study thoroughly two or three typical examples of each such flow "situation" and then to generalize to other regions. No good examples yet exist. As in most physical problems, the physical dynamics of continental margins are determined by boundary conditions around a control volume, initial conditions in the control volume, and constituent processes internal to the control volume. A regional regime may be such a control volume. The constituent processes may include those associated with flow over banks, around capes, etc. Once such processes are basically understood, models for the control volume can be constructed as an aggregate of all such constituent processes. The role of the purely geographic approach is then to locate, identify, and describe the constituent processes. The geographical work can be largely done by surveys; the road to understanding the processes sufficiently well for modeling or parametrization purposes generally requires field or laboratory experiments and theoretical investigation. For modeling the time-dependent circulation in the control volume, the boundary and initial conditions must be either prescribed or determined through monitoring. A few interior points need to be sampled for model verification. Again, good examples do not yet exist. These ideas may influence how geological and physical oceanographers will collaborate in process and modeling studies in the future.

OVERVIEW OF SUBSEQUENT PHYSICAL OCEANOGRAPHY CHAPTERS

The preceding material has served to introduce the reader to the milieu called the physical dynamics of continental margins. In subsequent chapters, the elements of four specific aspects of the physical dynamics of the continental margins are presented:

Chapter	Author	Title
3	A. Leetmaa	Some Simple Mechanisms for Steady Shelf Circulation
4	C. N. K. Mooers	Wind-Driven Currents on the Continental Margin
5	H. O. Mofjeld	Tidal Currents
6	O. S. Madsen	Wave Climate of the Continental Margins

The topics of these chapters are by no means all-inclusive, as can be judged from the earlier discussion, but they cover four representative classes of motion. Furthermore, the interactions of these classes of fluid motion are not examined. The classes of motion treated are distinguished by their space and time scales, their restoring forces, and their generation mechanisms.

In the final portion of this chapter, advantage is taken of these distinguishing features to derive a governing system of equations, or a single governing equation, and relevant boundary conditions for each class of motion from a common system of equations. In the course of this development some care is taken to indicate where various assumptions and approximations are introduced. In so doing, the reader is allowed some insight into where the theoretical foundations of the subject are weak and where they are strong. Also, some appreciation for neglected effects (terms) is developed, which may aid in recognizing areas of future research.

DEVELOPMENT OF SYSTEMS OF EQUATIONS

In subsequent chapters, recourse is had to simplified sets of equations. To lend the various treatments an element of unity, a fairly general subset of the Navier-Stokes equations for a rotating, density-stratified fluid is introduced. The reader is referred to the lists of commonly used symbols and differential and integral operator notation for physical and mathematical definitions (see p. 19). The momentum equations for x, y, and z directions, respectively, are

$$\rho_0\left(\frac{du}{dt} - fv\right) = -\partial_x p + \rho_0 \nu \nabla^2 u \tag{1}$$

$$\rho_0\left(\frac{dv}{dt} + fu\right) = -\partial_y p + \rho_0 \nu \nabla^2 v \tag{2}$$

and

$$\rho_0\left(\frac{dw}{dt} - b\right) = -\partial_z p + \rho_0 \nu \nabla^2 w \tag{3}$$

The first terms on the left-hand side of the momentum equations represent the total accelerations. The second terms in (1) and (2) represent the horizontal Coriolis forces, and in (3) the buoyancy force. The first terms on the right-hand side represent pressure gradient forces, and the second terms represent molecular dissipation. The effects of the earth's sphericity, the density fluctuations except in the buoyancy term, and the horizontal component of the Coriolis parameter have been neglected. Since the effects of compressibility, e.g., acoustic phenomena, are negligible for present purposes, the fluid is treated as incompressible, yielding the equation of continuity:

$$\partial_x u + \partial_y v + \partial_z w = 0 \tag{4}$$

To include thermohaline effects, heat and salt equations are introduced:

$$\frac{dT}{dt} = K_T \nabla^2 T \tag{5}$$

and

$$\frac{dS}{dt} = K_S \nabla^2 S \tag{6}$$

where the terms on the left-hand sides represent the total rates of change and those on the right-hand sides represent molecular dissipation.

To complete the system of equations, an equation of state is needed, i.e., $\rho = \rho(T, S, p)$. For present purposes, it is sufficient to consider a linearized equation of state and again neglect compressibility effects:

$$\rho = \rho_0(\alpha T - \beta S) \tag{7}$$

where α is the coefficient of thermal expansion and β is the coefficient of halal contraction.

To solve this system of equations, a particular problem must be posed which involves kinematic boundary conditions (e.g., vanishing normal velocity at rigid boundaries) and dynamic boundary conditions (e.g., continuity of pressure, momentum flux, and heat and salt fluxes at free surfaces or internal boundaries). Because it contains provisions for the "mean" motion, various classes of waves, and turbulence, the system of equations is still so complex that its solution is formidable. Because various time scales of circulation are known to exist, e.g., seasonal, storm, inertial-internal wave, surface gravity wave, and small-scale turbulence, one way to proceed is to partition the fields into mean and fluctuating components:

$$u = \bar{u} + u', \ldots$$

where

$$\bar{F} = \frac{1}{\mathfrak{I}} \int_0^{\mathfrak{I}} F \, dt \quad \text{and} \quad \bar{F}' = 0$$

The definition of the averaging time $\mathfrak{I}$ is left unspecified for now. The system of equations (1) through (7) is then averaged as above:

$$\frac{D}{Dt} \bar{u} - f\bar{v} = -\frac{1}{\rho_0} \partial_x \bar{p} - \nabla \cdot \bar{R}^{(x)} + \nu \nabla^2 \bar{u} \quad (1)'$$

$$\frac{D}{Dt} \bar{v} + f\bar{u} = -\frac{1}{\rho_0} \partial_y \bar{p} - \nabla \cdot \bar{R}^{(y)} + \nu \nabla^2 \bar{v} \quad (2)'$$

$$\frac{D}{Dt} \bar{w} - \bar{b} = -\frac{1}{\rho_0} \partial_z \bar{p} - \nabla \cdot \bar{R}^{(z)} + \nu \nabla^2 \bar{w} \quad (3)'$$

$$\partial_x \bar{u} + \partial_y \bar{v} + \partial_z \bar{w} = 0 \quad (4)'$$

$$\frac{D}{Dt} \bar{T} = -\nabla \cdot \bar{\mathbf{q}}_T + K_T \nabla^2 \bar{T} \quad (5)'$$

$$\frac{D}{Dt} \bar{S} = -\nabla \cdot \bar{\mathbf{q}}_S + K_S \nabla^2 \bar{S} \quad (6)'$$

and

$$\bar{\rho} = \rho_0 (\alpha \bar{T} - \beta \bar{S}) \quad (7)'$$

where

$$R = \begin{pmatrix} R^{(x)} \\ R^{(y)} \\ R^{(z)} \end{pmatrix} = \begin{pmatrix} uu & uv & uw \\ vu & vv & vw \\ wu & wv & ww \end{pmatrix}$$

is the Reynolds stress (momentum flux) tensor or matrix, $\mathbf{q}_T = (uT, vT, wT)$ is the heat flux vector, $\mathbf{q}_S = (uS, vS, wS)$ is the salt flux vector, and $(D/Dt)\bar{F} = \partial_t \bar{F} + \bar{u}\partial_x \bar{F} + \bar{v}\partial_y \bar{F} + \bar{w}\partial_z \bar{F}$. Note: Time variations of F that are slow compared to $\mathfrak{I}$ have been allowed for. A system of equations for a slowly varying "mean" circulation has now been obtained which takes into account the influence of the rapidly fluctuating fields in a precise fashion.

Before discussing and further reducing the system of mean equations, a system of equations for the fluctuating components is derived by simply subtracting the primed from the unprimed system of equations:

$$\frac{D}{Dt} u' + \nabla \cdot R^{(x)} + \mathbf{v}' \cdot \nabla \bar{u} - fv'$$
$$= -\frac{1}{\rho_0} \partial_x p' + \nabla \cdot \bar{R}^{(x)} + \nu \nabla^2 u' \quad (1)''$$

$$\frac{D}{Dt} v' + \nabla \cdot R^{(y)} + \mathbf{v}' \cdot \nabla \bar{v} + fu'$$
$$= -\frac{1}{\rho_0} \partial_y p' + \nabla \cdot \bar{R}^{(y)} + \nu \nabla^2 v' \quad (2)''$$

$$\frac{D}{Dt} w' + \nabla \cdot R^{(z)} + \mathbf{v}' \cdot \nabla \bar{w} - b'$$
$$= -\frac{1}{\rho_0} \partial_z p' + \nabla \cdot \bar{R}^{(z)} + \nu \nabla^2 w' \quad (3)''$$

$$\nabla \cdot \mathbf{v}' = 0 \quad (4)''$$

$$\frac{DT'}{Dt} + \nabla \cdot \mathbf{q}_T + \mathbf{v}' \cdot \nabla \bar{T} = \nabla \cdot \bar{\mathbf{q}}_T + K_T \nabla^2 T' \quad (5)''$$

$$\frac{DS'}{Dt} + \nabla \cdot \mathbf{q}_S + \mathbf{v}' \cdot \nabla \bar{S} = \nabla \cdot \bar{\mathbf{q}}_S + K_S \nabla^2 S' \quad (6)''$$

and

$$\rho' = \rho_0 (\alpha T' - \beta S') \quad (7)''$$

Thus, a system of equations for the fluctuating fields has been derived which exhibits several types of interactions: first, advection of fluctuations by the mean flow as represented by the terms $\bar{\mathbf{v}} \cdot \nabla F'$ in $(D/Dt)F'$; second, advection of mean fields by the fluctuating flow as represented by the terms $\mathbf{v}' \cdot \nabla \bar{F}$; and third, "self-advection" of the fluctuating field as represented by the terms $\nabla \cdot R^{(x)} = \mathbf{v}' \cdot \nabla u'$, etc. It is desirable to solve the primed and unprimed systems of equations jointly in some consistent fashion. That task is also formidable and further reductions are normally used.

Typically, reductions are achieved through elimination of terms by scaling the equations in nondimensional form, linearization of the equations, and parametrization of flux (stress) terms, often in a somewhat ad hoc manner. A typical parametrization of a momentum flux term involves the hypothetical relationship for the eddy coefficient, A_H

$$-\overline{u'v'} = A_H \, \partial_x \bar{v} \quad \text{or} \quad A_H = \frac{-\overline{u'v'}}{\partial_x \bar{v}}$$

There is no guarantee that A_H is actually positive or constant, although both are usually assumed. Though the development has been theoretical to this point, the above provides rich material for observational considerations. For example, the basis for determining $\mathfrak{I}$ could well be derived from observation of a "spectral gap," i.e., a significantly broad frequency band within which only a negligible fraction of the total energy lies and, thus, separating low-frequency (slowly varying "mean") and high-frequency ("turbulent") portions of the spectrum. Rather than employing ad hoc parametrizations, calculation of $\nabla \cdot \bar{R}$, $\nabla \cdot \bar{\mathbf{q}}_T$, and $\nabla \cdot \bar{\mathbf{q}}_S$ from observations is perfectly conceivable, though difficult to achieve. Alternatively, these calculations could be made from an approximate theoretical model for a limited set of idealized processes. The point to be conveyed is that the ability to model time-dependent shelf circulation meaningfully hinges on being able to determine the divergences of the turbulent stresses by one means or another. Some such stresses can be expected to be provided by meteorological responses as well as by tidal responses, instabilities of the fluid, and other effects.

For further purposes, use will be made of reduced systems of equations. [Examples of such reductions through introduction of scaling and nondimensional parameters can be found in the geophysical fluid dynamics literature (e.g., Fofonoff, 1962; Greenspan, 1969).] In the equations for the mean motion, time variations (steady-state assumption) and all nonlinear terms will be neglected, except in the buoyancy equation (to be introduced) and except for the adoption of the eddy coefficient parametrization of turbulent fluxes. The eddy coefficients are assumed to be much larger than the molecular viscosity and diffusivity, which can then be neglected. It will be further assumed that a single eddy viscosity and diffusivity coefficient exists in the horizontal and the vertical. Also, hydrostatic balance is assumed in the vertical momentum equation, an assumption that is well justified for slowly varying motions, i.e., motions on time scales large compared to N^{-1} (reciprocal of Väisälä-Brunt frequency, restoring time of displaced water parcel in stratified water column) and for small aspect ratios (D/L, ratio of vertical to horizontal scale) which are typical of the ocean, including the shelf. Finally, the heat and salt equations are combined to form a single equation for the buoyancy, using a linearized equation of state. Thus, the reduced system of equations for the mean motion is found from (1)′ through (7)′ to be

$$-f\bar{v} = -\frac{1}{\rho_0}\partial_x\bar{p} + \partial_x(A_H\,\partial_x\bar{u})$$
$$+ \partial_y(A_H\,\partial_y\bar{u}) + \partial_z(A_V\,\partial_z\bar{u}) \quad (8)$$

$$f\bar{u} = -\frac{1}{\rho_0}\partial_y\bar{p} + \partial_x(A_H\,\partial_x\bar{v})$$
$$+ \partial_y(A_H\,\partial_y\bar{v}) + \partial_z(A_V\,\partial_z\bar{v}) \quad (9)$$

$$-\bar{b} = -\frac{1}{\rho_0}\partial_z\bar{p} \quad (10)$$

$$\partial_x\bar{u} + \partial_y\bar{v} + \partial_z\bar{w} = 0 \quad (11)$$

and

$$\bar{u}\,\partial_x\bar{b} + \bar{v}\,\partial_y\bar{b} + \bar{w}\,\partial_z\bar{b} = \partial_x(K_H\,\partial_x\bar{b})$$
$$+ \partial_y(K_H\,\partial_y\bar{b}) + \partial_z(K_V\,\partial_z\bar{b}) \quad (12)$$

In Chapter 3, steady state thermohaline circulation is examined. There, constant eddy coefficients are adopted, horizontal diffusion is neglected, and only the horizontal pressure gradient is allowed to have alongshore (y) variation. Hence, the system of equations (8) through (12) reduces to

$$-f\bar{v} = \frac{-1}{\rho_0}\partial_x\bar{p} + A_V\,\partial^2_{zz}\bar{u} \quad (13)$$

$$f\bar{u} = \frac{-1}{\rho_0}\partial_y\bar{p} + A_V\,\partial^2_{zz}\bar{v} \quad (14)$$

$$\bar{b} = \frac{1}{\rho_0}\partial_z\bar{p} \quad (15)$$

$$\partial_x\bar{u} + \partial_z\bar{w} = 0 \quad (16)$$

and

$$\bar{u}\,\partial_x\bar{b} + \bar{w}\,\partial_z\bar{b} = K_V\,\partial^2_{zz}\bar{b} \quad (17)$$

The following are boundary conditions for this system:

1. Normal velocity vanishes at the sea bottom, $\bar{\mathbf{v}}\cdot\hat{n} = 0$ at $z = -h$, where $\hat{n}$ is the unit vector normal to the bottom.

2. Tangential velocity vanishes at the sea bottom, $\bar{\mathbf{v}}\cdot\hat{t} = 0$ at $z = -h$, where $\hat{t}$ is the unit vector tangential to the bottom.

3. Normal velocity vanishes at the sea surface, $\bar{w} = 0$ at $z = 0$.

4. Tangential stress equals wind stress at the sea surface, $A_V(\partial_z\bar{u}, \partial_z\bar{v}) = \boldsymbol{\tau}_w$.

5. Buoyancy flux vanishes at the sea bottom, $\nabla\bar{b}\cdot\hat{n} = 0$ at $z = -h$.

6. Buoyancy flux equals air-sea buoyancy transfer at the sea surface, $K_V\,\partial_z\bar{b} = B_0$ at $z = 0$.

A considerable reduction of the system of equations for the fluctuating fields is also made. All nonlinear and dissipative terms are neglected but vertical stress terms are included to provide for wind stress forcing. The advective accelerations provided by the mean flow and the advection of gradients of the mean fields by the fluctuations, except for the vertical variation of the mean buoyancy field in the buoyancy equation, are neglected. This neglect of terms involving the mean fields can be seriously in error on the continental margins; however, it is a useful simplification for tutorial purposes. Again, a buoyancy equation is also formed. Thus, the reduced system of equations for the fluctuating motion is found from $(1)''$ through $(7)''$ to be

$$\partial_t u - fv = -\frac{1}{\rho_0} \partial_x p + \frac{1}{\rho_0} \partial_z \tau^{(x)} \qquad (18)$$

$$\partial_t v + fu = -\frac{1}{\rho_0} \partial_y p + \frac{1}{\rho_0} \partial_z \tau^{(y)} \qquad (19)$$

$$\partial_t w - b = -\frac{1}{\rho_0} \partial_z p \qquad (20)$$

$$\partial_x u + \partial_y v + \partial_z w = 0 \qquad (21)$$

and

$$\partial_t b + wN^2 = \partial_z B \qquad (22)$$

where the primes have been dropped, $N^2 = \partial_z \bar{b}$, $\boldsymbol{\tau} = (\tau^{(x)}, \tau^{(y)}) = (\overline{u'w'}, \overline{v'w'})$ is the turbulent stress vector for the vertical flux of horizontal momentum, and $B = \overline{b'w'}$ is the vertical flux of buoyancy. This system of equations is examined further in Chapter 4, where wind-driven shelf circulation is considered. The boundary conditions used are strictly kinematic except for the parametrization of momentum and buoyancy fluxes across the sea surface.

In Chapter 6, surface gravity wave effects are considered. Since motions with frequencies that are large compared to the inertial and Väisälä-Brunt frequencies are considered there, rotational and buoyancy effects can be safely neglected in (18) through (22). Also, except when considering generation or dissipation problems, air-sea transfer and viscous damping terms can be neglected. Hence, the system (18) through (22) is reduced to

$$\frac{d}{dt} u = -\frac{1}{\rho_0} \partial_x p \qquad (23)$$

$$\frac{d}{dt} w = -\frac{1}{\rho_0} \partial_z p - g \qquad (24)$$

and

$$\partial_x u + \partial_z w = 0 \qquad (25)$$

where no analogs to (19) and (22) are required because x is the direction of horizontal propagation and effects of density stratification are neglected; also, the total time derivative (with nonlinear, advective accelerations) has been reintroduced. It immediately follows by cross-differentiation that $d\zeta/dt = 0$, where $\zeta \equiv \nabla \times \mathbf{v}$ is the vorticity and $\mathbf{v} = (u, w)$ is the velocity. Hence, if ζ is initially zero, it is always zero, and the fluid motion is irrotational. [This condition of persistent irrotationality generalizes to the three-dimensional case, though the arguments required are more subtle (cf. Batchelor, 1967).] Thus, there exists a velocity potential ϕ such

that $\mathbf{v} = \nabla \phi$. Then, from (25), the governing equation for surface gravity waves is

$$\nabla^2 \phi = 0 \qquad (26)$$

From (23) and (24), the Bernoulli equation is

$$p + gz + \tfrac{1}{2}[(\partial_x \phi)^2 + (\partial_z \phi)^2] + \partial_t \phi = F(t) \quad (27)$$

where $F(t)$ is a function of integration and depends on time only. Since the vertical velocity w must vanish at the bottom ($z = -h$),

$$\partial_z \phi = 0 \qquad \text{at} \qquad z = -h \qquad (28)$$

Since w must equal the time rate of change of the sea surface ($z = \eta$),

$$w = \frac{d\eta}{dt} \qquad \text{at} \qquad z = \eta \qquad (29)$$

By absorbing $F(t)$ into $\partial_t \phi$, imposing the dynamic boundary condition of continuous pressure at the sea surface, and setting the atmospheric pressure equal to a constant (zero), (27) becomes

$$g\eta + \tfrac{1}{2}[(\partial_x \phi)^2 + (\partial_z \phi)^2] + \partial_t \phi = 0 \qquad \text{at} \qquad z = \eta \ (30)$$

In linearized form, (29) and (30) combine to yield

$$g\, \partial_z \phi + \partial^2_{tt} \phi = 0 \qquad \text{at} \qquad z = 0 \qquad (31)$$

The problem to be solved for small-amplitude surface gravity waves is then (26) subject to boundary conditions (28) and (31).

The solutions to (18) through (22) generally have considerable vertical structure. These solutions can often be represented as a set of vertical modes; there is a countable infinity of such modes. With a free surface (i.e., a deformable sea surface), the simplest mode (i.e., with minimal vertical structure) is called the barotropic mode; all others are called baroclinic modes. The barotropic mode is interesting to analyze because of its simplicity, and because it is frequently encountered in observations. An approximate system of equations for the barotropic mode can be written by depth-averaging the system of equations for the fluctuating motion. First, the vertical acceleration in (20) and the buoyancy equation (22) are neglected, which is certainly safe for low-frequency (with respect to N) oscillations. Then (20) is integrated over depth to obtain

$$p(x, y, z) = p_a + \int_z^\eta \rho(z) g \, dz \qquad (32)$$

where η is the free surface elevation and p_a is the atmospheric pressure. Then $\nabla_H p = \nabla_H p_a + \rho(\eta) g \, \nabla_H \eta$, where ∇_H is the horizontal gradient. It is convenient to set $\rho(\eta) = \rho_s$ and to define $p_a = \rho_s g \eta_a$. Next, (18) and

(19) are integrated over the entire water column to obtain

$$\partial_t \tilde{u} - f \tilde{v} = -g \, \partial_x(\eta + \eta_a) + \frac{\tau_w^{(x)}}{\eta + h} \qquad (33)$$

and

$$\partial_t \tilde{v} + f \tilde{u} = -g \, \partial_y(\eta + \eta_a) + \frac{\tau_w^{(y)}}{\eta + h} \qquad (34)$$

where $\int_{-h}^{\eta} \rho_0 u \, dz \equiv (\eta + h)\rho_s \tilde{u}$ and $\int_{-h}^{\eta} \rho_0 v \, dz \equiv (\eta + h)\rho_s \tilde{v}$ are the depth-averaged horizontal velocities; $\tau_w^{(x)} = (\tau^{(x)}/\rho_0)(x, y, \eta)$ and $\tau_w^{(y)} = (\tau^{(y)}/\rho_0)(x \, y, \eta)$ are the wind stress components; and the bottom stress has been neglected. Now (21) is integrated over depth and kinematic boundary conditions are applied at the sea surface and sea bottom to obtain

$$\partial_x((\eta + h)\tilde{u}) + \partial_y((\eta + h)\tilde{v}) + \partial_t \eta = 0 \qquad (35)$$

Usually $n \ll h$; then it is safe to replace $\eta + h$ by h. Again, both wind stress and atmospheric pressure forcing are in the problem.

In Chapter 5, tidal phenomena are considered. The ordinary surface tides are basically barotropic motions. Hence, by neglecting the wind stress terms and replacing the atmospheric pressure term with an analogous term representing the tidal potential (η_F), (33) to (35) provide a simplified form of Laplace's tidal equations:

$$\partial_t \tilde{u} - f \tilde{v} = -g \, \partial_x(\eta + \eta_F) \qquad (36)$$

$$\partial_t \tilde{v} + f \tilde{u} = -g \, \partial_y(\eta + \eta_F) \qquad (37)$$

and

$$\partial_x(h \tilde{u}) + \partial_y(h \tilde{v}) + \partial_t \eta = 0 \qquad (38)$$

As a final note, all of the reductions given above were made on a heuristic basis. In a particular flow situation, where definite knowledge about scales is available, it is usually possible to proceed formally with a nondimensionalization and a scale analysis in such reductions.

SUMMARY

An introduction to the physical dynamics (physical oceanography and fluid dynamics) of the continental margin has been provided, thus acquainting the reader with some of the dynamic and geometric (geological) factors that must be borne in mind when considering the effects of continental margin circulation on sediment transport. Special emphasis has been placed on areas of interaction between geological properties of the continental margin and the margin's circulation. An attempt has been made to convey the present state of knowledge about the physical dynamics of the continental margin and the logical structure with which a fluid dynamicist/physical oceanographer addresses the problem. The amplitude, temporal, and spatial scales of motion, and their associated processes, have been identified tentatively. The concepts of transshelf and regional regimes have been introduced and explored for the purposes of localizing processes and of identifying control volumes for ecosystems and other models. The viewpoint that the physical dynamics of the margin incorporates a broad class of superposed and possibly interacting motions has been established within a general theoretical framework, from which the four classes of motion discussed in subsequent chapters have been isolated. These four classes of motion are surface gravity waves, tidal motions, responses to storms, and steady thermohaline circulation. They are four of the most important classes of motion, but they do not encompass all important classes. For example, the details of the turbulent surface and bottom boundary layers and of river plumes and oceanic fronts over the continental margin have not been dwelled upon. These latter areas have become subjects of contemporary research; in a few years more definitive knowledge may be available for review.

ACKNOWLEDGMENTS

Support for preparing this chapter was derived from National Science Foundation Grants NSF/DES 72-0147 A03 entitled "The Spatial and Temporal Structure of Physical Variability on Continental Shelves" and NSF/IDOE 75-09313 entitled "North American Participation in the Liège Colloquium on Continental Shelf Dynamics." Dr. John S. Allen, Mr. Thomas B. Curtin, and Mrs. Monica Abbott are thanked for their constructive technical and editorial criticisms.

SYMBOLS

A_H	horizontal eddy viscosity
A_V	vertical eddy viscosity
f	Coriolis parameter: $f = 2\Omega \sin \theta$
g	gravitational acceleration
h	mean water depth
K_H	horizontal eddy diffusivity of heat and salt
K_S	molecular salt diffusivity
K_T	molecular thermal conductivity
K_V	vertical eddy diffusivity of heat and salt
L	a horizontal scale length
b	buoyancy function: $b = -g\rho/\rho_0$, i.e., the effective gravitational acceleration in a density-stratified fluid

N Väisälä-Brunt frequency, or static stability:

$$N^2 = -\frac{g}{\rho_0}\left[\frac{d\rho}{dz}\bigg|_{\text{in situ}} - \frac{d\rho}{dz}\bigg|_{\text{adiabatic}}\right]$$

p fluid pressure

p_a atmospheric pressure

R Reynolds stress tensor or matrix

$\mathbf{q}_S$ salt flux vector

$\mathbf{q}_T$ heat flux vector

S salinity

T temperature

t time coordinate

$\mathfrak{J}$ averaging time

(u, v, w) fluid velocity vector; its positive direction is (east, north, up)

(x, y, z) position vector; its positive direction is (east, north, up)

η departure of free (sea) surface from mean position ($z = 0$)

θ latitude

ν molecular kinematic viscosity

ρ fluid density, i.e., a function of T, S, and p:
$$\rho = \rho(T, S, p)$$

ρ_0 time-averaged and horizontally averaged density, i.e., a function of depth (z) only:
$$\rho_0 = \rho_0(z)$$

$\boldsymbol{\tau}_w$ surface wind stress vector: $\boldsymbol{\tau}_w = (\tau_w^{(x)}, \tau_w^{(y)})$

ϕ velocity potential function

ψ stream function

Ω earth's rotation rate

DIFFERENTIAL AND INTEGRAL OPERATOR NOTATION

Note: F, a scalar, and $\mathbf{G} = (G^{(x)}, G^{(y)}, G^{(z)})$, a vector, are used as dummy functions for purposes of definition.

$(\partial_t F, \partial_x F, \partial_y F, \partial_z F)$ partial derivatives of F with respect to t, x, y, and z

$\dfrac{dF}{dt} = \partial_t F + u\,\partial_x F + v\,\partial_y F + w\,\partial_z F$ total time derivative of F

$\dfrac{DF}{Dt} = \partial_t F + \bar{u}\,\partial_x F + \bar{v}\,\partial_y F + \bar{w}\,\partial_z F$ time derivative of F with mean flow advection only

$\nabla F = (\partial_x F, \partial_y F, \partial_z F)$ three-dimensional gradient of F

$\nabla^2 F = \partial^2_{xx} F + \partial^2_{yy} F + \partial^2_{zz} F$ three-dimensional Laplacian of F

$\nabla \cdot \mathbf{G} = \partial_x G^{(x)} + \partial_y G^{(y)} + \partial_z G^{(z)}$ three-dimensional divergence of $\mathbf{G}$

$\nabla \times \mathbf{G} = (\partial_y G^{(z)} - \partial_z G^{(y)}, \partial_z G^{(x)} - \partial_x G^{(z)}, \partial_x G^{(y)} - \partial_y G^{(x)})$ three-dimensional curl of $\mathbf{G}$

$\bar{F} = \dfrac{1}{\mathfrak{J}}\displaystyle\int_0^{\mathfrak{J}} F\,dt$ time-averaged F

$F' = F - \bar{F}$ time perturbation of F

$\tilde{F} = \dfrac{1}{(\eta + h)}\displaystyle\int_{-h}^{\eta} F\,dz$ depth-averaged F

REFERENCES

Batchelor, G. K. (1967). *An Introduction to Fluid Dynamics*. London and New York: Cambridge University Press, 615 pp.

Bumpus, A. F. (1973). A description of the circulation on the continental shelf of the east coast of the United States. *Prog. Oceanogr.*, **6:** 111–156.

Cannon, G. (1972). Wind effects on currents observed in Juan de Fuca Submarine Canyon. *J. Phys. Oceanogr.*, **2**(3): 281–285.

Faller, A. J. and R. E. Kaylor (1966). A numerical study of the instability of the laminar Ekman boundary layer. *J. Atmos. Sci.*, **23**(5): 466–480.

Fofonoff, N. P. (1962). Dynamics of ocean currents. In M. N. Hill, ed. *The Sea*, Vol. 1. New York: Wiley-Interscience, pp. 323–395.

Greenspan, H. P. (1969). *The Theory of Rotating Fluids*. London and New York: Cambridge University Press, 328 pp.

Huyer, A., R. D. Pillsbury, and R. L. Smith (1975). Seasonal variation of the alongshore velocity field over the continental shelf off Oregon. *Limnol. Oceanogr.*, **20**(1): 90–95.

Huyer, A. and R. L. Smith (1974). A subsurface ribbon of cool water over the continental shelf off Oregon. *J. Phys. Oceanogr.*, **4**(3): 381–391.

Kaylor, R. and A. J. Faller (1972). Instability of the stratified Ekman boundary layer and the generation of internal waves. *J. Atmos. Sci.*, **29**(3): 497–509.

Lilly, D. K. (1966). On the instability of Ekman boundary flow. *J. Atmos. Sci.*, **23**(5): 481–494.

Niiler, P. P. (1975). A report on the continental shelf circulation and coastal upwelling. *Rev. Geophys. Space Phys.*, **13**(3): 609–614.

Orlanski, I. (1969). The influence of bottom topography on the stability of jets in a baroclinic fluid. *J. Atmos. Sci.*, **26**(6): 1216–1232.

Orlanski, I. and M. D. Cox (1973). Baroclinic instability in ocean currents. *Geophys. Fluid Dyn.*, **4**(4): 297–322.

Schmitz, W. J. (1974). Observations of low-frequency current fluctuations on the continental slope and rise near site D. *J. Mar. Res.*, **32**(2): 233–251.

Shepard, F. P. and N. F. Marshall (1973). Storm-generated current in La Jolla Submarine Canyon, California. *Mar. Geol.*, **15**(1): 19–24.

Shepard, F. P., N. F. Marshall, and P. A. McLaughlin (1974). Currents in submarine canyons. *Deep-Sea Res.*, **21**(9): 691–706.

Stefansson, U., L. P. Atkinson, and D. F. Bumpus (1971). Seasonal studies of hydrographic properties and circulation of the North Carolina shelf and slope waters. *Deep-Sea Res.*, **18**(4): 383–420.

Sturges, W. (1974). Sea level slope along continental boundaries. *J. Geophys. Res.*, **79**(6): 825–830.

Tareyev, B. A. (1971). Gradient-vorticity waves on the continental shelf. *Izv. Acad. Sci. USSR, Atmos. Oceanogr. Phys.*, **7**(4): 283–285 (English version).

Weatherly, G. L. (1972). A study of the bottom boundary layer of the Florida Current. *J. Phys. Oceanogr.*, **2**(1): 54–72.

Weatherly, G. L. (1975). A numerical study of time-dependent turbulent Ekman layers over horizontal and sloping bottoms. *J. Phys. Oceanogr.*, **5**(2): 288–299.

Wyatt, B., W. V. Burt, and J. G. Patullo (1972). Surface currents off Oregon as determined from drift bottle returns. *J. Phys. Oceanogr.*, **2**(3): 286–293.

Zimmerman, H. B. (1971). Bottom currents on the New England continental rise. *J. Geophys. Res.*, **76**(24): 5865–5876.

CHAPTER 3

Some Simple Mechanisms for Steady Shelf Circulation

ANTS LEETMAA

Atlantic Oceanographic and Meteorological Laboratories, Miami, Florida

An understanding of the mechanisms of sediment transport on the continental margins depends on a knowledge of the oceanic shelf circulations. This, in itself, is a complex phenomenon, and its study has just begun. Ideally, the geologist is generally not interested in the totality of the circulation pattern. Although sediment transport can occur at all levels, the primary interest of the geologist is in knowing the magnitude and direction of the flow close to the bottom; this can be obtained either from observations or theory. However, there is little of either. Bumpus (1973) describes what is known about the circulation on the continental shelf off the east coast of the United States. There is little theory to describe these observations. The motions appear to be complex and highly variable. Factors that determine the circulation on one shelf likely are not as important on another. Seasonal effects are dominant.

Some of the simplest models for shelf circulation are examined in this chapter. With these models, various concepts about forcing mechanisms and the nature of the dynamics can be explored. The relevance of the model for the real world depends on how well its results compare with observations. More complex models can be developed by using the simpler models whose dynamics are well understood and hopefully verified by observations, as building blocks. Realistic models ultimately permit calculating the nature of the flow close to the bottom, and determining which types of forcing are of importance for sediment transport.

The simplest models to treat theoretically are steady state ones. In this chapter attention is confined to these. Following chapters discuss wave effects, tidal flows, and other time-dependent phenomena. We start by exploring several models that differ only in their forcing mechanisms. The forces that are most important for steady motions on the shelf are thermohaline effects and wind stress. The former are caused by spatial differences in the temperature and/or salinity distributions. These produce pressure differences that drive water motions.

Results will depend strongly on the value of the frictional coefficients chosen. For very "viscous" models, the flows are little influenced by the rotation of the earth. However, for small values of viscosity, the solutions are strongly influenced by the earth's rotation. The magnitude of the viscosity is observationally extremely difficult to measure, and perhaps is best estimated from theoretical models.

THERMOHALINE FORCING

One example of thermohaline forcing is freshwater runoff from land. This produces fresher and consequently lighter water close to the coast than further offshore. This also is the dominant driving force in estuaries. To understand the general type of motion this creates, consider the following problem. We assume that the dynamics are governed by the following set of equations:

$$-f \frac{\partial v}{\partial z} = -g\beta \frac{\partial s}{\partial x} + A_v \frac{\partial^3 u}{\partial z^3} \qquad (1)$$

$$f \frac{\partial u}{\partial z} = A_v \frac{\partial^3 v}{\partial z^3} \qquad (2)$$

$$0 = \frac{\partial u}{\partial x} + \frac{\partial w}{\partial z} \qquad (3)$$

$$u \frac{\partial s}{\partial x} + w \frac{\partial s}{\partial z} = K_v \frac{\partial^2 s}{\partial z^2} \qquad (4)$$

For this right-handed coordinate system, x is perpendicular to the coast, y is parallel to it, and z is vertically upward; u, v, w are the x, y, z components of velocity, respectively; A_v and K_v are the vertical eddy mixing coefficients for momentum and salt; s is the salinity; and β is the coefficient of contraction for salt. The effects of the earth's rotation are contained in the terms fv_z and fu_z, where f is known as the Coriolis parameter and $f = 2\Omega \sin \theta$ where Ω is the angular velocity of the earth and θ is the latitude.

To arrive at these equations it is assumed that the effects of lateral mixing are small, and that the motions are slow and steady so that nonlinear effects and time dependence can be neglected. To simplify the problem further it is assumed that the motion does not vary in the direction parallel to the coast, i.e., $\partial(\ )/\partial y = 0$. The geometry is shown in Fig. 1.

At the coast there is a laterally distributed transport of fresh water (river runoff) denoted by T_R. This sets up a salinity gradient normal to the coast. What are the implications of this gradient? Equation 1 is

$$-fv_z = -g\beta s_x + A_v u_{zzz}$$

where

$$(\ )_z = \frac{\partial}{\partial z} (\ ), \qquad (\ )_{zz} = \frac{\partial^2}{\partial z^2} (\ ), \qquad \cdots$$

In an estuary where there are lateral side walls, v is either very small or absent. Similarly, on the shelf it can be shown that in shallow water, or for large values of A_v and K_v, this is also true. In such situations the term fv_z can be neglected and

$$v u_{zzz} = g\beta s_x$$

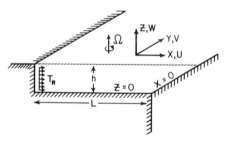

FIGURE 1. *Geometry for shelf models.*

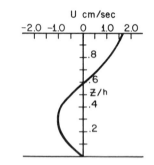

FIGURE 2. *Vertical profile of the offshore velocity for the following parameters:* $A_v = 10^2 \ sec^{-1}$; $g\beta s_x = 6.25 \times 10^{-8}$; $H = 5 \times 10 \ cm$; $T_R = 50 \ cm^2/sec$.

Situations exist on the shelf or in estuaries in which s_x is essentially independent of z. In these cases the solution for u is given by

$$u = -\frac{g\beta s_x h^3}{A_v} \left[\frac{1}{6}\left(\frac{z}{h}\right)^3 - \frac{5}{16}\left(\frac{z}{h}\right)^2 + \frac{1}{8}\left(\frac{z}{h}\right) \right] + 3\left[\frac{1}{2}\left(\frac{z}{h}\right)^2 - \frac{z}{h} \right]$$

As boundary conditions we have assumed that the stress is zero at the sea surface ($u_z = 0$ at $z = h$), the velocity is zero at the bottom ($u = 0$ at $z = 0$), and the net transport of water offshore is given by $\int_0^h u \, dz = T_R$. This solution is illustrated in Fig. 2. Although there is a net offshore transport of water, T_R, the magnitude of the flow toward the source of the river transport is much larger than T_R.

This perhaps is of importance for sediment or sewage transport onshore. Similar velocity profiles are obtained for more general and difficult problems. The circulations in estuaries are of this type. It should also be noted that the inflow velocity is at least an order of magnitude larger than the river outflow velocity T_R/h, 0.01 cm/sec. This type of problem becomes considerably more difficult when the distribution of s in x and z is solved for simultaneously.

The earth's rotation can play an important role in determining the nature of the flow. In the preceding example by the choice of geometry (in the case of estuaries) or by the assumption that the flow was very viscous, this effect was not present. If the value of the eddy coefficient is decreased or the water depth is increased, rotational effects become important. Then the solution, over most of the water column, becomes

$$u = 0, \qquad v = \left(\frac{g}{f}\beta \frac{\partial s}{\partial x}\right) z$$

Close to the surface and the bottom this solution is modified by friction. Note the differences between this solution and the preceding one. The primary flow is now parallel to the coast and northward. Only in thin regions close to the top and bottom is there on- or offshore transport (it is in these "boundary layers" that the offshore transport of fresh water occurs). The most likely direction for sediment transport is now parallel to the coast. The primary factor that determines whether this solution or the previous one applies is the value of A_v. Unfortunately this parameter is extremely difficult to measure directly, and within the range of physically possible A_v either solution can occur.

WIND FORCING

Consider a situation where a wind stress is applied at the surface, and $T_R = 0$ ($s_x = 0$). The governing equations are

$$-fv = -\frac{1}{\rho_0}p_x + A_v u_{zz}$$

$$fu = A_v v_{zz}$$

$$0 = -p_z - \rho_0 g$$

$$0 = u_x + w_z$$

where p is the pressure. At the sea surface we apply a stress $\rho_0 A_v(\partial v/\partial z) = \tau_y$. The boundary conditions are

at $z = h$, $\quad \rho_0 A_v \dfrac{\partial v}{\partial z} = \tau_y; \quad \dfrac{\partial u}{\partial z} = 0; \quad w = 0$

at $z = 0$, $\quad u = v = w = 0$

For small values of A_v the solution consists of two parts. In the interior of the fluid there is a "geostrophic" part where

$$u = 0, \quad v = \frac{1}{\rho_0 f}, \quad p_x = \frac{\tau_y}{\rho_0}\left(\frac{2}{A_v f}\right)^{1/2}$$

In addition, there are contributions that die away exponentially from the surface and bottom that match the interior solution to the boundary conditions. These are called Ekman layers. They have the property that the net transport in each layer is given by $T_{EL} = \tau_y f$, which is independent of the eddy coefficient. The velocities are of the order of a few centimeters per second for reasonable values of the wind stress.

The transport in the surface layer is to the right of the wind stress, or offshore in this example. The transport

in the lower layer is equal to that in the upper layer, but onshore. The detailed solutions for these layers are given in textbooks on dynamical oceanography.

We can understand this system of currents in the following way. When the wind begins to blow, the upper Ekman layer starts to transport water away from the coast. This lowers the sea surface next to the coast and creates a pressure gradient perpendicular to the coast. This is balanced by a flow parallel to the coast [$fv = (1/\rho_0)p_x$]. Close to the bottom, friction acting on this flow causes another Ekman layer to form with onshore transport. The pressure gradient continues to build until the transports in the upper and lower Ekman layers balance. A steady state is then obtained. This is the solution that was presented earlier.

In this problem the transports close to the bottom are again directed onshore and would support sediment transport in that direction.

The thickness (D_{EL}) of each Ekman layer is proportional to $(A_v/f)^{1/2}$, and this depends on the square root of the eddy coefficient. As the flow becomes more viscous (i.e., A_v increasing) D_{EL} increases, and the upper and lower Ekman layers merge.

When this happens the solution becomes (in the limit of $D_{EL} > h$)

$$u = 0, \quad v = \left(\frac{\tau_y}{h}\right)z$$

This is known as Couette flow. Rotational effects are no longer important and all the flow is parallel to the coast in the direction of the stress. The velocities close to the bottom are small.

In the examples presented so far, the magnitude of the eddy coefficient plays an important role in determining the nature of the solution. This is another reason for difficulty in formulating satisfactory models of shelf circulations. The criterion which determines the nature of the solution is the ratio of D_{EL}, the Ekman depth, to h, the depth of water. When D_{EL} is less than h the flows tend to be rotationally dominated. When D_{EL} is comparable to or larger than h, the effects of rotation diminish or disappear. Consequently, the nature of the solutions for a given value of A_v can depend also on the depth of the water. In deep water offshore, the solutions may be rotationally dominated, whereas inshore, where the water is shallower, they might become more Couette or estuarine in nature.

The simplest possible effects of two types of forcing in a simple model have now been briefly examined. As should be obvious the problems can become extremely complex even for this simple model, when the depth varies, when s_x is to be determined, and when rotational and viscous effects are equally important.

A MODEL OF CONTINENTAL SHELF CIRCULATION

The ideas explored in the preceding sections can be used to form a model of shelf circulations driven by freshwater runoff from land and by wind stress (Stommel and Leetmaa, 1972). As before a shelf of infinite length (y direction) and a semiinfinite width (extending from the deep ocean at $x = 0$ to negative x—infinity) is considered. The depth of the shelf is h. A mean flux of freshwater, T_R per unit length of coastline, flows toward the sea, due to the cumulative effects of river discharge along the coast. The steady wind stress components at the surface $z = h$ are τ_x and τ_y. The salinity and density are related by $\rho = \rho_0(1 + \beta s)$. Assume linear dynamics:

$$A_v \psi_{zzzz} - \beta g s_x - f v_z = 0$$

$$A_v V_{zz} + f \psi_z = 0$$

$$K_v s_{zz} + \psi_z s_x - \psi_x s_z = 0$$

where the motion is independent of y; x derivatives of diffusion terms are neglected because of the large ratio of horizontal to vertical scales; and the stream function ψ defines the velocity components $u = -\psi_z, w = \psi_x$. The boundary conditions in z are that

at $z = 0$, $\quad \psi = \psi_z = s_z = v = 0$

at $z = h$, $\quad \psi = -T_R; \quad -\rho_0 A_v \psi_{zz} = \tau_x;$

$$\rho_0 A_v V_z = \tau_y; \quad s_z = 0$$

This problem models wintertime conditions on the east coast continental shelf of North America. Winter is attractive because (1) density is primarily controlled by the salinity distribution and (2) the weak vertical density gradient in winter permits a simplification in the treatment of the third equation above, which is nonlinear. Even for this simple model, a complete solution is difficult. Instead the model is used to estimate the natural horizontal scale length $L = s_0 \nabla x / \nabla s$ where ∇x is the observed width of the shelf and ∇s is the decrease in salinity over that distance from the ocean value, s_0. The details of the solution are given by Stommel and Leetmaa (1972). The solution for L for various values of the wind stress and the eddy coefficient A_v is shown in Fig. 3, where E (the Ekman number) is equal to A_v/fh^2.

For these solutions, values of the parameters which are appropriate for the eastern U.S. continental shelf from Nantucket Shoals to Cape Hatteras have been chosen: $f = 0.7 \times 10^{-4}$ sec^{-1}, $h = 5 \times 10^3$ cm, $\beta g s_0 = 30$ cm/sec^2, $T_R = 50$ cm^2/sec, $A_v/K_v = 1$.

The upper curves of the diagram correspond to the purely wind-driven regime. The lowest curve, which is convex upward, is the pure density-driven model. For

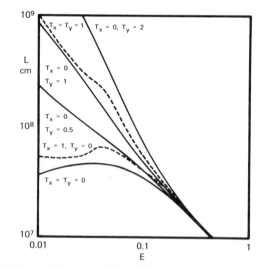

FIGURE 3. *Solutions for* L *for different values of the wind stress and the Ekman number* ($E = A_v/fH^2$).

large values of the mixing coefficient all curves coalesce and the motion is basically density driven and of the estuarine nature that was described earlier.

For small values of the vertical mixing coefficient, with y stress predominant, the Ekman transports which convect salt onshore and offshore are independent of A_v (as was pointed out earlier). However, vertical mixing, K_v, "short-circuits" these transports. This is proportional to A_v since we assumed that the Prandtl number, A_v/K_v, was unity. Thus small values of A_v correspond to small mixing between the upper and lower Ekman layers and large penetration of salt occurs (i.e., large L). When there is no applied wind stress, the Ekman transports are driven by stress associated with the shear produced by the horizontal salinity gradient, i.e., $A_v v_z = (A_v g \beta / f) s_x$. This diminishes as A_v becomes smaller, and despite a partial compensation because k is also smaller, the salt penetration diminishes. This accounts for the different behavior for L (A_v) for density as compared to wind forcing.

For large values of mixing, as pointed out earlier, the dynamics of the flow become nonrotational and also vertical mixing is enhanced. Thus all the curves coalesce in the purely salinity-driven case.

Attempts to compare this theory with the observations are difficult because a priori the appropriate values of A_v and k are unknown. It is assumed that $A_v/K_v = 1$. The observations in this area indicate that L is about 3.2×10^8 cm. Wintertime mean wind stress in this area can be estimated from Hellermann's (1967) world charts. These indicate that the magnitude of the x and y components of stress is about 1 dyne/cm^2. Thus with $\tau_x = \tau_y = 1$ and $L = 3.2 \times 10^8$, Fig. 3 indicates that A_v is about 37 cm^2/sec. This is consistent with the observations. It should also be noted that values of L as large as those observed imply that the shelf circulation, at least for this model, is basically wind-driven. As another test of the model the difference between the salinity at the top and bottom can be computed. This turns out to be 0.14‰. Again this is of the right order of magnitude according to the observations.

Despite these limited successes of the model, there is a serious discrepancy between the predicted v component of velocity and the observed value. All the observations indicate a negative v velocity of an order of 5 cm/sec. The theoretical v component is positive, about 20 cm/sec. If the observations are correct, this indicates that this simple model is not adequate to describe the observed shelf circulation.

There is some observational evidence to indicate that there is a northward rise in sea level along the coast (Sturges, 1974). If this feature is introduced into the model, the discrepancy in the direction of flow parallel to the coast can be resolved (Stommel and Leetmaa, 1972). However, the observations are not conclusive on this point. The theoretical flow close to the bottom then is in the right direction and is on the order of a centimeter or two per second.

CONCLUSIONS

As contradictions occur between model results and observations, more details can be added to the models. At some point, however, the question has to be asked as to how applicable steady state models are to shelf circulations and in particular to sediment transport. Examination of daily wind records at Nantucket Shoals light vessel shows that the wintertime root-mean-square wind stress is 5 to 10 times larger than the mean. Thus the transient fluxes are possibly an order of magnitude larger than the mean ones. For sediment transport this could be the dominant factor since the steady models give rather low near-bottom velocities.

Better observations are needed to indicate the direction that modeling should go. Long-term series of current and density measurements are needed to obtain an observational verification of the mean fields and their vertical structure. Time series of currents and density as functions of depth are needed; without these the more complex transient theories of shelf circulations cannot be adequately attained. Finally, for the results of the physical oceanographer to be of relevance to those interested in sediment transport we need to know whether the means or the transients are important in sediment transport.

In this chapter the reader is introduced to some of the problems facing a shelf modeler. Other more compli-

cated models exist that were not examined. Two of these are the models by Csanady (1974) and Pietrafesa (1973). Csanady discussed the barotropic (depth independent) response of a shelf to an imposed wind stress or external pressure gradient. Pietrafesa considers a steady state, nonlinear, wind-driven model of an eastern meridional coastal circulation. Both are considerably more complex analyses than the one presented here. A more extensive list of references can also be found in them.

SUMMARY

This chapter provides an introduction to steady state models of the oceanic circulation on the continental margin. Horizontal salinity gradients comprise a major forcing mechanism for shelf circulation. A gradient of seaward-increasing salinity will result in a seaward net transport of surface water and a larger landward net transport of bottom water, if the flow is relatively viscous (low values for the eddy coefficient and depth). With decreasing viscosity, the earth's rotation plays an increasingly important role in determining the nature of the flow. The primary flow tends to parallel the coast, while onshore and offshore transport is confined to the surface and bottom layers.

Wind forcing is effected by the application of wind stress to the sea surface. For small values of the eddy coefficient, the solution for the horizontal components of motion occurs in two parts. There is a coast-parallel "geostrophic" component of flow in the interior of the fluid. Additional flow components are experienced at the upper and lower boundaries, which die away exponentially toward the interior of the flow. Net transport in these *Ekman layers* is given by $T_{EL} = \tau_y f$, where τ_y is the component of shear stress parallel to the coast and f is the Coriolis parameter. Net transport is independent of the eddy coefficient. The thickness of each Ekman layer is proportional to the square root of the eddy coefficient.

As wind-driven flow becomes more viscous (because of increasing eddy coefficient or decreasing depth), the upper and lower Ekman layers merge and the vertical velocity gradient becomes linear in nature (Couette flow). Rotational effects are no longer important, and all flow is parallel to the coast in the direction of stress.

These relationships may be combined into a single steady state model for shelf circulation. When applied to the Middle Atlantic Bight of North America, the model predicts an eddy coefficient of 37 cm²/sec, and for the observed horizontal length scale, a primarily wind-driven circulation. However, it is necessary to postulate a northward rise in sea level along the coast, in order for the model to predict net flow to the south, as observed.

SYMBOLS

A_v	vertical eddy mixing coefficient for momentum
D_{EL}	thickness of the Ekman layer
E	Ekman number
f	Coriolis parameter: $f = 2\Omega \sin \theta$
h	depth of water
K_v	vertical eddy mixing coefficient for salt
L	natural horizontal length scale
p	pressure
s	salinity
T_R	river transport
u	x component of velocity
v	y component of velocity
w	z component of velocity
x	horizontal distance from origin perpendicular to coast
y	horizontal distance from origin parallel to coast
z	vertical distance upward from origin
β	coefficient of contraction for salt
θ	latitude
τ	shear stress
ρ	density
ψ	stream function
Ω	angular velocity of the earth

REFERENCES

Bumpus, D. F. (1973). A description of the circulation on the continental shelf of the East Coast of the United States. *Prog. Oceanogr.*, **6:** 111–158.

Csanady, G. T. (1974). Barotropic currents over the continental shelf. *J. Phys. Oceanogr.* 4(3): 357–371.

Hellermann, S. (1967). An update estimate of the wind stress on the world ocean. *Mon. Weather Rev.*, **95:** 607–626.

Pietrafesa, L. J. (1973). Steady baroclinic circulation on a continental shelf. Ph.D. Dissertation, Dept. of Oceanography and Geophysics Group, University of Washington, Seattle.

Stommel, H. and A. Leetmaa (1972). Circulation on the continental shelf. *Proc. Natl. Acad. Sci. U.S.A.*, **69**(11): 3380–3384.

Sturges, W. (1974). Sea level slope along continental boundaries. *J. Geophys. Res.*, **79**(6): 825–830.

CHAPTER 4

Wind-Driven Currents on the Continental Margin

CHRISTOPHER N. K. MOOERS

Rosenstiel School of Marine and Atmospheric Science, University of Miami, Miami, Florida

Currents on the continental margin are strongly influenced by meteorological forces. There are several components to the meterological forcing: momentum transfer through the wind stress; wind stirring of the water column; pressure adjustments; heat transfer through sensible, radiative, and evaporative heat fluxes; and mass (water) transfer through evaporation and precipitation. These forces operate on a host of space and time scales through various meteorological phenomena: for example, diurnal sea-land breeze systems; extratropical cyclones and anticyclones; tropical cyclones; midlatitude warm and cold fronts; and seasonal variations in intensity and location of air mass systems and their associated large-scale pressure and, thus, wind systems. In this chapter the emphasis is on transient circulations driven by the winds. In Chapter 3, the Ekman circulation produced by steady winds was noted.

The transient wind forcing is provided by atmospheric systems with an element of order and structure, having characteristic forms, intensities, and scales. For propagating disturbances, time scales imply space scales. Hence, a particular system or a set of systems of the same class can be considered. Some systems are "rare events," e.g., hurricanes, in the sense that the oceanic response to the event is perhaps completed before the next event occurs. In that case, the oceanic response to

an initial disturbance can be studied. The response will include wavelike "signatures" in physical fields. Such response problems are called initial value or adjustment problems. Other systems are recurrent or quasi periodic, e.g., atmospheric cyclones and cold fronts. In these cases, the oceanic response to a particular disturbance may be superimposed on that caused by preceding disturbances. Because of their quasi-periodic nature, though they may have random phase and amplitude distributions, the meteorological forces can be transformed into Fourier space, i.e., frequency and horizontal wave number space. The advantages of the latter are that (1) considerable knowledge exists about possible classes of wavelike motion which have natural representations in Fourier space; (2) possible near resonances can be identified; and (3) once the response function has been found, it is known once and for all and can be applied to any particular meteorological spectrum. The first viewpoint generally leads to Laplace transforms and the second to Fourier transforms. When applied to the same problem, the results are identical. From both viewpoints, consideration of the boundary conditions in a coastal region is an essential part of the problem, a part that makes the continental margin unique when compared with the open ocean. Of course, the geometric configuration of the continental margin and the properties of the water mass are vital ingredients too. The question of wind-driven currents on the continental margins has been placed in the following perspective: The currents realized on a continental margin are the response of the hydrodynamic system to at-

Contribution from the Rosenstiel School of Marine and Atmospheric Science, University of Miami, Miami, Florida 33149.

mospheric systems. Then it is immediately clear that something must be known of the continental margin's hydrodynamic system, as well as of the atmospheric systems providing the forcing.

In Chapter 2, the horizontal length scales (the Rossby radii of deformation) provided by the combined effects of density stratification and the earth's rotation were noted. Typical midlatitude values for the barotropic Rossby radius (L_0) and the baroclinic Rossby radii (L_n) were given as $L_0 = 10^3$ km and $L_n = 30/n$ km ($n = 1, 2, \ldots$). It was noted there that the ratio of the continental margin width to the L_n's could play a crucial role in determining the margin's response to meteorological forcing. Since margin widths vary from 50 to 500 km, a host of possibilities exists.

The horizontal scales of the meteorological forcing are also crucial. These scales vary over a broad range; for example, (1) several tens of kilometers for land-sea breeze systems; (2) a few hundred kilometers for mesoscale disturbances, e.g., atmospheric cold fronts; and (3) a few thousand kilometers for synoptic scale phenomena, e.g., midlatitude high-pressure cells. The corresponding time scales are (1) a few hours, (2) a few days, and (3) several months. The geological influence of the various scales of meteorological forces can be complex and subtle; for example, the circulation response to the vigorous small-scale phenomena can be effective for setting sediments into motion, whereas the response to the more sluggish large-scale phenomena is most important for net drift displacements.

Because of the proximity of the coastal boundary and the sea bottom, horizontal and vertical frictional effects can be expected to be important on the continental margin. Commonly used constructs for parametrizing turbulent dissipation are the eddy viscosity coefficients: A_H for the horizontal and A_V for the vertical. In a density-homogeneous fluid without bottom topography, a so-called horizontal Ekman layer will enter, whose width is given by $\delta_H = (A_H/f)^{1/2}$. Similarly, vertical Ekman layers may be found near the surface and bottom; if so, their thickness is given by $\delta_V = (A_V/f)^{1/2}$. Typical values used for A_H and A_V yield $\delta_H \sim 1$ to 10 km and $\delta_V \sim 3$ to 30 m in midlatitudes. The best that can be said for such estimates is that they seem reasonable when compared with observed circulation. Actually, the effects of density stratification and bottom topography provide a rich boundary layer structure, and the various effects usually enter in a combined fashion (Pedlosky, 1974a–c).

Before proceeding to specifics, it must be appreciated that no fully comprehensive, theoretical description yet exists that can take into account adequately all the features noted above. Though there are dubious ele-

ments in all treatments made, the general approach is to isolate a few specific phenomena for analysis.

LONG FREE WAVES

The time-dependent motions on a shelf may be locally driven (near-field case) by meteorological forces, by disturbances in an offshore current, or by fluctuations in river discharges; or they may be driven by disturbances generated at a distant source (far-field case) in the deep sea or at some other location along the coast. The near-field case encompasses the forced wave problem; the far-field case encompasses the free wave problem. Since the solution to the forced wave problem can be synthesized from solutions to the free wave problem, the free waves are considered first.

Attention is restricted to long waves, i.e., waves with space and time scales such that the effects of the earth's rotation play an essential role. In other words, pure gravity waves, including the coastally trapped Stokes edge waves, are neglected. Unlike the long waves, the Stokes edge waves are only weakly dependent on the Coriolis parameter and they can travel in either direction along a coast. Since they are forced by meteorological disturbances in a somewhat different manner than the long waves, they are not discussed further here. The several classes of free waves that can exist in a coastal region are progressively developed below.

Barotropic Waves

From (33), (34), and (35) of Chapter 2, the equations governing free (i.e., unforced) barotropic motions are in fully linearized form:

$$\eth_t u - fv = -g\,\eth_x\eta \tag{1}$$

$$\eth_t v + fu = -g\,\eth_y\eta \tag{2}$$

and

$$\eth_x(hu) + \eth_y(hv) + \eth_t\eta = 0 \tag{3}$$

where the tildes (for depth-averaged velocities) have been dropped for convenience. Multiplying (1) and (2) by h and solving for u and v in terms of η, equations for u and v are obtained:

$$D(hu) = -gh[\eth^2_{xt}\eta + f\,\eth_y\eta] \tag{4}$$

and

$$D(hv) = -gh[\eth^2_{yt}\eta - f\,\eth_x\eta] \tag{5}$$

where

$$D(\ ) \equiv \eth^2_t(\ ) + f^2(\ )$$

Then, by further cross-differentiation and substitution of (4) and (5) into (3), the governing equation for η is found:

$$\partial_t \left[\nabla^2_H \eta + \frac{\nabla_H h \times \nabla \eta}{h} \right] + \frac{f(\nabla_H h \times \nabla \eta)}{h} \cdot \hat{k} - \frac{\partial_t D(\eta)}{gh} = 0 \tag{6}$$

where $\hat{k}$ is the vertical unit vector. The solutions for the case of a long coastline located at $x = 0$ (x positive eastward) with no bathymetric variations in the y direction (y positive northward) are considered. Progressive wave solutions of the form $\eta = Z(x) \exp(i(\sigma t - ly))$ are admissible for (6), where the governing equation for Z is

$$Z'' + \frac{h'}{h} Z' + \left(\frac{-fh'}{\sigma h} l - l^2 + \frac{\sigma^2 - f^2}{gh} \right) Z = 0 \tag{7}$$

with $Z' = dZ/dx$, $h' = dh/dx$, σ is the wave frequency, and l is the y wave number. (For $\sigma > 0$, $l > 0$ implies northward phase progression.) Since the component of transport normal to the coastline must vanish, i.e., $hu = 0$ at $x = 0$, either $u = 0$ or $h = 0$ at $x = 0$. In the former case, it follows from (4) that

$$Z' - \frac{fl}{\sigma} Z = 0 \qquad \text{at} \qquad x = 0 \tag{8}$$

In the latter case, it follows from (3) that

$$h'Z' + \left(\frac{\sigma^2 - f^2}{g} - \frac{f}{\sigma} lh' \right) Z = 0 \qquad \text{at} \quad x = 0 \tag{9}$$

The boundary condition as $|x| \to \infty$ depends on whether coastally trapped or freely propagating waves are being treated.

The simplest examples involve cases of uniform depth, i.e., $h' = 0$. For these cases, (7) reduces to

$$Z'' + \left(-l^2 + \frac{\sigma^2 - f^2}{gh} \right) Z = 0 \tag{10}$$

First, the case of trapped waves, i.e., $Z \to 0$ as $|x| \to \infty$, is examined. With $k > 0$, the solution must be of the form $Z = Ae^{kx}$ as $|x| \to \infty$ for a west coast ($x \le 0$), and of the form $Z = Ae^{-kx}$ for an east coast ($x \ge 0$). Furthermore, (8) requires that $k = |f|l/\sigma$ for a west coast and $k = -|f|l/\sigma$ for an east coast of a land mass. Thus, in the northern ($f > 0$) hemisphere these waves can only propagate northward on a west coast and southward on an east coast. Similarly, in the southern ($f < 0$) hemisphere these waves can only propagate southward for a west coast and northward for an east coast. Since (10) requires that the propagation equation be

$$-k^2 + l^2 = \frac{\sigma^2 - f^2}{gh}$$

then $|l| = \sigma/c$, where $c = (gh)^{1/2}$ is the long gravity wave phase speed (~ 200 m/sec). Finally, $k = |f|/c = L_0^{-1}$; thus, the influence of these waves (barotropic Kelvin waves) is "confined" to a distance from the coast of L_0 (~ 1000 km). Also, $u = 0$ for *all* x and $v = (g/c) Z = (g/h)^{1/2} Z$. Thus, for a 20 cm sea-level perturbation at the coast, there would exist a 1 cm/sec flow parallel and adjacent to the coast. Also, the Kelvin waves are nondispersive, i.e., their phase speeds are independent of frequency, and they can exist for all frequencies.

Second, the case of freely propagating plane waves (called barotropic Poincaré waves) incident upon a coastal boundary at an arbitrary angle is examined, i.e.,

$$\eta_I = A_I \exp(i(\sigma t - kx - ly))$$

where $k^2 + l^2 = (\sigma^2 - f^2)/c^2$ is the propagation equation obtained from (10), k and l are real numbers, and A_I is the incident amplitude. As a consequence of the propagation equation, the Poincaré waves are limited to frequencies greater than the inertial, i.e., $\sigma > |f|$, and the inertial frequency serves as a low-frequency cutoff. A reflected wave of the form $\eta_R = A_R \exp(i(\sigma t + kx - ly))$ is generated at the coastal boundary. The boundary condition at the coast can only be satisfied by $\eta = \eta_R + \eta_I$; thus, (8) yields

$$-A_I \left(ik + \frac{fl}{\sigma} \right) + A_R \left(ik - \frac{fl}{\sigma} \right) = 0$$

or

$$A_R = -A_I \exp(i\theta)$$

where $\theta = 2 \tan^{-1}(\sigma k/fl)$. Altogether,

$$\eta = A_I \exp(i(\sigma t - ly))[\exp(-ikx) - \exp(i(kx + \theta))]$$

In other words, a phase shift θ is induced upon reflection. To examine the phase shift further, let ϕ equal the angle of incidence, measured with respect to the normal to the coastal boundary; then $k = K \cos \phi$ and $l = K \sin \phi$, where $K = (\sigma^2 - f^2)^{1/2}/c$. It follows that $\theta = 2 \tan^{-1}((\sigma/f) \cot \phi)$. For grazing incidence ($\phi = \pm\pi/2$), $\theta = 0$; for normal incidence ($\phi = 0$), $\theta = \pi$. For other angles of incidence, θ depends strongly on σ/f. A corollary of the above is that the interference pattern established by the reflected and incident wave patterns depends strongly on the angle of incidence and wave frequency. Poincaré waves can travel in either direction along a coast. They do carry energy in from the deep sea and carry it back after reflection. Thus, in a complete problem the sources and sinks of energy in the deep sea must be specified. For that matter, the Kelvin waves also carry energy from a distant coastal source to a distant coastal sink; such sources and sinks must also be specified in a complete problem.

For the case of stepped topography, e.g., $h = h_1$, $|x_1| \geq |x| \geq 0$; $h = h_2$, $|x| > |x_1|$; and $h_2 > h_1$, more general classes of coastally trapped waves are found. In the deep sea ($|x| > |x_1|$), the waves decay exponentially seaward and are Kelvin-like. On the continental margin ($|x| < |x_1|$), the waves can be either Kelvin-like or Poincaré-like, since the step in topography serves to refract or trap the Poincaré waves. Altogether, trapped Poincaré waves and the Kelvin waves emerge from such an analysis; in addition, a new wave mode, a so-called quasi-geostrophic mode, arises (Larsen, 1969). It depends for its existence only on the proportional discontinuity in depth, $\Delta h/h_2 = (h_2 - h_1)/h_2$, and the earth's rotation, f. All the trapped wave modes described are constrained to propagate in one direction as noted for the pure Kelvin waves. Because the imposed depth variation is discontinuous, only one such quasi-geostrophic mode occurs.

To examine this phenomenon under more realistic bathymetric conditions, the case of continuous bottom topography is considered next. To simplify the analysis of (10), the parameters of the problem are nondimensionalized in the following fashion:

$$s \equiv \sigma f^{-1}, \quad x' \equiv x L_c^{-1}, \quad \delta \equiv l L_c, \quad \text{and} \quad h' \equiv h h_0^{-1}$$

where L_c is the width of the zone of bathymetric variation and h_0 is a measure of the depth variation (bathymetric relief). Then (7) becomes

$$Z'' + \frac{h'}{h} Z' + \left[-\frac{\delta}{s} \frac{h'}{h} - \delta^2 + \epsilon \left(\frac{s^2 - 1}{h} \right) \right] Z = 0 \tag{11}$$

where $(\ )' = d/dx\ (\ )$ and $\epsilon \equiv f^2 L^2/gh_0$. (The original primes on x' and h' have been dropped.) Since $L_c \sim 100$ km and $h_0 \sim 4$ km, $\epsilon \sim 10^{-3}$; hence, there will exist solutions for which the term involving ϵ can be neglected in the first approximation. This approximation is called the nondivergent approximation; it "filters out" the Kelvin and Poincaré waves, i.e., all of the inertio-gravity waves, from the analysis. The approximation is also equivalent to writing (3) as

$$\partial_x(hu) + \partial_y(hv) = 0$$

which allows introduction of a stream function ψ, such that $\partial_y\psi = hu$ and $\partial_x\psi = -hv$. Then cross-differentiating (1) and (2) to eliminate η, and substituting for u and v in terms of $\psi = C(x) \exp(i(\sigma t - \delta y))$,

$$C'' - \frac{h'}{h} C' + \left[-\frac{h'}{h} \frac{\delta}{s} - \delta^2 \right] C = 0 \tag{12}$$

which is similar in form to (11). With (2), Z can be found from C:

$$Z = -\frac{1}{gh} \left[\frac{\sigma}{l} C' + fC \right]$$

The coastal boundary condition is simply that ψ vanishes, i.e., $C = 0$ at $x = 0$. Seaward of $|x| = 1$, the depth is assumed uniform. Hence, for $|x| > 1$, C must satisfy

$$C'' - \delta^2 C = 0 \tag{13}$$

For trapped waves, C must tend to zero as $|x| \to \infty$. At $|x| = 1$, C and C' must be continuous. In deep water, $C = C_2 = Ae^{x\delta}$; for a west coast $\delta > 0$ and for an east coast $\delta < 0$. Hence, these waves, like Kelvin waves, are constrained to propagate northward on the west coast and southward on the east coast in the northern hemisphere.

An exponential form for the bottom topography is chosen to facilitate the analysis. A west coast ($x \leq 0$) case is selected. Let $h = e^{-mx}$, $0 \geq x \geq -1$, where m is the proportional change in depth. (Then $h = e^m$ in the deep sea, $x \leq -1$.) Over the continental margin, (12) becomes

$$C'' + mC' + \left[m\frac{\delta}{s} - \delta^2 \right] C = 0 \tag{14}$$

Solutions to (14) of the form $C = e^{\alpha x} \sin(qx)$ exist, where $\alpha = -m/2$ and

$$q = \left[m\frac{\delta}{s} - \left(\delta^2 + \frac{m^2}{4} \right) \right]^{1/2} \tag{15}$$

From (15), qualitative information about δ can be obtained. For example, it is necessary that $m(\delta/s) \geq \delta^2 + m^2/4$ for q to be real, or

$$s \leq \frac{m\delta}{\delta^2 + m^2/4} \equiv s_u$$

where s_u is an upper bound on s. It is clear that s tends to zero if (1) m tends to zero (no bottom slope), (2) δ tends to zero (infinitely long waves), or (3) δ tends to infinity (infinitesimally short waves). From (2) and (3), there must exist a maximum value of s as a function of δ: max (s_u) occurs for $\delta = m/2$ and equals one. Thus, the inertial frequency is an upper bound on the frequency of these waves, a result that also follows for completely general bottom topographic variation. From the boundary conditions at $x = -1$, the dispersion relation is obtained:

$$\tan(q) = \frac{q}{(\delta + m/2)} \tag{16}$$

where δ can be evaluated in terms of q from (15). The dispersion relation yields a countable infinity of roots for q, the $\{q_n\}$, each of which corresponds to a distinct horizontal mode. Once q is determined from (16), then δ is found; i.e.,

$$\delta = \frac{m}{2s} \pm \left[\left(\frac{m}{2s} \right)^2 - \left(q_n{}^2 + \frac{m^2}{4} \right) \right]^{1/2}$$

Hence, for every permissible q_n, there are two values of δ for each frequency s such that

$$s < \frac{m^2}{(2q_n)^2 + m^2} \equiv s_{un}$$

the high-frequency cutoff for the nth mode. Thus, as q_n increases, s_{un} decreases. Similarly, as q_n increases, the value of δ at s_{un} increases (because s_{un} decreases). All of the qualitative features of the dispersion diagram have been captured without solving the problem quantitatively. A sketch of the dispersion diagram is given in Fig. 1.

These waves, commonly called continental shelf waves or topographic Rossby waves, are highly dispersive; i.e., their phase speed ($c = s/\delta$, the speed of phase propagation) is a function of frequency. Furthermore, the group speed ($c_g = \partial s/\partial \delta$, the speed of energy propagation) varies with frequency, too. In fact, at s_{un}, $c_g = 0$ for the nth mode; i.e., energy will not propagate alongshore to the present order of calculation. As a corollary, for $\delta < \delta(s_{un})$ where $\delta(s_{un})$ is the value of δ at $s = s_{un}$, the group and phase velocities have the same direction; conversely, for $\delta > \delta(s_{un})$, they have opposite directions. Typical phase speeds for the long-wave limit

(δ small) are of the order of 1 m/sec for the first (fastest) mode. Hence, these waves propagate more slowly than the common Kelvin waves; consequently, they have much shorter wavelengths for the same frequency.

These waves are vorticity waves. Thus, they appear as propagating circulation cells which look like elongated eddies if ψ is plotted. A perspective diagram of the first mode alongshore velocity field, V_1, is shown in Fig. 2 (Cutchin and Smith, 1973). Note the reversal in V_1 over the continental slope and that V_1 decays rapidly offshore. For the nth mode, V_n has n reversals over the continental margin. Also of significance to sediment transport concerns is the fact that the transisobath shelf wave flow, U_n, vanishes only at its nodal points (antinodes of V_n).

Baroclinic Waves

Since baroclinic waves involve the effects of density stratification, which is usually continuously distributed, a continuous model is considered. Alternatively, a two-layered or n-layered model of constant but different density in each layer could be considered. Such layered models can capture much of the essential physics of a problem, though it can often be difficult to relate them to the actual continental margin or ocean.

From the system of equations (18) to (22) of Chapter 2, the equations governing forced motions are established for discussion below. Through cross-differentiation of

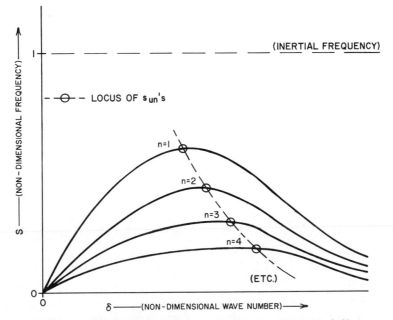

FIGURE 1. *Dispersion diagram for barotropic continental shelf waves. Key:* s_{un} *is high-frequency cutoff for the nth mode.*

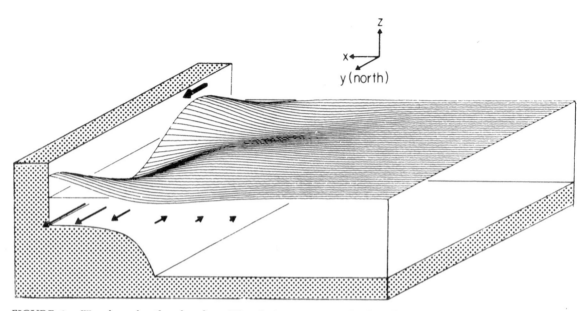

FIGURE 2. *Waveform for the alongshore* (V) *velocity component of a first (horizontal) mode barotropic continental shelf wave. From Cutchin and Smith (1973).*

(18) and (19) and of (20) and (22) of Chapter 2, the equations for the velocity components are obtained:

$$\rho_0(\partial^2_{tt}u + f^2u) = -(\partial^2_{xt}p + f\,\partial_y p) + \partial^2_{zt}\tau^{(x)}$$
$$+ f\,\partial_z\tau^{(y)} \tag{17}$$

$$\rho_0(\partial^2_{tt}v + f^2v) = -(\partial^2_{yt}p - f\,\partial_x p) + \partial^2_{zt}\tau^{(y)}$$
$$- f\,\partial_z\tau^{(x)} \tag{18}$$

$$\rho_0(\partial^2_{tt}w + N^2w) = -\partial^2_{zt}p + \rho_0\,\partial_z B \tag{19}$$

After τ is specified and p has been determined, u, v, and w can be found from (17) to (19). Through further cross-differentiation and substitution of (17) to (19) above into (21) of Chapter 2, the governing equation for pressure is obtained:

$$\partial_t\{(\partial^2_{tt}(\ ) + N^2)(\partial^2_{xx}p + \partial^2_{yy}p) + (\partial^2_{tt}(\ ) + f^2)\partial^2_{zz}p\}$$
$$= (\partial^2_{tt}(\ ) + N^2)\,\partial_z\{\partial_t\nabla\cdot\tau + f(\nabla x\tau)\cdot\hat{k}\}$$
$$+ (\partial^2_{tt}(\ ) + f^2)\rho_0\,\partial^2_{zz}B \tag{20}$$

where it has been assumed that N and f are constant, and $\hat{k}$ is again the vertical unit vector. The stress (τ) is ultimately provided by the surface wind stress. The solutions depend on the first vertical derivative of the curl and the divergence of τ and the second vertical derivative of the buoyancy flux; the curl of τ tends to be dominant for time scales much greater than the inertial period, i.e., f^{-1}. Upon application of the kinematic boundary condition at a coastline, say $u = 0$ at $x = 0$, τ itself is included in the forcing of the problem; cf. (17). Similarly, upon application of the dynamic boundary condition at the sea surface (i.e., $dp/dt = 0$, or in

linearized form, $\partial_t p_a + \rho_0 wg = 0$ at $z = 0$), the atmospheric pressure p_a is included in the forcing of the problem.

For free waves, the right-hand side of (20) vanishes. Oscillatory solutions of the form $\exp(i\sigma t)$ are admitted; then (20) reduces to

$$(N^2 - \sigma^2)(\partial^2_{xx}p + \partial^2_{yy}p) - (\sigma^2 - f^2)\,\partial^2_{zz}p = 0 \tag{21}$$

For a flat ocean bottom ($z = -h$), p can be decomposed into vertical modes; i.e.,

$$p = \sum_{n=0}^{\infty} \phi_n(z)Q_n(x, y)$$

where the $\{\phi_n\}$ each satisfy

$$\frac{d^2}{dz^2}\phi_n + \gamma^2_n(N^2 - \sigma^2)\phi_n = 0 \tag{22}$$

γ_n is the eigenvalue of the nth mode, and the associated boundary conditions are satisfied: $(d/dz)\phi_n = 0$ at $z = -h$ (because the vertical velocity w vanishes at the sea bottom) and $(d/dz)\phi_n + \gamma_n^2 g\phi_n = 0$ at $z = 0$ [because the pressure is continuous at the free (sea) surface]. The zeroth mode corresponds to the barotropic mode. For $n \geq 1$, the results can be closely approximated with a rigid sea surface. If N^2 is constant, the ϕ_n are trigonometric and $\gamma_n = (n\pi/h)/(N^2 - \sigma^2)^{1/2}$. Then the horizontal structure function Q_n satisfies

$$\partial^2_{xx}Q_n + \partial^2_{yy}Q_n + (\sigma^2 - f^2)\gamma^2_n Q_n = 0 \tag{23}$$

Each vertical mode admits a Kelvin wave solution. Analogous to the barotropic case, let $Q_n = \exp(k_n x)$

$\exp(-il_ny)$ on a west coast. Then, with $u = 0$ at $x = 0$, the unforced version of (17) requires that

$$k_n = \frac{fl_n}{\sigma} \qquad (24)$$

in the northern hemisphere. Then (23) requires that

$$l_n = \sigma\gamma_n \qquad (25)$$

For $n \geq 1$, a rigid sea surface, and N constant, (25) and (24) imply

$$l_n = \sigma\left(\frac{n\pi}{h}\right)\Big/(N^2 - \sigma^2)^{1/2}, \quad k_n = f\left(\frac{n\pi}{h}\right)\Big/(N^2 - \sigma^2)^{1/2}$$

For low frequencies ($\sigma^2 \ll N^2$), $k_n^{-1} = L_n$, the nth mode Rossby baroclinic radius of deformation, which decreases as n increases. For $n \geq 1$, these waves are called internal Kelvin waves; they can exist for all frequencies such that $\sigma < N$.

Similarly, each vertical mode admits a Poincaré solution. Analogous to the barotropic case, let $Q_n = \exp(-i(k_nx + l_ny))$. Then (23) requires that

$$(k^2_n + l^2_n) = (\sigma^2 - f^2)\gamma^2_n \qquad (26)$$

Again in the special case of $n \geq 1$, a rigid sea surface, and N constant, (26) implies

$$K_n \equiv (k^2_n + l^2_n)^{1/2} = \left(\frac{n\pi}{h}\right)(\sigma^2 - f^2)^{1/2}/(N^2 - \sigma^2)^{1/2}$$
$$(27)$$

where K_n is the magnitude of the horizontal wave number. These waves are called internal Poincaré waves; they can exist for all frequencies such that $f < \sigma < N$.

The existence of internal Kelvin and Poincaré waves vastly increases the species, and the vertical structure, of wave motions possible in the ocean and on the continental margin. These internal modes have very slow phase speeds, of the order of 1 m/sec or less. Such phase speeds are of the order of "mean" current velocities. Thus, these waves should interact with currents. When variable bottom topography is included, a rich set of reflection and refraction phenomena is encountered, a topic that is too technical to pursue here. Coastal trapping for internal modes over variable topography, analogous to the barotropic continental shelf waves, should be expected on general grounds. Because the baroclinic Kelvin waves and the barotropic continental shelf wave phase speeds are very similar, near-resonant modes have been found by Wang (1975) and Allen (1975) in two-layered models. The details are not discussed here but the crucial parameter does turn out to be the ratio of L_1 to the margin width. The dispersion

properties of the near-resonant modes are such that the coastal response to meteorological forcing may be profoundly influenced. Under near-resonant conditions, the barotropic continental shelf wave and internal Kelvin wave modes are mixed in such a way that they destructively interfere in the upper layer and constructively interfere in the lower layer, giving rise to alongshore motion trapped near the bottom over the outer shelf and the upper continental slope.

FORCED WAVES

Consideration of free waves is only valid at locations, or times, far removed from the source, or time of occurrence of a disturbance. To provide some appreciation for forced waves, a specific problem is considered. The forcing and geometry (west coast, northern hemisphere) of the problem are depicted in Fig. 3. A steady pattern of alternating, periodic (time scale assumed greater than an inertial period) atmospheric cyclones and anticyclones is hypothesized to propagate eastward in a zonal band. These "storms" strike a straight meridional coastline at normal incidence. The ocean is considered density homogeneous; thus, only barotropic waves are admitted. Below the surface Ekman layer, the ocean is treated as inviscid. The nondivergent approximation is made to eliminate Kelvin and Poincaré waves. For convenience, the "ocean basin" is taken to be a semi-infinite strip. Thus, the meridional coastline is of finite length. It has a continental margin whose depth increases exponentially offshore. The zonal coastlines are vertical barriers, i.e., they do not have continental

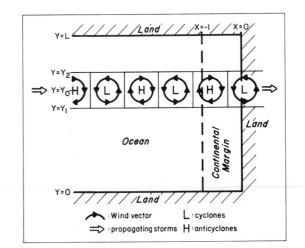

FIGURE 3. *Geometry of the forced wave problem. A system of atmospheric cyclones (L) and anticyclones (H), confined to a zonal band, are shown propagating from west to east across a deep ocean and a meridional (west coast) continental margin.*

margins. (This geometry is imposed to eliminate topographic long waves that would otherwise propagate along the zonal coastlines.) The problem is solved on an f plane; i.e., the meridional dependence of the Coriolis parameter is neglected. (This is done to eliminate planetary Rossby waves from the problem.) The several approximations and assumptions made yield a tractable problem which, as shall be seen, illustrates several physical effects. (Within this section, the symbols for physical quantities are the same as previously defined. Mathematical notation that is specific to this section is locally defined.)

The fully linearized, nondivergent, depth-integrated, and wind-stress-forced equations of motion and continuity are, from (33), (34), and (35) of Chapter 2,

$$\partial_t u - fv = -g\,\partial_x \eta + \frac{\tau^{(x)}}{\rho_0 h} \qquad (28)$$

$$\partial_t v + fu = -g\,\partial_y \eta + \frac{\tau^{(y)}}{\rho_0 h} \qquad (29)$$

$$\partial_x(hu) + \partial_y(hv) = 0 \qquad (30)$$

where the tilde over u and v and the subscript (w) on the wind stress have been dropped and the atmospheric pressure forcing has been neglected. From (28) to (30), a governing equation for the stream function (ψ, where $hu = -\partial_y\psi$ and $hv = \partial_x\psi$) is

$$\partial_t[\partial^2_{xx}\psi + \partial^2_{yy}\psi] + \frac{(dh/dx)}{h}[f\,\partial_y\psi - \partial^2_{xt}\psi]$$
$$= \nabla x \tau - \frac{(dh/dx)}{h}\tau^{(y)} \qquad (31)$$

Motions with a harmonic time dependence will be considered, i.e., $\sim \exp(i\sigma t)$. The problem is nondimensionalized as in (11), except, in addition, $\psi' \equiv \psi\psi_0^{-1}$, $\tau' \equiv \tau\tau_0^{-1}$, and $y' \equiv yL_c^{-1}$. Then (31) becomes

$$\partial^2_{xx}\psi + \partial^2_{yy}\psi - \mathcal{K}(x+1)b\left[\frac{-i}{s}\partial_y\psi - \partial_x\psi\right]$$
$$= \frac{1}{is}[\nabla x \tau + \mathcal{K}(x+1)b\tau^{(y)}]$$

where the primes have been dropped,

$$\mathcal{K}(x+1) = \begin{cases} 0, & x \le -1 \\ 1, & x \ge -1 \end{cases}$$

$$\frac{(dh/dx)}{h} = \begin{cases} 0, & x \le -1 \\ -b, & -1 \le x \le 0 \end{cases}$$

$b > 0$ and ψ_0 has been set equal to $(\tau_0 L_c)/\rho_0 f$. This relation between ψ_0 and τ_0 provides a means to estimate the strength of the circulation from the magnitude of the wind stress and the other parameters in the problem.

The factor $\mathcal{K}(x+1)$, the Heaviside step function, indicates that the governing equation differs between the deep ocean ($x \le -1$) and the continental margin ($-1 \le x \le 0$) because of the effects of sloping bottom topography. Measured in units of L_c, the basin extends from $y = 0$ to $y = L$.

Kinematic boundary conditions require that ψ vanish at $x = 0$, $y = 0$, and $y = L$. Similarly, ψ and $\partial_x\psi$ must be continuous between the deep ocean and the continental margin ($x = -1$). It is also necessary that ψ remain finite as $x \to -\infty$. With periodic boundary conditions in y, solutions are sought of the form $\psi = \sum_n A_n(x, y)\sin(l_n y)\exp(ist)$, where $l_n = n\pi/L$. It is to be understood that such summations are for $n = 1, 2, \ldots, \infty$. Solutions are composed of contributions from the homogeneous (free) and inhomogeneous (forced) problems. The solutions in the deep ocean and on the continental margin are different and must be matched at $x = -1$, which is the crux of the problem.

Consistent with Fig. 3, the wind stress is prescribed to have the form

$$\tau^{(x)} = [\mathcal{K}(y - y_1)$$
$$- \mathcal{K}(y - y_2)]\sin(k(y - y_0))\cos(st - k(x+1))$$

and

$$\tau^{(y)} = [\mathcal{K}(y - y_1)$$
$$- \mathcal{K}(y - y_2)]\cos(k(y - y_0))\sin(st - k(x+1))$$

where $y_1 = y_0 - M$, $y_2 = y_0 + M$, and $k = \pi/2M$. Then

$$\nabla x \tau = \{-2k[\mathcal{K}(y - y_1) - \mathcal{K}(y - y_2)]\cos(k(y - y_0))$$
$$- [\delta(y - y_1) - \delta(y - y_2)]\sin(k(y - y_0))\}$$
$$\times \cos(st - k(x+1))$$

where

$$\delta(y - y_1) = \begin{cases} \infty, & y = y_1 \\ 0, & \text{otherwise} \end{cases}$$

is the Dirac delta function. Already knowing the general form of ψ's y dependence, the forcing is given a Fourier sine series representation, i.e.,

$$\nabla x \tau + \mathcal{K}(x+1)b\tau^{(y)} = \sum \{F_n\cos(st - k(x+1))$$
$$+ \mathcal{K}(x+1)G_n\sin(st - k(x+1))\}\sin(l_n y)$$

where, for $k \neq l_n$,

$$F_n = \frac{4}{L}\frac{[l^2_n + k^2]}{[l^2_n - k^2]}\sin(l_n y_0)\cos(l_n m)$$

and

$$G_n = \frac{4}{L}\frac{bk}{[l^2_n - k^2]}\sin(l_n y_0)\cos(l_n m)$$

Thus, the prescribed forcing extends over all meridional harmonics [the $\sin(l_n y)$]; the strength of each harmonic of the forcing depends on the geometric parameters (L, b, M, and y_0) and the harmonic number (n). F_n and G_n are relatively large for n such that $l_n \sim k$. For $k = l_n$,

$$F_n = \frac{-2\pi}{L} \sin(l_n y_0)$$

and

$$G_n = \frac{b\pi}{l_n L} \sin(l_n y_0)$$

In the deep ocean, the forced solution is a zonally progressive wave:

$$\sum_n A_n \sin(st - k(x + 1)) \sin(l_n y)$$

where $A_n = -F_n/(s(k^2 + l_n^2))$. The free solution is a zonally damped standing wave:

$$\sum_n B_n \exp(l_n(x + 1)) \sin(st + \theta_n) \sin(l_n y)$$

which vanishes as $x \to -\infty$, and where the B_n and θ_n are real numbers to be determined by the boundary conditions at $x = -1$. Altogether, the deep-ocean solution ψ_D is

$$\psi_D = \sum_n \{A_n \sin(st - k(x + 1))$$
$$+ B_n \exp(l_n(x + 1)) \sin(st + \theta_n)\} \sin(l_n y)$$

For later purposes, ψ_D is written in an alternate form:

$$\psi_D = \sum_n \{Z(x_n)e^{ist} + Z^*(x_n)e^{-ist}\} \sin(l_n y)$$

where

$$Z_n = \frac{1}{2i} [A_n \exp(-ik(x + 1)) + C_n \exp(l_n(x + 1))]$$

$C_n = B_n \exp(i\theta_n)$, and Z_n^* is the complex conjugate of Z_n. Because of the complex conjugate relationship for the two parts of ψ_D, the solution for the continental margin is placed in a similar form. Then it will only be necessary to work with $Z_n(x)$ when applying the boundary conditions at $x = -1$.

Over the continental margin, the solution ψ_s is of the form

$$\psi_s = \exp(\alpha(x + 1)) \sum_n \{D_n(x) \cos(st - \beta y + \phi_n(x))$$
$$- D_n(0) \cos(\gamma_n x) \cos(st - \beta y + \phi_n(0))\} \sin(l_n y)$$

where $\cos(\gamma_n x)$ and $D_n(x)$ are solutions to the homogeneous and inhomogeneous problems, respectively. The solution is discussed more fully below, where α, β, and the γ_n are defined. The D_n and ϕ_n are real-valued

functions of x. Note that ψ_s vanishes at $x = 0$. Further, note the occurrence of meridional phase propagation through the βy term; since $\beta > 0$, the phase propagation is northward on a west coast. This term allows separation of variables and a Fourier sine series representation for the homogeneous problem. To facilitate solving the inhomogeneous problem and satisfying the boundary conditions at $x = -1$, it is advantageous to shift to the complex exponential form

$$\psi_s = \sum_n \{E_n(x, y)e^{ist} + E_n^*(x, y)e^{-ist}\} \sin(l_n y)$$

where

$$E_n(x, y) = \exp(\alpha(x + 1))X_n(x)e^{-i\beta y}$$

and

$$X_n(x) \equiv \tfrac{1}{2}[D_n(x) \exp(i\phi_n(x))$$
$$- D_n(0) \exp(i\phi_n(0)) \cos(\gamma_n x)]$$

The X_n are complex-valued functions; from their solutions, the D_n and ϕ_n can be found. With the form chosen for the E_n, the X_n satisfy

$$X_n'' + \gamma_n^2 X_n = \exp[-(\alpha + ik)(x + 1)] \frac{H_n}{s}$$

where

$$H_n \equiv \sum_m (-F_m + iG_m)P_{mn},$$

$$P_{mn} \equiv \frac{2i}{L} \int_0^L e^{i\beta y} \sin(l_m y) \sin(l_n y)\, dy$$

$$\alpha \equiv -\frac{b}{2}, \quad \beta \equiv \frac{b}{2s}, \quad \text{and} \quad \gamma_n^2 \equiv \beta^2(1 - s^2) - l_n^2$$

Note:

$$P_{mn} = \frac{4\beta l_m l_n[1 - (-1)^{m+n} \exp(i\beta L)]}{L[(\beta^2 - l_m^2 - l_n^2) + 2l_m l_n][(\beta^2 - l_m^2 - l_n^2) - 2l_m l_n]}$$

Hence, the βy phase factor serves to couple each of the continental margin's meridional harmonics to every meridional harmonic of the forcing. The corresponding meridional phase speed is $s/\beta = 2s^2/b$, which is dispersive and directly proportional to the width of the continental margin. There are possibilities of spatial resonance which will not be dwelled upon.

The solution for $X_n(x)$ which satisfies the inhomogeneous equation, *and* the boundary conditions $X_n(0) = 0$ and $X_n|_{x=-1} = X_n(-1)$, is

$$X_n(x) = \left[J_n(x) - J_n(0) \frac{\sin(\gamma_n(x + 1))}{\sin(\gamma_n)} \right]$$
$$- X_n(-1) \frac{\sin(\gamma_n x)}{\sin(\gamma_n)}$$

where

$$J_n(x) \equiv I_n(x) \frac{H_n}{s}$$

and

$$I_n(x) = [\exp(-(\alpha + ik)(x + 1))$$
$$+ \frac{(\alpha + ik)}{\gamma_n} \sin(\gamma_n(x + 1)) - \cos(\gamma_n(x + D))]$$
$$/[(\alpha^2 - k^2 + \gamma^2{}_n) + 2i\alpha k]$$

[It has been assumed that $\sin(\gamma_n) \neq 0$; otherwise, $X_n(x)$ must be defined in terms of $X_n'(-1)$ and $\cos(\gamma_n)$.] Note:

$$J_n(-1) = I_n(-1) = J_n'(-1) = I_n'(-1) = 0$$

It follows that

$$X_n'(-1) = - \frac{\gamma_n J_n(0)}{\sin(\gamma_n)} - \gamma_n X_n(-1) \cot(\gamma_n)$$

In the above γ_n has been treated as if it were real, which is only true if $\gamma^2{}_n > 0$, or $n < (L\beta/\pi)(1 - s^2)^{1/2} \equiv n_{c-0}$. For $n > n_{c-0}$, $\sin(\gamma_n x)$ and $\cos(\gamma_n x)$ are replaced by $\sinh(\gamma_n x)$ and $\cosh(\gamma_n x)$, respectively, in the expressions given above. This conversion from trigonometric to hyperbolic functions will be treated as understood in subsequent summations of the X_n.

Finally, there are the boundary conditions at $x = -1$ to be satisfied. First, $\psi_s = \psi_D$ at $x = -1$ implies

$$\sum_n X_n(-1)e^{-i\beta y} \sin(l_n y) = \frac{1}{i} \sum_n [A_n + C_n] \sin(l_n y)$$

$$(32)$$

Second, $\partial_x \psi_s = \partial_x \psi_D$ at $x = -1$ implies

$$\sum_n [\alpha X_n(-1) + X_n'(-1)]e^{-i\beta y} \sin(l_n y)$$
$$= \frac{1}{i} \sum_n [-ikA_n + l_n C_n] \sin(l_n y) \quad (33)$$

Since (32) and (33) must hold for $0 \leq y \leq L$, then

$$C_n = -A_n - \sum_m X_m(-1)P_{mn}{}^*$$

thus, with the A_n given, once the X_n are known the C_n are determined. Note that each free (trapped) harmonic in the deep ocean is excited by every continental margin harmonic.
Similarly,

$$\alpha X_n(-1) + X_n'(-1) = \sum_m [ikA_m + l_m \hat{B}_m]P_{mn}$$

Collecting results, it is found that

$$[\alpha - \gamma_n \cot(\gamma_n)]X_n(-1) = \frac{\gamma_n J_n(0)}{\sin(\gamma_n)}$$
$$+ \sum_m (ik - l_m)A_m P_{mn}$$
$$- \sum_m l_m [\sum_m X_p(-1)P_{pm}{}^*]P_{mn}$$

or

$$X_n(-1) + \sum_p \lambda_{pn} X_p(-1)$$
$$= \mu_n J_n(0) + \sum_p \nu_{pn}(ik - p)A_p \quad (34)$$

where

$$\nu_{pn} \equiv \frac{P_{pn}}{[\alpha - \gamma_n \cot(\gamma_n)]}, \quad \lambda_{pn} \equiv \sum_m l_m P_{pm}{}^* \nu_{mn},$$

and

$$\mu_n \equiv \frac{\gamma_n}{[\alpha \sin(\gamma_n) - \gamma_n \cos(\gamma_n)]}$$

A spatial resonance enters when $\tan(\gamma_n) = \gamma_n/\alpha$, the consequences of which are not pursued here. The right-hand side of (34) is determined by the forcing, whereas the left-hand side depends on the unknown $X_n(-1)$. Since (34) holds for all n, it represents an infinite set of equations. The solution can be written formally in matrix form:

$$X(-1) = (I + \lambda)^{-1}(\mu J(0) + \nu A)$$

where I is the identity matrix and the other symbols represent matrices with definitions obvious from (34). In practice, only a finite number (N) of the X_n would be determined approximately by using the first N equations with sums truncated after N terms. Convergence of the X_n to their true values would be tested by examining the solutions to the first $N + 1, \ldots ,$ equations.

In summary, the problem has been solved in principle, though not evaluated for a specific example, by reducing it to the problem of solving an infinite matrix. The physical problem is essentially a scattering problem, i.e., the scattering of forced waves into topographically trapped (free) waves. The P_{mn}, λ_{pn}, μ_n, and ν_{pn} are independent of the forcing and can be found once and for all when the geometry has been specified. The details of the forcing enter in the $J_n(0)$, A_n, and the $(ik - l_n)$. From inspection of the form of the solutions, the forcing generates a pattern of circulation cells on numerous meridional scales. In the deep ocean, the circulation cells have a zonal scale given by the forcing and they propagate onto the continental margin, where they excite continental margin and coastally trapped, deep-ocean circulation cells. The circulation cells over the continental margin have zonal scales given by the topography of the continental margin as well as by the forcing. The zonal scales of the trapped, deep-ocean circulation cells are given by the meridional geometry of the basin. The finite meridional dimension of the basin discretizes the meridional wave number; it, plus the sloping bottom topography, causes a northward phase progression, which is dispersive, over the continental margin. This phase progression is a modulation on the otherwise standing wave pattern in the meridional direction. The meridional phase progression also "leads" to the excitation of all continental margin meridional harmonics by each harmonic of the forcing, and to the

excitation of all trapped, deep-ocean harmonics by each continental margin harmonic. The resultant super-position of harmonics leads to a nonvanishing divergence of the time-averaged Reynolds stress. Thus, though the time-averaged wind stress vanishes, the continental margin's response to the fluctuating wind stress may be able to drive a mean current through the divergent Reynolds stress mechanism. It must be kept in mind that the problem considered involves a highly idealized geometry. The effects of the finiteness of the ocean basin most closely correspond to the effects of a large indentation in a coastline, e.g., a broad shelf region bounded by a pair of narrow shelf regions. With a more realistic enclosed basin, the alongshore wave number

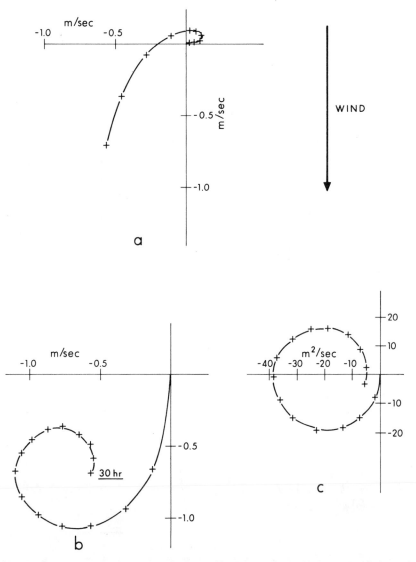

FIGURE 4. *Open basin.* (a) *Current profile after 30 hours; crosses give the tips of vectors at 10 m depth intervals.* (b, c) *Development of surface current and transport; crosses represent the tips of vectors at 2 hour intervals. From Forristall (1974).*

would still be discretized and there would also arise progressive, coastally trapped waves that propagate around the basin.

STORM SURGE

The subject of storm surge is one of great practical importance and has long been given at least some attention. Most studies have been preoccupied with the strong sea-level oscillations at the coast and the consequent runup and flooding. From the present point of view, storm surge is just an extreme case of meteorologically induced motions on the continental margin, a case that includes surface gravity wave effects. As a corollary then, strong current oscillations can be expected to be induced over the margin in conjunction with storm surge at the coast. Indeed, Forristall (1974) has considered a model for currents induced by a hurricane and finds large ($\sim$1 m/sec) currents induced on the shelf. Several results from his calculations are shown in Figs. 4, 5, and 6; at least some features of his calculations are said to be consistent with privileged data belonging to the petroleum industry. Frequently, numerical simulation models are employed for these problems; a recent example is found in the dissertation by Sloss (1972).

RECENT OBSERVATIONS

Over the past decade, especially because of the introduction of recording (Eulerian) current meters which are deployed on moored buoys, a considerable body of

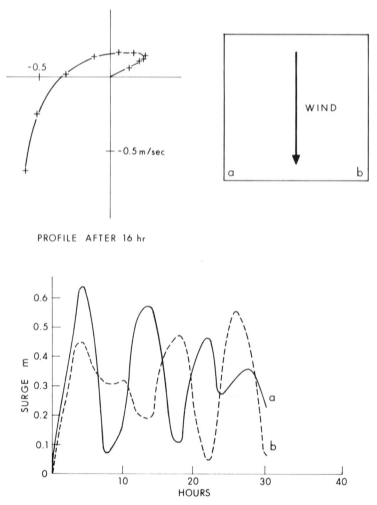

FIGURE. 5 *Closed basin. Profile vectors for 10 m depth intervals. The surge history is given for corners* a *and* b *of the basin in the upper diagram. From Forristall (1974).*

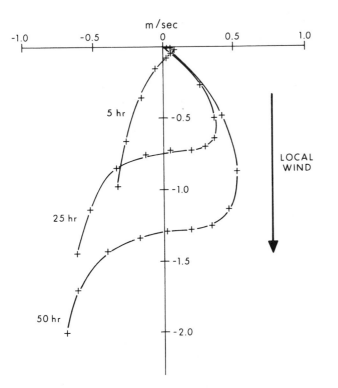

FIGURE 6. *Profiles for circular wind. Crosses show the tips of vectors at 10 m depth intervals. From Forristall (1974).*

observational knowledge has developed for the physical oceanography of continental margins. No attempt will be made to treat the entire set of observations. A few selected sets, which may be representative and provocative, will be noted. The continental shelf that has probably been most extensively and intensively studied is the one off northern Oregon. Examples of some of the subtidal frequency fluctuations occurring there during the coastal upwelling (summer) season are shown in Figs. 7, 8, and 9 (Smith, 1974). During the upwelling season, the isopycnals rise shoreward over the shelf. They tend to intersect the sea surface, though there can be considerable variability in their positions on a time scale of weeks, especially near the coast (Fig. 7). It can be seen that wind fluctuations on the time scale of a week are correlated with alongshore current fluctuations of the order of ±20 cm/sec throughout the water column (Figs. 8 and 9). There are phase shifts with time and depth in the current records with respect to the wind record. Similar fluctuations are shown in the coastal sea-level record, which has been adjusted isostatically for atmospheric pressure fluctuations. Similar types of records from the West Florida shelf are shown in Figs. 10, 11, and 12 (Price and Mooers, 1974); the amplitude of the oscillations is at least as great as in the Oregon case. The mean currents, computed for a month, exhibit currents, countercurrents, and undercurrents (Fig. 10). Again, the fluctuating currents appear to be associated with the fluctuating winds on the time scale of a week or more, have some vertical and horizontal correlation, and exceed the magnitude of the mean (Figs. 11 and 12). Other examples of the same phenomenon can be cited for other shelves, e.g., the New England shelf (Beardsley and Butman, 1974). Hence, it is reasonable to advance the speculation that wherever the winds exhibit an intermittent behavior on a time scale of several days to several weeks, the shelf currents will be dominated by intermittency on a similar

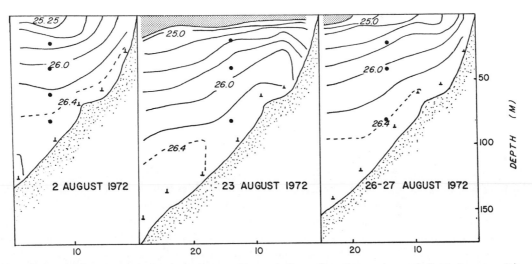

FIGURE 7. *Vertical sections of sigma-t from August occupations of Depoe Bay hydrographic line. Locations of operating current meters at DB-7 shown by large dots. Contour intervals (0.25 sigma-t units) not shown in shaded area (sigma-t is less than 24.75). From Smith (1974).*

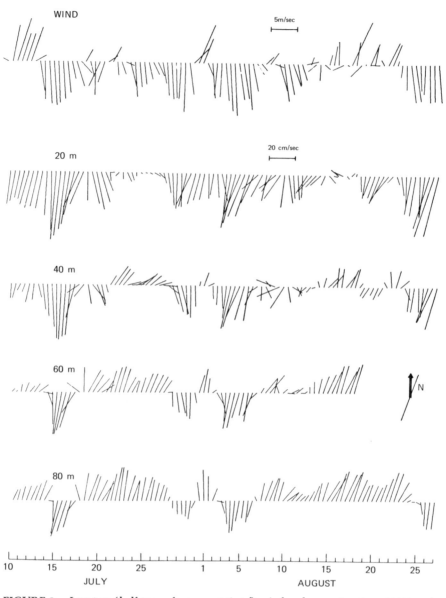

FIGURE 8. *Low-pass (half-power frequency 0.6 cpd) wind and current vectors at 0000 and 1200 UT daily from 1200 UT, July 10, to 0000 UT, August 27, 1972. Trend of local bathymetry indicated by line adjacent to northward arrow. From Smith (1974).*

time scale. Since this is the characteristic time scale of weather systems, this can be expected to be a rather universal phenomenon. The properties of these motions are reasonably consistent with the notion that they obey continental shelf wave dynamics, though definitive shelf wave studies have not yet been made. Motions on these time scales appear to be the most energetic motions on shelves. They are likely to be near-resonant responses to meteorological forcing and to play a central role in turbulent exchange processes. They should at least be significant for sediment transport.

With the introduction of the Cyclesonde, an autonomous current profiling system developed by Dr. John

Van Leer (University of Miami), a new dimension has been added to physical sampling, especially of continental margins. Figure 13 (Van Leer, 1974) shows time-depth plots of temperature, east-west velocity, and north-south velocity sampled by Cyclesondes on the West Florida shelf in a water depth of 200 m, i.e., at the outer shelf break. In these plots, considerable barotropic tidal and inertial structure is visible. Beginning on February 9, 1972 upon the passage of an atmospheric cold front, a rapid thickening of top and bottom boundary layers and the generation of strong near-surface inertial oscillations occurred. Thus, the shelf response to weather systems not only occurs at the weekly time scale but at

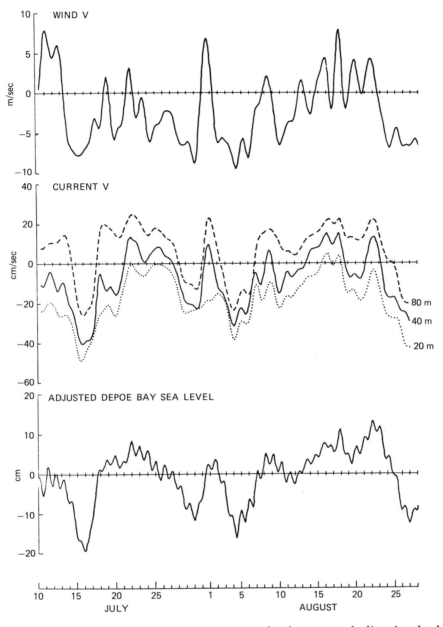

FIGURE 9. *North-south component of low-pass wind and currents and adjusted sea level. From Smith (1974).*

the inertial (order of a day) time scale. The response also involves thermodynamic effects, e.g., mixing, as well as dynamic effects. Clearly, storms can be expected to agitate bottom boundary layers, changing the velocity and the buoyancy. These results can be considered to be of rather universal applicability to conditions of physical variability on continental shelves.

Returning again to the Oregon coast, evidence of the wind and current variability in the tidal and inertial band is seen from the time-varying rotary spectra shown in Figs. 14 and 15 (Crew and Plutchak, 1974).

Again, the dominant time scale is of the order of weeks. Furthermore, the currents tend to be polarized in the clockwise sense whereas the winds are more rectilinear in this frequency band, which suggests that the ocean can be highly selective in the polarization of its response.

An entirely different type of current measurement, i.e., Lagrangian or drogue, shows several features off Oregon which are worthy of note in Figs. 16 through 19 (Stevenson et al., 1974). For instance, a system of currents and countercurrents (Fig. 16), a surface jet (Fig. 17), a system of current and undercurrent (Fig. 18),

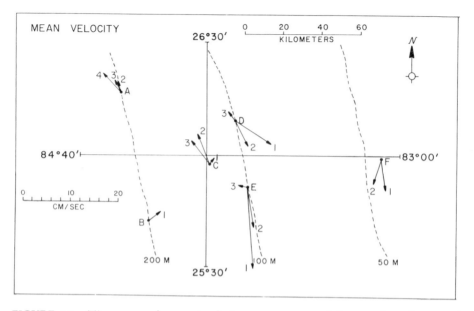

FIGURE 10. *Time-averaged current velocity vectors computed from 40-hour low-passed records sampled on the West Florida shelf. From Price and Mooers (1974).*

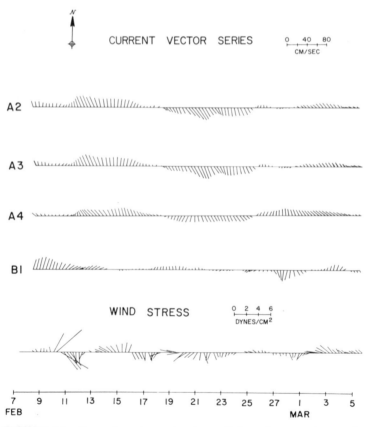

FIGURE 11. *Current vector series from 40-hour low-passed records sampled on the West Florida shelf. Wind stress vector series for 26°N, 84°W. Station symbols as in Fig. 10. From Price and Mooers (1974).*

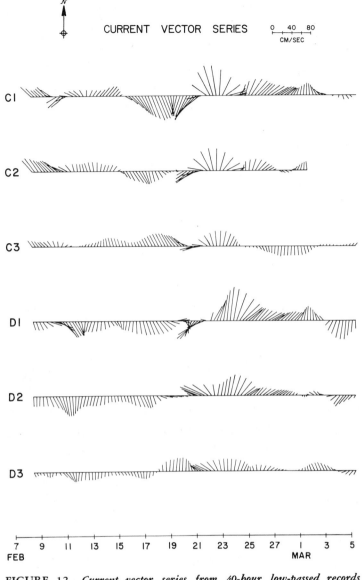

FIGURE 12. *Current vector series from 40-hour low-passed records sampled on the West Florida shelf. Station symbols as in Fig. 10. From Price and Mooers (1974).*

and a two-sided convergence at a surface front (Fig. 19) were documented. Most of the differences in current configurations were associated with different local wind conditions (including lack of wind), which had perturbed the seasonal state of coastal upwelling that had been set up much earlier. Oceanic fronts are commonly observed in coastal regions; meteorological forcing, offshore currents, and runoff can be influential in their formation, maintenance, and annihilation. The associated circulation is highly structured in the horizontal and vertical. The observed surface jet is a phenomenon that may be of fairly universal occurrence in coastal regions where upwelling-favorable winds are persistent

for several days or more. Last but not least, the existence of a (transient) undercurrent on the Oregon shelf serves to emphasize the fact that transient undercurrents over shelves and slopes are being widely reported. They are believed to be associated with the time-dependent wind forcing.

In summary, the observations mentioned here, as well as the aggregate of others, illustrate the importance of time-varying meteorological forces for the circulation on continental margins. They are consistent with the naive theories that exist today. The theories need substantial extension to include various interactions or nonlinear effects. The observations need to become

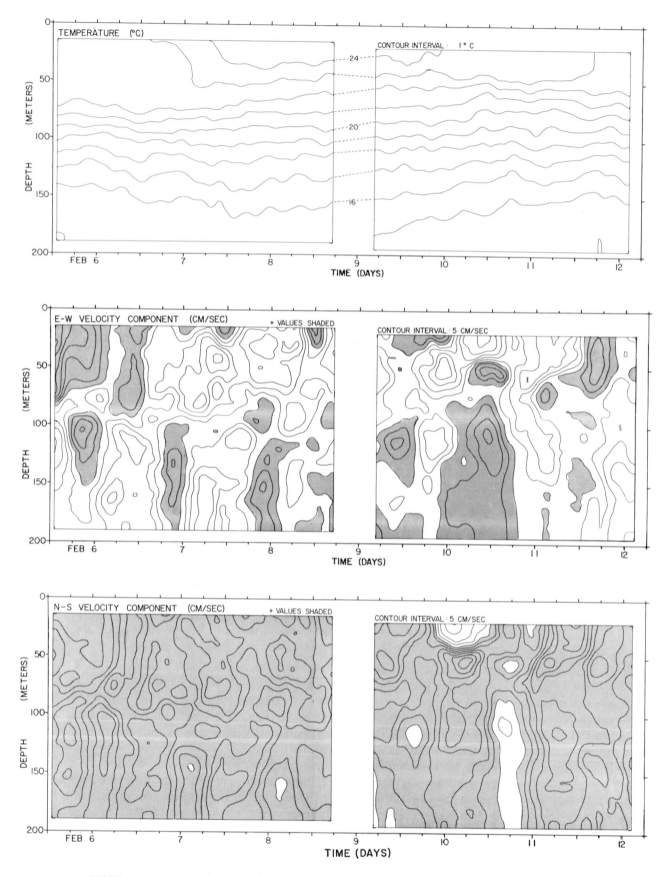

FIGURE 13. *Contour data from shelf break Cyclesonde station, February 1973. From Van Leer (1974).*

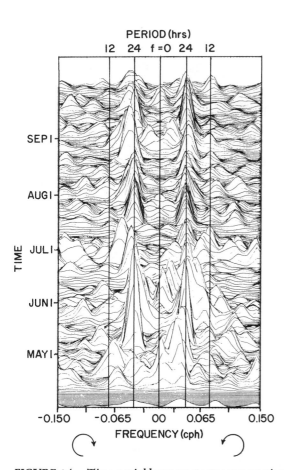

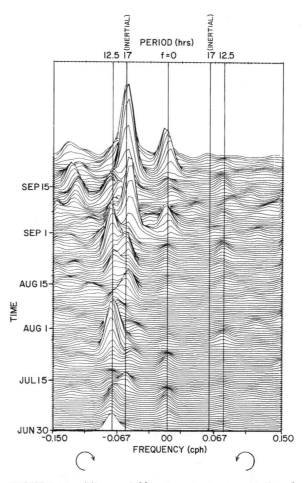

FIGURE 14. *Time-variable rotary spectra presentation of winds at the Oregon coast for April 20 through September 27, 1969. Negative frequencies represent clockwise rotation, positive frequencies anticlockwise rotation. Time increases along the ordinate where the separation of successive spectra is 15 hours. The beginning of each month is indicated on the ordinate. The salient frequencies are emphasized with lines and are labeled by period. From Crew and Plutchak (1974).*

FIGURE 15. *Time-variable rotary spectra presentation of currents for June 25 to October 2, 1969, at a location 35 miles off the Oregon coast at 45°N latitude. Negative frequencies represent clockwise rotation, positive frequencies anticlockwise rotation. Time increases along the ordinate where the separation of successive spectra is 15 hours. The beginning of each month is indicated on the ordinate. Salient frequencies are emphasized with lines and labeled by period. From Crew and Plutchak (1974).*

sharply focused and to take the form of process studies or experiments. As just one example, there is a need to examine the alongshore propagation of several-day to several-week time scale current oscillations to determine if they are coherent propagating free or forced waves, if they are topographically trapped incoherent eddies, or if they are a mixture of waves and eddies. The latter is probably true. If so, then major contributors to the Reynolds stress divergence may have been identified and the need to guess eddy coefficient values may be diminished.

ADDITIONAL READING

For further theoretical discussion of continental shelf waves, see Adams and Buchwald (1969), Buchwald (1973), Buchwald and Adams (1968), Caldwell et al.

(1972), Gill and Schumann (1974), Gill and Clarke (1974), Mysak (1967, 1968), and Niiler and Mysak (1971). For further discussion of continental shelf wave observations, see Mysak and Hamon (1969).

For further theoretical discussion of transient wind responses, see Birchfield (1972), Charney (1955), Heaps (1965), Jelesnianski (1965), Suginohara (1973, 1974), Walin (1972a), and Bennett (1974). For further discussion of observed coastal variability, see Collins and Pattullo (1970), Collar and Cartwright (1972), Csanady (1972a–c), Cragg and Sturges (1974), Csanady (1973, 1974), Walin (1972b), Vukovich (1974), Cartwright (1968, 1969), and Nowlin and Parker (1974).

For examples of numerical simulation models for coastal upwelling, see Hurlburt and Thompson (1973) and McNider and O'Brien (1973). For further theoretical

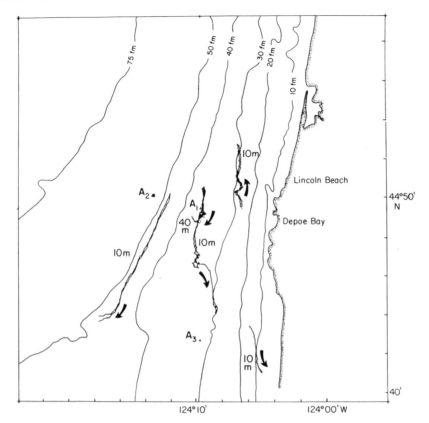

FIGURE 16. *Parachute drogue trajectories, August 23–24, 1972, as corrected for wind drag. The depth of the drogue pairs is indicated beside the trajectory. The arrows show the direction of motion. Location of the anchor buoys (A₁, A₂, A₃) used for radar references is shown. From Stevenson et al. (1974).*

and observational discussion of coastal upwelling, see Smith (1968), Allen (1973), Mooers et al. (1976), Muraki (1974), and Taylor and Stewart (1959).

For recent theoretical and observational discussions of upper layer wind mixing, see Niiler (1975) and Pollard and Millard (1970). For discussion of recent observations of bottom currents, see Smith and Hopkins (1972) and Murray (1970).

SUMMARY

Transient wind forcing of continental margin circulation is provided by atmospheric systems with an element of order and structure. Such systems have characteristic forms, intensities, and scales. They generate continental margin circulation with forms, intensities, and scales determined by the degree of correspondence between the characteristics of the forcing and those of intrinsic or free motions of the margin's waters.

Atmospheric disturbances occur with time scales of a few hours (land-sea breeze systems), a few days (cold fronts, low-pressure cells), and several months (seasonal phenomena). The corresponding space scales are several tens of kilometers, a few hundred kilometers, and a few thousand kilometers. The ratio of the baroclinic Rossby radius of deformation, approximately 10 to 30 km, to the width of the shelf plays a crucial role in determining the response of the margin's circulation to these various scales of forcing.

Some weather systems are "rare events" (e.g., hurricanes) in that the oceanic response is completed before the next event occurs. Other systems are quasi periodic (e.g., midlatitude lows) in that they reoccur at fairly regular and frequent intervals; the margin's response tends to consist of a superposition of responses to the succession of systems. In both cases, it is profitable to treat the oceanic response as a wavelike motion that can be characterized by its amplitude distribution in frequency–wave number (Fourier) space, even though the wave parameters may be randomly distributed.

The boundary conditions of continental margin flow are an essential part of the problem, which makes the continental margin unique when compared with the

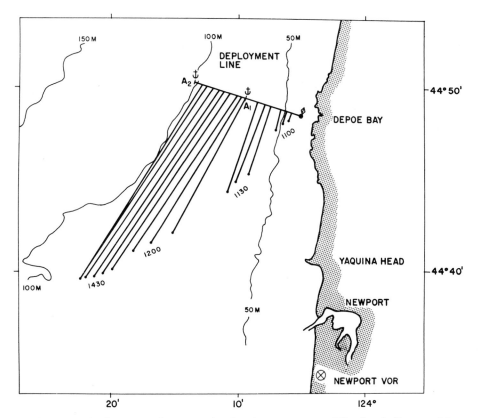

FIGURE 17. *Surface vane drogue motion for August 24, 1972. The dots indicate positions determined at a time near to that shown. Lines connect positions to those at deployment. From Stevenson et al. (1974).*

open ocean. Frictional effects leading to turbulent dissipation of energy are parametrized by eddy co-efficients. In a density-homogeneous fluid without variable bottom topography, a horizontal Ekman layer develops adjacent to the coastal boundary and has a width of the order of 1 to 10 km. Similarly, vertical Ekman layers develop near the top and bottom and have a thickness of the order of 3 to 30 m.

The time-dependent motions on a margin may be driven by local disturbances (near-field case), or by distant disturbances (far-field case). The near-field case encompasses the forced wave problem; the far-field case encompasses the free wave problem. In this chapter, attention has been restricted to long waves, with time scales such that the effects of the earth's rotation play an essential role. In Chapter 13, relatively short-period Stokes edge waves are considered; these waves are coastally trapped but do not strongly feel the effects of the earth's rotation.

Barotropic motions on a margin of uniform depth have been examined for both the trapped barotropic wave (Kelvin wave) and free barotropic plane wave (Poincaré wave) case. For Kelvin waves, the surface

elevation and the alongshore velocity component decay exponentially offshore, and there is no onshore-offshore velocity component. In the northern hemisphere, these waves may only propagate northward on a west coast, and southward on an east coast. Their influence is confined to a distance from the coast of about 1000 km. They are nondispersive, in that their phase speeds are independent of frequency, and they can exist for all frequencies. Their phase speed is that of deep-water gravity waves and is thus approximately 200 m/sec.

For Poincaré waves, the propagation equation shows that the wave frequency must be greater than the inertial frequency. These waves may travel in either direction along the coast. They are not coastally trapped and thus can strike the coast at an arbitrary angle of incidence. The waves that are then reflected, together with the incident waves, form complex interference patterns.

If the continental margin is topographically stepped, the waves may be either Kelvin-like or Poincaré-like, since the topographic step tends to refract or trap Poincaré waves. In addition a trapped "quasi-geostrophic" wave occurs, which depends for its existence only on the proportional discontinuity in depth and the

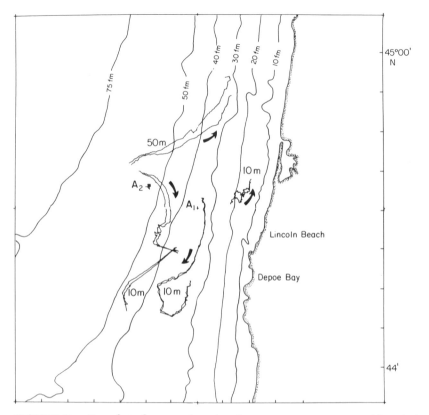

FIGURE 18. *Parachute drogue trajectories, August 7–9, 1972, as corrected for wind drag. The depth of the drogue pairs is indicated beside the trajectory. The arrows show the direction of motion. Location of the anchor buoys (A₁, A₂,) used for radar references is shown. From Stevenson et al. (1974).*

earth's rate of rotation. All of these types of waves trapped by stepped topography are constrained to propagate in one direction, as in the pure Kelvin wave case.

If the water depth is neither constant nor stepped, but varies continuously, topographic Rossby waves or "continental shelf waves" can exist. The inertial frequency is an upper bound on the wave frequency for these motions. They are highly dispersive (group and phase speeds are functions of frequency). Their phase speeds are determined by the shelf width and the Coriolis parameter. Typical phase speeds for the long-wave limit are small (approximately 1 m/sec); hence, these waves propagate much more slowly than do common Kelvin waves and have much shorter wavelengths for the same frequency. They are also constrained to propagate in the same direction as ordinary Kelvin waves. Their onshore-offshore particle velocities are not identically zero, in contrast to the ordinary Kelvin wave case.

Baroclinic waves can occur as internal Kelvin and Poincaré waves, and exist where there is significant density stratification over the water column. They also have very slow phase speeds, of the order of 1 m/sec or

less; hence, they may interact with "mean" currents and also with barotropic continental shelf waves. Over

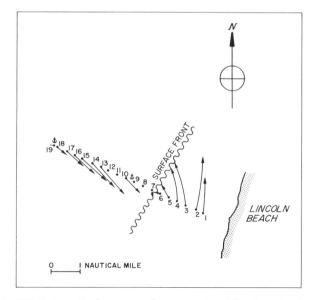

FIGURE 19. *Surface vane drogue motion following development from 1600 to 1845 local time August 7, 1972. Arrowheads indicate last position determined. From Stevenson et al. (1974).*

variable bottom topography, these waves are expected to experience complex reflection and refraction phenomena.

The present body of observations of transient wind-forced flows on the continental margin suggests that whenever the winds exhibit an intermittent behavior on a time scale of several days to several weeks, the current will be dominated by an intermittency on a similar time scale. The properties of these motions are reasonably consistent with the notion that they obey continental shelf wave dynamics. Motions on these scales appear to be the most vigorous motions on continental margins. They are likely to be near-resonant responses to meteorological forcing, and must play a significant role in sediment transport. Major questions confronting continental margin oceanography concern the extent to which transient components of wind-forced flows may be described as propagating free or forced waves, and the extent to which they may be described as large-scale eddies.

ACKNOWLEDGMENTS

Support for preparing this chapter was derived from National Science Foundation Grants NSF/DES 72-0147 A03 entitled "The Spatial and Temporal Structure of Physical Variability on Continental Shelves" and NSF/IDO 75-09313 entitled "North American Participation in the Liège Colloquium on Continental Shelf Dynamics."

REFERENCES

Adams, J. K. and V. T. Buchwald (1969). The generation of continental shelf waves. *J. Fluid Mech.*, **35**(4): 815–826.

Allen, J. S. (1973). Upwelling and coastal jets in a continuously stratified ocean. *J. Phys. Oceanogr.*, **3**(3): 245–257.

Allen, J. S. (1975). Coastal trapped waves in a stratified ocean. *J. Phys. Oceanogr.*, **5**(2): 300–325.

Beardsley, R. C. and B. Butman (1974). Circulation on the New England continental shelf: Response to strong wind storms. *Geophys. Res. Lett.*, **1**(4): 181–184.

Bennett, J. R. (1974). On the dynamics of wind-driven lake currents. *J. Phys. Oceanogr.*, **4**(3): 400–414.

Birchfield, G. E. (1972). Theoretical aspects of wind-driven currents in a sea or lake of variable depth with no horizontal mixing. *J. Phys. Oceanogr.*, **2**(4): 355–362.

Buchwald, V. T. (1973). On divergent shelf waves. *J. Mar. Res.*, **31**(2): 105–115.

Buchwald, V. T. and J. K. Adams (1968). The propagation of continental shelf waves. *Proc. Roy. Soc.*, **A305**: 235–250.

Caldwell, D. R., D. L. Cutchin, and M. S. Longuet-Higgins (1972). Some model experiments on continental shelf waves. *J. Mar. Res.*, **30**(1): 39–55.

Cartwright, D. E. (1968). A unified analysis of tides and surges round North and East Britain. *Phil. Trans. A,* **263**(1134): 1–55.

Cartwright, D. E. (1969). Extraordinary tidal currents near St. Kilda. *Nature,* **223**(5209): 928–930.

Charney, J. G. (1955). Generation of ocean currents by wind. *J. Mar. Res.*, **14**(4): 477–498.

Collar, P. G. and D. E. Cartwright (1972). Open sea tidal measurements near the edge of the Northwest European continental shelf. *Deep-Sea Res.*, **19**(10): 673–689.

Collins, C. A. and J. G. Pattullo (1970). Ocean currents above the continental shelf off Oregon as measured with a single array of current meters. *J. Mar. Res.*, **28**(1): 51–68.

Cragg, J. and W. Sturges (1974). *Wind-Induced Currents and Sea Surface Slopes in the Eastern Gulf of Mexico.* Technical Report, Dept. of Oceanography, Florida State University, Tallahassee, 51 numbered leaves.

Crew, H. and N. Plutchak (1974). Time varying rotary spectra. *J. Oceanogr. Soc. Japan,* **30**(2): 61–66.

Csanady, G. T. (1972a). The coastal boundary layer in Lake Ontario: Part I. The spring regime. *J. Phys. Oceanogr.*, **2**(1): 41–53.

Csanady, G. T. (1972b). The coastal boundary layer in Lake Ontario: Part II. The summer-fall regime. *J. Phys. Oceanogr.*, **2**(2): 168–176.

Csanady, G. T. (1972c). Frictional currents in the mixed layer at the sea surface. *J. Phys. Oceanogr.*, **2**(4): 498–508.

Csanady, G. T. (1973). Wind-induced baroclinic motions at the edge of the continental shelf. *J. Phys. Oceanogr.*, **3**(3): 274–279.

Csanady, G. T. (1974). Barotropic currents over the continental shelf. *J. Phys. Oceanogr.*, **4**(3): 357–371.

Cutchin, D. L. and R. L. Smith (1973). Continental shelf waves: Low frequency variation in sea level and currents over the Oregon continental shelf. *J. Phys. Oceanogr.*, **3**(1): 73–82.

Forristall, G. Z. (1974). Three-dimensional structure of storm-generated currents. *J. Geophys. Res.*, **79**(18): 2721–2729.

Gill, A. E. and A. J. Clarke (1974). Wind-induced upwelling, coastal currents, and sea level changes. *Deep-Sea Res.*, **21**(5): 325–345.

Gill, A. E. and E. H. Schumann (1974). The generation of long shelf waves by the wind. *J. Phys. Oceanogr.*, **4**(1): 83–90.

Heaps, N. S. (1965). Storm surges on a continental shelf. *Phil. Trans. A,* **257**(1082): 351–383.

Hurlburt, H. E. and J. D. Thompson (1973). Coastal upwelling on a β-plane. *J. Phys. Oceanogr.*, **3**(1): 16–32.

Jelesnianski, C. P. (1965). A numerical calculation of storm tides induced by a tropical storm impinging on a continental shelf. *Mon. Weather Rev.*, **93**(6): 343–358.

Larsen, J. C. (1969). Long waves along a single-step topography in a semi-infinite uniformly rotating ocean. *J. Mar. Res.*, **27**(1): 1–6.

McNider, R. T. and J. J. O'Brien (1973). A multi-layer transient model of coastal upwelling. *J. Phys. Oceanogr.*, **3**(3): 258–273.

Mooers, C. N. K., C. A. Collins, and R. L. Smith (1976). The dynamic structure of the frontal zone in the coastal upwelling region off Oregon. (To appear). *J. Phys. Oceanogr.*

Muraki, H. (1974). Poleward shift of the coastal upwelling region off the California coast. *J. Oceanogr. Soc. Japan,* **30**(2): 49–53.

Murray, S. P. (1970). Bottom currents near the coast during Hurricane Camille. *J. Geophys. Res.*, **75**(24): 4579–4582.

Mysak, L. A. (1967). On the theory of continental shelf waves. *J. Mar. Res.*, **25**(3): 205–227.

Mysak, L. A. (1968). Effects of deep-sea stratification and current on edgewaves. *J. Mar. Res.*, **26**(1): 34–42.

Mysak, L. A. and B. V. Hamon (1969). Low-frequency sea level behavior and continental shelf waves off North Carolina. *J. Geophys. Res.*, **74**(6): 1397–1405.

Niiler, P. P. (1975). Deepening of the wind-mixed layer. *J. Mar. Res.* (in press).

Niiler, P. P. and L. A. Mysak (1971). Barotropic waves along an eastern continental shelf. *Geophys. Fluid Dyn.*, **2**(4): 273–288.

Nowlin, W. D., Jr. and C. A. Parker (1974). Effects of a cold-air outbreak on shelf waters of the Gulf of Mexico. *J. Phys. Oceanogr.*, **4**(3): 467–486.

Pedlosky, J. (1974a). On coastal jets and upwelling in bounded basins. *J. Phys. Oceanogr.*, **4**(1): 3–18.

Pedlosky, J. (1974b). Longshore currents, upwelling and bottom topography. *J. Phys. Oceanogr.*, **4**(2): 214–226.

Pedlosky, J. (1974c). Longshore currents and the onset of upwelling over bottom slope. *J. Phys. Oceanogr.*, **4**(3): 310–320.

Pollard, R. T. and R. C. Millard, Jr. (1970). Comparison between observed and simulated wind-generated inertial oscillations. *Deep-Sea Res.*, **17**(4): 813–821.

Price, J. F. and C. N. K. Mooers (1974). *Current Meter Data Report from the Winter 1973 National Science Foundation Continental Shelf Dynamics Program.* UM-RSMAS-74020, University of Miami, Miami, Florida, 78 numbered leaves.

Sloss, P. W. (1972). Coastal processes under hurricane action: Numerical simulation of a free-boundary shoreline. Ph.D. Dissertation, Dept. of Geology, Rice University, Houston, Texas, 139 numbered leaves.

Smith, J. D. and T. S. Hopkins (1972). Sediment transport on the continental shelf off Washington and Oregon in the light of recent current measurements. In D. J. P. Swift, D. B. Duane, and O. H. Pikley, eds., *Shelf Sediment Transport: Process and Pattern.* Stroudsburg, Pa.: Dowden, Hutchinson & Ross, pp. 143–180.

Smith, R. L. (1968). Upwelling. In *Oceanography and Marine Biology Annual Review*, Vol. 6. London: Allen & Unwin, pp. 11–47.

Smith, R. L. (1974). A description of currents, winds, and sea level variations during coastal upwelling off the Oregon coast, July–August 1972. *J. Geophys. Res.*, **79**(3): 435–443.

Stevenson, M. R., R. W. Garvine, and B. Wyatt (1974). Lagrangian measurements in a coastal upwelling zone off Oregon. *J. Phys. Oceanogr.*, **4**(3): 321–336.

Suginohara, N. (1973). Response of a two-layer ocean to typhoon passage in the western boundary region. *J. Oceanogr. Soc. Japan*, **29**(6): 236–250.

Suginohara, N. (1974). Onset of coastal upwelling in a two-layer ocean by wind stress with longshore variation. *J. Oceanogr. Soc. Japan*, **30**(1): 23–33.

Taylor, C. B. and H. B. Stewart, Jr. (1959). Summer upwelling along the east coast of Florida. *J. Geophys. Res.*, **64**(1): 33–40.

Van Leer, J. C. (1974). *Progress Report on Cyclesonde Development and Use.* UM-RSMAS-74029, University of Miami, Miami, Florida, 77 numbered leaves.

Vukovich, F. M. (1974). The detection of nearshore eddy motion and wind-driven currents using NOAA 1 sea surface temperature data. *J. Geophys. Res.*, **79**(6): 853–860.

Walin, G. (1972a). On the hydrographic response to transient meteorological disturbances. *Tellus*, **24**(3): 169–186.

Walin, G. (1972b). Some observations of temperature fluctuations in the coastal region of the Baltic. *Tellus*, **24**(3): 187–198.

Wang, D.-P. (1975). Coastal trapped waves in a baroclinic ocean. *J. Phys. Oceanogr.*, **5**(2): 326–333.

CHAPTER 5

Tidal Currents

HAROLD O. MOFJELD

Atlantic Oceanographic and Meteorological Laboratories, Miami, Florida

In regions where they are sufficiently strong, tidal currents constantly rework bottom sediment. Weaker currents combine with storm-generated wave motion and currents to move sediment both at the water-bottom interface and in suspension. Tidal currents are especially effective agents of sediment transport because they persist throughout the year, whereas other types of water motion, particularly storm events, tend to be seasonal. They are the background upon which are superimposed other kinds of currents causing sediment transport.

The tides typically rise and fall twice a day (semi-daily tides), once a day (daily tides), or occur as a combination of daily and semidaily components. Figure 1 illustrates tides at different locations along the east coast of the United States and in the Gulf of Mexico. The tide is semidaily along the eastern seaboard and dominantly daily in the Gulf of Mexico (Pensacola and Galveston).

As the earth rotates about its axis, the forces producing the tides move across the earth's surface from east to west. The motion of the moon around the earth and the earth-moon system around the sun produce variations of the tides and tidal currents with periods of about two weeks, a month, six months, a year, and longer.

Approximately twice monthly the range of the tide is a maximum; that is, the difference in sea level between successive high tide and low tide is largest. These *spring tides* occur for both the daily and semidaily tides, although not necessarily on the same day. The term *neap tides* refers to the tides with minimum range.

If the ocean covered the entire earth to a constant depth, the pattern of sea-level changes caused by the tides would be simple. However, since the oceans have complicated shapes, these patterns in the real oceans are also complicated. The global distributions of the daily and semidaily tides are given in Figs. 2 and 3. The numbers in small type along the coast and at islands in Figs. 2 and 3 are the spring ranges averaged over a year.

The lines traversing the oceans in these figures give a general idea of the stage of the tide, i.e., when high, mean, and low water occur. For example, assume that in a particular region of the North Atlantic the semidaily high water is occurring along the cotidal line marked 0 hour. Then along the 3 and 9 hour lines, the semidaily tide is passing through mean sea level; and along the 6 hour line, the tide has reached low water. About 3 hours later, high water for the semidaily tide will occur along the 3 hour line, low water along the 9 hour line, and so forth. The pattern rotates counter-clockwise around a point in the North Atlantic where the range of the semidaily tide is zero. Such a rotating pattern is called an amphidromic system, and the point is called an amphidromic point. Amphidromic systems are a general feature of both the daily and semidaily tides in the deep ocean. Amphidromic systems also occur in shallow seas, such as the North Sea, as shown in Fig. 4.

The tides and tidal currents on the continental shelves and seas bordering the open seas are propagated as waves from the open oceans. These waves are partially

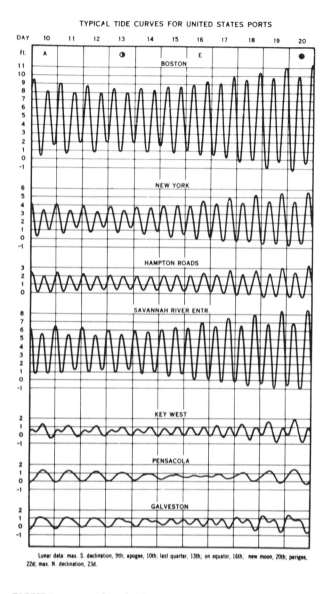

TYPICAL TIDE CURVES FOR UNITED STATES PORTS

Lunar data: max. S. declination, 9th; apogee, 10th; last quarter, 13th; on equator, 16th; new moon, 20th; perigee, 22d; max. N. declination, 23d.

FIGURE 1. *Predicted tides at selected ports along the U.S. east coast and in the Gulf of Mexico. The range and character of tides differ significantly along coasts and between distinct regions. From the U.S. Department of Commerce Tide Tables (1974).*

reflected back out to sea by the shoaling bottom which rises toward the beach. The combination of the incoming, or incident, wave and the reflected wave is called a standing tide; the semidaily tide on the continental shelf between Cape Hatteras and Long Island is an example of a standing tide. The wave may also propagate along the coast, in which case it is called a progressive tide; that is, the stages of the tide progress down the coastline. An example of a progressive tide is the daily tide along the U.S. east coast. Both of these examples can be seen in Figs. 2 and 3. The technical term for a tide that is generated elsewhere and propagates into a given region is cooscillation. The water on the continental shelf off the U.S. east coast cooscillates with the North Atlantic.

Coastal lagoons, bays, and estuaries cooscillate with the water on the continental shelves. The tides often enter these coastal bodies of water through inlets and over bars, both of which can significantly attenuate the tides so that the tides inside the embayment are smaller than the tides along the open coast.

EQUATIONS OF MOTION

To compute the tides and tidal currents within a region having a complicated shape and realistic bathymetry, computer programs have been developed which use information at a number of points to compute the motion at those or other points. How well the results of the calculations describe the motion depends on the spacing between points (how well the grid of points resolves depth variations), approximations to the fundamental equations, knowledge of the motions at the boundaries of the region, and estimates of the bottom stress coefficient determining the drag of the sediment on the water.

When idealized depth variations are assumed and when less significant forces are neglected, the tides and tidal currents can sometimes be described by simple formulas from which considerable insight can be gained into tidal phenomena. Historically, intensive research was done on the behavior of tides in channels having constant depth and vertical side boundaries. The channel theory of tides is used in the present discussion and then extended to open regions such as the continental shelves.

In a channel where bottom stress and the Coriolis effect due to the earth's rotation can be neglected, a tide causes the water to accelerate through the downchannel slope of the sea surface. Horizontal differences in the resulting tidal currents in turn cause sea level to change. The interplay between these two effects produces a wave that propagates down the channel, away from the source of the tide. A general theory of waves has been presented by Mooers in Chapter 2; tides are waves whose wavelengths are long compared with the water depth but whose amplitudes are small compared with the depth. With these assumptions, the tidal motion in a channel is described by the pair of equations

$$\frac{\partial u}{\partial t} = -g\,\frac{\partial \eta}{\partial x} \tag{1}$$

$$\frac{\partial \eta}{\partial t} = -h\,\frac{\partial u}{\partial x} \tag{2}$$

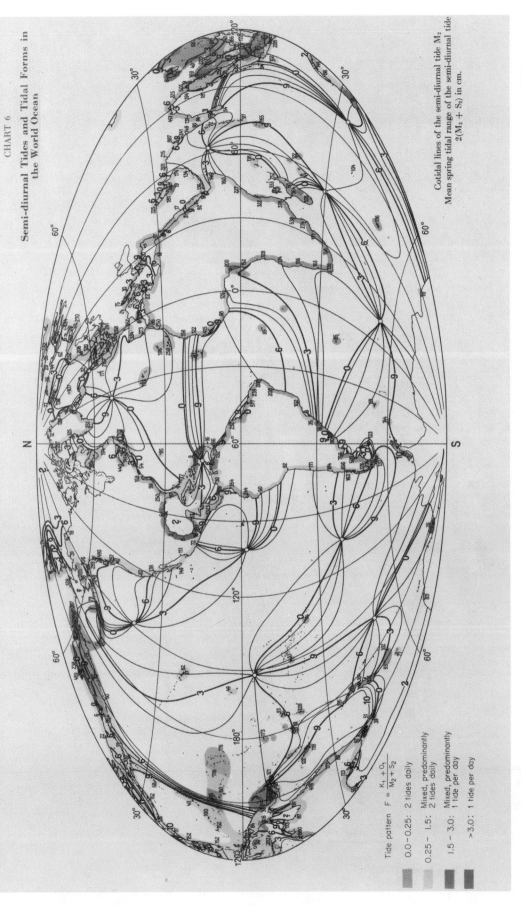

CHART 6

Semi-diurnal Tides and Tidal Forms in
the World Ocean

Tide pattern $F = \dfrac{K_1 + O_1}{M_2 + S_2}$

0.0 – 0.25: 2 tides daily

0.25 – 1.5: Mixed, predominantly
2 tides daily

1.5 – 3.0: Mixed, predominantly
1 tide per day

>3.0: 1 tide per day

Cotidal lines of the semi-diurnal tide M_2
Mean spring tidal range of the semi-diurnal tide
$2(M_2 + S_2)$ in cm.

FIGURE 2. *Global distribution of semidaily tides obtained from a theoretical model using coastal and island tide data as boundary conditions. The numbers in small type along the coasts and at islands are spring ranges in centimeters; the cotidal lines in the oceans are successive loci of high water as seen in hourly intervals. From Dietrich (1963).*

Diurnal Tides in the World Ocean

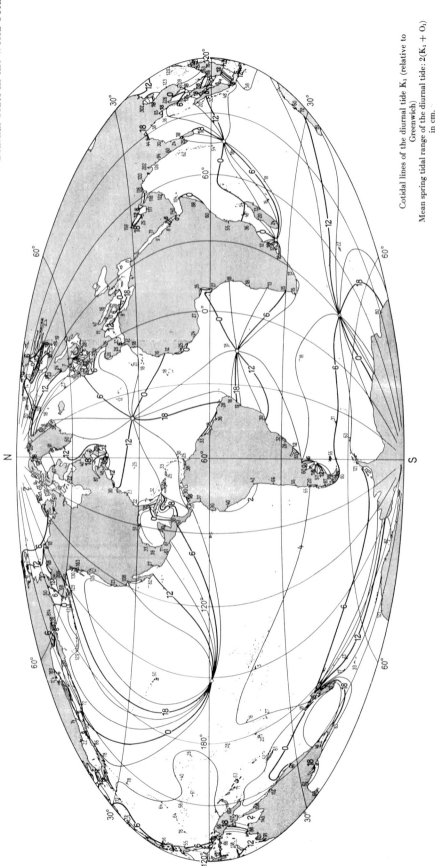

Cotidal lines of the diurnal tide K_1 (relative to Greenwich)
Mean spring tidal range of the diurnal tide: $2(K_1 + O_1)$ in cm.

FIGURE 3. *Global distribution of daily tides. From Dietrich (1963).*

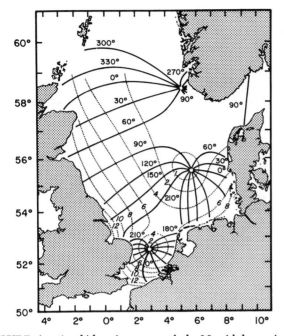

FIGURE 4. *Amphidromic systems of the* M_2 *tidal constituent (semidaily lunar tide) in the North Sea. The cotidal lines show the progress of the tide each constituent hour (30° phase change), the dotted corange lines show the decrease in feet of the M_2 tidal range away from the shore. From Doodson and Warburg (1941).*

Equation 1 states that the time rate of change of the downchannel horizontal velocity u is equal to the acceleration of gravity g multiplied by the downchannel slope of the sea surface, whose displacement above mean water is η; t is the time elapsed after high water at the source; and x is the distance away from the source of the tide as measured along the axis of the channel. Equation 2 states that the time rate of change of the sea surface displacement η is equal to the mean depth h multiplied

by the horizontal rate of change of the downchannel velocity u.

Assuming that the mean depth h is uniform throughout the channel, a tide with a period T and an amplitude a (one-half the range) would be described by

$$\eta = a \cos\left[\frac{2\pi}{T}\left(t - \frac{x}{c}\right)\right] \tag{3}$$

$$u = a\left(\frac{g}{h}\right)^{1/2} \cos\left[\frac{2\pi}{T}\left(t - \frac{x}{c}\right)\right] \tag{4}$$

where $c = (gh)^{1/2}$ is the speed of propagation at which the shape of the sea surface moves down the channel. For oceanic and shelf depths c is 200 and 31 m/sec, respectively.

If the depth h were representative of the open ocean ($h = 4000$ m) and the amplitude of the tide were $a = 0.5$ m, the maximum tidal current according to (4) would be 2.5 cm/sec, a relatively small speed. On the other hand, if the depth were representative of the continental shelves ($h = 100$ m), the corresponding current would be 15.8 cm/sec. For a given tidal amplitude, the maximum tidal currents are inversely proportional to the square root of depth. In very shallow water, where (4) would predict unrealistically large currents, the formula is not applicable since turbulent dissipation and bottom stress which would limit the currents have not been included.

In Fig. 5, the tide and tidal current are shown along a vertical section parallel to the channel axis; both are uniform across the channel. At any given time, the pattern repeats itself downchannel with a horizontal distance equal to the wavelength $\lambda = cT$ of the wave. The water velocity u is the same at every depth because no bottom stress is allowed in this idealized model.

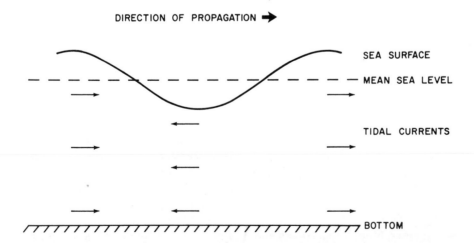

FIGURE 5. *Idealized progressive tide propagating in a narrow channel in which bottom stress effects on the tidal currents are neglected. The vertical scale is greatly exaggerated.*

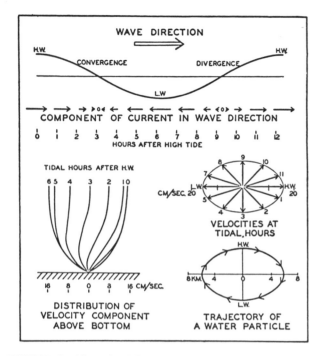

FIGURE 6. *Typical tidal currents on a continental shelf where bottom stress forces the currents to zero speed at the bottom. From Fleming and Revelle (1939, p. 130); after Sverdrup (1927).*

With turbulent stresses, a more realistic profile of u is shown in Fig. 6, in which u decreases within the bottom boundary layer to essentially zero at the water-sediment interface.

Since the tide propagates down the channel, it is a progressive tide. In this case, the maximum horizontal currents occur at high water and low water where the current is in the direction and opposite to the direction of propagation, respectively. As the sea level passes through mean sea level, $\eta = 0$, the current is momentarily zero.

As a water parcel moves in the channel, its position X is the integral in time of the horizontal velocity:

$$X = \int_0^t u \, dt + X_0 \qquad (5)$$

where X_0 is the position of the parcel at the initial time $t = 0$. Using (4), the horizontal displacement of the parcel is given by

$$X = \frac{a_0 T}{2\pi} \left(\frac{g}{h} \right)^{1/2} \sin \left[\frac{2\pi}{T} \left(t - \frac{x}{c} \right) \right] + X_0 \qquad (6)$$

Water subject to the tidal motion oscillates about an average position X_0 with an amplitude equal to half the total excursion. For a semidaily tide ($T = 12.5$ hours) and a tidal range $2a = 1$ m, the excursion for open sea

depths ($h = 4000$ m) is 0.35 km, whereas for shelf depths ($h = 100$ m) it is 2.25 km.

In (6), tidal currents would produce no net displacement of water or suspended particles. That is, if a water parcel were tagged using dye and observed throughout a tidal cycle, the parcel would return to the same location at the end of each tidal cycle.

STANDING TIDES

In bays and many estuaries, the incident progressive tide is reflected. The tide in this region is the combination of the incident and reflected tides. An analogous tide can be created in a channel by inserting a reflecting barrier. The sea surface displacement η above mean sea level and horizontal water parcel velocity for the standing tide are

$$\eta = a \, \cos\left(\frac{2\pi t}{T} \right) \cos\left(\frac{2\pi x}{\lambda} \right) \qquad (7)$$

$$u = a \left(\frac{g}{h} \right)^{1/2} \sin\left(\frac{2\pi t}{T} \right) \sin\left(\frac{2\pi x}{\lambda} \right) \qquad (8)$$

where a is the tidal amplitude at the reflecting barrier, x is the distance seaward from the barrier, and $\lambda = (gh)^{1/2}T$ is the wavelength of the tide.

The tide and tidal currents are out of phase. At mean water, the strongest flood tide current occurs. It brings water into the embayment whose sea level must then rise. The incoming current finally ceases when the tide outside the embayment has reached high tide. As sea level begins to drop outside, an ebb current develops which continues until low tide outside the embayment. On the next rising tide, the flood tide again refills the embayment.

An important parameter determining the characters of tides in bays, sounds (large bays), and estuaries is the ratio $R = L/\lambda$, of the distance L between the reflecting barrier and the mouth of the embayment to the wavelength λ of the tide. Where the ratio is small, such as a deep bay or a small indentation in the coastline, the tide as expressed in (7) has the same range as the tide outside the embayment. This is because the factor $\cos(2\pi x/\lambda)$ determining the variation of the sea surface displacement η over the embayment is equal to unity if $2\pi x/\lambda$ is always much less than unity. The tidal currents are small since the factor $\sin(2\pi x/\lambda)$ is equal to $2\pi x/\lambda$, a small number.

For example, a semidaily tide ($T \simeq 12.5$ hours) in a fjord with a depth h of 500 m and a length L of 100 km has a wavelength $\lambda = (gh)^{1/2}T$ of 3180 km and hence a maximum tidal current of 2.8 cm/sec at the mouth of the fjord ($x = L$) if the amplitude a of the tide is 1 m. The current decreases linearly from the mouth to the head where the reflecting barrier forces the horizontal tidal current to be zero. Usually, fjords are separated from the outside by a shallow sill which can strongly inhibit the tidal penetration into the fjord. The tidal currents over sills can reach several meters per second. The example above treats currents inside the sill.

For large, shallow embayments, areal variations in tides can occur, if the length L is comparable to one-fourth of the semidaily or daily tidal wavelengths. One measure of this variation is the ratio between amplitudes of the tide at the head of the embayment and at the mouth:

$$\frac{\eta(x = 0)}{\eta(x = L)} = \frac{1}{\cos(2\pi L/\lambda)} \qquad (9)$$

If L is close to $\lambda/4$, the ratio is large so that the tidal range inside the embayment is much larger than the tide outside. The embayment is near resonance with the tide.

The Gulf of Mexico is near resonance with the daily tides. In shallow bays, such as the Bay of Fundy, the Straits of Georgia, and Long Island Sound which are near resonance, bottom stress and other effects limit the motion. By assuming that the tides consist of progressive waves that diminish exponentially with the distance of propagation, Redfield (1950) has been able to reproduce most of the tidal characteristics in these bays. As the incident wave propagates up a bay, its amplitude diminishes; after reflection and return to the mouth of the bay, the wave is significantly smaller in amplitude than when it entered the bay. Near the mouth the tide is progressive, whereas near the closed end of the bay it is standing. The maximum amplification of the tide in naturally occurring bays is about four times the incident tide.

TIDES IN SHALLOW WATER

Where the tidal range is a significant fraction of the depth, processes that cause the waveform to propagate act in the deeper water under the wave crest to move that part of the waveform more rapidly than the waveform near the trough. The tide becomes distorted, with the slope of the sea surface greater on the leading side of the crest. Where this distortion is large, there is significantly more landward water discharge associated with the crest of the tidal wave than there is seaward discharge associated with the trough of the tidal wave. This landward net transport of water is known as Stokes drift. The difference between the distorted and

undistorted tides is a shallow water tide. The flood tidal current is also greater than the ebbing current.

Ebb and flood channels occur in shallow water in which the tidal currents in one direction are largely confined to one set of channels and the currents in the opposite direction are confined to other channels. Ebb-flood channel systems are described in detail in Chapter 10.

A shallow bar separating the shelf from an embayment offers considerable resistance to tidal flow. Large differences in sea level develop during the tidal cycles, which generate strong tidal currents. While the velocity of the water is large across the bar, the total amount of water that can flow in and out of the embayment is severely limited by the constriction. As a result, the tide in the embayment is less than it would be without the bar.

In an embayment having a complicated bathymetry, a water parcel wanders into a variety of tidal regimes, such as tidal flats and channels, bars, and shoals; the simple theory predicting a return of the parcel to its original position at the end of each tidal cycle does not apply in this case. The Stokes drift induced by the distinct tidal regimes is a net drift of the parcel which may be thought of as a steady current.

Under some circumstances, tides should cause a net transport of sediment through the Stokes drift. However, this phenomenon has not been adequately documented by field study.

In the frictional boundary layer near the bottom where the horizontal tidal currents increase with height above the bottom, the increase in wave tidal momentum with height produces a steady current. This steady current, which can advect suspended sediment, is driven by variations in tidal momentum and is limited by turbulent friction.

EFFECTS OF THE EARTH'S ROTATION

In larger bodies of water, the tides and tidal currents are subject to the Coriolis effect, caused by the earth's rotation. A moving water parcel experiences a force proportional to its speed which, looking down on the sea surface, is to the right in the northern hemisphere and to the left in the southern hemisphere. This Coriolis effect, when not counteracted by another force, drives the water in an elliptical path: the direction of the tidal current rotates clockwise in the northern hemisphere and counterclockwise in the southern hemisphere; the speed of the current is never zero. The semidaily tidal currents in the Middle Atlantic Bight are an example of this type of motion, shown schematically in

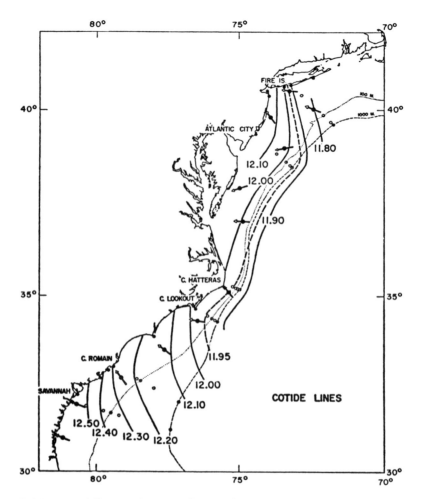

FIGURE 7. *Theoretical corange chart for the* M_2 *semidaily tide off the U.S. east coast. Ranges are in feet. From Redfield (1958).*

Fig. 6; the corange and cotidal charts are given in Figs. 7 and 8.

In regions where bottom stress can be neglected, the motion is determined approximately by the following equations:

$$\frac{\partial u}{\partial t} - fv = -g\frac{\partial \eta}{\partial x} \tag{10}$$

$$\frac{\partial v}{\partial t} + fu = -g\frac{\partial \eta}{\partial y} \tag{11}$$

$$\frac{\partial \eta}{\partial t} = -h\left(\frac{\partial u}{\partial x} + \frac{\partial v}{\partial y}\right) \tag{12}$$

The second terms, $-fv$ and $+fu$, in (10) and (11) represent effects of the earth's rotation.

There are two ways in which the Coriolis effect can alter tides and tidal current. In the case of a Poincare wave, the water parcel trajectories are ellipses whose major (larger) axis is in the direction of propagation; the ratio of the major to minor axis is the inertial period T_e divided by the period T of the tidal constituent. This type of tide can occur only where $T_e > T$ and is generally found in exposed regions such as continental shelves. The semidaily tide in the Middle Atlantic Bight (Fig. 7) is a standing Poincare wave (Redfield, 1958).

In restricted embayments such as the North Sea (Fig. 4) or on continental shelves where the direction of propagation of the tide is parallel to the coastline, a slope in the sea surface set up against the shore can balance the Coriolis effect. The result is a Kelvin wave. For a coastline parallel to the x direction and located at $y = 0$, a Kelvin wave has the form

$$\eta = ae^{-fy/c}\cos\left[\frac{2\pi}{T}\left(t - \frac{x}{c}\right)\right] \tag{13}$$

$$u = a\left(\frac{g}{h}\right)^{1/2}e^{-fy/c}\cos\left[\frac{2\pi}{T}\left(t - \frac{x}{c}\right)\right] \tag{14}$$

$$v = 0 \tag{15}$$

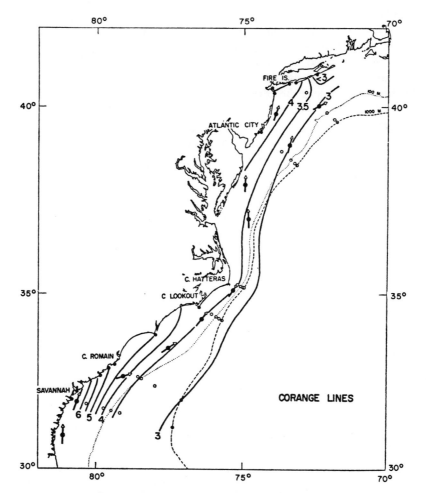

FIGURE 8. *Theoretical cotidal chart for the* M_2 *semidaily tide off the U.S. east coast. The cotidal lines are in hours after the Greenwich transit of the* M_2 *moon. From Redfield* (1958).

The tidal currents are parallel to shore ($v = 0$). At a latitude of 45° and with a depth of 50 m, a Kelvin wave decays to e^{-1} (36.8%) of its magnitude at the coast in a distance $y = c/f$ of 286 km. Conversely, a Kelvin wave propagating at 45°N along a continental shelf 150 km wide with a depth of 50 m has a tidal amplitude 59% of the amplitude at the coast.

A Kelvin wave propagating around a sea or ocean produces an amphidromic system. When a Kelvin wave enters an embayment in the northern hemisphere, such as the North Sea, it propagates counterclockwise around the embayment with the maximum tides and currents nearshore. Because the Kelvin waves do not decay rapidly away from their respective coasts, the motion at any given location is a combination of Kelvin waves. As a result, the tidal currents may not be colinear with the bathymetry. The sense of rotation of the tidal current direction is counterclockwise in this case, which is opposite to the direction for a Poincare wave on a continental shelf.

BOTTOM STRESS

Bottom stress modifies tides and tidal currents; its effect is greatest where strong tidal currents occur in shallow water. To model quantitatively the stress applied by the sediment on the water above, the flow is assumed to consist of a slowly varying tidal current superimposed on turbulence. The distribution of turbulent stress within the water determines the variation of tidal currents with distance above the bottom (velocity profile) and the dissipation of tidal energy. The details of flow near the bottom and estimates of bottom stress are central to the study of sediment transport.

The turbulent stress τ is often modeled as proportional to the rate of change of the current with increasing distance z above the bottom:

$$\frac{\tau}{\rho} = A_v \frac{\partial u}{\partial z} \qquad (16)$$

where the stress vector $\boldsymbol{\tau}$ is that part of the horizontal stress caused by vertical changes in the horizontal current $\mathbf{u}$ and A_v is the vertical eddy or turbulent viscosity. There are other terms caused by horizontal variations in $\mathbf{u}$ which could be added to the stress, but the term in (16) dominates turbulent processes in shallow water. A layer of water will produce a force opposite to the relative motion of the water just above the layer. The slower moving water near the bottom therefore acts as a drag on the water above. In general, A_v is determined by the spatial variations of currents, distance from boundaries, stratification of the water density, and the past history of the motion.

In turbulent boundary layers, a sublayer near the boundary layer exists where the stress is constant and the current speed increases logarithmically with distance from the boundary:

$$u = \frac{1}{k} \left(\frac{\tau_b}{\rho} \right)^{1/2} \ln \frac{30z}{z_0} \qquad (17)$$

where z_0 is the roughness length of the boundary which is determined by bottom irregularities, τ_b is the magnitude of the bottom stress, and k is von Karman's constant ($\simeq 0.4$). The effects of turbulence generally diminish with height above the water; the inertia of the water and the Coriolis effect become more important in balancing the pressure force due to the sea surface slope. In an oscillating tidal flow, the water farther from the bottom is moving faster and therefore has more inertia than the water near the bottom. In Fig. 6 the water farther from the bottom takes longer to respond to the pressure force and lags in time the motion near the bottom.

To model the attenuation of progressive tides due to bottom stress, an empirical formula is often used which relates the bottom stress τ_b/ρ to the vertically averaged tidal current U:

$$\boldsymbol{\tau}_b = -C_f \rho \mathbf{U}^2 \qquad (18)$$

The bottom stress is proportional to the square of the tidal current and opposite in direction to the current. The stress depends on the depth h through the current U, which is inversely proportional to $\sqrt{h}$. The stress is inversely proportional to h and hence is greater in shallower regions. The constant of proportionality C_f has been found from field studies to be about 0.0025. A number of such studies are described in Proudman (1952) for shallow regions around England.

INTERNAL TIDES

An internal tide is a wave with tidal period, associated with displacements within the water column and with very little displacement of the sea surface. Where there are two layers, the currents are in opposite directions in the two layers. The speed of propagation in this case is

$$c = \left(g \, \frac{\Delta \rho}{\rho} \cdot \frac{h_1 h_2}{h_1 + h_2} \right)^{1/2} \qquad (19)$$

where $\Delta\rho/\rho$ is the fractional change in water density between the lower and upper layers, g is the acceleration of gravity, and h_1 and h_2 are the thicknesses of the upper and lower layers. On a typical shelf with $\Delta\rho/\rho \simeq 0.002$, $g = 980$ cm/sec^2, $h_1 = 10$ m, and $h_2 = 50$ m, an internal wave would propagate with a speed c of 40.4 cm/sec, which is about 60 times slower than the surface tide's speed of propagation.

On the continental slope and at the shelf break, tidal currents interact with bathymetry to produce vertical displacements of density layers within the water column. The resulting undulations propagate both shoreward and seaward as internal tides.

As internal tides propagate inshore, the shoaling bottom thins the lower layer and hence slows the wave. Since the wave energy then becomes more concentrated, the amplitude of the currents increases as does the dissipation into turbulence. Sufficiently strong currents produce an internal bore in analogy to tidal bores in rivers. The internal tide becomes a series of pulses of waves with periods of several minutes, the pulses separated in time by the tidal period. The formation of internal bores occurs when the internal tidal currents equal the speed of propagation of internal waves.

Internal waves in a two-layered fluid cannot propagate shoreward of the intersection of the density interface with the bottom. Any internal waves that have not dissipated will lose the remainder of their energy to turbulence at the location where the water becomes unstratified. On some narrow shelves with strong stratification intercepting sharply rising bottom topography, internal tides are reflected back to sea, producing an internal standing tide.

A more realistic description of internal tides requires a continuously stratified water column and the Coriolis effect. The tides then propagate in the vertical as well as the horizontal direction. Whether an internal wave as it reflects off the bottom continues to propagate shoreward, or whether it is reflected seaward, is determined by the slope of the bottom and the direction of wave energy propagation (slope of the wave characteristic). A bottom slope steeper than the wave characteristic produces reflection seaward. Smaller slopes allow the wave to continue in the incident direction. A discussion of the reflection process may be found in Cacchione and Wunsch (1974).

Since the water density structure depends on the time of year, the existence and behavior of internal waves are

also seasonal. In summer when the shelf water is stratified, internal waves can exist over most of the shelf
regions; in winter, the lack of stratification precludes
occurrence of internal waves.

ADDITIONAL READING

This chapter was written to provide a qualitative introduction to the study of tidal currents. There is a large
literature on tidal phenomena; as in any scientific field,
the recent research is presented in succinct journal
articles which presuppose a knowledge of the field.

There are a number of texts which treat tides and tidal
currents in much more detail and more quantitatively
than was possible in this chapter. The general texts
by Sverdrup et al. (1942), Proudman (1952), and
Dietrich (1963) provide such treatments. The text by
Neumann and Pierson (1966) is more recent and more
advanced.

SUMMARY

Equations may be written to describe the propagation
of an idealized tidal wave down a straight-walled
channel. If bottom stress and the Coriolis effect are
neglected, the wave is seen to propagate as a result of
the interaction between water level displacement and
the flow of water induced by this displacement. The
speed of the tidal wave form (c) is equal to $(gh)^{1/2}$, where
g is the acceleration of gravity and h is water depth,
while the speed of the associated current (u) is proportional to this value. In very shallow water, u is reduced
by turbulent dissipation of energy and frictional loss of
energy to the bottom.

In nature, tides are propagated onto the continental
margin as waves from the open ocean. Such marginal
tides are said to cooscillate with the oceanic tide. Since
the incoming wave is rarely parallel to the coast, it
appears to propagate along the coast. Tidal waves
behaving in this fashion are referred to as *progressive
tidal waves*. The tidal wave may be partially reflected
back out to sea by the shoaling bottom and interact
with the next incoming wave so as to produce a *standing
tidal wave*. In a progressive tidal wave, maximum flood
velocity occurs at high water, while maximum ebb
velocity occurs at low water; in a standing tidal wave the
tide and tidal currents are out of phase, so that maximum
flood velocity occurs during the rising tide, and maximum ebb velocity occurs during the falling tide.

An important parameter determining the character
of tides in bays and estuaries is the ratio $R = L/\lambda$, where

L is the distance between the reflecting barrier and the
mouth of the embayment, and λ is the wavelength of
the tide. When the ratio is small, the tide within the bay
has the same range as outside, and tidal currents are
small. However, if L is comparable to one-fourth of the
semidaily or daily tidal wavelength, the embayment
resonates with the outside tide. Ranges are up to four
times higher, and currents are more intense.

When the tide range is a significant fraction of the
depth, the wave form becomes distorted, with the slope
of the sea surface becoming greater on the leading side
of the crest. The difference between the time–water
height curves of the undistorted and distorted tides is
called a shallow water tide. Where this distortion is
large, the velocity and discharge associated with the
crest are greater than those associated with the trough.
The resulting net transport of water is known as Stokes
drift.

In larger bodies of water, the tides and tidal currents
are subject to the Coriolis effect, caused by the earth's
rotation. A moving parcel of water experiences a force
proportional to its speed, which looking down at the sea
surface, is to the right in the northern hemisphere, and
to the left in the southern hemisphere. On open continental margins, the pressure force associated with the
passage of the tidal wave, together with the apparent
Coriolis force, results in a water parcel following an
elliptical trajectory with right-hand sense of rotation.
A tidal wave behaving in this fashion is a Poincare wave.
It occurs where the inertial period T_e is greater than the
period T of the tidal constituent. In restricted embayments such as the North Sea, or on continental shelves
where the tidal wave propagates parallel to the coastline,
coastward water flow induced by the Coriolis effect
is blocked by the coast, and there results a slope of the
sea surface up toward the coast. A tidal wave thus
modified is a Kelvin wave. A Kelvin wave propagating
around a sea or ocean is known as an amphidromic
system. The sense of rotation is counterclockwise.

The turbulent stress τ is often modeled as proportional
to the rate of change of the current with increasing
distance above bottom. The proportionality constant A_v
is the vertical eddy viscosity. It is determined by the
spatial variation of the currents, distance from boundaries, stratification of the water density, and the past
history of the motion. In turbulent boundary layers, a
sublayer near the boundary exists where stress is constant and the current speed increases logarithmically
with distance from the boundary. The slope of velocity
profile is determined in part by the degree of roughness
of the bottom, as measured by a bottom roughness
length Z_0.

An internal tide is a wave with a tidal period, associated with displacements within the water column, and with very little displacement of the sea surface. The wave may occur at the interface between fluids of two densities, or may occur in a continuously stratified fluid. On a typical shelf, an internal wave would propagate with a speed about 60 times slower than the surface tide's speed of propagation.

As the internal tide propagates inshore, the shoaling bottom thins the lower layer and hence slows the wave. Amplitude increases as does dissipation into turbulence; eventually the wave becomes a bore. Internal waves in a two-layered fluid cannot propagate shoreward of the intersection of the density interface with the bottom. At this point the waves lose their energy to turbulence, or if the bottom slope is steep enough, are reflected.

SYMBOLS

A_v vertical eddy coefficient

a amplitude

C_f drag coefficient

c phase velocity of tidal wave

g acceleration of gravity

h water depth

K a constant; von Karman's constant ($\simeq 0.4$)

L horizontal length scale

T period of tidal wave

t time

U vertically averaged tidal current

u current velocity

x horizontal distance

Z_0 roughness length

z vertical distance

λ wavelength

η vertical displacement of sea surface with respect to mean water level

ρ density

REFERENCES

Cacchione, D. and C. I. Wunsch (1974). Experimental study of internal waves over a slope. *J. Fluid Mech.*, **66:** 233–239.

Dietrich, G. (1963). *General Oceanography*. New York: Wiley-Interscience, 588 pp.

Doodson, A. T. and H. D. Warburg (1941). *Admiralty Manual of Tides*, London: HM Stationery Office, 270 pp.

Fleming, R. H. and R. Revelle (1939). Physical processes in the ocean. In P. D. Trask, ed., *Recent Marine Sediments*. New York: Dover, pp. 48–141.

Neumann, G. and W. J. Pierson (1966). *Principles of Physical Oceanography*. Englewood Cliffs, N.J.: Prentice-Hall, 545 pp.

Proudman, J. (1952). *Dynamical Oceanography*. New York: Dover, 409 pp.

Redfield, A. C. (1950). The analysis of tidal phenomena in narrow embayments. *Pap. Phys. Oceanogr. Meteorol.*, **11**(4): 1–36.

Redfield, A. C. (1958). The influence of the continental shelf on the tides of the Atlantic coast of the United States. *J. Mar. Res.*, **17:** 432–448.

Sverdrup, H. V. (1927). Dynamics of tides on the North Siberian shelf, results from the Maud Expedition. *Geofys. Publ.*, **4:** 5.

Sverdrup, H. V., M. W. Johnson, and R. H. Fleming (1942). *The Oceans*. Englewood Cliffs, N.J.: Prentice-Hall, 1087 pp.

U.S. Department of Commerce, National Oceanic and Atmospheric Administration, National Ocean Survey (1974). *Tide Tables, East Coast of North and South Americas, 1973*. National Ocean Survey, Rockville, Maryland, 288 pp.

CHAPTER 6

Wave Climate of the Continental Margin: Elements of its Mathematical Description

OLE SECHER MADSEN

Ralph M. Parsons Laboratory, Massachusetts Institute of Technology, Cambridge, Massachusetts

It should be evident to anyone who has spent more than a couple of hours at an ocean beach that waves are one of the dominating features of this environment. The energy associated with these waves is surprisingly large and it is to a large degree expended in the nearshore region through the process of wave breaking. The rate at which energy is expended in the surf when the breakers are about 1 m high is approximately equivalent to 3 kW/m of beach; i.e., the rate at which energy is dissipated along 400 km of this beach is equivalent to the rate of energy production of a nuclear power plant. It is therefore not surprising that wave action has the ability to alter the appearance of the shoreline drastically. The waves referred to in the preceding are the wind-generated waves of periods 3 to 20 seconds, which is the wave motion whose mathematical description is the topic of this chapter.

Along their way from the deep ocean to the shore the waves undergo changes because of the changing depth. At first, when the water is sufficiently deep, the waves are unaware of the presence of the bottom. As the depth decreases the waves start to feel the bottom and the bottom sediment starts to feel the waves. The oscillatory water motion associated with the waves exerts a shear stress on the bottom, which may be shown to be several times larger than the shear stress associated with a steady current of comparable magnitude. Since the ability of a water motion to move sediment generally is related to the magnitude of the shear stress exerted on the bottom the important role played by wave motion in the sediment transport process on the continental shelf is evident. The water velocity, immediately above the bed, associated with the waves is to the first approximation purely oscillatory; i.e., it moves back and forth without any net motion. Even if the shear stress associated with the wave motion were capable of setting the bottom sediment in motion, no net sediment motion would result. The addition of a slowly varying current superposed on the wave motion, however, brings out the importance of the presence of the wave motion. The large shear stress exerted on the bottom by the wave motion may stir up the sediment which may then be moved even by a current, which by itself would have been incapable of causing any sediment transport.

This qualitative description of the sediment transport processes taking place on the continental shelf emphasizes the necessity of having a better than superficial knowledge of the water movement on the continental shelf. Only through a thorough understanding of the quantitative description of the water movements and their contribution to the sedimentation on the continental shelf can this environment be effectively managed.

This and the preceding four chapters attempt to introduce the reader to the quantitative description of water movements on the continental shelf.

The mathematical description of a short-period wave motion is the topic of this chapter. For short-period wave motions the interaction between the fluid and a solid bottom (through the bottom shear stress) may to a first approximation be neglected and later accounted for by considering the influence of viscous effects in the bottom boundary layer. As previously mentioned this short-period wave motion may be regarded as a mechanism stirring up the bottom sediment. The currents acting as transporting agents once sediment is stirred up by the wave motion may be due to tidal forces, density differences, or wind. These water movements may be regarded as special types of waves, but their scales are generally of such magnitude that effects that may be neglected in the analysis of short-period waves have to be included in their mathematical description. The quantitative description of tides thus calls for accounting for the influence of the earth's rotation (Chapter 5) as well as bottom friction. In many cases the influence of varying density of the fluid may be neglected. Density differences do, however, induce large-scale circulations whose mathematical description is outlined in Chapter 3. Finally, the wind will induce a current in addition to generating the short-period waves treated in this chapter. Wind-driven currents are the topic of Chapter 4. The earlier chapters and the present chapter deal only with the fluid motion; the actual fluid-sediment interaction arising from the fluid motion is discussed in Part II of this book.

The motivation behind the present chapter is to attempt to introduce the reader to the simplest mathematical description of a wave motion. It is hoped that the presentation outlines the basic assumptions and concepts underlying the mathematical treatment in sufficient detail, so that it provides not only a set of formulas but also gives the reader a physical understanding of what a wave is and how it behaves. Numerical examples at the end of the chapter are included to illustrate the use of the formulas presented in the text. For a more detailed presentation of some of the topics touched upon in the following, the reader is referred to the textbook edited by Ippen (1966), which also treats topics not included in this presentation.

GOVERNING EQUATIONS

The equations governing fluid motions are derived from the basic principles of conservation of mass and momentum. For most water wave problems it is reasonable to

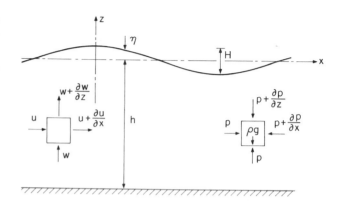

FIGURE 1. *Definition of coordinate system and symbols. Control volume and elementary particle of unit dimensions for the derivation of equations (1), (3), and (4).*

assume the fluid to be homogeneous and incompressible. In a Eulerian description, the conservation of mass principle then reduces to a statement of equality of the rate at which fluid enters and leaves a fixed control volume. In two dimensions, as shown in Fig. 1, this is expressed as

$$\frac{\partial u}{\partial x} + \frac{\partial w}{\partial z} = 0 \tag{1}$$

in which u is the horizontal and w the vertical component of the fluid velocity.

The conservation of momentum principle may be interpreted as an equilibrium of forces, inertia as well as body and surface forces, acting on an elementary fluid particle. Of body forces only gravity, acting in the negative z direction, will be considered. Surface forces generally consist of pressure forces and viscous forces. However, for most water wave problems the viscous forces are important only in the immediate vicinity of solid boundaries, within the boundary layer. Hence, treating the flow within the boundary layer separately, it is a reasonable assumption to regard the major portion of the fluid as being inviscid. With this assumption surface forces on an elementary fluid particle consist of pressure forces only, and force equilibrium in the horizontal direction becomes

$$\rho \frac{Du}{Dt} = - \frac{\partial p}{\partial x} \tag{2}$$

where ρ is the fluid density, t is time, and p is the fluid pressure. The acceleration term in (2) is to be interpreted as the time rate of change of horizontal velocity following the same fluid particle. Since a Eulerian description expresses the velocity as a function of space and time, i.e., $u = u(x, z, t)$, the substantial derivative, Du/Dt, can be expressed as

$$\frac{Du}{Dt} = \frac{\partial u}{\partial t} + u \frac{\partial u}{\partial x} + w \frac{\partial u}{\partial z} = - \frac{1}{\rho} \frac{\partial p}{\partial x} \tag{3}$$

where the first term on the left-hand side is termed the local acceleration and the two nonlinear terms are referred to as convective accelerations.

In a similar manner force equilibrium in the vertical direction yields

$$\frac{Dw}{Dt} = \frac{\partial w}{\partial t} + u\frac{\partial w}{\partial x} + w\frac{\partial w}{\partial z} = -\frac{1}{\rho}\frac{\partial p}{\partial z} - g \qquad (4)$$

in which g is the acceleration due to gravity.

Introducing the Velocity Potential

The three equations (1), (3), and (4) constitute the equations governing the three variables u, w, and p. Because of their nonlinearity, arising from the convective accelerations, these equations are not readily solvable. They may, however, be further simplified by assuming the existence of a *velocity potential* ϕ defined by

$$(u, w) = \left(\frac{\partial \phi}{\partial x}, \frac{\partial \phi}{\partial z}\right) \qquad (5)$$

The condition for the existence of a velocity potential is that the flow be irrotational, and a partial justification for making this assumption in the case of water waves is given by Lamb (1932). The ultimate justification for making this assumption is, however, that it leads to solutions that adequately describe the real problem.

Introducing the velocity potential defined by (5) in (1) leads to *Laplace's equation*:

$$\frac{\partial^2 \phi}{\partial x^2} + \frac{\partial^2 \phi}{\partial z^2} = 0 \qquad (6)$$

and when the velocity potential is introduced in the momentum equations, (3) and (4), these may be integrated to give *Bernoulli's generalized equation*:

$$\frac{\partial \phi}{\partial t} + \frac{1}{2}\left[\left(\frac{\partial \phi}{\partial x}\right)^2 + \left(\frac{\partial \phi}{\partial z}\right)^2\right] + gz + \frac{p}{\rho} = 0 \qquad (7)$$

where the right-hand side, strictly speaking, is an arbitrary function of time, which without loss of generality has been put equal to zero for the present analysis. Equation 7 is seen to reduce to Bernoulli's equation as it is used in hydraulics, if one assumes the problem to be steady, i.e., $\partial/\partial t = 0$.

It is seen that the introduction of the velocity potential has reduced the problem to the determination of two unknowns, ϕ and p, with the equation governing the kinematics of the problem, (6), being the well-known Laplace's equation. Once ϕ is determined the pressure is readily obtained from (7).

Boundary Conditions

To solve Laplace's equation certain boundary conditions must be specified. Taking the problem illustrated in Fig. 1 of the motion of a fluid with a free surface bounded below by a horizontal, impermeable bottom at $z = -h$, the boundary condition to be satisfied at $z = -h$ is that of no flow normal to the boundary:

$$\frac{\partial \phi}{\partial z} = 0 \qquad \text{at} \quad z = -h \qquad (8)$$

At the free surface the kinematic boundary condition states that a particle once on the free surface must remain there. Denoting the location of the free surface relative to the still water level, $z = 0$, by $\eta = \eta(x, t)$ the time rate of change of η following a particle on the free surface must equal the vertical velocity component of this particle. This leads to the *kinematic boundary condition*

$$\frac{D\eta}{Dt} = \frac{\partial \eta}{\partial t} + \frac{\partial \phi}{\partial x}\frac{\partial \eta}{\partial x} = \frac{\partial \phi}{\partial z} \qquad \text{at} \quad z = \eta(x, t) \qquad (9)$$

The difficulty in applying the kinematic boundary condition arises not only from its nonlinear form but also from the fact that it applies at the location of the free surface, $z = \eta$, which itself is part of the solution and hence an unknown. This calls for an additional boundary condition to be satisfied at the free surface. This boundary condition is obtained by expressing the dynamic condition that the pressure just below the free surface (surface tension effects being neglected) must equal the atmospheric pressure above the surface. For motions having a typical length scale, which is small relative to meteorological scales, the atmospheric pressure may without loss in generality be taken as constant and equal to zero. Hence, introducing $p = 0$ in (7) applied at the free surface leads to the *dynamic boundary condition*:

$$\frac{\partial \phi}{\partial t} + \frac{1}{2}\left[\left(\frac{\partial \phi}{\partial x}\right)^2 + \left(\frac{\partial \phi}{\partial z}\right)^2\right] + g\eta = 0$$
$$\text{at} \quad z = \eta(x, t) \qquad (10)$$

The solution of Laplace's equation subject to the boundary conditions stated is not readily performed. Although the governing equation itself is linear, the nonlinear boundary conditions (9) and (10) preclude a simple solution, unless further simplifications are introduced.

Linearized Governing Equations

Since the difficulties in obtaining a solution to the problem defined in the preceding sections are intimately related to the nonlinearity of the boundary conditions, the obvious simplification is to linearize these conditions. This linearization may be performed in a rigorous manner as done by Wehausen and Laitone (1960). For the present purpose, however, it suffices to state that the linearization of the boundary conditions (9) and (10) is justified when the wave height, H in Fig. 1, is small relative to the wavelength L and the water depth h. Furthermore, when small-amplitude waves are assumed, the free surface boundary conditions may with sufficient accuracy be applied at the known location, $z = 0$, rather than at the actual location of the free surface, $z = \eta$.

Thus, for infinitesimally small waves, the governing equation becomes

$$\frac{\partial^2 \phi}{\partial x^2} + \frac{\partial^2 \phi}{\partial z^2} = 0 \tag{11}$$

subject to the boundary conditions

$$\frac{\partial \phi}{\partial z} = 0 \quad \text{at} \quad z = -h \tag{12}$$

and

$$\frac{\partial^2 \phi}{\partial t^2} + g \frac{\partial \phi}{\partial z} = 0 \quad \text{at} \quad z = 0 \tag{13}$$

which is obtained from the linearized form of (9) by introducing the expression for the free surface profile obtained from (10):

$$\eta = -\frac{1}{g} \frac{\partial \phi}{\partial t} \quad \text{at} \quad z = 0 \tag{14}$$

Equations 11, 12, and 13 determine the velocity potential, and the corresponding surface profile is obtained directly from (14).

The pressure field associated with a solution of these equations is found from the linearized form of Bernoulli's generalized equation, (7):

$$p = -\rho \frac{\partial \phi}{\partial t} - \rho g z \tag{15}$$

where the term $\rho g z$ is identified as the hydrostatic pressure distribution, which would exist in the water in the absence of any wave motion. Thus, a convenient quantity p^+, expressing *the pressure caused by the wave motion*, is

$$p^+ = p + \rho g z = -\rho \frac{\partial \phi}{\partial t} \tag{16}$$

In spite of all the assumptions made in the preceding derivations of the linearized governing equations, these describe many observed features of water waves sufficiently accurately to be useful, which is the ultimate test of any mathematical description of a physical phenomenon.

SMALL-AMPLITUDE PROGRESSIVE WAVES

Velocity Potential and Surface Profile

Assume a velocity potential of the form

$$\phi = \frac{ga}{\omega} \frac{\cosh k(z + h)}{\cosh kh} \sin(kx - \omega t) \tag{17}$$

in which k and ω are constants of dimensions (length)$^{-1}$ and (time)$^{-1}$, respectively, and a has dimensions of length.

This velocity potential is readily seen to satisfy the governing equation, (11), and the boundary condition at the horizontal, impermeable bottom, (12). To get a feeling for the kind of fluid motion that is described by this velocity potential, the corresponding *free surface profile* η is obtained from (14) as

$$\eta = a \cos(kx - \omega t) \tag{18}$$

which shows the surface profile to be sinusoidal with maxima (wave crests) and minima (wave troughs) located a distance a above and below the still water level, respectively. The quantity a in (17) is consequently the wave amplitude or half the wave height, $H/2$, as defined in Fig. 1:

$$a = \frac{H}{2} = \text{wave amplitude} \tag{19}$$

Taking a snapshot view of the profile, say at $t = 0$, this is seen to vary sinusoidally in space with a *wavelength* L given by

$$k = \frac{2\pi}{L} = \text{wave number} \tag{20}$$

At any given location, say at $x = 0$, the free surface elevation varies sinusoidally with time having a *period* T determined by

$$\omega = \frac{2\pi}{T} = \text{radian frequency} \tag{21}$$

THE PHASE VELOCITY.   Imagine a comparison of two snapshots of the surface profile, one taken at $t = 0$ and the other at $t = \delta t$, where δt is a fraction of a wave period. The wave crest, $\eta = a$, which initially was located at $x = 0$, would at $t = \delta t$ be located at $x = \delta x$, where δx is determined by the constancy of the *phase*, $(kx - \omega t) = k \, \delta x - \omega \, \delta t = 0$. Hence,

$$\frac{\delta x}{\delta t} = c = \frac{\omega}{k} = \frac{L}{T} \tag{22}$$

which demonstrates that the surface profile (18) is that of a progressive wave propagating without change in form in the positive x direction at a constant speed, the *phase velocity c* given by (22). Had the phase $(kx + \omega t)$ been chosen, the argument above would have shown the wave to be propagating in the negative x direction.

The Dispersion Relationship

When the velocity potential is introduced in the remaining boundary condition, (13), the *dispersion relationship*

$$\omega^2 = kg \tanh kh \qquad (23)$$

is obtained. When this is written in the form

$$\frac{\omega^2 h}{g} \frac{1}{kh} = \tanh kh \qquad (24)$$

it is readily seen that (23) gives a unique determination of the wave number k for given wave period and water depth, since the left-hand side of (24) decreases monotonically with kh, whereas the right-hand side monotonically increases with kh.

The solution of (23) in terms of the wavelength as a function of wave period and water depth may be written as

$$L = \frac{gT^2}{2\pi} \tanh \frac{2\pi h}{L} \qquad (25)$$

which can also be rearranged to yield an expression for the phase velocity

$$c = \frac{L}{T} = \frac{gT}{2\pi} \tanh \frac{2\pi h}{L} \qquad (26)$$

The two expressions (25) and (26) contain the wavelength on both sides of the equations. However, in special cases they may be stated in explicit form.

SHORT WAVES (DEEP WATER). For wavelengths that are short compared to the water depth, i.e., kh large, $\tanh kh$ approaches unity and (25) and (26) may be written

$$L = L_0 = \frac{g}{2\pi} T^2 \quad \text{for} \quad \frac{h}{L} > 0.5 \qquad (27)$$

and

$$c = c_0 = \frac{g}{2\pi} T \quad \text{for} \quad \frac{h}{L} > 0.5 \qquad (28)$$

where the usual terminology of identifying deep-water values by the subscript zero has been introduced.

This result shows that the phase velocity in deep water is proportional to the wave period, i.e., the waves travel faster the longer their period, hence the term dispersion relationship. As indicated, these relationships are generally considered valid when $h/L > 0.5$.

LONG WAVES (SHALLOW WATER). For wavelengths that are long compared to the water depth the value of $\tanh kh$ approaches kh and (25) and (26) may be written

$$L = (gh)^{1/2} T \quad \text{for} \quad \frac{h}{L} < 0.05 \qquad (29)$$

and

$$c = (gh)^{1/2} \quad \text{for} \quad \frac{h}{L} < 0.05 \qquad (30)$$

These results show that long waves travel at the same speed independent of their period, so long as their wavelength satisfies the condition $h/L < 0.05$ as indicated. In this sense long waves are nondispersive.

INTERMEDIATE WAVES $(0.5 < h/L < 0.05)$. In the intermediate range of wavelengths, no simplification of (25) and (26) can be obtained. Introducing the deep-water wavelength, from (27), the general formula for the wavelength, (25) and (26) become

$$L = L_0 \tanh \frac{2\pi h}{L} \qquad (31)$$

and

$$c = c_0 \tanh \frac{2\pi h}{L} \qquad (32)$$

which show that, for a given period, i.e., L_0 and c_0 constant, the wavelength and the phase velocity decrease with decreasing water depth. As was shown previously the relationship expressed by (31) is unique and it is therefore possible to determine the value of, say, h/L as a function of the parameter h/L_0, which is readily evaluated for any given wave period and water depth. The tables by Wiegel (1954) are very useful in performing this task; a graphical solution is presented in Fig. 2.

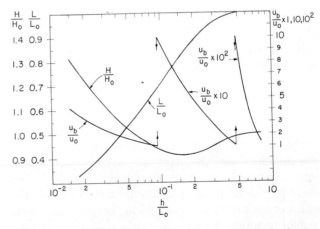

FIGURE 2. *Variation of wave parameters with depth. For* $h/L_0 < 0.015$ *the long-wave formulae apply (note: shoaling coefficient and near-bottom velocity are for* **normally incident waves**).

Wave-Induced Pressure

The pressure field associated with the wave motion described by the velocity potential given by (17) is expressed by (16),

$$p^+ = \rho g a \frac{\cosh k(z+h)}{\cosh kh} \cos(kx - \omega t) \qquad (33)$$

which shows that p^+, the pressure induced by the wave motion, not surprisingly is in phase with the surface profile; i.e., a positive pressure is induced under the crest, a negative pressure is induced under the trough.

The variation of the magnitude of the maximum wave-induced pressure with depth below the still water level is quite revealing. For short waves, large values of kh, the hyperbolic cosines may be approximated by exponentials and (33) may be written as

$$p^+ = \rho g a e^{kz} \cos(kx - \omega t) \qquad \text{for} \quad \frac{h}{L} > 0.5 \qquad (34)$$

which shows the wave-induced pressure to decay exponentially with depth (z is negative). At a depth corresponding to half a wavelength below the still water level the induced pressure variation is found to be only about 4% of the pressure variation immediately below the surface. For small values of kh, i.e., long waves, the hyperbolic cosines approach unity and (33) becomes independent of depth:

$$p^+ = \rho g a \cos(kx - \omega t) = \rho g \eta \qquad \text{for} \quad \frac{h}{L} < 0.05 \qquad (35)$$

Upon introducing (35) into the expression for the total pressure, (15), this becomes

$$p = \rho g(\eta - z) \qquad \text{for} \quad \frac{h}{L} < 0.05 \qquad (36)$$

which is nothing but a statement of hydrostatic pressure variation.

The statement of hydrostatic pressure for long waves is equivalent to a statement of the neglect of vertical fluid accelerations, i.e., setting the left-hand side of (4) equal to zero. If from the outset one assumes long waves and therefore takes the pressure distribution to be hydrostatic, a simple set of equations is obtained from the continuity and horizontal momentum equations. These are the equations used in Chapters 3, 4, and 5 which deal with motions of such large scales that $h/L \ll 0.05$. For these large-scale motions additional factors such as the effect of the earth's rotation and the bottom shear stress should be included.

The relationship between the free surface elevation and the wave-induced pressure is put to use when employing pressure transducers as a wave recording device.

Imagine, for example, a pressure transducer located near the bottom. The recorded pressure difference would then vary sinusoidally with time, as given by (33) with $z = -h$. In particular, the maximum pressure variation would be given by

$$\Delta p = p^+_{\max} - p^+_{\min} = \rho g \frac{2a}{\cosh kh} \qquad (37)$$

Hence, obtaining the wave period from the recorded variation of pressure with time would allow a determination of the wave period and hence the wavelength L, assuming a known depth. The wave height of the wave passing overhead can then be determined from (37) as

$$H = 2a = \frac{\Delta p}{\rho g} \cosh kh \qquad (38)$$

The requirement for accuracy of the recorded pressure when recording deep-water waves is evident from the discussion of the exponential decay with depth. In a way the depth at which a pressure transducer is placed may be used as a natural filter, filtering out short-period, i.e., high-frequency, oscillations.

Velocity Field and Accelerations

From the definition of the velocity potential, (5), the velocity field associated with a progressive wave given by (17) is readily evaluated. The *horizontal velocity component* is given by

$$u = \frac{\partial \phi}{\partial x} = a\omega \frac{\cosh k(z+h)}{\sinh kh} \cos(kx - \omega t) \qquad (39)$$

where the dispersion relationship (23) has been introduced to simplify the expression. It is noted that the horizontal velocity is in the direction of wave propagation under the crest and in the opposing direction under the trough. It is also seen that the horizontal water velocity component is of the order $a\omega = (ka)c$, where c is the phase velocity given by (22). Hence, the water velocity is much smaller than the phase velocity of the wave, since the basic assumption behind the linearized theory was that $a/L \ll 1$.

When investigating the fluid motion in the boundary layer, the maximum horizontal velocity component in the immediate vicinity of the bottom, u_b, is of particular interest. At $z = -h$, (39) gives

$$u_{\max} = u_b = \frac{a\omega}{\sinh kh} \qquad \text{at} \quad z = -h \qquad (40)$$

which shows the velocity at the bottom to vanish for deep-water waves, kh large, indicating that the bottom is hardly felt. For shallow-water waves, kh small, the velocity is constant throughout the depth and may be expressed in a particularly simple form:

$$u = \left(\frac{g}{h}\right)^{1/2} \eta \quad \text{for} \quad \frac{h}{L} < 0.05 \quad (41)$$

The *vertical velocity component* is found to be

$$w = \frac{\partial \phi}{\partial z} = a\omega \frac{\sinh k(z + h)}{\sinh kh} \sin(kx - \omega t) \quad (42)$$

which vanishes at the bottom, $z = -h$, as it should.

ACCELERATIONS. To the degree of accuracy consistent with the linear theory developed here the fluid accelerations are readily evaluated from (39) and (42) as $\partial u / \partial t$ and $\partial w / \partial t$.

Water Particle Orbits

So far a Eulerian description of the fluid motion associated with a progressive small-amplitude wave has been presented. An alternative description of the fluid motion is to determine the paths followed by individual water particles, a Lagrangian description. When observing, for example, the motion of a streak of dyed water particles one is in fact observing the motion in a Lagrangian sense. For linear progressive waves the transformation from a Eulerian description to a Lagrangian one is rather simple and is outlined in the following.

Denoting the mean position of a fluid particle by (X, Z) one may determine the position of this particle as a function of time (x_P, z_P) from

$$x_P = X + \int u_P \, dt \quad (43)$$

$$z_P = Z + \int w_P \, dt \quad (44)$$

where u_P and w_P are the Eulerian velocities evaluated at the unknown position (x_P, z_P) of the fluid particle. However, it is consistent with the linear approximation to evaluate u_P and w_P at the mean position of the particle. When this is done by introducing (39) and (42) and performing the integration the displacement of a water particle from its mean position may be expressed as

$$x_P - X = -a \frac{\cosh k(Z + h)}{\sinh kh} \sin(kX - \omega t) \quad (45)$$

for the *horizontal displacement*, and

$$z_P - Z = a \frac{\sinh k(Z + h)}{\sinh kh} \cos(kX - \omega t) \quad (46)$$

for the *vertical displacement*. As one would expect it is seen that the vertical displacement is identical to the expres-

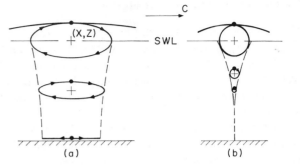

FIGURE 3. *Water particle orbits under progressive waves:* (a) *intermediate waves,* (b) *short waves.*

sion for the surface profile when $Z = 0$ is introduced into (46).

Equations 45 and 46 can be combined to yield the equation for the *particle orbits* relative to their mean position:

$$\frac{(x_P - X)^2}{A^2} + \frac{(z_P - Z)^2}{B^2} = 1 \quad (47)$$

in which

$$A = a \frac{\cosh k(Z + h)}{\sinh kh} \quad (48)$$

$$B = a \frac{\sinh k(Z + h)}{\sinh kh} \quad (49)$$

Equation 47 shows the particle orbits to be elliptical closed orbits with a *horizontal semiaxis A*, given by (48), and a *vertical semiaxis B*, given by (49). This is illustrated in Fig. 3.

In deep water the hyperbolic cosine and sine are identical, and (48) and (49) show that the horizontal and vertical axes of the elliptical orbit are identical. The water particles therefore describe circular orbits in deep water with radii

$$A = B = ae^{kZ} \quad \text{for} \quad \frac{h}{L} > 0.5 \quad (50)$$

decreasing exponentially with depth below the still water level.

In shallow water, long waves, the orbits become elongated with a vertical axis decreasing linearly with distance from the still water level where $B = a$ to zero at the bottom and with a constant length of the horizontal axis. In the general case, the horizontal semiaxis at the bottom $Z = -h$ becomes

$$A = A_b = \frac{a}{\sinh kh} \quad \text{at} \quad z = -h \quad (51)$$

which agrees with expression (40) derived for the horizontal particle velocity at the bottom, since these are necessarily related through $u_b = A_b \omega$.

Wave Energy and Energy Propagation

In the preceding sections the fluid motion associated with small-amplitude progressive waves has been analyzed in detail. The concept of the energy associated with a certain fluid motion is often helpful in the analysis of wave motions as it is, for example, in open channel hydraulics.

The energy associated with a wave motion consists of potential as well as kinetic energy. Since the fluid motion is time dependent it does not make sense to evaluate the instantaneous energy content of a fluid particle. However, an average energy associated with a progressive wave is useful. For a small-amplitude wave as considered here the average energy per unit surface area associated with the wave motion, *the specific energy* E_S, may be evaluated from

$$E_S = \frac{1}{L} \left(\int_0^L dx \int_{-h}^{\eta} \rho g(z + h)\, dz \right.$$

$$\left. + \int_0^L dx \int_{-h}^{\eta} \tfrac{1}{2}\rho(u^2 + w^2)\, dz - \tfrac{1}{2}\rho g h^2 L \right) \quad (52)$$

where the first term on the right-hand side expresses the potential energy, choosing the horizontal bottom as datum, the second term expresses the kinetic energy with u and w given by (39) and (42), and the last term expresses the energy of the fluid in the absence of any wave motion. Performing the integrations leads to the simple result that

$$E_S = \tfrac{1}{2}\rho g a^2 \quad (53)$$

with average potential energy and average kinetic energy per unit surface area being equal.

In addition to the specific energy the value of the rate at which energy propagates, the energy flux, is extremely useful. Again only a time-averaged value of this quantity makes sense and it is derived from the first law of thermodynamics, which states that the rate at which work is done on a body equals the rate of increase of energy of this body.

For a two-dimensional wave motion consider an imaginary "curtain" consisting of the water particles that are located at $x = 0$ at time $t = 0$ [see (45)]. Now consider the fluid, initially occupying the region $x < 0$, as doing work on the fluid on the other side of this curtain. During one wave period, work W_T is done on the fluid initially occupying the region $x > 0$, where

$$W_T = \int_0^T dt \int_{-h}^{\eta} p^+ u\, dz \quad (54)$$

where p^+ is given by (33) and u by (39). After the completion of one wave period the imaginary curtain has returned to its original position, $x = 0$, and the energy content of the fluid occupying the region $x > 0$ has increased by an amount equal to W_T. This indicates that an amount of energy equal to W_T has passed from the region $x < 0$ to the region $x > 0$ during one wave period. This energy propagation may be expressed as the average rate at which energy passes a certain point, the *energy flux*

$$E_F = \frac{1}{T} W_T = \frac{1}{2}\rho g a^2 \left[\frac{\omega}{k}\frac{1}{2}\left(1 + \frac{2kh}{\sinh kh}\right)\right] \quad (55)$$

where the energy flux is per unit length of wave crest, since the two-dimensional problem is being considered.

Comparison of (53) and (55) shows that the energy flux can be expressed as

$$E_F = E_S \left[c\,\frac{1}{2}\left(1 + \frac{2kh}{\sinh 2kh}\right)\right] = E_S\, c_g \quad (56)$$

where the bracketed term, which has the dimensions of a velocity, is referred to as *the group velocity* c_g:

$$c_g = \frac{1}{2}\left(1 + \frac{2kh}{\sinh 2kh}\right) c \quad (57)$$

For short waves $\sinh 2kh \gg 2kh$ and (57) shows that the group velocity is half of the phase velocity, i.e.,

$$c_g = \tfrac{1}{2}c_0 \qquad \text{for} \quad \frac{h}{L} > 0.5 \quad (58)$$

For long waves $\sinh 2kh \simeq 2kh$ and (57) shows that the group velocity equals the phase velocity:

$$c_g = c = (gh)^{1/2} \qquad \text{for} \quad \frac{h}{L} < 0.05 \quad (59)$$

A somewhat simplified physical interpretation of the group velocity is that it is the velocity at which the specific energy associated with the wave motion propagates. With this interpretation of the group velocity it follows that the energy associated with a group consisting of a finite number of identical periodic waves, and therefore the wave group itself, propagates at a speed equal to the group velocity. For deep-water waves, for which the group velocity is half the phase velocity, this gives rise to a peculiar feature, which may be observed in a wave flume when a finite number of short waves are generated. Individual waves are observed to travel up through the group disappearing at the front while new waves appear at the rear.

SUPERPOSITION OF SMALL-AMPLITUDE WAVES

The linearized governing equations possess the extremely convenient feature that any linear combination of particular solutions to this set of equations is itself a solution to the governing equations. To illustrate this *principle of superposition* assume a solution

$$\phi = \phi_1 + \phi_2 \qquad (60)$$

constructed by adding two velocity potentials, ϕ_1 and ϕ_2, each of which satisfies the linearized governing equations (11), (12), and (13). That ϕ also satisfies the governing equations is readily verified. Take, for example, the boundary condition (12):

$$\frac{\partial \phi}{\partial z} = \frac{\partial(\phi_1 + \phi_2)}{\partial z} = \frac{\partial \phi_1}{\partial z} + \frac{\partial \phi_2}{\partial z} = 0$$
$$\text{at} \quad z = -h \qquad (61)$$

which clearly is satisfied.

Any linear function of the velocity potential given by (60) would also be given as the linear combination of the two components corresponding to ϕ_1 and ϕ_2. Thus, the surface profile, η, corresponding to (60) would be given by (14) and be expressible as $\eta = \eta_1 + \eta_2$, showing that surface profiles may be superposed.

Waves Propagating in the Same Direction

Applying the principle of superposition to a large number of individual progressive waves of different frequencies ω_n and amplitudes a_n, all propagating in the positive x direction, gives a velocity potential

$$\phi = \sum_{n=1}^{N} \frac{g a_n}{\omega_n} \frac{\cosh k_n(z + h)}{\cosh k_n h} \sin(k_n x - \omega_n t + \psi_n) \qquad (62)$$

with a surface profile given by

$$\eta = \sum_{n=1}^{N} a_n \cos(k_n x - \omega_n t + \psi_n) \qquad (63)$$

where the phase angle ψ_n has been introduced to preserve generality. One phase angle can, however, be taken equal to zero by an appropriate choice of the origin, $x = 0$. The radian frequencies ω_n and wave numbers k_n of the individual components are related through the dispersion relationship (23), and the energy associated with the nth component is proportional to a_n^2, as (53) indicates.

Considering the temporal variation of the free surface given by (63) at a particular point, say $x = 0$,

$$\eta(t) = \sum_{n=1}^{N} a_n \cos(\omega_n t - \psi_n) \qquad (64)$$

it is readily imagined that a linear combination of different frequency oscillations can represent a rather complex temporal variation of the free surface, not unlike the confused conditions existing in the ocean.

THE CONCEPT OF THE ENERGY SPECTRUM. The left-hand side of (64) corresponds to the record obtained from a conventional wave gage and the problem generally facing the scientist is to extract useful information from such a record, which necessarily is of finite length.

Subjecting a finite record of the surface elevation as a function of time to a spectrum analysis produces information about the value of the square of the amplitude, a_n^2, associated with the wave motion of frequency ω_n. Since a_n^2 is proportional to the energy of this component the result of this analysis is referred to as the *energy spectrum*, which may be viewed as expressing the distribution of the energy associated with the observed sea state among the various frequencies.

It is beyond the scope of this presentation to go into the details of spectrum analysis, for which the reader is referred to Kinsman (1965) and Harris (1972); that the concept of the energy spectrum is useful should be evident, however, from the foregoing discussion. Imagine, for example, that the energy spectrum corresponding to a certain wave record shows that the wave energy is concentrated around two particular radian frequencies, say $\omega_L = 0.5 \text{ sec}^{-1}$ and $\omega_S = 1.0 \text{ sec}^{-1}$. This information may, for example, be interpreted as the presence of a long-period swell arriving from a distant storm of period $T_L = 2\pi/\omega_L \simeq 12$ seconds, and a short-period wave motion of period $T_S = 2\pi/\omega_S = 6$ seconds, which could be associated with locally generated waves. It should be quite evident that information of the type obtained above probably would have escaped the observer having only the raw record of the free surface variation as a function of time.

Reflected Waves

As yet another example of the principle of superposition imagine two progressive waves of the same radian frequency, one traveling in the positive x direction, one in the negative x direction, and of amplitudes a_i and a_r, respectively. The corresponding velocity potential would be given by

$$\phi = \frac{g}{\omega} \frac{\cosh k(z + h)}{\cosh kh} [a_i \sin(kx - \omega t) - a_r \sin(kx + \omega t)] \qquad (65)$$

and the surface profile

$$\eta = a_i \cos(kx - \omega t) + a_r \cos(kx + \omega t) \qquad (66)$$

where the subscripts are chosen to indicate one wave as being the incident wave, and the other the reflected wave.

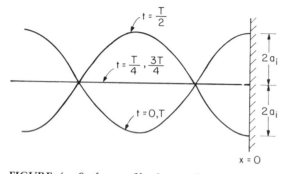

FIGURE 4. *Surface profile of a standing wave.*

STANDING WAVES.   Evaluating the horizontal velocity corresponding to the motion given by (65) at $x = 0$, one obtains

$$u = \frac{\partial \phi}{\partial x} = \frac{g}{\omega} \frac{\cosh k(z + h)}{\cosh kh} (a_i - a_r) \cos \omega t \qquad (67)$$

which shows that the horizontal velocity at $x = 0$ vanishes for $a_i = a_r$. The statement of $u = 0$ at $x = 0$ corresponds to introducing a vertical impermeable wall at $x = 0$, and the motion described by (66) for $a_i = a_r$ therefore corresponds to a fully reflected *standing wave*, whose surface profile is given by

$$\eta = 2a_i \cos kx \cos \omega t \qquad (68)$$

Inspection of (68) shows that the free surface appears completely flat, $\eta = 0$, whenever $\cos \omega t = 0$, i.e., at intervals of $T/2$. Also, it is seen that the surface elevation remains zero, at all times, at the points, the *nodes*, where $\cos kx = 0$, i.e., at intervals of $L/2$, whereas the total excursion of the free surface is a maximum at the *antinodes*, where $\cos kx = \pm 1$. This is illustrated in Fig. 4.

REFLECTION COEFFICIENT.   In the case of $a_i \neq a_r$ manipulation of (66) with use of trigonometric identities leads to an amplitude of the wave motion, denoted by $|\eta|$, which is a function of space

$$|\eta| = (a_i{}^2 + a_r{}^2 + 2a_i a_r \cos 2kx)^{1/2} \qquad (69)$$

This shows the resulting wave amplitude to vary between a maximum value of $|\eta|_{\max} = a_i + a_r$ and a minimum value of $|\eta|_{\min} = a_i - a_r$. Hence, knowledge of the spatial variation of the wave amplitude, or the wave height, associated with the combined motion enables the determination of the amplitudes of the incident and the reflected waves, since

$$a_i = \tfrac{1}{2}(|\eta|_{\max} + |\eta|_{\min}) \qquad (70)$$

and the *reflection coefficient* R, defined by

$$R = \frac{a_r}{a_i} = \frac{|\eta|_{\max} - |\eta|_{\min}}{|\eta|_{\max} + |\eta|_{\min}} \qquad (71)$$

THE ENERGY CONCEPT.   As an example of applying the energy concept, consider the problem treated in the preceding section, as arising from a wave incident on a partially reflecting beach.

The incident wave is associated with an energy flux in the positive x direction which per unit length of crest is given by (56):

$$E_{F,i} = \tfrac{1}{2}\rho g a_i{}^2 c_{g,i} \qquad (72)$$

whereas the reflected wave is associated with an energy flux in the negative x direction, which is

$$E_{F,r} = \tfrac{1}{2}\rho g a_r{}^2 c_{g,r} \qquad (73)$$

Since the water depth is constant and the waves have the same frequency the group velocity of the incident and reflected waves is identical, $c_{g,i} = c_{g,r} = c_g$, and the net rate at which energy propagates in the positive x direction per unit length of crest is given by

$$E_{F,i} - E_{F,r} = \tfrac{1}{2}\rho g a_i{}^2 c_g (1 - R^2) \qquad (74)$$

To conserve energy this net energy flux must equal the rate at which energy is dissipated per unit length of beach.

WAVE SHOALING AND REFRACTION

Normally Incident Waves on a Straight Beach

Consider an infinite train of normally incident periodic waves propagating from deep to shallow water over a gently sloping straight beach, i.e., the problem is two-dimensional.

CONSTANCY OF THE WAVE PERIOD.   That the wave period remains constant as the waves propagate over the gently sloping bottom may be seen from the following argument. Imagine an observer located at station I, where the water depth is h_I, and that this observer counts, say, N_I wave crests passing during a long time t_c. Another observer located at station II, where the water depth is h_{II}, counts the number of wave crests passing him, say N_{II}, during the same time interval, t_c. Now, the number of wave crests between stations I and II must necessarily remain constant and it follows that $N_I = N_{II}$. The wave period $T = t_c/N_I = t_c/N_{II}$ is consequently constant.

WAVE HEIGHT CHANGES DUE TO SHOALING.   It is assumed that the bottom slopes so gently that reflection can be ignored, also any energy dissipation is neglected. Considering the constancy of the wave energy in the region between stations I and II leads to the statement that the energy flux into and out of this region must be equal, i.e., with subscripts referring to stations I and II

$$E_{F,\text{I}} = \tfrac{1}{2}\rho g a_\text{I}^2 c_{g,\text{I}} = E_{F,\text{II}} = \tfrac{1}{2}\rho g a_\text{II}^2 c_{g,\text{II}} \quad (75)$$

where (56), which was derived for the constant-depth case, is assumed valid also for the case of a gently sloping bottom.

Equation 75 may be taken as an expression for the wave height variation due to the change in water depth between station I, $h = h_\text{I}$, and station II, $h = h_\text{II}$,

$$\frac{a_\text{II}}{a_\text{I}} = \frac{H_\text{II}}{H_\text{I}} \left(\frac{c_{g,\text{I}}}{c_{g,\text{II}}} \right)^{1/2} \quad (76)$$

If the waves may be characterized as long waves, i.e., $h/L < 0.05$, at both stations, (76) is particularly simple since (59) then gives

$$\frac{H_\text{II}}{H_\text{I}} = \left(\frac{h_\text{I}}{h_\text{II}} \right)^{1/4}, \qquad \frac{h_\text{I}}{L}, \frac{h_\text{II}}{L} < 0.05 \quad (77)$$

which is known as *Green's formula*.

If station I is assumed located in deep water, introduction of (58) gives

$$\frac{H}{H_0} = K_S = \left(\frac{c_0}{2c_g} \right)^{1/2} \quad (78)$$

where K_S is the *shoaling coefficient*, and the subscript II has been dropped. Since the general formula for c_g, (57), shows this to be a function of the parameter h/L, the shoaling coefficient may be presented as a function of the parameter h/L_0, as illustrated in Fig. 2. From Fig. 2 it is seen that the wave height as the waves propagate from deep to shallow water first decreases slightly, reaching a minimum value of $K_S = 0.91$, before increasing, apparently without limit, as the depth approaches zero. The increase of the wave height is ultimately limited by the breaking criterion, which is discussed in a later section.

When concerned with the effect of the wave motion on the bottom the maximum velocity immediately above the bottom is of major importance. This velocity is given by (40), and may conveniently be expressed as

$$\frac{u_b}{u_0} = \frac{u_b}{a_0\omega} = \frac{a}{a_0 \sinh kh} = \frac{K_S}{\sinh kh} \quad (79)$$

which as presented in Fig. 2 shows the rapid increase in near-bottom velocity as a wave shoals.

Wave Refraction

OBLIQUELY INCIDENT WAVES ON A STRAIGHT BEACH. As evidenced by the variation of wavelength with water depth presented in Fig. 2, it is seen that the phase ve-

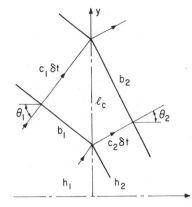

FIGURE 5. *Wave refraction. Wave crests, full lines; wave orthogonals, thin lines with arrows to indicate direction of propagation.*

locity, $c = L/T$, of a constant-period wave decreases with decreasing depth. A wave is assumed to propagate in a direction normal to its crest, in the direction of the wave orthogonal, and as a plane wave propagates from one region of constant depth, $h = h_1$, into another region where $h = h_2 < h_1$, its phase speed decreases from c_1 to c_2. As illustrated in Fig. 5, which shows two consecutive positions of a wave crest, the full line, an interval δt seconds apart, this change in phase velocity changes the direction of propagation from an angle of θ_1 with the x axis, which is perpendicular to the line along which the depth change occurs, to an angle θ_2.

Geometric considerations based on Fig. 5 yield

$$\sin \theta_1 = \frac{c_1 \delta t}{l_c} \qquad \text{and} \qquad \sin \theta_2 = \frac{c_2 \delta t}{l_c}$$

which may be combined to give

$$\frac{c_1}{\sin \theta_1} = \frac{c_2}{\sin \theta_2} \quad (80)$$

For a straight beach, bottom contours parallel with the y axis, one may regard the propagation of a wave train that in deep water has a direction of propagation at an angle θ_0 with the x axis, as a series of passages past small steps in the bottom. Each of these passages is associated with a directional change given by (80), and hence *Snell's law*,

$$\frac{c}{\sin \theta} = \frac{c_0}{\sin \theta_0} \quad (81)$$

is found to be valid for a beach with parallel contours.

Inspection of this result shows that the angle of propagation with the x axis decreases as the depth decreases. This means that the waves become more normally incident as the waves propagate into shallow water, a feature that may be observed at many beaches.

CHANGES IN WAVE HEIGHT DUE TO REFRACTION. To evaluate the change in wave height caused by refraction it is assumed that no energy propagates along a wave crest, that no reflection occurs, and that the energy dissipation is negligible. From these assumptions it follows that the energy flux between two neighboring orthogonals, a distance b apart, remains constant.

The variation of distance between two neighboring orthogonals may be found from Fig. 5, since this indicates that

$$\cos \theta_1 = \frac{b_1}{l_c} \quad \text{and} \quad \cos \theta_2 = \frac{b_2}{l_c}$$

or, by combining,

$$\frac{b_1}{b_2} = \frac{\cos \theta_1}{\cos \theta_2} \tag{82}$$

which again may be generalized for a straight beach with parallel bottom contours to read

$$\frac{b_0}{b} = \frac{\cos \theta_0}{\cos \theta} \tag{83}$$

with subscript zero referring to deep-water conditions.

The constancy of the energy flux between orthogonals therefore reads

$$E_{F,1}b_1 = \tfrac{1}{2}\rho g a_1{}^2 c_{g,1} b_1 = E_{F,2}b_2 = \tfrac{1}{2}\rho g a_2{}^2 c_{g,2} b_2 \tag{84}$$

or

$$\frac{a_2}{a_1} = \left(\frac{c_{g,1}}{c_{g,2}}\right)^{1/2} \left(\frac{b_1}{b_2}\right)^{1/2} \tag{85}$$

which by comparison with (76) shows that the effect of refraction is expressed by the factor $(b_1/b_2)^{1/2}$.

If station 1 is located in deep water, (85) may be written as

$$\frac{H}{H_0} = K_S \left(\frac{b_0}{b}\right)^{1/2} \tag{86}$$

where (78) has been introduced and the *refraction coefficient*

$$K_R = \left(\frac{b_0}{b}\right)^{1/2} \tag{87}$$

in the case of a straight beach with parallel bottom contours is given by (83).

Inspection of (83) shows that the effect of refraction of obliquely incident waves on a straight beach causes a decrease in wave height since, as (81) indicates, θ decreases with decreasing water depth.

WAVE REFRACTION DIAGRAMS. For a more complex bottom topography than the simple straight beach considered in the preceding section, formula (86) still holds, but the value of the refraction coefficient (87) must be obtained by graphical or numerical techniques.

Several graphical techniques are discussed by Johnson et al. (1948). These basically amount to applying Snell's law locally, in the form given by (80), thereby piecing together a graphical construction of a series of neighboring orthogonals. The distance b between these orthogonals is then measured and the refraction coefficient may be evaluated. This is a very tedious and time-consuming task and has for larger jobs been replaced by numerical methods utilizing high-speed computers and plotting routines.

The numerical methods are based on the pioneering work of Munk and Arthur (1952); for example, Dobson (1967) solves numerically the equation for the orthogonal separation factor along the orthogonal, thereby obtaining the value of the refraction coefficient.

For a discussion of the influence of the earth's curvature, which becomes of importance when calculating refraction diagrams over large distances, reference is given to Chao (1971) who also discusses the very complicated problem of what happens when neighboring orthogonals cross.

SOME NONLINEAR ASPECTS

The preceding sections have been limited to a discussion of waves of small enough amplitude to neglect the nonlinear terms in the governing equations (6), (8), (9), and (10). The solution obtained so far is simple and quite accurate; it predicts several useful features of water waves of amplitudes far exceeding the mathematical assumption on which the linearization was based, i.e., infinitesimally small waves. Although certain features are adequately described by the linear (first-order) wave theory, there are certain features that are not revealed by the linear solution.

Stokes Waves

From the linear (first-order) solution obtained for progressive small-amplitude waves, it is possible to justify the neglect of the nonlinear terms by substitution of the linear solution and evaluating the magnitude of the neglected terms relative to the terms that were retained. It turns out that the justification for the neglect of the nonlinear terms is, as previously stated, that a/L and a/h be small in some sense relative to unity.

One possible way of improving the solution is to introduce the linear (first-order) solution in the nonlinear governing equations in order to evaluate the nonlinear terms and then solve the resulting equations, which then would lead to a velocity potential ϕ', which essentially expresses the error committed by the neglect of the non-linear terms. Wehausen and Laitone (1960) outline the rigorous approach to be followed to obtain the corrections to any order of accuracy.

The first correction terms, which when added to the already stated linear terms, (17) and (18), produce a *second-order solution* for progressive Stokes waves, are for the velocity potential

$$\phi' = \frac{3}{8} a^2 \omega \frac{\cosh 2k(z+h)}{\sinh^4 kh} \sin 2(kx - \omega t) - \frac{a^2 \omega^2}{4 \sinh^2 kh} t \quad (88)$$

and for the surface profile

$$\eta' = a \frac{ka}{4} \frac{\cosh kh(2 + \cosh 2kh)}{\sinh^3 kh} \cos 2(kx - \omega t) \quad (89)$$

with the relationship between ω and k remaining unchanged.

WAVE ASYMMETRY. The surface profile of a second-order progressive wave is given by $\eta + \eta'$ with η given by (18) and η' by (89). As illustrated in Fig. 6, this second harmonic correction term, η', will cause the wave crest to become more peaked, whereas the wave trough will become flatter; i.e., an asymmetry that was not predicted by the linear solution appears. One can practically always tell what is up and down when looking at a wave record by using the result that crests are higher above the still water level than troughs are below.

This asymmetry is also present in the horizontal velocity component, which is evaluated from $u + u' = \partial\phi/\partial x + \partial\phi'/\partial x$ with ϕ and ϕ' given by (17) and (88), respectively. Under the crest the horizontal velocity in the direction of wave advance is larger than the horizontal velocity in the opposite direction under the trough.

This asymmetry is likely to have some effect on, for example, the net sediment transport associated with wave motion. A simple linear theory would clearly not produce a net sediment transport, since the motion near the bottom does not reflect the direction of propagation

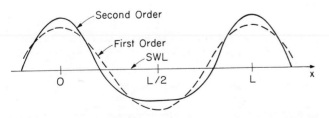

FIGURE 6. *Comparison of first- and second-order wave profile for* H = 0.6 m, h = 3.0 m, L = 30 m.

of the wave passing. However, the asymmetry associated with the second-order solution may change this.

STOKES MASS TRANSPORT. When evaluating the particle orbits to the accuracy consistent with a second-order theory it is found that these are no longer closed, i.e., there is a net motion of water associated with the wave motion, although the first-order approximation did not reveal this.

In fact, we may evaluate the average rate of volume transport per unit length of crest associated with a second-order wave by evaluating

$$q = \frac{1}{T} \int_0^T dt \int_{-h}^{\eta} u \, dz = \tfrac{1}{2} a^2 \omega \coth kh \quad (90)$$

which shows that a mass transport in the direction of wave advance is associated with the nonlinear wave motion. The mass transport associated with a wave motion is significantly influenced by viscous effects, as will be discussed later. However, its presence may have implications as far as any net transport of matter is concerned.

PHASE VELOCITY AND ENERGY. The energy and energy propagation associated with the second approximation of a progressive wave are identical to those previously obtained for the small-amplitude progressive waves, (53) and (55). Also, the phase velocity is unchanged to the second-order approximation. Advancing the approximation to third order, Wehausen and Laitone (1960) give the formula for the *phase velocity*

$$c = \left(\frac{g}{k} \tanh kh\right)^{1/2} \left(1 + (ka)^2 \frac{8 + \cosh 4kh}{8 \sinh^4 kh}\right)^{1/2} \quad (91)$$

where the first term on the right-hand side is identical to the result obtained from the linearized equations (23), and the second term expresses the *amplitude dispersion*, i.e., waves of larger amplitude travel faster.

Radiation Stresses

The averaged quantities of wave energy and energy flux were introduced based on linear theory. The equivalent to a momentum principle based on the *radiation stress* concept, as it was called by its discoverers Longuet-Higgins and Stewart (1960), has not been introduced. The reason for this is that a consistent expression for the radiation stresses, which are similar to Reynolds stresses, requires the use of the Stokes second-order solution. Consequently, the concept of radiation stresses is beyond the scope of this presentation of elementary wave theory. For a thorough discussion of the radiation stress concept the reader is referred to the summary given by Phillips

(1966) who also gives the references to the series of original articles by Longuet-Higgins and Stewart on this subject.

Some of the very interesting results obtained from the application of the radiation stress concept should be mentioned here, however, since they play an important role particularly in the nearshore zone. The radiation stress concept thus predicts an increase in mean water level inside the surf zone (Bowen et al., 1968) for normally incident waves. Also, Longuet-Higgins (1970a, b) used this concept to deduce the magnitude of a longshore current induced by obliquely incident waves on a straight beach. This *wave-induced longshore current*, combined with the wave excitation of the bottom sediment, is responsible for the longshore transport of sand (Longuet-Higgins, 1972). It is described in Chapter 13.

Limitations of Linear Theory

Besides providing insight into certain features of water wave motions, the second-order solution provides a means of establishing criteria for the validity of the linear wave theory. Clearly, if the linear wave theory is a good approximation the correction term, for example η', must be small compared to the linear term η. Comparison of the order of magnitudes of $|\eta'|$, (89), and $|\eta|$, (18), for long waves, kh small, gives

$$\frac{|\eta'|}{|\eta|} = \frac{3}{16\pi^2}\left(\frac{aL^2}{h^3}\right) \ll 1 \tag{92}$$

This reveals the parameter [Stokes parameter (aL^2/h^3)] as playing an important role when assessing the importance of nonlinearities for long-wave problems. The inequality given by (92) is rather loose and results presented by Madsen (1971) indicate that the applicability of a Stokes wave theory is limited to values of the Stokes parameter

$$\frac{aL^2}{h^3} < \frac{4\pi^2}{3} \tag{93}$$

This is not a severe restriction for relatively short waves; however, for long waves, say $h/L = 0.05$, (93) shows that only if $a/h < 0.03$ is it reasonable to assume that a Stokes wave theory describes the wave motion accurately.

Cnoidal Waves

The restriction on the application of a Stokes wave theory mentioned in the preceding section is obviously quite severe when dealing with relatively long nonlinear waves. By limiting one's attention to relatively long waves from the outset, it is possible to derive theories covering long waves of finite amplitude. For an excellent discussion of

the development of such theories, which is beyond the scope of the present chapter, the interested reader is referred to Peregrine (1972).

The wave theory that seems most appropriate when the inequality (93) is violated is the *cnoidal wave theory*, so named since its solution for the surface profile of a progressive wave of permanent form [first obtained by Korteweg and deVries (1895)] is expressed in terms of the Jacobian elliptic cosine function, generally designated by Cn.

The cnoidal waves show the features of a finite-amplitude wave, i.e., its speed of propagation depends on its height and the crest is peaked whereas the trough is long and flat. The ratio of the crest elevation above the still water level to the wave height increases with increasing values of the Stokes parameter, aL^2/h^3. In fact, a special solution with infinite wavelength, the *solitary wave*, consists of a waveform entirely above the still water level. Only in the special case of the solitary wave, shown in Fig. 7, can the surface profile be expressed in terms of the more familiar hyperbolic functions

$$\eta = H \cosh^{-2}\left[\left(\frac{3H}{4h}\right)^{1/2}\frac{x-ct}{h}\right] \tag{94}$$

where the phase speed c is given by

$$c = [g(h+H)]^{1/2} \tag{95}$$

In the general case the surface profile of a cnoidal wave may be determined from the tables prepared by Masch and Wiegel (1961).

It is evident that a wave propagating into shallower waters eventually would reach a point where the inequality (93) is violated, unless the wave broke prior to reaching this point. This essentially invalidates the further use of linear wave theory to predict the transformation of waves beyond this point. However, the complexity of the cnoidal wave solution has until recently precluded its use for such calculations, which out of necessity have been performed using linear wave theory. This problem has recently been resolved by Svendsen and Brink-Kjaer (1973) whose results may be used to obtain the shoaling characteristics of normally incident waves on a gently sloping beach based on the appropriate theory.

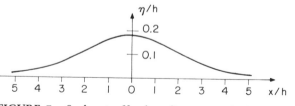

FIGURE 7. *Surface profile of a solitary wave* (**H/h** = *0.18*).

Wave Breaking

The ultimate limitation of any wave theory based on potential theory is given by the conditions at which the wave breaks.

However, theoretical studies of the limiting form of a progressive wave, based on potential theory, may be performed. It may be shown that the limiting form of a progressive wave of permanent form is symmetrical around the crest and exhibits a sharp angle of 120° at the crest (Wehausen and Laitone, 1960). This was used by Michell (1893) to deduce that the limiting form of a deep-water wave corresponds to a value of the wave steepness, *the limiting wave steepness,*

$$\left(\frac{H}{L}\right)_B = 0.142 \tag{96}$$

where the subscript B refers to breaking conditions. Since the limiting waveform clearly is nonlinear, its phase velocity and wavelength are a function of its height as well as of its period, as suggested by (91). In fact, Michell found that the wavelength L_B of a wave breaking in deep water is related to the wavelength determined from linear theory, L_0, given by (27), through

$$L_B \simeq 1.2 L_0 \tag{97}$$

Whereas the theoretical prediction of the limiting steepness based on an assumed symmetric profile seems reasonably realistic for deep-water waves, it is readily observed in a laboratory wave flume that waves approaching breaking as they propagate into shallower water exhibit a strong asymmetry, with the crest having a steep front slope and a flat back. Some waves approaching breaking show a relatively small degree of asymmetry and start to curl over only at the very tip of the crest (spilling breakers), whereas others rise with a practically vertical front slope from the trough of the preceding wave into which they plunge as they break (plunging breakers). For a description of various breaker types the reader is referred to Galvin (1972), who from laboratory experiments found that the parameter

$$B_0 = \frac{H_0}{L_0 \tan^2 \beta} \tag{98}$$

where H_0/L_0 is the deep-water wave steepness and β is the beach slope, may be useful in distinguishing between the different breaker types. Thus, Galvin found that for values of $B_0 > 4.8$ (large wave steepness or small slope) the breaker was of the spilling type. For values $0.09 < B_0 < 4.8$, the plunging type occurred. Clearly the breaker type is a continuous variable and the subjectiveness of the classification may cause a criterion such as the above to vary considerably from one observer to another.

In spite of the asymmetry exhibited by waves approaching breaking in shallow water attempts have been made to obtain theoretical estimates of the limiting wave form under the assumptions of a sharp-crested (120°) symmetrical profile. For waves of the solitary type such calculations have been carried out by several investigators with slightly varying results. The result by McCowan (1894) is the one most often quoted and gives the limiting height-to-depth ratio for a solitary wave as being

$$\left(\frac{H}{h}\right)_B = 0.78 \tag{99}$$

Munk (1949) argued that a periodic wave approaching breaking may be regarded as a solitary wave, since periodic waves in shallow water become increasingly nonlinear with peaked crests separated by long flat troughs. In addition, the generally gentle beach slope suggests the theoretical result (99), which was obtained for a horizontal bottom, to be approximately valid. This reasoning led Munk (1949) to adopt (99) as a breaking criterion for periodic waves in shallow water.

It should be mentioned that, although Munk found his criterion to be reasonably successful in representing experimental data, the reasoning behind it seems in error. Thus, Ippen and Kulin (1955) performed experiments with solitary waves on very gentle slopes (tan $\beta \simeq$ 0.023) and found the wave height to depth ratio at breaking to be 1.2, in poor agreement with (99).

Miche (1944) performed theoretical calculations for periodic waves in finite water depth and found his theoretical predictions for the limiting wave steepness to be reasonably well represented by the function

$$\left(\frac{H}{L}\right)_B = 0.140 \tanh\left(\frac{2\pi h}{L}\right)_B \tag{100}$$

This semitheoretical breaking criterion is quite general and is seen to be compatible with that of Michell (1893), (96), for deep-water waves ($h/L > 0.5$) and to lead to an expression similar to (99) for waves breaking in shallow water ($h/L < 0.05$) with the ratio $(H/h)_B = 0.88$.

The difficulty in applying (100), which may be adopted as an approximate general breaking criterion, is that it depends on the wavelength at breaking, L_B. However, it is reasonably accurate and certainly commensurate with the accuracy of this breaking criterion to take the wavelength at breaking as $1.2 \cdot$ (linear wavelength), as suggested by (97), for all $(h/L)_B$ ratios. This leads to a breaking criterion of the form

$$\frac{H_B}{L} = 0.17 \tanh\left(\frac{2\pi h_B}{1.2L}\right) \tag{101}$$

where L is the wavelength determined from linear wave theory.

The breaking criterion (101) is inherently approximate. Experimental results show that the height-to-depth ratio of waves breaking in shallow water depends on the beach slope and to some extent on the wave steepness. Thus, Galvin (1969) and Collins (1972) suggest empirical relationships for waves breaking in shallow water which incorporate the effect of beach slope. For small slopes (tan β < 0.07) their relationships are similar, although slightly different, and may be combined into

$$\left(\frac{H}{h}\right)_B = 0.72(1 + 6.4 \tan \beta)$$

$$\text{for} \quad \tan \beta < 0.1 \quad (102)$$

which yields values of $0.72 < (H/h)_B < 1.18$. For long waves and a reasonable beach slope of $1/30$, (102) yields the same results $[(H/h)_B = 0.88]$ as does (101).

THE BOTTOM BOUNDARY LAYER

When the governing equations used in the preceding sections were derived, it was mentioned that viscous forces were important only in the immediate vicinity of the solid boundary, within the boundary layer. The neglect of viscous forces gave a solution having a nonzero velocity at the bottom, whereas the boundary condition for a viscous fluid is a no-slip condition.

Imagine for the moment the fluid above a solid bed to be suddenly set in motion. The satisfaction of the no-slip condition at the bottom will initially be felt only by the fluid in the immediate vicinity of the boundary. However, as time goes on fluid further and further away from the boundary will feel the retarding effect of the solid boundary. Finally, the total depth of fluid will be affected by the solid boundary, the flow is fully developed, and a velocity profile as discussed in Chapter 7 results. The qualitative description of the boundary layer development given above may represent the conditions for very long-period waves for which the fluid velocity is unidirectional for sufficiently long time to achieve an essentially fully developed flow. This is, for example, often the case in the analysis of tides (Chapter 5) and the bottom friction may be approximately evaluated by conventional methods assuming the flow to be steady and unidirectional at any instant of time.

For short-period waves the fluid velocity changes direction before the boundary layer reaches the free surface. Therefore, the flow never becomes fully developed and viscous effects are for short-period waves limited to a region very close to the solid boundary, the boundary layer. Outside the boundary layer viscous effects are negligible, which explains the success of the wave theory developed in the preceding sections. Within the boundary layer the velocity is primarily horizontal, parallel to the bottom, and decreases in magnitude from its free stream value, given by (40), at the outer limit of the boundary layer to zero at the boundary. This introduces very large velocity gradients and therefore large shear stresses and associated energy dissipation.

Laminar Boundary Layer

The details of the fluid motion within the bottom boundary layer can be treated analytically only so long as the flow remains laminar. It is beyond the scope of this presentation to discuss the details of such an analysis; reference is given to any textbook on viscous flow, for example Schlichting (1960) or Batchelor (1967), where a derivation of the equations governing viscous flow may also be found.

The boundary conditions to be satisfied by the solution for the horizontal velocity within the boundary layer are that (1) the velocity approaches the *free stream velocity*, $u_b \cos(kx - \omega t)$, with u_b given by (40), as the distance from the boundary, $z + h$, increases; (2) the velocity vanishes at the boundary, i.e., $u = 0$ at $z = -h$.

The first-order solution to this problem neglects the influence of variation in the x direction, and the solution may be formally written as

$$u = u_b \left\{ \cos \omega t - \exp \left[- \left(\frac{\omega}{2\nu} \right)^{1/2} (z + h) \right] \right.$$

$$\left. \cos \left[\omega t - \left(\frac{\omega}{2\nu} \right)^{1/2} (z + h) \right] \right\} \quad (103)$$

where ν is the kinematic viscosity ($\nu \simeq 10^{-2} \text{ cm}^2/\text{sec}$ for water), and the x dependence has been omitted. This solution clearly satisfies the boundary conditions stated above, and it is seen that the difference between the free stream velocity and that given by (103), expressed by the exponentially decaying term, is less than 1% when the distance from the bottom is

$$\delta = z + h = 5 \left(\frac{2\nu}{\omega} \right)^{1/2} \quad (104)$$

which may be taken as a definition of the *boundary layer thickness* δ. This definition is clearly rather arbitrary. One could choose the distance where the difference was 5% or focus on the phase angle of the viscous term, to obtain different expressions for the boundary layer thickness. All such expressions would have the form given by (104), only the constant which here is chosen as 5 would be different. That the influence of viscous forces indeed is limited to a thin layer along the bottom can be seen by evaluating (104) for a wave of 3 second period, which gives $\delta \simeq 0.5$ cm.

MASS TRANSPORT.   It was mentioned that the solution given by (103) is a first approximation. When obtaining the second-order solution the spatial variation of the free stream velocity is considered, and this leads to the rather surprising result that a *steady streaming* is induced at the outer limit of the boundary layer. This streaming is in the direction of wave advance and causes viscous effects to diffuse into the main body of the fluid, so that the details of the fluid motion become affected by viscosity throughout the water depth. In particular, the mass transport associated with the wave motion is altered by this streaming.

Considering the viscous effects outlined above, the problem of mass transport in water waves was solved originally by Longuet-Higgins (1953) and later by Unluata and Mei (1970), who show the mass transport to be given by

$$q = \frac{1}{2} a^2 \omega \coth kh \left(1 + \frac{(kh)^2}{2} + \frac{3kh}{4 \sinh 2kh} \right) \quad (105)$$

which is considerably different from the inviscid result, Stokes mass transport, given by (90).

Water waves of normal incidence on an infinitely long straight beach could clearly not have a net mass transport in the direction of the wave advance. A return current must be generated to satisfy the physical condition of zero net mass transport. This condition of zero net transport also applies to waves observed in an experimental wave flume, where the Lagrangian velocity profile associated with a wave motion may be visualized by observing the motion of a dye streak. Expressions for the mass transport velocity distribution are presented in the references just cited. Of particular importance in the present context is the prediction that the mass transport velocity just outside the boundary layer is in the direction of wave advance and of a magnitude

$$U_{M,b} = \frac{5}{4} \frac{ka}{\sinh kh} u_b \quad (106)$$

with u_b given by (40). This unidirectional component introduced into an otherwise periodic problem may be of importance when considering the transport of matter associated with a wave motion.

BOTTOM SHEAR STRESS.   The fluid flowing above the bottom exerts a shear stress τ_b on the bottom. For a laminar boundary layer the velocity u is given by (103) and the shear stress on the bottom may be evaluated:

$$\tau_b = \rho \nu \left(\frac{\partial u}{\partial z} \right)_{z=-h} = \rho(\nu\omega)^{1/2} u_b \cos\left(\omega t + \frac{\pi}{4} \right) \quad (107)$$

Of particular interest in the present context is the maximum value of the bottom shear stress, which is seen to be $\pi/4$ out of phase with the velocity; i.e., maximum shear does not coincide with maximum velocity.

$$\tau_{b,\max} = \rho(\nu\omega)^{1/2} u_b \quad (108)$$

which shows the maximum shear stress associated with a wave motion of a given maximum velocity at the bottom to increase with decreasing wave period.

The Wave Friction Factor

In steady flow it is quite common to relate the bottom shear stress to the square of the velocity. The reason for this is that the shear stress tends to become proportional to the square of the velocity for *turbulent flow* which generally is the type of flow encountered in nature. For wave motion this form

$$\tau_{b,\max} = \frac{1}{2} f_w \rho u_b^2 \quad (109)$$

was chosen by Jonsson (1966) to relate the maximum bottom shear stress and the maximum free stream velocity, with the coefficient f_w being termed the *wave friction factor*.

Comparison of (108) and (109) shows that the wave friction factor corresponding to a laminar boundary layer may be expressed as

$$f_w = \frac{2(\nu\omega)^{1/2}}{u_b} = \frac{2}{[(u_b(u_b/\omega)/\nu]^{1/2}} = \frac{2}{(\mathrm{RE})^{1/2}} \quad (110)$$

where the particular form of the expression is chosen to bring out the fact that the denominator has the appearance of a Reynolds number, the *wave Reynolds number*, RE, which by use of (51) may be written as

$$\mathrm{RE} = \frac{u_b A_b}{\nu} \quad (111)$$

thus identifying the characteristic length scale of the viscous motion to be the maximum displacement of the water particles near the bottom.

As previously mentioned these formulas are valid only so long as the flow within the boundary layer remains laminar. For most wave conditions, outside of laboratory investigations, the flow in the boundary layer is *turbulent*.

TRANSITION FROM LAMINAR TO TURBULENT FLOW. Whether the flow remains laminar or turns turbulent depends on the value of the Reynolds number (111), and for rough boundaries such as a sand bottom, on the roughness of the boundary relative to the characteristic length scale of the motion, A_b. Denoting the *equivalent boundary roughness* by d_s with the subscript s indicating that it is the equivalent sand roughness, the flow remains laminar in the boundary layer, according to Jonsson (1966), only if the conditions

$$\text{RE} < 1.26 \cdot 10^4 \qquad (112)$$

and

$$\frac{A_b}{d_s} > \frac{4\sqrt{2}}{\pi} \, (\text{RE})^{1/2} \qquad (113)$$

are satisfied by the flow and boundary characteristics. The condition expressed by (112) clearly limits laminar flow to small-scale laboratory experiments.

WAVE FRICTION FACTOR DIAGRAM. For the case of turbulent flow within the boundary layer an analytical solution describing the details of the flow has been advanced by Kajiura (1968). It is beyond the present treatment to give the details of Kajiura's analysis. The wave friction factor concept introduced by Jonsson (1966) and expressed here by (109) may, however, be used as the basis for presenting empirical data obtained by various investigators.

Drawing on the conceptual analogy between the factors influencing the wave friction factor and those influencing its steady state analog, the Darcy-Weisbach friction factor, one would expect an empirical relationship of the type

$$f_w = f_w \left(\text{RE}, \frac{A_b}{d_s} \right) \qquad (114)$$

exhibiting the same features as those present in a Moody diagram (e.g., see Daily and Harleman, 1966, p. 275). That this is true is evident in Fig. 8, reproduced from Jonsson (1965), where the friction factor is seen to be a function only of the relative roughness, A_b/d_s, for conditions corresponding to rough turbulent flow in the boundary layer.

Figure 8 is a *wave friction factor diagram* and is in principle simple to use. If the wave conditions are known, the Reynolds number (111) and the value of the relative roughness, A_b/d_s, assuming d_s to be known, may be calculated. Use of Fig. 8 may therefore yield the corresponding value of the wave friction factor f_w, and the maximum shear stress exerted on the bottom may be evaluated from (109).

Energy Dissipation and Wave Attenuation

In addition to exerting a shear stress on the bottom, the viscous stresses cause an energy dissipation within the boundary layer. The rate at which energy is dissipated within the boundary layer must equal the rate at which the wave motion, as a whole, loses energy. Since it was found that the wave energy is related to the amplitude of the wave motion, it is seen that the energy dissipation taking place in the bottom boundary layer will cause a decrease in wave amplitude; i.e., the wave motion is attenuated because of viscous effects.

LAMINAR BOUNDARY LAYER. For laminar flow the problem of determining the rate of energy dissipation within the fluid may be solved analytically. As a general text the reader is referred to Eagleson and Dean (1966). Here it is stated that the result is an *exponential decay* of wave amplitude with distance traveled. As previously mentioned, the condition of laminar boundary layer applies only under laboratory conditions.

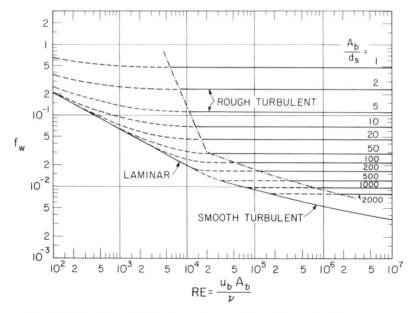

FIGURE 8. *Wave friction factor diagram. From Jonsson (1965).*

TURBULENT BOUNDARY LAYER. In the more general case of turbulent boundary layer one may generalize the expression for the maximum shear stress given by (109), by assuming it to hold at any instant of time, i.e.,

$$\tau_b = \tfrac{1}{2}\rho f_w u_b{}^2 |\cos \omega t| \cos \omega t \qquad (115)$$

which neglects the phase difference between shear stress and free stream velocity. The time dependence is taken as $|\cos \omega t| \cos \omega t$ to preserve the direction in which τ_b acts. For a case corresponding to rough turbulent flow it is seen from Fig. 8 that the insensitivity of f_w to the value of the Reynolds number indicates that an assumption of a constant friction factor equal to the wave friction factor f_w seems reasonable.

For steady, uniform flow in open channels (Henderson, 1966), the rate of energy dissipation per unit area is given as

$$P_D = \tau_b \bar{u} \qquad (116)$$

where $\bar{u}$ is a depth-averaged velocity. Adopting a similar relationship for the boundary layer flow, which clearly is an approximation, one obtains

$$P_D = \tau_b u_b \cos \omega t \qquad (117)$$

with τ_b given by (115). The time average of (117) gives the average rate of energy dissipation per unit area of the bottom, $\overline{P_D}$,

$$\overline{P_D} = \frac{2}{3\pi} \rho f_w u_b{}^3 \qquad (118)$$

which was shown by Putnam and Johnson (1949) to give a wave attenuation for periodic waves traveling in water of constant depth

$$\frac{a}{a_1} = \frac{1}{1 + K_f a_1 x} \qquad (119)$$

where a_1 is the amplitude at $x = 0$ and

$$K_f = \frac{2}{3\pi} \frac{f_w}{c_g g} \left(\frac{\omega}{\sinh kh}\right)^3 \qquad (120)$$

expresses the rate at which the wave amplitude attenuates.

NUMERICAL EXAMPLES

To illustrate the application of the elements of wave theory presented in the preceding sections, let a wave climate be represented in terms of the *deep-water conditions*

$$H_0 = 1 \text{ m}; \qquad T = 8 \text{ seconds}; \qquad \theta_0 = 45°$$

The wave is assumed to propagate from deep water into shallower water over a straight beach of gentle slope.

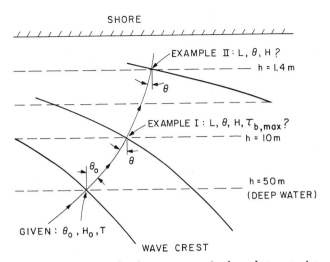

FIGURE 9. *Examples of wave propagation from deep water into shallower water: I, h = 10 m; II, h = 1.4 m. See text for explanation.*

In the following it is in addition assumed that small-amplitude wave theory applies and that energy dissipation caused by bottom friction may be neglected. The problem treated in the following examples is depicted in Fig. 9.

Example I, $h = 10$ m

When the wave reaches a water depth of 10 m we wish to estimate the wavelength L, the angle of incidence θ, the wave height H, and the maximum bottom shear stress $\tau_{b,\max}$, associated with the wave motion.

WAVELENGTH L. From (27) we obtain

$$L_0 = \frac{g}{2\pi} T^2 = 1.56 \cdot T^2 = 100 \text{ m}$$

and the parameter $h/L_0 = 10/100 = 0.1$ can be used with Fig. 2 to obtain $L/L_0 = 0.71$, or

$$\text{wavelength} = L = 0.71 L_0 = 71 \text{ m}$$

ANGLE OF INCIDENCE θ. Since the beach is straight, Snell's law, (81), applies and

$$\sin \theta = \frac{c}{c_0} \sin \theta_0 = \frac{L}{L_0} \sin \theta_0 = 0.502$$

results, since $c = L/T$ and T is constant. Therefore

$$\text{angle of incidence} = \theta \simeq 30°$$

WAVE HEIGHT H. The wave height is given by (86). The shoaling coefficient may be found from Fig. 2 as

$K_S = 0.93$, and the refraction coefficient $(b_0/b)^{1/2}$ is found from (87) to be

$$K_R = \left(\frac{\cos \theta_0}{\cos \theta}\right)^{1/2} = \left(\frac{\cos 45°}{\cos 30°}\right)^{1/2} \simeq 0.90$$

Equation 86 then gives

wave height $= H = K_S K_R H_0 = 0.93 \cdot 0.90 \cdot 1 = 0.84$ m

MAXIMUM BOTTOM SHEAR STRESS. To find the maximum bottom shear stress (109), the near-bottom velocity is needed. From (40) it is seen that

$$u_b = \frac{a\omega}{\sinh kh} = \frac{a}{a_0 \sinh kh} a_0 \omega = \left(\frac{K_S}{\sinh kh}\right) K_R a_0 \omega$$

where $a_0 = H_0/2 = 50$ cm. The value of $(K_S/\sinh kh)$ corresponds to the value of u_b/u_0, (79), and is found from Fig. 2 as $(u_b/u_0) \cdot 10 = 9.3$, i.e., $u_b/u_0 = 0.93$, and the near-bottom velocity becomes

$$u_b = 0.93 \cdot K_R \cdot a_0 \omega = 0.93 \cdot 0.9 \cdot 50 \cdot \frac{2\pi}{8} \simeq 33 \text{ cm/sec}$$

With (51)

$$A_b = \frac{u_b}{\omega} = \frac{33 \cdot 8}{2\pi} = 42 \text{ cm}$$

and $\nu = 10^{-2}$ cm²/sec, the wave Reynolds number becomes

$$\text{RE} = \frac{u_b A_b}{\nu} = 1.4 \cdot 10^5$$

If an equivalent sand roughness of the bottom is assumed, $d_s = 2$ cm (the large value reflects the possible presence of bed forms), the relative roughness becomes

$$\frac{A_b}{d_s} = \frac{42}{2} = 21$$

and the wave friction factor diagram (Fig. 8) gives

$$f_w = 4.6 \cdot 10^{-2}$$

The maximum bottom friction associated with the wave motion may finally be estimated, from (109), to be

maximum shear stress $= \tau_{b,\max} \simeq \frac{1}{2} f_w \rho u_b^2$

$$= \frac{1}{2} \cdot 4.6 \cdot 10^{-2} \cdot 1 \cdot 33^2 = 25 \text{ dynes/cm}^2$$

It is interesting to compare this result with that corresponding to the shear stress exerted on the bottom by an equivalent steady current of velocity $\bar{u} = u_b = 33$ cm/sec. For the same depth and $d_s = 2$ cm, use of the Moody diagram (Daily and Harleman, 1966, p. 275)

gives the value of the Darcy-Weisbach friction factor $f = 1.7 \cdot 10^{-2}$, and a corresponding shear stress

$$\tau_0 = \frac{f}{8} \rho \bar{u}^2 = \frac{1.7 \cdot 10^{-2}}{8} \cdot 1 \cdot 33^2 = 2.3 \text{ dynes/cm}^2$$

i.e., the shear stress associated with an equivalent current is in this case an order of magnitude smaller than that associated with the wave motion. The implication of this is that wave action may stir up sediment, which may then be moved by a current, although this current by itself would have been unable to cause any sediment transport.

Example II, $h = 1.4$ m

For the same deep-water conditions and with the conditions assumed in the preceding example, we now wish to determine the wave height when the waves reach a depth of 1.4 m.

The value of the parameter h/L_0 corresponding to this depth is $1.4 \cdot 10^{-2}$; i.e., it falls outside the range of values presented in Fig. 2. In this region, however, the long-wave approximation applies and (29) gives

$$L = (gh)^{1/2} T = (9.81 \cdot 1.4)^{1/2} 8 = 29.6 \text{ m} > 20 \cdot 1.4 \text{ m}$$

and application of Snell's law gives

$$\sin \theta = \frac{L}{L_0} \sin \theta_0 = \frac{29.6}{100} \sin 45° = 0.21$$

or $\theta \simeq 12°$, and the refraction coefficient becomes

$$K_R = \left(\frac{\cos \theta_0}{\cos \theta}\right)^{1/2} = \left(\frac{\cos 45°}{\cos 12°}\right)^{1/2} = 0.85$$

The value of the shoaling coefficient K_S cannot be obtained directly from Fig. 2. However, we may rewrite the definition of the shoaling coefficient in the following manner:

$$K_S = \frac{H}{H_0} = \left(\frac{H}{H_I}\right)\left(\frac{H_I}{H_0}\right)$$

and by choosing H_I corresponding to a water depth $h = h_I = 1.5$ m, the value of H_I/H_0 may be obtained from Fig. 2, corresponding to $h/L_0 = 1.5 \cdot 10^{-2}$, as $H_I/H_0 = 1.31$. The change in wave height from $h = h_I = 1.5$ m to the point of interest, $h = 1.4$ m, is given by Green's formula, (77), since the waves here may be categorized as long waves, i.e.,

$$K_S = \left(\frac{H}{H_I}\right)\left(\frac{H_I}{H_0}\right) = \left(\frac{1.5}{1.4}\right)^{1/4} 1.31 = 1.33$$

and the wave height is found from

wave height $= H = K_S K_R H_0 = 1.33 \cdot 0.85 \cdot 1 = 1.13$ m

If the breaking criterion (101) is adopted, it is seen that $(H/h)_B = 0.88$, whereas we have found $H/h = 0.81$ for $h = 1.4$ m. This indicates that the waves would break in a depth less than 1.4 m. In general, iterative use of the procedure just given would enable the determination of the depth where breaking occurs and the corresponding wave characteristics (in this example $h_B \simeq 1.3$ m). It should be noted that the application of linear theory up to the point of breaking is violating the limit of its applicability, (93), and that a more accurate determination of the breakpoint may be obtained by employing cnoidal wave theory.

SUMMARY

The mathematical description of progressive surface waves may be undertaken by means of a linearized form of the equations governing fluid motion. This simplification is valid if the waves are assumed to be of small amplitude relative to the wavelength and the water depth so that the nonlinear terms may be neglected (small-amplitude wave theory). The solutions to these equations permit the description of a progressive wave that propagates at a constant speed without change in form. They permit evaluation of the speed of wave propagation (phase velocity), u_{max} (maximum water particle velocity), the shape of the orbital trajectory of the water particles, and the wave-induced pressures as the wave passes overhead. General expressions for these parameters are presented for the case of short wavelengths (deep water), intermediate wavelengths, and long wavelengths (shallow water). The solutions show that the water particle orbits are closed; i.e., no net fluid motion is associated with a progressive sinusoidal wave.

For purposes of describing the changes in wave characteristics as waves propagate from deep to shallow water, it is important to be able to quantify the energy content of a surface wave and the energy flux occurring within a wave train. The average energy per unit surface area of a wave, partitioned equally between kinetic and potential forms, is the *specific energy E_S*, and is a function of the square of the wave amplitude. The energy flux E_F is the product of the *group velocity* and the specific energy. For short waves, the group velocity is half of the phase velocity; hence, individual waves tend to pass through a group of deep-water waves.

As waves propagate over a gently sloping bottom from deep to shallow water, they experience changes because of changing depth (the shoaling transformation) even if the effects of bottom friction are neglected. These changes are an expression of the principle of conservation of energy. The ratio of wave height to deep-water wave height, H/H_0, is equivalent to the *shoaling coefficient K_S*, which is a function of the ratio of water depth to deep-water wavelength, h/L_0. In nature, the height of waves propagating from deep to shallow water first decreases slightly, then increases until at $H/h \simeq 0.8$, the wave form becomes unstable and the wave breaks.

Certain aspects of surface wind waves may not be predicted by linear, first-order wave theory, based on the mathematical assumption of waves of small amplitude. Specifically, real waves tend to have crests that are narrow and peaked and troughs that are broad and flat. This characteristic may be described by the Stokes approximation, in which the linear, first-order solution is introduced into the nonlinear governing equations in order to evaluate the nonlinear terms. Correction terms are then added to the first-order governing equations. In Stokes waves, water particle orbits are no longer closed, and a net water transport (*Stokes mass transport*) in the direction of wave propagation is associated with the wave motion. Neither small-amplitude wave theory nor Stokes wave theory is adequate to describe the behavior of waves in very shallow water. Here *cnoidal wave theory* is more applicable, and a periodic wave approaching breaking behaves in some respects as a *solitary wave*, whose form is entirely above still water.

Wave motion is capable of entraining and transporting sediment, because the oscillatory flow associated with the wave motion penetrates to the seafloor and interacts with it. The significant portion of the flow for the problem of wave-sediment interaction is the flow in the immediate vicinity of the boundary, the boundary layer. Within the boundary layer the fluid velocity decreases rapidly from the value predicted by the wave theory to zero at the bed. The thickness of the boundary layer depends on the wave period, and the boundary layer associated with wave motion can never become fully developed beneath short-period waves. A second-order solution for horizontal variation within the boundary layer leads to the surprising result that a steady streaming (a unidirectional flow component) occurs in the upper portion of the boundary layer. Mass transport associated with wave motion is altered by this streaming which is in the direction of wave propagation.

The maximum bottom shear stress is related to the square of velocity by a *wave friction factor f_w*. For laminar flow, $f_w = 2/(\mathrm{RE})^{1/2}$, where RE is the *wave Reynolds number*, equal to $u_b A_b/\nu$, the product of the maximum bottom velocity and the maximum displacement of water particles near the bottom, divided by the kinematic viscosity of the fluid. For turbulent flow the wave friction factor is a function of the relative bottom roughness as well as the value of RE, and may be obtained from Jonsson's friction factor diagram, which is analogous to a Moody diagram.

Knowledge of the friction factor permits evaluation of the maximum bottom shear stress τ_b, which is proportional to the product of the friction factor and the square of the bottom velocity, and also permits evaluation of the average rate of energy dissipation per unit area of the bottom $\overline{P_D}$, which is proportional to the product of the friction factor and the cube of the bottom velocity. The maximum bottom shear stress associated with a wave motion is generally considerably larger than the shear stress associated with a steady current of comparable magnitude. This points out the importance of wind waves acting as a "stirring agent," which makes sediment available for transport even by very weak currents.

SYMBOLS

a	wave amplitude ($a = H/2$)
A	horizontal amplitude of water particle motion
b	distance between neighboring wave orthogonals
B	vertical amplitude of water particle motion
B_0	parameter identifying breaker type
c	phase velocity of wave
c_g	group velocity of wave
d_s	equivalent sand roughness of bottom
e	base of natural logarithm ($e = 2.71 \ldots$)
E_F	energy flux per unit length of wave crest
E_S	specific wave energy
f_w	wave friction factor
g	acceleration due to gravity ($g = 981 \text{ cm/sec}^2$)
h	water depth
H	wave height
k	wave number ($k = 2\pi/L$)
K_f	wave attenuation coefficient
K_R	refraction coefficient
K_S	shoaling coefficient
l	length
L	wavelength
n	integer
p	fluid pressure
p^+	fluid pressure due to wave motion
P_D	rate of energy dissipation per unit area of bottom
q	mass transport per unit length of crest
R	reflection coefficient
RE	wave Reynolds number
t	time
T	wave period
u	horizontal water particle velocity
u_0	horizontal water particle velocity at surface in deep water ($u_0 = a_0\omega$)
U_M	mass transport velocity
w	vertical water particle velocity
W_T	work done during one waveperiod
x	horizontal coordinate
X	mean horizontal position of water particle
z	vertical coordinate ($z = 0$ at still water level)
Z	mean vertical position of water particle
β	bottom slope
δ	boundary layer thickness
η	free surface elevation relative to still water level
θ	angle of wave incidence
ν	kinematic viscosity of fluid
π	3.14
ρ	fluid density
τ	shear stress
ϕ	velocity potential
ψ	phase angle
ω	radian frequency ($\omega = 2\pi/T$)
$(\ )_b$	refers to conditions at the bottom
$(\ )_B$	refers to conditions at breaking
$(\ )_i$	refers to conditions of incident wave
$(\ )_0$	refers to conditions in deep water
$(\ )_P$	refers to conditions of a certain water particle
$(\ )_r$	refers to conditions of reflected wave
$(\ )'$	second-order (small) correction
$(\overline{\ })$	indicates time-averaged value
$\vert\ \vert$	absolute value

REFERENCES

Batchelor, G. K. (1967). *An Introduction to Fluid Dynamics.* London and New York: Cambridge University Press, 615 pp.

Bowen, A. J., D. L. Inman, and V. P. Simmons (1968). Wave set-down and wave set-up. *J. Geophys. Res.,* **73:** 2569–2577.

Chao, Y. Y. (1971). An asymptotic evaluation of the gravity wave field near a smooth caustic. *J. Geophys. Res.,* **76:** 7401–7408.

Collins, J. I. (1972). Probabilities of breaking wave characteristics. *Proc. 12th Conf. Coastal Eng.,* **1:** 399–414.

Daily, J. W. and D. R. F. Harleman (1966). *Fluid Dynamics.* Reading, Mass.: Addison-Wesley, 454 pp.

Dobson, R. S. (1967). *Some Applications of a Digital Computer to Hydraulic Engineering Problems.* Technical Report No. 80, Dept. of Civil Engineering, Stanford University, Stanford, Calif.

Eagleson, P. S. and R. G. Dean (1966). Small amplitude wave theory. In A. T. Ippen, ed., *Estuary and Coastline Hydrodynamics.* New York: McGraw Hill, pp. 1–92.

Galvin, C. J., Jr. (1969). Breaker travel and choice of design wave height. *ASCE J. Waterw. Harbors Div.*, **99:** 175–200.

Galvin, C. J. (1972). Wave breaking in shallow water. In R. E. Meyer, ed., *Waves on Beaches*. New York: Academic Press, pp. 413–456.

Harris, D. L. (1972). Characteristics of wave records in the coastal zone. In R. E. Meyer, ed., *Waves on Beaches*. New York: Academic Press, pp. 1–52.

Henderson, F. M. (1966). *Open Channel Flow*. New York: MacMillan, 522 pp.

Ippen, A. T., ed. (1966). *Estuary and Coastline Hydrodynamics*, New York: McGraw Hill, 744 pp.

Ippen, A. T. and G. Kulin (1955). *Shoaling and Breaking Characteristics of the Solitary Wave*. Technical Report No. 15, M.I.T. Hydrodynamics Lab. Cambridge, 56 pp.

Johnson, J. W., M. P. O'Brien, and J. D. Isaacs (1948). *Graphical Construction of Wave Refraction Diagrams*. Technical Report No. 2, U.S. Naval Oceanographic Office, 35 pp.

Jonsson, I. G. (1965). *Friction Factor Diagrams for Oscillatory Boundary Layers*. Basic Research Report No. 10, Technical University of Denmark, Copenhagen, Denmark, pp. 10–21.

Jonsson, I. G. (1966). Wave boundary layers and friction factors. *Proc. 10th Conf. Coastal Eng.*, **1:** 127–148.

Kajiura, K. (1968). A model of the bottom boundary layer in water waves, *Bull. Earthquake Res. Inst.*, University of Tokyo, **46:** 75–123.

Kinsman, B. (1965). *Wind Waves*, Englewood Cliffs, N.J.: Prentice-Hall, 676 pp.

Korteweg, D. J. and G. deVries (1895). On the change of form of long waves advancing in a rectangular canal and on a new type of long stationary waves. *Phil. Mag. Edinburough*, **39**(5): 422–443.

Lamb, Sir H. (1932). *Hydrodynamics*, New York: Dover, 738 pp.

Longuet-Higgins, M. S. (1953). Mass transport in water waves. *Phil. Trans. Roy. Soc. London*, **A245:** 535–581.

Longuet-Higgins, M. S. (1970a). Longshore currents generated by obliquely incident sea waves 1. *J. Geophys. Res.*, **75:** 6778–6789.

Longuet-Higgins, M. S. (1970b). Longshore currents generated by obliquely incident sea waves 2. *J. Geophys. Res.*, **75:** 6790–6801.

Longuet-Higgins, M. S. (1972). Recent progress in the study of longshore currents. In R. E. Meyer, ed., *Waves on Beaches*, New York: Academic Press, pp. 203–248.

Longuet-Higgins, M. S. and R. W. Stewart (1960). Change in the form of short gravity waves on long waves and tidal currents. *J. Fluid Mech.*, **8:** 565–583.

McCowan, J. (1894). On the highest wave of permanent type. *Phil. Mag. Edinburough*, **38**(5): 351–358.

Madsen, O. S. (1971). On the generation of long waves. *J. Geophys. Res.*, **76:** 8672–8683.

Masch, F. D. and R. L. Wiegel (1961). *Cnoidal Waves Tables and Functions*. Council of Wave Research, The Engineering Foundation, University of California, Berkeley, 129 pp.

Miche, R. (1944). Forme limite de la houle lors de son déferlement. *Ann. Ponts Chaussées*, **114:** 131–164.

Michell, J. H. (1893). On the highest waves in water. *Phil. Mag. Edinburough*, **36**(5): 430–435.

Munk, W. H. (1949). Solitary wave theory and its application to surf problems. *Ann. N.Y. Acad. Sci.*, **51:** 376–424.

Munk, W. H. and R. S. Arthur (1952). Wave intensity along a refracted ray. In *Gravity Waves*. National Bureau of Standards, pp. 95–109.

Peregrine, D. H. (1972). Equations for water waves and the approximations behind them. In R. E. Meyer, ed., *Waves on Beaches*, New York: Academic Press, pp. 95–122.

Phillips, O. M. (1966). *Dynamics of the Upper Ocean*, London and New York: Cambridge University Press, 261 pp.

Putnam, J. A. and J. W. Johnson (1949). The dissipation of wave energy by bottom friction. *Trans. Am. Geophys. Union*, **30:** 67–74.

Schlichting, H. (1960). *Boundary Layer Theory*, 4th ed. New York: McGraw-Hill, 647 pp.

Svendsen, I. A. and O. Brink-Kjaer (1973). Shoaling of cnoidal waves. *Proc. 13th Conf. Coastal Eng.*, **1:** 365–384.

Unluata, U. and C. C. Mei (1970). Mass transport in water waves. *J. Geophys. Res.*, **75:** 7611–7618.

Wehausen, J. W. and E. V. Laitone (1960). Surface waves. *Handbuch der Physik*, Berlin and New York: Springer-Verlag, pp. 446–778.

Wiegel, R. L. (1954). Gravity waves. *Tables of Functions*. Council of Wave Research, The Engineering Foundation, University of California, Berkeley, 30 pp.

PART **II**

Sediment Entrainment and Transport

Part II deals with the interaction of fluid motion with the continental margin substrate that results in the entrainment and transport of sediment. This topic was mentioned in Chapter 6, by Madsen, who described the character of the bottom boundary layer associated with bottom wave surge. In Chapter 7, P. D. Komar describes the boundary layers that develop in water flow over a granular bottom in response to unidirectional currents, and the relationship between bottom shear stress, the drag coefficient, and the velocity profile. In Chapter 8, Komar analyzes the response of a sandy substrate to boundary flow. He describes the threshold criteria for sand entrainment and presents methods for estimating the transport of sand in response to unidirectional flows and reversing wave motions.

In Chapter 9, D. E. Drake provides a similar analysis of the entrainment and transport of cohesive fine sediment (mud) in response to boundary flow. The methodology of this relatively new field of study is also described, as are the mechanisms of transport of fine sediment across the continental shelf. This discussion provides a foundation for later chapters dealing with patterns of shelf sedimentation (Chapter 15) and the nature of fine sediment dispersal on the continental slope and rise (Chapters 16 to 18).

The remaining two chapters in Part II deal with further aspects of the response of substrates (primarily sand substrates) to boundary flow. In Chapter 10, D. J. P. Swift and J. C. Ludwick examine both textural response (grain-size frequency distributions) and morphologic responses (bed form arrays). They present a simple numerical model for determining the pattern of bed load transport, and the distribution of areas of erosion and sedimentation. In Chapter 11, G. V. Middleton and M. A. Hampton describe sediment gravity flows, a class of sediment transport mechanisms which occurs primarily at the shelf edge and on the continental slope (Chapters 16 and 17).

CHAPTER 7

Boundary Layer Flow Under Steady Unidirectional Currents

PAUL D. KOMAR

School of Oceanography, Oregon State University, Corvallis, Oregon

Whenever water or another fluid flows past a solid boundary there is a frictional drag that retards the motion, diminishing the velocity in proximity to the boundary. The water in immediate contact with the solid boundary adheres to it, and is therefore stationary relative to the surface even when the remainder of the water is flowing. The velocity of flow increases from the surface outward, at first increasing rapidly and then more slowly as the velocity reaches that of the freely flowing water at some distance away from the surface. This zone where the flow is appreciably retarded by friction against the surface is known as the *boundary layer*. A boundary layer is developed wherever water flows through a channel or over the open ocean bottom, or around some object immersed in the water. It also forms if the object moves relative to a still body of water such as the settling of a particle.

This chapter briefly reviews boundary layers that develop in water flow over a granular sediment bottom. Equations are presented that describe the velocity distribution through the boundary layer, the quadratic stress law that relates the stress directly to a single measure of the velocity, and the effects on the velocity profile from the presence of an appreciable suspended sediment load. This brief review is designed to provide the necessary background for the examination of sand transport in Chapter 8. More complete reviews can be found in Schlichting (1951), Monin and Yaglom (1971), and Clauser (1956).

LAMINAR AND TURBULENT FLOW STRESS

Consider a very thin plate moving in a viscous fluid (Fig. 1), the plate being parallel to a second stationary boundary and only slightly above it. Exerting a force F on the plate, we would note that the plate would accelerate until it reached a certain limiting velocity U which it would then maintain. No matter how small the force F might be, the plate would move; however, its velocity U would depend on the magnitude of the stress $\tau = F/A$, where A is the area of the plate. If the velocity were sufficiently low so that tubulence were not induced, it would be found that

$$U \propto \frac{hF}{A} \tag{1}$$

where h is the distance between the plate and boundary. Rearranging gives

$$\tau = \frac{F}{A} \propto \frac{U}{h} \tag{2}$$

Now suppose that at some instant of time t_0 we are able to place a vertical line of dots in the fluid between the plate and the boundary. At a small time Δt later it would be found that the dye dots had shifted position horizontally as shown in Fig. 1. We might first notice that the dot at the lower boundary had not moved at all during the time interval and that the dot adjacent to the moving plate had moved a distance equal to the move-

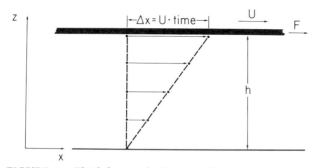

FIGURE 1. *Fluid shear and velocity profile produced by a plate moving with a velocity* U *relative to a parallel stationary boundary.*

ment of the plate. Dots at intermediate distances between the moving plate and the stationary bottom moved in a linear fashion according to their location between the plate and boundary. Since the velocity u is seen to increase linearly with distance z above the fixed bottom, and $u = 0$ at $z = 0$, and $u = U$ at $z = h$, it follows that

$$u = U \frac{z}{h} \tag{3}$$

Differentiating yields

$$\frac{du}{dz} = \frac{U}{h}$$

for the velocity gradient. Combining this with (2) yields

$$\tau = \mu \frac{du}{dz} \tag{4}$$

where μ is the constant of proportionality. This coefficient is called the *absolute viscosity* (or *dynamic viscosity*). It is a distinct property of the fluid and varies for different fluids. Table 1 gives values of water viscosity for a range of temperatures. The absolute viscosity is primarily a function of the temperature; pressure has only a small effect which can generally be ignored. The higher the value for the viscosity, the "thicker" will be the fluid. For example, at 20°C, water has an absolute viscosity of about 1 centipoise, heavy machine oil is 84 centipoise, and glycerine is 1490 centipoise.

The stress τ is a force per unit area and therefore in the cgs system it has units of dynes per square centimeter. The velocity gradient du/dz has units of cm/sec/cm, or sec^{-1}. The absolute viscosity therefore has units

$$\frac{\text{dyne} \cdot \text{sec}}{\text{cm}^2} = \frac{(\text{g} \cdot \text{cm/sec}^2) \cdot \text{sec}}{\text{cm}^2} = \frac{\text{g}}{\text{cm} \cdot \text{sec}}$$

One g/cm·sec is called a *poise*, but the poise is too large for most fluids so that absolute viscosities are generally given in *centipoise*. The equivalent in the English system of units is

$$1 \text{ poise} = \frac{1}{479} \text{ lbf} \cdot \text{sec/ft}^2$$

The shear stress of (4) is fundamentally a process of a transfer of momentum normal to the surface over which the stress acts. There is a continuous transfer or diffusion of momentum from faster to slower moving levels in the flow and thus toward the flow boundary. This momentum flux toward the boundary constitutes the frictional drag which saps the main flow of its total momentum. If no external force were applied, the flow would eventually lose all of its momentum and velocity to frictional drag. This momentum transfer aspect can be seen if we multiply numerator and denominator of (4) by the water density ρ to obtain

$$\tau = \frac{\mu}{\rho} \frac{d(\rho u)}{dz} \tag{5}$$

Since the quantity ρu is the momentum per unit volume of fluid, there is a gradient in horizontal momentum toward the boundary which gives rise to the diffusion of momentum and the fluid friction drag or stress. The ratio $\nu = \mu/\rho$ is the diffusivity of momentum and is known as the *kinematic viscosity*.

If the plate of Fig. 1 is moved at too great a velocity, it would be found that irregularities or disturbances

TABLE 1.   Density and Viscosity of Fresh Water and Seawater at a Pressure of 1 atm and a Range of Temperatures

Temperature (°C)	Fresh Water			Seawater, $S = 35\%$		
	ρ (g/cm³)	μ (centipoise)	ν (cm²/sec × 10⁻²)	ρ (g/cm³)	μ (centipoise)	ν (cm²/sec × 10⁻²)
0	1.0000	1.52	1.52	1.028	1.61	1.57
10	0.9997	1.31	1.31	1.027	1.39	1.35
15	0.9991	1.14	1.14	1.026	1.22	1.19
20	0.9982	1.005	1.00	1.025	1.07	1.05
30	0.9956	0.801	0.804	1.022	0.87	0.85

develop in the fluid flow. The motion would no longer be a simple linear increase in velocity with distance above the bottom. Dye placed in the moving fluid would quickly break up into eddies, diffusing across the flow as well as moving in the mean direction of flow. These two contrasting modes of fluid flow are designated as *laminar* and *turbulent*. In laminar flow the motion is characterized by the thin layers of fluid slipping over one another without mixing apart from a very slow molecular exchange. A trace of dye placed in a laminar flow forms a streak, losing its identity by molecular diffusion only after considerable time. Turbulent flow, on the other hand, is characterized by eddy motions where there is intense lateral mixing. Dye placed in a turbulent flow breaks up into eddies and quickly diffuses because of the eddy transfer. Turbulent flow may therefore be regarded as a motion where a complex secondary movement (the eddies) is superimposed upon the primary motion of translation.

Eddies are able to transfer momentum normal to the flow boundary at appreciably greater rates than the molecular transfer in laminar flow. This added transfer is incorporated in an *eddy viscosity* A_v so that the total stress becomes

$$\tau = (\mu + A_v)\frac{d\bar{u}}{dz} \qquad (6)$$

Since the eddy viscosity is much greater than the molecular dynamic viscosity $(A_v \gg \mu)$ the molecular viscosity is generally dropped altogether, giving

$$\tau = A_v \frac{d\bar{u}}{dz} \qquad (7)$$

The description of laminar flow is comparatively simple in that we need only give the three velocity components u, v, and w in the x, y, and z coordinate directions which when added vectorially give the flow direction and magnitude. On the other hand, turbulence has random velocity fluctuations that are superimposed over the average flow and which can be described today only with statistics. The instantaneous velocity component is then equal to the sum of a mean velocity $\bar{u}$, $\bar{v}$, and $\bar{w}$ plus a fluctuating component u', v', and w'. Thus for *xyz* coordinate directions, we write

$$u = \bar{u} + u', \quad v = \bar{v} + v', \quad w = \bar{w} + w' \qquad (8)$$

The mean velocities $\bar{u}$, etc., are obtained by time-averaging the measured u, v, and w values, the time being long in comparison to the time scale of the turbulence. Because the fluctuations u', v', and w' are both positive and negative, they average out when the mean is taken. The velocity fluctuations are therefore the random portion of the turbulent flow and are

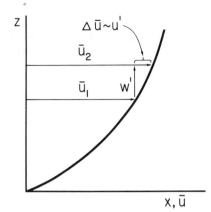

FIGURE 2. *Exchange of fluid elements of velocity $\bar{u}_1$ and $\bar{u}_2$ within the boundary layer due to the veritcal eddy velocity w'.*

found, in general, to follow the Gaussian distribution. It should be noted in (7) that with turbulent flow the stress is related to the gradient of the mean velocity $\bar{u}$, not to the instantaneous velocity u. No such distinction was of course required in steady laminar flow.

It is the eddy fluctuations u', v', and w' that are important in transferring momentum under turbulent flow. Consider the velocity profile of Fig. 2 with the fluid layers 1 and 2 moving with slightly different mean velocities. If the lower velocity fluid in layer 1 were to fluctuate with a w' velocity into layer 2, its velocity in the direction of the stream flow would be less than the velocity of the new environment. In the new environment it would appear to be deficient by an amount $\Delta u \simeq u'$. The drag of the faster moving surroundings would accelerate the element and increase its momentum. The flux crossing from level 1 to 2 would be $\rho w'$. Multiplying by u' gives the flow direction momentum change per second for this flux as $\rho u'w'$, or on the average over a time period, as

$$\tau = \overline{\rho u'w'} \qquad (9)$$

The rate of change of momentum represented by (9) is an effective resistance to motion and an effective shearing stress for fully developed turbulence.

It has been found empirically that for turbulent motion the stress is proportional to the square of the average velocity. The reason for this can be seen in part in (9). With increases in the fluctuating velocities in proportion to increases in the mean velocity, one has

$$u' \propto \bar{u} \quad \text{and} \quad w' \propto u' \propto \bar{u}$$

so that (9) becomes

$$\tau \propto \rho \bar{u}^2 \qquad (10)$$

We will return to this proportionality later.

Thus far we have not quantitatively specified the conditions under which the flow will become turbulent as opposed to remaining laminar. Of importance here are the forces that give rise to the instabilities, the turbulent eddies, versus the viscous forces that act to damp the disturbances. It is found that the dimensionless Reynolds number

$$\text{Re} = \frac{\rho u l}{\mu} \qquad (11)$$

is significant in this regard since it is proportional to the ratio of the inertial forces to the viscous force. l is a length parameter whose specification depends on the flow problem involved. For example, in river flow l becomes the hydraulic radius of the channel or the mean water depth if the channel is much wider than it is deep. For the river case u becomes the average velocity through the cross-sectional area of the channel. The generally accepted value for the critical Reynolds number in rivers is Re = 500. At Reynolds numbers lower than critical, local disturbances can initiate turbulent eddies but these do not persist for any distance away from the disturbance. The flow becomes laminar again unless new disturbances occur. At Reynolds numbers much above 500, irregularities develop and continue to grow so that the flow becomes fully turbulent. This is in part dependent on the number of obstacles on the flow boundary present to cause initial disturbances. In a smooth open channel, for example one made of smooth concrete, the flow may actually remain laminar until a critical Re = 2000 before becoming turbulent. The reason for the lower critical Re in natural channels is the great number of disturbance elements aiding the transition to turbulence.

Under natural conditions laminar flow is rare, occurring only for slow currents or for very small depths. In nearly all rivers and streams the flow is turbulent, unless the water is practically stationary or moves as thin sheet flow over a smooth boundary. Even in sheet flow local turbulence is common and dominates on account of the unevenness of the bottom. The only exception to this dominance of turbulent flow is very close to the bottom. If the bottom is sufficiently smooth without appreciable irregularities, there may develop a thin layer, the viscous sublayer, in which viscous stresses dominate over turbulent stresses.

BOUNDARY LAYER RELATIONSHIPS

It has been seen that when a fluid flows past a solid boundary, a boundary layer develops wherein the velocity decreases from its free value far from the boundary to zero at the wall itself. This boundary layer can be either laminar or turbulent, or a combination of

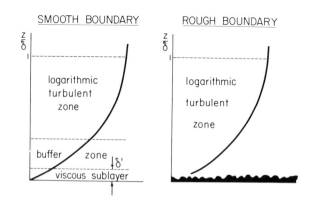

FIGURE 3. *Zones within the boundary layers above smooth and rough boundaries.*

both. This section reviews the equations that describe how the velocity $\bar{u}$ increases with distance above the bottom.

Along a *smooth boundary* the velocity profile is broken up into three distinct regions as illustrated in Fig. 3. No single equation will describe the velocity profile over its entire thickness. At the wall where the velocities are low there is a *viscous sublayer* in which the mean shear stress is controlled by the dynamic molecular viscosity μ and where the velocity gradient du/dz is very nearly constant. We therefore have

$$\tau = \mu \frac{du}{dz} = \text{const}$$

Defining

$$u_* = \left(\frac{\tau}{\rho}\right)^{1/2} \qquad (12)$$

which has units of velocity and is therefore called the *frictional drag velocity*, gives

$$u_*^2 = \frac{\tau}{\rho} = \nu \frac{d\bar{u}}{dz}$$

and integration yields

$$\bar{u} = \frac{u_*^2}{\nu} z + \text{const}$$

The constant of integration equals zero since $u = 0$ at $z = 0$, leaving

$$\frac{\bar{u}}{u_*} = \frac{u_* z}{\nu} \qquad (13)$$

which shows that the velocity increases linearly with increasing distance above the bottom, the rate of increase depending mainly on u_*.

It is found that (13) applies over the range of z given by

$$0 < \frac{u_* z}{\nu} < 5$$

from which we can evaluate the thickness of the viscous sublayer to be ($z = \delta'$)

$$\delta' = 5\,\frac{\nu}{u_*} \qquad (14)$$

The viscous sublayer is generally so thin that measurements to investigate its properties are very difficult to carry out. The sublayer, while dominated by viscous transfer, is observed not to be laminar, but is accompanied by considerable irregular fluctuations.

Above the viscous sublayer is found a second narrow zone in which the viscous and turbulent transfer of momentum are comparable. This is sometimes called the *buffer zone*. Farther out into the boundary layer where

$$\frac{u_* z}{\nu} > 30 \text{ to } 70$$

turbulent transfer is found to dominate over viscous transfer. Above the buffer zone the velocity takes the form

$$\frac{\bar{u}}{u_*} = \frac{1}{\kappa} \ln \frac{u_* z}{\nu} + 5.5 \qquad (15)$$

where κ is von Karman's constant (approximately 0.4 for many flows). The 5.5 constant is empirical, determined experimentally by Nikuradse (1933) and others.

Large roughnesses at the boundary will disrupt the viscous sublayer so that the turbulence extends to the wall. This is commonly the case where the flow occurs over sand and coarser beds where the sediments may also develop ripples and other roughness elements. In the equations above the velocity distribution depended only on the magnitude of the frictional drag velocity u_*. With rough surfaces the size and form of the roughness elements become additional important parameters. For rough boundaries the equation for the velocity profile is

$$\frac{\bar{u}}{u_*} = \frac{1}{\kappa} \ln \frac{z}{k_s} + 8.5, \quad \begin{array}{c} \dfrac{u_* z}{\nu} > 50 \text{ to } 100, \\[1.5ex] \dfrac{z}{\delta} < 0.15 \end{array} \qquad (16)$$

where k_s is a measure of the roughness called the *roughness element* and δ is the boundary layer thickness. The constant 8.5 is again empirical, obtained in pipes covered with a uniform layer of sand. Under such conditions k_s is equal to the diameter of the sediment grains since that is the distance they project upward into the flow. Comparative flume experiments have shown that with a complete sediment grain-size distribution, the sieve size, of which 75% of the mixture by weight is finer, should be taken as a measure of k_s (Einstein, 1950). This is because the larger particles project upward into the flow while the smallest particles are hidden within the crevices between the large grains. The bed roughness is also affected by the presence of ripples and bars on the bottom.

An alternative relationship to (16) for a rough boundary, one that is commonly used, is

$$\frac{\bar{u}}{u_*} = \frac{1}{\kappa} \ln \frac{z + z_0}{z_0} \qquad (17)$$

where z_0 is the so-called *roughness length* but in fact is the notional height within the sediment bed at which $\bar{u} = 0$ and $z = 0$. In practice the velocity distribution $\bar{u}$ is plotted against $\ln(z)$ to obtain a first approximation intercept value of z_0, which then in turn is used to plot $\bar{u}$ versus $\ln(z + z_0)$ to obtain a second approximation value for z_0, and so on (Inman, 1963). The slope of the line yields u_*. Equation 17 becomes equivalent to (16) if $z_0 < z$ and

$$z_0 = \frac{k_s}{30} \qquad (18)$$

This shows the relationship between the zero-velocity intercept z_0 and the roughness k_s which in turn is a function of the sediment grain size and the presence of bed forms. However, it will be seen later that the value of z_0 is also affected by the presence of a suspended sediment load transporting close to the bed.

On what basis does one decide to use (15) for flow over a smooth boundary versus either (16) or (17) for a rough boundary? The distinction was made by Nikuradse (1933) who, as a result of his experiments on the flow over a sandy bed, concluded that three types of flow could be distinguished according to the ratio of the grain size or roughness k_s to the thickness of the viscous sublayer δ' given by (14). These flows are as follows:

1. *Smooth flow*

$$0 < \frac{k_s}{\delta'} < 1 \qquad \left(0 < \frac{u_* k_s}{\nu} < 5 \right)$$

All of the sand grains or other irregularities lie completely within the viscous sublayer. The resistance to flow is not affected by the grains and the surface therefore exhibits the same resistance to flow as does a completely smooth surface.

2. *Transitional flow*

$$1 < \frac{k_s}{\delta'} < 14 \qquad \left(5 < \frac{u_* k_s}{\nu} < 70 \right)$$

The most exposed sand grains now project through the viscous sublayer and shed eddies into the main flow.

There is now a form resistance that increases the overall resistance to the fluid flow.

3.   *Fully rough flow*

$$14 < \frac{k_s}{\delta'} \qquad \left(70 < \frac{u_* k_s}{\nu}\right)$$

All of the irregularities project up through the viscous sublayer so that there is no longer a coherent sublayer. The grains shed eddies and the resistance to the flow is predominantly due to form resistance of the sand grains or other irregularities.

For the transitional flow the velocity distribution is given by

$$\frac{\bar{u}}{u_*} = \frac{1}{\kappa} \ln \frac{z}{\alpha k_s} + 8.5 \qquad (19)$$

where α is a correction factor with a value that depends on the ratio k_s/δ' as given in Fig. 4. The empirical α is again based on the experimental results of Nikuradse (1933).

One difficulty with (15), (16), and (19) is that they are found to apply only to the lower portion of the total boundary layer thickness, i.e., to approximately $z/\delta < 0.15$, where δ is the thickness of the boundary layer. In this region the stress τ is nearly constant and equal to the value at the boundary.

In the early days of velocity profile investigations, it was found that the velocity reduction or defect, $\bar{u}_\infty - \bar{u}$, where $\bar{u}_\infty$ is the velocity outside the boundary layer, was almost entirely dependent on the magnitude of the shear stress velocity u_*. In Fig. 5 the ratio of the velocity defect to u_* is seen to bear a constant relationship to z/δ, the distance above the bottom relative to the thickness of the boundary layer. The correlation is surprisingly

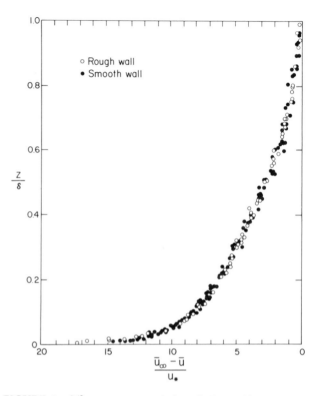

FIGURE 5.   *The convergence of the velocity profiles over rough and smooth boundaries when the velocity defect $\bar{u}_\infty - \bar{u}$ is related to the relative distance above the bottom, z/δ, where δ is the thickness of the boundary layer. From Daily and Harleman (1966).*

good over the turbulent portion of the boundary layer; it does not hold within the viscous sublayer. Empirically the boundary layer is divided into two zones where the equations

$$\frac{\bar{u}_\infty - \bar{u}}{u_*} = -5.6 \log \frac{z}{\delta} + 2.5 \qquad \left(\frac{z}{\delta} < 0.15\right) \quad (20)$$

$$\frac{\bar{u}_\infty - \bar{u}}{u_*} = -8.6 \log \frac{z}{\delta} \qquad \left(\frac{z}{\delta} > 0.15\right) \qquad (21)$$

apply. Equations 20 and 21 are known as the *universal velocity defect law*. It can be shown that (20) is directly derivable from (16) with some modification to the empirical coefficients. Therefore, the two equations are not independent.

In truth, none of the equations above is particularly good for describing the velocity profile adjacent to a boundary such as found in rivers and streams where ripples and dunes cause local boundary layer separation. It is partly because of this deficiency that progress in the understanding of sediment transport and deposition has been hindered.

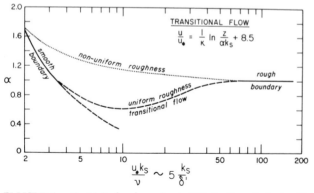

FIGURE 4.   *Empirical values of α in (19) for the velocity profile in transitional flow conditions.*

EKMAN BOUNDARY LAYER

For both the atmosphere and ocean the main flow above the boundary layer is approximately geostrophic. Therefore the Coriolis force is clearly significant within the boundary layer, producing a rotation of the velocity vector with increasing distance above the bottom. In the northern hemisphere the velocity vectors rotate to the right (clockwise) with increasing height while also increasing their magnitudes. Such a rotational velocity profile is called an *Ekman boundary layer* after V. W. Ekman who considered the theory of such flows with application to the wind-induced drift in the surface waters of the ocean. The relationship of the Ekman boundary layer to the regional flow field has been described in Chapter 3.

This Ekman boundary layer is considerably thicker than the logarithmic boundary layer already considered. Typical values are given in Table 2 for flows on the continental shelf over a rough bottom, for a deep ocean flow over a smooth bottom, and for the atmospheric boundary layer. In all cases the logarithmic boundary layer is only a small portion of the Ekman boundary layer thickness. Within this small thickness the amount of rotation is minor and can be neglected when considering the logarithmic boundary layer alone. The total Ekman layer has a thickness of approximately $\kappa u_*/f$ where $f = 2\Omega \sin\phi$ is the Coriolis parameter, Ω is the angular velocity of the earth's daily rotation (0.729×10^{-4} sec^{-1}), and ϕ is the geographic latitude (Csanady, 1967; Wimbush and Munk, 1970).

TABLE 2.   Representative Boundary Layer Parameters

	Continental Shelf (Rough)	Deep Sea (Smooth)	Atmosphere
$\bar{u}_\infty$	30 cm/sec	3 cm/sec	15 m/sec
u_*	1.0 cm/sec	0.1 cm/sec	50 cm/sec
Thickness			
Viscous sublayer	1 mm	2 cm	0.5 cm
Constant stress layer	1 m	10 cm	50 m
Logarithmic layer	10 m	1 m	100 m
Ekman layer	50 m	5 m	2 km

Source. From Wimbush and Munk (1970) and a seminar presented by Wimbush (November 1974).

It is significant that in the northern hemisphere the free velocity above the Ekman layer will trend to the right of the mean direction of flow in the Ekman layer. Associated with currents of this nature is the generation of upwelling and downwelling currents along a coast (see Chapter 14, pp. 261–264). In addition, the bottom stress of the fluid flow will not necessarily be in the direction of the free flow above the boundary layer. Therefore, the sand transport and bed forms oriented with the near-bottom flow would be at an angle to the free flow above the Ekman layer.

The Ekman layer has been found in the atmospheric boundary layer, but it is generally complicated by various instabilities. It has also been produced in the laboratory in rotating tanks and ocean models, but its existence in the ocean has not been clearly demonstrated.

BOUNDARY LAYER THICKNESS

If a flow of uniform velocity $\bar{u}_\infty$ passes over a flat plate, as shown in Fig. 6, a boundary layer will develop, thickening gradually downstream from the leading edge. Over some initial distance from the edge the flow will remain laminar and the boundary layer thickness will be (Schlichting, 1951, p. 104)

$$\delta_{\text{lam}} = 5 \left(\frac{\mu x}{\rho \bar{u}_\infty} \right)^{1/2} \qquad (22)$$

where x is the distance from the leading edge of the plate. The laminar boundary layer thickens according to the square root of the distance from the start of boundary layer development. Beyond some critical distance on the plate the inertial forces will predominate and the flow then becomes turbulent. This transition takes place at approximately

$$\text{Re}_x = \frac{\bar{u}_\infty x}{\nu} = 1 \times 10^5$$

For a turbulent boundary layer the thickness may be expressed by the equation

$$\delta = 0.38 x \left(\frac{\nu}{\bar{u}_\infty x} \right)^{1/5} \qquad (23)$$

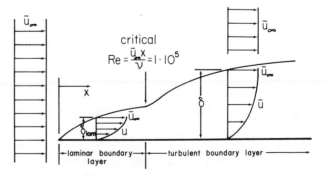

FIGURE 6. *The growth of the boundary layer over a plate, the boundary layer being first laminar and then turbulent.*

which also predicts an increase in the boundary layer thickness with increasing distance x along the plate. According to (23) the boundary layer should continue to grow until the entire fluid field is affected. This condition is encountered in pipes and in open channels such as rivers where the boundary layer thickness approximately corresponds to the flow depth (actually slightly depressed below the water surface because of the development of cross-channel secondary circulations). In the open ocean the boundary layer does not continue to grow as indicated by (23). According to Csanady (1967), a turbulent Ekman layer, such as would be found in the ocean, differs from the two-dimensional turbulent boundary layer considered above in that it may exist with its thickness remaining constant in the direction of flow. If outside the boundary layer the pressure gradient and Coriolis force balance, then within the boundary layer the pressure gradient is much the same as without but the Coriolis force is less since the velocity is reduced. This excess pressure gradient is then available for "reenergizing" the fluid flow and thus acts against the flow velocity reduction and growth of the boundary layer. Under such conditions, Csanady envisions that the boundary layer thickness may remain constant.

Above the boundary layer the flow remains nearly free of turbulence and shear so that the field of mean velocities conforms closely to potential flow conditions. Actually there is some turbulence but it is negligibly small relative to the much higher state of turbulence inside the boundary layer. Although the boundary layer thickness δ can be defined in terms of the mean velocity $\bar{u}$, at any instant the border zone consists of fingers of turbulence extending up into the nonturbulent free stream flow and fingers of nonturbulence extending deep into the turbulent region of the boundary layer. Daily and Harleman (1966, p. 229) discuss one case in which they measured such turbulent fingers extending above δ to distances as great as 1.2δ and nonturbulent zones protruding into the boundary layer as far as 0.4δ, more than halfway down through the boundary layer. The average position of the turbulent versus nonturbulent interface was at 0.78δ, 22% short of the total boundary layer thickness based on the disappearance of the mean shear stress.

STRUCTURE OF THE VISCOUS SUBLAYER

When present, the viscous sublayer is not a simple shearing of fluid layers over one another as in purely laminar flow. Instead, it displays instabilities involving periodic events of fluid inrush followed by ejection.

Associated with this vertical exchange is a horizontal structure of high- and low-velocity streaks extending in the flow direction with spacings on the order of $100v/u_*$ (Einstein and Li, 1958; Kline et al., 1967; Corino and Brodkey, 1969). The streaks can be detected with dye injection and can also produce lineations of fine sand and mica flakes on an otherwise smooth sediment bottom. Extensive work has been conducted on the ejection or "bursting" of fluid from the viscous sublayer (Rao et al., 1971; Willmarth and Lu, 1972; Offen and Kline, 1974; Gordon, 1974). The sweep and ejection events occupy only a relatively short period of time but are important in that they are responsible for most of the vertical transport of momentum and consequently for most of the horizontal shear stress. The bursting phenomenon has been recorded in marine boundary layers where such events contributed 80% of the total stress while occurring only 20% of the total flow time.

BOUNDARY LAYER SEPARATION

The boundary layer may become separated from the boundary itself if the bottom is sufficiently irregular on either a small or large scale (Fig. 7). Separation may occur behind individual sediment grains or behind ripples, dunes, or other obstacles (see Chapter 10). In boundary layer separation the layer actually detaches from the solid surface and enters the main body of fluid flow as a *free shear layer*. The point S of Fig. 7 is called the *separation point*, and the point A, where the boundary layer again attaches to the bed, is called the *attachment point*. On the downside of the barrier there is a zone of sluggish recirculating flow in the form of a large eddy. If the object is sufficiently blunt, separation may also occur on the upstream side (Fig. 7).

Boundary layer separation occurs because of adverse pressure gradients that oppose the fluid flow. Where the streamlines over the object become closer together because the velocity is increased, the pressure described by Bernoulli's equation is lower than where the streamlines are more widely separated and the velocity is lower (Fig. 8). Therefore, on the lee side there is a pressure gradient opposing the flow. If sufficiently strong, this pressure gradient may actually cause the flow to stagnate ($\bar{u} = 0$) in that zone, turn back upon itself, and eventually form the lee eddy as shown. The tendency for boundary layer separation is greater when the boundary layer is laminar than when it is turbulent. This is because with turbulent flow the free flow above the object is better able to add more momentum to the boundary layer adjacent to the object, momentum that enables the flow to better overcome the opposing

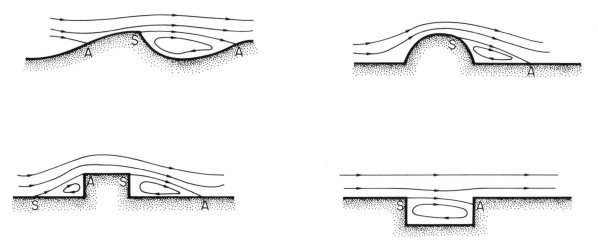

FIGURE 7. *Boundary layer separation over a variety of bottom irregularities.*

pressure gradient. With a *streamlined* object, the added momentum is able to overcome any adverse pressure gradient so that no separation occurs.

One important consequence of boundary layer separation is that it gives rise to *form resistance* or *form drag*. With separation there is a significant net force from the balance of the normal forces acting on the body, the net force being in the direction of flow, felt as an additional drag above the tangential stresses of viscous shear.

QUADRATIC STRESS LAW

Employing $u_* = (\tau/\rho)^{1/2}$, algebraic manipulation of (17) leads to

$$\tau = \frac{\kappa^2}{\{\ln\,[(z + z_0)/z_0]\}^2}\,\rho\bar{u}^2 \qquad (24)$$

which gives a proportionality between τ and $\bar{u}^2$ which was first indicated in (10). The proportionalty factor of (24) contains z, the distance above the bottom, but if it is specified that u is measured at $z = 100$ cm, then

$$\tau = \frac{0.16}{\{\ln\,[(100 + z_0)/z_0]\}^2}\,\rho(\bar{u}_{100})^2 \qquad (25)$$

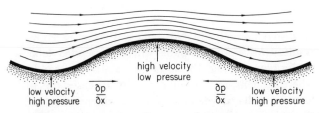

FIGURE 8. *Pressure gradients resulting from variations in velocity over an irregular bottom. On the lee side of the hump the pressure gradient $\partial p/\partial x$ opposes the velocity and if sufficiently great may cause boundary layer separation.*

having set $\kappa = 0.40$. Substituting

$$C_{100} = \frac{0.16}{\{\ln\,[(100 + z_0)/z_0]\}^2} \qquad (26)$$

where C_{100} is called the *drag coefficient*, gives

$$\tau = C_{100}\rho(\bar{u}_{100})^2 \qquad (27)$$

There is considerable precedence for a relationship such as (27), a proportionality between the stress τ and $\bar{u}^2$. One example is in the flow of water through a pipe. In that case, for fully developed turbulence the drag coefficient depends on the wall roughness of the pipe (Nikuradse, 1933), analogous to the C_{100} dependence on z_0 in (26). Under low flow conditions the drag coefficient depends on the Reynolds number as well as on the roughness factor.

There has been considerable interest in (27), called the *quadratic stress law* (Sternberg, 1968, 1972; McCave, 1973; Ludwick, 1975). It is apparent that if one properly evaluates the drag coefficient C_{100}, then only a measure of the velocity $\bar{u}_{100}$ is needed at 100 cm above the bed to evaluate the stress τ, rather than measuring the complete velocity profile and determining u_* and hence τ by fitting the profile to a relationship such as (17). Sternberg (1968, 1972) has obtained many measurements in Puget Sound, Washington, measuring the complete profile to determine τ and then calculating the drag coefficient

$$C_{100} = \frac{\tau}{\rho(\bar{u}_{100})^2} = \left[\frac{u_*}{\bar{u}_{100}}\right]^2 \qquad (28)$$

His results are shown in Fig. 9 and include areas that are rocky (A), gravelly (B), rippled sand (D, F), and indistinctly roughened sand (E, C). All of the data are compiled in Fig. 10, indicating that

$$C_{100} = 3.1 \times 10^{-3}$$

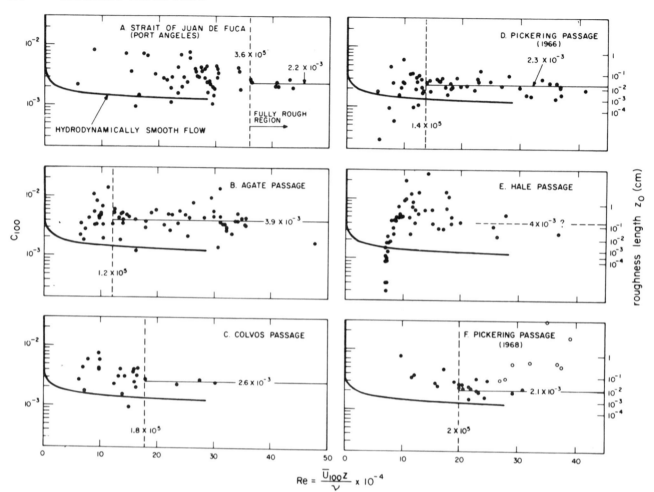

FIGURE 9. C_{100} *values from Sternberg (1968, 1972), obtained in various locations within Puget Sound.*

is a reasonable average value for the drag coefficient for fully developed turbulent conditions ($\bar{u}_{100}z/\nu > 1.5 \times 10^5$). It can be seen in Fig. 9 that the average values of C_{100} for the various bed types range from about 2×10^{-3} to 4×10^{-3}. This difference is of the same order of magnitude as the variance in C_{100} values from one particular location. The relationship between the average C_{100} values and z_0 is shown schematically in Fig. 11.

With a relatively constant value of C_{100}, (27) becomes

$$\tau = 3.1 \times 10^{-3}\rho(\bar{u}_{100})^2 \qquad (29)$$

which would simplify the evaluation of the bottom stress since one would need only a single current meter to measure the velocity at 100 cm above the bed. This would also simplify the calculations of sand transport under such conditions (Chapter 8). However, there are two severe limitations on the employment of this average drag coefficient and (29). Although the data of Sternberg (1968, 1972) cover a wide range of bed

roughness conditions, it does not include large-scale ripples, dunes, or sand waves. As pointed out by Sternberg, such appreciable bed forms (>5 cm in height) would offer greater resistance to the flow than other bed types and would therefore affect the value of C_{100}. The other limitation is with regard to the presence of an appreciable suspended load. Most of the measurements of Sternberg were for conditions of negligible suspended load. When an important suspended load was present, as shown by the open circles of Fig. 9F, the C_{100} values increased significantly. This is also shown by the results of McCave (1973) who made his measurements in the English Channel and North Sea. Figure 12 shows his results over a shingle bank where suspended load was negligible, and Fig. 13 shows the data from a sand bank where suspension is significant. In all cases the turbulence is fully developed (C_{100} is independent of the Reynolds number). It is seen in Figs. 12 and 13 that the C_{100} values obtained by McCave tend to be higher than those reported by Sternberg (1968, 1972). One

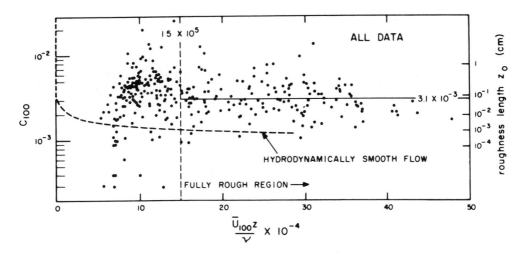

FIGURE 10. *All of the* C_{100} *values of Sternberg (1968, 1972) giving a mean value of 3.1 × 10⁻³ under fully rough conditions.*

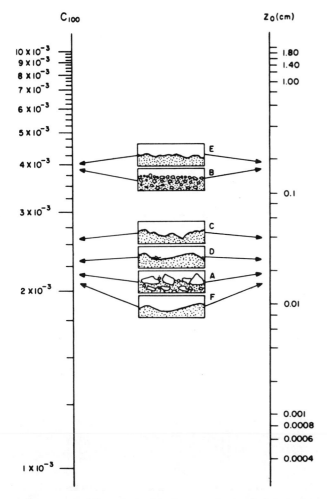

FIGURE 11. *The relationship between* C_{100} *and the roughness length* z_0. *The letters A through F refer to the locations in Fig. 9. From Sternberg (1972).*

unusual feature of the McCave data is that he obtains different C_{100} values between flood and ebb tide (note the changes in current direction). The locations are approximately the same although there was a slight relocation of the current meters required by the ship relocation, the ship being moored by a single bow anchor (a distance change of about 175 m between measuring sites of flow at flood and ebb tide). Some of the difference between the drag coefficient values may have resulted from this change of location, however small. However, there was no noticeable difference between the two areas as seen by divers.

Ludwick (1975) found an even greater variation in drag coefficient values, obtained at the entrance to Chesapeake Bay, Virginia, C_{100} ranging over more than two orders of magnitude. There is some question with regard to his evaluations of the bottom stress since many of his velocity profiles show maxima and minima, and density stratification may be present in the water column. However, this would not account for all of the variability found in the C_{100} values. That there is a

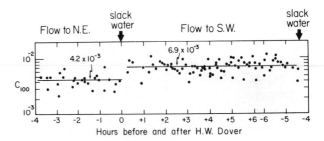

FIGURE 12. C_{100} *values from McCave (1973) obtained over a gravel bank in the English Channel.*

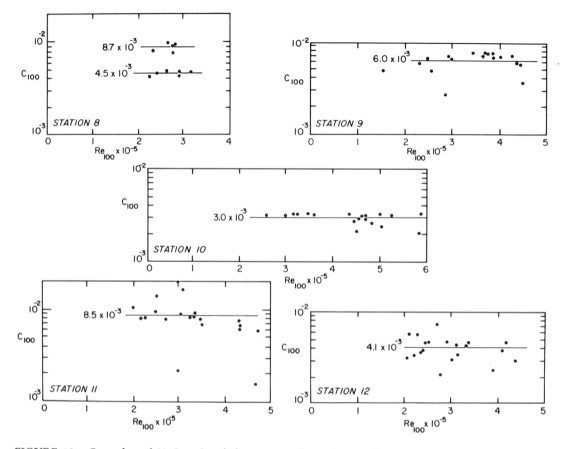

FIGURE 13. C_{100} *values of McCave (1973) for stations on the sandy Hinder Bank in the North Sea. The values have been corrected for sediment suspension* *effects placing them on a "suspension-free" basis. Two distinct drag coefficient values at station 8 correspond to two positions of the ship at anchor.*

systematic change in the drag coefficient over the two orders of magnitude is shown in Fig. 14 where C_{100} is plotted against u_*. Ludwick attributes this increase in C_{100} with increasing u_* at least in part to the hierarchy of bed forms and their sequence. This change in C_{100} with varying u_* and changing bed form is shown in

Table 3 for the flume data of Raudkivi (1963), also plotted in Fig. 14 as the dashed curve. In the measurements of Ludwick (Fig. 14) C_{100} does not take on a constant value as in the data of Sternberg or McCave, even at high flow rates. Nor does it reach a maximum and then decrease as do the Raudkivi data. Ludwick

TABLE 3. Calculated Values of C_{100} for Various Bed Forms*

u_* (cm/sec)	Bed Form	f_b	C_{100}
1.219	Threshold	0.02	4.22×10^{-3}
2.316	Ripples, dunes	0.06	2.07×10^{-2}
3.413	Ripples, dunes	0.10	5.30×10^{-2}
4.237	Ripples, dunes	0.12	7.87×10^{-2}
4.206	Sand waves	0.05	1.54×10^{-2}
3.658	Transition	0.02	4.22×10^{-3}

Source. From Ludwick (1975).

* Calculated from Raudkivi (1963) who gives the resistance f_b for 0.14 and 0.40 mm sands. In Ludwick's calculations it is assumed that $\delta = 10$ m.

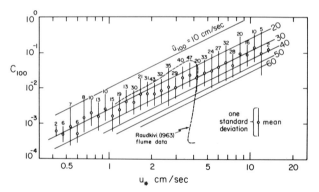

FIGURE 14. C_{100} *values of Ludwick (1975) obtained at the entrance to Chesapeake Bay, showing a progressive increase with increasing* u_*. *Also shown is a curve based on the flume data of Raudkivi (1963), summarized in Table 3. After Ludwick (1975).*

hypothesized that this continued increase in C_{100} is because of the presence in the Chesapeake Bay entrance of megaripples, large sand waves, and shoals not found in the laboratory flume studies such as those of Raudkivi. The enhancement of the suspended load at high $u*$ may also, in part, explain the increase in the C_{100} values.

SEDIMENT SUSPENSION

As the fluid velocity increases an appreciable suspended sediment load may develop which will modify the velocity profile. Bagnold (1966) has suggested that for sediment of settling velocity w_s to remain in suspension the upward-directed components of the turbulent eddy velocity fluctuation, w_{up}', must exceed w_s. On this basis Bagnold derived the suspension criterion

$$\theta = \frac{\tau}{(\rho_s - \rho)gD} > 0.4 \, \frac{w_s^2}{gD} \qquad (30)$$

for quartz sand ($\rho_s = 2.65$ g/cm³), D being the grain diameter. This relationship is plotted in Fig. 15 along with the curve for the threshold of sediment motion. Equation 30 is also shown plotted in Fig. 15 with a proportionality coefficient of 0.19 rather than 0.4. This value is based on considerations of McCave (1971) who argued that the suspension criterion should be determined relative to the lower levels of the flow where w' is larger rather than taking a mean value for the boundary layer as Bagnold did. It is seen in Fig. 15 that according to (30) material of $D < 0.17$ mm will be transported largely in suspension immediately after the threshold is exceeded. Using the 0.19 coefficient of McCave, material finer than 0.23 mm would be transported in suspension. As the relative stress θ increases it is seen that even coarse grain sediment can be transported in suspension.

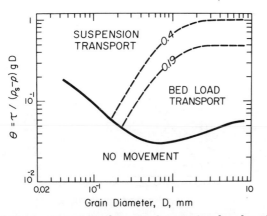

FIGURE 15. *Curves for the onset of suspension, based on (30). The 0.4 curve is that given by Bagnold (1966), and the 0.19 curve is that suggested by McCave (1971). Also shown is the curve for threshold of sediment motion.*

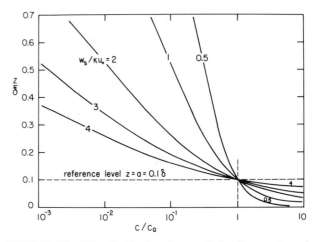

FIGURE 16. *The distribution of suspended sediment above the bottom according to (32), referred to a concentration* c_a *at the level* $z = a$. *The distribution is strongly dependent on the value of* $w_s/\kappa u*$, *the exponent in (32).*

The elementary treatment of the distribution of suspended sediment above the bottom follows the same principles as that of turbulent mixing and diffusion of fluid momentum. There is a tendency for an upward diffusion transport of sand, carried by the turbulent eddies, since there is a concentration gradient with highest concentrations closest to the bed. This upward diffusion of sand is opposed by the downward settling of the grains due to their settling velocity w_s. At equilibrium the upward diffusion balances the downward settling, giving the equation

$$cw_s + \nu_s \frac{\partial c}{\partial z} = 0 \qquad (31)$$

where c is the concentration at level z, and ν_s is the sediment diffusion rate, analogous to the kinematic eddy viscosity which specifies a fluid momentum diffusion. Under certain assumptions the solution of (31) leads to

$$\frac{c}{c_a} = \left[\frac{\delta - z}{z} \frac{a}{\delta - a} \right]^{w_s/\kappa u_*} \qquad (32)$$

(Rouse, 1938) where c_a is the reference concentration at $z = a$. With (32) it is possible to calculate the concentration of material of a particular grain size at any level in the flow, provided the concentration is known as some reference level. Figure 16 shows plots of c/c_a versus z/δ for a range of exponents $w_s/\kappa u*$, according to (32). It is seen that in each case $c = 0$ at $z = \delta$, and increases downward toward the bed, becoming $c = c_a$ at $z = a$. According to the relationship the concentration continues to increase still closer to the bed, becoming infinite when $z = 0$. The suspension equation obviously does not pertain to such close proximity to the bed as the sand will transport as bed load there.

The rate of concentration decrease with increasing height above the bottom as seen in Fig. 16 is heavily dependent on the exponent of (32). It is seen that the greater the value of the exponent the more rapid the decrease in c with distance above the bed. This means that for a given boundary stress u_*, grains of larger settling velocity w_s will decrease upward at a faster rate than grains with smaller w_s. For a given w_s, the larger the value of the stress velocity u_*, the greater the distance above the bed will significant concentrations of that grain size be found. This dependence on w_s and u_* thus fits one's intuition. Now suppose the coefficient has some value ξ such that

$$\frac{w_s}{\kappa u_*} = \xi \tag{33}$$

Substituting $u_* = (\tau/\rho)^{1/2}$, this can easily be shown to be equivalent to

$$\frac{\tau}{(\rho_s - \rho)gD} = \frac{1}{\kappa^2 \xi^2} \frac{\rho}{(\rho_s - \rho)} \frac{w_s^2}{gD} \tag{34}$$

This is the same as (30) above determined by Bagnold (1966). Equivalence of the two relationships gives $\xi = 3.12$ for the 0.4 coefficient of Bagnold, and $\xi = 4.46$ for the coefficient proposed by McCave. It is seen in Fig. 16 that with these exponents in (32) very little suspended load would exist above the bottom.

Measurements in rivers and in flumes have shown that (32) gives a good representation of the actual distribution of suspended load. The relationship is for a state of equilibrium, but divergences will occur under many natural conditions so that the equation must be used with a certain amount of caution. Other formulations have been derived for the distribution of suspended load. These are reviewed by Raudkivi (1967) and Yalin (1972).

Einstein and Chien (1952, 1955), Vanoni (1953), and Vanoni and Nomicos (1959) have shown that the high concentrations of suspended sediment found near the bed modify the velocity profile and the von Karman coefficient κ from the clear-water case. Detailed theoretical analyses of the effects of suspended sediment particles on the turbulent eddies are given by Hino (1963) and Yalin (1972). Both conclude that the presence of the suspension causes an appreciable reduction of the amplitude of the turbulent fluctuations but little reduction of the fluctuating velocities.

Einstein and Chien (1955) performed experiments with high concentrations of suspended sand in a flume with an otherwise fixed sand bed. An example of the observed effects on the velocity profile is shown in Fig. 17. It is seen that there is an erroneously high indication of z_0, based on the intercept of the extrapolated logarithmic velocity profile. In addition, the slope of the straight-line plot is increased, and since the slope as given by (17) is u_*/κ and u_* remains the same in the two cases in Fig. 17, κ must decrease as a result of the presence of the suspended sediment. Figure 18 gives the variations in κ with k_s/z_0 where k_s is the mean boundary roughness as before, equal to the sand grain diameter in the fixed bed experiments of Einstein and Chien (1955). At the upper end of the line of Fig. 18 $\kappa = 0.40$

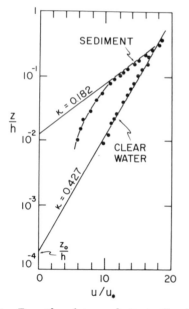

FIGURE 17. *Examples of two velocity profiles from Einstein and Chien (1955) showing the effects of the presence of suspended sediment. The conditions of bottom slope, depth of water, and discharge remain the same so that the current u_* is the same for the two cases. The change in slope of the curve therefore is due to the effects on the value of κ of the presence of the suspension.*

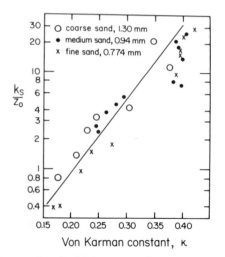

FIGURE 18. *Graph of k/z_0 versus the von Karman constant showing the effects of the presence of suspended sediments. From Einstein and Chien (1955).*

at $k_s/z_0 = 30$, the values found by Nikuradse (1933) with clear water flowing over a fixed sand roughness. McCave (1973) compared the value of the von Karman κ with the suspended grain concentration in the lower 10 cm of flow, obtaining his measurements from West Hinder Bank in the southern North Sea. At a mean sediment concentration of about 20 g/l, $\kappa = 0.40$, its clear-water value, but progressively decreased to $\kappa = 0.25$ at a sediment concentration of around 60 g/l. Such deviations did not occur over shingle banks because of the absence of appreciable suspended loads. Based on Fig. 18, calculations indicate that at the higher concentration found by McCave the drag coefficient C_{100} would be increased by about 40% over its clear-water value. This increase depends on the size of the roughness factor k_s: the smaller the value of k_s, the greater the percentage increase due to suspension (the 40% increase is for $k_s = 0.1$ cm, which is the roughness implied by the clear-water profile of Fig. 17). Thus the presence of a suspended sediment load may have a significant effect on the drag coefficient but apparently cannot alone account for the two orders of magnitude variation found by Ludwick (1975).

SUMMARY

A boundary layer is developed whenever water flows past a solid wall, whether through a channel or over the open ocean bottom. A boundary layer also develops in the atmosphere when winds blow over the ground. The boundary layer is characterized by a progressive decrease in velocity as the solid wall is approached, becoming zero at the wall itself. The description of the velocity profile is important in considerations of the fluid flow, and in application to the evaluation of sediment transport.

The velocity profile depends on the relative sizes of the viscous sublayer and any bottom irregularities such as sediment grains, ripples, or other bed forms. If the bottom irregularities are small in comparison, then the viscous sublayer may persist. Under such smooth conditions, above the viscous sublayer is a buffer or transitional zone, above which in turn is a fully turbulent zone where the velocity increases logarithmically with distance above the bottom. The viscous sublayer is not a simple laminar flow, but shows complex secondary motions of "bursting" or "sweeps and ejections." When the bottom is rough the viscous sublayer does not persist, and the logarithmic zone extends to the bottom.

The boundary layer may become separated from the boundary itself if the bottom is sufficiently irregular, such as behind individual sediment grains or behind ripples, dunes, or other obstacles. A description of the velocity profile under such circumstances is more difficult.

The quadratic stress law was also examined which relates the stress exerted by the flow to the velocity at some level (usually 100 cm) above the bottom. The drag coefficient, a proportionality factor, shows only small variations over a wide range of flow conditions and small bed forms, but varies considerably in the presence of an appreciable suspended sediment load or where large bed forms are present (dunes and larger).

SYMBOLS

A	area
A_v	vertical eddy viscosity
c	sediment concentration
c_a	reference sediment concentration at $z = a$
C_{100}	drag coefficient determined by measurements 100 cm off bottom
D	sediment grain diameter
f	Coriolis parameter: $f = 2\Omega \sin \phi$
g	acceleration of gravity (980 cm/sec^2)
k_s	bottom roughness element size
l	length parameter
p	fluid pressure
Re	Reynolds number
U	velocity of a plate
u	horizontal water velocity
$\bar{u}$	time-averaged horizontal water velocity
u'	horizontal eddy velocity in turbulent flow
$\bar{u}_\infty$	time-averaged velocity at the top of the boundary layer
u_{100}	velocity at 100 cm off the bottom
u_*	friction velocity
v'	eddy velocity component in y direction
w'	eddy velocity component in z direction
w_s	settling velocity of sediment grains
w'_{up}	upward component of vertical eddy velocity
x	horizontal distance in flow direction
y	horizontal distance across flow direction
z	vertical distance, zero at bottom
α	a correction factor
δ	boundary layer thickness
δ'	viscous sublayer thickness
κ	von Karman's constant ($\cong 0.4$)
μ	absolute or dynamic viscosity

ν kinematic viscosity

ν_s sediment diffusion rate

ρ density of water

ρ_s density of sediment grains

τ stress exerted by fluid flow

ξ coefficient for suspended sediment concentration
       equation

ϕ latitude

Ω angular velocity of earth's rotation

REFERENCES

Bagnold, R. A. (1966). An approach to the sediment transport problem from general physics. *U.S. Geol. Surv., Prof. Pap. 422*, 37 pp.

Clauser, F. H. (1956). The turbulent boundary layer. *Adv. Appl. Mech.*, **4**: 1–51.

Corino, E. R. and R. S. Brodkey (1969). A visual investigation of the wall region in turbulent flow. *J. Fluid Mech.*, **37**: 1–30.

Csanady, G. T. (1967). On the "resistance law" of a turbulent Ekman layer. *J. Atmos. Sci.*, **24**: 467–471.

Daily, J. W. and D. R. F. Harleman (1966). *Fluid Dynamics*. Reading, Mass.: Addison-Wesley, 454 pp.

Einstein, H. A. (1950). *The Bed-Load Function for Sediment Transportation in Open Channel Flow*. U.S. Dept. of Agriculture, Soil Conservation Service, T.B. No. 1026.

Einstein, H. A. and N. Chien (1952). *Second Approximation to the Solution of the Suspended Load Theory*. University of California Institute of Engineering Research and U.S. Army Corps of Engineers, Missouri River Division, M.R.D. Sediment Series, No. 3, 30 pp.

Einstein, H. A. and N. Chien (1955). *Effects of Heavy Sediment Concentration near the Bed on the Velocity and Sediment Distribution*. University of California, Institute of Engineering Research and U.S. Army Corps of Engineers, Missouri River Division, M.R.D. Sediment Series, No. 8, 76 pp.

Einstein, H. A. and H. Li (1958). The viscous sublayer along a smooth boundary. *ASCE Trans.*, **123**: 293–317.

Gordon, C. M. (1974). Intermittent momentum transport in a geophysical boundary layer. *Nature, Phys. Sci.*, **248**: 392–394.

Hino, M. (1963). Turbulent flow with suspended particles. *Proc. ASCE J. Hydraul. Div.*, **89**(HY 4): 161–185.

Inman, D. L. (1963). Sediments: Properties, mechanics of sedimentation. In F. P. Shephard, ed., *Submarine Geology*, 2nd ed. New York: Harper & Row, pp. 101–151.

Kline, S. L., W. C. Reynolds, F. A. Schraub, and P. W. Runstadler (1967). The structure of turbulent boundary layers. *J. Fluid Mech.*, **30**: 741–773.

Ludwick, J. C. (1975). Variations in boundary drag coefficient in the tidal entrance to Chesapeake Bay, Virginia. *Mar. Geol.*, **19**: 19–28.

McCave, I. N. (1971). Sand waves in the North Sea off the coast of Holland. *Mar. Geol.*, **10**: 199–225.

McCave, I. N. (1973). Some boundary-layer characteristics of tidal currents bearing sand in suspension. *Mém. Soc. Roy. Sci. Liège, 6e Série*, **6**: 107–126.

Monin, A. S. and A. M. Yaglom (1971). *Statistical Fluid Mechanics: Mechanics of Turbulence*. Cambridge, Mass.: M.I.T. Press, 769 pp.

Nikuradse, J. (1933). *Laws of Flow in Rough Pipes*. Natl. Advisory Comm. Aeronautics Tech. Memo. 1292 (translation from German, 1950).

Offen, G. R. and S. J. Kline (1974). Combined dye-streak and hydrogen-bubble visual observations of a turbulent boundary layer. *J. Fluid Mech.*, **62**: 223–239.

Rao, K. N., R. Narasimha, and M. A. B. Narayanan (1971). The "bursting" phenomenon in a turbulent boundary layer. *J. Fluid Mech.*, **48**: 339–352.

Raudkivi, A. J. (1963). Study of sediment ripple formation. *ASCE J. Hydraul. Div.*, **89**(HY 6): 15–33.

Raudkivi, A. J. (1967). *Loose Boundary Hydraulics*. Oxford: Pergamon, 331 pp.

Rouse, H. (1938). Experiments on the mechanics of sediment suspension. *Proc. 5th Int. Congr. Appl. Mech., Cambridge, Mass.*

Schlichting, H. (1951). *Boundary-Layer Theory*. New York: McGraw-Hill, 747 pp.

Sternberg, R. W. (1968). Friction factors in tidal channels with differing bed roughness. *Mar. Geol.*, **6**: 243–260.

Sternberg, R. W. (1972). Predicting initial motion and bedload transport of sediment particles in the shallow marine environment. In D. J. P. Swift, D. B. Duane, and O. H. Pilkey, eds., *Shelf Sediment Transport*, Stroudsburg, Pa.: Dowden, Hutchinson & Ross, pp. 61–82.

Vanoni, V. A. (1953). Some effects of suspended sediment on flow characteristics. *Proc. 5th Hydraul. Conf., Studies in Eng. Bull. 34*. Iowa State University, Ames, pp. 134–158.

Vanoni, V. A. and G. N. Nomicos (1959). Resistance properties of sediment-laden streams. *Proc. ASCE J. Hydraul. Div.*, **85**(HY 5): 77–107.

Willmarth, W. W. and S. S. Lu (1972). Structure of the Reynolds stress near the wall. *J. Fluid Mech.*, **55**: 65–92.

Wimbush, M. and W. Munk (1970). The benthic boundary layer. In A. E. Maxwell, ed., *The Sea*, Vol. 4, Part 1. New York: Wiley-Interscience, pp. 731–758.

Yalin, M. S. (1972). *Mechanics of Sediment Transport*. Oxford: Pergamon, 290 pp.

CHAPTER 8

The Transport of Cohesionless Sediments on Continental Shelves

PAUL D. KOMAR

School of Oceanography, Oregon State University, Corvallis, Oregon

There has been an increase of interest within the past few years concerning the processes of sediment transport on continental shelves. This is partly in response to such practical applications as the building of offshore nuclear power plants. The question has arisen as to what effect such construction would have on the shelf processes and the sediment transport. In addition, a variety of large dunes and other sedimentary features have been discovered in various parts of the world and we are seeking an understanding of their formation. Finally, the deep-sea oceanographer is interested in how sediments bypass the continental shelves, going from the land to the deep ocean basin. The purpose of this chapter is to review the transport of coarse sediments on the continental shelf, mainly sands, perhaps as fine as silts: those that remain close to the bottom and are transported by water motions interacting with the sediment bottom. Chapter 9 by Drake considers transport of the finer materials in suspension.

On continental shelves we must be interested in both unidirectional and oscillatory flows. In this chapter the only oscillatory flow considered is that of surface wave motions. As discussed by Madsen in Chapter 6, an orbital motion of individual water particles is associated with the surface waves. In deep water ($h/L > 0.5$) the orbits are nearly circular, with the orbital diameter equaling the wave height at the surface and decreasing exponentially with depth. In intermediate water depths the orbits are elliptical, again decreasing in size with depth, the ellipses also becoming flatter. At the bottom itself the elliptical motion degenerates to a straight-line to-and-fro motion of the water, the periodicity of the motion being the same as that of the surface wave profile variations. In contrast, true unidirectional currents have no periodicity, for example, normal ocean currents driven by water density differences, turbid-water currents (nepheloid layers) produced by the density contrast between clear ocean water and sediment-laden water, and currents caused by winds blowing over the water surface. We shall also consider tidal currents and the flow associated with internal waves as being "unidirectional." This is because the periods are so large that they behave and give results that are better grouped with the true unidirectional currents than with the short-period ($T < 20$ seconds) ocean waves. Finally, as discussed by Madsen in Chapter 6, the waves themselves can induce unidirectional currents in addition to their oscillatory motions.

There is some disagreement as to whether unidirectional currents or wave orbital motions are most important to sediment transport on the continental shelf. This disagreement is actually more apparent than real. The evidence indicates that surface wave action can cause bottom sediment stirring to depths of 125 m or more on high-energy ocean coasts. Some of this evidence is reviewed later. Of course on low wave energy coasts this depth of sediment movement due to waves is considerably less. However, in general sediment transport theories that consider unidirectional currents alone probably apply only at greater depths on the outer shelf where the wave motions become negligible. At shallower depths the sediment transport becomes much more complicated in that we must consider the effects of combined wave action plus superimposed unidirectional currents. Transport under these conditions is generally viewed as the more effective to-and-fro wave motions placing the sediment in motion but producing little or no *net* transport of sediment due to the orbits being closed, the water (and sediment) returning approximately to its starting point. However, if a unidirectional current were superimposed upon this to-and-fro orbital motion, a *net* transport of water and sediment would result. The unidirectional current could be very weak in magnitude and still cause a transport since the sediment is already in motion under the wave action. In regard to area and quantity, this type of transport is probably the most important on continental shelves. Unfortunately, at present we are not in a position to make a quantitative prediction of sediment transport under these conditions.

This chapter attempts a brief summary of the mechanics of sediment transport on continental shelves. Derivations are not presented; these can be found in the original papers. Instead, the conclusions are summarized and placed in perspective. The review begins with a summary of what is known concerning the threshold of sediment motion, under both unidirectional and oscillatory water motions. Obviously sediment must be placed in motion before it can be transported. Next we shall examine sediment transport, first under unidirectional flow alone and then under combined waves and unidirectional flow. Incorporated into this review is a summary of the relationship between the water flow and the sedimentary structures produced on the sediment substrate. This is necessary since the sedimentary features influence the sediment transport. Also, of course, geologists are interested in these sedimentary features in their own right and would like to use their presence as an indication of the water flow environment.

SEDIMENT THRESHOLD

As the velocity of fluid flow over a bed of sediments is increased, there comes a stage when the fluid exerts a force or stress on the particles sufficient to cause them to move from the bed and be transported. This stage is generally known as the threshold of sediment movement, or as the critical stage for erosion or entrainment.

Considerable attention has been given to the threshold of sediment movement under unidirectional currents

such as are found in rivers, with blowing winds, and in certain circumstances on continental shelves. Nearly all the data are from laboratory flumes, although Sternberg (1967, 1971) has measured the conditions for initial movement within the tidal channels of Puget Sound, Washington. Within the limits of experimental error the various studies agree with one another. The results have led to a variety of threshold curves that form the basis for predicting initial sediment motion, knowing some combination of sediment grain diameter and density or water velocity and stress. Figure 1 shows the curve of Shields (1936), modified to include the results of White (1970). It relates the critical entrainment function (relative stress)

$$\theta_t = \frac{\tau_t}{(\rho_s - \rho)gD} \tag{1}$$

to a Reynolds number $u_* D/\nu$. The stress τ exerted by the water flow over the sediment bed is given by

$$\tau = C_{100}\rho(\bar{u}_{100})^2 \tag{2}$$

(Chapter 7) where C_{100} is the frictional drag coefficient and $\bar{u}_{100}$ is the velocity that is measured at 100 cm above the bottom. Since $\bar{u}$ varies with the distance above the bottom through the boundary layer, it is necessary to use its magnitude at that 100 cm level in conjunction with the empirical value of C_{100} to obtain the correct bottom stress. Sternberg (1968, 1972) has shown that for hydrodynamically rough flows the drag coefficient C_{100} assumes a constant value of about 3×10^{-3}, ranging from about 2×10^{-3} to 4×10^{-3} (Chapter 7). This occurs on naturally sorted sand and gravel sediments at Reynolds numbers greater than about 1.5×10^5 (equivalent to $\bar{u}_{100} = 15$ cm/sec). For these conditions the bottom stress becomes approximately

$$\tau = 3 \times 10^{-3}\rho(\bar{u}_{100})^2 \tag{3}$$

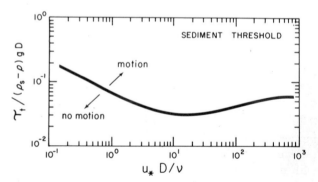

FIGURE 1. *Threshold of sediment motion under unidirectional currents. From Shields (1936), modified after results of White (1970).*

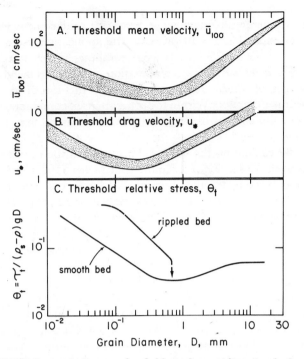

FIGURE 2. *Sediment threshold under unidirectional flow, relating the threshold* $\bar{u}_{100}$, $u_* = (\tau/\rho)^{1/2}$ *and* θ_t *to the sediment grain size* D. *From Bagnold (1963) and Inman (1963).*

In terms of the friction velocity

$$u_* = \left(\frac{\tau}{\rho}\right)^{1/2} \qquad (4)$$

this can also be expressed as

$$u_* = 5.47 \times 10^{-2}\bar{u}_{100} \qquad (5)$$

It is apparent that there are practical difficulties in employing Fig. 1 to evaluate the threshold. Sediment grain parameters (diameter D and density ρ_s) and water flow parameters (τ, u_*, ρ, ν) are contained within both dimensionless quantities θ_t and u_*D/ν. For example, given D and ρ_s it is not a simple procedure to arrive at the fluid stress τ_t necessary for threshold. Because of this, Bagnold (1963) replotted θ_t as a function of the grain diameter D alone as shown in Fig. 2. Also given are plots of D versus the threshold friction velocity u_* and the threshold value of the velocity $\bar{u}_{100}$ 1 m above the bed. For application purposes any of the curves of Fig. 2 are easier to employ than the original Shields curve.

The conditions for threshold under unidirectional currents are fairly well established and should not change markedly with the addition of more data. This is not the case for the threshold under oscillatory wave motions. There have been fewer studies of oscillatory threshold and the agreement between the several studies is not always very good. The findings of Komar and Miller

(1973, 1975a), who have recently reviewed most of the available data, are summarized briefly here.

Many equations have been proposed for the threshold of sediment motion under waves; Silvester and Mogridge (1971) present 13 different equations gathered from the literature. The available data are from oscillating bed experiments where a cradle holding the sediment is oscillated harmonically in still water (Bagnold, 1946; Manohar, 1955), from an oscillating-flow water tunnel (Rance and Warren, 1968), and from laboratory wave channels (Horikawa and Watanabe, 1967).

We found in our review (Komar and Miller, 1973, 1975a) that for grain diameters less than about 0.5 mm (medium sands and finer) the threshold is best related by the equation

$$\frac{\rho u_m{}^2}{(\rho_s - \rho)gD} = 0.21\left(\frac{d_0}{D}\right)^{1/2} \qquad (6)$$

where u_m and d_0 are the near-bottom threshold velocity and orbital diameter of the wave motion such as are obtained in linear Airy wave theory, being related to the water depth h, the wave period T, wave height H, and the wavelength L through

$$u_m = \frac{\pi d_0}{T} = \frac{\pi H}{T \sinh(2\pi h/L)} \qquad (7)$$

Equation 6 for the threshold is modified after an empirical equation obtained by Bagnold (1946).

For grain diameters greater than 0.5 mm (coarse sands and coarser) the threshold is best predicted with an empirical curve relating d_0/D to $\rho u_m/(\rho_s - \rho)gT$, first presented by Rance and Warren (1968). This latter dimensionless number represents the ratio of the acceleration forces to the effective gravity force acting on the grains. The empirical curve can be expressed as the equation

$$\frac{\rho u_m{}^2}{(\rho_s - \rho)gD} = 0.46\pi\left(\frac{d_0}{D}\right)^{1/4} \qquad (8)$$

which is similar to (6) for the finer grain sizes.

Equation 6 can be used to evaluate the threshold for grain sizes at least as fine as the lower silt range (where cohesive effects can be expected to cause departures from the established relationship), and (8) gives good results for grain sizes larger than 0.5 mm and as coarse as 5 cm (the coarsest material studied). For a given grain density and diameter, the threshold under waves can therefore be established by a certain wave period T and orbital velocity u_m or diameter d_0. Since $u_m = \pi d_0/T$, only two of these three parameters need be established in defining the threshold. Utilizing (6) and (8), Komar and Miller (1975a) produced Fig. 3 for the threshold values of T and u_m. This graph assumes $\rho_s =$

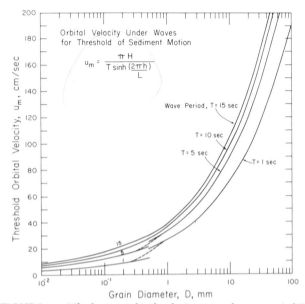

FIGURE 3. *The bottom orbital velocity* u_m *and wave period* T *necessary for the threshold of sediment motion under waves. From Komar and Miller (1975a).*

2.65 g/cm³ for the sediment density and so applies only to ordinary quartz sands. Given a grain diameter D, the required combinations of T and u_m for sediment motion are easily selected. It is seen that there are multiple combinations of T and u_m for any given grain size; the higher the wave period T, the greater is the required orbital velocity u_m. At low wave periods and grain sizes around 0.5 mm there is disagreement between (6) and (8) so that the curves of Fig. 3 do not join smoothly. In this transition region we have drawn smooth dashed curves as a compromise.

Once the threshold wave period and orbital velocity u_m are determined, there are of course many combinations

of water depth h and wave height H that could yield the required u_m. To further expedite application, Komar and Miller (1975b) prepared a computer program that for a given D and ρ_s combination prints incremented values of T, H, and h which define the threshold of sediment motion. Within the program itself combinations are eliminated which specify wave conditions that are impossible.

By evaluating the drag coefficients with the results of Jonsson (1967, Fig. 6), Komar and Miller (1975a) were able to compare the threshold under waves with the threshold curves for unidirectional steady currents. This comparison is shown in Fig. 4 where the curve is the same as that in Fig. 2, from Bagnold (1963). The comparison is very good, especially on the coarser grained limb of the curve. The scatter on the fine-grained end is due mainly to the systematic difference in the results between the Bagnold (1946) and the Manohar (1955) data. It should be pointed out that the unidirectional data upon which the curve is based show a comparable scatter. Madsen and Grant (1975) independently arrived at results similar to ours in the comparison between the unidirectional and oscillatory threshold.

Data on the development of oscillatory ripple marks under waves can be utilized as a further check on the equations of sediment threshold. If the threshold curve is correct, the ripple data should lie in a stress field above the threshold curve. In Fig. 5, from Komar and Miller (1975c), it is seen that this is largely the case, although some data points do fall below the curve. The field data of Inman (1957), denoted by the + symbol, especially extends below what should be the threshold limit. This is in part due to the existence of wide wave spectra under field conditions, which may give high instantaneous orbital velocities over short periods of time. Every

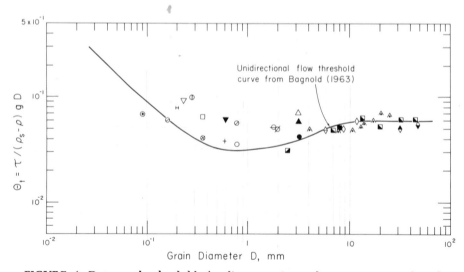

FIGURE 4. *Data on the threshold of sediment motion under waves compared to the Bagnold (1963) curve for unidirectional flow sediment threshold. See Komar and Miller (1975a) for identification of symbols.*

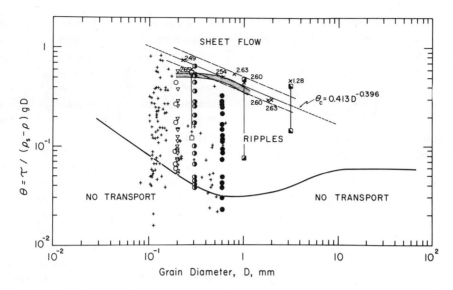

FIGURE 5. *Use of oscillatory ripple mark data under waves to further define the threshold limit of sediment motion. Also shown is the transition from a rippled bed to flat bed sheet flow. The × symbol denotes data on this transition to sheet flow. The + symbol is the field data of Inman (1957) on the occurrence of oscillatory ripples. The remaining symbols are laboratory data on oscillatory ripples from Carstens et al. (1969) and Manohar (1955). From Komar and Miller (1975c).*

hour or so the waves at that particular ripple location combine or sum to form unusually strong orbital motions which exceed the threshold and produce the ripples. There might be no further motion for a considerable span of time until another such summation occurs. It should be recognized that all the other data on threshold and ripple occurrence come from laboratory tests with a single wave period (a very narrow spectrum). The oscillating-flow water tunnel ripple data of Carstens et al. (1969), shown in Fig. 5, are also conservative in that a semicircular rod was used as a disturbance element. The conclusion is that in the field, oscillatory ripple marks may be formed under conditions that are defined by the threshold relationships as being below threshold. It might be argued that the threshold curve should be lowered to incorporate such conditions. However, this seems undesirable as the sediment may be in motion for only, say, 5 minutes out of each hour. Komar and Miller (1975c) conclude that the threshold relationships should be retained as given above, but if one is interested in the evaluation of conditions for the generation of oscillatory ripple marks, then the coefficient of (6) should be reduced from 0.21 to 0.11.

These equations for threshold under waves refer to a sinusoidal wave motion. Very often in deep water under ocean swell wave conditions this is an adequate description of the orbital motions. It has been seen above that with a wide wave spectrum the orbital motions are irregular, being the summation of many sinusoidal motions (see also p. 72, Chapter 6). In addition, as waves

enter intermediate and shallow water the wave crests peak and the troughs become wide and flat. Under such waves the orbital motions give strong onshore velocities under the crests and weaker offshore velocities under the troughs (see Fig. 8). With such orbital motions the threshold equations are inadequate except for perhaps a rough estimate as to whether threshold is reached some time during the wave cycle. Under such circumstances, the approach of Ippen and Eagleson (1955), Eagleson et al. (1958), and Johnson and Eagleson (1966) may be attempted. They relate the stresses exerted on the sediment particle to the instantaneous near-bottom water motions. These stresses include hydrodynamic drag, lift, the virtual or "apparent" mass force, a force caused by the instantaneous pressure gradients under the waves, the inertia force, and the force of gravity. The analysis is much the same as that of White (1940) who performed a similar analysis for grain threshold under steady flows, threshold occurring when the moment of the forces due to the water flow exceeds the moment of gravity so that the particle is tipped out from among the other particles that form the bed. This leads to an equation of the form

$$(C_D + C_L \tan \phi)\rho \left(\frac{\pi D^2}{4}\right) u^2$$

$$+ (1 + C_M)\rho \left(\frac{\pi D^3}{6}\right) \dot{u} = \frac{\pi}{6}(\rho_s - \rho)gD \tan \phi \quad (9)$$

where u is the instantaneous velocity and $\dot{u} = du/dt$ is the acceleration. C_D is a drag coefficient, C_L is a lift coefficient, C_M is the coefficient of virtual mass, and ϕ is the angle of contact of the particle with the particles below. Giving u the sinusoidal motion

$$u = u_m \sin\left(\frac{2\pi t}{T}\right) \qquad (10)$$

where T is the period, t is time, and u_m is the maximum velocity reached in the orbital motion when $\sin(2\pi t/T) = 1$, (9) becomes

$$k_1 \rho D^2 u_m^2 \left[\sin\left(\frac{2\pi t}{T}\right)\right]^2 + k_2 \rho D^3 \left(\frac{2\pi u_m}{T}\right) \cos\left(\frac{2\pi t}{T}\right)$$
$$= \frac{\pi}{6}(\rho_s - \rho)gD^3 \quad (11)$$

where the coefficients k_1 and k_2 incorporate the several coefficients C_D, C_L, C_M, and $\tan\phi$. The coefficients k_1 and k_2 are also introduced because the velocity u of (9) signifies the velocity at the edge of the grain that is to be moved, somewhere within the boundary layer under the wave orbital motions. The introduction of k_1 and k_2 is therefore also an attempt to ignore the boundary layer and to relate the threshold to the orbital motions calculated from the potential flow equations that describe the wave motions. The u_m of (11) is now the maximum orbital velocity given by (7), generally viewed as the velocity at the top of the boundary layer, rather than the velocity at some distance within the boundary layer.

Rearrangement of (11) gives

$$\frac{6k_1}{\pi}\frac{\rho u_m^2}{(\rho_s - \rho)gD}\sin^2\left(\frac{2\pi t}{T}\right)$$
$$+ 12k_2\frac{\rho u_m}{(\rho_s - \rho)gT}\cos\left(\frac{2\pi t}{T}\right) = 1 \quad (12)$$

which shows the importance of the dimensionless parameters $\rho u_m^2/(\rho_s - \rho)gD$ and $\rho u_m/(\rho_s - \rho)gT$ obtained in the analysis that led to (6) and (8). This suggests that a comparison of (12) with (6) and (8), which are based on the available data, could lead to an evaluation of the coefficients k_1 and k_2. These values could then be used in motions other than the simple sinusoidal motions assumed in obtaining (12) from (9), necessary in the comparison with (6) and (8) and because the available data on threshold are for simple sinusoidal motions.

The approach to a comparison between (11) and the semiempirical equations (6) and (8) would be for a given grain diameter D and wave period T. Either (6) or (8) could be used to determine the maximum orbital velocity u_m needed for threshold. This threshold u_m could then be used in (11) to determine the necessary values of k_1 and k_2 required to obtain that u_m as the threshold

for the given D and T. Since there are two coefficients to be determined, it is necessary to fit the coefficients to a range of values for u_m and T for a given grain size D. For example, for a grain size $D = 0.10$ cm and wave period $T = 5$ seconds, $k_1 = 0.075$ and $k_2 = 5.5$ give a reasonable threshold condition. But even here there is some subjective decision as to how much movement is required for threshold, similar to the decision in a laboratory experiment on threshold.

The best approach to the threshold problem would be to utilize (9) with evaluations of the various coefficients and considerations of the boundary layer under wave motions. The drag and lift coefficients depend on the instantaneous Reynolds number uD/ν and so continuously change as the velocity u varies in the orbital motions. Although this approach is much more difficult, it could be used to examine the threshold under irregular wave motions and also the threshold under combined wave and unidirectional flows, both conditions being important to sediment transport considerations.

To what water depths can normal surface waves be expected to cause sediment motion? Figure 6 gives the water depth of movement for waves of period $T = 15$ seconds and a range of grain diameters and wave heights. The curves are based on the threshold equations (6) and (8), with (7) used to relate the threshold u_m to the water depth and wave parameters. It is seen that reasonable wave conditions could produce sediment movement down to depths of 125 m and more. As pointed

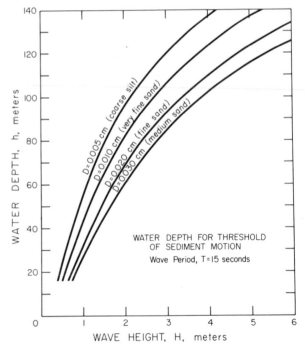

FIGURE 6. *Expected water depth of sediment movement due to surface waves for period* T = 15 *seconds and a range of sediment grain sizes and wave heights. From Komar and Miller (1975a).*

out above, these calculations are conservative as sediment threshold might be achieved at still greater depths than those indicated by the equations because interactions of wave trains of slightly differing period can generate higher instantaneous velocities, velocities may be higher than given by (7) from first-order Airy wave theory, and because small protuberances on the bed could cause sediment motion at lower velocities than implied by the analysis. However, there is basic agreement on the depth of movement curves of Fig. 6 and the water depths to which oscillatory ripple marks are observed on continental shelves. For example, Komar et al. (1972) found such ripples in bottom photographs on the Oregon continental shelf in water depths as great as 125 m and sometimes considerably greater (Fig. 14).

The results also indicate that waves are important to considerable water depths in the consideration of sediment transport. This should not come as a surprise, as it was recognized as early as the turn of the century. There were many stories such as gravel and large stones being washed into lobster pots by the wave surge [see the summary by Johnson (1919, pp. 78–83)]. On this basis and on theoretical calculations by Sir George Airy (of linear Airy wave fame), the belief was established that there is appreciable oscillatory motion to depths of 600 ft (183 m) and perhaps deeper. More recently, studies such as those by Curray (1960) and Draper (1967) considered wave-induced bottom surge and the frequency with which the bottom sediment would be stirred. Curray (1960) suggested that off the Texas coast hurricane waves are capable of moving sediment for the entire range of depths across the continental shelf. Draper's (1967) data indicate that on the shelf edge west of Britain wave action should be sufficiently intense to stir fine sand at a depth of 600 ft (183 m) for more than 20% of the year. McCave (1971) has also demonstrated the effectiveness of surface waves in controlling the sediment distribution and pattern of bed forms in the North and Celtic seas.

SEDIMENT TRANSPORT DUE TO UNIDIRECTIONAL CURRENTS

Before examining the sediment transport under combined wave action and superimposed unidirectional currents, let us review the approaches to evaluating the transport when steady unidirectional flow alone is involved. The approaches are similar to considerations of sediment transport in rivers and draw heavily on that literature. The studies of Kachel and Sternberg (1971) and Sternberg (1967, 1968, 1972) have contributed much to our ability to predict bed load sediment trans-

port under continental shelf conditions. They obtained their data from tidal channels in Puget Sound, Washington, the bed load being evaluated from the migration rates of the current ripples observed in bottom photographs and the water velocity profiles measured by means of a large instrumented tripod (Sternberg and Creager, 1965). In their analysis they also include transport data from rivers as summarized by Guy et al. (1966). Their transport relationships are based on the theoretical developments of Bagnold (1963) who relates the rate of mass transport of sediment as bed load to the power exerted by the fluid moving over the boundary. On this basis Bagnold deduces the relationship

$$\frac{\rho_s - \rho}{\rho_s} g j = K \omega \tag{13}$$

where j is the mass discharge of sediment (g/cm·sec), ω is a measure of the power expended on the bed by the fluid, and K is a dimensionless proportionality coefficient. The left side of the equation is the immersed weight sediment transport rate, having the same units as the power ω. When the immersed weight transport rate is multiplied by $(\tan \theta - \tan \beta)$, the difference between the friction angle (comparable to a coefficient of friction) and the bottom slope, one obtains the power utilized in transporting the sediment. Since ω is the total available power to perform work (sediment transport), it is apparent that the dimensionless K includes an efficiency factor. The sediment transport process is therefore likened to a machine in which one compares power available and utilized to determine the efficiency of the machine.

Bagnold (1963) expresses the fluid power ω as

$$\omega = \tau \bar{u} \tag{14}$$

where τ is the boundary shear stress and $\bar{u}$ is the fluid velocity near the boundary. Inman et al. (1966) use the friction velocity $u_* = (\tau/\rho)^{1/2}$ rather than the mean velocity $\bar{u}$ to give

$$\omega = \rho u_*^3 \tag{15}$$

Equations 14 and 15 can be shown to differ only by the inclusion of the drag coefficient. The Bagnold sediment bed load transport equation then becomes

$$\frac{\rho_s - \rho}{\rho_s} g j = K \rho u_*^3 \tag{16}$$

Kachel and Sternberg (1971) relate their field measurements to (16). They find that the value of the coefficient K is empirically related to the grain size D of the sediment to be transported and to the excess boundary shear stress $(\tau - \tau_t)/\tau_t$, where τ_t is still the threshold stress for initial sediment movement. This empirical relation-

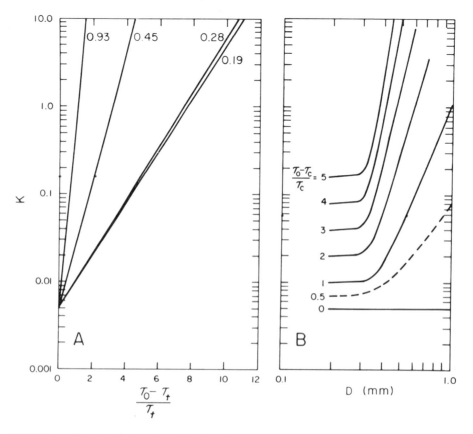

FIGURE 7. *Sternberg's (1972) empirical values of* K *in (16) versus the grain diameter* D *and the excess stress* $(\tau - \tau_t)/\tau_t,$ *where* τ_t *is the threshold stress. From Sternberg (1972).*

ship is given in Fig. 7. The procedure for evaluating the transport is then the following:

1. If $\bar{u}_{100}$ is measured, the velocity 1 m above the bottom, then the stress may be determined as

$$\tau = 3 \times 10^{-3}\rho(\bar{u}_{100})^2$$

as already seen, and

$$u_* = 5.47 \times 10^{-2}\bar{u}_{100}$$

2. Knowing a value of the mean sediment diameter D, Fig. 2 is then employed to determine whether the evaluated fluid stress τ exceeds the critical stress to initiate sediment movement.

3. Compute the excess shear stress $(\tau - \tau_t)/\tau_t$ and together with D use Fig. 7 to estimate the magnitude of the coefficient K.

4. Use this K value in (16) and the previously evaluated u_* to calculate the mass transport j of bed load sediment.

As pointed out by Sternberg, the approach is applicable to mean sediment sizes between approximately 0.20 and 1 to 2 mm. For the finer grain sizes there are special effects of shape, packing, and finally cohesiveness that

must then be considered. Application to grain sizes greater than 2 mm would require unwarranted extrapolation from the empirical results. The equation calculates only the bed load transport, not the finer grained suspended load. Not only is the suspended load not included, if it is large it might also affect the velocity profile sufficiently to introduce appreciable error into the evaluation of the bed load (Chapter 7). One way to reduce this error would be to measure the velocity profile, and determine $\bar{u}$ at three or more levels above the bottom rather than just the single $\bar{u}_{100}$ value. With the velocity profile one can determine τ directly rather than using (3) with its empirical drag coefficient obtained with little or no suspended load present (Chapter 7). Such an approach of obtaining τ from the measured velocity profile was undertaken, for example, by Ludwick (1974) in his calculations of sediment transport under tidal currents at the entrance to Chesapeake Bay.

Attempts to evaluate shelf sediment transport quantitatively have almost entirely relied on Bagnold's theoretical considerations rather than on kinematic methods. One exception is the study of Smith and Hopkins (1972) who were interested in the movement of sediments from the Columbia River across the Washington shelf. They developed a kinematic approach to calculate the transport of suspended sediment but the results require the measurement of the suspended load concentration at some level close to but above the bottom. Since such measurements were not available and there is no method for its theoretical prediction, Smith and Hopkins were unable to make the final step of actually evaluating the transport. Instead, they resorted to employing three different empirical equations that are based on steady flow sediment transport in rivers. Such an application to shelf sediment transport of equations that are empirically fitted to river data alone is not particularly satisfactory. However, since the transport estimates will be only very crude no matter what the approach, such a technique is certainly acceptable at this stage in our understanding.

BED FORMS UNDER UNIDIRECTIONAL FLOW

Unidirectional currents give rise to a characteristic series of bed forms. On most shelves where the currents are relatively weak the structures consist only of the usual varieties of small current ripple marks. More impressive are the large-scale features that develop on shelves that are dominated by strong tidal currents. One such area, the one that has received the most study, is the North Sea, Irish Sea, Celtic Sea, and English Channel, surrounding Britain and northern Europe. There the near-surface tidal velocities have values of 100 to 150 cm/sec over wide areas and reach values of as much as 250 cm/sec locally. The many studies point to the bed load transport being in response to these strong tidal currents. A variety of sedimentary bed forms are developed in the sediments, including sand ribbons, sand waves, megaripples, sandbanks and ridges, and longitudinal furrows. The characteristics of these bed forms are reviewed in Chapter 10. Many of these bed forms can be used to determine paths of bed load transport through these seas (Stride, 1963; Kenyon and Stride, 1970); see Chapter 15. The migrations of sand waves have been utilized to estimate sand transport rates. For example, Stride and Cartwright (1958) estimated a rate of advance of 10 cm/day which would give a displacement of 36 m in a single year. These sand waves had a height of 3 m so the volume sand transport rate over the 64 km front of the field of sand waves would be 4×10^6 m³/year. Other estimates of rates are comparable or somewhat smaller. It is apparent that appreciable quantities of net sediment transport can occur under unidirectional currents on the shelf.

SEDIMENT TRANSPORT DUE TO WAVES

As discussed in the introductory paragraphs, on most continental shelves waves are an important agent in sediment transport, so one cannot consider unidirectional currents alone. In this section we review the mechanics of this sediment transport process, first by waves alone and then by waves with a superimposed unidirectional current. We also briefly examine the relationship between the wave orbital motions and the oscillatory ripple marks produced in the bottom sediments.

It is generally believed that in shallow water the waves alone can produce an onshore sediment transport because of the asymmetry of their orbital motions (Cornish, 1898). As illustrated in Fig. 8, this is because in shallow water the wave crests are sharp and separated by long flat troughs. The orbital motion under the crest is of high speed but low duration, whereas the offshore re-

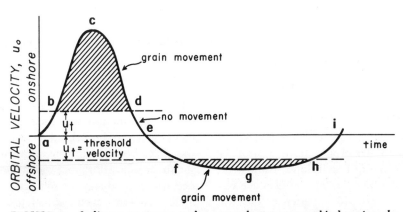

FIGURE 8. *Sediment movement under a near-bottom wave orbital motion, the velocity at* c *under the wave crest reaching a much higher value that at* g *under the trough. This causes a net onshore movement of certain sediment sizes, the grain movement being represented by the cross-hatched areas. However, the degree of transport is not directly proportional to the areas of cross-hatching.*

turn flow under the troughs is of lower speed but longer duration. Some grain sizes may be of sufficient size that they are transported only by the stronger onshore orbital motion and are not moved at all by the return offshore flow. It is readily apparent that each crest passage would shift the grain onshore and it would progressively hop toward shore with no intervening offshore motion. A somewhat finer grain size might be moved both during the onshore motion and offshore flow but would only shift offshore a small distance during the return orbit since most of the current of the return orbit is not sufficient to move the grain. The particle would zigzag back and forth but with a slow net movement in the direction of wave advance. Still finer grains would move equally back and forth, being in transit throughout the entire wave orbital motions so that there would be no net transport, presuming there is no unidirectional current superimposed on this system. In this way the waves may selectively drive pebbles and cobbles toward the beach but not produce a shoreward transport of the finer sand. Johnson (1919, p. 93) summarizes some dramatic examples of this. For example, Murray (1853) demonstrated that shingle and chalk ballast dropped into the sea off Sunderland, England, at a distance of 10 to 15 km from the land, were ultimately thrown on the beach by storm waves. Gaillard (1904) quotes Robinson as an authority for pig lead washing up at Madras, India, during a violent storm, the lead having come from a vessel wrecked more than 2 km offshore. A wave tank study by Bagnold (1940, pp. 32–33) provided better confirmation that the bigger the particle the more pronounced is the onshore creep. The biggest particles therefore tend to collect on the beaches.

Now if the bottom is sloping offshore there would be a component of gravity acting on the grains that might oppose this onshore shift because of the inequality of the orbital velocities. It is sometimes envisioned that a balance could be achieved between the two forces for a certain grain size so that that grain size would be at a null-point position and would move back and forth in equilibrium but with no net transport. This null point is an unstable equilibrium, however, as that grain size in slightly deeper water would move offshore and in slightly shallower water would move onshore. In addition, according to the model, all grains coarser than this critical equilibrium grain size would have a stronger offshore component because of gravity and would tend to shift offshore while the finer grains would move onshore. This obviously conflicts with experience. This hypothesis of an equilibrium oscillation point of a certain grain size is the null-point hypothesis as formulated by Cornaglia [see the discussion in Munch-Peterson (1950)]. The studies of Ippen and Eagleson (1955), Eagleson et al.

(1958), and Eagleson and Dean (1961) have placed the null-point hypothesis into mathematical terms; they conducted experiments in the laboratory wave tank which appear to verify the theory at least in part. Miller and Zeigler (1958) compared onshore variations in medium grain size with a theoretical onshore increase based on these studies. Although both theory and observation agreed that there should be an increase in grain size and an improvement in sorting in the onshore direction as the breaker zone is approached, this cannot be taken as confirmation of the null-point hypothesis as other mechanisms could be responsible for the sorting. Miller and Zeigler (1964) made a further study of this and concluded that there was little agreement with the theoretical distribution based on the null-point hypothesis.

In Komar (in press, Chapter 11) I present a more thorough review of the null-point hypothesis. The conclusion is that because it does not consider unidirectional currents, even those induced by the waves, and because it probably overrates the importance of the offshore gravity component acting on the grains, the null-point hypothesis does not present a valid model for sediment transport or equilibrium nontransport under wave action.

Bagnold (1963, p. 518) developed a model of sediment transport produced by a coupling of wave action with superimposed linear currents. According to the model the stress exerted by the wave motion supports and suspends sediment above the bottom but without causing a net transport since the wave orbits are closed. Superimposed on this to-and-fro motion is any unidirectional current that produces a net transport of the sediment, the direction of transport being the same as the current. Since the waves have already supplied the power to put the sand into motion, the unidirectional current can cause a net transport no matter how weak the flow, even if the current taken alone is below the threshold of sediment transport. According to this model, i_ζ, the immersed weight sediment transport rate per unit bed width, is given by the relationship

$$i_\zeta = K'\Omega \frac{\bar{u}_\zeta}{u_0} \qquad (17)$$

where Ω is the power the waves expend in placing the sediment in motion near the bottom; u_0 is the horizontal component of the orbital velocity at the bottom such that Ω/u_0 becomes the stress exerted by the waves; $\bar{u}_\zeta$ is the superimposed unidirectional current flowing near the bed which produces the net drift of sediment; and K' is a dimensionless coefficient of proportionality that contains efficiency factors of the transport system just as before.

The model of (17) has been successfully applied to sediment transport along beaches where $\bar{u}_\zeta$ becomes the longshore current in the surf zone (see the review in Chapter 13). In shelf sediment transport $\bar{u}_t$ becomes unidirectional currents such as those driven by winds, tidal currents, normal ocean thermohaline currents, as well as mass transports induced by the waves themselves. Inman and Bowen (1963) made a wave tank study of the applicability of (17) to this problem and found that matters are complicated in deep water when a current flows over a rippled sediment bottom. They made measurements of the sand transport caused by combined waves and currents traveling over a horizontal sand bed in water 50 cm deep. The waves had heights of 15 cm and periods of 1.4 or 2.0 seconds. The superimposed currents in all cases flowed in the direction of wave travel (onshore) with steady uniform velocities of 2, 4, and 6 cm/sec. With low superimposed currents the behavior was as expected; the sand transport along the tank increased with increasing velocity. However, at the higher velocities they obtained the peculiar results in which the sediment transport actually decreased with an increase in current; in one case the sand even moved upchannel opposite to the direction of current and wave motion. This resulted from the complex periodic movement of the sand above the ripples. At low superimposed current velocities the wave-induced ripple marks remained symmetrical with eddies forming between the ripples during the onshore and offshore semiorbital motions (Fig. 9). At times of reversal in the direction of orbital motion the eddies were thrown upward off the bottom, the sand being placed in suspension and therefore most susceptible to the current and a net transport. The sand was thrown somewhat higher off the bottom just preceding the crest passage, probably because the near-bottom water motions are vertical at that stage. The effect of an increase in the superimposed current was to reduce the symmetry of the ripples by increasing the effective onshore orbital velocity and decreasing the offshore velocity. The resulting asymmetrical ripples had flattened upwave faces and steepened onshore faces. The vortex generated on the downwave face during onshore semiorbits (Figs. 9 and 10) became by far the stronger and when thrown upward from the bottom was directed in the offshore direction, accounting for the reduction in transport and the one case of an offshore net transport. Such a reversal would be only temporary; with still stronger superimposed onshore currents the wave motions would have become relatively less important and the net transport would again have been strongly in the direction of current and wave travel. Even with just this pilot study of Inman and Bowen (1963) it is apparent that onshore-offshore transport of sediments under combined wave and current systems is going to be very complex. It is conceivable that with an offshore current

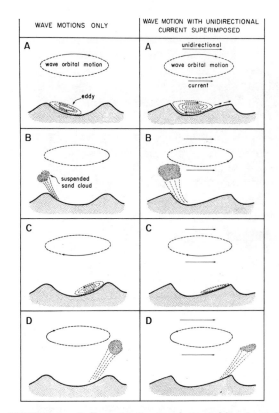

FIGURE 9. *Sediment suspension due to eddies formed in protective lee of ripple crests, without and with unidirectional currents superimposed on the wave motions. With a superimposed onshore current the ripples become asymmetric so that more sediment is thrown into suspension in an offshore direction, as shown, giving an unexpected offshore sediment transport. From Komar (in press).*

superimposed on wave motion, fine sediments that are thrown entirely into suspension will be carried offshore while at the same time coarser sediments that remain near the bottom are carried onshore (Komar et al. 1972; Inman and Tunstall, 1973).

Cook and Gorsline (1972) investigated the net transport of sand by waves through the use of directional sediment traps placed within the bed. They found that the net sand transport rate correlated best with the ratio of onshore to offshore orbital velocity strengths, simultaneously measured near the bottom. Some of their data showed a net transport in the offshore direction where the offshore orbital velocity exceeded the onshore velocity. They also provide measurements of the vertical distribution of suspended sand quantity above the bottom.

Studies of the transport of sands under waves have also been undertaken at the University of California, Berkeley. A good summary of their findings are presented by Einstein (1972). They have made significant contributions to our understanding of the nature of the

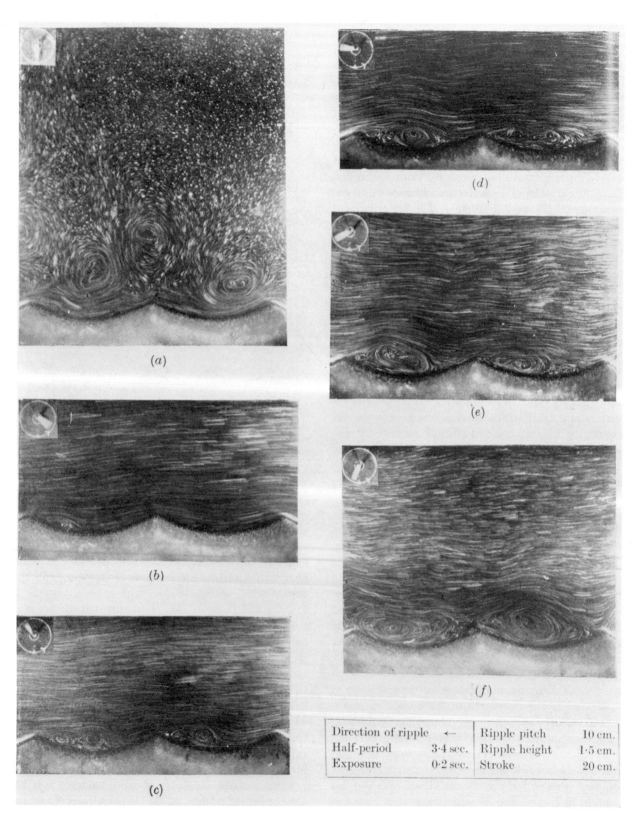

Direction of ripple	←	Ripple pitch	10 cm.
Half-period	3·4 sec.	Ripple height	1·5 cm.
Exposure	0·2 sec.	Stroke	20 cm.

FIGURE 10. *Vortices developed above oscillatory ripple marks. The series shows half of a complete oscillation (T/2), the pointer giving the phase, where the end of the stroke is at* the bottom. The ripple troughs are occupied during most of the orbital motion by eddies which develop from the crest lee, growing in size. From Bagnold (1946).

boundary layer under wave motions, and provided data on threshold of sediment motion under waves and on the characteristics of oscillatory ripple marks. However, their models for sediment transport under waves do not include adequate considerations of the role of the ripple marks in the process, which, as we have seen from the results of Inman and Bowen (1963), is extremely important.

Kennedy and Locher (1972) summarize the work carried out on the mechanisms of sediment suspension conducted at the University of Iowa Institute of Hydraulic Research. Their research, although still in its initial stages, makes a good attempt at evaluating the role of ripple marks in causing sediment suspension. As an example, Fig. 11 shows the time variations of sediment concentration over a wave period at two different stations above a ripple profile. One can immediately see that there are important time variations, with peaks and troughs in the concentration level, and that there are differences between the ripple crest position and ripple flank position. Also, the concentration falls off very rapidly with distance above the bottom. In the upper diagram of Fig. 11 the maximum concentration is much greater than in the lower diagram of Fig. 11, the former figure extending down into the eddy formed in the lee of the ripple crest. That peak concentration occurs at $t/T = 0.5$ when the horizontal orbital velocity is strongest and the lee eddy is fully developed. The second largest peak in Fig. 11 (upper), at about $t/T = 0.25$, results from suspension at a more distant ripple being swept past the sensor by the horizontal velocity. Figure 11 (lower), measured at the ripple crest, shows a small peak at $t/T = 0.5$, an influence of the adjacent lee eddy, and stronger peaks at $t/T = 0.25$ and 0.75 resulting from the sweeping past of eddies thrown upward out of the ripple lees as the wave crest or trough approaches. Other concentration peaks occur from clouds from more distant ripples or a second pass of the cloud produced by the adjacent ripples.

Figure 12, from Kennedy and Locher (1972), presents the time-averaged concentration of sediment distribution above the bottom at various ripple positions. It is seen that there is a rapid falloff in average concentration with distance above the bottom. Above $z/h = 0.08$, where z is the distance above the bottom and h is the total water depth, the average concentration decreases upward at an exponential rate, and is independent of the ripple position. Below $z/h = 0.08$ the concentration depends largely on the ripple position, being greatest on the shoreward lee flank of the ripple where the strongest eddies develop.

Kennedy and Locher (1972) present theoretical considerations aimed toward the evaluation of the distribution of suspended sediment concentration above the bed. They considered separately two suspension mechanisms, turbulent diffusion and wave-induced orbital motions. Unfortunately, the data on sediment concentration are inadequate at present for a precise determination of the roles played by the two mechanisms.

Jensen and Sørensen (1973) obtained field measurements of sediment concentration distribution under combinations of waves and tidal currents. The results did not show any systematic variation of concentration arising from the variation of the tidal current velocity so that the suspension must primarily have been due to the wave action. The field measurements were supplemented with tests carried out in an oscillatory water tunnel. The concentration decreased with distance above the bottom (Fig. 12) much as found in the study of Kennedy and Locher (1972).

It is apparent from the results of the studies of Inman and Bowen (1963) and Kennedy and Locher (1972) that the rate at which sediment is placed in suspension is governed in large part by the ripple geometry, the spacing and height, since that is important in determining the intensity of lee vortices downcurrent from the ripple crests. If predictions of sediment transport under waves are ultimately to be made, we must first have a better understanding of the relationship between the sediment ripple geometry and the wave orbital motions. There have been many descriptive studies of oscillatory ripples but few good quantitative investigations. Probably the two best are the field study of Inman (1957) and the laboratory study of Mogridge and Kamphuis (1973). Komar (1974) has reviewed much of the available published data on this relationship.

The investigation of Inman (1957) was conducted through the use of underwater breathing apparatus, and extended from the surf zone to depths of about 50 m off the California coast. Figure 13, based on his results and supplemented with laboratory measurements, yields an empirical relationship between the ripple spacing λ and the orbital diameter d_0 next to the bottom, given by

$$d_0 = \frac{H}{\sinh(2\pi h/L)} \tag{18}$$

It is seen in Fig. 13 that for sands with median diameters in the range 88 to 177 μ, as the wave orbital diameter increases the ripple spacing would first increase with approximately $\lambda = d_0$ (more exactly $\lambda = 0.8d_0$), the ripple length ultimately reaching a maximum of about $\lambda = 20$ cm beyond which λ decreases with a further increase in d_0. Such a relationship between λ and d_0 is also demonstrated by the wave channel study of Mogridge and Kamphuis (1973). With very large d_0 a stress is finally reached at which ripples cease to exist and

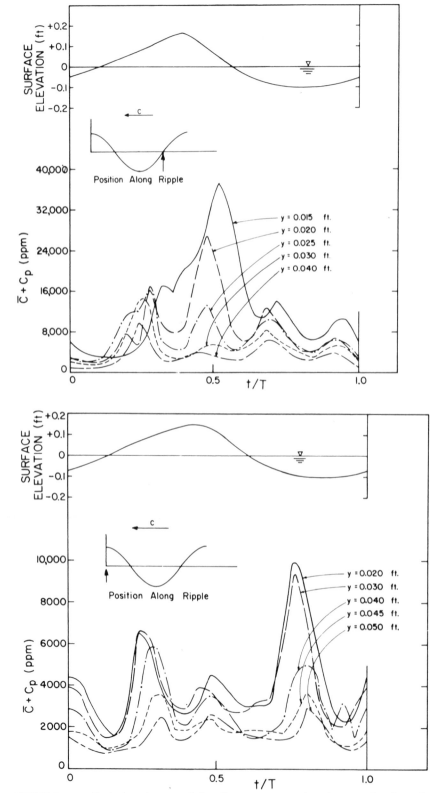

FIGURE 11. *Variations in suspended sediment concentration above ripples through one wave period. The corresponding surface elevation of the waves with respect to the sediment suspension is also shown. From Kennedy and Locher (1972).*

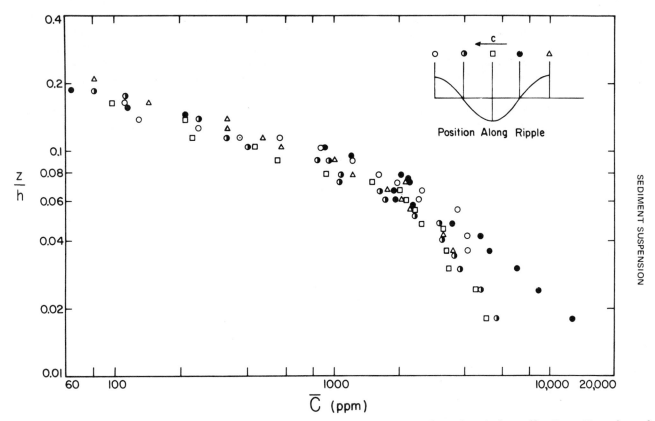

FIGURE 12. *Time-averaged suspended sediment concentration at various positions along the ripple profile. From Kennedy and Locher (1972).*

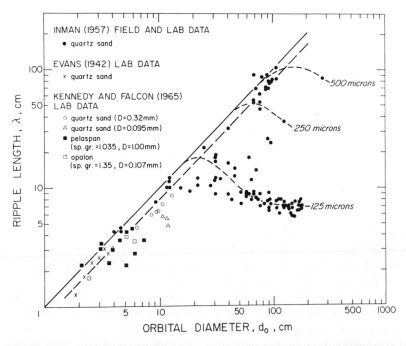

FIGURE 13. *Oscillatory ripple mark spacing related to the near-bottom wave orbital diameter for a range of sediment grain sizes. From Inman (1957) and Komar (1974).*

121

to-and-fro sheet sand movement occurs; the λ versus d_0 curve of Fig. 13 terminates on the right. This might suggest that in an onshore direction of decreasing water depth and therefore increasing d_0, the ripple spacing would first increase, reach a maximum, and then decrease. This is not generally the case, however, as was discussed by Komar et al. (1972) and at greater length by Komar (1974). For the conditions in which $\lambda = 0.8d_0$, threshold of sediment motion cannot be achieved unless the wave periods are short ($T < 3$ seconds, approximately) and the water depths are small. Therefore the $\lambda = 0.8d_0$ line of Fig. 13 is generally applicable only to short-period waves found on lakes and confined bays of limited fetch, and in the laboratory wave tank. Under ocean conditions the right limb of Fig. 13 is applicable where $\lambda \ll d_0$. This can be seen in the distribution of the data; the laboratory data form the $\lambda = 0.8d_0$ line while the field data of Inman (1957) form the $\lambda \ll d_0$ curve.

One consequence of this is that in lakes and bays where $\lambda = 0.8d_0$, the ripple spacings will tend to increase onshore, whereas under ocean conditions the opposite is true and the ripples tend to decrease in spacing onshore (Komar, 1974). Because of the complex nature of waves reaching the coast, however, actual patterns of ripple lengths are seldom this systematic. Komar et al. (1972), using an underwater stereo camera system, found oscillatory ripples in an onshore-offshore transit across the Oregon continental shelf at depths up to 125 m in which the ripple length first increased in the offshore direction, reached a maximum spacing, and then decreased in length at still greater depths (Fig. 14). We attributed this variation to waves arriving at the coast from a distant storm in a dispersed state, the longest period waves arriving first with progressively shorter period waves following. The level of competence of sediment movement gradually shifts onshore under such conditions, leaving ripples preserved in deeper water from an earlier, higher wave period phase of the storm-produced waves.

This discussion applies only to sands with median diameters in the range 88 to 177 μ for which data are available. For coarser sands it is not yet possible to define the dependence of λ on d_0 but it may be similar to that depicted by the curves labeled 250 μ and 500 μ in Fig. 13. The few available measurements in coarser sands indicate that the sand size is the single most important factor in determining the size of the ripples. For a given orbital diameter of water wave motion, the ripple lengths produced in 500 μ sands could be more than a factor of 10 greater than ripples formed in fine sands (Cook and Gorsline, 1972). More data are required on this.

As the bottom orbital diameter increases, the ripple heights first increase, reach a maximum, and then decrease until the ripples disappear altogether. As a corollary, the ripple steepness, the ratio of the ripple height to the length, first increases to a maximum and then decreases until the ripples disappear. This sequence is shown by the studies of Bagnold (1946), Carstens et al. (1969), Yalin and Russell (1963), and Mogridge and Kamphuis (1973). The same experience can be gained in swimming shoreward in water of progressively shoaling depth. In deep water the ripples are sharp-crested with pronounced heights. As the breaker zone is approached the ripple heights progressively decrease and finally disappear within the intense shear under the breaking waves. As the ripple heights decrease there is less development of lee vortices during the wave orbital motions and therefore less suspended sand thrown upward off the bottom. When the ripples disappear under the intense shear, the sand moves as a thin carpet of high concentration close to the bottom rather than as clouds of sand thrown up into the water.

SUMMARY

Sediment transport on continental shelves can be divided into three types according to the processes as (1) resulting from unidirectional currents alone, (2) produced by wave action alone, and (3) caused by unidirectional currents superimposed on the wave action. Of the three, only in the first case are we in a position to make quantitative estimates of sediment transport rates, given the required current measurements. Even there we are only able to estimate the bed load transport, not the suspended load transport. Waves alone apparently can cause an onshore transport of coarse sediment that moves only during wave-crest passage. This has not been adequately studied nor have any attempts been made at quantifying the process. Most commonly, case (3) of waves with superimposed unidirectional currents is important in shelf sediment transport. We have seen that this depends heavily on the ripple geometry, and the net transport is a complex function of the interaction of this geometry with the wave and current motions. Because of these complications, in spite of many studies on transport of sand under waves, we are nowhere near a state of art that permits quantitative predictions of sediment transport.

Sediment transport by steady, uniform flows in open channels (rivers) has been the subject of intense study for more than a century. After all these investigations there is still no consensus of agreement as to an analytical framework, and quantitative estimates may be in error

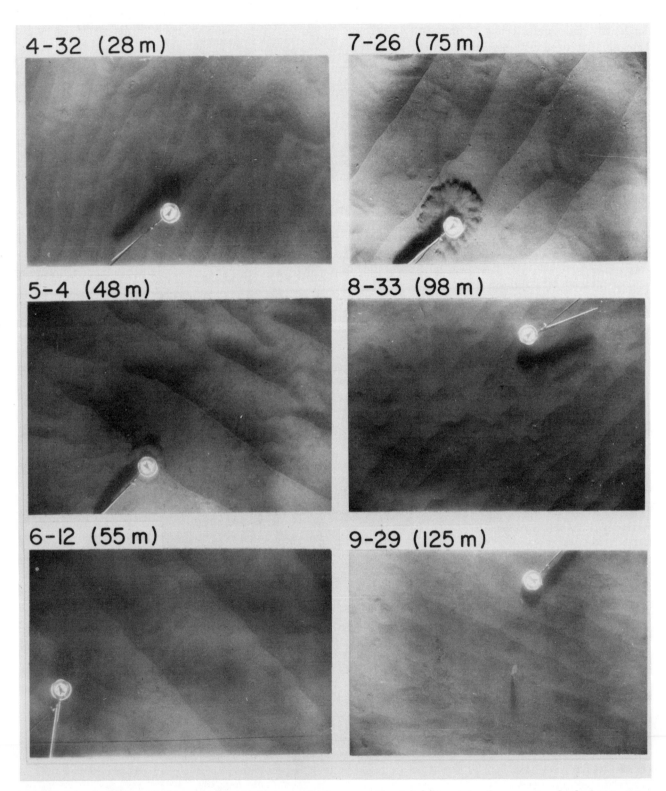

FIGURE 14. *Oscillatory ripple marks in a profile across the Oregon continental shelf, ranging from a water depth of 28 to 125 m. Ripple statistics are given by Komar et al. (1972).*

by as much as a factor of 10. This rather bleak appraisal can hardly give one a feeling of optimism concerning the evaluation of sand transport on continental shelves, under far more complicated conditions. This should be a fruitful line of research for many years to come.

ACKNOWLEDGMENTS

This review was undertaken while supported by the Oceanography Section, National Science Foundation, NSF Grant GA-36817. I thank I. N. McCave for his helpful suggestions and for reviewing the manuscript.

SYMBOLS

C_{100} drag coefficient determined from measurements 100 cm off bottom

d_0 bottom orbital diameter under waves

D sediment grain diameter

g acceleration of gravity (981 cm/sec^2)

h water depth

H wave height

j mass discharge of sediment transport

K dimensionless coefficient

L wavelength

T wave period

u water velocity

$\bar{u}_{100}$ time-averaged velocity measured 100 cm off bottom

u_* friction velocity

u_m maximum near-bottom orbital velocity under waves

θ relative stress

λ sediment ripple spacing

ν kinematic viscosity

ρ density of water

ρ_s density of sediment grains

τ stress exerted by water flow

τ_t threshold stress

τ_0 bottom stress

ω power expended by unidirectional water flow

REFERENCES

Bagnold, R. A. (1940). Beach formation by waves; some model experiments in a wave tank. *J. Inst. Civil Eng.*, **15**: 27–52.

Bagnold, R. A. (1946). Motions of waves in shallow water: Interaction between waves and sand bottoms. *Proc. Roy. Soc. London*, series A, **187**: 1–15.

Bagnold, R. A. (1963). Mechanics of marine sedimentation. In M. N. Hill, ed., *The Sea*, Vol. 3, *The Earth Beneath the Sea*. New York: Wiley-Interscience, pp. 507–582.

Carstens, M. R., F. M. Neilson, and H. D. Altinbilek (1969). *Bed forms generated in the laboratory under an oscillatory flow: analytical and experimental study.* U.S. Army Corps of Engineers, Coastal Engr. Res. Center, Tech. Memo. No. 28, 39 pp.

Cook, D. O. and D. S. Gorsline (1972). Field observations of sand transport by shoaling waves. *Mar. Geol.*, **13**: 31–55.

Cornish, V. (1898). On sea beaches and sand banks. *Geogr. J.*, **11**: 528–559, 628–647.

Curray, J. R. (1960). Sediments and history of Holocene transgression continental shelf, northwest Gulf of Mexico. In Shepard, Phleger, and van Andel, eds., *Recent Sediments, Northwest Gulf of Mexico*. Tulsa, Okla.: Am. Assoc. Pet. Geol., pp. 221–266.

Draper, L. (1967). Wave activity at the sea bed around northwestern Europe. *Mar. Geol.*, **5**: 133–140.

Eagleson, P. S. and R. G. Dean (1961). Wave-induced motion of bottom sediment particles. *Trans. ASCE*, **126**: 1162–1189.

Eagleson, P. S., R. G. Dean, and L. A. Peralta (1958). *The Mechanics of the Motion of Discrete Spherical Bottom Sediment Particles Due to Shoaling Waves.* U.S. Army Beach Erosion Board, Tech. Memo. No. 104, 41 pp.

Einstein, H. A. (1972). A basic description of sediment transport on beaches. In R. E. Meyer, ed., *Waves on Beaches*, New York: Academic Press, pp. 53–93.

Gaillard, D. D. (1904). Wave action in relation to engineering structures. *U.S. Army Corps of Engineers, Prof. Pap. No. 31.*

Guy, H. P., D. B. Simons, and E. V. Richardson (1966). Summary of alluvial channel data from flume experiments, 1956–61. *U.S. Geol. Surv., Prof. Pap. 462–I,* 96 pp.

Horikawa, K. and A. Watanabe (1967). A study on sand movement due to wave action. *Coastal Eng. Japan,* **10**: 39–57.

Inman, D. L. (1957). *Wave-Generated Ripples in Nearshore Sands.* U.S. Army Corps of Engineers, Beach Erosion Board, Tech. Memo. No. 100, 66 pp.

Inman, D. L. (1963). Sediments: Physical properties and mechanics of sedimentation. In F. P. Shepard, ed., *Submarine Geology*, 2nd ed. New York: Harper & Row, Chapter 5, pp. 101–151.

Inman, D. L. and A. J. Bowen (1963). Flume experiments on sand transport by waves and currents. *Proc. 8th Conf. Coastal Eng.*, 137–150.

Inman, D. L., G. C. Ewing, and J. B. Corliss (1966). Coastal sand dunes of Guerrero Negro, Baja California, Mexico. *Geol. Soc. Am. Bull.*, **77**: 787–802.

Inman, D. L. and E. B. Tunstall (1973). Phase dependent roughness control of sand movement. *Proc. 13th Conf. Coastal Eng.*, 1155–1171.

Ippen, A. T. and P. S. Eagleson (1955). *A Study of Sediment Sorting by Waves Shoaling on a Plane Beach.* U.S. Army, Beach Erosion Board, Tech. Memo. No. 63, 83 pp.

Jensen, J. K. and T. Sørensen (1973). Measurements of sediment suspension in combinations of waves and currents. *Proc. 13th Conf. Coastal Eng.*, 1097–1104.

Johnson, D. W. (1919). *Shore Processes and Shoreline Development.* New York: Hafner, 584 pp. (1965 facsimile).

Johnson, J. W. and P. S. Eagleson (1966). Coastal processes. In A. T. Ippen, ed., *Estuary and Coastline Hydrodynamics*. New York: McGraw-Hill, pp. 404–492.

Jonsson, I. G. (1967). Wave boundary layers and friction factors. *Proc. 10th Conf. Coastal Eng.* 127–148.

Kachel, N. B. and R. W. Sternberg (1971). Transport of bedload as ripples during an ebb current. *Mar. Geol.*, **19:** 229–244.

Kennedy, J. F. and F. A. Locher (1972). Sediment suspension by water waves. In R. E. Meyers, ed., *Waves on Beaches*. New York: Academic Press, pp. 249–295.

Kenyon, N. H. and A. H. Stride (1970). The tide-swept continental shelf between the Shetland Isles and France. *Sedimentology*, **14:** 159–173.

Komar, P. D. (1974). Oscillatory ripple marks and the evaluation of ancient wave conditions and environments. *J. Sediment. Petrol.*, **44:** 169–180.

Komar, P. D. (in press). *Beach Processes and Sedimentation*. To be published by Prentice-Hall, Englewood Cliffs, N.J.

Komar, P. D. and M. C. Miller (1973). The threshold of sediment movement under oscillatory water waves. *J. Sediment. Petrol.*, **43:** 1101–1110.

Komar, P. D. and M. C. Miller (1975a). Sediment threshold under oscillatory waves. *Proc., 14th Conf. Coastal Eng.*, 756–775.

Komar, P. D. and M. C. Miller (1975b). On the comparison of the threshold of sediment motion under waves and unidirectional currents with a discussion of the practical evaluation of the threshold. *J. Sediment. Petrol.*, **45:** 362–367.

Komar, P. D. and M. C. Miller (1975c). The initiation of oscillatory ripple marks and the development of plane-bed at high shear stresses under waves. *J. Sediment. Petrol.*, **45:** 697–703.

Komar, P. D., R. H. Neudeck, and L. D. Kulm (1972). Observations and significance of deep-water oscillatory ripple marks on the Oregon Continental Shelf. In D. J. P. Swift, D. B. Duane, and O. H. Pilkey, eds., *Shelf Sediment Transport*, Stroudsburg, Pa.: Dowden, Hutchinson & Ross, pp. 601–619.

Ludwick, J. C. (1974). Tidal currents and zig-zag sand shoals in a wide estuary entrance. *Geol. Soc. Am. Bull.* **85:** 717–726.

McCave, I. N. (1971). Sand waves in the North Sea off the coast of Holland. *Mar. Geol.*, **10:** 199–225.

Madsen, O. S. and W. D. Grant (1975). The threshold of sediment movement under oscillatory waves. A discussion. *J. Sediment. Petrol.*, **45:** 360–361.

Manohar, M. (1955). *Mechanics of Bottom Sediment Movement Due to Wave Action*. Beach Erosion Board, Tech. Memo. No. 75, 121 pp.

Miller, R. L. and J. M. Zeigler (1958). A model relating dynamics and sediment pattern in equilibrium in the region of shoaling waves, breaker zone, and foreshore. *J. Geol.* **66:** 417–441.

Miller, R. L. and J. M. Zeigler (1964). A study of sediment distribution in the zone of shoaling waves over complicated bottom topography. In R. L. Miller, ed., *Papers in Marine Geology*. New York: Macmillan, pp. 133–153.

Mogridge, G. R. and J. W. Kamphuis (1973). Experiments on bed form generation by wave action. *Proc. 13th Conf. Coastal Eng.*, 1123–1142.

Munch-Peterson (1950). Littoral drift formula. *U.S. Army Beach Erosion Board Bull.*, 4(4): (1938 speech).

Murray, J. (1853). On movement of shingle in deep water. *Minutes Proc. Inst. Civil Engr.*, **12:** 551.

Rance, P. J. and N. F. Warren (1968). The threshold movement of coarse material in oscillatory flow. *Proc. 11th Conf. Coastal Eng.*, 487–491.

Shields, A. (1936). *Anwendung der Ahnlichkeits Mechanik und der Turbulenzforschung auf die Geschiebe Bewegung: Preuss*. Versuchanstalt fur Wasserbau und Schiffbau, Berlin, 20 pp.

Silvester, R. and G. R. Mogridge (1971). Reach of waves to the bed of the continental shelf. *Proc. 12th Conf. Coastal Eng.*, 651–667.

Smith, J. D. and T. S. Hopkins (1972). Sediment transport on the continental shelf off of Washington and Oregon in light of recent current measurements. In D. J. P. Swift, D. B. Duane, and O. H. Pilkey, eds., *Shelf Sediment Transport*. Stroudsburg, Pa.: Dowden, Hutchinson & Ross, pp. 143–180.

Sternberg, R. W. (1967). Measurements of sediment movement and ripple migration in a shallow marine environment. *Mar. Geol.*, **5:** 195–205.

Sternberg, R. W. (1968). Friction factors in tidal channels with differing bed roughness. *Mar. Geol.*, **6:** 243–261.

Sternberg, R. W. (1971). Measurements of incipient motion of sediment particles in the marine environment. *Mar. Geol.*, **10:** 113–119.

Sternberg, R. W. (1972). Predicting initial motion and bedload transport of sediment particles in the shallow marine environment. In D. J. P. Swift, D. B. Duane, and O. H. Pilkey, eds., *Shelf Sediment Transport*. Stroudsburg, Pa.: Dowden, Hutchinson and Ross, pp. 61–82.

Sternberg, R. W. and J. S. Creager (1965). An instrument system to measure boundary-layer conditions at the sea floor. *Mar. Geol.*, **3:** 475–482.

Stride, A. H. (1963). Current-swept sea floors near the southern half of Great Britain. *Q. J. Geol. Soc. London*, **119:** 175–199.

Stride, A. H. and D. E. Cartwright (1958). Sand transport at southern end of the North Sea. *Dock Harbour Authority*, **38:** 323–324.

White, C. M. (1940). The equilibrium of grains on the bed of a stream. *Proc. Roy. Soc. London*, series A, **174:** 322–338.

White, S. J. (1970). Plane bed thresholds of fine grained sediments. *Nature*, **228**(5267): 152–153.

Yalin, M. S. and R. C. H. Russell (1963). Similarity in sediment transport due to waves. *Proc. 8th Conf. Coastal Eng.*, 151–167.

CHAPTER 9

Suspended Sediment Transport and Mud Deposition on Continental Shelves

DAVID E. DRAKE*

Atlantic Oceanographic and Meteorological Laboratories, Miami, Florida

An attempt to synthesize our knowledge of the distribution, transport, and deposition of fine sediment on continental shelves is in certain respects a rather frustrating undertaking. The reason is that our collection of field data on shelf water circulation and suspended matter is meager and of very uneven quality and density. Less than 10% of the world's shelf areas have been studied for suspended matter and many of these studies were restricted to surface water sampling; shelf circulation is only slightly better known. One might suggest that the reason for our lack of data is that there has been no compelling need to study suspended matter over continental shelves. In fact, there is some truth to this (if one limits his thinking to marine geology) for two reasons:

1. We have been content to simply describe end products (bottom sediments) and speculate briefly about processes of transport.

2. For economic reasons we have been preoccupied with sand on continental margins.

Consequently, looking back over the literature, it is difficult to find more than a few articles prior to 1960 dealing with processes of fine sediment movement seaward of the coast. Much of the early work was pioneered by European sedimentologists. The situation

began to change in the 1960s in large part as a spinoff from concern over environmental crises. Owing to increased development of the coastal zone and the need to assess man's impact on the ecology of the continental shelf, it became apparent that the dispersal and fate of fine sediment was of great importance, in terms of both biological and chemical processes (e.g., see Aston and Chester, 1973; Riley and Chester, 1971; Osterberg et al., 1964; Gross, 1972; Gordon, 1970). For completeness, it should be mentioned that thorough studies of suspended solids in the sea have been neglected because of the considerable problems in recovering and analyzing samples (see section on methods). In short, suspended solids research is time-consuming and expensive; and commonly neat, watertight conclusions are elusive because of the problems associated with separating short-term and long-term processes.

Before launching into the body of this chapter, let us state some definitions:

1. *Suspended load* refers to the solid materials that are fully enclosed and supported by the surrounding fluid.

2. *Suspended sediment, suspended particulate matter, particulates, suspended solids, suspended matter,* and *seston* have all been used more or less interchangeably to denote the solid fraction enclosed within a fluid medium. The terms *suspended solids* and *suspended matter* are used in this chapter.

* Present address: U.S. Geological Survey, Menlo Park, California 94025.

3. *Turbidity* and related terms should be used carefully since a fluid can obviously be turbid because of dissolved materials.

4. *Nepheloid layer* is a special term that is applied to the near-bottom zone of increased suspended solids concentration in the deep sea. This is an unfortunate restriction because there are "nepheloid" layers (water layers exhibiting high light scattering and absorption) at intermediate depths and over the continental shelf.

METHODS AND INSTRUMENTS

As is typical of any fairly new research area, there have been a variety of methods and instruments applied to suspended solids experiments with a slow evolution toward standardization (Gibbs, 1974).

Water Sample Recovery and Processing

WATER "BOTTLES." This includes any container that is lowered from the sea surface, closed at depth, and hauled up to recover a sample of suspended solids. The amount of seawater needed is dictated by the analyses to be performed. Generally, 5 to 10 liters will be sufficient for simple gravimetric analysis and some chemical tests. For x-ray diffraction studies up to 50 to 60 liters may be required over the outer parts of some shelves (Meade et al., 1975).

Because particle concentrations are commonly below 100 μg/l at more than 30 km from the coast, great care must be exercised to ensure that the sampling gear is clean. Internal parts which contact the sample should be minimized and replaced frequently since they can act as particle generators. The Niskin bottle is probably used most for suspended matter collections, with the new "top-drop" model showing promise for reducing contamination from internal surfaces.

PUMPS AND HOSES. Although in situ pumping systems can be designed to rapidly deliver large volumes of water directly to deck collection equipment, they are generally only useful in shallow water or for surface water sampling while under way. Unfortunately, pumping systems are difficult to keep clean (free of metallic contamination and algal growth) and in almost all cases they drastically change the physical characteristics of the particles by breaking down loosely bound organic and inorganic aggregates. The author is skeptical of suspended solids samples taken with pumping systems for virtually any type of study.

IN SITU PARTICLE SEPARATIONS. In principle, separation of the particles from the seawater in situ by pulling water through a small pore-size filter is an ideal solution to problems of contamination (Spencer and Sachs, 1970). However, such systems are expensive and time-consuming in the field. In fact, it seems likely that they would not sample representatively because of the distinct possibility that water drawn through the filter once will be drawn through again, giving artificially low particle concentrations.

Schubel and Schiemer (1972) described an in situ water freezing technique that seems especially well suited for microscopic particle size analyses in an area where concentrations are fairly high (>3 mg/l).

PARTICLE SEPARATIONS. Most water samples for suspended solids analysis are recovered with some type of retrievable container and the particles are then separated on shipboard using vacuum or pressure filtration or high-speed centrifugation (Spencer and Sachs, 1970; Lisitzin, 1972; Manheim et al., 1970). It is quite important that samples be withdrawn from the water containers immediately after recovery to avoid particle settling and sample bias. In addition, the water should be processed within a few hours after collection. This requirement generally precludes the use of centrifuges if many samples are being taken.

On the other hand, membrane filtration is a fully developed and accepted technique for separation of particles from fluid media (Banse et al., 1963; Manheim et al., 1970; Meade et al., 1975). The advantages are many:

1. Cost. Inexpensive multiple filtration units are available or can be easily constructed to handle an unlimited number of samples at once with either vacuum or pressure as the driving mechanism.

2. Filters. A wide variety of filter materials, diameters, and pore sizes are readily available to suit the type of analysis to be carried out.

In summary, most researchers have concluded that membrane filtration techniques are superior for particle separation and subsequent laboratory work. However, there are several aspects of filtration techniques that can cause problems and are easily overlooked.

1. Pore sizes and filter diameter. Membrane filters can be obtained with mean pore sizes ranging from 0.025 to 14 μm and standard diameters from 13 to 293 mm. The pore sizes most commonly used are 0.8 and 0.45 μm; the latter being preferred if studies are to be performed in relatively clean ocean water. In areas where particle content is known to be high (>1 mg/l) there is probably no significant difference in retention characteristics of these two filters because of the rapid accumulation of a filter cake which greatly enhances

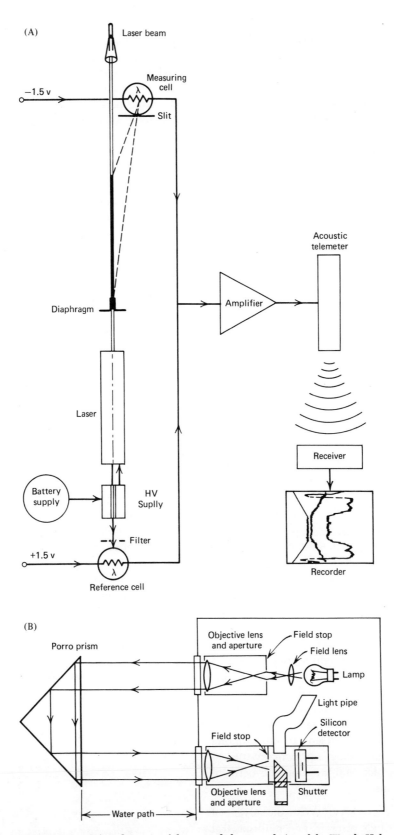

(A) Laser beam

Measuring cell

−1.5 v

Slit

Diaphragm

Laser

Battery supply

HV Suplly

Filter

+1.5 v

Reference cell

Amplifier

Acoustic telemeter

Receiver

Recorder

(B)

Porro prism

Objective lens and aperture

Field stop

Field lens

Lamp

Light pipe

Silicon detector

Field stop

Objective lens and aperture

Shutter

Water path

FIGURE 1. (A) *Schematic of laser nephelometer designed by Woods Hole Oceanographic Institution scientists. From Meade at al. (1975). (B) Schematic of 1 m path-length beam transmissometer designed by the Visibility Laboratory, Scripps Institution of Oceanography. From Petzold and Austin (1968).*

small particle entrapment (Sheldon and Sutcliffe, 1969).

Because of this filter-cake action, attempts to separate suspended matter according to size using membrane filters are not valid. In particular, a reviewer of this chapter has drawn attention to the fact that some filters exhibit a rather wide range of pore sizes below the nominal pore size. Thus a 0.8 μm filter will trap many smaller particles even without a filter cake.

The diameter of filter to be used depends on the subsequent analyses. The smaller 47 mm and 25 mm filters are excellent for gravimetric work, whereas larger filters are not conveniently weighed but allow large volumes of water to be processed.

2. A source of considerable errors can be traced to incomplete rinsing of filters after sample processing. Care must be taken to ensure that the *entire* filter has been freed of sea salt with distilled water. Filters should be dried at low temperature (50°C) or frozen on shipboard to minimize the decomposition of organic fractions.

PARTICLE ANALYSES. Table 1 presents a series of analyses that can be undertaken on suspended solids. The list could be easily expanded but only the most fundamental analyses are within the scope of this chapter.

OPTICAL SENSORS. Suspended solids can be indirectly studied by means of light scattering and absorption measurements (Jerlov, 1968). Table 1 includes the most common methods and references. "Nephelometers" (scattering meters) and transmissometers (Fig. 1) can be configured to continuously record optical properties from the sea surface to oceanic depths. By recording in real time, water sampling efficiency is greatly increased. Although optical measures can be related to particle concentrations and characteristics, the relationships are not unique because of the great natural variability of particle refractive indices, sizes, and shapes. The most promising optical instruments for suspended solids research are those that measure light scattered out of a collimated or coherent (laser) beam of light because scattering is almost exclusively caused by particulate matter. Application of optical techniques to problems of interest to marine geologists is a rather new and rapidly expanding field of research. At present, in situ profiling of light attenuation and scattering is principally used to aid in water sample location and interpolation between discrete depth data points.

SATELLITE AND AERIAL IMAGERY. With the advent of earth satellites, high-altitude aircraft, and multispectral scanners and image enhancement techniques, remote sensing of the surficial layers of the ocean and shallow bottom topography has become a major new oceano-graphic tool. Stevenson and Uchupi (1969), Mairs (1970), and Charnell et al. (1974) have discussed satellite imagery data for several coastal areas.

Figure 2 shows the results of an image enhancement technique performed on the energy sensed within the blue-green spectral band by the first Earth Resources Technology Satellite (Charnell et al., 1974). The enhancement technique, termed "contrast stretching," amplifies small differences in energy sensed over the ocean by the multispectral scanner, thus making barely perceptible features easily recognizable. An amazing variety of intriguing features have been found in these images (see Fig. 2).

PHYSICAL ASPECTS OF SUSPENDED SEDIMENT TRANSPORT

Particulate matter is maintained in suspension if the upward components of sediment diffusion equal or exceed the settling velocities of the particles. This principle can be expressed in equation form as

$$cw_s + \nu_s \frac{\partial c}{\partial z} = 0 \qquad (1)$$

(Komar, Chapter 7), where c is the sediment concentration at level z, ν_s is the sediment diffusion rate, and w_s is the particle settling velocity. The hydraulic characteristics of the suspended particles will depend on the energy expended by the buoyant forces. Lateral transport of the sediment will be a function of the diffusive and advective motions of the fluid. It follows that, provided we can specify the concentrations and hydraulic characteristics of the suspended particles along with the diffusive and advective components of the shelf water motions, we should be able to describe satisfactorily the dispersal of the particulate matter. The dispersion of a substance in a turbulent fluid may be fully described by the three-dimensional convective-diffusive conservation of mass equation (neglecting molecular diffusion):

$$\frac{\partial \bar{c}}{\partial t} + \bar{u}\frac{\partial \bar{c}}{\partial x} + \bar{v}\frac{\partial \bar{c}}{\partial y} + \bar{w}\frac{\partial \bar{c}}{\partial z}$$

$$= \frac{\partial}{\partial x}\left(k_x \frac{\partial \bar{c}}{\partial x}\right)$$

$$+ \frac{\partial}{\partial y}\left(k_y \frac{\partial \bar{c}}{\partial y}\right)$$

$$+ \frac{\partial}{\partial z}\left(k_z \frac{\partial \bar{c}}{\partial z}\right) - Kc \qquad (2)$$

TABLE 1. Suspended Solids Analyses

Parameter	Methods	References
Total suspended solids (gravimetric study)	1. Scattering of light (nephelometry)—Forward scattering over some angular span gives approximate particle concentrations (see Fig. 1A)	1. Jerlov (1968), Beardsley et al. (1970), Meade et al. (1975)
	2. Beam attenuation—Scattering and absorption of light from a coherent or collimated beam. Can be influenced by dissolved substances and is not as sensitive as nephelometers (see Fig. 1B). Gives approximate particle content	2. Jerlov (1968), Petzold and Austin (1968), Drake et al. (1972)
	3. Secchi disk	3. Manheim et al. (1972)
	4. Filtration of water—25 or 47 mm diameter, 0.8 or 0.4 μm pore size filters available from several manufacturers. *Nuclepore* polycarbonate filters have proved to be very stable and are preferred for open shelf and deep-sea work	4. Banse et al. (1963), Sheldon and Sutcliffe (1969), Spencer and Sachs (1970), Manheim et al. (1970)
	5. Centrifugation and weighing—Superspeed centrifuge is required to separate finest material. Quantitative transfer is difficult and time-consuming. Not recommended in areas of low particle concentrations	5. Jacobs and Ewing (1965), Lisitzin (1972)
Organic components		
Combustible matter	1. Combust sediment at 450–500°C in platinum crucibles or foil for 2 hours. Standard Millipore and Nuclepore filters leave negligible ash weight	1. Manheim et al. (1970), Meade et al. (1975)
Organic carbon	1. Precombusted glass fiber filters are used to separate particles from water. The filter is then combusted at 500°C in oxygen and the CO_2 is collected and passed through an infrared absorption gas analyzer	1. Menzel and Vaccaro (1964), Strickland and Parsons (1968), Meade et al. (1975)
Carbonate carbon	1. Dissolve CO_3 in *dilute* acid using glass or other resistant filters. Determine weight loss of filter or analyze CO_2 with infrared absorption gas analyzer	
Mineral matter	1. Ash residue after combustion at 450°C is composed of terrigenous and biogenic minerals	1. Manheim et al. (1970)
	2. Terrigenous fraction can be estimated by microscope counts of strongly birefringent grains. This is very difficult if suspension is aggregated	2. Bond and Meade (1966)
	3. Chemical analyses of major ions can give estimate of relative abundance of some siliceous minerals	3. Spencer and Sachs (1970)
Mineralogy	1. X-Ray diffraction using silver membrane filters. Results are strongly influenced by the amount of material on filter. Quantitative work is very difficult	1. Hathaway (1972), Pierce et al. (1972)

TABLE I (Continued)

Parameter	Methods	References
Texture	1. Microscope — Standard petrographic microscope can be used to particle diameters of $\sim$4 μm. Electron microscopy is useful for studies of agglomeration and grains smaller than 10 μm. A variety of industrial particles sizers are available to speed this work. Small amounts of water are passed through membrane filters to produce a one-grain layer. The filter can then be cleared with mineral oil	1. Bond and Meade (1966), Schubel (1969), Manheim et al. (1972)
	2. Electronic sizing — Raw water samples are passed through an electric field; the changes in electrical resistance of the seawater electrolyte are related to the volume of the particles in suspension. This method is very rapid but tells one nothing about the shape and density of particles and, hence, their probable settling characteristics	1. Strickland and Parsons (1968), Carder et al. (1971)
Settling rates	1. Settling determinations require relatively large amounts of suspended matter and cannot be done on shipboard. Electronic sedimentation balances (Cahn Instruments, California), which can handle low sample concentrations are available. The greatest problem is in maintaining the suspension close to a natural state prior to analysis. However, direct settling rates studies would yield much useful data and such methods should receive more attention	1. Schubel (1971), Lisitzin (1972)

where the terms on the left (with the exception of $\partial \bar{c}/\partial t$) represent advective flow and those on the right describe the horizontal and vertical diffusion. The term Kc must be added to account for settling or other losses of suspended particles. Unfortunately, the analytical complexity of the full conservation of mass equation is such that it generally cannot be applied to sediment dispersal problems. It is common to reduce (2) to its one-dimensional or two-dimensional forms for which reasonable assumptions can be made and analytical solutions are feasible (Harleman, 1966). Nevertheless, even in these simplified cases, we still require precise knowledge of the advective and diffusive fluid motions, the settling rates of the particles, and the rates of particle addition and loss from the system. In fact, all of these requisite variables are poorly known (Okubo and Pritchard, 1969). Schubel and Okubo (1972) attempted to determine the long-term seaward flux of suspended matter off Chesapeake Bay using a two-dimensional "transport diffusion" equation. Because there were no available data, they were forced to estimate values for the average particle settling velocity

and the sediment exchange rate at the seafloor. Their results "support the concept of the by-passing of the shelf by fine-grained terrigenous sediment" (Schubel and Okubo, 1972, p. 345). However, in the same volume and elsewhere, Meade (1972) points out that the overwhelming bulk of qualitative and semiquantitative data regarding the fate of silt and clay on the Atlantic shelf are in favor of little bypassing of terrigenous suspended sediment. Which viewpoint are we to believe? For reasons that are discussed later, I am inclined to believe that the state of the art does not allow application of mathematical models of dispersion. We know far too little about *actual* particle settling rates, deposition and erosion, and shelf circulation (see Komar, Chapters 7 and 8; Leetma, Chapter 3; and Mooers, Chapter 4).

Whereas complete description of suspended particle dispersion over shelves is not now possible, study of the transport caused by unidirectional currents is feasible. To perform such a computation one needs to know the velocity and dimensions of the flow and the concentration and distribution of particles in the current. As reviewed by Komar (Chapter 7), the distribution of

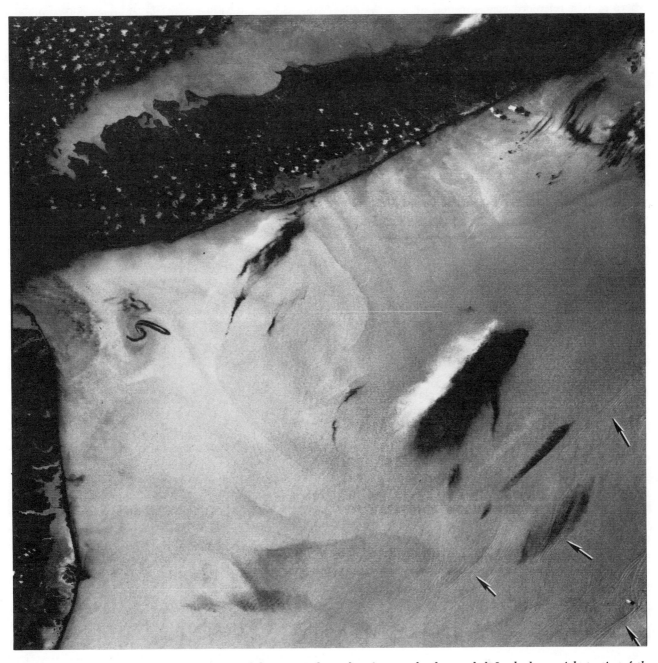

FIGURE 2. *ERTS-1 satellite image of the apex of the New York Bight (August 16, 1972). "Contrast stretched" data received in MSS-4 (0.5–0.6 μm) band. Long Island extends along top of figure. The surface plume of turbid water leaving Raritan Bay and distinct dark patches in the bight apex caused by ocean dumping are clearly revealed. In the lower right portion of the figure one can detect the surface manifestations of packets of internal waves originating at the shelf break. From Charnell et al. (1974).*

material of a given particle size in a turbulent, unidirectional, steady current is a function of the settling rate and upward mixing. In equation form this is expressed as

$$cw_s + v_s \frac{\partial c}{\partial z} = 0 \qquad (3)$$

where c is concentration, w_s is settling rate, v_s is upward diffusion rate, and z is the level above the bottom. Equation 3 can be solved to produce

$$\frac{c}{c_a} = \left[\frac{\delta - z}{z} \frac{a}{\delta - a} \right]^{w_s / \kappa \mu_*} \qquad (4)$$

where c_a is a reference concentration at $z = a$, κ is von Karman's constant (0.4), δ is the flow depth, and μ_* is the shear velocity $[\mu_* = (\tau/\rho)^{1/2}]$. Equations 3 and 4 are well founded in theory and have been verified in flumes and rivers. Unfortunately, application of these relations in the ocean requires extensive (and expensively elaborate) simultaneous sampling of the *turbulent* velocity profile, depth of flow, sediment concentration, and particle settling rates. One difficult problem is the proper determination of the reference sediment concentration and settling rate, c_a and w_s.

The concentration cannot be measured accurately with optical devices. On the other hand, the problem of obtaining a statistically-adequate series of reference values as close to the sea floor as possible (usually within 1 to 5 cm of the bed) using mechanical devices has not been solved. This sampling problem is a major obstacle to meaningful application of (4) on the continental shelf (Smith, in press).

In contrast to coarse silt and sand, precise determination of the settling rates of grains smaller than about 31 μm is complicated by the formation of grain aggregates which tend to exhibit a wide range in density. As discussed later, it appears that a significant part (perhaps the bulk) of the fine silt and clay minerals on shelves is aggregated with organic matter to form larger, low density, irregularly shaped settling units (McCave, 1975). Optical measures give little or no reliable textural data and direct sampling will doubtless disturb the fragile grain aggregates. Even if the suspension can be sampled with minimal disturbance, standard techniques for settling rate analyses are not extended easily to the range of concentrations typical over continental shelves (<50 mg/l). Because (4) is so strongly dependent on the value $w_s/\kappa\mu_*$, the calculation will be a gross estimate, at best, without precise settling velocity data.

EROSION OF COHESIVE SEDIMENTS

Komar (Chapter 8) presents a thorough discussion of the general approach to quantitative estimates of bed load (sand) transport. He concludes that after "intense study for more than a century . . . quantitative estimates may be in error by as much as a factor of 10" in the relatively simple cases presented by river flows. The situation for beds of fine-grained sediment is even more complicated owing to the cohesiveness of particles smaller than about 100 μm (Fig. 7). Flume studies have shown that the threshold shear stress for compact fine silt or clay beds can exceed that necessary to move gravel-sized material. This cohesiveness is produced by

the same unsatisfied particle charges that induce a heavy suspension to flocculate; the effectiveness of these forces will be inversely related to water content (compaction). However, the smooth curves of Fig. 7 are misleading because the cohesiveness, and hence the threshold shear stress, varies markedly with subtle changes in mineral composition, organic matter content, and the sediment bulk properties.

Graf (1971) summarized the primitive state of our knowledge of mud bed erosion. Attempts to quantitatively relate cohesive bed scour to water content, shear strength, plasticity index, mineral composition, and textural measures have met with varying, but generally only fair, degrees of success. For the present, the most honest statement that can be made is that mud beds will be eroded at current velocities on the order of 10 to 30 cm/sec (measured 100 cm over the bed) provided the water content exceeds 80%. At lower water contents, the threshold increases and can be determined precisely only through actual scour experiments. While field experiments over sand bottoms have been carried out (Sternberg, 1972), no comparable studies have been published for muddy bottoms on the continental shelf.

FINE PARTICLE DEPOSITION

It is intuitively comprehensible that we should have difficulty in predicting cohesive bed scour. Surprisingly, recent work has shown that our ability to predict deposition from a steady, uniform flow may be equally poor. The difficulty arises because we cannot precisely describe the fluid motions close to the seafloor under turbulent flow conditions. Consequently, it is not yet possible to relate vertical suspending current energy to the mean velocity measured some distance away from the boundary. In fact, recent studies suggest that such a relationship may hold only in a most general sense (Gordon and Dohne, 1973; Gordon, in press). The fluid motions within the boundary layer are dependent on the boundary form and the flow characteristics (i.e., steady versus fluctuating velocity). For a thorough description of boundary layer relationships, refer to Komar (Chapter 7). His Fig. 3 (Chapter 7) depicts the mean velocity curves for turbulent fluid flow near smooth and rough boundaries. Recently developed techniques for observing the details of turbulent boundary layer mixing (Corino and Brodkey, 1969) have led to the definition of a so-called bursting phenomenon. The mechanism involves disruption of the near boundary zone (viscous sublayer) as slow moving fluid moves away and is rapidly replaced by an "inrush of high speed fluid from the outer boundary layer" (Gordon, in press).

Sediment entrainment is closely related to the rate of energy expended during bursting (Grass, 1971). Our primitive understanding of bottom stress and vertical momentum transport in boundary layers is clearly illustrated by the surprising findings of Gordon (in press). He has found that turbulent velocity fluctuations (which are important in sediment entrainment) are more frequent and vigorous during *decelerating* phases of tidal currents. As emphasized by Komar (Chapter 8), shelf currents are rarely steady and uniform and extrapolation of relationships derived from carefully controlled laboratory flume experiments should be treated cautiously.

MUD SOURCES AND DEPOSITIONAL SITES

Although locally the present supply of terrigenous material can be dominated by eolian transport or shoreline or shelf erosion, the bulk of the fine-grained inorganic matter entering the ocean gets there in rivers. Holeman (1968) estimated an annual river runoff of 18×10^9 metric tons of suspended solids. The quantities of material contributed by shore erosion and eolian transport are both less than 2% of the river transport (Gilluly, 1955; Judson and Ritter, 1964; Judson, 1968; Prospero and Carlson, 1972). The contribution of shelf erosion to suspended solids in the ocean and to the development of mud deposits is unknown. However, it is evident that the reworking of fine materials delivered during glacial epochs is an important process on shelves that now receive minor amounts of fluvial sediment (for example, the Atlantic shelf of the United States and the Wadden Sea) and resuspension of Holocene sediment contributes greatly to near-bottom increases in suspended matter.

TABLE 2a. Fluvial-Sediment Discharge to Oceans

Continent	Drainage Area (% of total)	Short tons/ mile²/year	10^9 Short tons/year
North America	20	245	1.96
South America	19	160	1.20
Africa	19	70	0.54
Australia	5	115	0.23
Europe	9	90	0.32
Asia	25	1530	15.91
		Total	20.16 (18 metric tons)

Source. From Holeman (1968).

TABLE 2b. Suspended Sediment Discharge to Ocean for Selected Major Rivers

River	Location	Annual Suspended Solids Discharge (10^9 short tons)	Percentage of World Total
Yellow	China	2.08	10.0
Ganges	India	1.60	8.0
Bramaputra	East Pakistan	0.80	4.0
Yangtze	China	0.55	2.7
Indus	West Pakistan	0.48	2.4
Amazon	Brazil	0.40	2.0
Mississippi	United States	0.34	1.7
Irrawaddy	Burma	0.33	1.7
Mekong	Thailand	0.19	0.8
Colorado	United States	0.15	0.7
Red	North Vietnam	0.14	0.6
	Total	7.06	34.6

Source. From Holeman (1968).

It is essential to realize that more than one-third of the world's fluvial sediment load is carried by about one dozen major rivers (Lisitzin, 1972); the Ganges and Yellow rivers alone carry nearly 20% of the total (Table 2). Clearly the major input points for suspended solids to the world ocean are the large rivers, most of which have built deltas and most of which are located in southern Asia (Holeman, 1968).

The fate of this material in the ocean is our ultimate concern and it is of interest to examine rates of inorganic sedimentation in the oceans as a first approximation to this problem. McCave (1972) and Lisitzin (1972) have briefly summarized our knowledge of terrigenous sedimentation rates (Table 3). These data suggest the following:

1. Less than 10% of modern riverborne suspended solids reach the deep sea (basin and ocean ridge systems). Most inorganic sediment is deposited on continental margins (including shelf, slope, and rise) or in marginal seas.

2. Of the approximately 1.1×10^9 tons of terrigenous sediment that accumulates annually in the deep sea, between 10 and 40% may be transported in the atmosphere (Prospero and Carlson, 1972) and another few percent may be cosmic and volcanogenic. In fact, the proportion of terrigenous deep-sea sediment that is derived from atmospheric transport may be substantially greater than one-half. A thorough review of research in this area is beyond the scope of this chapter but the interested reader is directed to the

TABLE 3. Terrigenous Sedimentation Rates

Location	Rate per 10^3 years (g/cm²)	Area (km²)	Mass (metric tons)
Deep-sea basins and ocean ridge system	~0.4	2.7×10^8	1.1×10^9
Continental rise	~3-7	2.0×10^7	1.0×10^9

Source. From McCave (1972).

works of Rex and Goldberg (1958), Griffin and Goldberg (1963), Biscaye (1965), Windom (1968; 1975), Heath (1969), and Mullen et al. (1972) in addition to the paper by Prospero and Carlson (1972) and the references cited therein.

3.   Assuming that 50% of the river supply is deposited in estuaries and coastal wetlands, approximately 8×10^9 tons must be deposited on the continental slope and rise. However, only approximately 10^9 tons can be accounted for in these provinces. This means that about 40% of the total annual riverborne load is accumulating off perhaps two dozen major input sites.

In summary, processes that tend to move fine sediment to the deep sea appear to be rather ineffective relative to those factors that tend to hold material on the continental margins or in marginal seas. This conclusion is in agreement with the geographic distribution of Holocene sediments on the seafloor and the rather clear association of sediment composition with latitudinally controlled sources (Lisitzin, 1972). The latter discovery means that processes of mixing in the oceans are not sufficiently vigorous to homogenize the suspended solids contributed from continents. This is surprising in light of the very slow settling rates of fine silt and clay particles (<4 μm diameter) which comprise the bulk of the terrigenous sediment that does reach the deep sea.

Settling of small particles follows Stokes' law:

$$w_s = \frac{2}{9} \frac{\rho_1 - \rho_2}{\eta} g r^2$$

where ρ_1 is particle density, ρ_2 is fluid density, η is the viscosity of the fluid, g is the acceleration due to gravity (980 cm/sec²), and r is the particle radius in centimeters. In 20°C water a 4 μm quartz sphere will settle at about 1.4×10^{-3} cm/sec and a clay particle (<2 μm) will settle at 0.3×10^{-3} cm/sec or less. On the average, particles between 4 and 2 μm will settle only 86 cm in one day in *still, vertically homogeneous water.* If such particles are introduced near the ocean surface, they should remain in suspension for weeks or months over

shelves and for years over deep-sea areas, even if the water was motionless. It is difficult to explain the known distribution of bottom sediments if fine particles settle individually. For example, sediment with a median grain diameter of <2 μm accumulates within 20 km of the Mississippi Delta at depths of 20 to 40 m (Shepard, 1960). It seems evident that these particles do not settle individually and, consequently, the usual laboratory methods of fine sediment textural analysis (involving peptizing agents) will yield little information on transport processes and vertical flux rates.

We will now turn to those processes that may operate to limit the seaward escape of fine-grained terrigenous sediments. The preceding discussion suggested that perhaps one-half (and possibly much more) of the sediment delivered by rivers is deposited in estuaries, coastal lagoons, and embayments. Clearly if this is the case, the composition and distribution of suspended solids in shelf waters off such coasts will be significantly affected by nearshore processes. The following section summarizes the considerable evidence to support entrapment of river sediment in estuaries and concludes with a discussion of the factors that have been called upon to explain this entrapment.

SEDIMENTS IN ESTUARIES

Composition

Recent mineralogical studies have demonstrated striking differences between the composition of sediments in several estuaries and in the rivers which now flow into these estuaries. Pevear (1972) has shown that estuarine muds between Florida and South Carolina contain substantial amounts of montmorillonite, whereas present-day rivers arising in the Piedmont carry small amounts of this clay species. Furthermore, bottom muds taken along the axes of the estuaries show compositional gradients consistent with entrapment of fluvial silt and clay in the landward portions of each estuary; the montmorillonite (plus other material) is evidently moved into the estuaries from the adjacent shelf. In agreement with this conclusion, Pevear and Pilkey (1966) and Pilkey and Field (1972) present evidence for present-day onshore transport of shelf sands off the southeastern United States. Transport in some instances is thought to have involved sand deposits up to 20 km offshore. Similarly, mineral composition in the Neuse estuary of North Carolina (Griffin and Ingram, 1955), the Rappahannock estuary, Virginia (Nelson, 1960), and Chesapeake Bay (Powers, 1954) supports a model involving retention of fluvial sediment in the upper reaches of estuaries and admixture with inner shelf sediment near the sea. Meade (1969, 1972) and Hathaway (1972) have summarized much of these data for the Atlantic coastal plain and added much new information for the entire eastern seaboard.

Sediment composition data from other areas in agreement with this conclusion have been presented by Van Straaten (1960) for the Ems estuary; Postma (1967) for the Wadden Sea; Kulm and Bryne (1967) for the Yaquina Bay estuary; Greer (1969) for the Hampton Harbor estuary (Massachusetts); Guilcher (1956) for the Kapatchez River estuary (West Africa); and Rajcevic (1957) for the Seine estuary.

Rates of Accumulation

Rates of sediment accumulation in certain estuaries and the amounts of material delivered by tributary rivers are "fairly" well known because of the navigational problems associated with harbor shoaling. The rates of shoaling are especially pronounced in "partially mixed" estuaries for reasons that are discussed later. Principal sources for the mud shoals are the rivers, erosion (cannibalism) of estuary shores, and the ocean. As pointed out by Meade (1969) and Schubel (1971), in those estuaries where precise records have been kept on channel dredging, it is clear that subtantially more sediment must be dredged than is currently delivered by fluvial sources (examples are Savannah Harbor, Hudson Harbor, and Charleston Harbor). Meade (1969, 1972) has reviewed the evidence in some detail and has concluded that on the order of only 10% (*at most*) of the riverborne suspended solids now entering estuaries ever reaches the sea. In fact, as we have seen the estuarine environment may also be an excellent sink for sediments moving over the inner parts of adjacent continental shelves.

Processes of Transport and Deposition

The evidence for nearly perfect retention of fluvial sediment in many estuaries and coastal wetlands is uncontestable. Schubel (1971) shows that bottom muds of Chesapeake Bay, when texturally analyzed using standard peptizing methods, contain up to 50% clay particles (settling rate 10^{-4} to 10^{-3} cm/sec). For the same estuary, Pritchard (1956) estimates an upward water flow rate of 10^{-3} cm/sec to account for the distribution of salinity. Clearly, this sediment should not have accumulated in the bay unless it settled as larger *composite particles*. In addition to this there are several other factors which promote entrapment of fine sediments in estuarine, tidal flat, and lagoon environments.

Chief among these are the circulation pattern in shallow marine waters and the unique depositional and erosive characteristics of muds. Because these factors are fundamental to the formation of mud deposits at any location in the ocean and also control the distribution of particles in the water column, they will be discussed in some detail.

Formation of Composite Particles

The subject of particle agglomeration and flocculation is controversial not because there is any question that several mechanisms can and do produce composite particles in the sea, but because we do not know the relative importance of the various mechanisms or their significance relative to water circulation factors. Although there is quite a spectrum of potentially important mechanisms for the formation of composite particles, only two general types have been studied in detail: biological mechanisms and physicochemical flocculation.

There are many animals in the sea that obtain their food by filtering suspended solids from the water; some examples are mollusks, barnacles, and copepods. Filter-feeders are capable of separating particles from 1 μm to about 50 μm and ejecting them as fecal pellets which range in length from a few tens of microns to about 3000 μ (Haven and Morales-Alamo, 1972; Smayda, 1969). The voided pellets have a density of about 1.2 and will contain the undigested portions of the suspended solids (terrigenous minerals, biogenic minerals such as silica tests, and unused organic matter). Depending on size and density, fecal pellets have been found to settle at rates ranging from 10^{-2} to 2 cm/sec, rates typical of coarse silt and fine sand grains (Fig. 3).

Those few studies that have quantitatively assessed the role of filter-feeders in sedimentation have produced some surprising data. Early work in the Clyde Sea by Moore (1936) indicated that during the spring, the pellets from barnacles and euphausids were deposited at up to 33.4 mg/cm² each week! Verwey (1952) suggested that Wadden Sea mussels could deposit as much as 150,000 metric tons of sediment each year, and Haven and Morales-Alamo (1966) showed that small oysters from Chesapeake Bay could deposit about 1.6 grams per individual per week.

Schubel (1971) reported several preliminary attempts to estimate the effects of copepods on suspended solid agglomeration and settling in Chesapeake Bay. Schubel split a sample of bay water into two 2-liter beakers. Fifty copepods were added to one beaker and the other beaker was used as a control. After 74 hours the mean volume diameter of the suspended solids in each beaker

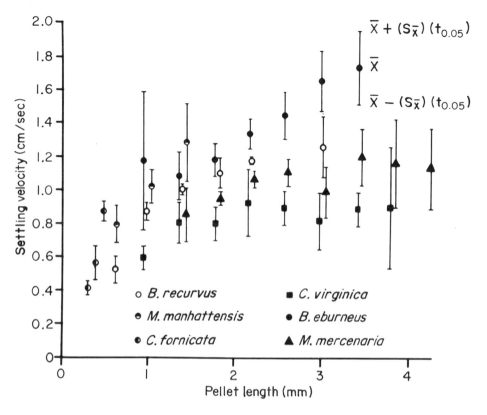

FIGURE 3. *Settling rates of fecal pellets produced by bottom deposit and filter-feeders. From Haven and Morales-Alamo (1972).*

was determined and compared to the same parameter determined prior to the start of the experiment. The mean size of the grains in the *control* beaker showed a fourfold increase, whereas the copepod beaker showed a tenfold size increase (from 6.8 to 70 μm). It is likely that copepod densities rarely are as high as 25 liter^{-1} in most estuaries, and certainly not over continental shelves, although 5 to 15 individuals per liter are sometimes present over the shelf during summer months off southern California and Long Island, New York (Drake, unpublished data). Nevertheless, Schubel's experiments indicate the possible importance of biologic agglomeration by organisms that are widely dispersed throughout estuarine waters. In addition, the size increase in the *control* beaker, while less dramatic, suggests that abiotic production of composite grains may be significant.

Biological production of aggregates and fecal pellets is not restricted to the suspended solids. Recent studies in shallow bays near Cape Cod suggest that aggregation by deposit-feeders totally changes the physical character of the surface sediments (Johnson, 1974). Johnson reported that 26 sediment samples contained an average of slightly more than 40% (by number) organic-mineral aggregates in the >5 μm size fractions. In fact, the state of aggregation was so high that Johnson concluded that nearly all of the fine silt and clay mineral grains were incorporated in aggregates. When these sediments are resuspended they remain aggregated. Consequently, the transport of silt and clay along or across the shelf, its distribution in the water column, and its ultimate deposition can be understood only if one recognizes that the bulk of these grains are probably incorporated in large, relatively rapidly settling composite particles.

Any sedimentologist who has determined the size distribution of <62 μm sediment by settling techniques is well aware of the process of particle flocculation in the presence of organic and inorganic salts. Excellent theoretical discussions of flocculation are presented by Krone (1962, 1972), and Verwey and Overbeek (1948). The process depends on the fact that small particles have unbalanced charges near their boundaries. If immersed in an electrolyte, the particles will attract ions to satisfy this charge imbalance and the attracted ionic layer will attract a second outer layer of oppositely charged ions. The result is a double layer of ions clustered about the particle. The boundary charge depends on the nature and concentration of electrolyte as well as the nature of the particle; when this charge is above a critical value no flocculation occurs even if particle concentrations are as high as 150 g/l (Krumbein and Pettijohn, 1938). However, an increase in electrolyte concentration decreases the boundary charge and causes particle attraction, provided the particles collide.

In a suspension of particles that, by virtue of the salinity of the fluid, are capable of adherence, the extent of flocculation will depend on the rate of collisions. Collisions can be produced by random Brownian motion of submicron grains, by settling of grains in polydisperse systems, and by fluid shearing (Krone, 1972). As floccules form by these mechanisms, the larger composite grains will tend to sweep out smaller grains, and adherence of separate flocs will also begin to occur. Krone (1962) found that "aggregation of previously formed flocs" involved progressively weaker bonds and an overall loss in floccule shear strength. Therefore, in a turbulent (shearing) fluid, collisions are produced, but beyond some critical point an increase in turbulence actually leads to disaggregation. In this way the low-density flocs are culled out of the population in favor of smaller, more compact composite particles (Krone, 1972).

Some results of laboratory studies of sediment floccu-

lation are presented in Fig. 4. Using pure clay suspensions, Whitehouse et al. (1960) found that flocculation begins abruptly at chlorinities of 1 to 2‰. Chlorinities above 2‰ produced additional flocculation in montmorillonite but had little effect on kaolinite and illite. Their data show that kaolinite and illite form large aggregates which settle at about 0.014 and 0.018 cm/sec, respectively. The substantial difference between these settling rates and that for montmorillonite has been invoked to explain mineral distributions on the Niger Delta (Porrenga, 1966).

The experiments of Krone (1962) are of special interest because they demonstrate the strong dependence of flocculation on the particle concentration (Fig. 4A). This is because all three processes that cause collision proceed at rates that are directly proportional to particle content. Natural muds from San Francisco Bay were used for his studies; therefore they should be more representative than the pure clay suspensions used by Whitehouse and his colleagues. We may conclude that at concentrations below 200 mg/l physical flocculation is *not* an effective means of composite particle formation. As we shall see, particle concentrations commonly

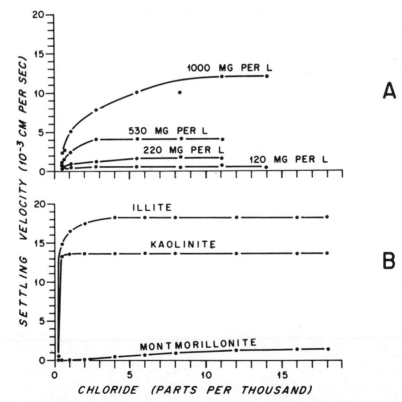

FIGURE 4.   (A) *Variation in clay particle settling rates with chlorinity and sediment concentration. Data based on Krone (1962); figure from Meade (1972). (B) Settling rates of clay floccules with changing chlorinity. Data based on Whitehouse et al. (1960); figure from Meade (1972).*

exceed 200 mg/l in rivers and estuaries where conditions of increasing salinity and internal shearing are also appropriate for the flocculation process. Nevertheless, field measurements and analyses of suspended solids samples from estuaries do not lend strong support to the widespread notion that physicochemical flocculation is the predominant process for the formation of composite particles (Schubel, 1971).

Let us look at some of the evidence. An excellent example of changes in suspended solids concentration over a tidal cycle at a point in Savannah Harbor is presented in Fig. 5; these data are representative of many estuaries. The striking aspect of the suspended solids curves is the rapid loss of material as the flood tide current wanes (from 1500 to <100 mg/l in only 2.5 hours). By solving (4), which defines the vertical distribution of suspended solids under conditions of steady, turbulent shear flow, for the settling velocity, Krone (1972) found that settling rates in Savannah Harbor ranged from about 0.3 to 4.5 cm/sec. In other words, the Savannah Harbor *clay* was settling at rates equivalent to fine- and medium-grained quartz sand. Unfortunately, Krone (1972) did not attempt to determine directly the characteristics of the composite particles. However, it seems unlikely, in light of the data presented in Fig. 4, that physicochemical flocculation can account for such tremendously high settling velocities. Velocities in excess of 0.1 cm/sec are in much better agreement with the settling rates of fecal pellets (Smayda, 1969; Schubel, 1971; Haven and Morales-Alamo, 1972).

Manheim et al. (1972) found that aggregates were common within a transitional zone between the turbid shelf water near the Mississippi Delta and the clear waters of the Gulf of Mexico. An order of magnitude estimate suggested that shelf water copepods might be significant particle generators on a long-term basis. However, no quantitative estimates of the proportions of aggregated versus discrete grains were presented.

Circumstantial evidence in support of physicochemical flocculation in natural waters can be drawn from Drake et al.'s (1972) analysis of the fate of flood sediment off southern California. In 1969, more than 50×10^6 tons of sand, silt, and clay were delivered to a segment of the mainland shelf near Santa Barbara by the Santa Clara River; 25 to 35% of this sediment was fine silt and clay. During peak river flow suspended solids concentrations a few kilometers upstream from the coast were as high as 50,000 mg/l (Drake, 1972).

The flood sediment was discharged directly to the littoral zone and, using box cores, Drake (1972) showed that more than 80% of the river sediment could be accounted for on the shelf at depths of less than 50 m and distances less than 20 km from the river mouth. Yet, currents on this shelf average 10 to 20 cm/sec with measured high velocities of 30 to 35 cm/sec. Furthermore, the shelf is exposed to moderate wave action from the North Pacific. The nearly complete retention of even fine silt and clay on the inner shelf after this flood does not prove physicochemical flocculation, but such rapid deposition is difficult to explain in any other way (deposition was too rapid to be attributed to biological mechanisms).

Composite particles ranging from a few tens of microns to several millimeters in length are common constituents of marine waters. Schubel (1971) found that by number the agglomerates make up less than 10% of most samples from Chesapeake Bay but *by volume* they contain more than 85% of the mineral grains. The larger particles usually occur in forms (spheroids, rods, etc.) which are demonstrably the result of filter-feeding organisms, whereas smaller composite particles ($<62 \mu m$) are irregularly shaped agglomerates of mineral grains and amorphous organic matter that may be floccules or fragments of fecal pellets. Composite particles that one can unquestionably ascribe to purely physicochemical flocculation are very rare. But this problem may be caused simply by the large amount of amorphous organic matter that is always available to interact with mineral floccules. Electron microscope studies of suspended solids from New York Bight and southern California shelf waters suggest that composite particles incorporating terrigenous grains are significant in both areas (Drake, unpublished data). These studies are in preliminary stages but the results thus far are similar to those reported by Schubel (1971). Generally, a shelf water sample will contain one or more coarse silt or very fine sand size aggregates for each 300 to 500 grains counted. Off southern California, the composite particles contain from 5 to 35% terrigenous minerals predominantly in the size range of less than 4 to 16 μm. The great significance of these particles is illustrated by a simple calculation. Assuming a mineral content of 25% and a mean grain size of 4 μm within a spherical aggregate of 62 μm diameter, one such composite grain will include nearly 1000 mineral particles.

In summary, it is evident that much fine silt and clay do not settle as single grains in marine waters. At present the data suggest that physicochemical flocculation can and does produce settling rate increases which range up to about one order of magnitude, whereas biological processes can account for rate increases of up to several orders of magnitude. Both mechanisms are geologically important although their effects in a particular area may be highly variable in time and space.

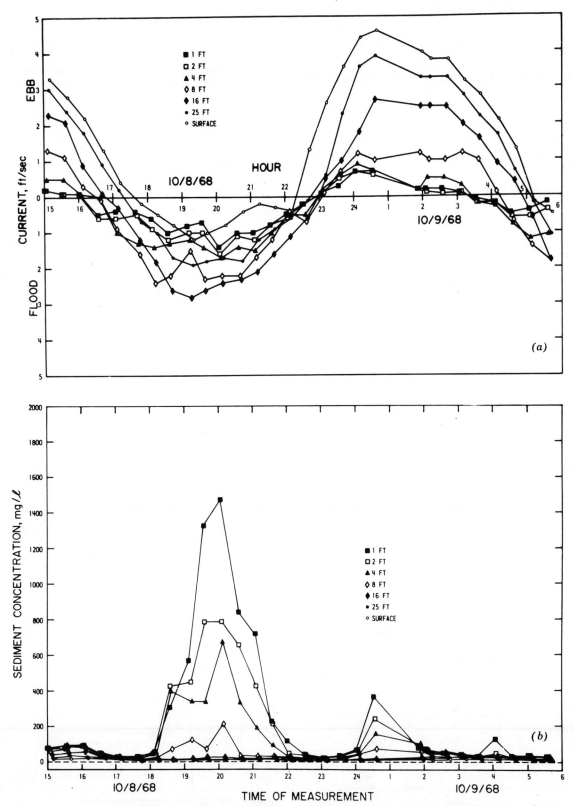

FIGURE 5. *Currents and suspended solids at one location in Savannah Harbor. Sample levels are referenced to bottom. Note that flood and ebb currents are stronger at near-bottom and near-surface levels, respectively. From Krone (1972).*

Estuarine and Inner Shelf Circulation

Recent advances in our knowledge of shelf circulation patterns suggest that residual currents over the inner portions of some shelves are similar to the nontidal currents in stratified and partially stratified estuaries. In estuaries, circulation is controlled by freshwater flow and frictional and tidal mixing (Fig. 6). If freshwater flow is large compared with the tidal range, salt water is prevented from entering the estuary (or enters as a sharply defined salt wedge) and the bulk of the fluvial sediment is carried directly out over the inner shelf; [examples are the Amazon River (Gibbs, 1972) and the Mississippi River at high river stages]. However, most estuaries fall in the partially mixed class described by Pritchard (1956). These estuaries are characterized by a landward flow of salt water along the bottom and seaward flows of fresher water near the surface; in such cases, the freshwater head is not sufficient to prevent saltwater density flow into the estuary. Because of the tides and the surface layer flow, there is mechanical mixing of water between the two layers which necessitates a net landward flow of shelf water to make good the losses from the lower layer (Fig. 6). The result is a nontidal circulation typified by more vigorous and persistent flood tide currents within the intruding bottom layer. Fluvial sediment entering this type of system and fine particles resuspended by current and wave action along the estuary shores will move seaward with the "fresh" water flow. As this current encounters the saline underflow the coarser materials will settle into the lower layer to be carried back upstream. Fine-grained particles will begin to flocculate as the salinity reaches 2 to 5‰ or will be incorporated in fecal pellets by estuarine zooplankton; thus the finest sediments will

also settle into the landward-flowing salt layer. Entrainment of water from the bottom layer carries fine sediments back into the surface layer, and the system "recycles." One result of this recirculation is the well-known "turbidity maximum" located near the landward limit of salt water (Postma, 1967; Schubel, 1971). It must be stressed that this phenomenon is not caused by flocculation, but it probably promotes flocculation and agglomeration of fine particles by increasing particle collisions and particle residence time.

Although partially stratified estuaries effectively trap suspended matter, the recent work of Milliman et al. (1975) on the Amazon River plume demonstrates a similar tendency in a classical salt wedge estuary. Milliman et al. (1975) suggest flocculation to help explain the rapid clearing of surface waters seaward of the river.

Mud Deposits and Scour Resistance

Although we have shown that sediments in estuaries settle rapidly owing to formation of composite particles and that the two-layer circulation promotes retention of these materials within the estuarine water column, it is clear that these processes alone are not enough to explain the permanent deposition of clay in nearshore areas and the short "life span" of estuaries. Sooner or later concentrations in suspension would reach such high levels that seaward diffusion would balance new fluvial sediment contributions.

In order to understand mud accumulation fully we must consider the unique ability of muds to resist current scour. Flume studies by Sundborg (1956), Krone (1962), Kuenen (1965), and Postma (1967) have demonstrated that beds composed of mineral particles

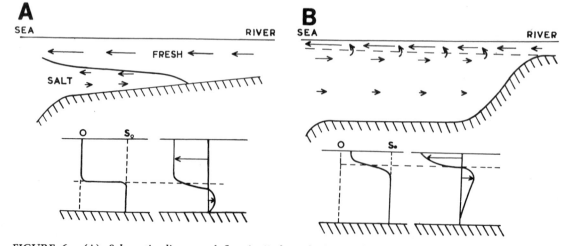

FIGURE 6.   (A) *Schematic diagram of flow in "salt wedge" type of estuary.* (B) *Flow in a partially stratified estuary. From Bowden* (1967).

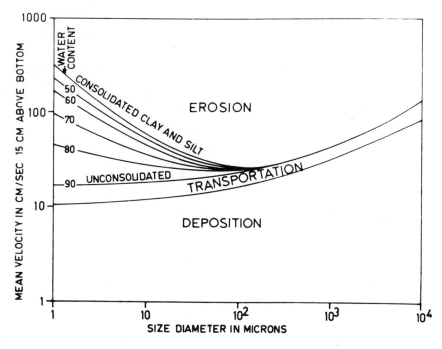

FIGURE 7. *Current velocities required to transport particulate materials. From Postma (1967).*

smaller than about 100 μm develop a cohesiveness that depends on composition and water content (degree of consolidation). Stated another way, there is a difference between the current velocity required to transport clay and silt and the velocity (boundary shear stress) required to resuspend the same sediment (Fig. 7). Thus, fine sediments that are able to settle to the bottom in areas where average tidal velocities are reduced or during slack tide periods will accumulate unless much higher current velocities are reached later.

Because the cohesion of fine sediment is a complex function of mineral composition, sediment texture, water content, and organic matter content, there is currently no reliable means of predicting sediment entrainment as a function of boundary shear stress. Experimental studies of mud scour in flumes are hampered by the necessity of disturbing the natural sediment. Yet the entrainment rate of fine particles from the bottom is critical to any computation of sediment dispersion in the ocean (unless it is assumed that the particles diffuse with the fluid and settling and deposition are neglected, an assumption not supported by our knowledge of suspended solids distributions, apparent settling rates, and depositional sites). Probably the most promising means for determining rates of bed/water sediment exchange involves in situ flume studies. Such a flume, capable of use at all ocean depths, is under development by J. B. Southard and R. Young at the Massachusetts Institute of Technology.

Rates of mud accumulation will depend on the balance between supply (suspended solids concentrations) and current energy (including both fair-weather and storm conditions). Consequently, estuarine shoaling is normally most rapid near the turbidity maximum at the landward tip of the saline underflow.

The same processes of agglomeration and accelerated settling, estuarine-like circulation, and mud cohesion also operate over the inner portions of continental shelves. The estuarine-like circulation is especially well documented on the eastern shelf of the United States through the work of Bumpus (1965) and Harrison et al. (1967); see Fig. 8. The increased intensity of landward bottom water flow near the mouths of eastern seaboard estuaries demonstrates a genetic relationship to the seaward mixing of low-salinity surface waters. A similar flow pattern is suggested by salinity data off the Po River Delta (Nelson, 1970). However, it appears that estuaries and fluvial discharge are not required for estuarinelike inner shelf circulation. An additional mechanism which may drive water shoreward over the inner shelf floor is the continuous net transport associated with progressive surface waves (see Chapter 13). It appears that the net shoreward drift of bottom waters on the Northeast Pacific shelf (Gross et al., 1969) begins at depths where prevailing long-period waves should "feel bottom" (about 40 m). Furthermore, progressive waves will produce a net advective flow landward near the sea surface and increased vertical

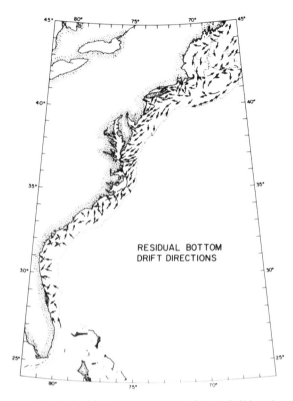

RESIDUAL BOTTOM
DRIFT DIRECTIONS

FIGURE 8. *Residual bottom currents on Atlantic shelf based on drifter studies by Bumpus (1965) and Harrison et al. (1967). From Hathaway (1972).*

mixing throughout the shallow, inner shelf water column. The net result of these landward water motions is the "inner shelf/littoral zone" turbidity maximum present over virtually all shelf areas (Figs. 9–11). Sediment is maintained in suspension within this zone at concentrations ranging from a few milligrams per liter to more than 100 mg/l, depending on supply and wave activity. Even in areas of limited terrigenous sediment supply, the particles over the inner shelf are predominantly noncombustible at 500°C (Drake, 1972; Manheim et al., 1970). Because advective currents over the inner shelf are principally coast-parallel and onshore, sediment that manages to escape from estuaries or directly from rivers will move along the coast some distance before being deposited, swept into a downstream estuary, or carried offshore.

The littoral/inner shelf turbid zone is analogous to a river flowing parallel to the coast (McCave, 1972). Suspended solids concentrations are maintained at high levels by additions from land, wave action, and the slow but important drift of water toward shore. Losses of sediment from this "river" occur through seaward-flowing rip currents (Shepard and Inman, 1951) that extend mostly only a few hundred meters from the beach; advective plumes trending offshore because of

the convergence of opposing coastal currents; diffusive mixing with open shelf waters; and, if possible, through sedimentation on the inner shelf and in coastal wetlands. Because of relatively rapid settling rates of composite particles (formed by flocculation off major muddy rivers and through the life processes of filter-feeding organisms in estuaries and on the shelf), much of the sediment carried seaward by these mechanisms will settle to the near-bottom waters within 10 to 20 km of the shore. Thereafter, as in estuaries, some of it will be slowly transported back to the inner shelf to reenter the nearshore turbid zone (see Fig. 11). Obviously, the seaward loss of fine sediment is dependent on the width of the shelf, the energy of marine transport processes, the settling rates of the suspended solids, and the availability of estuaries and coastal wetlands which will provide suitable deposition sites. As pointed out by Meade et al. (1975) all of the factors are optimal to impede the seaward loss of sediment from the Atlantic shelf of the United States. At the other extreme are the narrow shelves off the western United States (Drake et al., 1972) and West Africa (Emery et al., 1973, 1974). Nevertheless, it is apparent from an earlier discussion of sedimentation rates in the deep sea, that even those fine particles which do manage to bypass the shelf are principally trapped in marginal basins and seas, accreted to the fronts of deltas, or deposited on the continental rise (discussion in Chapter 18).

SUSPENDED MATTER ON CONTINENTAL SHELVES

We have already noted that present-day rates of terrigenous sedimentation in the deep sea are small compared to the annual fluvial sediment supply. We can also advance several reasons for the present entrapment of fine sediment either on shelves or a short distance from them.

1. Holocene sea-level rise has produced wide shelves, increased the distance from supply points to the shelf edge, and created numerous nearshore sites suitable for mud accumulation. Thus, much of the material in suspension over the continental slopes and rise may be resuspended sediment delivered during lowered sea level (see the discussion by Southard and Stanley, Chapter 16).

2. Many of the large rivers that supply one-third to one-half of the present terrigenous sediment enter marginal seas (Mississippi, Yellow, Colorado, Irrawaddy, Mekong, and Po rivers) or onto relatively wide shelves (Amazon and Orinoco rivers).

3. Circulation patterns over the inner portions of many shelves oppose the seaward transport of fine

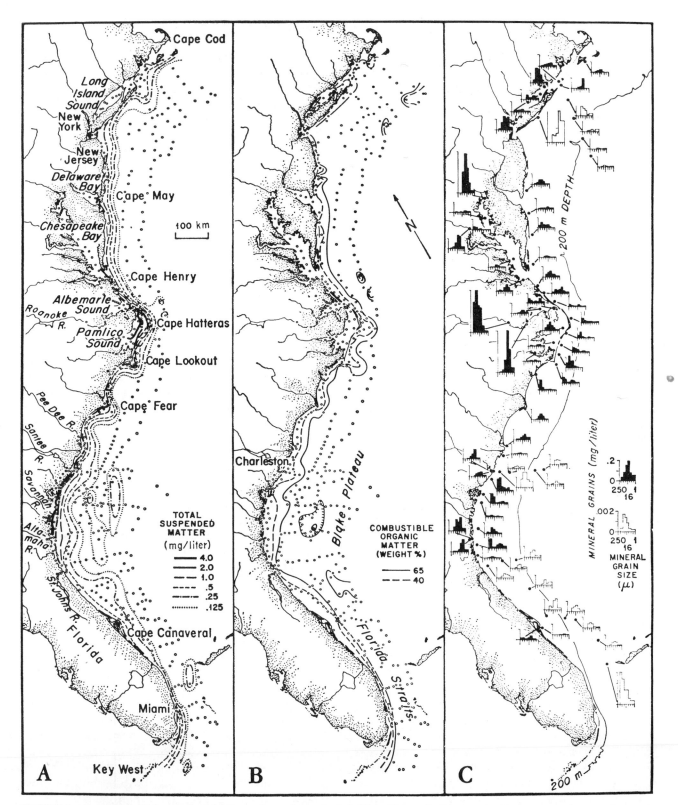

FIGURE 9. *Suspended solids over the Atlantic continental margin of the United States. Surface water only. From Manheim et al.* (1970).

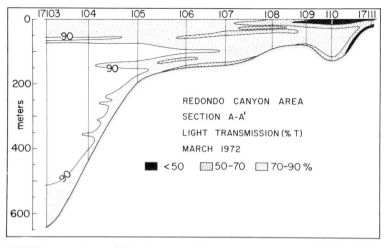

FIGURE 10. *Vertical distribution of light transmission values (% T/m) over mainland shelf off Redondo Beach, California. High values correspond to clear waters. Values of 50% T/m correspond to approximately 1 mg/l of suspended matter. From Drake and Gorsline (1973).*

sediment. Consequently, mud delivered by major rivers may be moved along the shore and accreted to the coast.

4. Flocculation and biological agglomeration are important processes that accelerate the settling of silt and clay by factors of 2 to more than 10 times. Biologic agglomeration may be most important in estuaries, whereas flocculation by purely physicochemical means may be significant where rivers are directly tributary to the coast [for example, the clay deposits at depths of 10–40 m off North Pass of the Mississippi Delta (Shepard, 1960) which could not have been deposited as single grains].

5. Flume studies suggest that mud deposition occurs at current velocities that are surprisingly high—up to 10 to 15 cm/sec measured 1 m above the bed (Kuenen, 1965; Partheniades et al., 1969). Einstein and Krone (1962) and McCave (1970, 1972) suggest that the deposi-

tion of silt and clay at such current velocities is related to the presence of the viscous sublayer in turbulent boundary flows. It is suggested that fine particles entering this near-bed layer of laminar flow are deposited and an initially heavy suspension will eventually clear even though the "free stream" velocity is greater than 10 cm/sec.

6. After deposition mud is rather easily resuspended provided no dewatering has occurred (Postma, 1967). Resuspension of rapidly deposited, loose surficial mud appears to occur at velocities of approximately 20 to 25 cm/sec measured 1 m over the bed. However, if deposition takes place slowly so that less water is incorporated in the bed or if a sufficient thickness of material is deposited such that the bed is compacted under its own weight, the critical erosion velocity will increase dramatically to over 50 cm/sec. Clearly, if

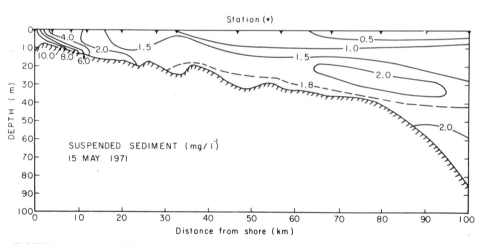

FIGURE 11. *Suspended solids distribution over the continental shelf off Chesapeake Bay. From Schubel and Okubo (1972).*

deposition is not soon followed by high current velocity events, a mud bed has a good chance of resisting all but the most intense flows.

This last factor, that of bed cohesiveness and resistance to resuspension, is complicated by the presence of benthic (and nektonic) organisms that stir, burrow, and otherwise disturb bottom sediments. Like biological agglomeration processes, the abilities of the benthos to resuspend particles directly and to prevent compaction of muds (by introducing water) has not been examined in sufficient detail. The few studies that have been reported suggest that sediment reworking by organisms may be extremely significant. For example, Rhoads (1963) notes that a clam, *Yoldia limatula*, is able to rework sediment in Buzzards Bay faster than it accumulates. During the feeding of this clam, sediment is stirred into suspension several centimeters above the seafloor. Stanley et al. (1972) have observed similar activities of both benthos and fish near the shelf edge off the Middle Atlantic States, and Barnes (1970) found that intense bioturbation of the upper 25 cm of box core samples in the Santa Cruz Basin off southern California was the rule rather than the exception. Clearly, studies of all aspects of the interaction of organisms with fine sediment should be among our highest research priorities.

Cross-Shelf Transport

Cross-shelf transport of sediment can occur through diffusion (wave and tidal mixing) and advection (e.g., mud streams off deltas, current convergences, turbid layer density flow, and cold water cascades across the shelf edge). The effectiveness of these processes will depend on shelf width, coastal morphology, and the energy level of marine erosion and transport mechanisms (i.e., storm frequency, wave activity). Some examples in the form of suspended solids distributions over shelves will illustrate some of these factors. Unfortunately, data collected during or immediately following storms are *very* sparse. Nevertheless, we do know that wind- and wave-driven currents associated with the passage of weather fronts and hurricanes can exceed 50 cm/sec 1 m above the seafloor at water depths of 50 to 70 m (Smith and Hopkins, 1972). This deficiency will be difficult to overcome; however some evidence is available from moderate storms (see Chapter 10).

The vertical distribution of suspended solids has been found to be quite similar on all shelves, whereas the composition and concentrations encountered are highly variable, depending on the geologic setting (Table 4). Figures 9 to 14 present examples from several shelf areas. Two particle-rich zones are evident in each area:

a near-surface layer and a near-bottom layer. At shallow depths these layers tend to merge such that the entire water column is turbid. Over the middle and outer shelf one or more midwater layers of higher particle concentrations may be interspersed with relatively clear layers.

The surface turbid layer generally correlates well with the wave-mixed layer or with low-salinity plumes over the inner portions of the shelf. Farther from the coast an increase in particle concentration within the thermocline is commonly observed as gradients in the thermocline become steeper and the settling of minerals and plankton is retarded (Drake, 1971; Drake et al., 1972; Proni et al., in preparation).

The composition of the particles in the surface turbid layer depends on the relative magnitudes of biogenic and terrigenous supply. Off the Atlantic coastal plain terrigenous supply is limited by the entrapment of fluvial sediment in coastal wetlands and the estuarinelike inner shelf water motions. Consequently, the ratio of combustible to noncombustible materials increases rapidly away from the coast (Manheim et al., 1970; Meade et al., 1975). On the other hand, off southern California where the fluvial sediment load is only slightly greater, noncombustible matter comprises more than 50% of the total suspended solids over the entire nearshore shelf and well out over the basins of the California borderland. This difference is attributed to the fact that much more of the riverborne load of California streams is introduced directly to the ocean (Drake et al., 1972) and the nearshore shelf is relatively narrow. In fact, Emery (1960) found good agreement between annual river supply and nonbiogenic sediment accumulation in the basins of the California borderland; the narrow shelf and general lack of nearshore sites for mud deposition are important factors in this seaward transport of suspended solids.

Although terrigenous sediment transport in near-surface waters can be significant off major rivers (see Fig. 14), the composition and high concentrations of material in suspension near the bottom demonstrate that material introduced at the shore quickly settles to form the shelf nepheloid layer (Table 4). Even off coasts with a modest supply of sediment from land, the near-bottom layer is predominantly composed of noncombustible solids (Drake, 1972, 1974; Meade et al., 1975). In general, the particles near the bottom are somewhat coarser grained and the concentrations and texture respond rapidly to the passage of storms. For example, Rodolfo et al. (1971) observed a fourfold increase in suspended loads near the bottom off Cape Hatteras shortly after Hurricane Gerda had moved through the area. Similarly, Drake (in press) estimated that suspended solids concentrations in the apex of the New

TABLE 4. Suspended Solids Statistics

Location	Water Layer	Suspended Solids Concentration	Composition	Reference
1. Rivers	—	Highly variable; depends on discharge, drainage basin lithologies, and climate. Range from <10 mg/l to >50 g/l	5–15 % combustible organics	Scruton (1960), Holeman (1968), Nelson (1970), Schubel (1971), Drake (1972), Lisitzin (1972), Meade (1972), Curtiss et al. (1973)
2. Estuaries	Surface	Varies with river discharge and location in estuary. Range from ~5 to 500 mg/l	20–40% combustible organics	Patten et al. (1963, 1966), Postma (1967), Meade (1969, 1972), Schubel (1971), Krone (1972), Drake (1974)
	Near bottom	Varies extremely with tidal flow and location. Range from 5–10 to >5000 mg/l	20–30% combustible organics	
3. Shelf southern California	Surface	2–5 mg/l within 3 km of shore; <1 mg/l 10 km offshore. Both values can increase by an order of magnitude during extreme winter floods	10–20% combustible organics over inner shelf; 20–80% organics over outer shelf 10–20 km from coast	Rodolfo (1964), Beer and Gorsline (1971), Felix and Gorsline (1971), Drake (1972), Drake et al. (1972), Drake and Gorsline (1973)
	Near bottom	Wide range depending on proximity to supply, bottom sediment type, and currents. Range from ~5 to 50 mg/l at depths <30 m. <1 to ~10 mg/l at depths >30 m	<10 to 30% combustible organics	
Atlantic shelf (United States)	Surface	Range from <1 to 15 mg/l near the shore to <0.5 mg/l 50 km from shore. Steep gradient decrease within 10–20 km of coast	20–40% organics within 20 km of shore; 60–90% organics over middle and outer shelf	Manheim et al. (1970), Buss and Rodolfo (1972), Pierce et al. (1972), Schubel and Okubo (1972), Meade et al. (1975)
	Near bottom	Varies with current resuspension. 2–20 mg/l within 10 km of shore and <30 m depth. Less than 0.5 to 5 mg/l at >30 m	Predominantly noncombustible matter to equal proportions of combustible and noncombustible	
Gulf of Mexico (northern shelf)	Surface	Ranges from <1 mg/l over Florida shelf to >64 mg/l near Mississippi Delta. Concentrations of up to 100 mg/l (or higher) near the delta distributaries. Concentrations decrease to <0.25 mg/l 100 km from Mississippi River	50–80% terrigenous matter to shelf edge near delta. Mostly organic matter more than 100 km from Mississippi Delta	Scruton and Moore (1953), Manheim et al. (1972), Griffin and Ripy (unpublished data)

Location	Position	Description	Notes	Reference
Gulf of Mexico	Near bottom	Up to 800–900 mg/l in delta distributaries. Concentrations of >50 mg/l are present within 10 km of shore off central Louisiana, decreasing to 5–10 mg/l at depths >20 m. Probably decreases to values of 2–10 mg/l over western shelf and to <1–2 mg/l over Florida shelf	Controlled by Mississippi River and other large fluvial sources	Emery et al. (1973)
Southwest Africa	Surface	>8 mg/l near rivers decreasing to <0.5 mg/l at 100 km from coast. Terrigenous fraction decreases more rapidly to <0.25 at shelf edge. Advective plumes off major rivers and narrow shelf indicate important cross-shelf transport	Combustible fraction (as %w) ranges from 30 off Congo River to 90% beyond shelf break	Lisitzin (1972)
Bering Sea	Surface over shelf	Up to 7 mg/l	—	McManus and Smith (1970)
	Near bottom	3–11 mg/l or higher near rivers	—	
Gulf of Maine	Surface	<0.5 mg/l in central parts of area	25–75% combustible matter depending on season	Spencer and Sachs (1970)
	Near bottom	Increases to ~1.0 mg/l, with one value up to 3.3 mg/l	10–20% combustible matter	
North Sea	Surface	Up to about 2 mg/l at shelf depths of 30–40 m	—	Joseph (1955)
	Near bottom	Ranges from 3 to 10 mg/l, depending on relatively strong tidal currents	—	Joseph (1955)
N. Adriatic near Po River Delta (depths of <50 m extend >100 km south of delta)	Surface (Po River)	Ranges from ~40 to 300 mg/l, depending on river stage		Nelson (1970)
	Surface (over shelf)	>60 mg/l within 5 km of delta distributaries. Decreases rapidly to 8 mg/l 20 km from delta and to <2 mg/l 50 km offshore (see Fig. 14)		Nelson (1970)
	Near bottom	Up to 30–50 mg/l at depths <10 m near delta; decreases to <1 mg/l at depths >30 m		Nelson (1970)

York Bight increased an average of approximately 0.5 mg/l (10–80%) throughout the entire water column during a short storm that produced 1 to 2 m seas (depths are from 10 to 40 m).

The composition of the shelf nepheloid layer suggests that it is principally supplied by resuspension of bottom sediment. In some areas such as the Gulf of Maine (Spencer and Sachs, 1970) and the North Sea (Postma, 1967), it is clear that the resuspended sediment represents ongoing reworking of Pleistocene glacial outwash and fluvial deposits. However, over other shelf areas such as the Gulf and Pacific coasts of the United States there seems little reason to doubt that the nepheloid layer is mostly composed of recently contributed fluvial material. Sediment near the bottom is alternately deposited and resuspended in response to fluctuations in tidal and wind- and wave-driven currents. If sediment supply is abundant (such as off the Mississippi, Amazon, and Orinoco rivers), cohesive nearshore mud beds capable of withstanding hurricanes will develop. If the sediment supply is not sufficient to form scour-resistant mud beds, the silt and clay (principally in flocculant or agglomerated form) will be repeatedly deposited, resuspended, and transported until it reaches a low-energy area (cf. Swift, Chapter 15). As we have seen, current patterns over the inner shelf will tend to concentrate resuspended sediment in the nearshore zone. If estuaries are present, much of this material will be funneled into the estuaries and deposited. If no estuaries are present, suspended solids concentrations will still be maintained at high levels over the inner shelf and the sediment must then diffuse seaward or move seaward in advective currents to form middle shelf, outer shelf, or slope mud deposits (Curray, 1960). Deposition of silt and some clay will begin at shelf depths where wave surge currents fall below about 5 to 10 cm/sec; this depth will depend on the exposure of the coast but generally ranges from 20 to 50 m. This material is not a static accumulation of fine sediment. Rather it is a reservoir of silt and clay which is in dynamic equilibrium with supply from the inner shelf turbid zone and diffusive or storm current losses to deeper shelf areas (Smith and Hopkins, 1972).

The results of bottom drifter and current-meter measurements over the outer portions of continental shelves along with theoretical tidal current calculations suggest that much of the fine sediment that reaches the outer shelf will eventually be bypassed to the continental slope. Bumpus (1965) and Harrison et al. (1967) on the Atlantic coast and Gross et al. (1969) and Smith and Hopkins (1972) working on the Oregon and Washington shelf have concluded that net bottom water transport at depths greater than 50 m is directed offshore. Figures 10 and 11 show that near-bottom concentrations of suspended solids remain relatively high to the shelf edge. However, with the exception of areas near major deltas, the concentrations of sediment over the outer shelf

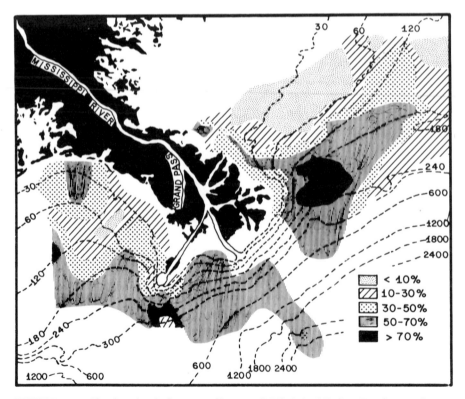

FIGURE 12. *Clay fraction in bottom sediments off Mississippi Delta. Depths are shown in feet. From Shepard (1960).*

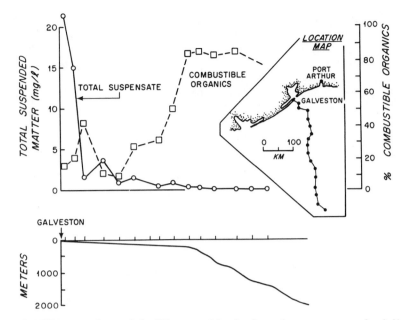

FIGURE 13. *Suspended solids composition in the surface waters over the shelf off Texas. From Manheim et al. (1972).*

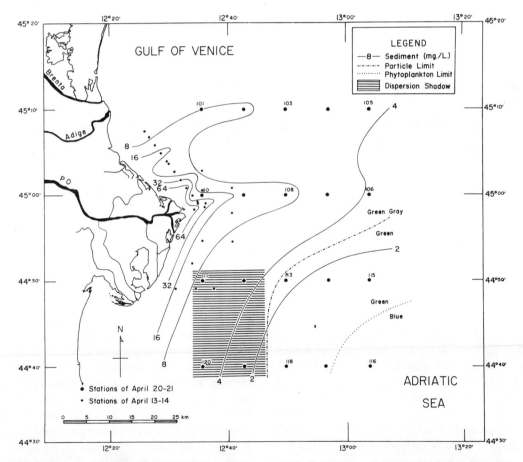

FIGURE 14. *Suspended solids in the surface water off the Po River Delta. From Nelson (1970).*

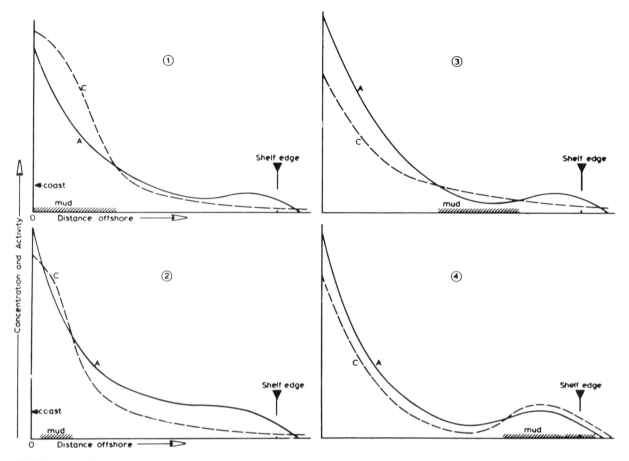

FIGURE 15. *Conceptual model to explain mud deposition sites on continental shelves. A denotes current activity and* *C denotes suspended load concentrations near the seafloor. From McCave (1972).*

rarely exceed 5 mg/l (Meade et al., 1975; Pierce, Chapter 18; Drake and Gorsline, 1973; Stanley et al., 1972; Lisitzin, 1972). McCave (1972) points out that the outer shelf concentrations are too low to supply more than a few tens of milligrams per square centimeter to the seafloor each year. The burrowing and stirring activities of organisms should easily prevent the formation of a cohesive mud layer, and storm currents of the magnitude measured by Smith and Hopkins (1972) would easily remove these thin outer shelf deposits.

A very useful conceptual explanation of mud deposition on shelves has been presented by McCave (1972) and is reproduced in Fig. 15. The diagram emphasizes the fact that the location and rate of mud accumulation are controlled by the balance between sediment supply (concentration in the shelf nepheloid layer) and marine transport ability (the sum of tidal, wave, wind, and density currents near the seafloor). Our knowledge of suspended solids in shelf waters demonstrates that there is always fine sediment available for deposition at all points on the shelf. But a permanent mud accumulation at any shelf position will only occur if sediment supply

is sufficient to form a cohesive bed that will withstand the disturbing effects of benthic organisms and storm current erosion. As more data on bottom current shear, suspended solids concentrations, and settling rates are collected, it should be possible, using models of the type presented by McCave (1970), to determine whether existing mud deposits on shelves are expanding or contracting.

Submarine Canyons and Suspended Sediment Transport

It is well known that submarine canyons which incise the continental shelf act as sumps for sand moving along the coast (Gorsline and Emery, 1959). It has also been suggested that canyons act as important transport conduits for fine sediments which either settle into canyons or are drawn into canyons through a mechanism termed "low density turbid layer flow" (Moore, 1969; McCave, 1972). The turbid layer flow mechanism is essentially a very low-density turbidity current driven by the excess density imparted to the fluid by the solid material. As noted by Drake (1971) and McCave (1972) suspended solids concentrations over gently sloping continental shelves are generally not high enough to produce significant downslope flows. In fact, the density increases caused by temperature stratification in shelf waters are generally sufficient to prevent downslope migration of water containing as much as 100 to 1000 mg/l of suspended solids (10 mg/l of mineral matter produces a density increase roughly equal to a 0.1°C temperature decrease). Furthermore, downslope density flow is unlikely over the shelf because currents generated in this manner will be influenced by Coriolis force and turned to roughly parallel bottom contours. However, the flow velocities necessary to maintain the particles in suspension as the current moves along the shelf could only be achieved by excessively high concentrations of suspended solids, on the order of several hundred milligrams per liter.

Because of these problems, McCave (1972) suggested that turbid layer density flow may only be significant in submarine canyons where flow and dispersion are confined by the canyon walls. In order to test this possibility, Drake and Gorsline (1973) and Drake (1974) measured suspended sediment concentrations in several southern California canyons during periods when F. P. Shepard of Scripps Institution of Oceanography measured near-bottom currents. Shepard found that the predominant currents flow alternately up- and down-canyon and tend to reverse in agreement with tidal and internal wave periods. In all of the canyons examined the *net* flow is downcanyon at rates of about 1 cm/sec to a few centimeters per second (Shepard and Marshall, 1973). All canyons contained a near-bottom nepheloid layer with from 0.5 to 6 mg/l of suspended solids. Using the equation for turbidity current flow developed by Middleton (1966), Drake and Gorsline (1973) found reasonable agreement between measured and computed net downcanyon currents, lending strength to the argument that turbid layer flow is important in canyons.

However, subsequent research in another California canyon has *not* supported this conclusion (Drake, 1974). In this later work, currents in Santa Cruz Canyon (a canyon with very little fine sediment supply and a coarse sand and rocky bottom) were found to be decidedly downcanyon at net rates of several centimeters per second. Yet, suspended solids concentrations were well below 0.2 mg/l near the canyon floor throughout the current measurement period. It would be impossible for such small concentrations to produce a measurable density current. Thus, the results from this canyon suggest that other mechanisms, wholly unrelated to suspended solids, were causing the net downcanyon flow.

Regardless of the origin of downcanyon flow, it is of considerable interest to compute the rate of seaward

suspended solids transport in southern California canyons (Table 5). We know the approximate rates at which fine sediment must cross the mainland shelf in order to account for measured deposition in the nearshore basins of the California borderland (Rodolfo, 1964). Assuming that the suspended solids distributions measured by Drake and Gorsline (1973) are typical of all canyons cutting the mainland shelf of southern California, and that a bottom layer 20 m in thickness flows downcanyon at an average rate of 2 cm/sec, the total annual water transport would average about 6.3 × 10⁹ m³ for each of the 11 major canyons (see Emery, 1960). Assuming this water contains an average of 1 mg/l of sediment (1 g/m³), then all 11 canyons would carry about 69,000 metric tons/year under fair-weather conditions. Emery (1960) determined that 1 to 1.5 million tons of terrigenous sediment accumulates yearly in the nearshore basins to which the 11 canyons are tributary. Therefore, only about 4 to 7% of this sediment is carried by the slow downcanyon drift of canyon bottom waters. This estimate omits storm conditions which might increase the transport by a factor of 2 or greater. It seems, however, that the bulk of fine sediment moved seaward is carried by full-scale, high-density turbidity currents in canyons or by other mechanisms operating along other pathways (see discussion in Chapter 17 by Kelling and Stanley).

TABLE 5.   Submarine Canyon Suspended Sediment Transport

Canyon Nepheloid Layer	
Average thickness	20 m
Average width (canyon axis)	500 m
Downcanyon flow rate	2 cm/sec
Annual water transport (one canyon)	6.3 × 10⁹ m³
Total canyon bottom layer flow (11 canyons)	6.9 × 10¹⁰ m³
Suspended solids concentration	1 mg/l (1 g/m³)
Total annual sediment flux	6.9 × 10⁴ tons

Source. Based on canyon morphology studies by Emery (1960); canyon current measurements by Shepard and Marshall (1973); and suspended solids studies of Drake and Gorsline (1973).

SUMMARY

Problems related to the dispersal of silt and clay have been studied for many years in rivers and estuaries and, as man's impact spreads to the continental shelf and beyond, we have become much more aware of the importance of fine particle movement in the open ocean. Small particles scavenge many pollutants ranging from heavy metals to pesticides and hydrocarbons. A wide variety of planktonic and benthonic organisms feed on detritus and, in this manner, potentially dangerous pollutants are introduced near the base of the marine food chain. In addition, abnormally high concentrations of suspended solids will limit the productivity of phytoplankton through reductions in light penetration and changes in sedimentation patterns can be disastrous for benthic organisms.

Environmental problems are present to some degree in all parts of the ocean, but nowhere are they more threatening and severe than in the highly productive coastal wetlands and shallow waters bordering the continents. Unfortunately, the continental shelf is also the most complex oceanic environment, demanding well-organized, multidisciplinary, repeated surveys in order to quantify fundamental mechanisms and processes.

Terrigenous (including pollutants), biogenic, cosmogenic, and authigenic particles comprise the sediments found in suspension or on the bottom. The last two sources are negligible on nearly all shelf areas. Terrigenous sediments are principally carried to the sea by approximately two dozen of the world's rivers; coastal erosion and atmospheric transport are only locally important. Annual production of biogenic particles exceeds terrigenous sediment supply (approximately 18×10^9 tons) by a factor of 10 or 20, yet these siliceous and calcareous particles are relatively rapidly decomposed and, generally, constitute a minor part of shelf deposits. Consequently, surface waters may contain more than 90% biogenic detritus while the bottom sediments contain less than 10% biogenic material.

Estimates of the annual deposition of silt and clay minerals in various parts of the world ocean indicate retention of the bulk of terrigenous particles near continental sources in deltaic deposits and marginal basins. Sedimentation rates in the deep sea are anomalously low. The factors that may explain this phenomenon are entrapment of sediment in estuaries and on wide shelves produced by the latest transgression; patterns of advective currents which tend to retard seaward transport of suspended particles; and flocculation and biologic aggregation of small grains into large, rapidly settling units. The continental shelf off the eastern United States receives relatively small amounts of terrigenous sediment and apparently retains at least 90% of the present-day input. Disposal of wastes in the nearshore waters on this shelf guarantees larger waste-handling problems at some future time. Unfortunately, similar, but perhaps less severe, situations are found on comparatively narrow and higher energy coastlines.

The two factors that are characteristic of fine-grained particulate matter and which tend to greatly increase shelf retention of muds are grain aggregation and interparticle cohesive forces. Microscopic studies of shelf suspended sediments almost always show a high degree of particle aggregation. Aggregates with average diameters in the medium to coarse silt range commonly contain hundreds of very fine silt and clay particles bound together by amorphous organic matter and by van der Waals forces. Typically 50 to over 90% of the suspended mineral grains are incorporated in larger aggregates. Although the densities of the aggregates probably are in the range 1.2 to 1.7 g/cm³, they will settle significantly more rapidly than single clay and fine silt particles. The increased settling rate results in a rapid loss of suspended matter from seaward-diffusing surface waters into near-bottom waters that, in many cases, appear to migrate onshore slowly (e.g., toward estuaries). Thus, size increases through aggregation will oppose shelf bypassing and allow clay deposition in comparatively "high energy" locations.

The presence of a near-bottom turbid zone even over "clean" sand shelf deposits is evidence for rapid particle settling and for resuspension of sediments. We are currently not able to model adequately the flux of sediments at the sediment-water interface because of our lack of understanding of the response of cohesive deposits to turbulent currents, our poor knowledge of biological activities, and our inability to properly measure the critical variables in the equations which describe the distribution of suspended solids.

We have only recently moved from studies of shelf deposits to studies of processes. We must continue to increase our efforts to combine shelf water circulation investigations with long-term measurements of suspended sediment distributions, mechanisms and rates of particle aggregation, and sediment-water interface dynamics. Because fine-grained particles are relatively mobile, provide food for many marine animals, and carry much of man's less noteworthy products, these studies should be among our highest priorities.

SYMBOLS

c suspended sediment concentration

$\bar{c}$ time-averaged sediment concentration

g acceleration due to gravity: $g = 980$ cm/sec²

k_x eddy diffusivity in x direction

k_y eddy diffusivity in y direction

k_z eddy diffusivity in z direction

r particle radius

t time coordinate

$\bar{u}$ mean current velocity in x direction

$\bar{v}$ mean current velocity in y direction

v_s verticle sediment diffusion rate

$\bar{w}$ mean current velocity in z direction

w_s settling velocity

δ flow depth

κ von Karman's constant: $\kappa = 0.4$

μ_* shear velocity: $\mu_* = (\tau/\rho)^{1/2}$

η fluid viscosity

τ shear stress

REFERENCES

Aston, S. R. and R. Chester (1973). The influence of suspended particles on the precipitation of iron in natural waters. *Estuarine Coastal Mar. Sci.*, **1**: 225–231.

Banse, K., C. P. Falls, and L. A. Hobson (1963). A gravimetric method for determining suspended matter in sea water using Millipore filters. *Deep-Sea Res.*, **10**: 639–642.

Barnes, P. W. (1970). Marine geology and oceanography of Santa Cruz Basin off Southern California. Unpublished Ph.D. Thesis, University of Southern California, Los Angeles, 170 pp.

Beardsley, G. F., Jr., H. Pak, K. Carder, and B. Lundgren (1970). Light scattering and suspended particles in the eastern equatorial Pacific Ocean. *J. Geophys. Res.*, **75**: 2837–2845.

Beer, R. M. and D. S. Gorsline (1971). Distribution, composition and transport of suspended sediment in Redondo submarine canyon and vicinity (California). *Mar. Geol.*, **10**: 153–175.

Biscaye, P. E. (1965). Mineralogy and sedimentation of recent deep-sea clay in the Atlantic Ocean and adjacent seas and oceans. *Bull. Geol. Soc. Am.*, **76**: 803–832.

Bond, G. C. and R. H. Meade (1966). Size distributions of mineral grains suspended in Chesapeake Bay and nearby coastal waters. *Chesapeake Sci.*, **7**: 208–212.

Bowden, K. F. (1967). Circulation and diffusion. In G. H. Lauff, ed., *Estuaries*. AAAS, Publ. 83, Washington, D.C., pp. 15–36.

Bumpus, D. F. (1965). Residual drift along the bottom on the continental shelf in the Middle Atlantic Bight area. *Limnol. Oceanogr.*, **10**: R50–53.

Buss, B. A. and K. S. Rodolfo (1972). Suspended sediments off Cape Hatteras, North Carolina. In D. J. P. Swift, D. B. Duane, and O. H. Pilkey, eds., *Shelf Sediment Transport: Process and Pattern*. Stroudsburg, Pa.: Dowden, Hutchinson & Ross, pp. 263–279.

Carder, K. L., G. F. Beardsley, and H. Pak (1971). Particle size distributions in the eastern equatorial Pacific. *J. Geophys. Res.*, **76**: 5070–5077.

Charnell, R. L., J. R. Apel, W. Manning, and R. H. Qualset (1974). Utility of ERTS-1 for coastal ocean observation: The New York Bight example. *Mar. Technical Soc.*, **8**: 42–47.

Corino, E. R. and R. S. Brodkey (1969). A visual investigation of the wall region in turbulent flow. *J. Fluid Mech.*, **50**: 233–255.

Curray, J. R. (1960). Sediments and history of Holocene transgression, continental shelf, northwest Gulf of Mexico. In F. P. Shepard, F. B. Phleger, and Tj. H. Van Andel, eds., *Recent Sediments, Northwest Gulf of Mexico*. Tulsa: Am. Assoc. Pet. Geol., pp. 221–266.

Curtiss, W. F., J. K. Culbertson, and E. B. Chase (1973). *Fluvial-Sediment Discharge to Oceans from the Conterminous United States*. U.S. Geol. Surv. Circ. 670, 17 pp.

Drake, D. E. (1971). Suspended sediment and thermal stratification in Santa Barbara Channel, California. *Deep-Sea Res.*, **18**: 763–769.

Drake, D. E. (1972). Suspended matter in Santa Barbara Channel, California. Unpublished Ph.D. Thesis, University of Southern Calitornia, Los Angeles, 357 pp.

Drake, D. E. (1974). Distribution and transport of suspended solids in submarine canyons. In R. Gibbs, ed., *Suspended Solids in Sea Water*. New York: Plenum Press, pp. 133–153.

Drake, D. E. (in press). Suspended particulate matter in the New York Bight apex, fall 1973. *J. Sediment Petrol*.

Drake, D. E. and D. S. Gorsline (1973). Distribution and transport of suspended particulate matter in Hueneme, Redondo, Newport and La Jolla submarine canyons, California. *Geol. Soc. Am. Bull.*, **84**: 3949–3968.

Drake, D. E., R. L. Kolpack, and P. J. Fischer (1972). Sediment transport on Santa Barbara–Oxnard Shelf, Santa Barbara Channel, California. In D. J. P. Swift, D. B. Duane, and O. H. Pilkey, eds., *Shelf Sediment Transport: Process and Pattern*. Stroudsburg, Pa.: Dowden, Hutchinson & Ross, pp. 307–332.

Einstein, H. A. and R. B. Krone (1962). Experiments to determine modes of cohesive sediment transport in salt water. *J. Geophys. Res.*, **67**: 1451–1561.

Emery, K. O. (1960). *The Sea Off Southern California*. New York: Wiley, 360 pp.

Emery, K. O., J. D. Milliman, and E. Uchupi (1973). Physical properties and suspended matter of surface waters in the southeastern Atlantic Ocean. *J. Sediment. Petrol.*, **43**: 822–837.

Emery, K. O., F. Lepple, L. Toner, E. Uchupi, R. H. Rioux, W. Pople, and E. M. Hurlburt (1974). Suspended matter and other water properties of surface waters of the northeastern Atlantic Ocean. *J. Sediment. Petrol.*, **44**: 1087–1110.

Felix, D. W. and D. S. Gorsline (1971). Newport submarine canyon, California: An example of the effects of shifting loci of sand supply on canyon position. *Mar. Geol.*, **10**: 177–198.

Gibbs, R. J. (1972). Amazon River estuarine system. In B. W. Nelson, ed., *Environmental Framework of Coastal Plain Estuaries*. Geol. Soc. Am. Mem. 133, pp. 85–88.

Gibbs, R., ed. (1974). *Suspended Solids in Water*. New York: Plenum Press, 320 pp.

Gilluly, J. (1955). Geologic contrasts between continents and ocean basins. *Geol. Soc. Am. Spec. Paper 62*, 7–18.

Gordon, C. M. (in press). Sediment entrainment and suspension in a turbulent tidal flow. *J. Sediment. Petrol*.

Gordon, C. M. and C. F. Dohne (1973). Some observations of turbulent flow in a tidal estuary. *J. Geophys. Res.*, **78**: 1971–1978.

Gordon, D. C. (1970). Some studies on the distribution and composition of particulate organic carbon in the North Atlantic Ocean. *Deep-Sea Res.*, **17**: 233–244.

Gorsline, D. S. and K. O. Emery (1959). Turbidity-current deposits in San Pedro and Santa Monica Basins off Southern California. *Bull. Geol. Soc. Am.*, **70**: 279–290.

Graf, W. H. (1971). *Hydraulics of Sediment Transport*. New York: McGraw-Hill, 513 pp.

Grass, A. J. (1971). Structural features of turbulent flow over smooth and rough boundaries. *J. Fluid Mech.*, **50**: 233–255.

Greer, S. A. (1969). Sedimentary mineralogy of the Hampton Harbor estuary, New Hampshire and Massachusetts. In *Coastal Environments*. N.E. Mass. and New Hampshire, Field Trip, Eastern section of Soc. of Econ. Paleontologists and Mineralogists, University of Massachusetts, pp. 403–414.

Griffin, G. M. and R. L. Ingram (1955). Clay minerals of the Neuse River estuary. *J. Sediment Petrol.*, **25**: 194–200.

Griffin, J. J. and E. D. Goldberg (1963). Clay-mineral distributions in the Pacific Ocean. Chap. 26, pp. 728–741. In M. N. Hill, ed., *The Sea*, Vol. 3. New York: Wiley-Interscience, Chapter 26, pp. 728–741, 963 pp.

Gross, M. G. (1972). Geologic aspects of waste solids and marine waste deposits, New York metropolitan region. *Geol. Soc. Am. Bull.*, **83**: 3163–3176.

Gross, M. G., B. A. Morse, and C. A. Barnes (1969). Movement of near-bottom waters on the continental shelf off the northwestern United States. *J. Geophys. Res.*, **74**: 7044–7047.

Guilcher, A. (1956). L'envasement du Rio Kapatchez et ses causes. *XVIII Congr. Int. Géog.*, Rio de Janeiro, **2**: 241–247.

Harleman, D. R. F. (1966). Pollution in estuaries. In A. T. Ippen, ed., *Estuary and Coastline Hydrodynamics*. New York: McGraw-Hill, pp. 630–647. pp.

Harrison, W., J. J. Norcross, N. A. Pore, and E. M. Stanley (1967). Circulation of shelf waters off the Chesapeake Bight. *Environ. Sci. Serv. Admin. Prof. Pap. 3*, 82 pp.

Hathaway, J. C. (1972). Regional clay mineral facies in the estuaries and continental margin of the United States East Coast. In B. W. Nelson, ed., *Environmental Framework of Coastal Plain Estuaries*. Geol. Soc. Am. Mem. 133, pp. 293–316.

Haven, D. S. and R. Morales-Alamo (1966). Aspects of biodeposition by oysters and other invertebrate filter-feeders. *Limnol. Oceanogr.*, **11**: 487–498.

Haven, D. S. and R. Morales-Alamo (1972). Biodeposition as a factor in sedimentation of fine suspended solids in estuaries. In B. W. Nelson, ed., *Environmental Framework of Coastal Plain Estuaries*. Geol. Soc. Am. Mem. 133, pp. 121–130.

Heath, G. R. (1969). Mineralogy of Cenozoic deep-sea sediments from the equatorial Pacific Ocean. *Bull. Geol. Soc. Am.*, **80**: 1997–2018.

Holeman, J. N. (1968). The sediment yield of major rivers of the world. *Water Resour. Res.*, **4**: 737–747.

Jacobs, M. D. and M. Ewing (1965). Mineralogy of particulate matter suspended in sea water. *Science*, **149**: 179–180.

Jerlov, N. G. (1968). *Optical Oceanography*. Amsterdam: Elsevier, 194 pp.

Johnson, R. G. (1974). Particulate matter at the sediment-water interface in coastal environments. *J. Mar. Res.*, **32**: 313–330.

Joseph, J. (1955). Extinction measurements to indicate distribution and transport of water masses. *Proc. UNESCO Symp. Phys. Oceanogr.*, 59–75.

Judson, S. (1968). Erosion of the land. *Am. Scientist*, **56**: 356–374.

Judson, S. and D. F. Ritter (1964). Rates of regional denudation in the United States. *J. Geophys. Res.*, **16**: 3395–3401.

Krone, R. B. (1962). *Flume Studies of the Transport of Sediment in Estuarial Shoaling Processes, Final Report*. Hydraulic Eng. Lab. and Sanitary Eng. Res. Lab., University of California, Berkeley, 110 pp.

Krone, R. B. (1972). *A Field Study of Flocculation as a Factor in Estuarial Shoaling Processes*. Tech. Bull. 19, U.S. Army Corps of Engineers, Committee on Tidal Hydraulics, 62 pp.

Krumbein, W. C. and F. P. Pettijohn (1938). *A Manual of Sedimentary Petrography*. New York: Appleton, 549 pp.

Kuenen, Ph. H. (1965). Experiments in connection with turbidity currents and clay-suspensions. In W. F. Whittard and R. Bradshaw, eds., *Submarine Geology and Geophysics*. Colston Research Society, Butterworths, London, pp. 47–71.

Kulm, L. D. and J. V. Bryne (1967). Sedimentary response to hydrography in an Oregon estuary. *Mar. Geol.*, **4**: 85–118.

Lisitzin, A. P. (1972). *Sedimentation in the World Ocean*. Soc. Econ. Paleontologists and Mineralogists Spec. Publ. 17, 218 pp.

McCave, I. N. (1970). Deposition of fine-grained suspended sediment from tidal currents. *J. Geophys. Res.*, **75**: 4151–4159.

McCave, I. N. (1972). Transport and escape of fine-grained sediment from shelf areas. In D. J. P. Swift, D. B. Duane, and O. H. Pilkey, ed., *Shelf Sediment Transport: Process and Pattern*. Stroudsburg, Pa.: Dowden, Hutchinson & Ross, pp. 225–248.

McCave, I. N. (1975). Vertical flux of particles in the ocean. *Deep-Sea Res.*, **22**: 491–502.

McManus, D. A. and C. S. Smith (1970). Turbid bottom water on the continental shelf of the northern Bering Sea. *J. Sediment. Petrol.*, **40**: 869–873.

Mairs, R. L. (1970). Oceanographic interpretation of Apollo photographs. *Photogramm. Eng.*, **36**: 1045–1058.

Manheim, F. T., J. C. Hathaway, and E. Uchupi (1972). Suspended matter in surface waters of the northern Gulf of Mexico. *Limnol. Oceanogr.*, **17**: 17–27.

Manheim, F. T., R. H. Meade, and G. C. Bond (1970). Suspended matter in surface waters of the Atlantic continental margin from Cape Cod to the Florida Keys. *Science*, **167**: 371–376.

Meade, R. H. (1969). Landward transport of bottom sediments in estuaries of the Atlantic coastal plain. *J. Sediment. Petrol.*, **39**: 222–234.

Meade, R. H. (1972). Transport and deposition of sediments in estuaries. In B. Nelson, ed., *Environmental Framework of Coastal Plain Estuaries*. Geol. Soc. Am. Mem. 133, pp. 91–120.

Meade, R. H., P. L. Sachs, F. T. Manheim, J. C. Hathaway, and D. W. Spencer (1975). Sources of suspended matter in waters of the Middle Atlantic Bight. *J. Sediment. Petrol.*, **45**: 171–188.

Menzel, D. W. and R. F. Vaccaro (1964). The measurement of dissolved and particulate organic carbon in sea water. *Limnol. Oceanogr.*, **9**: 138–142.

Middleton, G. V. (1966). Experiments in density and turbidity currents. *Can. J. Earth Sci.*, **3**: 627–637.

Milliman, J. D., C. P. Summerhayes, and H. T. Barretto (1975). Oceanography and suspended matter off the Amazon River February–March 1973. *J. Sediment. Petrol.*, **45**: 189–206.

Moore, D. G. (1969). Reflection profiling studies of the California continental borderland: Structure and Quaternary turbidite basins. *Geol. Soc. Am. Spec. Pap. 107*, 142 pp.

Moore, H. B. (1936). The muds of the Clyde Sea area. III. Chemical and physical conditions; rate of sedimentation; and infauna. *Mar. Biol. Assoc. U.K. J.*, **17**: 325–358.

Mullen, R. E., D. A. Darby, and D. L. Clark (1972). Significance of atmospheric dust and ice rafting for Arctic Ocean sediment. *Bull. Geol. Soc. Am.*, **83**: 205–212.

Nelson, B. W. (1960). Clay mineralogy of bottom sediments, Rappahannock River, Virginia. In *Clays and Clay Minerals*, Natl. Conf. on Clays and Clay Minerals, 7th. New York: Pergamon, pp. 135–147.

Nelson, B. W. (1970). Hydrography, sediment dispersal and recent historical development of the Po River Delta, Italy. In J. P. Morgan and R. H. Shaver, eds., *Deltaic Sedimentation Modern and Ancient*. Tulsa, Okla.: Soc. Econ. Paleontologists and Mineralogists, pp. 152–184.

Okubo, A. and D. W. Pritchard (1969). *Summary of Our Present Knowledge of Physical Processes of Mixing in the Ocean and Coastal Waters*. Report No. NYo-3109-40, U.S. Atomic Energy Commission, 159 pp.

Osterberg, C., A. G. Carey, and H. Curl (1964). Acceleration of sinking rates of radionuclides in the ocean. *Nature*, **200:** 1276–1277.

Partheniades, E., R. H. Cross, and A. Ayora (1969). Further results on the deposition of cohesive sediments. *Proc. 11th Conf. Coastal Eng.*, London, ASCE, **1:** 723–742.

Patten, B. C., D. K. Young, and M. Roberts (1963). *Suspended Particulate Matter in the Lower York River, Virginia, June 1961–July 1962*. Virginia Institute Marine Sci. Spec. Rep. 44, 19 pp.

Patten, B. C., D. K. Young, and M. Roberts (1966). Vertical distribution and sinking characteristics of seston in the lower York River, Virginia. *Chesapeake Sci.*, **7:** 20–29.

Petzold, T. H. and R. W. Austin (1968). *An Underwater Transmissometer for Ocean Survey Work*. Scripps Inst. Oceanogr. Technical Report, Ref. 68-9, 5 pp.

Pevear, D. R. (1972). Sources of recent nearshore marine clays, southeastern United States. In B. W. Nelson, ed., *Environmental Framework of Coastal Plain Estuaries*. Geol. Soc. Am. Mem. 133, pp. 317–335.

Pevear, D. R. and O. H. Pilkey (1966). Phosphorite in Georgia continental shelf sediments. *Bull. Geol. Soc. Am.*, **77:** 849–858.

Pierce, J. W., D. D. Nelson, and D. J. Colquhoun (1972). Mineralogy of suspended sediment off the southeastern United States. In D. J. P. Swift, D. B. Duane, and O. H. Pilkey, eds., *Shelf Sediment Transport: Process and Pattern*. Stroudsburg, Pa.: Dowden, Hutchinson & Ross, pp. 281–306.

Pilkey, O. H. and M. E. Field (1972). Onshore transportation of continental shelf sediment: Atlantic southeastern United States. In D. J. P. Swift, D. B. Duane, and O. H. Pilkey, eds., *Shelf Sediment Transport: Process and Pattern*. Stroudsburg, Pa.: Dowden, Hutchinson & Ross, pp. 429–446.

Porrenga, D. H. (1966). Clay minerals in recent sediments of the Niger Delta. In *Clays and Clay Minerals, Natl. Conf. on Clays and Clay Minerals, 14th.* New York: Pergamon, pp. 245–249.

Postma, H. (1967). Sediment transport and sedimentation in the estuarine environment. In G. H. Lauff, ed., *Estuaries*. Washington, D.C.: Am Assoc. Adv. Sci., pp. 158–179.

Powers, M. C. (1954). Clay diagenesis in the Chesapeake Bay area. In A. Swineford and N. V. Plummer, eds., *Clay and Clay Minerals, Natl. Conf. on Clays and Clay Minerals, 2nd.* Natl. Res. Council Publ. 327, pp. 68–80.

Pritchard, D. W. (1956). The dynamic structure of a coastal plain estuary. *J. Mar. Res.*, **15:** 33–42.

Prospero, J. M. and T. N. Carlson (1972). Vertical and areal distribution of Saharan dust over the western equatorial North Atlantic Ocean. *J. Geophys. Res.*, **77:** 5255–5265.

Rajcevic, B. M. (1957). Etudes des conditions de sédimentation dans l'estuaire de la Seine. *Ann. Inst. Techn. Batiment Trav. Publ.*, Paris, **117:** 745–775.

Rex, R. N. and E. D. Goldberg (1958). Quartz contents of pelagic sediments of the Pacific Ocean. *Tellus*, **10**(1): 153–159.

Rhoads, D. C. (1963). Rates of sediment reworking by *Yoldia limatula* in Buzzards Bay, Massachusetts, and Long Island Sound. *J. Sediment. Petrol.*, **33:** 723–727.

Riley, J. P. and R. Chester (1971). *Introduction to Marine Chemistry*. New York: Academic Press, 465 pp.

Rodolfo, K. S. (1964). Suspended sediment in southern California waters. Unpublished M.S. Thesis, University of Southern California, Los Angeles, 135 pp.

Rodolfo, K. S., B. A. Buss, and O. H. Pilkey (1971). Suspended sediment increase due to Hurricane Gerda in continental shelf waters off Cape Lookout, North Carolina. *J. Sediment. Petrol.*, **41:** 1121–1125.

Schubel, J. R. (1969). Size distributions of the suspended particles of the Chesapeake Bay turbidity maximum. *Netherlands J. Sea Res.*, **4:** 283–309.

Schubel, J. R. (1971). Estuarine circulation and sedimentation. In J. R. Schubel, ed., *The Estuarine Environment: Estuaries and Estuarine Sedimentation*. Am. Geol. Inst. Short Course Lecture Notes, Am. Geol. Inst., Washington, D.C. pp. VI–117.

Schubel, J. R. and A. Okubo (1972). Some comments on the dispersal of suspended sediment across continental shelves. In D. J. P. Swift, D. B. Duane, and O. H. Pilkey, eds., *Shelf Sediment Transport: Process and Pattern*. Stroudsburg, Pa.: Dowden, Hutchinson & Ross, pp. 333–346.

Schubel, J. R. and E. W. Schiemer (1972). A device for collecting in situ samples of suspended sediment for microscopic analysis. *J. Mar. Res.*, **30:** 269–273.

Scruton, P. C. (1960). Delta building and the deltaic sequence. In F. P. Shepard, F. B. Phleger, and Tj. H. Van Andel, eds., *Recent Sediments, Northwest Gulf of Mexico*. Tulsa: Am. Assoc. Pet. Geol., pp. 82–102.

Scruton, P. C. and D. G. Moore (1953). Distribution of surface turbidity off Mississippi Delta. *Bull. Am. Assoc. Pet. Geol.*, **37:** 1067–1074.

Sheldon, R. W. and W. H. Sutcliffe, Jr. (1969). Detention of marine particles by screens and filters. *Limnol. Oceanogr.*, **14:** 141–144.

Shepard, F. P. (1960). Mississippi Delta: Marginal environments, sediments and growth. In F. P. Shepard, F. B. Phleger, and Tj. H. Van Andel, eds., *Recent Sediments, Northwest Gulf of Mexico*. Tulsa, Okla.: Am. Assoc. Pet. Geol., pp. 56–81.

Shepard, F. P. and D. L. Inman (1951). Nearshore circulation. *Proc. First Conf. Coastal Eng.*, Council on Wave Research, pp. 50–59.

Shepard, F. P. and N. F. Marshall (1973). Currents along the floors of submarine canyons. *Am. Assoc. Pet. Geol.*, **54:** 244–264.

Smayda, T. J. (1969). Some measurements of the sinking rates of fecal pellets. *Limnol. Oceanogr.*, **14:** 621–626.

Smith, J. D. (in press). Modeling of sediment transport on continental shelves. In *The Sea*, Vol. 6. New York: Interscience.

Smith, J. D. and T. S. Hopkins (1972). Sediment transport on the continental shelf off of Washington and Oregon in light of recent current meter measurement. In D. J. P. Swift, D. B. Duane, and O. H. Pilkey, eds., *Shelf Sediment Transport: Process and Pattern*. Stroudsburg, Pa.: Dowden, Hutchinson & Ross, pp. 143–180.

Spencer, D. W. and P. L. Sachs (1970). Some aspects of the distribution, chemistry, and mineralogy of suspended matter in the Gulf of Maine. *Mar. Geol.*, **9:** 117–136.

Stanley, D. J., P. Fenner, and G. Kelling (1972). Currents and sediment transport at Wilmington Canyon shelfbreak, as observed by underwater television. In D. J. P. Swift, D. B. Duane, and O. H. Pilkey, eds., *Shelf Sediment Transport: Process and Pattern*. Stroudsburg, Pa.: Dowden, Hutchinson & Ross, pp. 621–644.

Sternberg, R. W. (1972). Predicting initial motion and bedload transport of sediment particles in the shallow marine environment. In D. J. P. Swift, D. B. Duane, and O. H. Pilkey, eds., *Shelf Sediment Transport: Process and Pattern*. Stroudsburg, Pa.: Dowden, Hutchinson & Ross, pp. 61–82.

Stevenson, R. E. and E. Uchupi (1969). The dispersal of suspended sediments off southeastern United States. *Geol. Soc. Am. Abstr. Programs*, **1**: 216.

Strickland, J. D. H. and T. R. Parsons (1968). *A Practical Handbook of Seawater Analysis*. Fisheries Res. Bd. Canada, Bull. 167, Ottawa, 311 pp.

Sundborg, A. (1956). The River Klarälven: A study of fluvial processes. *Geogr. Ann.*, **38**: 127–316.

Van Straaten, L. M. (1960). Transport and composition of sediments. In *Symposium Ems-Estuarium, Nordsee. Verhandel. Koninkl. Ned. Geol. Mijnbouw. Genoot. Geol. Ser.*, **19**: 279–292.

Verwey, J. (1952). On the ecology of distribution of cockle and mussel in the Dutch Wadden Sea, their role in sedimentation and the source of their food supply, with a short review of the feeding behavior of bivalve mollusks. *Arch. Néeland. Zool.*, **10**: 171–239.

Verwey, E. J. W. and J. Th. G. Overbeek (1948). *Theory of the Stability of Lyophobic Colloids*. Amsterdam: Elsevier, 205 pp.

Whitehouse, N. G., L. M. Jeffrey, and J. D. Debrecht (1960). Differential settling tendencies of clay minerals in saline waters. In A. Swineford, ed., *Clay and Clay Minerals*. Oxford: Pergamon, pp. 1–79.

Windom, H. L. (1968). Atmospheric dust records in glacial snowfields: Applications to marine sedimentation. Ph.D. Dissertation, California University, Scripps Institution of Oceanography, San Diego, 105 pp.

Windom, H. L. (1975). Eolian contributions to marine sediments. *J. Sediment. Petrol.*, **45**: 520–529.

CHAPTER 10

Substrate Response to Hydraulic Process:
Grain-Size Frequency Distributions and Bed Forms

DONALD J. P. SWIFT

Atlantic Oceanographic and Meteorological Laboratories, Miami, Florida

JOHN C. LUDWICK

Institute of Oceanography, Old Dominion University, Norfolk, Virginia

Chapters 8 and 9 dealt with the entrainment of sand and mud, respectively, on the continental shelf. In addition, Chapter 8 discussed the most ubiquitous shelf bed form, the sand ripple formed by bottom wave surge, since it plays a critical role in the entrainment and transport of sand on the continental shelf.

This chapter explores further the response of the shelf floor to the hydraulic climate. Two key responses that are used to infer regional patterns of sediment transport are grain-size frequency distributions and substrate bed forms. The chapter also describes a numerical model for estimating sediment transport and areas and rates of erosion and deposition.

GRAIN-SIZE FREQUENCY DISTRIBUTIONS

Krumbein (1934) was the first to bring to popular attention the concept that the size frequency distribution of sand samples tends to be log-normally distributed. It has become a tenet of conventional wisdom that this distribution, as defined by its mean and standard distribution, is the signature of the depositional event, and that deviations from log normality, as measured in terms of

standard deviation, skewness, and kurtosis, reflect both the provenance and subsequent hydraulic history of the sediment (see Inman, 1949; Friedman, 1961; Visher, 1969).

Genesis of the Normal Curve

Recent theoretical studies (Middleton, 1968; Swift et al., 1972b) have attempted to present this hypothesis in a more rigorous manner by consideration of probability theory. The reader is referred to these papers for the mathematical foundation of the following discussion.

The probability model for the genesis of a log-normally distributed grain population considers a flow over a sand substrate in which the total load is adjusted to flow conditions. If deposition is to occur, there must be a decrease in bottom shear stress $(-\partial \tau_0/\partial x)$ and discharge $(-\partial q/\partial x)$ down the transport path. The distribution of grain sizes in the load undergoing transport down this shear stress gradient and the absolute value of the gradient is such that for each grain-size class, an upstream portion of the path is experiencing supercritical stress, and a downstream portion is experiencing subcritical stress. We are concerned with the central portion of the transport path, where a series of transition points for critical shear stress occur, with each successive downstream transition point being associated with a successively finer grain-size class. The grains are assumed to travel down the transport path in a series of discrete hops as a consequence of the turbulent structure of the flow, and as a consequence of a larger scale cycle of flow events separated by periods of quiescence. The model is thus a stochastic model, with an inherently random aspect to its behavior, and the problem may be dealt with in terms of probability theory.

Under these conditions, it is conceptually possible to define the grain-size frequency distribution at each point as the product of two probability vectors, an admittance vector and a retention vector (Fig. 1). The admittance vector is the sequence of probabilities of entrance of the size classes present, ordered in sequence of decreasing grain size. The retention vector is similarly the sequence of probabilities of retention of successively finer grain sizes.

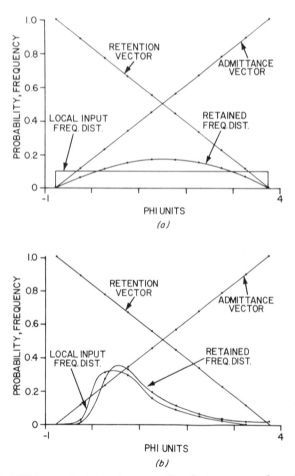

FIGURE 1. *Grain-size frequency distributions as a product of a retention vector and an admittance vector. See text for explanation. From Swift et al. (1972b).*

If P_{jn} is an element in an admittance vector, where j denotes the jth station in the transport path and n denotes one of n grain-size classes, and if P'_{jn} is a corresponding element in a retention vector for the same station, then the product of the two probabilities, $P_{jn}(1 - P'_{jn})$ gives the probability that the particle in the local input enters but does not leave the station. The product of the input vector with all corresponding elements in the admittance and retention vectors for a station gives the frequency distribution for that station (Fig. 1). This is a restatement, in probabilistic terms, of the intuitively apparent fact that the modal diameter of a deposit is that grain size most likely to arrive and least likely to be carried away from the place of deposition under prevailing flow conditions; progressively coarser sizes are progressively less frequent because they are less likely to arrive, and progressively finer sizes are progressively less frequent because they are more likely to be carried away.

In Fig. 1a, the two linear numerical filters (admittance and retention vectors) are applied to a local input frequency distribution that is uniform in nature and a symmetrical retained frequency distribution results. If, however, the local input has a skewed distribution (Fig. 1b), then the retained distribution is still skewed, although it has been modified by the station probabilities. If the filters are not linear, then further modification of the input vector occurs.

In Fig. 2, various hypothetical input distributions are subjected to sorting down the stations of a hypothetical transport path according to the probabilistic algorithm described above. In column A, an initially rectangular distribution is seen to evolve into a distribution with a distinct mode, and the mode is seen to shift toward the finer end of the distribution at successive stations. The coarse flank of the mode becomes visibly sigmoid (S-shaped) as is characteristic of the side of the normal distribution frequency curve. The increasingly sigmoid shape is the consequence of the multiplication of successive admittance vectors in order to obtain the coarse admixture of the local input frequency distribution. For instance, if the admittance vector has the form

$$0.100, \quad 0.200, \quad 0.300, \quad 0.400, \quad 0.500, \quad 0.600,$$
$$0.700, \quad 0.800, \quad 0.900, \quad 1.00$$

and retains this form from station to station, then at the third station the size frequency distribution of the coarse admixture will be determined primarily by the third power of the admittance vector,

$$0.001, \quad 0.008, \quad 0.027, \quad 0.064, \quad 0.125, \quad 0.216,$$
$$0.343, \quad 0.512, \quad 0.729, \quad 1.00$$

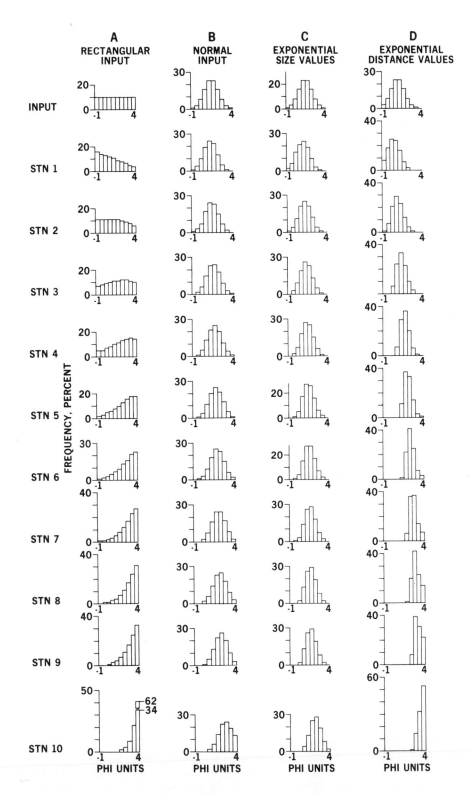

FIGURE 2. *Grain-size frequency distributions along sediment transport paths under different conditions. See text for explanation. From Swift et al.* (1972b).

The initially small probabilities have decreased more than the initially large ones, and the resulting curve of frequency against size class will be exponential in form.

The modal shift is the phenomenon of *progressive sorting* (Russell, 1939) whereby the deposit becomes finer down the transport path, as a consequence of the steady depletion of the transported material in coarser particles. In physical terms, this means that the coarsest particles tend to get left behind whenever the bottom is eroded by a flow event that is weaker than the one that preceded it.

If it is assumed that the input distribution is normal to begin with (Fig. 2, column *B*), and if it is assumed that the probabilities of admittance and retention vary linearly with grain size, then the mode shifts toward the fine end of the distribution as the sediment is traced down the transport path with no change in the shape of the normal curve. However, in a more realistic case, the probabilities of admittance and retention are assumed to vary exponentially with grain size; in other words, the transport rate varies exponentially with grain size. As a consequence, vector multiplication acts on the two sides of the frequency curve in a dissimilar fashion (column *C*). The greater range of transport probabilities assigned to the coarser sands results in greater efficiency of sorting on that side of the curve, and progressive steepening of that side, as the sediment is traced down the transport path. The sediment becomes increasingly enriched in the fine admixture at the expense of the coarse admixture (becomes fine-skewed), as well as becoming finer down the transport path. This effect is particularly marked where the intensity of the flow field is made to decrease down the transport path (column *D*).

Size Frequency Subpopulations and Flow Regimes

Thus there are at least theoretical reasons supporting the concept that the size frequency distribution of fluid-deposited sands constitutes hydraulic signatures of the flow process. Attempts to interpret these signatures have in general generated more heat than light (Emery and Uchupi, 1972, p. 375). However, the analysis of the subpopulations constituting sand samples has proved more fruitful. The basic work has been undertaken by Moss (1962, 1963, 1972). He notes that most grain-size frequency distributions of sand deposited from fluid flow do not plot as a straight line on probability paper as they should if they are normally distributed. Instead the curves are Z-shaped (Fig. 3). He has demonstrated that these Z-shaped curves are composite distributions and are the consequence of the presence of three or more log-normally distributed subpopulations, and that these subpopulations are an outcome of the manner in which

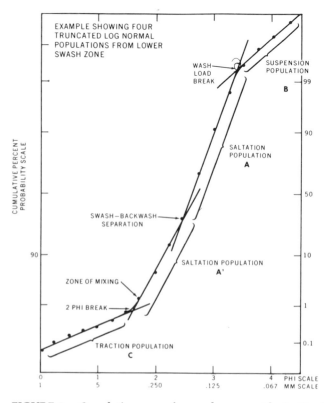

FIGURE 3. *Cumulative curve of a swash zone sample. In Moss' terminology, A is framework population, B is interstitial population, and C is contact population. From Visher (1969).*

the bed is built (Fig. 4). A *framework population* (A population) constitutes the bulk of the sample. Its modal diameter is a function of the average dimensions of the relatively large spaces between grains on the aggrading surface. There is a strong feedback in this system between deposit grain size and bed load grain size; the dimensions of grains selected from bed load for deposition in such holes depend on the dimensions of grains already deposited, which in turn depend on the dimensions of available grains in the bed load, and ultimately on the dimensions of the hydraulic parameters of the flow.

A fine *interstitial subpopulation* (B population: fine tail of the size frequency distribution curve) consists of grains that are small enough to filter into the interstices of the grain framework of the deposit. Their average diameter is not that of the bowl-shaped openings on the bed surface but the smaller average diameter of the interstices within the deposit.

A coarse *contact subpopulation* (C population: coarse tail of the frequency curve) consists of grains that are too coarse to fit into or through the surface openings as do the grains of the A and B populations. Instead they accumulate as slowly moving to stationary clogs of mutually interfering coarse grains on the bed surface. When a critical area of these rejected coarse particles has ac-

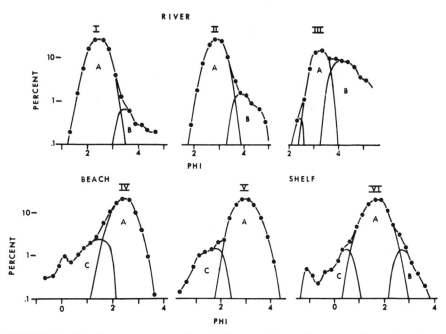

FIGURE 4. *Size-frequency curves of sands from various environments. Curves have been dissected to indicate subpopulations. From McKinney and Friedman (1970).*

cumulated, it will be buried beneath further layers of A population grains.

It is important to assess the relationship between Moss' rather sophisticated theory of subpopulation genesis, and the prevailing equation of transport modes with subpopulation characteristics. It has been generally assumed [see the review by Visher (1969)] that the contact population represents particles moved by dragging or rolling, the framework population represents particles moving by saltation, and the interstitial population represents particles traveling in suspension. There *is* a correlation between these differing modes of transport and the per-

centage of respective subpopulations in the deposit, because each of these modes is most likely to carry the appropriate size of material for the subpopulations with which they have been correlated. The relative percentages of subpopulations, however, are a *direct* consequence of mechanisms of bed construction, and only indirectly reflect modes of transport. Moss has shown, for instance, that both B and A subpopulations may be generated from saltative transport alone.

The percentages of these three populations in a given deposit will vary, within limits set by grain geometry and grain interaction processes, according to the regional

TABLE 1. Nomenclature and Grain-Size Characteristics of Sediment Flow Regimes

Moss (1972)	Southard and Boguchwal (1973)	Population			Mean Diameter (Moss, 1972)
		A	B	C	
Fine ripple stage	Ripples	Dominant	Abundant	Scarce	0.07–0.25 mm (3.75–2.00ϕ)
Coarse ripple stage	Ripples	Dominant	Scarce	Scarce	0.25–0.92 mm (2.00–0.25ϕ)
Dune stage	Dunes	Dominant	Scarce	Scarce	0.25–2.2 mm (2.0 to −1.1ϕ)
Rheologic stage	Transition Upper flat bed Antidunes	Dominant	Abundant	Abundant	0.17–4.8 mm (2.6 to −2.3ϕ)

availability of the three populations, and also according to the hydraulic microclimate of the bed. Moss (1972), on the basis of flume studies and studies of river deposits, has described five bed regimes. These may be correlated with the flow regimes described by Southard and Boguchwal (1973, Fig. 23). Each tends to form a characteristic admixture of subpopulations (Table 1). Moss (1972) notes that in the *fine ripple* stage, grains do not protrude through the laminar sublayer of the bottom boundary layer of the flow and microturbulence is absent from the bed surface. Fine particles can become concentrated near the bed, and can pass copiously into the interstices. Hence the fine ripple regime is characterized by an abundant B population.

In the *coarse ripple stage* and *dune stage*, grains protrude through the lamina sublayer. Fluid dynamic lift and bed grain turbulence operate to keep fine particles from being concentrated near the bed, and the interstitial (B) population is normally a minor bed constituent.

In the *rheologic stage*, flow is supercritical, and bed load particle behavior is dominated by the dispersive pressure associated with grain collisions (Bagnold, 1954). These pressures force the particles against and into the bed. This effect is evidently dominant over the lift forces which act at the bed, and the interstitial B population again passes copiously into the bed. The rheologic stage is furthermore the only stage in which Moss observed an abundant contact (C) population.

Moss' theory may thus be used to infer flow regime from the grain-size distribution. It must be applied with caution, however, as it was developed for quasi-steady flows, and the continental margin environment tends to be subjected to an additional oscillatory flow component because of wave surge. Grain-size distributions consequently tend to indicate more intense unidirectional flows than actually exist (Stubblefield et al., 1975).

BED FORMS

In this section it is necessary to deal with more varied and larger bed forms than the wave ripples described in Chapter 7. Sand wave fields and sand ridge fields may generate bed form spacings of a kilometer or more, and bed form amplitudes of up to 30 m. Such large-scale bed form arrays become significant storage elements in continental margin sediment budgets, and such budgets cannot be understood without an awareness of bed form mechanics. Furthermore, large-scale bed forms impact directly on human usage of the continental margin. Large tankers navigate the Thames estuary channels (Langhorne, 1973) with scant meters of clearance over sand wave crests. Sewage outfalls and nuclear power

plants are planned or are being constructed in the inner shelf ridge fields of the Atlantic shelf. Seafloor well heads are subject to burial by migrating bed forms.

CONCEPTS. A bed form is an irregularity in the particulate substrate of a fluid flow. This definition includes the subaqueous sand wave and sand ridge fields of the earth's shelves, the subaerial dune fields of the earth's deserts and those photographed on Mars, and the bed forms of the base surge deposits surrounding the lunar craters, sedimented out of a transient fluid of gas, dust, and debris generated by the impact of meteors. Bed forms are not independent phenomena; they are equilibrium configurations of the interface between a mobile, usually cohesionless substrate, and an overlying flow field, and tend to occur in repetitive arrays rather than alone. They are the product of feedback between flow structure and substrate structure. The three-dimensional pattern of flow does not "cause" the bed form to arise, nor does the bed form "cause" the deformation of the boundary layer of the flow field; instead, strictly speaking, these two elements of a flow-substrate system interact to cause each other.

Wilson (1972, p. 204) notes that when a fluid is sheared, either against another fluid, against itself, or against a rigid boundary, there are many situations in which secondary flows develop. Secondary flows are regularly repeated patterns of velocity variation superimposed on the mean flow. The primary flows satisfy the three continuity laws (of mass, energy, and momentum), but in such a way that any small disturbance is initially self-aggravating; in other words, the flow is an unstable system. In sheared flows, this usually involves the development of any combination of such secondary flows as transverse internal waves, or transverse or flow-parallel vortices. Such secondary flows may occur simultaneously at several scales.

Wilson further notes that sheared fluids may become unstable in response to almost any sort of strong gradient in velocity, pressure, viscosity, temperature, or density in the direction normal to the shear force. These may arise over completely plane beds. Eventually, however, as the perturbed flow and the bed deform in response to each other, a new stable state is attained.

The theory of fluid instability has been outlined by Lin (1955), Chandrasekhar (1961), Rosenhead (1963), and Yih (1965), and these authors have discussed many cases to which it has been applied. Allen (1968a, p. 50) has summarized their computational approach. The algorithm requires that equations of motion be set up to describe the fluid motion of interest. These equations are solved to discover whether a small sinusoidal disturbance of one variable will be damped or amplified under the chosen limits for other variables. The motion is stable if the disturbance is damped, but unstable if it is amplified. In nature the unstable disturbance is amplified until the other variables of the system set some limiting condition on the amplification and a new state of quasi equilibrium is attained. Stability analysis has been successfully applied to the problem of ripple and sand wave formation (e.g., Smith, 1969) and it seems likely that all bed forms will ultimately prove susceptible to this mode of attack.

BASIC MODES OF BED FORM BEHAVIOR. Most bed forms fall into two basic categories: those that are oriented across the flow direction, such as sand waves and ripples, and those that are oriented parallel to the flow direction, such as sand ribbons. These two basic patterns must correspond to two basic patterns within the flow field itself, a transverse pattern in which zones of scour and aggradation alternate down the flow path, and one in which zones of scour and deposition alternate across the flow path (Figs. 5 and 6). There is considerable evidence to indicate that this is the case, although the basic mechanisms are far from clear.

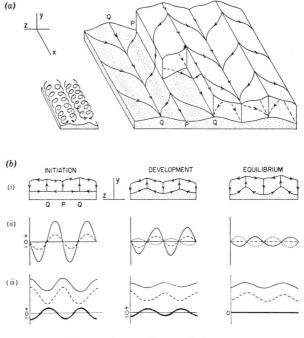

FIGURE 6. *The development of a longitudinal bed form.* (a) *The pattern of secondary flow over longitudinal bed form elements: PP, flow attachment lines along ridge trough; QQ, flow separation lines along crests.* (b) *Development of longitudinal elements in vertical cross section perpendicular to mean flow direction.* (i) *Form and flow components;* (ii) *components in z direction;* (iii) *components in x direction. Numbers as in Fig. 4. Note alternate notation of coordinate axes. From Wilson (1972).*

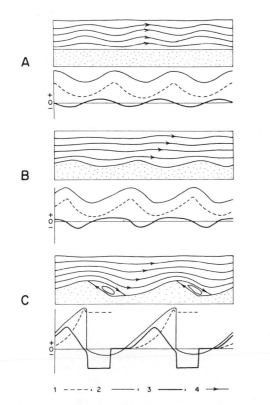

FIGURE 5. *The development of a transverse bed form.* (A) *Initiation;* (B) *growth;* (C) *equilibrium.* (1) *Sand transport rate;* (2) *shear velocity at bed;* (3) *erosion rate;* (4) *streamlines. From Wilson (1972).*

Transverse Bed Forms

MODE OF FORMATION. As noted by Wilson (1972), most transverse bed forms are probably caused by transverse wave perturbations in the flow. The problem is a complex one, and the solutions offered to date have not been altogether satisfactory. Summaries are presented by Allen (1968a, pp. 130–149), Kennedy (1969, p. 151), and Smith (1970, p. 5928).

Smith points out that many of these studies are unnecessarily restrictive; they assume an eddy viscous mean flow but neglect the inertial terms in the equation of motion (Exner, 1925, in Raudkivi, 1967) or assume inviscid irrotational flow (Kennedy, 1969). These assumptions require an a priori phase shift in the velocity field relative to the interface disturbance in order for instability to occur. Smith (1970) has undertaken a stability analysis employing inertial terms in the equations of motion. His results indicate that the interface is unstable with respect to infinitesimal perturbations of wavelength greater than the wavelength for which the inertia of the grains is important (wavelengths less than 10 times mean grain diameter). Smith utilizes the sediment continuity equation, which may be presented in its simplest two-dimensional form as

$$\frac{\partial \eta}{\partial t} = -\kappa \frac{\partial q}{\partial x} \qquad (1)$$

where η is height of the interface above a datum, t is time, κ is a constant, q is sediment discharge at a level near the bed, and x is horizontal distance. In physical terms, the time rate of change of the height of the interface at a point above the datum is proportional to the horizontal discharge gradient at that point, assuming saturation of the boundary flow with sediment; a decrease in discharge across the point $(-\partial q/\partial x)$ must result in aggradation, while an increase $(\partial q/\partial x)$ must result in erosion. Smith has rewritten the equation in terms of boundary shear stress and discharge:

$$\frac{\partial \eta}{\partial t} = -\left(\frac{1}{c_0}\frac{\partial q}{\partial \tau_0}\right)\frac{\partial \tau_0}{\partial x} \qquad (2)$$

where c_0 is the boundary concentration of sediment, q is the mean volume flux of sediment per unit width, and τ_0 is the local mean shear stress on the bed.

Smith's analysis divides the nonuniform horizontal velocity along the waveform interface into an in-phase component and an out-of-phase component. The in-phase component consists of accelerating flow over crests with maximum shear stress at those points, as required by flow continuity. Since boundary erosion varies directly with $\partial \tau_0/\partial x$, this in-phase component simply causes upstream erosion of the interface perturbation and downstream deposition; the perturbation moves downstream with neither growth nor decay. However, the inertia of the high-velocity water of the upper part of the water column causes it to converge with the rising bed on the upstream side of an interface perturbation, and there is as a consequence an out-of-phase component

of velocity and bottom shear stress which attains its maximum value at this zone. This maximum persists when the components are added; hence, since deposition is proportional to $-\partial \tau_0/\partial x$, some sand must always be deposited on the crest of the perturbation.

Smith (1970) cites Exner's (1925) earlier stability analysis as qualitatively correct, despite neglect of the inertial terms in the momentum equations. Exner had shown that when downcurrent spacing is wide, the crests of perturbations move faster than the troughs between them, resulting in oversteeping of the downcurrent slope to the angle of repose (30° underwater), and consequently, in the formation of a horizontal roller eddy (wake, flow separation bubble) downstream of the crest. The perturbation is now a mature ripple or sand wave.

The generation of a wake behind a growing bed form results in propagation of interface instability in the downstream direction. Smith (1970) cites Schlichting (1962, p. 200) who has studied the development of a turbulent wake. Behind a negative step such as the avalanche slope of a growing transverse bed form, flow accelerates downstream of the attachment line (Fig. 7) and at the same time a boundary layer is initiated that grows in height downstream. Shear increases downstream because of flow acceleration, then decreases as the effects of boundary layer growth dominate over the effects of flow acceleration. Here, in a zone where $\partial \tau_0/\partial x < 0$, sand is deposited and a new ripple grows, which in turn deforms, develops a wake, and triggers a third. Smith's stability analysis does not specify wavelength for growing bed form perturbations, and it is apparent that this parameter must be defined by spatial adjustments in the turbulent velocity field. As downstream ripples grow in height and their separation bubbles in width, they must grow in length, which is accomplished by the smaller ripples moving faster, and stretching out the ripple field.

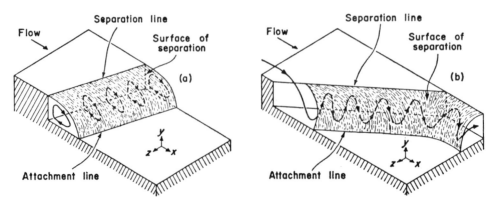

FIGURE 7. *Three-dimensional separated flows.* (a) *Roller;* (b) *vortex.* **Note** *alternative notation of coordinate axes. From Allen (1970).*

Smith's scheme of transverse bed form formation by the spontaneous deformation of the interface into a moundlike perturbation, its increasing asymmetry, the formation of a separation bubble, and the downstream propagation of the instability, has been strikingly confirmed in experimental work by Southard and Dingler (1971). Their work suggests that if the critical flow threshold is approached slowly, preexisting bed irregularities may trigger downstream ripples in the interval of metastability before the threshold is attained. However, if the threshold is passed rapidly, or if marked preexisting irregularities do not occur, mounds will spontaneously appear and transform themselves into regular ripples.

Other schemes for the formation of transverse bed forms have been proposed, in which the wavelike perturbation of flow precedes bed deformation, rather than arising from interaction with the bed. Cartwright (1959) has proposed that the shelf-edge sandwave field of La Chapelle Bank in the Celtic Sea are responses to stationary internal waves (tidal lee waves) in the stratified water column. Furnes (1974) has analyzed the formation of sand waves in response to internal waves of a fluid whose density stratification is a consequence of its suspended load. While compatible with the field evidence, these models for sand wave formation remain unconfirmed. They are important contributions, however, if only in that they reduce the bias toward the results of experimental laboratory work. The space and time scales and the internal structure of shelf flows are qualitatively different than those of laboratory flumes and there is no reason to assume that such further modes of sand wave formation do not exist in nature.

TYPES OF TRANSVERSE BED FORMS.   Field and laboratory observations show that there tend to be two overlapping populations of transverse bed forms: *ripples*, with wavelengths up to 0.6 m, and *sand waves*, with wavelengths in excess of 0.6 m (Fig. 8). Sand waves commonly bear ripples on their backs. The two populations appear to be responses to two distinct genetic mechanisms. As small forms grow up through the velocity gradient of the boundary layer, the zone of maximum stress on the upcurrent flank shifts to the crest, at which point the entire upcurrent slope is erosional and the lee slope depositional; upward growth is stopped, and the ripple migrates at constant speed (Wilson, 1972, p. 200). Transverse bed forms of larger wavelengths are insensitive to the boundary layer velocity gradient and their upflank zone of maximum shear stress shifts to the crest only when the whole flow is significantly deformed by their upward growth. As a consequence, the equilibrium height of sand waves in shallow flows is proportional to flow depth, while the equilibrium height of ripples is

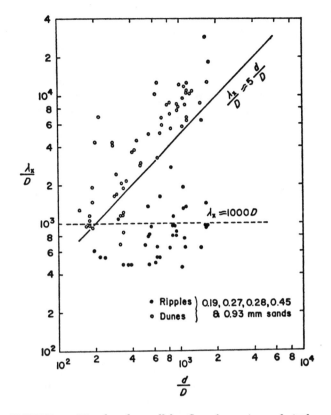

FIGURE 8.   *Wavelength parallel to flow of experimental ripples and sand waves in relation to flow depth. Data of Guy et al. (1966). From Allen (1970).*

depth independent for all flow depths (Allen, 1967); see Fig. 8. Expressions for the equilibrium heights of ripples and sand waves have been considered by Kennedy and used by McCave (1971).

Stride (1970) has plotted measurements of height versus depth for sand wave fields of the North Sea at depths of 90 to 60 m, and found no correlation. Large bed forms grow slowly, and equilibrium heights may be rarely obtained in such shallow tidal seas subject to strong periodic storm surges. Deep-sea sand waves (Lonsdale and Malfait, 1974) can obviously never equilibrate with total water depth, although the significant flow depth may be only a small fraction of total depth, because of density stratification.

The distinction between small- and large-scale bed forms may be due to more than interaction of wavelength with the velocity gradient. Kennedy (1964) has suggested that small transverse bed forms represent perturbations of the traction and saltation loads that move very near to or on the bed, and hence must react quickly to changes of flow speed. Larger transverse bed forms, on the other hand, could reflect a perturbation of the suspended load, which will tend to respond slowly, and therefore over a large distance to a change in flow speed.

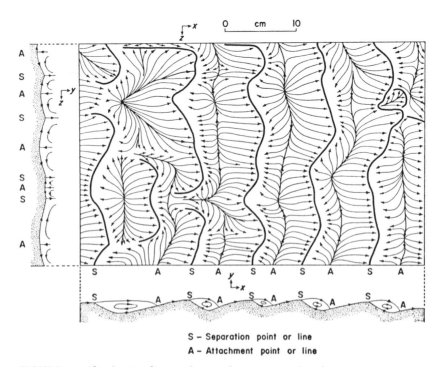

S – Separation point or line
A – Attachment point or line

FIGURE 9. *Skin friction lines and streamlines associated with a portion of a bed of experimental ripples in fine-grained quartz sand. Mean flow velocity 22 cm/sec from left to right. Mean flow depth 4.5 cm. Note alternative notation of coordinate axes. From Allen (1970).*

In continental margin sand wave fields, there are often three orders of transverse bed forms: current ripples, sand waves, and larger sand waves. McCave (1971) suggests that the two classes of sand waves may be the consequence of Kennedy's two categories of substrate response.

Because of the turbulent diffusion of sand normal to the flow direction, an initially equant interface perturbation will tend to extend itself normal to flow, hence the quasi-two-dimensional nature of ripples and sand waves (ripple profile does not change down the length of the ripple crest). However, at increasing values of mean velocity, and therefore of turbulent instantaneous velocity component, transverse bed forms tend to become three-dimensional (Znamenskaya, 1965). As crests become locally inclined to the mean flow direction, the horizontal "roller eddy" of the separation bubble becomes a horizontal helical vortex (Fig. 7) and irregular patterns of skin flow result (Fig. 9). Under yet more intense flows, the irregularities may take on ordered patterns (Fig. 10). Bagnold (1956) attributes one particularly common pattern, that of the lingoid ripple, to "... the partial diversion of grain flow ... and its funneling into channels between existing ripples; deposition (of a new lingoid ripple) would take place immediately downstream of such a funnel." A diagonal or diamondlike pattern of lingoid ripples results.

TRANSVERSE BED FORMS AND FLOW REGIMES. It has long been known that as a shallow flow over a noncohesive substrate intensifies, a sequence of bed configurations transpires (Simons et al., 1961; Simons and Richardson, 1963; Guy et al., 1966). The flow variables governing this sequence are h, depth of flow; $\bar{u}$, mean velocity of flow; ρ, density of fluid; ρ_s, density of sediment; μ, viscosity of fluid; and D, mean diameter of sediment.

The critical parameters are fluid power (proportional to $\bar{u}^3$; see Chapter 8) and grain size (Fig. 11). Grain density is variable to the extent that heavy minerals may be present; and fluid density and viscosity vary somewhat with temperature and salinity. Flow depth determines whether or not the flow is subcritical or supercritical as expressed by the dimensionless Froude number $F = \bar{u}/(gh)^{1/2}$, where $(gh)^{1/2}$ is the celerity of a shallow water wave. In supercritical flows ($F > 1$), surface waves couple with substrate perturbations (antidunes) that tend to migrate upcurrent. On the continental margin supercritical flows are confined to the swash and breakpoint zones of the surf, and to tidal flats; and the antidunes and rhomboid ripples that form in these zones are ephemeral.

Southard (1971) and Southard and Boguchwal (1973) have argued that bed configuration diagrams such as Fig. 11 should be presented in terms of dimensionless depth, velocity, and grain size to eliminate the overlap-

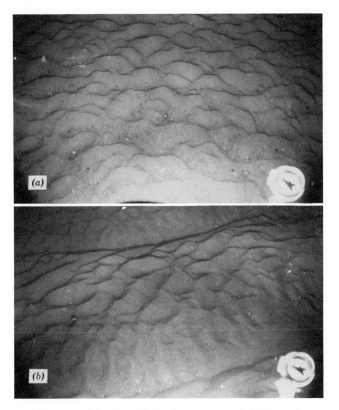

FIGURE 10. (a) *Lingoid ripple pattern on shelf floor off Cape Hatteras, North Carolina.* (b) *Lingoid ripples on back of sand wave, straight ripples in trough, same area.*

ping of fields that occurs in diagrams utilizing fluid power or bed shear stress.

Figure 11 shows that dunes (sand waves) occur at higher values of fluid power than do ripples by themselves. This fact is consonant with Kennedy's suggestion that sand wave formation involves suspended load transport, which requires higher values of fluid power than does bed load transport.

TRANSVERSE BED FORMS AND TIDAL FLOWS. Tidal flows, which reverse every 6 hours, generate transverse bed forms in a cohesionless substrate. Tidal current ripples are no different than ripples generated by unidirectional currents, except that their sense of asymmetry is reversed as the tide changes. Small sand waves (height of 1 m or less) may have their asymmetries partly or wholly reversed by strong reversing tidal currents (Klein, 1970). Larger sand waves tend to display a time-integrated response to reversing tidal flows, maintaining an ebb or flood asymmetry in accord with the dominant flow component residual to the semidiurnal cycle. "Cat-backed" sand waves are large sand waves that have a sloping upcurrent side, a flat top, and (in profile) an "ear" perched on the edge of the downcurrent slope (Van Veen,

1936). The ear is a response to the subordinate portion of the tidal cycle. Tide-formed sand waves in areas of equal ebb and flood flow are commonly symmetrical.

As distance from shore increases, the tidal current is no longer reversing but rotary (Chapter 5). The advent of midtide cross flow tends to inhibit the formation of sand waves large enough to survive through the tidal cycle (McCave, 1971). Under such circumstances longitudinal bed forms are favored (Smith, 1969).

Longitudinal Bed Forms

Wilson (1972) comments that practically all longitudinal bed form elements, whether formed in wind or water, are initiated by regular helical vortices with axes parallel to flow. His reasons for his admittedly sweeping assertion are as follows:

1. Longitudinal helical flow cells occur in many different kinds of situations. They are the only kind of flow perturbations known to fluid mechanics whose wavelength is measured normal to the mean flow direction.

2. With the exception of alternating parallel lanes of fast and slow flow, the double helical pattern is the only one that meets the theoretical requirements, namely bilateral symmetry parallel to flow, regular repetition normal to flow, and conformity with the law of continuity.

3. Many investigations of flow over longitudinal bed forms resulted in some evidence for the occurrence of helical flow over the longitudinal elements, for instance, model ripples and dunes (Allen, 1968a); in river channels (Gibson, 1909); over tidal sand ridges (Houbolt, 1968); and over desert dunes (Hanna, 1969).

The theory of longitudinal flow perturbation is less well developed than the theory of transverse flow perturbation. Such perturbations are not as obvious in laboratory flumes as transverse (streamwise) flow perturbations, and many occur at scales far beyond those of laboratory flumes. As in the case of transverse bed forms, longitudinal bed forms appear to be able to form in response to perturbations of boundary flow, or in response to perturbations of the whole flow field. As in the case of transverse bed forms, they appear to form during the course of flow-substrate interaction, and also in response to the preexisting internal structure of a sheared flow.

Preexisting flow structures appear to be more important than in the case of transverse perturbations. Perhaps the most general statement that can be made is that in a sheared flow that is wide relative to its depth, a significant portion of flow energy must be diverted to an ordered secondary flow component, in order to maintain lateral flow continuity. At least three basic varieties of such secondary flow structure exist.

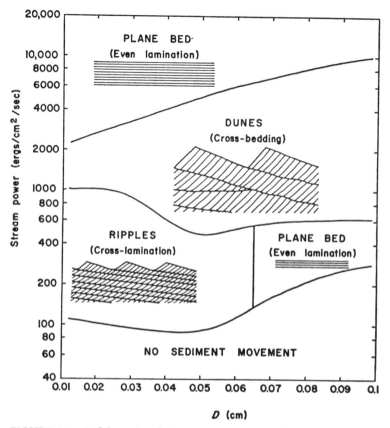

FIGURE 11. *Bed forms in relation to stream power and grain size. Data of G. P. Williams, H. P. Guy, D. B. Simons, and E. V. Richardson. From Allen (1970).*

MICROSCALE LONGITUDINAL BED FORMS (PARTING LINEATION). It has been repeatedly suggested that the logarithmic boundary layer tends to be so patterned, although an adequate analytic model has not yet been devised (Schlichting, 1962, pp. 500–509). Kline (1967) and Kline et al. (1967) have conducted dye experiments in flumes which suggest that the laminar sublayer and the lower part of the buffer sublayer of the turbulent boundary layer have a structure characterized by vigorous transverse components of flow (see discussion, p. 94). Dye introduced into the boundary layer forms into bands that are more slowly moving than those in the intervening water zones. Although the streaks are randomly generated, they have a mean transverse spacing of $\lambda_z = 100\nu/u_*$ in which ν is the kinematic viscosity and u_* is the shear velocity (Kline, 1967). The response to helically structured boundary flow over a cohesionless particulate substrate is, however, well known; it is the ubiquitous parting lineation (Sorby, 1859), so named for the tendency of flagstones (silty sandstones with strong bedding fissility) to exhibit lineations on bedding planes. Closer examination reveals a waveform bedding surface whose undulations parallel flow direction; ridges are a few grain diameters high and are up to several

centimeters apart (Allen, 1964; 1968a, pp. 31–32); see Fig. 12. There is clear evidence for the divergence and convergence of bottom flow in that the azimuths of long grains are bimodal, although this evidence does not resolve the secondary flow pattern. A similar structure has been reported from mud beds (Allen, 1969). Here the notches are frequently narrower than the ridges.

Coupling probably occurs between bed and flow structure, in that the grain ridges localize flow cells. Also, the sand of the ridges is coarser (Allen, 1964) and the resulting roughness would tend to slow crestal flow. This feature would cause downstream growth in the retarded wake of the grain ridges, and would perhaps induce upward ridge growth until ridge crests reach a level whose flow is rapid enough to counteract growth.

MESOSCALE LONGITUDINAL BED FORMS (CURRENT LINEATIONS). "Current lineations" (McKinney et al. 1974) is a generic term for low-amplitude strips of sand resting on a coarser substrate (sand ribbons) and for strips of coarse sand or gravel flooring and elongate depression of slight depth (longitudinal furrows). Current lineations are a larger scale of longitudinal bed form, with spacings ranging from a few meters to many hun-

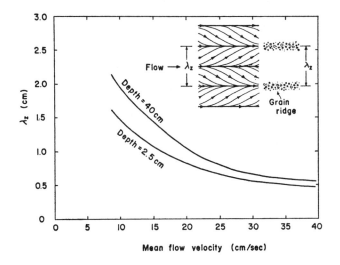

FIGURE 12. *The mean transverse spacing of parting lineations as a function of mean flow velocity and flow depth. From Allen (1970).*

dreds of meters (Allen, 1968a). They are best observed by means of sidescan sonar (Figs. 13 and 14). The large-scale patterns are characteristic of shelves with strong tidal flows (Kenyon, 1970); see Fig. 15. Sand ribbons and longitudinal furrows are probably the most common mesoscale bed form on the continental shelf, being widely distributed on both shelves dominated by tidal flows (Kenyon, 1970; Belderson et al., 1972) and storm-dominated shelves (Newton et al., 1973; McKinney et al., 1974); see Fig. 14. Unpublished data of the Atlantic Oceanographic and Meteorological Laboratories, Miami, Florida, show them to be characteristic of large sectors of the Middle Atlantic Bight. Relief is negligible relative to width. Kenyon would restrict the term "sand ribbon" to features having length-to-width ratios of 1:40 and refers to shorter, broader features as elongate sand patches, but the distinction seems arbitrary. Unlike parting lineation ridges, sand ribbons tend to consist of streamers of finer sand in transit over a coarser substrate which may, in fact, be a gravel. A continuum may exist between a sand ribbon pattern of sand and gravel streets of equal width, to a "longitudinal furrow" pattern (Stride et al., 1972; Newton et al., 1973) in which widely spaced, elongate erosional windows in a thin sand sheet reveal a coarser substrate. Ribbon width relative to the width of the interribbon zone does not appear to be simply a function of the height of a sinusoidal surface of sand layer over a coarser substrate, since the windows in profile are notchlike affairs separated by flat plateaulike zones (Fig. 16). Furthermore, the ribbons are commonly rather asymmetrical, as though the sand sheet occurs at minimum thickness on one side, and increases very slowly to maximum thickness on the other side. Such asymmetrical ribbons could be interpreted as

degraded sand waves, but the sharpness of the contacts plus the lack of relief suggests instead asymmetrical helical flow cells (Fig. 17).

Small sand ribbons may be large-scale analogs of the responses described in the preceding section that involve the entire logarithmic boundary layer. However, most shelf sand ribbon patterns have spacings of tens or hundreds of meters, and as noted by Allen (1970, p. 69), can only be responses to the entire depth of flow.

Theoretical studies (Faller, 1971; Faller and Kaylor, 1966; Brown, 1971; Lilly, 1966) and experimental studies (Faller, 1963) show that there is a mechanism by which a helical flow pattern may be induced in the large-scale flows of the continental margin. When such flows are in geostrophic balance (pressure term balanced by Coriolis term in the equation of motion; see p. 25). The lower portion of the flow is an Ekman boundary layer (see p. 97). The basal meter behaves as a logarithmic boundary layer in that flow speed decreases rapidly to a zero value or nearly so at the seafloor. Flow direction (in the northern hemisphere) is to the left of the free-stream direction, however, since the Coriolis term is reduced along with mean velocity; the equation of motion more nearly constitutes a balance between friction and pressure terms. With increasing height off the bottom, flow is more nearly geostrophic and its direction is more nearly parallel to the isobars, until free stream conditions are reached. Thus velocity vectors at successively higher levels constitute a left-handed spiral. On the continental shelf, this lower Ekman boundary layer may extend to the base of the mixed layer, if it exists, or to the surface, where it is overprinted with a right-handed Ekman spiral (upper Ekman boundary layer) because of direct wind stress (Ekman, 1905, Plate 1).

Above a critical Reynolds number, this Ekman layer is unstable. However, because the instability transpires in an Ekman field subject to the Coriolis effect, the instability does not result in random turbulence, but instead in a regular pattern of secondary flow (Faller, 1971, pp. 223–225). In this pattern zones of surface convergence, downwelling, and bottom divergence alternate with zones of surface divergence, upwelling, and bottom convergence. The resulting flow structure consists of horizontal helical cells with alternating right- and left-hand senses of rotation (Fig. 6). Angles of convergence and divergence (pitch) are generally a few degrees; in other words, the secondary component of flow is weak, relative to the main geostrophic component. The flow cells may occur at several scales (Faller and Kaylor, 1966). In laboratory studies (Faller, 1963), smaller scale cells have a spacing of approximately $11D$, where D is a characteristic depth of the Ekman layer, and tend to be oriented up to 14° to the left of the mean flow. They occur at Reynolds numbers above 125. Larger

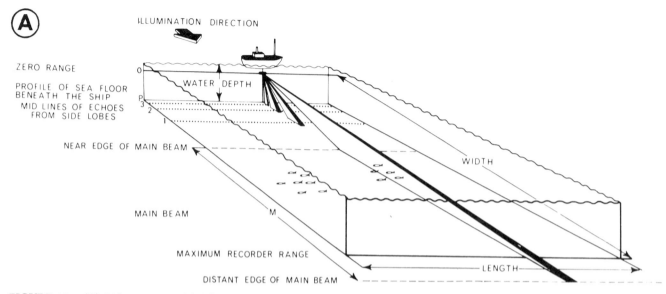

FIGURE 13. (A) *Sidescan sonar.* (B) *The resulting record. A, Bottom of seafloor; B, turbulence in water column due to ship's wake; C, zigzag pattern is due to refraction of sound in density-stratified water; D, main lobe (see above); E, side lobe. From Belderson et al. (1972).*

scale cells have wavelengths much greater than $11D$ and are oriented to the right of geostrophic flow. They occur at much lower Reynolds numbers.

Helical flow structure may occur in the upper Ekman layer where its wind-driven stirring creates the mixed layer above the thermocline (Faller, 1971), or may occur in the lower layer (Faller, 1963). In surface helical flow the downwelling zones that collect the high-velocity wind-driven surface water are more sharply defined than the upwelling zones (Langmuir, 1925). In bottom helical flow, downwelling zones deliver higher velocity water to the seafloor, and may also be more sharply defined than the upwelling zones. During intense flows, when stratification breaks down and the layers partly or completely overlap, a compound top-to-bottom helical flow structure might be expected.

Observational and theoretical studies required to link this scheme to the observed shelf sand ribbon patterns have not been undertaken; however, there are obvious points of compatibility. The ribbons tend to be parallel, or oriented at a small angle to the regional trend of shelf contours, and presumably to the mean geostrophic flow direction. The greater intensity of downwelling zones would explain the dissimilar width of ribbons and intervening erosional windows. The Reynolds numbers required are not excessive for either tide- or wind-driven shelf flows.

LONGITUDINAL SAND RIDGES.   Large-scale longitudinal bed forms of the continental margin, with relief of up to

10 m and spacings measured in kilometers, are called sand ridges (Off, 1963; Swift et al., 1974); see Fig. 18. They are comparable in scale to the seif dunes, and the yet larger "draas" of the sand seas of the world's deserts (Wilson, 1972), except that as befits submarine sand bodies, their side slopes are much lower, usually being measured in fractions of a degree.

Sand ridges appear to form in two basic types of situations. They are characteristic of the reversing flows of tidal estuaries and bays, where they tend to form in complex arrays parallel to the estuary axis (Figs. 18A,B). They also appear on inner shelves of coasts undergoing erosional retreat (Figs. 18C,D), where they appear to be specific responses to the coastal boundary of the shelf flow field (Duane et al., 1972; Robinson, 1966; Swift et al., 1972a); the mechanism is discussed in detail in Chapter 14. On the inner shelf, either tidal or storm flows may be the forcing mechanism (Duane et al., 1972; Swift, in press). The ridges tend to extend obliquely seaward from the shoreface. Like sand ribbon patterns, the generally larger scale sand ridge fields tend to comprise discontinuous sheets of finer sand over a coarser substrate. However, where sand ridges build up into the wave-agitated zone on open coasts, their crests tend to be coarser than their flanks although generally not as coarse as the substrate exposed in trough axes (Houbolt, 1968; Swift et al., 1972b; Stubblefield et al., 1975).

SAND RIDGES AS RESPONSES TO WIND-DRIVEN FLOWS. Sand ridges are found on continental shelves seaward

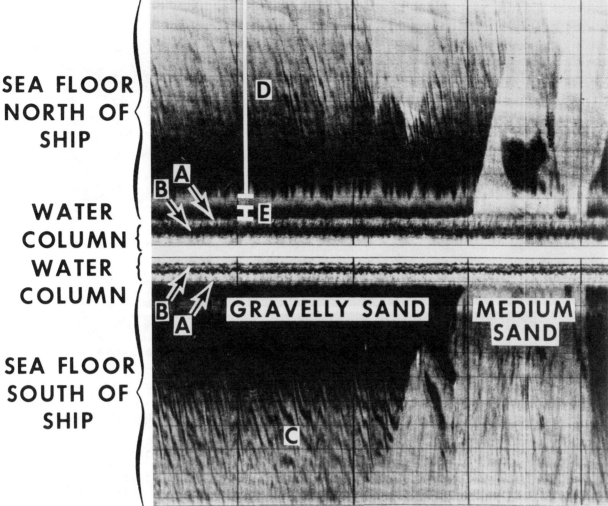

SEA FLOOR
NORTH OF
SHIP

WATER
COLUMN

WATER
COLUMN

SEA FLOOR
SOUTH OF
SHIP

GRAVELLY SAND   MEDIUM SAND

FIGURE 13—*Continued.*

of active inner shelf-generating zones, and occur as well on some shelves whose inner margins are not actively forming them (Swift et al., 1974). It appears that shelf flow fields can continue to maintain these ridges of coastal origin after the retreating shoreline has abandoned them, and can even locally generate them afresh (see discussion in Chapter 14). Without a maintaining mechanism, shelf flows might be expected to degrade sand ridges by leveling crests and filling in troughs. In fact, however, the ridges of the Atlantic continental shelf tend to expose compact clays or lag gravels in their troughs, indicating continuing trough scour (Swift et al., 1972a; McKinney et al., 1974).

There are a variety of competing hydraulic mechanisms that may serve to explain the formation and maintenance of large-scale sand ridges on the continental margin, none of which is clearly understood. On the open shelf, cellular flow structure in storm flows, as described in the preceding section, may couple with the shelf floor. Such cellular flow structure might generate ridges along the coastal boundary (see discussion in Chapter 14) where wind-driven flows are frequent and intense and there is an abundant supply of sand. As these ridges have been left behind by the retreating shoreline during the Holocene transgression, the same cellular flow structure may be continuing to maintain them on the outer shelf.

If this analysis is correct, then sand ribbons and sand ridges may differ in that sand ribbons represent responses to one flow event or a flow season while sand ridges represent time-averaged responses to repeated flow events, whose emerging relief tends to localize the position of large-scale flow cells. Events capable of forming such large-scale flow cells would presumably be peak storm

FIGURE 14. *Sand ribbon patterns from the Spanish Sahara shelf. Light is sand; dark is coarser sand and gravel. Distortion ellipse with scales on first record. (a) 1, 2: Sharply defined ridges of asymmetrical ribbons; 3: pinna (pelecypod) bed. (b) 1: Symmetrical ribbon; 2: sand waves. From Newton et al. (1973).*

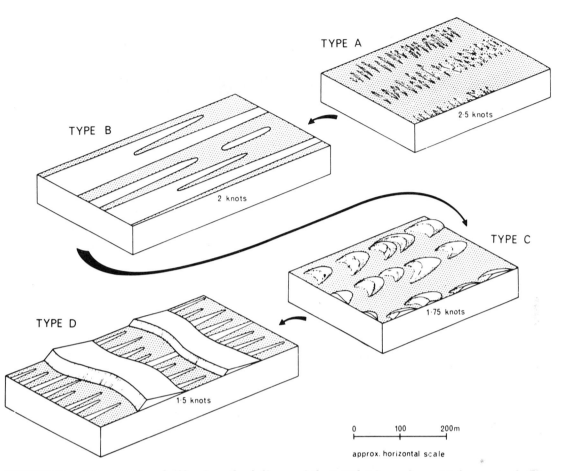

FIGURE 15. *Categories of sand ribbon from the shelf around the British Isles, and associated current velocities. From Kenyon (1970).*

or tidal flows, in which secondary circulation involves the entire water column.

The most problematic aspect of shelf ridge fields is the depth-to-width ratios of the troughs, which range from 1:10 to 1:150. The smaller ratios are compatible with the "type I" flow cells of Faller's (1963) experimental work, whereas the large ratios may derive from Faller's "type II" cells which have "much greater" di-

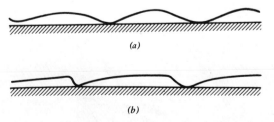

FIGURE 16. (a) *Wide sand ribbons alternating with narrow streets of coarse sand (erosional windows) due to intersection of sinusoidal surface of sand sheet with horizontal substrate.* (b) *Same pattern due to notchlike incision of erosional windows. The latter pattern is a common one.*

mensions. It is perhaps easier to conceive of such flattened cells if it is remembered that the central downwelling zone is the only sharply defined portion of a double helical flow cell; the marginal zones of diffuse upwelling may take up much of the "stretch," serving to complete flow continuity in a fashion analogous to the role of "ground" in electrical circuitry.

The advent of appreciable relief in a growing system of sand ridges may bring other hydraulic mechanisms into play. Secondary flow cells appear to be an innate response to channeled flow. It has long been known (Gibson, 1909; Jeffreys, 1929; Einstein and Li, 1958; Leopold et al., 1964, pp. 251–284; Wilson, 1972) that driftwood or ice in a river tends to move toward the center, whose surface is elevated slightly above that of the margins, and that the thread of maximum velocity tends to be depressed below the surface. The result is a double helical flow cell, in which bottom water spreads, rises along the margins, converges over the center, and sinks there. Flume and theoretical work (Kennedy and Fulton, 1961; Gessner, 1973) indicates that in flumes of

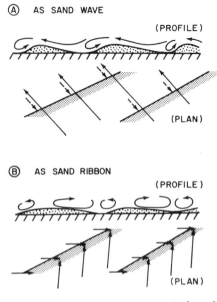

FIGURE 17. *Interpretation of asymmetrical sand ribbons; B is more probable.*

square cross section the unequal distribution of turbulent (Reynolds) stresses will result in secondary flow from the center toward the corners. The resulting multiple flow cells do not form the double helical pattern postulated for natural channels, however, and their applicability to the natural situation is uncertain. Bagnold (1966, pp. I12–I15) offers an independent explanation. He suggests that there is asymmetrical exchange of momentum between the bottom boundary layer of a river and the overlying flow in that tongues of boundary water abruptly penetrate the overlying flow, to be compensated by a general sinking of the latter (see Chapter 7, p. 98). This results in elevation of the water surface over the channel axis where this exchange is most intense. The ensuing pressure head, he suggests, drives the secondary component of flow.

The preceding discussion has dwelt on double helical flow cells as mechanism for generating a large-scale sand ridge topography. An attempt has been made to match

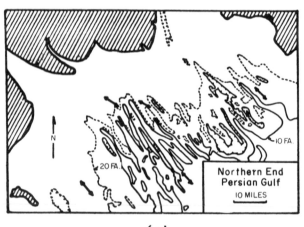

(a)

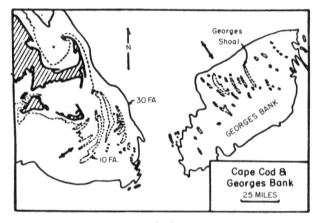

(c)

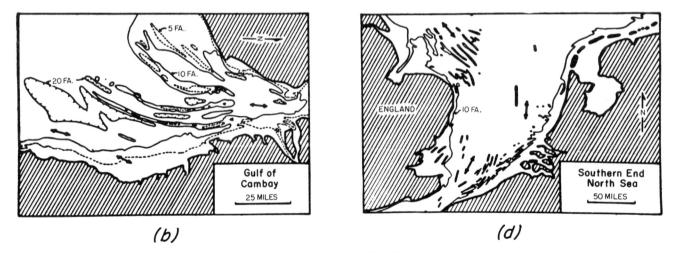

(b)

(d)

FIGURE 18. *Patterns of sand ridges on tide-dominated shelves. From Off (1963).*

theoretical and experimental studies with characteristics of shelf ridge fields. However, such large-scale coupling of flow with substrate has not yet been observed in the field. It is worth noting that there is an independent mechanism that is theoretically capable of maintaining a ridge topography, either by itself or in conjunction with other mechanisms. The mechanism described by Smith (1969) requires that ridges be aligned with mean flow direction and that the variance in flow direction be high, either because the flow is a rotary tidal flow; or because it is storm-driven, and the direction of flow varies during a storm and also among storms (see Chapter 4). As a consequence, most flows intense enough to entrain sand will be aligned at an oblique angle with the ridges during most of their duration. Flow across the ridge can be treated two-dimensionally according to slender body theory (Smith, 1969) and the stability analysis of Smith (1970) applies (see the preceding section). First one flank then the other flank of the ridge will be eroded, with sand transferred to the crest and far flank each time.

SAND RIDGES IN RESPONSE TO TIDAL FLOWS. The reversing nature of nearshore tidal flows adds another mechanism capable of maintaining a ridged topography. The velocity of the tidal wave is a function of water depth, and flow over a step or across a sill in a cohesionless substrate will result in a phase discontinuity between the behavior of the tidal wave on either side of the sill

(Fig. 19). Thus, when the tide is in the last stages of ebb on one side, it may be already beginning to flood on the other, so that there is an opposing sense of flow over the crest of the sill. If the flow is broad relative to its depth, and if the sill is a relatively large-scale feature, then this is an inherently unstable situation. Slight irregularities in the seafloor on either side of the sill will result in inequalities in the rate of propagation of the tidal wave, and during the brief period of opposing flow the two water masses on either side of the sill will tend to interpenetrate along a zigzag front. The tongues of flow on either side of the sill crest will tend to scour its channels until the crestline of the sill has also become zigzag (Fig. 19). A channel that is on the side of the sill facing the oncoming tidal wave and opens in that direction is called a *flood sinus* (Ludwick, 1973). It experiences an excess of flood over ebb discharge (is flood-dominated). A channel on the other side of the sill is called an *ebb sinus*, and is ebb dominated.

Scour in the interdigitating channels of such an ebb-flood channel complex is matched by aggradation of the interchannel shoals. This transfer is perhaps aided by the secondary circulation mechanisms described in the preceding section. As a consequence of the residual current pattern, net vectors of bottom flow integrated over the tidal cycle meet obliquely head-on over crests (Fig. 20), with the result that each ridge becomes a sand circulation cell, or closed loop in the sand transport pattern. Mean $\partial \tau_0 / \partial x$ is negative along these vectors toward the

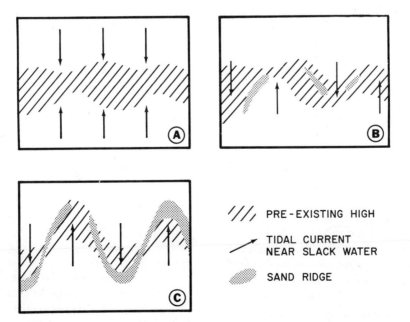

/// PRE-EXISTING HIGH

⟍ TIDAL CURRENT
NEAR SLACK WATER

SAND RIDGE

FIGURE 19. *Hypothetical scheme of development of an ebb-flood channel topography as a consequence of the phase lag experienced by the tidal wave in its passage across a submarine sill.*

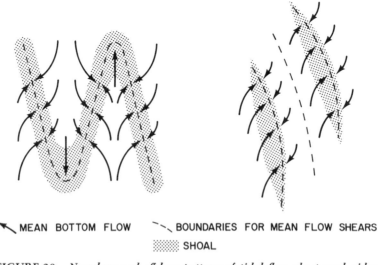

MEAN BOTTOM FLOW   BOUNDARIES FOR MEAN FLOW SHEARS
SHOAL

FIGURE 20. *Nearshore and offshore patterns of tidal flow about sand ridges. Based on Ludwick (1970b) and Caston and Stride (1970).*

zone of residual current shear at the ridge crest. The ridges therefore tend to aggrade toward the intertidal zone where they become "drying shoals" or swash platforms dominated by wave processes (Oertel, 1972). On open coasts, however, wave surge erosion may balance tidal current construction when the crests are still subtidal.

Tidal sand ridges that partition ebb- and flood-dominated flows usually experience a stronger residual current on one side than on the other, and tend to migrate away from that side (Fig. 21). In such cases, where the cross-ridge component of flow is strong, a ridge may itself deform into a sigmoid pattern, and eventually into two or three separate ridges (Caston, 1972); see Fig. 21.

Sills with interdigitating ebb and flood channel systems occur at the mouths of most tidal estuaries (Fig. 22), as a consequence of frictional retardation of the tidal wave within the estuary, and the resultant phase lag. On the Bahama Banks, they occur on the inner sides of islands, where the two wings of the tidal wave meet as they refract around the island (Fig. 23). The evolution of such a system portrayed in Fig. 19 probably rarely occurs in nature; the channel systems form simultaneously with such sills, not afterward. For instance, the ebb-flood channel system of the Chesapeake Bay mouth shoal appears to have formed during the Late Holocene reduction in the rate of sea-level rise (Ludwick, 1973). It can be inferred from the present morphology that the sill prograded south across the bay mouth, fed by the littoral drift discharge of the Delmarva coastal compartment. The ridges would have developed in zigzag fashion, alternately and progressively segregating the flow into ebb-dominated and flood-dominated channels (Fig. 24).

Tidal flows often occur in the presence of salinity stratification so intense as to persist for part or all of the tidal cycle despite the powerful mixing effect of flow turbulence. Flow structure may be yet more complicated as a result. In Fig. 25, the residual circulation over the Hudson estuary mouth shoal (New York Harbor entrance) is seen to be a resultant response to flow interdigitation due to the phase lag effect (Fig. 25C) and to estuarine (two-layer) circulation (Fig. 25B).

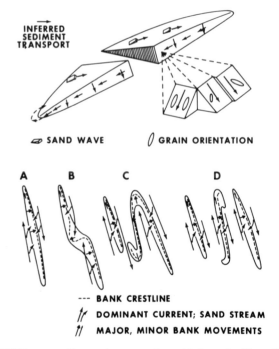

INFERRED SEDIMENT TRANSPORT

SAND WAVE   GRAIN ORIENTATION

A   B   C   D

--- BANK CRESTLINE
DOMINANT CURRENT; SAND STREAM
MAJOR, MINOR BANK MOVEMENTS

FIGURE 21. *Above: Anatomy of a tidal sand ridge. From Houbolt (1968). Below: Evolution of a tidal sand ridge. From Caston (1972).*

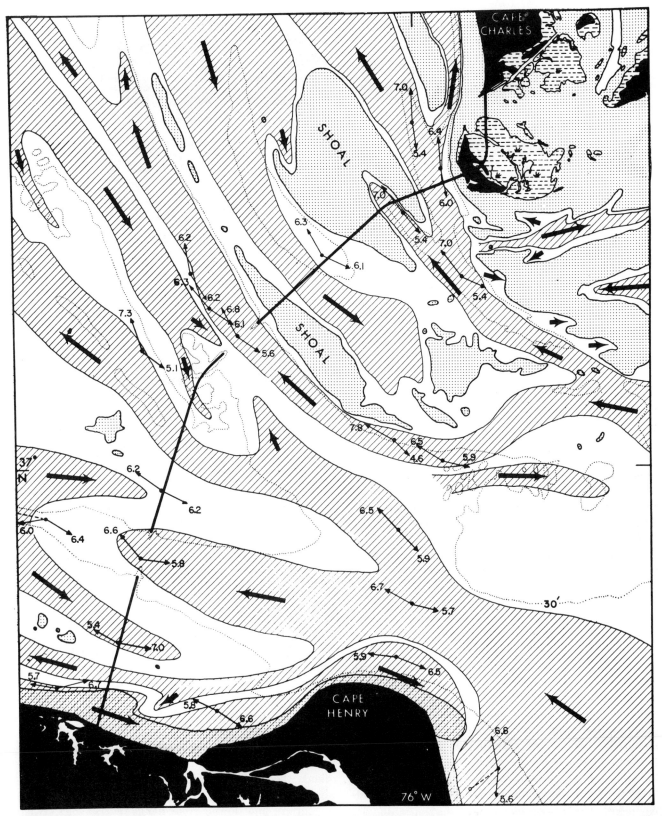

FIGURE 22. *A hydraulic and geomorphic interpretation of the net nontidal (residual) flow pattern at the bottom in the entrance to Chesapeake Bay. Numbers are measured flood and ebb durations at the bottom in hours; small arrows show measured direction of near-bottom currents. Stippled areas are shoaler than 18 ft. Ruled areas show where there is an ebb or a flood flow predominance. From Ludwick (1970a).*

FIGURE 23.   *Ebb-flood channel pattern on the Great Bahama Bank. Altitude 3000 ft. Photo: Charles True.*

Stratification may also play a role in the formation of ridge topography within the estuary. Weil et al. (in press) describe the formation of subtidal levees in Delaware Bay as the consequence of the penetration of subsurface saline tongues up the channels during flood tide, resulting in an internal pressure head that can drive channel axis downwelling (Fig. 26), and as a consequence of the overriding of the tongues by fresher water during the ebb tide, with similar effects. One of us (Ludwick, in press) has mapped near-bottom convergences and divergences of flow in the Chesapeake Bay mouth during flood tide. These are absent during the more thoroughly stratified ebb. Here stratification appears to inhibit channel axis downwelling and bottom current divergence (see Fig. 31). Velocity profiles of Chesapeake Bay mouth tidal flows tend to be parabolic but with markedly sigmoidal perturbations (Ludwick, 1973), and may imply the presence of standing internal waves or wakes from shoals.

Tidal flows in confined estuary mouths thus tend to develop an interdigitating pattern of ebb- and flood-dominated channels, whose sequence of partitioning ridges tends to alternate between clockwise and counterclockwise current flows (Fig. 20). On the offshore shelf, however, the tide becomes rotary rather than reversing and a different pattern tends to appear (Caston and Stride, 1970). Ridges appear in free-standing sets rather than in continuous zigzag arrays. Residual current shears occur in channel axes as well as on ridge crests, and successive ridges experience residual flows with the same sense of rotation.

Huthnance (1972) attributes this open shelf flow pattern to interaction of the ridges with the shelf tide. His model considers a rectilinear reversing tide whose flow directions make an oblique angle with the ridge axis. The cross-ridge component of flow must accelerate over the ridge crest for continuity reasons. The ridge-parallel component of flow must decrease up the upcurrent flank as the water column shoals, and influence of friction becomes proportionately greater. However, because high-velocity fluid is being transported into the shoal region, the decrease in the ridge-parallel flow component lags behind the decrease in depth. On the downcurrent flank, the restoration of the ridge-parallel flow to ambient velocity is similarly lagged. When the tide changes, upcurrent and downcurrent flanks reverse roles. When flow is averaged over the tidal cycle, a clockwise pattern of residual flow around the ridge results (or counter-

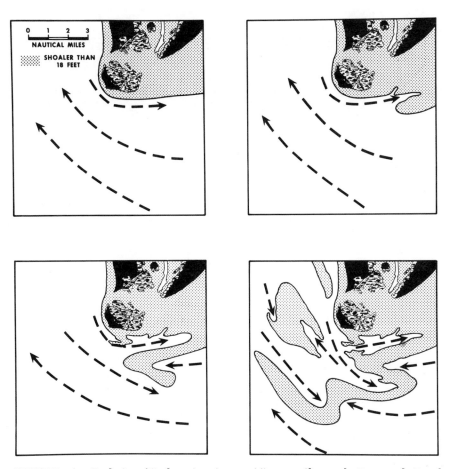

FIGURE 24. *Evolution of "submarine zigzag spit" across Chesapeake Bay mouth. Based on Ludwick* (1972).

clockwise, depending on whether the oblique, reversing tidal stream is sinistral or dextral with respect to the ridge). Huthnance proposes a second mechanism whereby in the northern hemisphere, Coriolis force also results in clockwise circulation.

Huthnance's mechanisms are interesting, but the requirement that there be a significant angle between the axis of the tidal stream and that of the ridge presents a problem. The ridges are a response element within the flow field–substrate system, not an independent forcing element. It seems doubtful that ridges of cohesionless sand could maintain a significant angle with the tidal stream for any length of time, unless it were somehow an equilibrium response to flow. Smith (1969) notes that tidal sand ridges might be expected to orient themselves parallel to the long axis of the tidal ellipse, as the sand body would then be at a small angle of attack throughout most of the high-velocity part of the tidal cycle. According to slender body theory the cross-shoal component of flow during this period can be considered to

be two-dimensional and driven by the cross-shoal pressure gradient. It would thus sweep sand first up one side and then up the other as the tide rotated.

Possibly the dilemma is resolved by the lag effect cited by Postma (1967) and Stride (1974); see Fig. 27. Because of a lag in the entrainment of sand, the period of maximum sand transport is believed to lag behind maximum flood flow, and again behind maximum ebb flow. The result should be to align the response element (sand ridge) obliquely across the major axis of the tidal ellipse. It also seems likely that the large-scale, unbounded tidal flow field of the open shelf might at least locally generate Ekman flow structure during midtide, and couple with inner shelf ridge fields in the fashion that has been suggested for wind-driven flows.

Limiting Conditions of Bed Form Formation on the Continental Margin

In attempting to apply the elements of bed form theory presented on the preceding pages to analyses of conti-

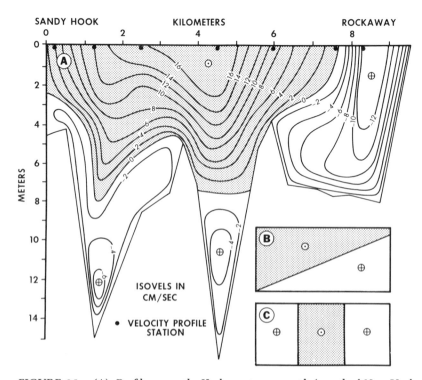

FIGURE 25. (A) *Profile across the Hudson estuary mouth (mouth of New York Harbor) contoured for velocity residual to the tidal cycle. The flow pattern is a resultant response to component flow patterns shown in (B) and (C). (B) Schematic diagram of two-layered, estuarine flow pattern. (C) Schematic diagram of component of flow pattern resulting from phase lag of tidal wave. From Duedall et al. (in press), after Kao (1975).*

nental margin sedimentation patterns, it is useful to keep in mind some generalizations presented by Allen (1966; 1968a, pp. 50–53; 1968b). Allen, following Bagnold (1956), notes that the grain-fluid system is a decidedly multivariate one, and that we should expect to find co-existing instabilities of several different modes and scales. Flows that experience both transverse and streamwise perturbations may develop *bed form associations* consisting of two different bed form types, for instance a reticulate pattern with sand waves overprinted on sand ridges. Likewise, flows tend to experience one or more instability modes at several different spatial scales, resulting in a *bed form hierarchy* as, for instance, in the case of the Diamond Shoals sand wave field (Hunt et al., in press), where photos show that current ripples are superimposed on sand waves (Fig. 10) and sidescan sonar records show in turn that sand waves are superimposed on giant sand waves (Fig. 28). Elaborate hierarchical associations of bed forms occur over vast areas of the earth's surface, in subaerial sand seas (Wilson, 1972), and also in widely disparate environments on the continental margin (compare Fig. 29 with Fig. 30).

The physical scale at which bed forms occur affects their response characteristics, and in turn the flow fre-quency to which they are tuned. For instance, on the crests of the drying sand ridges of the Minas Basin, current ripples reflect radial drainage at the last stages of ebb, sand waves are oriented with slip faces seaward as responses to peak ebb flow, while larger dunes locally are landward facing, reflecting a stronger flood than ebb flow (Swift and McMullen, 1968; Klein, 1970; Dalrymple, 1973).

The largest scale transverse and longitudinal bed forms have had to readjust to continuous environmental change associated with Holocene deglaciation and the accompanying transgression of the continental margins. In some cases, they appear to have taken nearly the duration of the Holocene to form. Sand ridges on the central New Jersey shelf have basal strata containing 11,000-year-old shells (Stubblefield et al., 1975). These features and many other shelf ridge fields appear to have been formed by shoreface ridge formation and detach-ment (Swift, in press) during the Holocene transgression; see Fig. 28, Chapter 14. Plan geometry and internal structure of Atlantic Shelf ridge fields suggest that ridge spacing has increased by ridge migration or coalescence as the water column deepened (Swift et al., 1974).

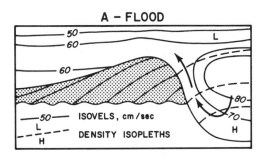

A – FLOOD

ISOVELS, cm/sec
DENSITY ISOPLETHS

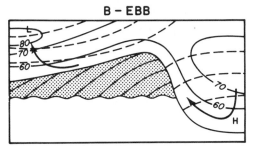

B – EBB

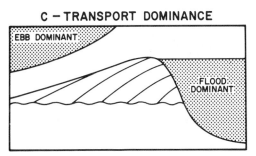

C – TRANSPORT DOMINANCE

EBB DOMINANT

FLOOD DOMINANT

FIGURE 26. *Tidal sand ridge as a submarine levee, formed in response to stratified flow. From Weil et al. (in press).*

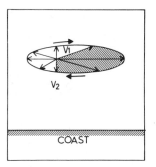

COAST

FIGURE 27. *Lag effects in a rotating tide. Radial arrows are vectors of tidal current velocity at intervals through the tidal cycle. Sand entrainment starts at velocity V_1 and continues to velocity V_2. Net sand transport is to right and onshore. From Postma (1967).*

The response of larger bed forms tends to lag beyond the peak flow event or may comprise an average response to repeated events. Allen (1973) notes that maximum sand wave height in the Fraser River and Gironde estuary is lagged behind peak tidal flow by as much as a quarter of the tidal period. Ludwick (1972) notes that tidal sand waves are symmetrical over portions of the Chesapeake Bay mouth where the tidal cycle is symmetrical, but are asymmetrical when there is flood or ebb asymmetry in the tidal cycle. Thus their response to reversing tidal flow is time-averaged in a manner entirely analogous to the response of oscillation ripples to wave surge (Chapter 8). Tidal sand waves in Chesapeake Bay mouth attain their greatest height and slopes during the summer months when wave activity is

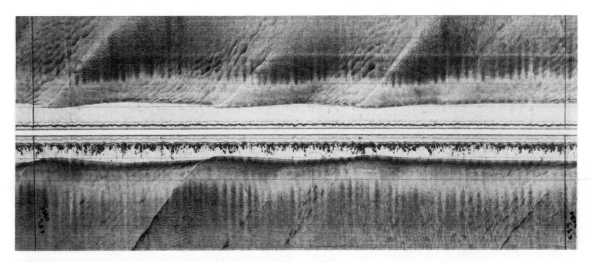

FIGURE 28. *Sidescan sonar record of sand waves on the back of giant sand waves, Cape Hatteras, North Carolina. Sand waves are larger in coarse sand of trough than on finer sand of giant waves. Giant sand waves are 120 m apart, 7 m high. Unpublished data of Swift and Hunt.*

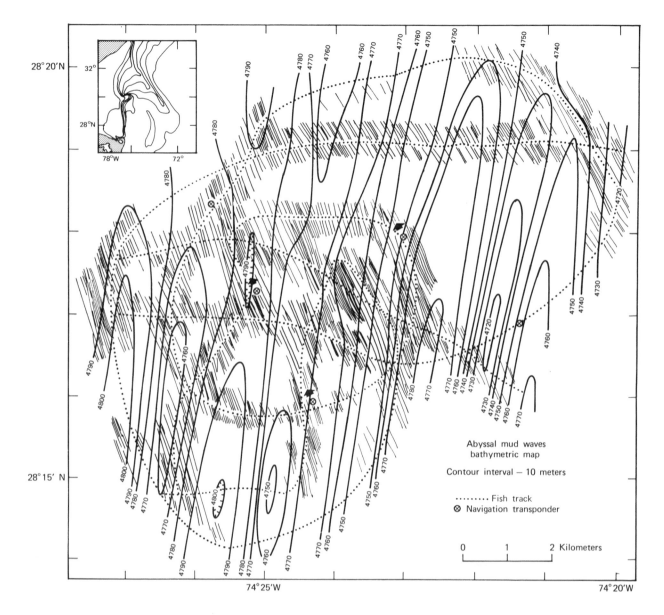

FIGURE 29. *Erosional furrows and large-scale silt ridges on the Blake-Bahamas outer ridge, 4700 m depth. From Hollister et al. (1974).*

at a minimum; they are degraded by the more intense wave activity of winter months (Ludwick, 1970a,b, 1972). The Platt Shoals sand wave field on the open Virginia shelf appears to be induced by storm flows, hence it may have the opposite behavior pattern; sand waves would be highest in the winter months and would tend to be degraded by fair-weather wave surge and burrowing organisms during the summer (Swift et al., 1974).

ESTIMATES ON SEDIMENT TRANSPORT

A Numerical Model

The ultimate goal of dynamic sedimentology is to define the sources, transport paths, and sinks of sediment trans-

port systems, and to determine the rates of erosion, transport, and sedimentation associated with these elements. Much of the material in the following chapters is devoted to available information of sediment sources, pathways, and sinks on the continental margin. However, there have been very few attempts to estimate rates of sediment transport. It should be possible to measure the time history of a marine flow by means of a current-meter array, then employ the empirical relationships developed by hydraulic engineers to estimate the time history of sediment transport. The difficulties however, are formidable. In situ recording current meters are expensive and difficult to maintain. Data processing is complicated and expensive, and the records of meters deployed in shallow water are distorted by the effects of

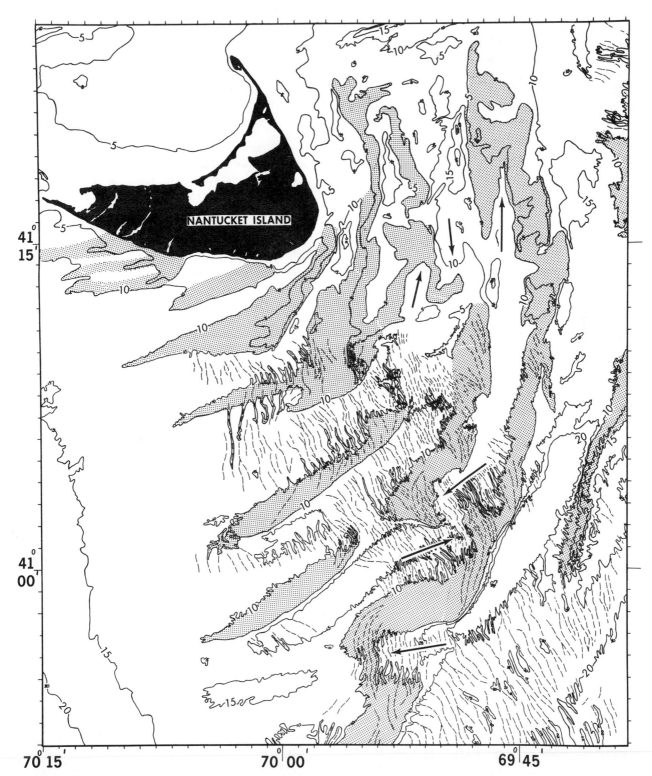

FIGURE 30. *Pattern of sand waves (dark lines) and sand ridges at Nantucket Shoals. Significant highs are stippled and ebb-flood channel couplets are indicated by arrows. Ebb and flood sinuses as inferred from morphology, are indicated by arrows. The ridges* *appear to have initially formed as shoreface-connected ridges similar to those attached to the shoreface of modern Nantucket Island, and to have been stranded on the shelf floor as the shoreface underwent erosional retreat. From Swift (1975).*

wave surge. There is no general agreement on the most satisfactory transport equation, or on the applicability of equations developed under laboratory conditions to the complex deep-water flow fields of the continental margin.

A simple numerical model for estimating rates and patterns of sediment transport in areas of tidal flow has been devised by one of us (Ludwick, in press). It is summarized below.

STRUCTURE OF THE MODEL.   The model requires determination of the distribution across the study area of a sediment transport index $\tau_0 u_{100}$ over a tidal cycle. The index is derived from Bagnold's (1956) work (see p. 113), in which sediment discharge q is set proportional to fluid power ω defined as

$$\omega = \tau_0 \bar{u}$$

so that

$$q = K \tau_0 \bar{u}$$

where τ_0 is bottom shear stress and $\bar{u}$ is the depth-averaged flow velocity. For convenience of measurement, Ludwick substitutes u_{100}, the velocity measured 100 cm off the bottom. With this information, it is possible to use the sediment continuity equation (p. 166) to determine the distribution and relative rates of aggradation and erosion along streamlines of sediment transport.

DETERMINATION OF τ_0.   In order to determine the distribution of τ_0, current velocities were measured over 27 hour intervals at 24 stations in the mouth of Chesapeake Bay. A Kelvin Hughes direct reading current meter was employed from an anchored ship. At each station the current meter was used successively through 11 different depth levels. Hourly profiles with 4 minute observation periods were obtained at each level.

These speed values were then reduced to pseudo-synoptic data sets for standard times and depths at each station (see Ludwick, in press). Each data set was fitted to Hama's (1954) parabolic velocity defect law (see the discussion in Chapter 7, p. 96). This empirical function pertains to outer boundary flow, at distances greater than $0.15h$, where h is the thickness of a turbulent boundary layer, or water depth in the case of fully developed flow in a uniform channel. The equation is

$$\frac{\bar{u}_\infty - u}{u_*} = 9.6 \left(1 - \frac{z}{h}\right)^2$$

where $\bar{u}_\infty$ is the free stream velocity, u is velocity at distance z above the bed, and u_* is the friction or shear velocity.

An estimate of u_* on the bottom is then obtained by least squares curve fitting. The value can be converted to an estimate of τ_0, the boundary shear stress, through the relationship $u_* = (\tau_0/\rho)^{1/2}$, where ρ is fluid density.

This measurement, obtained by observation of the entire water column, provides a far more reliable estimate for $\tau_0 u_{100}$, the fluid power, than does τ_0 determined simply from the product $C_{100}\rho(u_{100})^2$, due to uncertainties in determining C_{100} (see Chapter 7, p. 99).

MAPS OF BED SEDIMENT TRANSPORT.   Values of the sediment transport index obtained for 24 stations must be converted to maps of near-bottom streamlines of sediment transport. The values are adjusted to the mean tidal range, a process described by Ludwick (1973). They are further corrected by subtracting 150 dyne·cm/sec cm², a threshold value for the initiation of sediment movement (Fig. 31). The value at each station is integrated separately over each flood and each ebb half-cycle, and the results are averaged for ebb and flood. After averaging and integrating, the units of the sediment transport index are dyne·cm/cm² per average ebb (or flood) half-cycle.

The values obtained at points on the field grid of 24 irregularly placed stations must then be redistributed over a systematic grid. This is a problem in vector interpolation. The first step is to prepare separate maps of the north-south and east-west components of $\tau_0 u_{100}$ for the flood cycle. Each map is contoured. The flood component maps are superimposed. Resultant vectors may now be calculated at any point, if the contour interval is sufficiently small. The density of resultant vectors may be increased in areas of complex flow. Finally, streamline maps may be prepared by drawing lines that are everywhere tangent to the vectors (Fig. 32A). The process is repeated for the ebb half-cycle and the vector sum of ebb and flood (Figs. 33A and 34A).

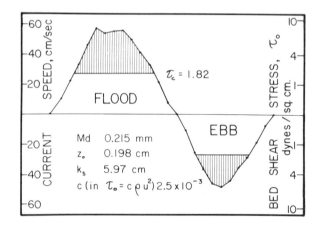

FIGURE 31. *Tidal current speed and bottom shear stress at a flood channel station, Chesapeake Bay mouth. Speed values are for a distance of 18.5 ft off the bottom. Total depth, 56 ft. Observed speeds were corrected from mean tidal range and averaged over six cycles. z_0 is the roughness length estimated from vertical velocity profiles, k_s is the height of bottom roughness elements, and τ_c is the critical shear stress, calculated from the Shields entrainment diagram. From Ludwick (1970a).*

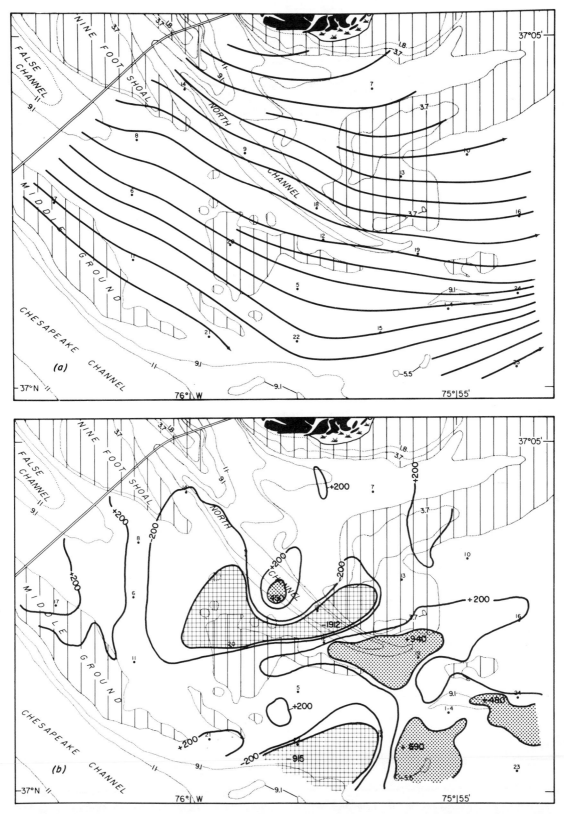

FIGURE 32. *Ebb-directed sediment transport at the bed.* (a) *Streamlines of the sediment transport vector* $\tau_0 u_{100}$; *depths are in meters; vertically ruled areas are shoaler than 5.5 m.* (b) *Erosion-deposition chart on which erosion is positive* (+) *and deposition is negative* (−); *units are dyne–cm/cm² per ebb half-tidal cycle per 463 m of transport* × 10^{-4}; *cross-hatched areas indicate erosion intensity greater than* −400 *units; stippled areas indicate deposition intensity greater than* +400 *units. From Ludwick (in press).*

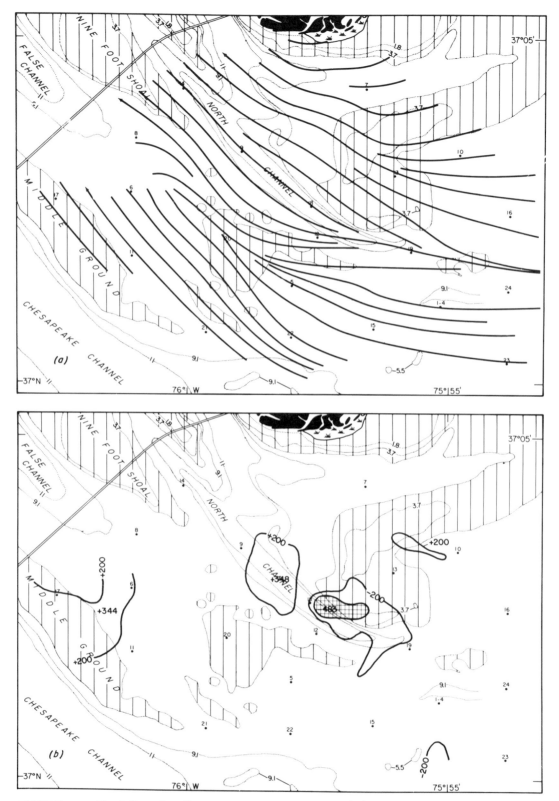

FIGURE 33. *Flood-directed sediment transport at the bed. (a) Streamlines of the sediment transport vector $\tau_0 u_{100}$; depths are in meters; vertically ruled areas are shoaler than 5.5 m. (b) Erosion-deposition chart on which erosion is positive (+) and deposition is negative (−); units are dyne–cm/cm² per flood half-tidal cycle per 463 m of transport × 10⁻⁴; cross-hatched areas indicate erosion intensity greater than −400 units; stippled areas indicate deposition intensity greater than +400 units. From Ludwick (in press).*

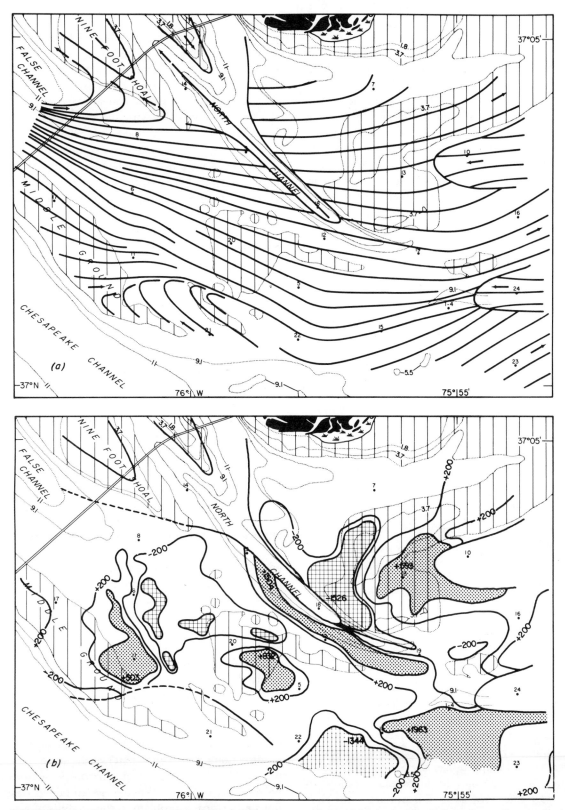

FIGURE 34. *Vector sum of ebb and flood sediment transport at the bed.* (a) *Streamlines of the resultant sediment transport vector* $\tau_0 u_{100}$; *depths are in meters; vertically ruled areas are shoaler than 5.5 m.* (b) *Erosion-deposition chart on which erosion is positive* (+) *and deposition is negative* (−); *units are dyne–cm/cm² per tidal cycle per 463 m of resultant transport* $\times$ 10^{-4}; *cross-hatched areas indicate erosion intensity greater than* −400 *units; stippled areas indicate deposition intensity greater than* +400 *units. From Ludwick (in press).*

It is important to realize the limitations of the stream-lines of bottom sediment transport so determined. The redistribution of information has not in any way increased the accuracy or resolution of the original data. Sediment input in a stream tube does not necessarily equal sediment output, since deposition or erosion may occur. There is no underlying stream function in the method, and the spacing of the streamlines is not a measure of transport rates. It is assumed, however, that transport is confined to a path of unit width that conforms to the bathymetry of the seafloor, and that the streamline is the center line of this path. It is further assumed that conditions are steady and nonuniform for the entire pattern.

NET SEDIMENTATION MAPS. As a separate and ensuing procedure, it is possible to estimate the extent of areas of erosion and deposition, and also the rates at which these processes occur. The estimate utilizes the sediment continuity equation written in terms of discharge:

$$\frac{\partial \eta}{\partial t} = -\epsilon \frac{\partial q}{\partial x}$$

where η is bed elevation relative to a datum plane, t is time, ϵ is a dimensional constant related to sediment porosity, q is the weight rate of bed sediment transport per unit width of streamline path, and x is distance along the streamline.

Discharge (q) may be taken as proportional to $\tau_0 u_{100}$ and the right-hand partial derivative may be approximated by a finite difference:

$$\frac{\partial q}{\partial x} \simeq \frac{\Delta q}{\Delta x} = K \frac{(\tau_0 u_{100})_2 - (\tau_0 u_{100})_1}{x_2 - x_1}$$

The term $x_2 - x_1$ is held constant arbitrarily at a value of 465 m, hence

$$\frac{\partial \eta}{\partial t} \propto -\Delta \tau_0 u_{100}$$

Thus a decrease in transport rate along a transport pathway induces deposition; an increase causes erosion.

The resultant vector map for a half-cycle is superimposed on the equivalent streamline map. The magnitude of $\tau_0 u_{100}$ is determined at equispaced points along each streamline; $\Delta \tau_0 u_{100}$ is determined as a positive or negative value, and mapped over the area of study as an estimate of relative erosion and deposition intensity. In Figs. 32B, 33B, and 34B, net sedimentation maps have been prepared for the ebb and flood half-cycles and for the vector sum of ebb and flood.

UTILITY OF THE MODEL. Such a manipulation of the data from 24 current-meter stations extracts a surprising amount of information from them. Streamlines of bed sediment transport associated with the ebb tidal jet are seen to pass over the bay mouth shoal in parallel fashion. Flood streamlines, however, form a pattern in which flow divergence and flow convergence alternate across the flow in sympathy with the topographic pattern of interdigitating ebb and flood channels.

The vector sum map shows a complex pattern of flow dominance that is also correlated with bottom morphology. Patterns of net sedimentation do not correlate as closely with the topography, probably because they do not indicate the areas of maximum relief, but instead areas undergoing maximum change. In particular, the parabolic shoals that envelop each ebb or flood sinus are seen to be subject to a systematic pattern of sedimentation. The sides of shoal segments that face the dominant flow, however obliquely, are eroding. The crest and downcurrent sides, however, are undergoing aggradation. Thus, the processes that Smith (1970) has inferred to cause sand waves (see p. 165) appear also to be applicable to ebb-flood channel topography.

The model can be generalized for portions of the shelf dominated by storm flows if each flow event is treated in the same fashion as Ludwick treated a tidal half-cycle, or sediment transport can be integrated over an arbitrary period of observation.

Transport Estimates from Tracer Dispersal Studies

One of the main stumbling blocks in divising quantitative estimates of sediment transport has been the limited applicability of empirical relationships based on laboratory observations to the complex flows of the marine environment. The model partially circumvents this problem by recourse to the sediment transport index, based on Bagnold's generalized evaluation of fluid power (Chapter 8, p. 113). In doing so, it provides only a relative answer. Sediment transport is proportional to fluid power, and the proportionality constant remains unevaluated. Despite this sacrifice, the model has not resolved the problem of adequately treating the complex time and space scales of marine flows. In particular, it fails to deal with the vexing problem of the role of bottom wave surge in "lubricating" bottom sediment transport by reducing the effective transport threshold for a unidirectional flow component (see discussion, Chapter 8, p. 115). This wave surge factor becomes part of the proportionality constant. Wave-surge-amplified transport is not that critical a problem in the analysis of a primarily tide-built topography. It becomes critical, however, in open shelf transport, where wind-driven unidirectional flows attain their maximum intensities just as the wave regime does.

It is clear that the best resolution of a marine sediment transport system will be obtained when a model such as the one presented above is employed together with an independent method for evaluating the proportionality constant. The most promising method to date is the deployment of radioisotope tracers. Fluorescent tracers have been widely used (see Ingle and Gorsline, 1973; Inman and Chamberlain, 1959). However, since counting of labeled particles must be done in the laboratory, the analysis is tedious, and it is generally not possible to watch the development of dispersal patterns in real time. Furthermore, fluorescent tracers have a very limited applicability seaward of the surf, as a consequence of the limited sensitivity of the method and the difficulties of hand sampling. Tracer dispersal can usually be observed in an area 50 m in diameter or less, under fair-weather conditions. After a storm, when a major displacement of sediment has occurred, the tracer grains are liable not to be there at all.

Radioisotope tracers avoid much of this difficulty. The RIST (Radio Isotope Sand Tracer) system, developed by Oak Ridge National Laboratories and the Coastal Engineering Research Center (Duane, 1970) detects radioisotope-labeled tracers by means of a towed scintillometer. The data logging system provides for real time readout, which greatly aids mapping of the dispersal pattern. A relatively long-lived isotope such as ruthenium-103, with a half-life of 40 days, permits effective tracing for three times that duration, or an entire storm season.

A numerical estimate of sediment transport may be fine-tuned by quantitative analysis of radioisotope tracer dispersal patterns. The procedure requires not only the mapping of successive outlines of the tracer pattern but the ability to account for all of the labeled particles at each stage. In order to establish such a mass balance, it is necessary to know the depth of reworking, which is the depth to which labeled particles have penetrated during dispersal. This depth can be calculated from the known ability of the sediment to absorb radiation, if it is assumed that the tagged particles are mixed into the reworked layer in a homogeneous fashion. If tracer particles can be accounted for through successive mappings of the dispersal pattern, then the rate of sediment transport as indicated by dispersal of tracers may be checked against the rate of transport as estimated from current-meter records in one of two ways. Transport rates may be determined directly from the dispersal pattern and compared with estimates based on current-meter records. Or current-meter records may be used to simulate tracer dispersal patterns, and these ideal patterns may be compared with observed dispersal patterns.

Figure 35 shows a series of radioisotope dispersal patterns from an experiment conducted by J. W. Lavelle and his associates, Atlantic Oceanographic and Meteorological Laboratories, Miami, Florida. Water-soluble bags of labeled sand were released along a line in 20 m of water on the south shore of Long Island during April and May of 1974. Over a period of 69 days, a typical fair-weather dispersal pattern formed (panels *B–F*). The data in these panels have been corrected for decay, but in the last panel, the corrected values on the margins of the pattern are so much weaker than background, that they were lost when smoothed background values were subtracted. The mild summer storms during this period only briefly generated flows strong enough to transport sand, and much of the labeled sand remained in close proximity to the drop line, where it was not readily resolved by the towed scintillometer. It will be necessary to apply a statistical smoothing function to the data, in order to undertake a mass balance calculation for the dispersed tracer sand. If continued experimentation leads to improved field techniques and data processing, then radioisotope tracers should prove a fruitful method for calibrating numerical models of sediment transport.

SUMMARY

The size frequency distribution of marine sand samples tends to be log-normally distributed. This distribution, as defined by its mean and standard deviation, is the "signature" of the depositional event, and deviations from log normality, as measured in terms of skewness and kurtosis, may be taken to reflect both the provenance and the hydraulic history of the sediment.

The modal diameter of a sand deposit is that grain size most likely to arrive and least likely to be carried away from the place of deposition; progressively coarser sizes are progressively less frequent because they are progressively less likely to arrive, and progressively finer grain sizes are progressively less frequent because they are more likely to be carried away. This intuitively apparent concept can be explained in terms of probability theory.

As sand progresses down a transport path by intermittent hops, it tends to leave its coarser grains behind, and the deposits are progressively finer in the direction of transport (have undergone progressive sorting). They also will tend to be fine-skewed, particularly if the intensity of hydraulic activity also declines down the transport path.

Moss has shown that the size distributions of marine sands tend to be made up of three log-normal popula-

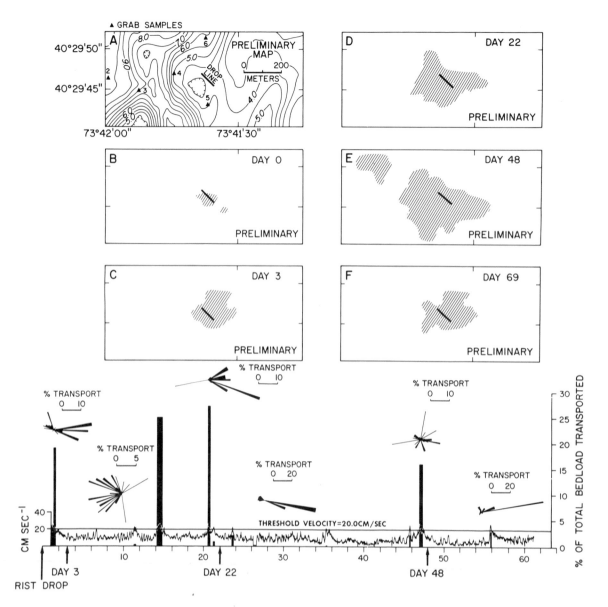

FIGURE 35. *Time sequence of dispersal of radioisotope-labeled sand, south shore of Long Island, April 22–July 2, 1974. (A) Background radioactivity in arbitrary units. Heavy black line is line of emplacement of sand labeled with ruthenium-103. (B–F) Maximum extent of detectable signal after removal of background on successive mapping days. Data have been corrected for decay. Bottom panel: time-velocity record (jagged line). Height of vertical bars indicates percentage of total bed load transported, as determined from a normalized sediment transport index, $V_{100} - V_T$. Width of vertical bars indicates duration of transport event. Rose diagrams indicate direction of transport. Length of radial bar indicates percentage of transport during that event; width is proportional to direction and intensity. Velocity was sampled every 10 minutes. Unpublished data of Lavelle et al., Atlantic Oceanographic and Meteorologic Laboratories, Miami, Florida.*

tions as a consequence of the fashion in which the bed is built; the main subpopulation (A population) comprises the framework of the deposit. A fine B population is interstitial; a coarse C population is the consequence of "traction clogs." The A:B:C ratio varies with the flow regime.

Bed forms are irregularities in the particulate substrate of a fluid flow. Sheared flow is innately unstable, and tends to develop repeated patterns of velocity variation, either parallel or normal to the flow direction. Such instabilities tend to interact with the bed so as to cause rhythmic variations in elevation. Flow and bed perturbation amplify each other until equilibrium is attained.

Bed forms occur in *associations* (more than one genetic type present), in *hierarchies* (successive scales of bed forms of similar genesis), and in hierarchical associations. *Transverse bed forms* interact with wavelike perturbations of flow transverse to the flow direction. *Current ripples* are small-scale transverse bed forms that appear to result from boundary layer instability; their wavelength is independent of depth. *Sand waves* result from perturbations of the whole flow field, or a density-homogeneous portion of it. Several scales of sand waves may occur together; the smaller scale may perhaps be a response to primarily tractive transport, whereas the larger scale may be a response to primarily suspensive transport.

Longitudinal bed forms are caused by velocity perturbations parallel to flow. In some cases, the perturbation takes the form of horizontal, flow-parallel vortices whose sense of rotation is alternately right and left handed, and this may be true for all cases. *Parting lineations* are small-scale longitudinal bed forms. They are sand ridges a few grain diameters high and a few centimeters apart. *Current lineations* have wavelengths ranging from a few meters to hundreds of meters; their heights are negligible relative to width.

In a characteristic pattern, *sand ribbons* occur on a gravel substrate. In *longitudinal furrow* patterns, the lows are more sharply defined than the intervening highs.

Sand ridges may have wavelengths of hundreds of meters to several kilometers, and amptitudes of 10 m or more. They are induced by tidal flows at estuary mouths, by tidal or wind-driven flows on the shelf, and perhaps by boundary undercurrents on the continental rise. They appear to be time-averaged responses to intermittent flow, and in many cases have survived successive environmental transitions associated with the Holocene transgression.

A simple numerical model for estimating bed load transport on the continental margin requires as input current-meter measurements. Streamlines of bottom sediment transport may be based on the sediment transport index. The index is derived from Bagnold's energetics, in which sediment discharge is set proportional to fluid power, equal to bottom shear stress times the depth-averaged velocity. The sediment continuity equation is used to predict areas and relative intensities of erosion and deposition. In this equation, the discharge gradient along a streamline is related to the time rate of change of bottom height above a datum by a dimensional constant.

It may be possible to evaluate Bagnold's proportionality constant for sediment transport by means of mass balance assessments of radioisotope dispersal patterns.

However, improvements in field tracer techniques and data processing are required before such an evaluation is possible.

ACKNOWLEDGMENTS

We thank J. R. L. Allen and R. L. Miller for critical review of the manuscript.

SYMBOLS

C_{100}	drag coefficient determined from measurements 100 cm above the bottom
D	grain diameter
h	water depth
K	a constant
k_s	height of bottom roughness elements
q	sediment discharge
t	time
u	velocity
$\bar{u}$	depth-averaged velocity
$\bar{u}_\infty$	time-averaged free stream velocity
u_{100}	velocity 100 cm off the bottom
u_*	shear velocity; shear stress with velocity units
x	distance downstream
y	distance above the bed
z	distance transverse to flow
z_0	roughness length
ϵ	dimensional constant related to sediment porosity
λ	wavelength
ρ	fluid density
ρ_s	sediment density
η	elevation of a surface above a datum
τ	shear stress
τ_0	shear stress at the bed
τ_c	critical bed shear stress
ν	kinematic viscosity
ω	fluid power

REFERENCES

Allen, J. R. L. (1964). Primary current lineation in the lower old red sandstone (Devonian) Anglo-Welsh Basin. *Sedimentology*, **3:** 89–108.

Allen, J. R. L. (1966). On bedforms and paleocurrents. *Sedimentology*, **6:** 153–190.

Allen, J. R. L. (1967). Depth indicators of clastic sequences. *Mar. Geol.*, **5**: 429–446.

Allen, J. R. L. (1968a). *Current Ripples*. Amsterdam: North-Holland Publ., 433 pp.

Allen, J. R. L. (1968b). The nature and origin of bedload hierarchies. *Sedimentology*, **10**: 161–182.

Allen, J. R. L. (1969). Erosional current markings of weakly cohesive mud beds. *J. Sediment. Petrol.*, **34**: 607–623.

Allen, J. R. L. (1970). *Physical Processes of Sedimentation*. New York: American Elsevier, 248 pp.

Allen, J. R. L. (1973). Phase difference between bed and flow in natural environments and their geologic relevance. *Sedimentology*, **20**: 323–329.

Bagnold, R. A. (1954). Experiments on a gravity-free dispersion of large solid in a Newtonian fluid under shear. *Proc. Roy. Soc. London*, **A225**: 49–63.

Bagnold, R. A. (1956). The flow of cohesionless grains in fluids. *Proc. Roy. Soc. London*, **A249**: 235–250.

Bagnold, R. A. (1966). An approach to the sediment transport problem from general physics. *U.S. Geol. Surv. Prof. Paper 422-I*, 37 pp.

Belderson, K. H., N. A. Kenyon, A. H. Stride, and A. R. Stubbs (1972). *Sonographs of the Sea Floor*. Amsterdam: Elsevier, 185 pp.

Brown, R. A. (1971). A secondary flow model for the planetary boundary layer. *J. Atmos. Sci.*, **27**: 742–757.

Cartwright, D. E. (1959). On submarine sand waves and tidal lee waves. *Proc. Roy. Soc. London*, **A253**: 218–241.

Caston, V. N. D. (1972). Linear sand banks in the southern North Sea. *Sedimentology*, **18**: 63–78.

Caston, V. N. D. and A. H. Stride (1970). Tidal sand movement between some linear sand banks in the North Sea off northeast Norfolk. *Mar. Geol.*, **9**: M38–M42.

Chandrasekhar, B. (1961). *Hydrodynamic and Hydromagnetic Stability*. New York and London: Clarendon Press, 654 pp.

Dalrymple, R. W. (1973). Preliminary investigations of an intertidal sand body, Cobequid Bay, Bay of Fundy. *Marit. Sediments*, 4: 21–28.

Duane, D. B. (1970). Synoptic observations of sand movement. *Proc. 12th Coastal Eng. Conf., Washington, D.C.*, 799–813.

Duane, D. B., M. E. Field, E. P. Meisburger, D. J. P. Swift, and S. J. Williams (1972). Linear shoals on the Atlantic inner shelf, Florida to Long Island. In D. J. P. Swift, D. B. Duane, and O. H. Pilkey, eds., *Shelf Sediment Transport: Process and Pattern*. Stroudsburg, Pa.: Dowden, Hutchinson & Ross, pp. 447–498.

Duedall, I., H. B. O'Connors, J. B. Parker, R. Wilson, W. Miloski, and G. Hulse (in press). The seasonal and tidal variation of nutrients and chlorophyl in the New York Bight Apex. Marine Sciences Research Center, State University of New York.

Einstein, H. A., and H. Li (1958). Secondary currents in straight channels. *Trans. Am. Geophys. Union*, **39**: 1085–1088.

Ekman, U. W. (1905). On the influence of the earth's rotation on ocean currents. *Ark. Mat. Astron. Fysik*, **2**: 1–53.

Emery, K. O. and E. Uchupi (1972). Western North Atlantic Ocean; topography, rocks, structure, water, life, and sediments. *Am. Assoc. Pet. Geol. Mem.*, **17**: 532.

Exner, F. M. (1925). Über die wechselwirking zwischen wasser und geschiebe in flussen. *Setzungsberichte der Academie der Wissenschaften*. Vienna Heft 3–4. pp. 165–180.

Faller, A. J. (1963). An experimental study of the instability of the laminar Ekman boundary layer. *J. Fluid Mech.*, **15**: 560–576.

Faller, A. J. (1971). Oceanic turbulence and the Langmuir cir culations. *Ann. Rev. Ecol. System.*, **2**: 201–233.

Faller, A. J. and R. B. Kaylor (1966). A numerical study of the instability of the laminar Ekman boundary layer. *J. Atmos. Sci.*, **23**: 466–480.

Friedman, G. M. (1961). Distinction between dune, beach, and river sands from their textural characteristics. *J. Sediment. Petrol.*, **31**: 514–529.

Furnes, G. K. (1974). Formation of sand waves on unconsolidated sediments. *Mar. Geol.*, **16**: 145–160.

Gessner, F. B. (1973). The origin of turbulent secondary flow along a corner. *J. Fluid Mech.*, **58**: 1–25.

Gibson, A. H. (1909). On the depression of the filament of maximum velocity in a stream flowing through an open channel. *Proc. Roy. Soc. London*, **A82**: 149–159.

Guy, H. P., D. B. Simons, and E. V. Richardson (1966). Summary of alluvial channel data from flume experiments 1956–1961. *U.S. Geol. Surv. Prof. Pap. 462-J*, 96 pp.

Hama, F. R. (1954). Boundary characteristics for smooth and rough surfaces. *Trans. Soc. Naval Architects and Marine Engineers*, **62**: 333–358.

Hanna, S. R. (1969). The formation of longitudinal sand dunes by large helical eddies in the atmosphere. *J. Appl. Meteorol.*, **88**: 874–883.

Hollister, C. D., R. D. Flood, D. A. Johnson, P. Lonsdale, and J. B. Southard (1974). Abyssal furrows and hyperbolic echo traces on the Bahama outer ridge. *Geology*, **2**: 395–400.

Houbolt, J. J. H. C. (1968). Recent sediments in the southern bight of the North Sea. *Geol. Mijnbouw*, **47**: 245–273.

Huthnance, J. M. (1972). Tidal current asymmetries over the Norfolk sandbanks. *Estuarine Coastal Mar. Sci.*, **1**: 89–99.

Ingle, J. C. and D. S. Gorsline (1973). Use of fluorescent tracers in the nearshore environment. In *Tracer Techniques in Sediment Transport*. Vienna: Int. At. Energy Agency, pp. 125–148.

Inman, D. L. (1949). Sorting of sediments in the light of fluid dynamics. *J. Sediment. Petrol.*, **19**: 51–70.

Inman, D. L. and T. K. Chamberlain (1959). Tracing beach sand movement with irradiated quartz. *J. Geophys. Res.*, **64**: 41–47.

Jeffreys, H. (1929). On the transverse circulation in streams. *Proc. Cambridge Phil. Soc.*, **25**: 20–25.

Kao, A. (1975). A study of the current structure in the Sandy Hook–Rockaway Point transect. M.S. Research Paper, Marine Sciences Research Center, State University of New York, Stony Brook.

Kennedy, J. F. (1964). The formation of sediment ripples in closed rectangular conduits and in the desert. *J. Geophys. Res.*, **69**: 1517–1524.

Kennedy, J. F. (1969). The formation of sediment ripples, dunes and antidunes. *Ann. Rev. Fluid Mech.*, **1**: 147–168.

Kennedy, R. J. and J. F. Fulton (1961). The effect of secondary currents upon the capacity of a straight open channel. *Trans. EIC*, **5**: 12–18.

Kenyon, N. H. (1970). Sand ribbons of European tidal seas. *Mar. Geol.*, **9**: 25–39.

Klein, G. D. (1970). Depositional and dispersal dynamics of intertidal sand bars. *J. Sediment. Petrol.*, **40**: 1095–1127.

Kline, S. J. (1967). Observed structure features in turbulent and transitional boundary layers. In G. Sovran, ed., *Fluid Mechanics of Internal Flow*. Amsterdam: Elsevier, pp. 27–68.

Kline, S. J., W. C. Reynolds, F. A. Schraub, and P. W. Runstler (1967). The structure of turbulent boundary layers. *J. Fluid Mech.*, **30**: 741–773.

Krumbein, W. C. (1934). Size-frequency distributions of sediments. *J. Sediment. Petrol.*, **4**: 65–77.

Langhorne, D. N. (1973). A sandwave field in the outer Thames estuary. *Mar. Geol.*, **14**: 129–143.

Langmuir, I. (1925). Surface motion of water induced by the wind. *Science*, **87**: 119–123.

Leopold, L. B., M. G. Wolman, and J. P. Miller (1964). *Fluvial Processes in Geomorphology*. San Francisco: Freeman, 522 pp.

Lilly, D. K. (1966). On the instability of Ekman boundary flow. *J. Atmos. Sci.*, **23**: 481–494.

Lin, C. C. (1955). *The Theory of Hydrodynamic Stability*. London and New York: Cambridge University Press, 155 pp.

Lonsdale, P. and B. Malfait (1974). Abyssal dunes of foraminiferal sand on the Carnegie Ridge. *Geol. Soc. Am. Bull.*, **85**: 1697–1712.

Ludwick, J. C. (1970a). Sand waves in the tidal entrance to Chesapeake Bay: Preliminary observations. *Chesapeake Sci.*, **11**: 98–110.

Ludwick, J. C. (1970b). *Sandwaves and Tidal Channels in the Entrance to Chesapeake Bay*. Technical Report 1, Inst. of Oceanography, Old Dominion University, Norfolk, Va., 7 pp.

Ludwick, J. C. (1972). Migration of tidal sandwaves in Chesapeake Bay entrance. In D. J. P. Swift, D. B. Duane, and O. H. Pilkey, eds., *Shelf Sediment Transport: Process and Pattern*. Stroudsburg, Pa.: Dowden, Hutchinson & Ross, pp. 377–410.

Ludwick, J. C. (1973). *Tidal Currents and Zig-Zag Sand Shoals in a Wide Estuary Entrance*. Technical Report 7, Inst. of Oceanography, Old Dominion University, Norfolk, Va., 23 pp.

Ludwick, J. C. (in press). Tidal currents, sediment transport, and sand banks in Chesapeake Bay entrance, Virginia. In M. O. Hayes, ed., *Second International Estuarine Conf. Proc. Myrtle Beach, South Carolina, Oct. 15–18, 1973*.

McCave, I. N. (1971). Sandwaves in the North Sea off the coast of Holland. *Mar. Geol.*, **16**: 199–225.

McKinney, T. F. and G. M. Friedman (1970). Continental shelf sediments of Long Island, New York. *J. Sediment. Petrol.*, **40**: 213–248.

McKinney, T. F., W. L. Stubblefield, and D. J. P. Swift (1974). Large scale current lineations on the central New Jersey shelf: Investigations by sidescan sonar. *Mar. Geol.*, **17**: 79–102.

Middleton, G. V. (1968). The generation of log normal size frequency distributions in sediments. In *Problems of Mathematical Geology*. Leningrad: Science Press, pp. 37–46.

Moss, A. J. (1962). The physical nature of common sandy and pebbly deposits (Part I). *Am. J. Sci.*, **260**: 337–373.

Moss, A. J. (1963). The physical nature of common sandy and pebbly deposits (Part II). *Am. J. Sci.*, **261**: 297–343.

Moss, A. J. (1972). Bedload sediments. *Sedimentology*, **18**: 159–219.

Newton, R. S., E. Seibold, and F. Werner (1973). Facies distribution patterns on the Spanish Sahara continental shelf mapped with sidescan sonar. *Meteor. Forsch. Ergebnisse*, **15**: 55–77.

Oertel, G. F. (1972). Sediment transport of estuary entrance shoals and the formation of swash platforms. *J. Sediment. Petrol.*, **42**: 857–863.

Off, T. (1963). Rhythmic linear sand bodies caused by tidal currents. *Am. Assoc. Pet. Geol. Bull.*, **47**: 324–341.

Postma, H. (1967). Sediment transport and sedimentation in the marine environment. In G. H. Lauff, ed., *Estuaries*. Washington, D.C.: Am. Assoc. Adv. Sci., pp. 158–180.

Raudkivi, A. J. (1967). *Loose Boundary Hydrodynamics*. Oxford: Pergamon, 331 pp.

Robinson, A. H. W. (1966). Residual currents in relation to shoreline evolution of the East Anglian coast. *Mar. Geol.*, **4**: 57–84.

Rosenhead, L. (1963). *Laminar Boundary Layers*. London and New York: Oxford University Press (Clarendon), 687 pp.

Russell, R. D. (1939). Effects of transportation on sedimentary particles. In P. P. Trask, ed., *Recent Marine Sediments*. London: Thomas Murphy, pp. 32–47.

Schlichting, H. (1962). *Boundary Layer Theory*. New York: McGraw-Hill, 647 pp.

Simons, D. B. and E. V. Richardson (1963). Forms of bed roughness in alluvial channels. *ASCE Trans.*, **128**: 284–302.

Simons, D. B., E. V. Richardson, and M. L. Albertson (1961). Flume studies using medium sand (0.45 mm). *U.S. Geol. Surv. Water Supply Pap. 1498A*, 76 pp.

Smith, J. D. (1969). Geomorphology of a sand ridge. *J. Geol.*, **77**: 39–55.

Smith, J. D. (1970). Stability of a sand bed subjected to a shear flow of low Froude number. *J. Geophys. Res.*, **30**: 5928–5940.

Sorby, H. C. (1859). On the structures produced by the current present during the deposition of stratified rocks. *Geologist*, **2**: 137–147.

Southard, J. B. (1971). Representation of bed configurations in depth-velocity-size diagram. *J. Sediment. Petrol.*, **41**: 903–915.

Southard, J. B. and L. A. Boguchwal (1973). Flume experiments on the transition from ripples to lower flat bed with increasing grain size. *J. Sediment. Petrol.*, **43**: 1114–1121.

Southard, J. B. and J. R. Dingler (1971). Flume study of ripple propagation behind mounds on flat sand beds. *Sedimentology*, **16**: 257–263.

Stride, A. H. (1970). Shape and size trends for sand waves in a depositional zone of the North Sea. *Geol. Mag.*, 469–477.

Stride, A. H. (1973). Interchange of sand between coast and shelf in European tidal seas (abstract). In *Abstracts Symposium on Estuarine and Shelf Sedimentation, Bordeaux, France, July 1972*, p. 97.

Stride, A. H. (1974). Indication of long term tidal control of net sand loss or gain by European coasts. *Estuarine Coastal Mar. Sci.*, **2**: 27–36.

Stride, A. H., R. H. Belderson, and N. H. Kenyon (1972). *Longitudinal Furrows and Depositional Sand Bodies of the English Channel*. Mémoire Bureau Recherches Géologique et Minières, No. 79, pp. 233–244.

Stubblefield, W. L., J. W. Lavelle, T. F. McKinney, and D. J. P. Swift (1975). Sediment response to the present hydraulic regime on the central New Jersey shelf. *J. Sediment. Petrol.*, **15**: 227–247.

Swift, D. J. P. (1975). Tidal sand ridges and shoal retreat massifs. *Mar. Geol.*, **18**: 105-133.

Swift, D. J. P., D. B. Duane, and T. F. McKinney (1974). Ridge and swale topography of the Middle Atlantic Bight, North America: Secular response to the Holocene hydraulic regime. *Mar. Geol.*

Swift, D. J. P., B. Holliday, N. Avignone, and G. Shideler (1972a). Anatomy of a shoreface ridge system, False Cape, Virginia. *Mar. Geol.*, **12**: 59–84.

Swift, D. J. P., J. C. Ludwick, and W. R. Boehmer (1972b). Shelf sediment transport: A probability model. In D. J. P. Swift, D. B. Duane, and O. H. Pilkey, eds., *Shelf Sediment Transport: Process and Pattern.* Stroudsburg, Pa.: Dowden, Hutchinson & Ross, pp. 195–223.

Swift, D. J. P. and R. M. McMullen (1968). Preliminary studies of intertidal sand bodies in the Minas Basin, Bay of Fundy, Nova Scotia. *Can. J. Earth Sci.*, **5**: 175–183.

Van Veen, J. (1936). *Underzoekingen in der Hookden.* The Hague: Algemene Landsdrukkerij, 252 pp.

Visher, G. S. (1969). Grain size distribution and depositional processes. *J. Sediment. Petrol.*, **39**: 1074–1106.

Weil, C. B., R. D. Moose, and R. E. Sheridan (in press). A model for the evolution of linear tidal built sand ridges in Delaware Bay, U.S.A. In G. Allen and A. Klingbiel, eds., *Symposium International: Relations Sédimentaires entre estuaires et plateaux continentaux, University of Bordeaux, July 9–14, 1973.*

Wilson, I. G. (1972). Aeolian bedforms—their development and origins. *Sedimentology*, **19**: 173–210.

Wilson, I. G. (1973). Equilibrium cross-section of braided and meandering rivers. *Nature*, **241**: 393–394.

Yih, C. S. (1965). *Dynamics of Non-homogeneous Fluids.* New York: MacMillan, 235 pp.

Znamenskaya, N. S. (1965). The use of the laws of sediment dune formation in computing channel formation. *Soviet Hydrology Selected Papers (Am. Geophys. Union 1966)*, **5**: 415–432.

CHAPTER 11

Subaqueous Sediment Transport and Deposition by Sediment Gravity Flows

GERARD V. MIDDLETON

Department of Geology, McMaster University, Hamilton, Ontario, Canada

MONTY A. HAMPTON*

Department of Geology, University of Rhode Island, Kingston

Much of the coarse detritus removed from the continental margins is transported into the deep-ocean basins by "mass flows" rather than by the action of waves, tides, and ocean currents. In this chapter we discuss the definition and classification of such flows, and we describe what is known or inferred about the way in which they transport and deposit sediment.

We propose that flows consisting of sediment moving downslope under the action of gravity be called *sediment gravity flows*, synonymous with mass flows, and we distinguish four main types of such flows. We believe that the mechanisms that characterize these four types of flows are theoretically distinct, but that only one type of flow (turbidity currents) has previously been demonstrated beyond reasonable doubt to be an effective mechanism for moving coarse sediment into deep water. We further suggest some probable characteristics of sand or conglomerate beds deposited from each of the four hypothetical flow types. These suggestions are, of course, somewhat speculative in nature, but we hope that they may provide a starting point for the interpretation of some of those perplexing groups of sands and conglomerates whose deep-water environment of deposition is strongly indicated in many parts of the stratigraphic record.

DEFINITIONS

During sediment transport the average path of a grain is parallel to the bed. Movement of grains parallel to the bed, however, is not possible unless the particles are first lifted up above the stationary grains or other obstructions that are always present on the bed. It is possible to distinguish the many different transport mechanisms from one another according to the origins of the forces involved, i.e., those that cause movement of grains parallel to the bed, and those that support the grain above the bed so that movement parallel to the bed is possible.

We base our definitions of sediment gravity flows on these two fundamental aspects. First, sediment gravity flow is defined as sediment transport in which movement parallel to the bed is provided directly by the pull of gravity. Thus, sediment gravity flow is distinguished from fluid gravity flow by the relative importance of sediment and fluid in driving the flow: In a fluid gravity flow (e.g., a river or an ocean current) the fluid is moved by gravity and drives the sediment parallel to the bed, but in a sediment gravity flow it is the sediment that is moved by gravity, and the sediment motion moves the

* Present address: Marine Geology Branch, U.S. Geological Survey, Menlo Park, California 94025.

interstitial fluid. [Generally, movement of the sediment-fluid dispersion also drags along the fluid above the sediment gravity flow to produce an entrained layer (see Fig. 3) and this layer may contain a lower concentration of sediment, derived by mixing between the entrained layer and the stationary fluid above. So in practice, there is no absolute distinction between sediment gravity flows and fluid gravity flows.]

It should be noted that mechanisms such as suspension (by turbulence), saltation (by hydraulic lift forces and drag), and traction (by dragging or rolling of particles on the bottom) may all operate in some types of sediment gravity flows, as they do in some types of fluid gravity flows. In addition, it is probable that there are mechanisms such as upward intergranular flow, direct interaction between grains, and support of grains by a "strong" matrix, which are important for some sediment gravity flows but play only a minor role in the movement of sediment by most fluid flows.

Four individual sediment gravity flow types are distinguished on the basis of how grains are supported above the bed (Fig. 1): (1) *turbidity currents*, in which the sediment is supported mainly by the upward component of fluid turbulence, (2) *grain flows*, in which the sediment is supported by direct grain-to-grain interactions (collisions or close approaches), (3) *fluidized sediment flows*, in which the sediment is supported by the upward flow of fluid escaping from between the grains as the grains are settled out by gravity, and (4) *debris flows*, in which the larger grains are supported by a "matrix," that is, by a

mixture of interstitial fluid and fine sediment, which has a finite yield strength. The flows may be either subaerial or subaqueous; most of the phenomena of geological interest are submarine.

This classification is a genetic one, and we believe it is sound because it is based on the fundamental aspects of sediment transport. It could therefore be extended to a unified classification of all sediment transport processes.

Sediment gravity flow is distinguished from gravity sliding or slumping on the basis of the degree of internal deformation [extensive in flows, slight in slides, intermediate in slumps; see Varnes (1958) and Dott (1963)]. In slides, large blocks or slabs of material move on a few relatively well-defined slippage planes, but in slumps the material may break up into many blocks, and is generally internally deformed. Care must be taken to distinguish slumps from tectonic folds (Helwig, 1970) and from deformation produced by loading (Anketell et al., 1970).

It is very important to realize that in most real sediment gravity flows, more than one grain support mechanism will be important. Our definitions above are of conceptual end members, whereas real flows exist throughout a continuum between these end members. Furthermore, other mechanisms (such as traction) may operate during the last stages of deposition and produce or modify some of the textures and structures observed in the sediment bed that is finally deposited from the flow. Thus, it will be very difficult to distinguish between deposits formed from the different types of flows. Only by recognizing the full spectrum of mechanisms that may

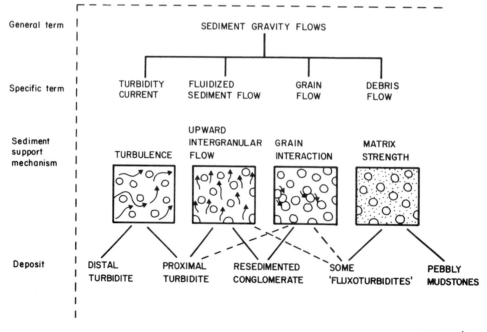

GVM, Jan, '73

FIGURE 1. *Classification of subaqueous sediment gravity flows.*

be operating during transport and deposition, however, can further progress in recognizing and interpreting the different facies of deposits be assured.

TURBIDITY CURRENTS

The concept of turbidity currents dates back at least to observations by Forel (1885) on the undercurrent formed by the Rhone River on entering Lake Geneva. Later, after construction of large reservoirs in the southwest United States and elsewhere, engineers were concerned with density currents formed by rivers entering the reservoirs. These density currents resulted from the higher density of the river water, which was due to differences in temperature or suspended sediment content (or, more usually, a combination of the two). Experimentation to determine the hydraulics of such density currents was stimulated by these occurrences and by the problem of saltwater intrusion in estuaries and locks [for reviews, see Middleton (1969, 1970)].

In geology, the concept of turbidity currents first attracted widespread interest after the suggestion by Daly (1936) that turbidity currents produced by wave action on the continental shelves during the Pleistocene low sea-level stands might have flowed down the continental shelves and eroded submarine canyons [the name was not coined by Daly, but by Johnson (1938)]. Kuenen's early experiments were performed to test the hypothesis that turbidity currents could erode canyons.

The interest soon changed, however, from the erosive capacity of dilute turbidity currents to the capacity of relatively high-concentration currents to transport sandy and pebbly sediments into deep water and to form graded beds. Kuenen (1950) produced graded beds in laboratory experiments with mixtures of sand and mud, Kuenen and Migliorini (1950) used the experiments to explain graded beds in the Apennines, and Heezen and Ewing (1952) used the results to explain cable breaks and sand layers in the Atlantic Ocean.

Following a period of intense field investigation of both recent and ancient sediments, much of the knowledge about sedimentary structures and turbidite facies was summarized by Bouma (1962). The "Bouma sequence" (Fig. 2) was interpreted in terms of laboratory flow regimes by Harms and Fahnestock (1965) and Walker (1965) and was made the basis of a detailed discussion of turbidite facies by Walker (1967); see Walker (1973) for a full discussion of the historical development of the turbidity current "paradigm."

Building on the earlier results of hydraulic engineers, renewed experimentation during the late 1960s has led to a clearer understanding of the hydraulics of turbidity currents. It has also become clear from field investigation of the coarser facies that other mechanisms besides that of the "classical" turbulent turbidity current are necessary to explain all the features observed.

Mechanics of Flow

In the oceans, most turbidity currents appear to be *surges*, that is, they are initiated by some event (presumably more or less catastrophic in nature—see below) and they move downslope away from their source. Most tur-

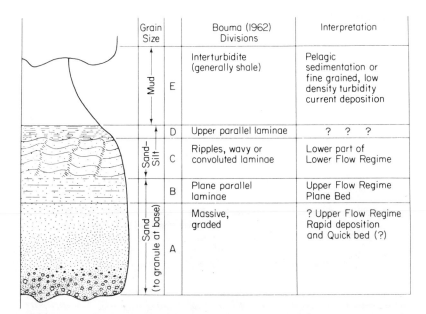

FIGURE 2. *Ideal sequence of structures in a turbidite bed (the Bouma sequence), From Bouma (1962) and Blatt et al. (1972).*

bidity currents appear to originate in submarine canyons: At first they are confined to channels [either the canyon itself or channels on the surface of the fan; see Nelson and Kulm (1973)] but later, on the distal parts of submarine fans and on the basin floor, it appears that the current is unconfined and, presumably, spreads laterally.

The current lasts only so long as there is a continued supply of sediment-water mixture, and the supply produced by the initial "catastrophe" lasts only for a short period of time. Such surges have been modeled in the laboratory by suddenly releasing a volume of denser fluid (saltwater, clay-suspension or sediment-water mixture) into a channel filled with less dense fluid, usually fresh water (see Fig. 3). Kuenen (1950) poured sediment-water mixtures into a channel from a bucket. Keulegan (1957, 1958) released salt water from a lock at the end

of a horizontal channel, and Middleton (1966a, 1967) modeled his experiments with sediment suspensions on Keulegan's experiments in order to compare the hydraulics of concentrated, coarse sediment turbidity currents with the hydraulics of saltwater (or clay-suspension) density flows. Numerical simulation by Daly and Pracht (1968) confirmed and refined the results obtained by Keulegan and Middleton. Later experimental work is reviewed by Middleton (1970).

Most surges, as they travel away from the source, develop into flows that may be divided into three main parts: the head, the body, and the tail (Fig. 3). The head is generally thicker than the rest of the flow and it has a characteristic shape and hydraulic behavior (see below), but behind the head there is a region (the body) where the flow is almost uniform in thickness. For a large surge, flow in the body changes only slowly with

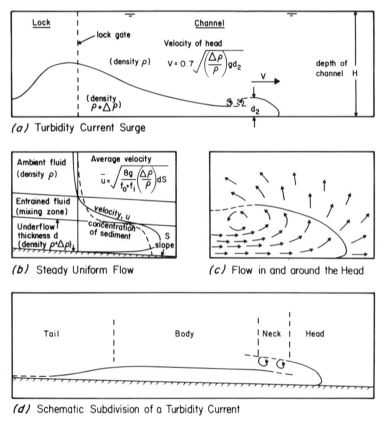

(a) Turbidity Current Surge

(b) Steady Uniform Flow

(c) Flow in and around the Head

(d) Schematic Subdivision of a Turbidity Current

FIGURE 3. *Hydraulics of turbidity currents.* (A) *Turbidity current surge, as observed in a horizontal channel after releasing suspension from a lock at one end. The velocity of the head* v, *is related to the thickness of the head* d_2, *the density difference between the turbidity current and the water above* $\Delta\rho$, *the density of the water* ρ, *and the acceleration due to gravity* g. (B) *Steady, uniform flow of a turbidity current down a slope* s. *The average velocity of flow* u *is related to the thickness of the flow* d, *the density difference, and the frictional resistance at the bottom* f_0 *and upper interface* f_i. (C) *Flow pattern within and around the head of a turbidity current.* (D) *Schematic division of a turbidity current into head, neck, body, and tail.*

time or space and so approximates a steady, uniform flow (see below). At the end of the current there must be a region (the tail) where the flow thins rapidly and becomes more dilute. Mixing between the body of the current and the water above produces an entrained layer, with a low concentration of sediment, dragged along by the current below. This layer may continue to flow, because of inertia and low friction, after the body of the current has passed by, and it is possible that it deposits some sediment and reworks the upper part of the sediment deposited by the body and tail of the turbidity current.

A second type of turbidity current that might be important in nature, and that has been extensively investigated theoretically and in the laboratory [see Middleton (1966b) for a review], is the *steady, uniform flow* of the denser fluid in a channel below a less dense fluid (Fig. 3). To maintain such a flow there must be a sloping channel and a steady supply of denser fluid from upstream. Probably these conditions are not often fully met in nature, though it seems likely that rivers heavily laden with sediment may occasionally produce steady turbidity currents in the oceans as they do in lakes.

Keulegan (1957, 1958) and Middleton (1966a) found that the *heads* of both saltwater density currents and high-density turbidity currents were characterized by a well-defined shape, flow pattern, and hydraulics (Fig. 3). Fluid and sediment sweep forward and upward through the head and circulate round to the back of the head, where a part is lost as eddies torn away by separating flow. The coarser sediment presumably falls back into the flow (to be circulated around again) but the finer sediment is incorporated into a dilute "cloud" that trails behind the head and is swept along by the entrained flow. The flow pattern has several important consequences: (1) the head may be a region of erosion (and formation of scour marks), even when deposition is taking place from the body of the current behind the head; (2) denser fluid (sediment-water mixture) must be continuously supplied into the head, to replace that lost in eddies; consequently it may be expected that the coarsest sediment will be progressively concentrated in the head; (3) the head moves somewhat more slowly than the body of the current, and, in the steeper parts of the channels, will generally be at least twice as thick [though Komar (1972) has pointed out that, on very low slopes—about 0.002—the body might be expected to become as thick as the head, so that spill over the channel levees would be from the body rather than from the head of the current].

Allen (1971a) noted that the head of experimental turbidity currents is lobate in form and has an "overhanging" front, which implies that some of the water in front of the current is trapped beneath it, particularly in "tunnels" between the lobes of the head. This trapped water must ultimately be mixed into the flow, leading to dilution of the sediment-water mixture. The overhang of the front is caused by viscous resistance at the bottom. Experiments by Simpson (1972) have shown that the ratio between the height of the overhang, h, and the thickness of the head, d_2, is a strong function of the Reynolds number (size) of the flow, and decreases from about 0.2 for a Reynolds number of 100 to 0.075 for a Reynolds number of 10,000. Further reduction in relative importance of this effect is expected at the large Reynolds numbers characteristic of natural flows.

The lobate nature of the head may have other important consequences, particularly for the formation and orientation of sole marks, such as flutes and grooves, most of which must be cut by the head and then rapidly buried by sediment deposited from the body of the flow. At least some of the divergent and crossing sole-mark directions may be explained by local divergence of flow directions within a strongly lobate turbidity current head [see Simpson (1969, 1972) for further details on the lobate nature of heads of experimental flows].

Deposition

In both rivers and turbidity currents, the main mechanism supporting sediment is turbulence. Turbulence is generated mainly at the boundaries of the flow (at the bed, in a river, and at both bed and upper boundary in the case of a turbidity current). But in a river, the movement that causes turbulence is produced by gravity acting downslope on the fluid (water), whereas in a turbidity current the water would not move downslope unless it contained suspended sediment, which makes it denser than the overlying water. There must be an initial input of energy to form the suspension before flow of a turbidity current can take place, but once flow has begun, it is possible that it may be continued almost indefinitely. In the hypothetical state of dynamic equilibrium, called "autosuspension" by Bagnold (1962), sediment is maintained in suspension by turbulence, which is generated by flow caused by the downslope action of gravity acting on a density difference that is due to sediment suspended by turbulence, i.e., there is a complete feedback loop:

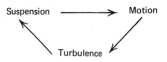

and all that is necessary to keep the system going is that the energy loss caused by friction should be compensated by the input of gravitational (potential) energy as the current moves downslope.

In this theoretical model, therefore, it is possible to have a turbidity current that neither erodes nor deposits sediment. In reality, complications arise: Sediment is lost from the current at the rear of the head or by mixing across the upper interface, or it is lost by deposition of the coarser sizes in a current that is still competent to carry the finer sizes; the current becomes more dilute by incorporating water into the head or by mixing across the upper interface of the body of the current; and not all parts of a turbidity current surge can be in the state of autosuspension at the same time. From experiments, we expect that erosion can take place at the head at the same time as deposition takes place from the body or tail of the same current.

The rate of deposition from turbidity currents must also be highly variable. In many cases it seems to have been very rapid (see below in discussion of structures). Rapid deposition from suspension may be expected in proximal regions because of decay of intense turbulence produced by whatever mechanism generated the current in the first instance. Rapid deposition may also result from rapid flattening of the slope, for example at the mouth of a submarine canyon. The energy of part of the current may be suddenly dissipated by overbank flow, or by spreading out of the current as it moves from the channel onto the basin floor or the lower part of the fan. In all these cases, there should not be an extended phase of traction, because sediment is buried almost as soon as it is deposited: The lower part of the bed should be massive or show only poorly developed faint lamination.

But in other circumstances (generally in more distal regions), where there is a much closer approach to autosuspension, there may be a relatively extended period of scouring of the bed by the head of the flow, followed by relatively slow deposition from the body and tail of the flow. Under these circumstances one expects the sole of the bed to have well-developed scour marks (e.g., flutes) and the bed itself to display traction structures such as lamination and cross-lamination.

Textures and Structures

The diagnostic features of an idealized turbidite bed were systematized by Bouma (1962) in the now familiar Bouma sequence (Fig. 2). The most characteristic features are the sole markings [well described and illustrated in the book by Dzulynski and Walton (1965)], the size grading, and the sequence of "intervals" [renamed divisions by Walker (1965)]: massive, plane laminated, cross-laminated or convoluted, and again plane laminated. Not all turbidites show all of these divisions,

but it is certainly rare to find them in a reverse order. Many different varieties of grading and sedimentary structures have been described, so the sections that follow do not pretend to be complete, but only to discuss the genetic significance of a few of the main structures in the light of modern experimental and theoretical investigations.

Sole Marks

The most common varieties of sole markings fall into four categories: scour marks (of which flutes are the most important), tool marks (grooves, striations, prod marks, bounce marks, etc.), organic markings, and load structures. All four types may be present on the same sole. Allen (1971b) has written an exhaustive treatise on flutes; he has shown convincingly from experimental studies that flutes are formed by erosion of a cohesive bed by flow separation; that is, the flute is scoured by a captive eddy rotating about a near-horizontal axis transverse to the flow. A wide variety of different forms have been described by many workers and analyzed in terms of the flow separation theory by Allen (1971b). Pett and Walker (1971) found that most flutes fall into a spectrum between deep bulbous flutes (most characteristic of A beds—those beginning with a massive division) and shallow pointed flutes (associated mainly with B or C beds—those beginning with laminated or cross-laminated divisions). The single most important conclusion to be drawn from the presence of flutes on the sole of a sandstone bed is that the flow that cut the flutes was turbulent and capable of at least minor erosion of a mud bed. It also seems to be necessary for deposition to follow erosion rapidly to ensure preservation of the flute; it has never yet been conclusively demonstrated that the bed overlying a flute was deposited by a different current from that which cut the flute. This does not mean that the current could not change flow direction; a progressive change in direction of as much as 90° has been demonstrated for some beds by Parkash and Middleton (1970).

The most common types of tool marks are grooves. Frequently more than one set is found preserved on a single sandstone bed. The most common tools appear to be shale clasts (poorly consolidated mud fragments, presumably eroded from the banks or bed of the channel only a short distance upflow). Grooves are generally straight but may be somewhat curved. Their characteristics can be explained by the large size and inertia of the tool, and its presence in the lowest part of the flow. It does not seem necessary to suppose that the flow was laminar rather than turbulent.

Prod, bounce, and brush marks are formed by impacting of small tools on the bottom. They are generally found together and covered by a bed showing well-developed lamination or cross-lamination. They imply relatively slow deposition from suspension with saltation of the larger particles over the bed.

Load Structures

Load structures form when a reversed density gradient is set up, that is, when the sand bed is denser than the mud on which it is deposited, and when the viscosity and strength of the sand and mud are low enough to permit deformation of the interface between them. Anketell et al. (1970) indicate that the form of the structure depends on the ratio of the viscosities of the two layers: If the viscosities are almost equal, the loading will produce symmetrical sinuous deformations, but if the viscosity of the lower layer is much greater (as usually seems to be the case), there will be broad, rounded loads of sand and sharp-crested "flames" of mud extending upward between them. Loading seems to be localized by initial irregularities in the mud surface, or in the sand deposited on the mud surface (many flutes show some loading also, and "load-casted ripples" are formed as the sand ripple sinks down into the mud as the ripple is being formed). An extreme form of load structure results when a sand layer liquefies and has such a low strength and viscosity that it forms isolated "pseudonodules" in the soft mud beneath. Such structures were produced experimentally by Kuenen (1958).

Structures Produced by Liquefaction

Load casting may be expected to be most prominent in sand beds deposited rapidly on a soft mud substratum. The sand will be denser than the mud, and rapid deposition may lead to an unstable fabric that breaks down either during deposition or shortly afterward, with consequent "liquefaction," i.e., loss of strength and reduction in viscosity for a period until some pore water is lost and the sand assumes a more stable fabric. Partial liquefaction of very fine to fine sand (the fine grain size and presence of some mud laminae greatly reduce permeability) seems to be the main cause of the "convolute lamination" found in many turbidite beds (particularly in the C division). Many examples of convolute lamination show a sequence of deformed internal erosion surfaces that shows that liquefaction took place at several intervals during the deposition of the bed (e.g., Sanders, 1965). Orientation of fold axes and overturning in a downcurrent direction suggest that the current drag was itself at least one cause of liquefaction.

The massive division found at the base of many turbidites is probably also to be explained in terms of very rapid deposition from suspension, which results in a period of liquefaction (or more correctly, a period when the bed behaves as though it were liquefied, because it is still compacting and consequently the grains are still partly supported by water escaping from the pores). Middleton (1967) showed experimentally that beds deposited rapidly from very high-concentration flows behaved for a short period after deposition like viscous fluids (or pseudoplastics) and were "churned" by waves moving along the interface between the "quick" bed and the entrained layer still flowing above the bed. Rapid deposition from suspension prevents the formation of traction structures such as lamination, and "churning" by interfacial waves destroys whatever traces of lamination may have been formed.

Grading

Middleton (1967) distinguished two types of size grading in experimental turbidites, both types are also found in nature. In *distribution grading* the whole size distribution shifts to finer sizes progressively from the bottom to the top of the bed. This type of grading was characteristic of graded beds deposited layer by layer from relatively dilute experimental turbidity currents. In *coarse tail grading* only the coarsest percentiles of the size distribution show the grading; in experimental flows this type of grading (and reverse grading confined to the base of the bed) was shown only by high-concentration flows. Grading is probably caused in part by decay of initial turbulence, in part by decreasing competency in the tail of the turbidity current, and in part by the concentration of coarse grains in the head of the current by the flow circulation described earlier.

Reverse grading seems to be a characteristic of high-concentration flows and is well shown by many conglomeratic sediment gravity flows (Fisher, 1971). It has been demonstrated experimentally but its cause is still not well understood.

Walker (1975) has examined many examples of conglomerates deposited by sediment gravity flows. He finds that it is possible to distinguish three basic models on the basis of the internal sequence of structures: (1) inverse-to-normally graded, (2) graded-stratified, and (3) disorganized-bed. Beds showing inverse grading at the base almost always also show normal grading at the top, but lack internal plane stratification. Plane stratification, however, *is* commonly shown by finer grained graded conglomerate beds that lack a basal division of reverse grading.

Plane Lamination

The lamination seen in the B division of many turbidites is generally not marked by clear size, shape, or density contrast between the grains. In many cases it is very diffuse, does not display good parting lineation, and although the grains generally show good alignment parallel to the current, they do not show the bimodal orientation typical of parallel lamination formed by traction over an aggrading upper flow regime plane bed.

Walker (1975) has shown that, in conglomerate beds showing inverse-to-normal grading, the pebbles are oriented with the long axis parallel to the flow and dipping upstream. In these beds, therefore, the fabric was *not* formed by "normal" traction (rolling and sliding of pebbles) which is known to produce a transverse or bimodal orientation of pebble long axes.

There is still very little information on pebble orientations in conglomerate beds showing well-developed plane stratification—in beds where the stratification is diffuse the fabric seems to be the same as in the inverse-to-normal graded beds.

Therefore, the significance of plane lamination in turbidites is not clear; some of it probably is true "upper flow regime" lamination, but some of it is probably simply the result of deposition that was rapid enough to prevent the formation of bed forms but not so rapid that all segregation of grain sizes was prevented or that a "quick" bed resulted.

Recently, Hand (1974; also Hand et al., 1972) has shown that antidunes formed by a turbidity current must have a wavelength at least 12 times the thickness of the current. Most large turbidity currents are probably more than a meter thick, so one might expect that antidunes, if formed by turbidity currents, would have wavelengths in the tens of meters. Probably the amplitude would be small, so that lamination formed by such antidunes (which might be expected to be below the B division) would be almost indistinguishable in the field from lamination formed on a plane bed. Even in flumes and rivers, lamination formed by antidunes is very faint and diffuse. A good example of antidunes has recently been described from the base of some unusual turbidites in the Cloridorme Formation by Skipper (1971).

Ripple-Cross Lamination

Most turbidite sections show at least some ripple cross-lamination (forming the C division). Some contain beds with well-developed climbing ripple drift, which results from a combination of traction (to form ripples) and rapid fallout of sediment from suspension. The different types formed depend on the relative importance of fallout and traction, as thoroughly discussed by Jopling and Walker (1968), Walker (1969), and Allen (1971c,d).

Conclusions

In summary, most of the structures observed in turbidites can be explained by a combination of erosion at the head of the current and more or less rapid deposition from the body and tail of the current. Very rapid deposition gives rise to a quick bed that compacts to form a structureless division A. Somewhat less rapid deposition may form diffuse parallel lamination, and (at the appropriate grain size) climbing ripple drift or convolute lamination. If deposition is slow enough, more extended traction takes place with the formation of typical "upper flow regime" plane lamination or ripple cross-lamination. Generally, the Bouma sequence indicates both a decreasing flow intensity (or flow regime) and a decreasing rate of deposition from suspension and increasing importance of traction as one goes from the bottom to the top of the bed. The sequence seems to fit well with the sequence that might be expected to be formed by deposition from a turbidity current surge, starting some distance behind the head and continuing as the body and tail of the current move by.

GRAIN FLOWS

The concept of grain flow came about mainly as a result of the experimental and theoretical work of R. A. Bagnold. Bagnold (1954) claimed to have measured an upward supportive stress acting on grains within flowing sediments, resulting from grain-to-grain interaction rather than fluid turbulence. This *dispersive pressure* is proportional to the shear stress transmitted between grains, and it counteracts the tendency for grains to settle out of the flow. Bagnold applied his analysis to the case of *grain flow*, or sand avalanching, where concentrated dispersions of cohesionless sediment move downslope in response to the pull of gravity, and grains remain in a dispersed state essentially by bouncing upward off one another.

F. P. Shepard and R. F. Dill have reported numerous occurrences of small-scale grain flows in the upper reaches of submarine canyons (Shepard, 1961; Dill, 1966; Shepard and Dill, 1966). They observed "rivers of sand" cascading down floors of canyons where slopes exceed the angle of repose of sand, and remarked on the ability of these flows to erode bedrock of the canyons.

Sanders (1965) speculated that grain flows may be a companion to some turbidity currents. He envisaged sediment gravity flows that consist of two parts: an upper turbidity current with grains supported by turbulence, and a lower, faster moving grain flow with grains supported by dispersive pressure. Sanders used this model to interpret sedimentary structures of some turbidites.

Inferred large-scale grain flows preserved in ancient rocks were described by Stauffer (1967). His careful description of the morphology, texture, and sedimentary structures has served as the basic model for recognition of grain flow deposits (e.g., see Klein et al., 1972). Caution must be used in applying Stauffer's model, because the transport mechanism (grain flow) is highly inferential. Link (1971) and Van der Kamp et al. (1973) have reinterpreted the Matilija sandstones (Stauffer's type "grain flow deposits") as "normal" proximal turbidites. There is a serious lack of data on the features of unequivocal grain flow deposits.

Mechanics of Flow

Bagnold's experiments consisted of shearing cohesionless sediments in the annular space between two concentric drums. He rotated the outer drum and measured shear and dispersive stresses produced on the inner drum. Bagnold found that grain flow can occur in a viscous regime, where fluid viscosity is important in producing dispersive pressure, or in an inertial regime, where viscosity is insignificant and grain inertia dominates.

In both regimes, dispersive pressure is produced by momentum change associated with grain interaction, when grains "bounce off" one another. In the viscous regime, momentum change results from interaction of grains and interstitial fluid as grains come into close proximity. In contrast, under inertial conditions the influence of the interstitial fluid is insignificant and dispersive pressure is generated as grains undergo elastic solid-to-solid collisions. A transition region, where both fluid viscosity and grain inertia are important, exists between the two regimes.

The relationship between shear stress T and dispersive pressure P, expressed as $T/P = \tan \phi$, has distinct constant values for the inertial and the viscous regimes. The angle ϕ is interpreted as a "dynamic angle of internal friction," and values of 37° for viscous flow and 18° for inertial flow were determined from Bagnold's experiments. The magnitude of ϕ varies with sediment type, however, and the experimental values above cannot be generally applied to sediments. In particular, inertial ϕ probably exceeds 18° for most natural sediments. For example, Bagnold (1966) measured a value of 32° for a beach sand and concluded that inertial ϕ should be somewhere between 0 and 8° less than the static angle of internal friction, which typically is 30 to 35° for natural sediments. Predictions about viscous ϕ are not currently possible, except that it is larger than inertial ϕ.

Analogous to using the static angle of internal friction for sand (approximately 30°) to calculate the angle of repose, the dynamic angle of internal friction can be used to calculate the slope angle necessary for sustained movement of grain flows. For the case of a subaqueous grain flow with plain water as the interstitial fluid, the minimum slope angle necessary for movement is equal to the angle of internal friction. Thus, viscous regime grain flows (which occur at slow flow velocities, for example) using Bagnold's experimental "sand" require a rather steep slope of 37° to sustain movement, whereas an inertial regime flow would move on a somewhat gentler, though still steep slope of 18°. Grain flows in real sand typically would require slopes greater than 18°, up to 30°+. Of course, such slopes are rare on a regional scale on the ocean floor, so pure grain flows probably do not account for significant long-distance transport of marine sediments.

Bagnold (1954) and Middleton (1970) suggested, however, that some grain flows might move on relatively gentle slopes. If the density of the interstitial fluid is greater than that of plain water, the effective weight of the solid grains is reduced by buoyancy which therefore reduces the magnitude of dispersive pressure necessary to support the grains. Dispersive pressure ultimately is generated by the downslope pull of gravity on the grains, which varies directly with the slope angle. Thus, smaller dispersive pressures require smaller slope angles. The added weight of a high-density fluid also increases the downslope driving force.

Middleton (1970) calculated that if the interstitial fluid is mud with a density of 2.0 g/cm³, the slope angle necessary for viscous grain flow is 10.5° and for inertial flow is only 4.5°. (This calculation assumed a solid grain density of 2.6 g/cm³ and a solid grain concentration of 0.55.) However, mud (clay and water) is a cohesive mixture, so the theory developed for grain flow does not strictly apply to this sediment. Debris flow theory, discussed on the following pages, describes the flow of cohesive sediments.

Some question exists as to the role of turbulence in grain flows. Some workers have described grain flow as "laminar shear of cohesionless sediment" implying the absence of turbulence (e.g., Stauffer, 1967, p. 502). Also Bagnold (1954, p. 60; 1956, p. 288) concluded from his experiments that high solid-grain concentrations suppress small-scale turbulence in interstitial fluids, and thus that the role of turbulence in grain flow is negligible. Probably, though, at high-energy conditions flow might become turbulent. In this event, turbulence would help support the solid grains, in addition to the support provided by dispersive pressure. Thus, a continuum may exist between grain flows and turbidity currents and some flows may be combinations of the two types.

Deposition

Deposition from grain flows is fundamentally by mass emplacement. In contrast to traction-current deposition, typical of rivers and ocean currents, in which grains are laid down particle by particle upon the nonmoving bed, mass emplacement ideally involves a sudden "freezing" of the flow resulting in simultaneous deposition of a layer several grains thick. "Freezing" occurs when the driving stress becomes less than that necessary to propel the flow. Reduction of driving stress typically occurs by a reduction in slope angle, and deposition occurs when slope angle becomes less than the value of ϕ.

Textures and Structures

Simple experimental grain flows can be produced by causing sand to slide down an inclined chute or down the slope of a sand pile. An obvious feature of these flows is lateral and vertical inverse grading. Large grains migrate to the top and front of the flow. Inverse graded bedding should therefore be a feature of grain flow deposits (although not necessarily restricted to them). Fisher and Mattinson (1968) reported inverse grading in Wheeler Gorge conglomerates in California and suggested a high-density, highly fluid emplacement mechanism, perhaps grain flow.

Two ideas have been proposed for the mechanism by which inverse grading forms. Bagnold (1954, p. 62; 1968, p. 50) concluded that dispersive pressure acts to size-sort grains, pushing larger grains to the top of the flow (the area of least shear strain rate), and smaller grains move to the bottom, producing reverse grading. According to Bagnold's theory, this sorting effect decreases as flow becomes more viscous, for example at

low velocities or for small grain sizes. Middleton (1970) disagreed with the dynamic sorting mechanism proposed by Bagnold and suggested that reverse grading is the product of a kinetic sieve mechanism whereby small grains fall downward between large grains during flow, displacing the large grains upward. Shake a box of popcorn or a container of dry sand to see this effect.

Stauffer (1967) described supposed grain flow beds in Tertiary rocks of California as being thick and sharply bounded (Fig. 4). Soles of the beds are either flat and featureless or they have one or more "peculiar sole markings." Included as sole markings are frondescent marks, slide marks, load structures, and "ropy" sole marks. Internally, the grain flow beds described by Stauffer typically are massive and ungraded, but diffuse parallel lamination and dish structures occur in some beds. Stauffer also reported the presence of large, outsized lutite and siliceous clasts in grain flow beds. These clasts occur at any level within a bed, commonly floating in a sandy or muddy matrix. Occasionally, clasts are abundant enough for the rocks to be classed as conglomerates.

Rees (1968) showed that grain flow in sands (and probably also in coarser sediments) produces grain orientation parallel to the flow and imbricated upflow. This is similar to the grain orientation produced in sands by deposition from suspension (with minor traction). At present there is no real information about grain orientation in fluidized sediment flows; possibly it would be of the same type.

Stauffer suggested that grain flows are further characterized by the absence of traction current or turbidity current features, such as erosional sole markings (e.g., flute casts), convolute lamination, cross-lamination, ripple marks, and current scours.

Positive identification of grain flow deposits is still problematical. All features proposed to be characteristic of grain flows probably can be produced by other transport processes. Dish structure and flute casts occur together in Cambrian sandstones in Quebec. In addition, debris flow deposits typically contain floating outsized clasts (Johnson, 1970, p. 461). Slide marks are characteristic of certain turbidity current and debris flow deposits (Hampton, 1970).

Another point to consider is that, in the absence of excess pore pressures, relatively large slopes (18–37°) are required for true grain flow. Slopes of this magnitude are rare in the ocean, suggesting that most grain flow deposits are localized features.

Conceptually, the mechanics of true grain flow are quite clear—gravity flow of cohesionless sediments with grains supported by dispersive pressure—but the manifestations of this mechanism as preserved in sedimentary

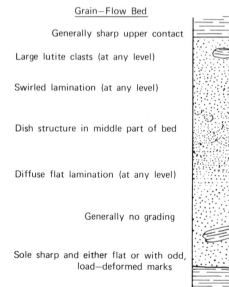

Grain–Flow Bed

Generally sharp upper contact

Large lutite clasts (at any level)

Swirled lamination (at any level)

Dish structure in middle part of bed

Diffuse flat lamination (at any level)

Generally no grading

Sole sharp and either flat or with odd, load–deformed marks

FIGURE 4. *Typical sequence of structures in supposed "grain flow" sandstones, Tertiary of California. From Stauffer (1967).*

deposits are speculative at the present time. The point to remember is that grain flow basically is an end member of a spectrum of sediment gravity flow processes. A given flow may have combined characteristics of grain flow, debris flow, and turbidity current flow. Furthermore, the nature of a given flow may vary in time and space, giving rise to deposits with combined characteristics of several processes.

FLUIDIZED SEDIMENT FLOWS

The term "fluidization" has been used by chemical engineers to describe the process of expanding a granular bed by an upward flow of fluid through it, to the point where the expanded bed is a concentrated dispersion that behaves in some ways like a viscous fluid. Certain sands have an unstable fabric that may be destroyed by a sudden shock (or series of shocks) with consequent loss of strength or "liquefaction" (Casagrande, 1936; Terzaghi, 1956; Seed, 1968). A liquefied (or "quick") sand is in a state similar to that of a fluidized bed, except that the state is transient and the bed gradually solidifies from the base up as fluid is lost upward through the pores and consolidation takes place. The concept of fluidization was applied to volcanic pipes by Reynolds (1954) and to ash flows by McTaggart (1960). Andresen and Bjerrum (1967) described several examples of sediment gravity flows caused by liquefaction and Van der Knaap and Eijpe (1969) suggested that liquefaction might be the cause of some turbidity currents. The application to sediment gravity flows and to deposition from high-concentration turbidity currents was also discussed by Middleton (1969, 1970) and Hendry (1973).

Mechanics

Not all sands or gravels are subject to spontaneous liquefaction. Shepard and Dill (1966, p. 48) investigated the sands accumulating in the head of Scripps Canyon and concluded that they would not be subject to liquefaction. Densely packed sands resist liquefaction because it is necessary to expand the fabric (increase the porosity) in order to shear the sand. Sands subject to liquefaction are those that are loosely packed or that have textural properties (at present little understood) that are favorable to breakdown of the fabric [for a recent experimental investigation see Castro (1969)].

Once the fabric is destroyed, the grains no longer form a rigid framework to the deposit, but must be supported at least in part by the pore fluids: consequently pore fluid pressures rise much above normal hydrostatic pressures ("excess pore pressures"). So long as grains are supported by the pore fluid, the sand has little strength and behaves like a fluid with a viscosity on the order of 1000 times that of water. Thus liquefied or fluidized sand can flow rapidly down relatively gentle slopes (3–10°); see Middleton (1969).

The main problem with fluidization as a mechanism for sediment transport is that the excess pore pressures are rapidly dissipated by loss of pore fluids: Van der Knaap and Eijpe (1969) and Middleton (1969) calculate times of the order of a few hours for persistence of liquefaction of a fine sand layer 10 m thick, and the times become much shorter as the grain size increases or the thickness decreases.

The real significance of fluidization, however, is its combined occurrence with other support mechanisms, such as turbulence of dispersive pressure. Grain flows, in particular, can be expected to be aided by fluidization effects, because during flow the grains, by necessity, are in a relatively dispersed state, but deposition involves consolidation with consequent upward expulsion of pore water. The upward fluid drag is a fluidization effect, helping to support grains, and thereby prolonging the life of a flow.

The "fluidization effect" may even be felt after flow ceases, if deposition is rapid enough to trap excess pore water, which is eventually expelled by consolidation. Upward drag may not be strong enough to fluidize the bed completely in these instances, but it might mobilize enough grains to affect textures and structures significantly.

Deposition

Deposition from a fluidized sediment flow takes place because of loss of pore fluids. It may be expected, therefore, that the flow will gradually "freeze" from the bottom up, though it is also possible that the whole flow will come to rest on a slope too low to overcome internal friction.

Textures and Structures

Some of the structures produced by liquefaction phenomena have already been discussed under turbidity currents. One must add quicksand phenomena such as sand sills and dykes and related injection phenomena. Some of these may be formed some time after deposition of the sand and may be caused, perhaps, by a combination of supply of water from compacting clays and strong vibration by earthquake shocks.

Fluidized sediment flows may be expected to be so concentrated that it will be impossible for normal traction structures to form. However, the pore-fluid expulsion that causes fluidization can cause segregation of grains to occur by dragging some grains to the upper levels of the sediment. This process is especially effective while the sediment is flowing but also occurs after deposition. Upward movement of a particular grain takes place when the pore-fluid expulsion drag exceeds the grain's terminal settling drag and when the grain is small enough to make its way between the surrounding, nonrising grains. Thus, both settling velocity and grain size are important in controlling the mobility of a particular grain (Schwab and Hampton, 1975).

The enigmatic "dish structure," as named by Wentworth (1967) and described by Stauffer (1967), may possibly result from consolidation of a fluidized bed. Lowe and LoPiccolo (1974) recently presented detailed descriptions of dish structures from several localities. The structure consists of a series of concave-upward laminations, generally in the middle part of a bed but occurring at any and all levels of some beds. They are found most typically in massive to faintly laminated, medium- to coarse-grained sands, but also in beds of other grain sizes and in beds with ripples or convolute laminations. The dishes are generally about 4 to 50 cm wide and 1

or 2 cm deep. They are typically flatter near the bottom of a division and grade up into an anastomosing pattern and, further upward, into tightly curved dishes. Pillar structures, which are approximately vertical crosscutting columns of sand, commonly occur with dishes.

Dish structure is known from the Jackfork Group of Oklahoma (Lowe and LoPiccolo, 1974), the Eocene of California (Stauffer, 1967), Cretaceous-Paleocene of California (Chipping, 1972), several Paleozoic formations in Quebec—Cap Enragé at Bic (Hubert et al., 1970), Cap Enragé at St. Simon [unpublished theses by Mathey (1970) and Davies (1972)], St. Damase formation (?), St. Roch formation, and Kamouraska formation—and from the Marnoso-Arenacea formation of Italy (Ricci Lucchi, personal communication, 1971) and several other localities. (See Fig. 5.)

Lowe and LoPiccolo (1974) suggest that dishes and pillars are formed by pore-fluid expulsion during consolidation of rapidly deposited sediments. Semipermeable laminations within the bed retard the upward-moving pore fluid, forcing some of it to migrate horizontally to "weak" points where it can continue to escape upward. Fine sediment and planar and low-density grains are

FIGURE 5. *Dish structure, Cap Enragé formation. Scale in centimeters.*

filtered out by the semipermeable laminations which, when deformed, define the dish shape. Relatively violent, abrupt expulsion, breaking through laminations, forms pillars.

DEBRIS FLOW

Subaerial debris flow has been recognized as a significant sediment transport process in arid and semiarid regions since Blackwelder's (1928) classic paper on the subject. More recent discussions of debris flow mechanics and deposits are those of Hooke (1967) and Johnson (1965, 1970). Johnson developed a model for debris flow that can be used to characterize this process.

Debris flows are a concern in some urban and suburban areas, where they pose a threat to life and property (e.g., Kenny, 1969; Jahns, 1969).

Subaqueous debris flows are only now receiving consideration as a major transporter of ocean sediments. Hampton (1970, 1972a,b, 1975) and Fisher (1971) have published studies of submarine debris flow. Mountjoy and Playford (1972) and Stanley (1974) have described suspected debris flow deposits in ancient and modern sediments. Hampton has emphasized the possible importance of fine-grained debris flow activity in the ocean. Previously debris flow, or some allusion to it, was called on only to account for rather uncommon muddy, coarse-grained marine conglomerates, such as pebbly mudstone (Crowell, 1957; Shepard, 1963, p. 328; Peterson, 1965). Interpretation of these deposits was based mainly on textural comparison with subaerial debris flow deposits, which typically consist of large clasts set in a muddy, typically clay-rich matrix. Hampton (1970, 1972a, 1975) has demonstrated that sandy debris flows should be common in the ocean, and that sandy submarine debris flow deposits can be low in clay similar to sands emplaced by other processes.

Mechanics of Flow

Debris flow refers to the sluggish downslope movement of mixtures of granular solids (e.g., sand grains, boulders), clay minerals, and water in response to the pull of gravity. Rates of flow are assumed to be greater than those of soil creep. Debris flow essentially resembles flow of wet concrete. Most observations of debris flow have been made on land, but they can be readily extrapolated to include submarine debris flow.

Debris flows on land typically are episodic events; they are more frequent in the spring season, when water is exceptionally abundant. An individual flow moves in a series of waves, or surges. Large boulders, several feet in diameter, are commonly present. They are rafted along rather gently in the flow, slowly shifting position but not being violently tossed about. Debris flows can transport large objects while moving sluggishly. They have been observed to travel over slopes as low as 1 or 2° (Curry, 1966; Sharp and Nobles, 1953).

Granular solids in a debris flow are more or less "floated" during transport. Recall that debris is composed of clay minerals, other granular solids, and water. The clay minerals and water combine together to act as a single fluid, and this fluid has finite cohesion (strength), a property not possessed by plain water. Yield strength provides a major amount of support for solid grains in a debris flow. Thus, grains are supported much as the floor supports a chair. In addition, clay-water fluids have greater densities than plain water; thus they provide greater buoyancy. Support of grains by cohesion of the fluid phase distinguishes true debris flow from grain flow and turbidity current flow.

Johnson (1965; 1970, Chapter 14) developed a rheological model that describes the flow behavior of debris, as well as sediment transport by debris flow. Equations derived from his model predict the velocity profile within a debris flow and the critical conditions necessary for maintaining debris flow.

Johnson's model, referred to as the Coulomb-viscous model, has the form

$$T = c + \sigma_\mu \tan \phi + \mu\dot{\epsilon}, \qquad T > c + \sigma_\mu \tan \phi$$

where T is internal shear stress, σ_μ is internal normal stress, ϕ is the angle of internal friction, μ is viscosity, and $\dot{\epsilon}$ is the rate of shear strain (velocity gradient). The inequality on the right means that flow occurs if the shear stress exceeds the total yield strength of the debris. (The total yield strength has components of cohesion c and intergranular friction $\sigma_\mu \tan \phi$.) The statement on the left says that within a steadily moving debris flow, the driving stress is equal to the resistance supplied by cohesion c plus friction $\sigma_\mu \tan \phi$ plus viscosity $\mu\dot{\epsilon}$. Note that if the total strength is zero, this equation reduces to the Newtonian equation, $T = \mu\dot{\epsilon}$, which describes the flow of water.

Competence of a true debris flow is controlled by the strength and density of the clay-water fluid. Grains are supported by strength and buoyancy, rather than by turbulence, upward escape of fluid, or dispersive pressure. The largest grain suspended in a debris flow has a downward-acting weight that just equals the upward-acting support of fluid strength and buoyancy. Larger grains overcome the strength, sink downward, and thus settle out of the flow.

Competence of a clay-water fluid is approximated by the following expression:

$$D_{max} = \frac{8 \cdot 8c}{g(\rho - \rho_f)}$$

where D_{max} is the diameter of the largest grain suspended in the fluid, c is the cohesion of the fluid, g is the acceleration of gravity, ρ is the density of the solid grain, and ρ_f is the density of the fluid. This equation assumes spherical grains. Thus competence, expressed as D_{max}, increases with cohesion and with decreasing density contrast between the grains and the fluid.

Hampton (1970, 1972a, 1975) applied this equation to hypothetical submarine debris flows and demonstrated experimentally that the amount of clay, relative to solid grains, necessary to support sand-size material is on the order of 2 to 20%. He concluded, therefore, that the textures of sandy debris flow deposits might be similar to those of sands deposited by other processes. Identification of sandy debris flow deposits on the basis of texture (grain-size distribution) might be difficult or impossible.

Subaerial debris flows have been observed to have surprising competence, even with low clay content. Flows carrying boulders several feet in diameter have been reported (Jahns, 1949, 1969; Johnson, 1970, p. 486; Curry, 1966). Curry, for example, reported 2.5 ft boulders in a flow with only 1.1% clay-size material in the matrix. Water content of the matrix was low—9.1%—and density of the matrix was high—2.53 g/cm³.

Transport of solid grains by debris flow, as considered above, is in dispersion. Grains also may be transported by traction, being dragged along as bed load, but the significance of bed load transport in debris flow has not been evaluated.

Another point is that dispersive pressure probably is significant in most debris flows. Bagnold (1954) measured dispersive pressure in sediment flows with as little as 9% volume concentration of solid grains. Debris flows with this concentration of grains, or greater, should generate dispersive pressure that would help to support grains. However, strength of the fluid phase affects the role of dispersive pressure. In grain flow the dispersive pressure must be great enough only to lift the weight of the grains. But in the presence of a clay-water fluid, if dispersive pressure is to lift grains it must also overcome the strength of the overlying fluid. On the other hand, once grains become dispersed the strength of the fluid phase tends to keep them so, and therefore the role of dispersive pressure might be reduced.

The mobility of a debris flow, in terms of the minimum slope angle and thickness of debris required to maintain flow, is determined by the magnitudes of matrix strength, matrix density, and dispersive pressure. Pure debris flow, with a fluid matrix strong and dense enough to support all grains and with no dispersive pressure, will move on any slope, no matter how gentle, if the thickness of the debris is great enough. Dispersive pressure effects introduce the necessity of some minimum slope angle, regardless of thickness (Hampton, in preparation).

Once again, it should be emphasized that true debris flow, with support of grains by strength, is best used as a conceptual model, one end member of the complete spectrum of sediment gravity flow. Debris flow in nature might involve dispersive pressure, and perhaps even turbulence, but strength and buoyancy are the main support mechanisms.

Deposition

Deposition from debris flow occurs by mass emplacement, when the driving stress of gravity decreases below the strength of the debris and "freezing" occurs. Johnson (1970, p. 435) recognized *lateral* and *medial* deposits from debris flows. Lateral deposits form on the sides of the channel in which the flow moved, as terraces, or they form adjacent to the channel, as levees. Medial deposits form in the center of the channel.

Lateral and medial deposits are explained as follows. Johnson calculated that debris flowing in channels must have certain regions of nondeformation, where the applied stress is less than the strength. These regions typically exist along the edge of the channel (dead regions of Fig. 6) and in the central portion of the flow (rigid plug of Fig. 6). Material in the dead regions forms the lateral deposits and in some shapes of channels (e.g., triangular) they do form medial deposits.

The rigid plug is characteristic of debris flow and plays an important role in movement and deposition.

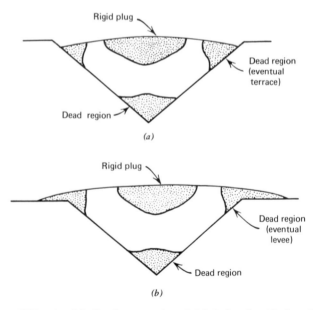

FIGURE 6. *Idealized cross sections of debris flow in a V-shaped channel.* (a) *Flow is contained within the channel.* (b) *Flow overtops the channel. Modified from Johnson (1970).*

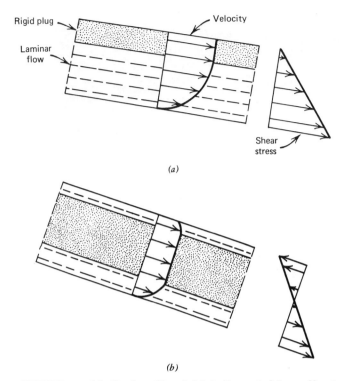

(a)

(b)

FIGURE 7. *Idealized profiles of debris flows. At left: profile of debris flow, showing rigid plug and zones of flow. Velocity profile also is shown. At right: internal shear stress distribution (driving stress). (a) Flow with no shear stress at upper surface of flow (e.g., subaerial flow). Internal shear stress at any level represents downslope component of weight of overburden. Rigid plug is rafted along on top of the flow. (b) Flow with significant shear stress at the upper surface (e.g. subaqueous debris flow). Internal shear stress at any level is the difference between the downslope component of the overburden and the interface stress (generated by frictional resistance of the overlying water). Rigid plug floats within the flow, with zones of shear above and below.*

The rigid plug is rafted along on top of debris that is in a state of flow. In some subaqueous flows—those that have a significantly large stress imposed at the upper interface by the overlying water—the plug floats below the surface (Fig. 7).

The thickness of the plug varies directly with strength and inversely with density and slope angle. If for some reason the thickness of a debris flow decreases, or if the thickness of the plug increases (because of a decrease in slope angle, for instance), the flow comes to a stop when the thickness of the plug equals the thickness of the flow. This forms medial deposits.

The terminus of a debris flow deposit typically has a blunt, steep-sided profile similar to the meniscus formed at the edge of a drop of water. This form, in contrast to a wedge-shaped or tapered profile, is due to the strength properties of the debris.

The effect of continued debris flow deposition in an area is a general smoothing of the topography (Johnson,

1970, p. 513; Hooke, 1967, p. 452). Surface irregularities with relief somewhat less than the thickness of the rigid plug will not be reflected at the surface of an overlying debris flow deposit.

Johnson's explanation of subaerial debris flow deposition should apply generally to subaqueous debris flow.

Textures and Structures

Textures of bouldery debris flow deposits resemble those of tillites, which has resulted in misinterpretation of some ancient deposits (Dott, 1961). As previously pointed out, sandy debris flow deposits can have textures similar to other sandstones, with regard to low clay content. Judging from the competence equation, however, debris flow should have positive-skewed size distributions (fine tail). Assuming a source material with an approximate normal distribution of particle sizes, all those grains too large to be supported by the fluid should be left near the source while the rest of the material is carried to the site of final deposition. The resulting size distribution would be abruptly truncated at some maximum size. This skewness might be obscured by a strength sufficient to carry almost all material available, or by participation of other support mechanisms, or by a late-stage period of bed load transport.

Internally, bouldery debris flow deposits are typically massive. Large boulders are set at random in a fine-grained matrix. Hampton (1975) speculates that fine-grained debris flow deposits might contain grain-size lamination, perhaps with normal or inverse grading. He determined experimentally that fluid matrix strength, and hence matrix competence, decreases when the matrix is sheared, so that finer grain sizes should be carried in the zones of flow than in the rigid plug (Fig. 7). Since debris flow deposition preserves these zones, coarse-fine laminations should result.

Bedding-plane structures of debris flows should be similar to those of grain flow deposits with load casts (?) and slide marks. Johnson and Hampton (1969) and Hampton (1970) observed that debris flows typically do not abrade the underlying bed unless pull-apart and rigid-block sliding occur, in which case slide marks can be produced. Pull-apart involves tensile separation of a debris flow due to loss of friction at the base of the flow, for example when the channel is smooth and wet. Blocks produced by separation slide along rigidly ahead of the main flow.

Pull-aparts have been observed in natural subaerial debris flows and in experimental subaqueous debris flows. The experimental subaqueous debris flows produced slide marks (low-relief, parallel grooves or striations) on the channel (Fig. 8). This is the type of feature

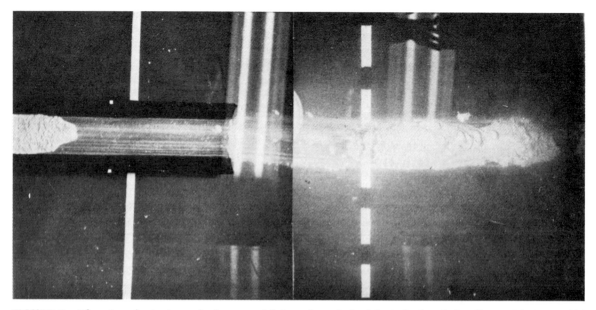

FIGURE 8.   *Plan view of experimental subaqueous debris flow, showing pull-apart and parallel grooves. Main body of the flow, at left, is contained in a 6 in. semicircular*   *channel. At right is the detached pull-apart block. Parallel grooves, created by the rigidly sliding pull-apart block, are evident in the channel ahead of the main body of the flow.*

described by Stauffer at the base of grain flow deposits. Hampton has observed similar features in supposed sediment gravity-flow deposits in the Cretaceous Venado formation of California and the Eocene Tyee formation of Oregon.

Pull-aparts (and slide marks) can form in mass emplacements other than debris flows. They may account for displaced blocks showing evidence of previous flow (and therefore not landslide or slump deposits) that are found in ancient rocks (Hendry, 1972; Mountjoy et al., 1972).

A summary of the types of sedimentary structures that might be found in beds deposited by hypothetical ("pure end-member") turbidity currents, grain flows, fluidized flows and debris flows, is shown in Figure 9.

INITIATION OF SEDIMENT GRAVITY FLOWS

There are three main aspects to the problem of initiation of sediment gravity flows: (1) What causes mass movement to begin? (2) How do submarine mass movements, such as slides and slumps, turn into sediment gravity flows? (3) What transitions are possible between one type of sediment gravity flow and another, and what causes them to take place? Answers to these questions can only be tentative at present. The whole question of initiation remains one of the least studied and least understood aspects of sediment gravity flows.

It has generally been held that one of the main causes of slumping and sliding is earthquakes. The association of slumping with earthquakes is certainly well documented, the best known example being the Grand Banks earthquake, slump, and turbidity current of 1929 (Heezen and Ewing, 1952; Emery et al., 1970). Other examples of major submarine slumps are described by Menard (1964). Not all slumps and slides are triggered by earthquakes: In many cases slumping is probably due mainly to rapid deposition of sediment on a slope, leading to underconsolidation of muds, formations of large excess pore fluid pressures, and consequent low internal angles of friction (Morgenstern, 1967). Spontaneous liquefaction of sensitive muds or sands may be produced by relatively minor triggering events.

One sequence of events suggested by many authors is the gradual filling of the head of a submarine canyon with sediment moved into it by longshore drift, followed by the periodic emptying of the canyon head by mass movement or gravity flow. This appears to be the case for Scripps Canyon, though Shepard and Dill (1966) and Dill (1964, 1967, 1969) have demonstrated that earthquakes do not always result in emptying of the canyon-head fill and that emptying is not triggered by dynamite blasts. They have also shown that the sand fill does not have soil mechanical properties that would make spontaneous liquefaction probable. Sand movement down the very steep upper part of the canyon takes place by slow creep and by local grain flow ("sand

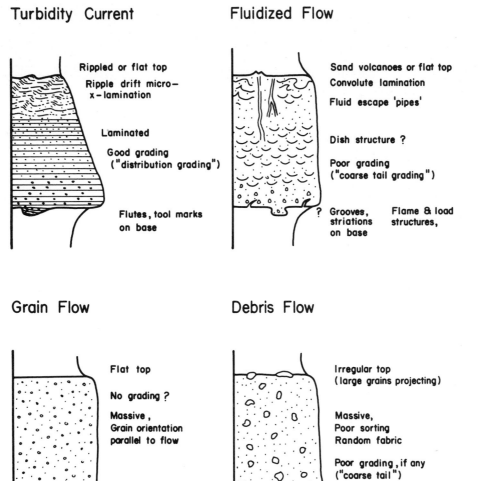

FIGURE 9. *Sequence of structures in hypothetical single-mechanism deposits.*

flows"). Sand may also be moved by currents, with speeds of as much as 50 cm/sec, that move alternately up and down the canyon (Shepard and Marshall, 1973), but there is still no satisfactory explanation for the sudden disappearance of sand from the head of the canyon that has been observed to take place.

Hampton (1972b) has observed the subaerial generation of debris flows by remolding of slides and slumps and incorporation of water. It seems probable that submarine sliding and slumping would be much more likely to give rise to debris flows, because the main requirement besides mechanical shock and strain (which breaks down the fabric of unconsolidated sediment) is the addition of a small quantity of water. Mixing in only a small percentage of water is generally enough to reduce the

strength of unconsolidated sediment many times, to the point where flow is possible.

Hampton (1972b) also observed some cases where flows were generated by spontaneous liquefaction. Other, submarine, examples have been described by Andresen and Bjerrum (1967) and Middleton (1969).

There remains the major topic of the conversion of one form of sediment gravity flow into another and particularly the generation of turbidity currents from other, higher density, types of flows. The basic problems are how to mix enough water into the flow and generate the turbulence necessary to produce a turbidity current.

Turbulence will be generated in a flow by friction at the upper or lower interface or by flow separation phenomena, provided that the scale and speed are large enough to overcome the damping effects of viscosity and strength. For a flow without strength (a fluidized sediment flow or a grain flow) the appropriate criterion is a Reynolds number. Flow will become turbulent only when a critical value (about 2000) of the Reynolds number is exceeded (Re = thickness × speed/kinematic viscosity). For a flow with both viscosity and strength (such as a debris flow) the criterion is also a Reynolds number, but the critical value of the Reynolds number is a function of the Bingham number, which depends on the strength (Hampton, 1972b). In general, high concentrations of sediment increase the size and speed of flow that can remain viscous by several orders of magnitude, as compared with clear-water flows. Nevertheless, it seems probable that very large highly concentrated sediment flows moving down a relatively steep slope would accelerate rapidly to the point where they would become turbulent.

Mixing of water with a concentrated sediment gravity flow is most likely to take place in three ways: (1) Turbulent mixing across the upper interface. This has already been discussed with reference to turbidity currents; the degree of mixing probably depends mainly on the Froude number, and is likely to be most effective on steep slopes and for low concentrations. High concentrations (and therefore high density contrasts between the flow and the water above) tend to reduce mixing across the interface. (2) Mixing into the flow of water (or mud) trapped beneath the flow, as suggested by Allen (1971a) and already discussed above. (3) Mixing by flow separation at the head of the flow. In a turbidity current, flow separation takes place in the lee of the head, so that the current trails behind a dilute cloud of sediment torn away from the lee of the head. In sub-aqueous debris flows, Hampton (1972b) has shown that flow separation takes place on the upper part of the "snout" of the debris flow and that such separation is very effective in eroding away the snout and producing a cloud of less concentrated suspension, which then moves away as a turbidity current, independent of the debris flow which gave rise to it.

In conclusion, it is probable that there will be transitions between the various possible highly concentrated sediment gravity flows, partly because of the probability that two or more sediment support mechanisms will operate simultaneously (for example, dispersive pressure and fluidization, or matrix strength and dispersive pressure), and partly because of the instability of large, high-speed flows, which tend to become turbulent as the flow accelerates and then to cease being turbulent as the flow decelerates just before the main phase of deposition. Probably it will be difficult to mix enough water into such a flow, across either the upper or the lower interface, to cause a gradual transition to a much less concentrated flow. Possibly such a transition might be produced in some types of flows by progressive deposition of sediment out of the flow. Generally, however, it seems more probable that mixing processes, and particularly mixing at the head of the flow as described by Hampton, would produce a much more dilute suspension. It might be expected, in other words, that most sediment gravity flows would be of two main types: (1) highly concentrated flows, where sediment is supported by a range of different mechanisms, including turbulence, and (2) relatively low concentration flows, with sediment supported by turbulence (turbidity currents). The densities of the highly concentrated flows might be expected to be only slightly less than those of unconsolidated sediment (per-

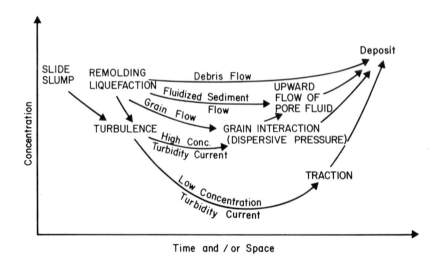

Figure 10. *Hypothetical evolution of a single flow, either in time or in space. See text for explanation.*

haps 1.5–2.4 g/cm³) whereas the density of the relatively low-concentration turbidity currents might be expected to be in the range 1.03 to 1.2 g/cm³ (Fig. 10).

SUMMARY

Sediment gravity flows are flows in which sediment is moved parallel to the bed directly by gravity. In fluid gravity flows, for example rivers, gravity moves the fluid, and the moving fluid causes sediment movement. In sediment gravity flows, gravity acts directly on the sediment grains, and sediment movement causes movement of intergranular fluids. In all sediment gravity flows, sediment must not only be moved parallel to the bed, but it must be held in dispersion, i.e., some mechanism must act to prevent sediment from settling onto the bed. Four different types of sediment gravity flows are distinguished on the basis of how grains are supported: (1) turbidity currents (grains supported by turbulence), (2) grain flows (grains supported by direct grain interactions), (3) fluidized sediment flows (grains supported by upward escape of intergranular fluid), (4) debris flows (grains supported by matrix strength).

Most turbidity currents are surges, in which a head, body, and tail may be distinguished (see Fig. 3). The head is thicker than the body and has a characteristic shape and hydraulic behavior. It moves at a speed which, on slopes greater than about 1 in 1000, tends to be less than that of the flow behind the head. Sediment is swept from the body of the flow into the head, and lost in turbulent eddies that separate from the rear of the head. There is generally erosion of the bed and little deposition of sediment from the head: the coarsest sediment in the flow tends to become concentrated into the head. The body of the current approaches uniform conditions. Friction at the upper interface causes mixing with and entrainment of the fluid above the turbidity current. In the tail of the current, thinning and dilution by mixing and deposition cause deceleration and eventual dissipation of the current. The characteristic sequence of structures observed in turbidites—erosional sole marks, massive (A) division, laminated (B) division, cross-laminated (C) division, and size grading—may be explained by deposition progressively from the head to the tail of such a current. It seems, however, that the formation of the massive (A) division may be due to a transient stage, resulting from rapid deposition, when the bed, which is still partly fluid, shows some of the characteristics of grain flow, fluidized sediment flow, or debris flow.

Grain flows generally are only possible on very steep slopes (about 20° or more); however, grain interactions (dispersive pressure) may act in combination with other sediment gravity flow mechanisms. Deposition from grain flows is by mass emplacement and probably produces inverse grading and grain orientation parallel to the flow.

Fluidized sediment flows can move very rapidly on slopes as low as 1 in 100 but the support mechanism is rapidly dissipated as intergranular fluid escapes. Fluid escape structures, such as dish structures, pillars, and vertical sheet structures, may be formed during deposition.

Debris flows can move at relatively low speeds down slopes as low as 1 in a 100. Matrix strength prevents settling of all but the largest grains and restricts internal deformation to regions outside a central rigid plug. The deposits may be coarse, poorly sorted, and internally massive, with poorly developed sole marks and an irregular upper surface. It is possible, however, that some sands, with only moderate amounts of mud matrix, move into deep water as sandy debris flows.

Real sediment gravity flows probably combine, at various stages of their evolution, two or more of the theoretical "end member" flow mechanisms described above. It is probable that large-scale grain or fluidized sediment flows become turbulent, thus changing into turbidity currents. Conversely, in many high-density turbidity currents turbulence is probably lost as a dense sediment-water mixture separates to form a "quick bed" that displays many of the features of grain, fluidized sediment, or debris flows.

ACKNOWLEDGMENTS

We thank Arvid M. Johnson and Roger G. Walker for discussion. Financial support was provided by U.S. Geological Survey, U.S. National Science Foundation, and National Research Council of Canada.

This chapter is a revised version of one originally published in *Turbidites and Deep Water Sedimentation*—lecture notes for a short course sponsored by the Pacific Section of S.E.P.M. and given in Anaheim, May 12, 1973.

SYMBOLS

D_{max}	diameter of largest grain
H	depth of channel
P	dispersive pressure, transmitted by grain interaction
Re	Reynolds number
S	slope
T	internal shear stress
U	velocity

V	velocity of head
c	cohesion
d	depth (of density current)
d_2	thickness of head of density current
g	acceleration due to gravity
h	height of overhang at the head of a density current
f_i	coefficient of friction at the upper interface of a density current
f_0	coefficient of friction at the bottom of a density current
g	acceleration of gravity (980 cm/sec^2)
$\Delta\rho$	difference in density between two fluids
$\dot{\epsilon}$	rate of shear strain (velocity gradient)
μ	viscosity
ρ_f	density of fluid
ρ	density of solid grains
σ_μ	internal normal stress
ϕ	dynamic angle of internal friction

REFERENCES

Allen, J. R. L. (1971a). Mixing at turbidity current heads, and its geological implications. *J. Sediment. Petrol.*, **41**: 97–113.

Allen, J. R. L. (1971b). Transverse erosional marks of mud and rock: their physical basis and geological significance. *Sediment. Geol.*, **5**(3/4, special issue): 165–385.

Allen, J. R. L. (1971c). A theoretical and experimental study of climbing-ripple cross-lamination, with a field application to the Uppsala esker. *Geogr. Ann.*, **A53**: 157–187.

Allen, J. R. L. (1971d). Instantaneous sediment deposition rates deduced from climbing-ripple cross-lamination. *J. Geol. Soc. London*, **127**: 553–561.

Andresen, A. and L. Bjerrum (1967). Slides in subaqueous slopes in loose sand and silt. In A. F. Richards, ed., *Marine Geotechnique*, Urbana: University of Illinois Press, pp. 221–239.

Anketell, J. M., J. Cegla, and S. Dzulynski (1970). On the deformational structures in systems with reversed density gradients. *Ann. Soc. Géol. Pologne*, **40**: 3–30.

Bagnold, R. A. (1954). Experiments on a gravity-free dispersion of large solid spheres in a Newtonian fluid under shear. *Roy. Soc. London Proc.*, **A225**: 49–63.

Bagnold, R. A. (1956). The flow of cohesionless grains in fluids. *Roy. Soc. London Phil. Trans.*, **A249**: 235–297.

Bagnold, R. A. (1962). Auto-suspension of transported sediment; turbidity currents. *Roy. Soc. London Proc.*, **A265**: 315–319.

Bagnold, R. A. (1966). The shearing and dilatation of dry sand and the "singing" mechanism. *Proc. Roy. Soc. London*, **A295**: 219–232.

Bagnold, R. A. (1968). Deposition in the process of hydraulic transport. *Sedimentology*, **10**: 45–56.

Blackwelder, E. (1928). Mudflow as a geological agent in semi-grid mountains. *Geol. Soc. Am. Bull.*, **39**: 465–484.

Blatt, H., G. Middleton, and R. Murray (1972). *Origin of Sedimentary Rocks*. Englewood Cliffs, N.J.: Prentice-Hall, 634 pp.

Bouma, A. H. (1962). *Sedimentology of some Flysch Deposits*. Amsterdam: Elsevier, 168 pp.

Casagrande, A. (1936). Characteristics of cohesionless soils affecting the stability of slopes and earth fills. *J. Boston Soc. Civil Eng.*, **23**: 13–32.

Castro, G. (1969). *Liquefaction of Sands*. Harvard Soil Mech. Ser. No. 81, 112 pp.

Chipping, D. H. (1972). Sedimentary structure and environment of some thick sandstone beds of turbidite type. *J. Sediment. Petrol.*, **42**: 587–595.

Crowell, J. C. (1957). Origin of pebbly mudstones. *Geol. Soc. Am. Bull.*, **68**: 993–1009.

Curry, R. R. (1966). Observations of alpine mudflows in the Tenmile Range, central Colorado. *Geol. Soc. Am. Bull.*, **77**: 771–776.

Daly, B. J. and W. E. Pracht (1968). Numerical study of density-current surges. *Phys. Fluids*, **11**: 15–30.

Daly, R. A. (1936). Origin of submarine canyons. *Am. J. Sci., 5th Ser.*, **31**: 410–420.

Dill, R. F. (1964). Sedimentation and erosion in Scripps submarine canyon head. In R. L. Miller, ed., *Papers in Marine Geology (Shepard Commemorative Volume)*. New York: Macmillan, pp. 23–41.

Dill, R. F. (1966). Sand flows and sand falls. In R. W. Fairbridge, ed., *Encyclopedia of Oceanography*. New York: Rheinhold, pp. 763–765.

Dill, R. F. (1967). Effect of explosive loading on the strength of seafloor sands. In A. F. Richards, ed., *Marine Geotechnique*. Urbana: University of Illinois Press, pp. 291–299.

Dill, R. F. (1969). Earthquake effects on fill of Scripps submarine canyon. *Geol. Soc. Am. Bull.*, **80**: 321–328.

Dott, R. H., Jr. (1961). Squantum "tillite," Massachusetts—Evidence of ancient glaciation or subaqueous mass movements? *Geol. Soc. Am. Bull.*, **72**: 1289–1306.

Dott, R. H., Jr. (1963). Dynamics of subaqueous gravity depositional processes. *Am. Assoc. Pet. Geol. Bull.*, **47**: 104–128.

Dzulynski, S. and E. K. Walton (1965). *Sedimentary Features of Flysch and Greywackes*. Amsterdam: Elsevier, 274 pp.

Emery, K. O., E. Uchupi, J. D. Phillips, C. O. Bowin, E. T. Bunce, and S. T. Knott (1970). Continental rise off eastern north America. *Am. Assoc. Pet. Geol. Bull.*, **54**: 44–108.

Fisher, R. V. (1971). Features of coarse-grained, high-concentration fluids and their deposits. *J. Sediment. Petrol.*, **41**: 916–927.

Fisher, R. V. and J. M. Mattinson (1968). Wheeler Gorge turbidite-conglomerate series, California; inverse grading. *J. Sediment. Petrol.*, **38**: 1013–1023.

Forel, F. (1885). Les ravins sous-lacustres des fleuves glaciaires. *C. R. Acad. Sci. Paris*, **101**: 725–728.

Hampton, M. A. (1970). Subaqueous debris flow and generation of turbidity currents. Unpublished Ph.D. Thesis, Stanford University, Stanford, Calif., 180 pp.

Hampton, M. A. (1972a). Transport of ocean sediments by debris flow (abstract). *Am. Assoc. Pet. Geol. Bull.*, **56**: 622.

Hampton, M. A. (1972b). The role of subaqueous debris flow in generating turbidity currents. *J. Sediment. Petrol.*, **42**: 775–793.

Hampton, M. A. (1975). Competence of fine-grained debris flows. *J. Sediment. Petrol.*, **45**: 834–844.

Hand, B. M. (1974). Supercritical flow in density currents. *J. Sediment. Petrol.*, **44**: 637–648.

Hand, B. M., G. V. Middleton, and K. Skipper (1972). Discussion: Antidune cross-stratification in a turbidite sequence,

Cloridorme Formation, Gaspe, Quebec. *Sedimentology*, **18**: 135–138.

Harms, J. C. and R. K. Fahnestock (1965). Stratification, Bed Forms, and Flow Phenomena (with an Example from the Rio Grande). In G. V. Middleton, ed., *Primary Sedimentary Structures and their Hydrodynamic Interpretation*. Soc. Econ. Paleontologists and Mineralogists Spec. Publ. 12, pp. 84–115.

Heezen, B. C. and M. Ewing (1952). Turbidity currents and submarine slumps, and the 1929 Grand Banks earthquake. *Am. J. Sci.*, **250**: 849–873.

Helwig, J. (1970). Slump folds and early structures, northeastern Newfoundland Appalachians. *J. Geol.*, **78**: 172–187.

Hendry, H. E. (1972). Breccias deposited by mass-flow in the Breccia Nappe of the French pre-Alps. *Sedimentology*, **18**: 277–292.

Hendry, H. E. (1973). Sedimentation of deep water conglomerates in Lower Ordovician rocks of Quebec—Composite bedding produced by progressive liquefaction of sediment. *J. Sediment. Petrol.*, **43**: 125–136.

Hooke, R. L. (1967). Processes in arid-region alluvial fans. *J. Geol.*, **75**: 438–460.

Hubert, C., J. Lajoie, and M. A. Leonard (1970). Deep sea sediments in the lower Paleozoic Quebec Supergroup. In J. Lajoie, ed., *Flysch Sedimentology in North America*. Geol. Assoc. Can. Spec. Pap. 7, pp. 103–125.

Jahns, R. H. (1949). Desert floods. *Eng. Sci. Newslett., Calif. Inst. Technol.*, **May**: 10–14.

Jahns, R. H. (1969). California's ground-moving weather. *Eng. Sci. Mag., Calif. Inst. Technol.*

Johnson, A. M. (1965). A model for debris flow. Unpublished Ph.D. Thesis, Pennsylvania State University, University Park, 205 pp.

Johnson, A. M. (1970). *Physical Processes in Geology*. San Francisco: Freeman, 577 pp.

Johnson, A. M. and M. A. Hampton (1969). *Subaerial and Subaqueous Flow of Slurries*. Final Report to U.S. Geol. Surv., Branner Library, Stanford University, Stanford, Calif.

Johnson, D. (1938). Origin of submarine canyons. *J. Geomorphol.*, **1**: 111–129, 230–243, 324–340; **2**: 42–60, 133–158, 213–236.

Jopling, A. V. and R. G. Walker (1968). Morphology and origin of ripple-drift cross-lamination, with examples from the Pleistocene of Massachusetts. *J. Sediment. Petrol.*, **38**: 971–984.

Kenny, N. T. (1969). Southern California's trial by mud and water. *Natl. Geogr.*, **136**(Oct.): 552–573.

Keulegan, G. H. (1957). *Thirteenth Progress Report on Model Laws for Density Currents. An Experimental Study of the Motion of Saline Water from Locks into Fresh Water Channels*. U.S. Natl. Bur. Stand. Rep. 5168, 21 pp.

Keulegan, G. H. (1958). *Twelfth Progress Report on Model Laws for Density Currents. The Motion of Saline Fronts in Still Water*. U.S. Natl. Bur. Stand. Rep. 5831, 29 pp.

Klein, G. deV., U. deMelo, and J. C. D. Favera (1972). Subaqueous gravity processes on the front of Cretaceous deltas, Reconcavo basin, Brazil. *Geol. Soc. Am. Bull.*, **83**: 1469–1492.

Komar, P. D. (1972). Relative significance of head and body spill from a channelized turbidity current. *Geol. Soc. Am. Bull.*, **83**: 1151–1156.

Kuenen, Ph. H. (1950). Turbidity currents of high density. *18th Int. Geol. Congr., London, Rep.*, Pt. 8, pp. 44–52.

Kuenen, Ph. H. (1958). Experiments in geology. *Trans. Geol. Soc. Glasgow*, **23**: 1–28.

Kuenen, Ph. H. and C. I. Migliorini (1950). Turbidity currents as cause of graded bedding. *J. Geol.* **58**: 91–127.

Link, M. H. (1971). Sedimentology and environmental analysis of the Matilija Sandstone north of the Santa Ynes fault, Santa Barbara County, California. M.A. Thesis, University of California, Santa Barbara, 106 pp.

Lowe, D. R. and L. D. LoPiccolo (1974). The characteristics and origins of dish and pillar structures. *J. Sediment. Petrol.*, **44**: 484–501.

McTaggart, K. C. (1960). The mobility of *nuées ardentes*. *Am. J. Sci.*, **258**: 369–382.

Menard, H. W. (1964). *Marine Geology of the Pacific*. New York: McGraw-Hill, 271 pp.

Middleton, G. V. (1966a). Experiments on density and turbidity currents. I. Motion of the head. *Can. J. Earth Sci.*, **3**: 523–546.

Middleton, G. V. (1966b). Experiments on density and turbidity currents. II. Uniform flow of density currents. *Can. J. Earth Sci.*, **3**: 627–637.

Middleton, G. V. (1967). Experiments on density and turbidity currents. III. Deposition of sediment. *Can. J. Earth Sci.*, **4**: 475–505.

Middleton, G. V. (1969). Turbidity currents and grain flows and other mass movements down slopes. In D. J. Stanley, ed., *The New Concepts of Continental Margin Sedimentation*. Am. Geol. Inst. Short Course Notes, pp. GM-A-1 to GM-B-14.

Middleton, G. V. (1970). Experimental studies related to problems of flysch sedimentation. In J. Lajoie, ed., *Flysch Sedimentology in North America*. Geol. Assoc. Can. Spec. Pap. 7, pp. 253–272.

Morgenstern, N. (1967). Submarine slumping and the initiation of turbidity currents. In A. F. Richards, ed., *Marine Geotechnique*. Urbana: University of Illinois Press, pp. 189–220.

Mountjoy, E. W., H. E. Cook, P. N. McDaniel, and L. C. Pray (1972). Allochthonous carbonate debris flows—Worldwide indicators of reef complexes, banks, or shelf margins. *24th Int. Geol. Congr., Montreal*, Sect. 6, pp. 172–189.

Mountjoy, E. W. and P. E. Playford (1972). Submarine megabreccia debris flows and slumped blocks of Devonian of Australia and Alberta—a comparison (abstract). *Am. Assoc. Pet. Geol. Bull.*, **56**: 641.

Nelson, C. H. and L. D. Kulm (1973). Submarine fans and deepsea channels. In G. V. Middleton and A. H. Bouma, eds., *Turbidites and Deep-water Sedimentation*. S.E.P.M. Short Course Notes, Anaheim, Calif., pp. 39–78.

Parkash, B. and G. V. Middleton (1970). Downcurrent textural changes in Ordovician turbidite greywackes. *Sedimentology*, **14**: 259–293.

Peterson, G. L. (1965). Implications of two Cretaceous mass transport deposits, Sacramento Valley, California. *J. Sediment. Petrol.*, **35**: 401–407.

Pett, J. W. and R. G. Walker (1971). Relationship of flute cast morphology to internal sedimentary structures in turbidites. *J. Sediment. Petrol.*, **41**: 114–128.

Rees, A. I. (1968). The production of preferred orientation in a concentrated dispersion of elongated and flattened grains. *J. Geol.*, **76**: 457–465.

Reynolds, D. L. (1954). Fluidization as a geological process and its bearings on the problem of intrusive granites. *Am. J. Sci.*, **252**: 577–614.

Sanders, J. E. (1965). Primary Sedimentary Structures Formed by Turbidity Currents and Related Resedimentation Mechanisms. In G. V. Middleton, ed., *Primary Sedimentary Structures and their Hydrodynamic Interpretation*. Soc. Econ. Paleontologists and Mineralogists Spec. Publ. 12, pp. 192–219.

Schwab, W. C. and M. A. Hampton (1975). *Some Textural Effects of Pore-Fluid Expulsion*. Geol. Soc. Am. Abstracts with Programs, Northeastern Section, 7: 116–117.

Seed, H. B. (1968). Landslides during earthquakes due to soil liquefaction. *Proc. ASCE, J. Soil Mech. Found. Div.*, 94(SM5): 1053–1122.

Sharp, R. P. and L. H. Nobles (1953). Mudflow of 1941 at Wrightwood, Southern California. *Geol. Soc. Am. Bull.*, 64: 547–560.

Shepard, F. P. (1961). Deep-sea sands. *21st Int. Geol. Congr. Rep.*, Pt. 23, pp. 26–42.

Shepard, F. P. (1963). *Submarine Geology*, 2nd ed., New York: Harper & Row, 557 pp.

Shepard, F. P. and R. F. Dill (1966). *Submarine Canyons and other Sea Valleys*. Chicago: Rand McNally, 381 pp.

Shepard, F. P. and N. F. Marshall (1973). Currents along floors of submarine canyons. *Amer. Assoc. Pet. Geol. Bull.*, 57: 244–264.

Simpson, J. E. (1969). A comparison between laboratory and atmospheric density currents. *Q. J. Roy. Meteorol. Soc.*, 95: 758–765.

Simpson, J. E. (1972). Effects of the lower boundary on the head of a gravity current. *J. Fluid Mech.*, 53: 759–768.

Skipper, K. (1971). Antidune cross-stratification in a turbidite sequence, Cloridorme Formation, Gaspé, Quebec. *Sedimentology*, 17: 51–68.

Stanley, D. J. (1974). Pebbly mud transport in the head of Wilingmton Canyon. *Marine Geol.*, 16: M1–M8.

Stauffer, P. H. (1967). Grain-flow deposits and their implications, Santa Ynes mountains, California. *J. Sediment. Petrol.*, 37: 487–508.

Terzaghi, K. (1956). Varieties of submarine slope failures. *Proc. 8th Texas Conf. Soil Mech. Found. Eng.*, 41 pp. See also (1957). *Teknisk Ukeblad*, No. 43–44, pp. 1–16.

Van der Kamp, P. C., J. D. Harper, J. J. Conniff, and D. A. Morris (1973). Facies relations and paleontology in Eocene-Oligocene, Santa Ynes Mountains, California (abstract). *Am. Assoc. Pet. Geol. Bull.*, 57: 809–810.

Van der Knaap, W. and R. Eijpe (1969). Some experiments on the genesis of turbidity currents. *Sedimentology*, 11: 115–124.

Varnes, D. J. (1958). Landslide types and processes. In *Landslides and Engineering Practice*. Highway Res. Board Spec. Rep. 29, U.S. Natl. Res. Council Publ. 544, pp. 20–47.

Walker, R. G. (1965). The origin and significance of the internal sedimentary structures of turbidites. *Yorkshire Geol. Soc. Proc.*, 35: 1–32.

Walker, R. G. (1967). Turbidite sedimentary structures and their relationship to proximal and distal depositional environments. *J. Sediment. Petrol.*, 37: 25–43.

Walker, R. G. (1969). Geometrical analysis of ripple-drift cross-lamination. *Can. J. Earth Sci.*, 6: 383–391.

Walker, R. G. (1973). Mopping-up the turbidite mess. A history of the turbidity current concept. In R. N. Ginsberg, ed., *Evolving Concepts in Sedimentology*. Baltimore: Johns Hopkins Press, pp. 1–37.

Walker, R. G. (1975). Generalized facies models for resedimented conglomerates of turbidite association. *Geol. Soc. Am. Bull.*, 86: 737–748.

Wentworth, C. M. (1967). Dish structure, a primary sedimentary structure in coarse turbidites (abstract). *Am. Assoc. Pet. Geol. Bull.*, 51: 485.

Patterns of Sedimentation in Space and Time

Part III examines the sedimentary environments of the continental margin, using concepts of water circulation and substrate response developed in the preceding parts. In Chapter 12, H. R. Wanless describes the nature of sedimentation in the intracoastal zone of estuaries, lagoons, and deltas. In Chapter 13, P. D. Komar analyzes the wave-driven nearshore circulation system of beaches facing the open shelf and the response of the beach and surf zone substrate to this circulation pattern. In Chapter 14, D. J. P. Swift considers the interaction of the wave-driven nearshore circulation with the coastal boundary of the shelf flow field, and the resulting character of the coastal sediment budget. In Chapter 15, Swift describes the role of the coastal zone in modulating the input of sediment onto the shelf surface, and the several shelf sedimentary regimes that may result. In Chapter 16, J. B. Southard and D. J. Stanley assess water flow and substrate response at the second major discontinuity of the continental margin transport system, the shelf edge. In Chapter 17, G. Kelling and D. J. Stanley apply the principles of gravity-induced transport that were presented in Chapter 11 to problems of sedimentation in the outer margin environments, the continental slope and rise. Finally, in Chapter 18, J. W. Pierce extends the analysis of fine sediment transport presented in Chapter 9 by considering its dispersal over the continental margin as a whole.

CHAPTER 12

Intracoastal Sedimentation

HAROLD R. WANLESS

Rosenstiel School of Marine and Atmospheric Science, Miami, Florida

The coastal transition from terrestrial upland to open sea may be simple—a broad gently sloping beach or sheer rocky headland—or tortuously complex. Where preexisting topography or evolving coexisting topography has produced an intricate shoreline pattern, there results a complex interweaving of nearshore sedimentary environments *within* the coastal maze.

This intracoastal zone consists of four elements:

1. Preexisting coastal topography (the framework onto which present coastal sedimentary environments are superimposed),
2. coast-constructing sedimentary bodies (sand barriers and spits, deltas, sand and mud belts and bars, reefs),
3. resultant holes (estuaries and lagoons), and
4. hole and shadow fillings (tidal flats, storm flats, and tidal marshes and swamps).

These elements may combine to form a simple pattern of coastal environments (barrier beach with resultant lagoon marginally filled by marsh) or a coastal complex (resulting from repetitive, slightly offset, development of coast-constructing bodies and resultant holes and hole fillings).

Coastal urbanization and industrial, shipping, and recreational development can and have profoundly modified both the sedimentary environments and the controlling processes of many intracoastal systems.

There are many approaches toward an understanding of intracoastal sedimentation. Early studies focused on morphologic attributes, still recognized as providing a basic guide to the system's dynamics. More recent studies have examined coastal sedimentary dynamics by aerial photographic documentation of historical coastal evolution (see Shepard and Wanless, 1971), detailed analyses of surface sediment texture, composition, and fabric and biotic communities (Ginsburg, 1956; Rhoads, 1963; Howard et al., 1972), core borings through the Holocene sedimentary accumulates (Kraft, 1971; Wanless, 1969), direct measurements on sediment and fluid dynamics (Postma, 1967; Visher and Howard, 1974), laboratory modeling (McKee and Sterrett, 1961), and computer modeling (Komar, 1973). Howard (1972) has recently revitalized the concept that physical and biological sedimentary structures, used in comparative analysis with those of ancient rock sequences, can be very valuable in deciphering the effective dynamics of coastal environments.

Perhaps the prime concept to be derived from recent coastal research is that a workable understanding (one allowing valid modeling and forecasting) of intracoastal sedimentary environments can be achieved only when the spectrum of interacting controls of deposition is recognized and resolved (Nelson, 1972, p. 5).

The purpose of this chapter is to provide the reader with an awareness of the spectrum of depositional controls in the intracoastal environment, and to give examples of how these depositional controls interact to form, maintain, and modify the resultant intracoastal sedimentary environments.

DEPOSITIONAL CONTROLS

Attributes controlling sedimentation can be divided into primary controls and dependent controls. Primary depositional controls are preexisting topography, climatic-hydrographic setting, character of sea-level history, and rate of terrestrial sediment influx. These primary attributes are essentially independent of one another. Dependent controls are resultant waves and currents (patterns, intensity, frequency), rates and patterns of sediment influx and dispersal, evolving coexisting topography and bathymetry, water chemistry, and organism influences. Dependent controls may have interacting influence. Resultant coexisting topography, for example, both modifies waves and currents and is changed by them.

Primary Controls

Climate includes wind regime, temperature, precipitation, humidity, and solar radiation. The intensity and frequency distribution of these attributes determine many characteristics of the intracoastal environment (local wave climate, coastal vegetation, precipitation: evaporation balance, storm frequency and intensity, flooding frequency, and intensity from storm surges and terrestrial runoff). Fundamental climatic divisions are tropical, temperate, arctic, rainy, and arid.

Hydrodynamic setting includes general offshore wave climate, tidal range and distribution, and offshore circulation. Variations in the combined climatic-hydrographic setting produce fundamental differences in resultant intracoastal sedimentary environments. End members of the resultant aqueous setting are "low energy" (subjected to small day-to-day physical processes but intense sporadic storm action) and "high energy" (dominated by moderate to intense day-to-day physical processes). Between these atmospheric and hydrodynamic extremes the resultant effect of climatic and hydrodynamic attributes depends on their intensity, frequency, and temporal relationships.

Preexisting topography, through its effect on wave action, tidal currents, and wind-driven circulation, must be considered the basic attribute of the local setting. The subaqueous offshore profile and bathymetric plan modify offshore wave climate and dampen or amplify tidal current flow. Local submerged and emergent relief controls wave and current patterns and circulation. Mainland and offshore emergent topographies define the initial coastal pattern.

Preexisting topography may be either rock or unconsolidated sediment. The former provides an essentially unyielding, rigid framework. The latter may be both modified by coastal processes and reworked to supply sediment to present environments.

Sea-level history has a major influence on the resultant pattern of sedimentation. During rapid sea-level rise, coastal environments must (1) play catch-up sedimentation, (2) rapidly migrate landward (transgress), grow upward, or be overridden by the rising sea, and (3) constantly change in pattern in response to changing topographic-bathymetric setting. With slow sea-level rise or a steady state and sufficient sediment nourishment, coast-constructing environments may maintain a stationary position or even expand seaward (regress), and intracoastal holes will tend to fill in toward some equilibrium condition. With insufficient nourishment, mobile constructional coasts may migrate landward or dissipate.

Sea level has been rising during the past 10,000 years as glacial melt has gradually replenished the oceans (Fig. 1). Sea level rose with decreasing rapidity from about −100 m 10,000 years ago to −2 m about 3500 years ago, and then at a much slower rate to the present near-stationary level. The effects of this sea-level rise are accentuated in areas of active tectonic downwarping or sediment loading and diminished in areas of tectonic uplift and glacial unloading.

Effects of river discharge are examined with deltaic and estuarine environments.

Dependent Controls

Resultant waves, currents, circulation, sediment transport, and source of sediment are examined in other chapters.

EVOLVING COEXISTING TOPOGRAPHY AND BATHYMETRY. A sedimentary body, once formed, becomes an integral part of the coastal bathymetry and, if emergent, topography. Resultant coastal morphology is a combination of the remaining exposed preexisting topography-bathymetry plus the superimposed topography-bathymetry of present sedimentary accumulates. This resultant evolving coexisting topography-bathymetry is constantly changing, and the relationship of coastal morphology to processes may not be apparent, especially during or following periods of rising sea level.

ORGANISM ACTIVITY. Organism communities can play a major influence in the construction, maintenance, and modification of intracoastal sedimentary environments. Organism communities create pericoastal constructional accumulates such as coral-algal reefs, oyster reefs, and sabillarid worm rock "reefs." Skeletal remains are a source of sediment, and commonly comprise nearly the entire sediment accumulate of

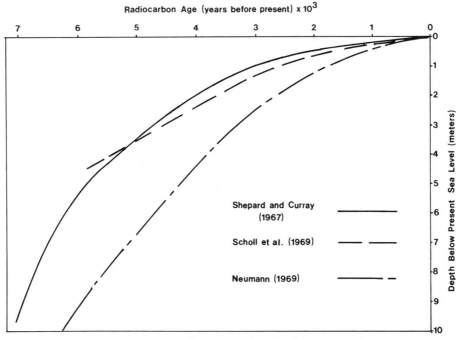

FIGURE 1. *Curves of the Late Holocene rise of sea level from areas of apparent tectonic stability (diagram by E. R. Warzeski).*

tropical and subtropical shallow marine environments unless diluted by terrigenous sediment influx. Organic remains provide much or all of the sediment accumulating in marginal marine marshes and swamps and protected lagoons and ponds.

Organism communities also modify the substrate through bottom stabilization (marsh and swamp communities, sea grasses, algal mats), sediment entrapment (vegetative current baffles, filter-feeding, mucilaginous films), and textural modification (pellet aggregation and biocorrosion).

CONSTRUCTIONAL COASTS, BELTS, BARS, AND REEFS

Sand barriers, sand spits, and deltas are active constructional sedimentary environments that produce emergent coasts and effectively isolate adjacent intracoastal lagoons and ponds. Other active constructional sedimentary environments, coral-algal reefs and sand and mud bars and belts, do not form emergent coasts but do shoal sufficiently to separate adjacent landward lagoonal environments from the open sea. Inlets through these accumulates provide the main pathway for sediment influx and water exchange.

Sand Barriers and Spits

SAND BARRIERS. Barrier beaches are single, elongate, slightly emergent sand ridges extending generally parallel to the coast but separated from it by a lagoon; barrier islands consist of multiple sand ridges and commonly have dunes and vegetation on the island and marsh, swamp, or sand washover flats along the lagoonward margin.

Sand barriers are characteristic sediment bodies formed over low coastal plains during the late stages of the present postglacial rise of sea level. Sand barriers border about 47% of the U.S. coastline (Shepard and Wanless, 1971) and nearly 13% of the world's continental coastline (Schwartz, 1973).

The mechanism for generating sand barriers has remained a focus of repeated controversy during the past century. DeBeaumont (1845; see Schwartz, 1973) pioneered the theory that barriers were thrown up from the sea as a wave-generated offshore submarine bar and gradually built up to a barrier beach and then island, sand being derived by reworking the drowned coastal plain into a new profile of equilibrium (Fig. 2A). Price (1963) has observed that offshore bars formed by exceptionally high storm tides may become emergent as water returns to normal, providing the transition to emergent beach. A second mechanism, originally proposed by Gilbert (1885), suggests that barriers form by longshore-drift-generated sand spits extending from headlands and are subsequently segmented into islands by tidal inlet breaches (Fig. 2B). Hoyt (1967) precipitated a recent flare of controversy in proposing that barrier islands form by gradual submergence of pre-

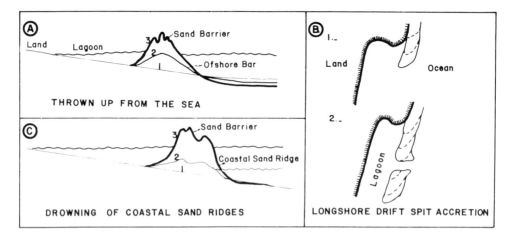

FIGURE 2. *Basic theories of barrier island formation.* (A) *Idealized cross sections of barrier island* (3) *evolving from an offshore bar* (2), *the sediment being supplied during modification of the offshore profile.* (B) *Idealized map of sand barrier formation by longshore drift spit accretion* (1) *and subsequent inlet dissection into barrier islands* (2). (C) *Idealized cross sections of barrier island* (3) *forming by submergence of coastal sand ridge* (2). *Modified from Hoyt* (1967).

existing mainland coastal beach-dune ridges, a lagoon forming as the landward area is flooded (Fig. 2C). His principal argument is the absence of ocean beach and neritic sediments in the Holocene coastal sedimentary sequence.

Shepard (1960) has effectively argued that barriers may form only in times of slow submergence or a steady state. Leontyev and Nikiforov (1965, 1966) offer a contrary position.

Much of the controversy just mentioned is the result of insufficient preserved evidence. Sand barriers, once formed, may respond to gradual changes in sea level, sediment supply, wave climate, or subaqueous offshore profile by landward (erosional onlap, transgression) or seaward (constructional offlap, regression) migration (Kraft, 1971). Many islands have had a partly transgressive history during the latter part of the present postglacial rise of sea level, destroying the barrier's early history. Shepard (1960) and Schwartz (1971) propose a multiple causality of sand barriers, as evidenced by the observed processes of maintenance and Late Holocene stratigraphic history.

"Cheniers," elongate narrow sand barriers or ridges bordering the seaward margins of some delta plains, are generally considered to be tossed up as waves rework the seaward face of a prograded delta lobe. Littoral drift may concentrate the sand away from the area of erosion (Shepard, 1960; Hoyt, 1969). Sand barrier formation, maintenance, and modification are discussed at greater length in Chapter 14.

SAND SPITS. Barrier spits are barriers connected with the mainland at one end, and extending part way across the seaward margin of a bay or estuary (Fig. 2B). Longshore drift of sand provides nourishment for their growth and maintenance. Spit accretion may extend to the next preexisting topographic headland, producing a nearly continuous "bay barrier" across the enclosed bay (see Shepard and Wanless, 1971, Fig. 2.17).

Deltas

Deltas are sediment piles protruding from the coast in areas where rivers carry in more detritus than can be removed by ocean processes. As this definition implies, delta morphology and dynamics are the result of the interaction of fluvial effluent and coastal reworking processes, set within the particular climatic, topographic, and sea-level history framework (Morgan, 1970a,b; Scott and Fisher, 1969).

Six general concepts provide a basis for understanding deltaic dynamics. First, sediment commonly accumulates extremely rapidly along and just seaward of active deltaic distributary channels, resulting, unless checked by coastal processes, in rapid extension of both the emergent and submerged portions of the delta fan (progradational or constructive phase). Second, with rapid progradation, distributary channel profiles may overextend, causing the river to seek a new more direct course to the sea by abandoning the previous delta fan. Following this, the sediment of the abandoned delta lobe will be subjected to reworking (destructive phase). A delta accumulate is thus an orderly, imbricate stack of individual subdelta packages, each recording a constructional and destructional phase (Fig. 3). Third, deltaic morphology and the relative dominance of

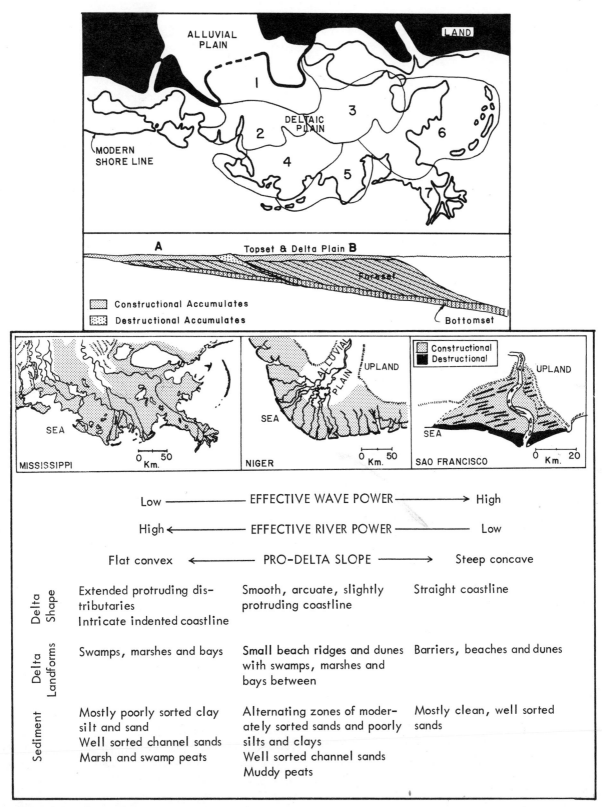

FIGURE 3. *Top: Map of imbricating subdelta lobe sequence formed by Mississippi River during the past 5000 years. Modified from Scruton (1960). Middle: Idealized cross section across imbricate delta accumulate showing relationship of constructional and destructional deposits. Modified from Scruton (1960). Bottom: Table and sketch of delta form-process relationships. Modified from Wright and Coleman (1973).*

constructional sedimentary environments are determined by the interaction of effective river discharge and effective wave-current climate. Fourth, deltas are gradational to estuarine and tidal inlet systems. Fifth, the specific character of deltaic subenvironments is strongly influenced by climatic setting. Sixth, deltas are subjected not only to eustatic sea-level change but also to rapid subsidence caused both by delta mud compaction and local crustal downwarping.

DELTA-BUILDING CYCLE. Prograding subdelta lobes are composed of one to several distributary channels (floored by sand or silt) bordered by natural levees (sandy at the channel margin crest but rapidly fining down the backslope) which isolate shallow interdistributary ponds or swamps (containing fine muds and organic detritus or peats) (Russell and Russell, 1939; Frazier and Osanik, 1969). Floods locally crevasse the channel margin levees and spread coarser detritus across interdistributary ponds. Seaward of the emergent delta plain is a shallow, constructional submarine plain (topset beds) yielding seaward to a marginal prodelta slope (foreset beds) rapidly advancing over a deeper area of mixed deltaic and shelf sedimentation (bottomset beds) (Scruton, 1960). Prodelta slopes range from less than 1° off the low-energy Mississippi River Delta to as much as 25° off small coarse-grained deltas affected by waves (Guilcher, 1963). More gently sloping prodelta slopes are commonly finer than adjacent topset or bottomset beds, whereas steep slopes are generally sandier and deeply gullied. Sedimentation rates increase upward on the prodelta slope and drop abruptly at the emergent delta plain. Scruton (1960) measured maximum accumulation rates of 0.4 m/year for the active Mississippi subdelta.

An active delta plain is maintained by contemporaneous deposition of floodwater detritus (dominant adjacent to channel banks) and swamp or marsh peats (dominant away from channel margins). Where peat accumulation does not keep up with subsidence, marshes give way to shallow interdistributary ponds which expand by continuing subsidence and wind-driven waves (Frazier and Osanik, 1969).

As a subdelta lobe is abandoned by the river, growth ceases and ocean waves and currents rework the accumulate, creating new subenvironments. Waves winnow finer silts and clays from the prodelta face and delta plain causing rapid retreat of the shoreline. Sands concentrate shoreward to form low beach ridges (cheniers) against the eroding delta plain fringe. Swamp and marsh growth may keep up with subsidence on the abandoned delta lobe, or the delta plain may be slowly inundated by the sea (Shepard and Wanless, 1971, p. 204). When inundated, sand cheniers remain as offshore barriers and, unless a continued supply of sand is provided from offshore, will quickly migrate landward (Guilcher, 1963; Morgan et al., 1958) or dissipate as their sand is spread across the drowned delta plain behind (Frazier and Osanik, 1969, p. 70). Longshore drift may concentrate cheniers away from areas of active erosion, as illustrated by a broad chenier plain west of the Mississippi Delta complex (see Fig. 3; Gould and McFarlan, 1959; Shepard, 1960) and may even serve as a primary nourishment source for downdrift barrier island chains (Shepard, 1960; Wright, 1970).

The resultant delta is an imbricate stack of subdelta lobe packages each containing a constructive and destructive sedimentary record (see Fig. 3; Scruton, 1960; Frazier and Osanik, 1969).

FLUVIAL-MARINE INTERACTION. The interaction of fluvial (river) effluent and physical ocean processes determines both the resultant morphology of the delta and the extent and timing of destructional processes. Basic parameters of fluvial influence are mean annual discharge, discharge intensities (peaks), discharge strength (discharge per foot of river mouth width), and the amount and textural distribution of material being supplied. Parameters of ocean influence are effective wave power (wave climate attenuated by the subaqueous bottom profile), tidal range and currents, and littoral drift.

Where river influence dominates, deltas have highly irregular convex shorelines in which the active subdelta may have a birdfoot pattern. Earlier subdelta lobes are distinguishable at the surface and have received minor modification during destructive processes. Rapid progradation of river-dominated deltas produces a low-angle convex prodelta slope which further dampens the influence of offshore waves (Mississippi River Delta: Wright and Coleman, 1973).

Under the dominance of coastal processes, the delta will have an essentially straight coast and a steep concave prodelta profile offering little attenuation of the offshore wave climate. Destructive reworking of both the abandoned subdelta systems and the contemporaneous subdelta deposit is extensive, and the resultant accumulate consists primarily of reworked sands (Senegal River Delta: Wright and Coleman, 1973). Subdelta lobes will be nearly vertically stacked and difficult to differentiate in cross section.

Between these two extremes deltas respond to the temporal and spatial distribution of the interacting processes. Wright and Coleman (1972, 1973) have attempted to provide a quantitative first-order resolution of process-form interaction using a discharge effectiveness index (total discharge strength divided by attenuated wave power per unit of wave crest) to compare delta systems (Fig. 3).

Higher resolution is qualitative at present. Peak wave power, coinciding with peak discharge, will distribute detritus as it emerges, causing smooth delta margins and rather simple patterns of progradation (Ebro River Delta, Spain). When the two parameters are out of phase, waves seasonally rework bulges of effluent accumulation, resulting in more complex progradational patterns of successive shore ridges separating delta plain marshes or alluvium (Sao Francisco Delta, Brazil: Wright and Coleman, 1973).

Focused river discharge will tend to generate more lobate deltas (Mississippi River Delta), whereas active discharge spread among many channels will generate a smoother delta form though commonly with an intricate internal pattern of delta plain subenvironments (Niger River Delta, Nigeria). The latter, having steeper prodelta profiles providing less attenuation to offshore waves, will be more prone to modification.

Texture of river effluent may determine the extent and character of both constructional and destructional sedimentary accumulates. Rivers, such as the Mississippi, releasing mostly fine silt and clay, build broad delta front mud aprons well beyond the delta plain. Little sand will be released during reworking, and the resultant regressive deposits should be of lesser importance. With coarser textured river effluent, material settles out nearer shore, yielding steeper prodelta profiles, and more sand is provided (for a given amount of reworking) during destructive phases.

The primary effect of littoral drift on deltas entering low- to moderate-energy coasts is to skew destructive phase sedimentation downdrift. Littoral drift along higher energy coasts may divert the debouching distributary channel by near-coast formation of "peri-littoral" levees (Senegal River Delta, Senegal).

A large tidal range dissipates or impedes delta growth both by flushing much of the river effluent from the potential delta plain and by repeatedly exposing the accumulate to ocean processes.

CLIMATIC INFLUENCE. Delta response to climatic variation is primarily in modification of the internal dynamics and sedimentary environments of the delta plain. Eolian redistribution of sediment is common in zones of strong onshore winds (generally also associated with high degree of wave reworking) and arid climates (lack of vegetation; Senegal River Delta). Hurricane-generated tides and floods sweep subtropical delta plains, causing drastic modification of the delta plain

and often change in channel course (North Mahavavy Delta, Madagascar: Guilcher, 1963). Deltas from rivers draining arid lands may be largely tidal for much of the year with interdistributary ponds becoming salt pans or sabkhas and the delta plain subjected to eolian reworking (Senegal River Delta).

Arctic periglacial deltas are frozen and inactive for most of the year and may have a low-energy morphology, since near the coast pack ice inhibits wave action even in the summer (Shepard and Wanless, 1971). Meltwaters and breakup of river ice from the more southerly drainage basin may reach the delta while the latter is still frozen. These often catastrophic floods overtop the frozen delta, annually creating new channels (Lena River Delta, Siberia: Guilcher, 1963). The resultant delta plain is commonly a confused braided pattern subjected to intense permafrost action (Shepard and Wanless, 1971).

SUBSIDENCE. Subsidence, through compaction of the rapidly accumulated subdelta packages and local crustal downwarping in response to the increasing sediment overburden, is an integral part of deltaic dynamics. In both cases it is a time-lagged response continuing well after a subdelta lobe is abandoned or after a drop in the sediment load of the stream through postglacial vegetation of upland areas, damming entrapment of river load, or major diversion of river course.

Rapidly prograding delta fronts may be quite unstable (especially along low-energy coasts and where extending out into deep water). Earthquake shocks can cause major portions of the delta front to slump (outwash delta of Valdez Glacier, Alaska, 1964 earthquake) (Wilson and Torum, 1968; Shepard and Wanless, 1971).

Marine Sand and Mud Belts and Bars

Linear belts of sand are common near the margins of many shallow marine platforms. These belts parallel the coast and are somewhat landward of the slope break (seaward threshold of a shallow platform beyond which water depth increases) which intensifies and directs incoming wave and tidal currents (Ball, 1967). These sand belts are most common in tropical climates and are commonly composed of ooid sands produced on the sand belt with some skeletal admixture. Elsewhere sand-bar belts may be formed of skeletal or silicic-clastic sand from either reworking of the underlying substrate or littoral drift (Southwest Florida coast: Ball, 1967). The rippled surface of the sand belt may extend into the intertidal zone. Internal cross-bedding from spillover lobes indicates sand transport across the belt with net landward movement (Ball, 1967). These belts effectively separate the landward "bank" or lagoon from offshore wave action but provide only minor restriction to circulation.

Smaller slope breaks on the platform interior and nearshore may also produce linear belts of sand and mud ("Safety Valve" mud bar belt, Biscayne Bay, Florida) (Wanless, 1969, 1970).

Tidal bar belts are common along arcuate shelf breaks (south end of the Tongue of the Ocean and north end of Exuma Sound, Bahamas) or in association with lateral constrictions in the topography (entrance to estuaries and bays) that amplify tidal flow (Ball, 1967; Off, 1963). These bars, composed of locally generated ooid-skeletal carbonate sands or terrigenous silicic-clastic sands, may form belts paralleling the coast or slope break (see discussion of tidal sand ridges, Chapter 10). Tidal bar belts are composed of numerous sandbars extending perpendcular to the belt's axis, each separated by tidal channels. Although only locally emergent, tidal bar belts may form effective seaward barriers to nearshore marine environments.

Organically Constructed Coasts

REEFS. Coral-algal reef growths may generate peritidal shelf margin barriers in tropical and subtropical areas of moderate to high wave energy not affected by turbid water outflow. Broad backreef rubble flats and high intertidal coralline algae pavements on the reef crest (especially in the Pacific) provide a highly effective separation of the interior lagoon. Coastal reef growth may also bridge coastal embayments in much the same configuration as sand spits.

Turbid or low-salinity water outflow from rivers (east side of Andros Island, Bahamas) or bays (South Florida reef tract) may locally inhibit reef growth. In subtropical climates, winter air temperature inhibits growth of coral in the intertidal zone.

Tidal Inlets

Tidal inlets across constructional accumulates provide circulation to the contained lagoons and estuaries and are an important conduit for sediment influx and exchange. Inlets cutting sand barriers commonly form as storm surge or river flood breaches. Inlets associated with estuaries or in areas of rock substrate are commonly inherited from preexisting topography; others have an uncertain origin.

Some tidal inlets are comparatively permanent features being maintained by tidal exchange with the lagoon supplemented by sporadic peaks of river outflow or storm-generated currents (Shepard, 1960; Pierce, 1970).

Flood tidal deltas commonly form just inside an inlet as current velocity drops beyond the topographic constriction. Tidal deltas may form rapidly from storm breach inlets (El-Ashry and Wanless, 1965) draining sand from the littoral platform, partially filling the lagoon, and restricting flow through the inlet. In areas of active littoral drift, the shoreface may eventually block the inlet on the seaward margin (Dickinson et al., 1972). Bruun and Gerritsen (1955, p. 84) note that tendency toward inlet closure is related to the ratio between littoral drift and tidal flow bypassing of sand past the inlet (see also O'Brien and Dean, 1972).

Spit accretion on the updrift side of a tidal inlet commonly induces longshore migration of the inlet mouth. With drift, inlets may become overextended and recut a new more direct course across the barrier (Nauset Beach, Massachusetts; Fire Island Inlet, New York); see Shepard and Wanless (1971) and Chapter 14. A broad zone of tidal delta lobes extending into the lagoon may result if repetitive recutting is from storm breaching of the extended inlet mouth.

Tidal deltas also form on the seaward side of ebb-dominated inlets, draining lagoons and estuaries along low-energy coasts (Shepard and Wanless, 1971, p. 156). Sand and mud for ebb-dominated tidal deltas are provided by (a) storm discharge of lagoon-derived sediment (Caesars Creek delta draining the south end of Biscayne Bay, Florida), (b) seaward shift of long shore drifting sand, or (c) estuarine discharge of river effluent. Ebb tidal deltas are stunted or dissipated by wave energy and littoral drift in higher energy settings.

Tidal channels across mud and sandbar belts are formed and maintained by sporadic storm processes in moderate- to low-energy settings and day-to-day currents in areas of very strong tidal action (Ball, 1967). Change in channel configuration appears to be a mixture of migration and recutting, the latter dominating in low-energy settings.

Channels across coral-algal reef barriers may be the result of freshwater or turbid water outflow inhibiting coral growth. Shipping channels across many coral reefs are man-made. Tidal inlets are considered further in Chapter 14.

INTERCOASTAL HOLES

Lagoons

Lagoons are marginal marine water bodies protected and partially isolated from the open sea or shelf by constructional coastal accumulates and/or by partly drowned preexisting topography. Lagoons, as defined here, do not contain significant river input. Lagoons are characteristic of coastal plains, occurring as linear water bodies paralleling the coast. However, they may take on a variety of shapes. To understand the dynamic processes in a lagoon one must consider how the framework elements (topography, hydrodynamic setting, climate and character of sea-level rise) interact to produce the resultant wave patterns, circulation, chemical parameters, biota, and rates and patterns of sediment influx, production, dispersal, and accumulation. A few basic concepts are emerging from this potential maze of interaction.

CARBONATE VERSUS SILICIC-CLASTIC LAGOONS. Fundamental differences may exist between tropical carbonate lagoons and silicic-clastic lagoons. Sediment in many tropical lagoons is largely calcium carbonate skeletal and nonskeletal remains produced in or near the area of accumulation. Further, tropical carbonate lagoons commonly form on top of earlier calcium carbonate sediment accumulates that were lithified during the previous glacial lowering of sea level. The underlying topography thus is rock. Limestone ridges may also form the seaward rim to the coastal lagoon as illustrated by the fossil coral limestone ridge bordering lagoons from Biscayne Bay to Florida Bay, southern Florida. The limestone topography is essentially rigid and not modifiable, and slight changes in sea level cause dramatic changes in the patterns of circulation and sedimentation within (Ball, 1967; Wanless, 1969, in press).

Temperate lagoons and some tropical lagoons, in contrast, are dominated by terrigenous silicic-clastic sediment detritus transported into the lagoon with only minor additions of locally produced skeletal remains. The preexisting substrate in coastal plain areas is predominantly unlithified sands, muds, and gravels that can be modified by and reworked into the present sedimentary environment. Seaward barriers to coastal plain lagoons are generally mobile sand barriers. Mobile barriers are responsive to slow sea-level rise and commonly evolve to maintain the enclosed lagoonal environments with little change (though major shifts in position may occur). A mobile seaward barrier, however, can undergo dramatic short-term changes that may cause rapid modification to the adjacent lagoon (storm inlets, tidal deltas, sand washovers) (El-Ashry and Wanless, 1965; Dickinson et al., 1972). In other areas rocky headlands provide some rigidity to the lagoonal setting.

WAVES AND SEDIMENT DISTRIBUTION. Lagoons along the Atlantic coast of North America average 2 m in depth (Emery and Uchupi, 1972). Bottom sediments move by wave agitation and storm currents; day-to-day

tidal currents are generally small except near tidal inlets. Two types of textural zonation are observed in lagoons (Fig. 4). Silt and clay-sized particles are concentrated on the shallow margins of about half of the lagoons and in the deeper portions of the rest (Emery and Uchupi, 1972).

The former appears characteristic of shallow lagoons in higher energy settings where day-to-day wave agitation reaches the lagoon floor, and fine sediment is carried by wind- and tidal-generated circulation to accumulate in protected coastline irregularities, tidal marshes, and flats and along leeward coastlines. High-energy lagoons will tend to fill by continuous lateral accretion of margins.

The latter is characteristic of (a) shallow lagoons in low-energy settings, such as Biscayne Bay, Florida (Wanless, 1969) and Rockport Bays, Texas (Shepard and Moore, 1960) (b) deep lagoons, and (c) lagoons with stratified water columns, such as Harrington Sound, Bermuda (Neumann, 1965). In these lagoons, day-to-day wave agitation affects only the shallower margins, allowing fine sediment to accumulate in the deeper central portion. These lagoons will fill by gradual shallowing. Sporadic major storm agitation may redistribute this vulnerable reservoir of fine sediment, producing sediment bodies similar to those of high-energy settings (marginal flats or marshes) or quite different from these (muddy tidal deltas, bars, and belts, storm levees). Shepard and Moore (1960) conclude that such lagoons along the Gulf coast of North America are accumulating sediment at a rate of 20 to 40 cm/100 years. For arid Laguna Madre, Rusnak (1960, 1967) has found rates of only 10 cm/100 years.

LAGOON PARTITIONING. A conspicuous feature of many shallow lagoons is their partitioning by the formation of mud or sand cuspate spits into subcircular bays (Fig. 5). Wind-generated circulation cell patterns appear responsible (Zenkovitch, 1959, 1967; Taft and Harbaugh, 1964; Wanless, 1969). Observations from a variety of settings indicate that sea-level change, tidal currents, standing tidal waves, and preexisting shoreline irregularities are not responsible. Cuspate spit accretion is important in determining the pattern of accumulation in both narrow elongate and wide shallow lagoons and ponds.

ORGANISM INFLUENCE. Biogenic activity is an important dynamic influence in nearly all lagoonal systems. Lagoonal bottom sediments are intensely reworked by a variety of infaunal and grazing benthic organisms (Rhoads, 1963). An important by-product is the packaging of fine silt and clay detritus and organic debris into sand-sized fecal pellets. Although having a much less

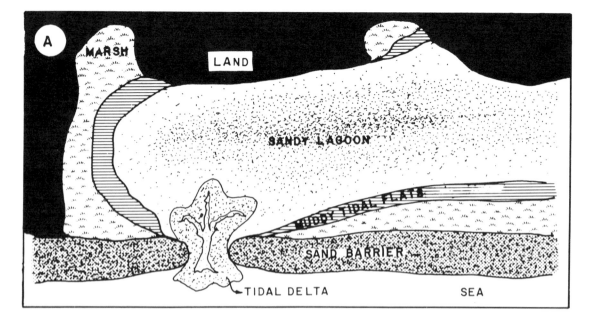

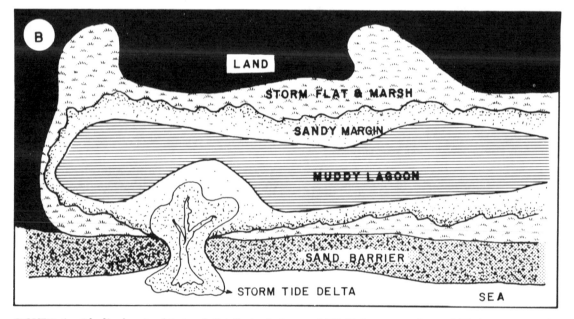

FIGURE 4. *Idealized maps of textural distribution in lagoon of* (A) *high-energy setting and* (B) *low-energy setting.*

effective excess density than comparable mineral grains, the pellet packages (0.1 to over 1 mm in size) will respond to the hydrodynamic environment as sand-sized particles rather than the fine silt, clay, and organic debris of which they are composed (Haven and Morales-Alamo, 1972; see the discussion in Chapter 9 (p. 137). Pellets of silicic-clastic detritus are mostly loosely bound and often easily broken. Carbonate pellets are commonly much firmer and in certain areas may even have a lightly cemented crust (Bathurst, 1971). Where pellets are important, sediment analysis of disaggregated bottom

sediment is of little value in understanding the hydrodynamic behavior of bottom sediment in lagoons, estuaries, and mud flats.

Vegetation may also be extremely important in determining the pattern and character of lagoonal sedimentation (Ginsburg and Lowenstam, 1958; Wanless, 1969; Scoffin, 1970). Sea grasses such as *Thalassia* are common in lagoons with sufficient light penetration. Grass blades, acting as a current baffle and having dense epiphytic growth attached, aid in the entrapment of mud. Dense rhizome and rootlet systems

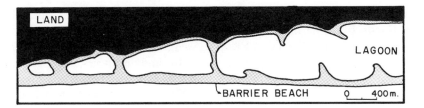

FIGURE 5. *Dissection of lagoon into subcircular bays by cuspate spit accretion. Sketch from aerial photograph in Zenkovitch (1967).*

provide highly effective substrate stability during storms (Fig. 6). Dense sea-grass beds can effectively protect the bottom from 2 knot tidal currents (Hay, 1967). Wanless (1969) calculated that sea-grass beds in Biscayne Bay sufficiently modify bottom currents and substrate to permit lime mud accumulation up to a depth of 3 m, whereas from purely physical considerations this sediment should be restricted to depths greater than 5 m in that particular wave climate and bathymetric setting.

Gelatinous subtidal algal mats may also be important to bottom sediment entrapment and stabilization (Bathurst, 1967; Neumann et al., 1970).

CLIMATE. In arid regions, lagoons having slow exchange with the ocean may become very saline, reducing organism activity and generating precipitation of evaporitic minerals (southern Laguna Madre, Texas: Rusnak, 1960; Miller, in press). Floors of tropical lagoons having above-normal salinity may be loosely to firmly bound by carbonate cements (Bathurst, 1971; Logan et al., 1970).

Sediment influx to lagoons is largely restricted to channel passes in wet climates. In dry climates, however, sand barriers are not stabilized by vegetation and much of the sediment influx may be by wind and storm tide transport across the barrier as illustrated by southern

FIGURE 6. *Dense rhizome (horizontal) and rootlet (vertical) system of the sea grass* Thalassia testudinum *exposed in the wall of a storm-eroded sand pocket. Base of dark portion of blades at top marks prestorm sediment surface. Taken off Eleuthera Island, Bahamas following passage of a small hurricane in 1973 (photo by N. P. James).*

Laguna Madre, Texas. Broad wind sediment flats extend well across the lagoon from the highly mobile, unvegetated Padre Island barrier (Rusnak, 1960; Miller, in press).

Estuaries

Estuaries are active river valleys inundated by the sea. Cut by downslope drainage, estuarine embayments mostly trend at a large angle to the general coastline and commonly digitate along tributary river valleys. Sediments in the estuary may have a river or ocean source. Estuaries have stratified flow patterns (estuarine circulation, described below) that tend to retain supplied sediment and draw inward marine-transported detritus (Guilcher, 1967; Morton, 1972). Estuaries are mostly somewhat deeepr than lagoons, having an average depth of 4 m along the Atlantic coast of North America (Emery and Uchupi, 1972). As a result, estuarine sedimentation is dominated more by tidal flow and estuarine circulation than by wave action. The mouth of the estuary may be wide and quite open, but in areas of strong tidal flow, tidal sand deltas and bars may constrict the seaward opening (see the discussion in Chapter 14, p. 296). Estuarine mouths are generally also constricted along coasts of active littoral drift by a combination of coexisting sand barriers and bars and preexisting coastal accumulates. Modern and relic littoral drift accumulates have skewed the mouth of Chesapeake Bay far south of its ancestral opening (Harrison et al., 1965).

ESTAURINE CIRCULATION.   Superimposed on the periodic tidal motion and sporadic wind-driven circulation is an "estuarine circulation" generated by density differences between freshwater outflow and seawater (see Chapter 3, p. 24; Chapter 9, p. 142). Denser seawater flows inland as a wedge under the freshwater discharge producing a stratified water column. Water from the saline wedge decreases in salinity with vertical mixing and is gradually recycled seaward drawing more seawater into the wedge. Estuaries with small tidal influence may maintain stratification over most of the estuarine basin or channel and throughout much of the tidal cycle, but have only slow movement within the saline wedge (Southwest Pass, Mississippi River Delta: Meade, 1972). During times of flood peaks and in areas of moderate tidal action, stratified estuarine circulation will be restricted to the inner part of the estuary, the outer being a "mixed" or vertically homogeneous estuary in which tidal and wind-wave-generated processes dominate (Savannah River, Georgia; Suison Bay, California; Chesapeake Bay, Maryland: Meade,

1972; Postma, 1967). Upper flow regime conditions may occur during flood tides at the base of these saline wedges (Visher and Howard, 1974).

In sandy estuaries, sand ridges may separate ebb and flood channels (Gironde estuary, France: Allen et al., 1973), reducing the extent and modifying the pattern of stratification (see Chapter 10, p. 177).

TURBIDITY MAXIMUM.   Suspended detritus commonly has a "turbidity maximum" near the inner margins of stratified estuarine circulation (Fig. 7). Concentrations are considerably higher than either the associated river water or seawater (Postma, 1967). Early studies attributed this to flocculant aggregation of riverborne material upon contact with seawater (Luneburg, 1939). Postma and Kalle (1955) demonstrated that the following hydrodynamic processes alone are sufficient to produce this maximum. In the mixed boundary zone between the seaward-moving river water and upstream-moving seawater wedge is a plane of no horizontal motion, averaged over the tidal cycle. Sediment particles from river outflow may settle through the mixed layer and into the saline wedge. As a particle settles into the mixed boundary zone, its seaward transport will be slowed or nulled, providing time for settling. Once in the saline wedge, the particles will be carried landward, then eventually mixed into the river water and returned seaward, and so on. Similarly, particles entering from the seaward end may be trapped in this cycle many times before release or deposition.

An aggrading mud or sandbar may form at the landward edge of the saline wedge both as river flow separates from the bottom dropping its sand load and as silt is forced out of the suspension turbidity maximum (Van Veen, 1950; Postma, 1967); see Chapter 14.

Clay-sized particles may settle too slowly to be effectively concentrated by estuarine circulation, but little work has been done on the grain sizes effectively trapped by turbidity maxima (Meade, 1972); see

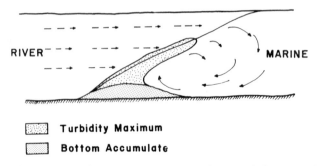

Turbidity Maximum

Bottom Accumulate

FIGURE 7. *Schematic cross section of suspended material transport (arrows), suspended turbidity maximum, and bottom sediment accumulation associated with estuarine circulation. Modified from Meade (1972).*

Chapter 9 (p. 142). Visher and Howard, studying the marsh- and mud flat-bordered Altamaha estuary, Georgia, show estuarine-tidal circulation to be a highly effective mechanism for size segregation of sediment into differing populations by "bedload transport, suspension and recycling during successive tidal cycles" (Visher and Howard, 1974, p. 302).

FILLING TOWARD AN EQUILIBRIUM. Estuarine processes result in an accumulation of sediment that will slowly fill the estuary. With shoaling, tidal and wind-wave scour and circulation become effective in redistributing sediments. Filling will also reduce the tidal prism, promoting further sedimentation. Allen et al. (1973) have theorized that the estuary may attain an equilibrium configuration in which there is a linear upstream decrease in the volume contained in the tidal prism. This equilibrium estuary (Chapter 14) may be contained within the estuarine valley or, with large supply of river effluent, may emerge as a delta.

HOLE AND SHADOW FILLINGS

Intracoastal depressions and energy shadows generated by preexisting and constructional coexisting topography may be partly or totally filled by peritidal flats of sand and mud (tidal flats and wind or storm flats) and/or peat and peaty mud (tidal marshes and swamps). Peritidal flats may also form along leeward open coasts in very low-energy settings. Flats, marshes, and swamps form and persist where sediment influx and/or production is greater than can be removed by waves and currents. Sediment is entirely provided by the sea for peritidal flats in lagoons, sounds, and open coast energy shadows. Sediment may be partly riverborne on delta margin flats and within estuaries, but even that is mostly reworked into its final configuration by marine tides, waves, and currents. Many flats, swamps, and marshes depend on an interaction of physical processes and vegetative communities.

Sand and Mud Flats

ENERGY SETTING. It is meaningful to define two types of detrital sediment flats to distinguish the temporal distribution of effective processes. "Tidal flats" are those detrital accumulates built up into the intertidal zone, maintained and modified primarily by day-to-day processes (Fig. 8A). Tidal flats are characteristic of higher energy settings. "Wind" or "storm flats" are those detrital accumulates whose nourishment and effective dynamics depend primarily on sporadic storm-generated "tides" (Fig. 8B). These are characteristic of low-energy settings.

CHANNELED SYSTEMS. Sand and mud flats may be either dissected by channels or essentially channel free. Channel systems digitate inland, resembling an inverse delta with a seaward source (Van Straaten, 1954; Shinn et al., 1969). Channels on tidal flats are mostly bordered by broad intertidal sand and mud flats. On storm flats, channel margins may have natural levees extending well above normal high-water level but fading inland (Shinn et al., 1969). Leveed channels separate inter-channel ponds. The marine margin of the flat, marsh, or swamp may also have a high natural storm levee, further restricting circulation within.

Several mechanisms pump sediment into channeled flat systems. Offshore sediment is provided to tidal flats by both landward asymmetry in the offshore tidal rotation (Postma, 1967; McCave, 1972) and shoreward increase in wave agitation with its attendant net land-ward flow (Chapter 8, Fig. 8). Residual landward transport also occurs because of the "settling lag" phenomenon; suspended particles continue to drift landward after tidal flow has decelerated below threshold velocity, until they reach the bottom (Postma, 1967). The net result of these mechanisms is to cause tidal flat sediments to move landward over successive tidal cycles. Onshore-directed storm surges of agitated sediment-laden offshore waters appear to be the main source of nourishment to storm flats (Wanless, 1973; Gebelein, 1973).

Within channeled flats, tidal (or surge) currents decrease in strength gradually in a landward direction along channel axes and rapidly across flats. In addition, tidal flat channels frequently experience their most intense currents immediately before and after low water, rather than at midebb and midflood. This time-velocity asymmetry is a consequence of the retardation of the leading edge of the tidal wave in shallow water by friction. Both the time-velocity asymmetry of the tidal cycle and the spatial velocity contrasts favor the settling of particles on the flats at high tide over their settling in the channels at low tide (Postma, 1967). Decreasing velocity will also decrease competency (capability to carry a certain grain size), the larger particles being deposited near the mouth and only the finest particles reaching the landward edge. Settling from waters ponded on flats can also be an efficient trapping mechanism, especially on leveed storm flats.

Wind-generated waves and currents are very important in reworking and distributing sediment on some tidal flats. On shallow flats, as the Wadden Zee, Netherlands, grain size increases with decreasing depth and

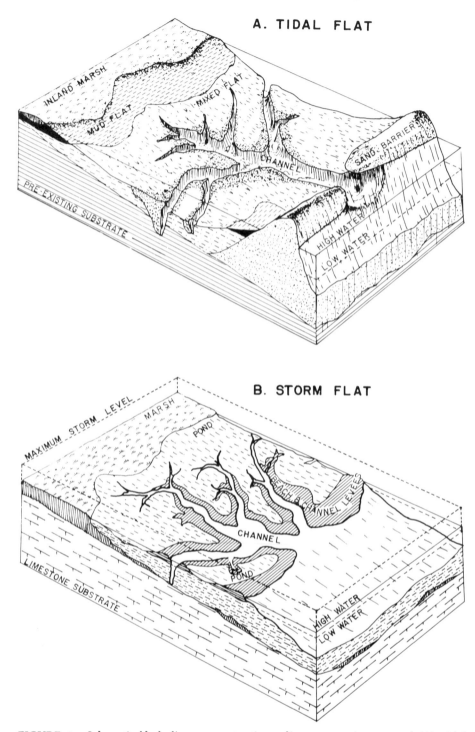

FIGURE 8. *Schematic block diagrams contrasting sedimentary environments of (A) tidal flats (as those bordering the North Sea) and (B) storm flats (as that bordering northwestern Andros Island, Bahamas).*

increasing wave fetch (Postma, 1967). Wave-swept carbonate tidal flats in Shark Bay, Australia, show a similar relationship (Davies, 1970).

Tidal flats, in areas of large tidal range, are characterized near low-water level by channel margin sand flats continually reworked by strong tidal currents (Fig. 8*A*). The midtide zone is dominated by mixed flats composed of an alternation of rippled sand strata (midtide deposits) and draping clay strata (slack tide deposits). Oyster bars may cover mixed flats. Meandering tidal channels cut the sand flats and mixed flats and rework the accumulate by point bar migration. Upper mud flats, near high-water level, are a finely laminated accumulate from slack water settleout (Van Straaten, 1954; Reineck, 1967, 1972; Knight and Dalrymple, in press). Algal mats may cover the surface, helping to trap and bind the sediment (Evans, 1965). The landward margin of the flat may be a narrow to broad marsh underlain by carbonaceous muds which tend to become peatier landward.

Channeled storm flats (Fig. 8*B*) are only well documented from environments of carbonate sedimentation, especially the flats bordering northwestern Andros Island, Bahamas (Shinn et al., 1969; Ginsburg et al., 1970; Wanless, 1973; Ginsburg and Hardie, in press). Here subtidal channels contain rippled skeletal sands and mud lumps. Laminated supratidal levees are muddy skeletal or pellet sands becoming muddier on the gradual backslope into the ponds. Levees build during storm-tide overbank flooding by traction-load sand and algal stick-on-mud laminae (Wanless, 1973). Pond muds are intensely reworked by a restricted suite of burrowing and grazing organisms. Pond cores contain a few thick mud layers recording settleout accumulation from ponded storm waters. Inland from the channeled storm flat is a supratidal algal marsh. This seasonally wet and dry marsh receives sediment only during major storms. Hurricane Betsy (1965) covered this flat and marsh with nearly 3 m of water (Perkins and Enos, 1968). The underlying sedimentary record is an alternation of thick storm sediment layers and algal-sediment mush.

Vegetative influences are commonly intense on storm flats if climatic conditions permit. The laminated levee surface of channeled carbonate storm flats in South Florida and the Bahamas is bound by filamentous blue-green algae, and the surface when ripped up has a leathery character. The algal-bound surface inhibits erosion during storm flooding, aids in the entrapment of storm-derived muds on the levees, and provides effective stabilization to channel margins so that migration is primarily through current undercutting of the cohesive mud.

In dry tropical settings, evaporite precipitation and cementation become very important (Laguna Madre, Texas: Dickinson et al., 1972; Shark Bay, Australia: Logan et al., 1970; Persian Gulf: Kinsman, 1966) and algal influence may extend into the subtidal zone (Logan et al., 1970).

UNCHANNELED SYSTEMS. Rapidly prograding flats in areas of low tidal range may not be channeled. The carbonate storm flat bordering southwestern Andros Island, Bahamas, appears to build seaward in pulses by repeated storm growth emergence of linear mud bars on the very shallow offshore platform. The linear bars isolate elongate ponds that are subsequently dissected into subcircular "bays" and gradually filled by ponding of sediment-laden storm water (Gebelein, 1973).

The "wind flats" bordering southern Laguna Madre blanket much of the area between Padre Island and the Texas mainland. Much of the sediment is blown across from the unvegetated island barrier but is redistributed and mixed with lagoonal muds by sporadic wind tides which spread across the flats. Gypsum precipitates from water ponded in the subtle topographic irregularities of the semiarid flat (Miller, in press).

Tidal Swamps and Marshes

Tidal swamps are marginal marine (paralic) flats densely vegetated by trees and shrubs; marshes by sedges, reeds, or algae. The accumulate may be totally *autochthonous* (the result of organic production within the marsh or swamp), *allochthonous* (composed of organic and mineral detritus transported in), or a mixture of the two. Marshes and swamps may dominate flats that (1) have built to the upper part of the tidal level or have low tidal range, (2) do not have excessive allochthonous sediment influx, and (3) have minimal reworking of the substrate.

Tidal marshes dominate in temperate climates (see Teal and Teal, 1969). Two species of the sedge *Spartina* dominate in North America. Tall *Spartina alternaflora* occurs well out into the intertidal zone; *S. patens* is restricted to the very upper part of the tidal zone. Dense blade growths baffle currents aiding in sediment entrapment, and dense root systems stabilize the sediment surface. Tidal channels digitating into many marshes have undergone little or no migration in the past 50 years.

Sediments in *Spartina* marshes are usually carbonaceous mud, grading into purer autochthonous peat inland away from detrital sediment influx.

Marshes generally yield to tidal mangrove swamps in subtropical and tropical climates (Craighead, 1964).

Mangrove swamp accumulates are commonly nearly pure autochthonous peat, even though receiving storm influxes of sediment-laden water from adjacent carbonate lagoons. Detrital carbonate sediment spread across the swamp may be entirely leached by acidic (tannic acid) waters from the mangroves. In areas of high detrital influx (as marginal storm levees or where silicic-clastic sediment is abundant) sediment is a sandy or muddy (marly) peat (Spackman et al., 1969).

Mangroves propagate by floating seedlings and, once established, spread by arching prop roots (red mangrove) and runners (black mangrove). Red mangroves may occur throughout the tidal range and slightly below it. Black mangroves are restricted to near high-water level and generally occur on the landward margin of the swamp.

In contrast to early thoughts, mangrove swamps are an essentially passive sedimentary environment offering some resistance to erosion, which will be self-maintaining only if the area is protected from strong physical agitation and excessive detrital sediment influx. Once an area of mangrove swamp is eroded to a depth of only 0.5 m below low water, the mangrove cannot build back. A mangrove shoreline can thus only recede during a period of rising sea level and in the presence of intermittent wave energy, unless shallow mud or sand flats form over which the red mangrove can again take root (Wanless, 1969, in press).

Mangrove swamps may separate vast freshwater marshes or supratidal flats from the marine environment. Along the rainy South Florida coast a nearly continuous mangrove fringe separates the freshwater Everglades marsh from the coastal marine bays and lagoons. In arid areas, such as Madagascar, mangroves may separate supratidal evaporitic flats (Guilcher, 1963).

Mangrove root systems are a partial baffle to currents. Perhaps more important are the sponges on the proproots that filter much of the suspension from flood tide water. During hurricanes, mangroves are an important energy buffer to inland environments, though the mangrove fringe itself may be extensively damaged (Spackman et al., 1969).

SUMMARY

Intracoastal sedimentary environments consist of four elements: (1) rigid or mobile preexisting coastal topographic framework; (2) coast-constructing sedimentary bodies (sand barriers, spits, and deltas) or pericoast-constructing bodies (sand and mud bars and belts, reefs); (3) resultant holes (estuaries and lagoons); and (4) hole and energy shadow fillings (tidal flats, storm flats, tidal marshes, and swamps). The character, patterns, and complexity of intracoastal sedimentary deposits are determined by the resultant interacting influence of primary (independent) and dependent depositional controls.

Sand barriers, constructional slightly emergent sand ridges extending generally parallel to the coast, effectively separate intracoastal lagoons. Sand barriers are formed by spit accretion, reworking of drowned coastal plain into a new profile of equilibrium, or gradual submergence of mainland beach dune ridges.

Delta formation potential is determined by fluvial effluent within a framework of preexisting topography, climate, and rate of sea-level change. Spatial and temporal interaction of fluvial effluent and physical ocean processes determines both resultant morphology of the delta and extent and timing of destructional reworking.

Marine reefs and sand or mud belts and bars, generated in association with breaks in slope or lateral topographic constrictions, may effectively separate a landward "bank" or lagoon from offshore wave action but provide only minor restriction to circulation.

Tidal inlets across constructional accumulates provide circulation to the contained lagoons and estuaries and are important conduits for sediment influx and exchange.

Lagoons, marginal marine water bodies protected and partially isolated from the open sea by constructional accumulates or preexisting topography, tend to parallel the coast and be sufficiently shallow that the wind-generated wave energy and circulation determine resultant patterns of sediment accumulation and lagoon partitioning. Fundamental differences may exist between silicic-clastic lagoons and tropical carbonate lagoons—the latter being characterized by local sediment production, intense organism influences, and a rigid limestone preexisting topographic framework.

Estuaries, active river valleys partially inundated by the sea, tend to be deeper than lagoons and oriented at a large angle to the coast. Sedimentation processes are dominated by stratified flow patterns (estuarine circulation) and tidal motion.

Tidal flats, maintained by day-to-day processes, and wind or storm flats, nourished by sporadic storm processes, are detrital fillings of intracoastal holes or energy shadows. Organism or evaporitic influences may be intense where respective climatic conditions permit.

Tidal swamps are marginal marine flats densely vegetated by trees and shrubs; marshes by sedges, reeds, or algae. Swamp and marsh accumulates may be totally locally produced or partly detrital and dominate intertidal environments that do not have excessive sediment influx or substrate reworking.

ACKNOWLEDGMENTS

Manuscript preparation was supported by Sea Grant. Critical reviews were provided by Drs. N. P. James and J. van de Kreeke, and Mr. E. R. Warzeski.

REFERENCES

Allen, G. P., P. Carbonel, P. Castaing, J. Gayet, J. M. Jouanneau, A. Klingebiel, C. Latouche, Ph. Legigan, M. Pojos, and G. Vernette (1973). *Guidebook on Environments and Sedimentary Processes of the North Aquitaine Coast.* Bordeaux: Institut de Géologie du Bassin d'Aquitaine, 106 pp.

Ball, M. M. (1967). Carbonate sand bodies of Florida and the Bahamas. *J. Sediment. Petrol.*, **37:** 556–591.

Bathurst, R. G. C. (1967). Subtidal gelatinous mat, sand stabilizer and food, Great Bahama Bank. *J. Geol.*, **75:** 736–738.

Bathurst, R. G. C. (1971). Carbonate sediments and their diagenesis. *Developments in Sedimentology*, Vol. 12. Amsterdam: Elsevier, 620 pp.

Bruun, P. and F. Gerritsen (1955). *Stability of Coastal Inlets.* Amsterdam: North-Holland, 123 pp.

Craighead, F. C. (1964). Land, mangroves and hurricanes. *Bull. Fairchild Trop. Gard.*, **19:** 1–28.

Curray, J. R. and F. P. Shepard (1972). Some major problems of Holocene sea levels (abstr. with fig.). *Am. Quat. Assoc., 2nd Natl. Conf., Miami, Fla.* pp. 16–18.

Davies, G. R. (1970). Carbonate bank sedimentation, eastern Shark Bay, western Australia. In B. W. Logan et al., eds., *Carbonate Sedimentation and Environments, Shark Bay, Western Australia.* Am. Assoc. Pet. Geol. Mem. 13, pp. 85–168.

deBeaumont, E. (1845). *Leçons de Géologie Pratique.* Paris, pp. 223–252.

Dickinson, K. A., H. L. Berryhill, Jr., and C. W. Holmes (1972). Criteria for recognizing ancient barrier coastlines. In J. K. Rigby and W. K. Hamblin, eds., *Recognition of Ancient Sedimentary Environments.* Soc. Econ. Paleontologists and Mineralogists Spec. Publ. 16, pp. 192–214.

El-Ashry, M. T. and H. R. Wanless (1965). Birth and early growth of a tidal delta. *J. Geol.*, **73:** 404–406.

Emery, K. O. and E. Uchupi (1972). *Western North Atlantic Ocean: Topography, Rocks, Structure, Water, Life, and Sediments.* Am. Assoc. Pet. Geol. Mem. 17, 532 pp.

Evans, G. (1965). Intertidal flat sediments and their environments of deposition in the Wash. *Q. J. Geol. Soc. London*, **121:** 209–245.

Frazier, D. E. and A. Osanik (1969). Recent peat deposits—Louisiana coastal plain. In E. C. Dapples and M. E. Hopkins, eds., *Environments of Coal Deposition.* Geol. Soc. Am. Spec. Pap. No. 114, pp. 63–85.

Gebelein, G. D. (1973). Sedimentology and stratigraphy of recent shallow marine and tidal-flat sediments, southwest Andros Island, Bahamas. *Am. Assoc. Pet. Geol. Bull.*, **57:** 780–781.

Gilbert, G. K. (1885). *The Topographic Features of Lake Shores.* U.S. Geol. Surv. 5th Ann. Rep., pp. 69–123.

Ginsburg, R. N. (1956). Environmental relationships of grain size and constituent particles in some south Florida carbonate sediments. *Am. Assoc. Pet. Geol. Bull.*, **40:** 2384–2427.

Ginsburg, R. N. and L. A. Hardie (in press). Tidal and storm deposits, northwestern Andros Island, Bahamas. In R. N. Ginsburg, ed., *Tidal Deposits, a Source Book of Recent Examples and Fossil Counterparts.* Berlin and New York: Springer-Verlag.

Ginsburg, R. N. and H. A. Lowenstam (1958). The influence of marine bottom communities on the depositional environment of sediments. *J. Geol.*, **66:** 310–318.

Ginsburg, R. N., O. P. Bricker, H. R. Wanless, and P. Garrett (1970). Exposure index and sedimentary structures of a Bahama tidal flat. Discussion Paper. *Geol. Soc. Am. Abstr. Program*, **2**(7): 744–745.

Gould, H. E. and E. McFarlan, Jr. (1959). Geologic history of the chenier plain, southwestern Louisiana. *Gulf Coast Assoc. Geol. Soc. Trans.*, **9:** 1–10.

Guilcher, A. (1963). Estuaries, deltas, shelf and slope. In M. N. Hill, ed., *The Sea*, Vol. 3. New York: Wiley-Interscience, pp. 620–654.

Guilcher, A. (1967). Origin of sediments in estuaries. In G. H. Lauff, ed., *Estuaries.* Washington, D.C.: Am. Assoc. Adv. Sci., Publ. No. 83, pp. 149–157.

Harrison, W., R. J. Malloy, G. A. Rusnak, and J. Terasmae (1965). Possible late Pleistocene uplift, Chesapeake Bay entrance. *J. Geol.*, **73:** 201–229.

Haven, D. S. and R. Morales-Alamo (1972). Biodeposition as a factor in sedimentation of fine suspended solids in estuaries. In B. W. Nelson, ed., *Environmental Framework of Coastal Plain Estuaries.* Geol. Soc. Am. Mem. 133, pp. 121–130.

Hay, W. W. (1967). Bimini lagoon: Model epeiric sea. *Am. Assoc. Pet. Geol. Bull.*, **51:** 468–469.

Howard, J. D. (1972). Nearshore sedimentary processes and geologic studies. In D. J. P. Swift, D. B. Duane, and O. H. Pilkey, eds., *Shelf Sediment Transport: Process and Pattern.* Stroudsburg, Pa.: Dowden, Hutchinson & Ross, pp. 645–648.

Howard, J. D., et al. (1972). Georgia coastal region, Sapelo Island, U.S.A.: Sedimentology and biology. *Senckenbergiana Marit.*, **4:** 1–223.

Hoyt, J. H. (1967). Barrier island formation. *Geol. Soc. Am. Bull.*, **78:** 1125–1135.

Hoyt, J. H. (1969). Chenier versus barrier, genetic and stratigraphic distinction. *Am. Assoc. Pet. Geol. Bull.*, **53:** 299–306.

Kinsman, D. J. J. (1966). Gypsum and anhydrite of Recent age, Trucial coast, Persian Gulf. *2nd Symposium on Salt, Northern Ohio Geol. Soc., Cleveland, Ohio.* **1:** 302–326.

Knight, R. J. and R. W. Dalrymple (in press). Intertidal sediments from the south shore of Cobequid Bay, Bay of Fundy, Nova Scotia. In R. N. Ginsburg, ed., *Tidal Deposits, a Source Book of Recent Examples and Fossil Counterparts.* Berlin and New York: Springer-Verlag.

Komar, P. D. (1973). Computer models of delta growth due to sediment input from rivers and longshore transport. *Geol. Soc. Am. Bull.*, **84:** 2217–2226.

Kraft, J. C. (1971). Sedimentary facies patterns and geologic history of a Holocene marine transgression. *Geol. Soc. Am. Bull.*, **82:** 2131–2158.

Leontyev, O. K. and L. G. Nikiforov (1965). On the reasons of world wide spread of shore barriers. *Oceanology* **4:** 653–661.

Leontyev, O. K. and L. G. Nikiforov (1966). An approach to the problem of the origin of barrier bars. *Int. Oceanogr. Congr., 2nd Abstr. Papers*, pp. 221–222.

Logan, B. W., G. R. Davies, J. F. Read, and D. E. Cebulski

(1970). *Carbonate Sedimentation and Environments, Shark Bay, Western Australia*. Am. Assoc. Pet. Geol. Mem. 13, 223 pp.

Luneburg, H. (1939). Hydrochemische untersuchungen in der elbmundung mittels elektrokolorimeter. *Arch. Dtsch. Seewarte*, **59:** 1–27.

McCave, I. N. (1972). Transport and escape of fine grained sediment from shelf areas. In D. J. P. Swift, D. B. Duane, and O. H. Pilkey, eds., *Shelf Sediment Transport: Process and Pattern*. Stroudsburg, Pa.: Dowden, Hutchinson & Ross, pp. 245–252.

McKee, E. D. and T. S. Sterrett (1961). Laboratory experiment on form and structure of longshore bars and beaches. In J. A. Peterson and J. C. Osmond, eds., *Geometry of Sandstone Bodies*. Am. Assoc. Pet. Geol., pp. 13–28.

Meade, R. H. (1972). Transport and deposition of sediments in estuaries. In B. W. Nelson, ed., *Environmental Framework of Coastal Plain Estuaries*. Geol. Soc. Am. Mem. 133, pp. 91–120.

Miller, J. A. (in press). Facies characteristics of Laguna Madre wind-tidal flats. In R. N. Ginsburg, ed., *Tidal Deposits, A Source Book of Recent Examples and Fossil Counterparts*. Berlin and New York: Springer-Verlag.

Morgan, J. P., ed. (1970a). *Deltaic Sedimentation, Modern and Ancient*. Soc. Econ. Paleontologists and Mineralogists Spec. Publ. 15, Tulsa, Okla., 312 pp.

Morgan, J. P. (1970b). Depositional processes and products in the deltaic environment. In J. P. Morgan, ed., *Deltaic Sedimentation, Modern and Ancient*. Soc. Econ. Paleontologists and Mineralogists Spec. Publ. 15, Tulsa, Okla., pp. 31–47.

Morgan, J. P., L. G. Nichols, and M. Wright (1958). *Morphological Effects of Hurricane Audrey on the Louisiana Coast*. Tech. Ser., Rep. 10, Louisiana State University, Coastal Studies Inst., 53 pp.

Morton, R. W. (1972). Spatial and temporal distribution of suspended sediment in Narragansett Bay and Rhode Island Sound. In B. W. Nelson, ed., *Environmental Framework of Coastal Plain Estuaries*. Geol. Soc. Am. Mem. 133, pp. 131–141.

Nelson, B. W., ed. (1972). *Environmental Framework of Coastal Plain Estuaries*. Geol. Soc. Am. Mem. 133, 619 pp.

Neumann, A. C. (1965). Processes of Recent carbonate sedimentation in Harrington Sound, Bermuda. *Bull. Mar. Sci.*, **15:** 987–1035.

Neumann, A. C., C. D. Gebelein, and T. P. Scoffin (1970). The composition, structure and erodability of subtidal mats, Abaco, Bahamas. *J. Sediment. Petrol.*, **40:** 274–297.

O'Brien, M. P. and R. G. Dean (1972). Hydraulics and sedimentary stability of coastal inlets. In *Proc. 13th Coastal Eng. Conf. ASCE, New York*, pp. 761–780.

Off, T. (1963). Rhythmic linear sand bodies caused by tidal currents. *Am. Assoc. Pet. Geol. Bull.*, **47:** 324–341.

Perkins, R. D. and P. Enos (1968). Hurricane Betsy in the Florida-Bahama area—geologic effects and comparison with Hurricane Donna. *J. Geol.*, **76:** 710–717.

Pierce, J. W. (1970). Tidal inlets and washover fans. *J. Geol.*, **78:** 230–234.

Postma, H. (1967). Sediment transport and sedimentation in the estuarine environment. In G. H. Lauff, ed., *Estuaries*. Washington, D.C.: Am. Assoc. Adv. Sci., Publ. No. 83, pp. 158–179.

Postma, H. and K. Kalle (1955). Die entstehung von trubungszonen im unterlauf der flosse, speziell im hinblick auf die verhaltnisse in der unterelbe. *Dtsch. Hydrograph.*, **8:** 137–144.

Price, W. A. (1963). Origin of barrier chain and beach ridge. In *The Geological Society of America, Abstracts for 1962*. Geol. Soc. Am. Spec. Pap. 73, p. 219.

Reineck, H. E. (1967). Layered sediments of tidal flats, beaches and shelf bottom. *Am. Assoc. Adv. Sci.*, **83:** 191–206.

Reineck, H. E. (1972). Tidal flats. In J. K. Rigby and W. K. Hamblin, eds., *Recognition of Ancient Sedimentary Environments*. Soc. Econ. Paleontologists and Mineralogists Spec. Publ. 16, pp. 146–159.

Rhoads, D. C. (1963). Rates of sediment reworking by *Yoldia limatula* in Buzzards Bay, Massachusetts, and Long Island Sound. *J. Sediment. Petrol.*, **33:** 723–727.

Rusnak, G. A. (1960). Sediments of Laguna Madre, Texas. In F. P. Shepard, F. B. Phleger, and T. H. van Andel, eds., *Recent Sediments, Northwest Gulf of Mexico*. Tulsa, Okla.: Am. Assoc. Pet. Geol., pp. 153–196.

Rusnak, G. A. (1967). Rate of sediment accumulation in modern estuaries. In G. H. Lauff, ed., *Estuaries*. Washington, D.C.: Am. Assoc. Adv. Sci., Publ. 83, pp. 180–184.

Russell, J. R. and R. D. Russell (1939). Mississippi River Delta sedimentation. In P. D. Trask, ed., *Recent Marine Sediments, a Symposium*. Tulsa, Okla.: Am. Assoc. Pet. Geol., pp. 153–177.

Scholl, D. W., F. C. Craighead, and M. Stuiver (1969). Florida submergence curve revised: Its relation to coastal sedimentation rates. *Science, 163:* 562.

Schwartz, M. L. (1971). The multiple causality of barrier islands. *J. Geol.*, **79:** 91–94.

Schwartz, M. L., ed. (1973). *Barrier Islands*, Stroudsburg, Pa.: Dowden, Hutchinson & Ross, 451 pp.

Scoffin, T. P. (1970). The trapping and binding of subtidal carbonate sediments by marine vegetation in Bimini Lagoon, Bahamas. *J. Sediment. Petrol.*, **40:** 249–273.

Scott, A. J. and W. L. Fisher (1969). Delta systems and deltaic deposition. In *Delta Systems in the Exploration for Oil and Gas, A Research Colloquium*. University of Texas, Austin, 78 pp.

Scruton, P. C. (1960). Delta building and the deltaic sequence. In F. P. Shepard, F. B. Phleger, and T. H. van Andel, eds., *Recent Sediments, Northwest Gulf of Mexico*. Tulsa, Okla.: Am. Assoc. Pet. Geol., pp. 82–102.

Shepard, F. P. (1960). Gulf coast barriers. In F. P. Shepard, F. B. Phleger, and T. H. van Andel, eds., *Recent Sediments, Northwest Gulf of Mexico*. Tulsa, Okla.: Am. Assoc. Pet. Geol., pp. 197–220.

Shepard, F. P. and D. G. Moore (1960). Bays of central Texas coast. In F. P. Shepard, F. B. Phleger, and Tj. H. van Andel, eds., *Recent Sediments, Northwest Gulf of Mexico*. Tulsa, Okla.: Am. Assoc. Pet. Geol., pp. 117–152.

Shepard, F. P. and H. R. Wanless (1971). *Our Changing Coastlines*. New York: McGraw-Hill, 579 pp.

Shinn, E. A., R. M. Lloyd, and R. N. Ginsburg (1969). Anatomy of a modern carbonate tidal flat, Andros Island, Bahamas. *J. Sediment. Petrol.*, **39:** 1202–1228.

Spackman, W., W. L. Riegel, and C. P. Dolsen (1969). Geological and biological interactions in the swamp-marsh complex of southern Florida. In E. C. Dapples and M. E. Hopkins, eds., *Environments of Coal Deposition*. Geol. Soc. Am. Spec. Pap. No. 114, pp. 1–35.

Taft, W. H. and J. W. Harbaugh (1964). Modern carbonate sediments of southern Florida, Bahamas, and Espirito Santo Island, Baja California: A comparison of their mineralogy and chemistry. *Stanford Univ. Publ., Geol. Sci.*, **8**(2): 133 pp.

Teal, J. and M. Teal (1969). *Life and Death of the Salt Marsh.* Boston: Little, Brown, 278 pp.

Van Straaten, L. M. J. U. (1954). Composition and structure of recent marine sediments in the Netherlands. *Leidse Geol. Mededel.*, **13:** 1–110.

Van Veen, S. (1950). Eb-en vloedschaarsystemen in de Nederlandse getijwateren. "*Waddensymposium*," *Tijdschr. Kon. Ned Aardrijksk. Genoot.*, 43–65.

Visher, G. S. and J. D. Howard (1974). Dynamic relationship between hydraulics and sedimentation in the Altamaha estuary. *J. Sediment. Petrol.*, **44**(2): 502–521.

Wanless, H. R. (1969). *Sediments of Biscayne Bay: Distribution and Depositional History.* Inst. Marine Sci., University of Miami Tech. Rep. 69–2, 260 pp.

Wanless, H. R. (1970). Influence of preexisting bedrock topography on bars of "lime" mud and sand. Biscayne Bay, Florida. *Am. Assoc. Pet. Geol. Bull.*, **54:** 875.

Wanless, H. R. (1973). Cambrian of the Grand Canyon—A reevaluation of the depositional environment. Ph.D. Dissertation, Johns Hopkins University, Baltimore, 115 pp.

Wanless, H. R. (in press). Mangrove sedimentation in geologic perspective. In P. Gleason, ed., *South Florida Environments: Present and Past.* Miami, Fla.: Miami Geol. Soc., 452 pp.

Wilson, B. W. and A. Torum (1968). *The Tsunami of the Alaskan Earthquake, 1964: Engineering Evaluation.* U.S. Army Corps of Engineers, Coastal Eng. Res. Cent., Tech. Mem. 25, 401 pp.

Wright, L. D. (1970). The influence of sediment availability on patterns of beach ridge development in the vicinity of the Shoalhaven River delta, N.S.W. *Aust. Geogr.*, **3:** 336–348.

Wright, L. D. and J. M. Coleman (1972). River delta morphology: Wave climate and the role of subaqueous profile. *Science*, **176:** 282–284.

Wright, L. D. and J. M. Coleman (1973). Variations in morphology of major river deltas as functions of ocean wave and river discharge regimes. *Am. Assoc. Pet. Geol. Bull.*, **57**(2): 370–398.

Zenkovitch, V. P. (1959). On the genesis of cuspate spits along lagoon shores. *J. Geol.*, **67:** 267–277.

Zenkovitch, V. P. (1967). *Processes of Coastal Development.* New York: Wiley-Interscience, 738 pp.

CHAPTER 13

Nearshore Currents and Sediment Transport, and the Resulting Beach Configuration

PAUL D. KOMAR

School of Oceanography, Oregon State University, Corvallis, Oregon

With respect to magnitudes of physical processes, the nearshore zone of breaking waves and surf is the most intense within the continental margin area. In this nearshore zone the waves expend most of their energy, causing coastal erosion, generating longshore currents, and producing sediment transport. Considerable progress has been made in the past decade in understanding these processes. The purpose of this chapter is to review briefly our present understanding of the generation of longshore currents and the resulting sediment transport. This chapter also examines the response of the beach configuration to this sediment transport, controlling the shoreline shape in the longshore direction. The variety of rhythmic shoreline features found on beaches, including beach cusps and rhythmic topography, are briefly reviewed. This examination is an abstract of a more thorough treatment given elsewhere (Komar, in press).

NEARSHORE CURRENTS

Two systems of nearshore currents can be distinguished: (1) a cell circulation of rip currents, and (2) longshore currents caused by waves breaking at an angle to the shoreline. Very often the two systems combine and so cannot be treated separately, but as a first examination we shall treat them as being independent. The basic difference between the two involves the driving forces responsible for the currents. The cell circulation is caused principally by longshore variations in the water level within the surf zone produced either by waves dumping water shoreward of a bar or by wave setup being higher shoreward of the larger breaking waves. When waves arrive at an angle to the shoreline there is a longshore component to the wave-associated momentum flux that generates longshore currents.

Cell Circulation

The cell circulation (Fig. 1) consists of rip currents and associated longshore currents. To replace water leaving the surf zone by way of the rips, there must also be a much slower shoreward mass transport.

Waves moving toward the shore carry with them a momentum flux as well as energy. Longuet-Higgins and Stewart (1964) have formulated the theoretical basis for this momentum flux caused by the presence of waves, and have coined the term *radiation stress*. Radiation stress has proved to be a very powerful tool in treating many wave-associated problems including, as we shall see, the generating of nearshore currents, both the cell circulation and the currents caused by an oblique wave approach.

When the waves arrive at the shore with their crests parallel to the shoreline, this radiation stress (momentum flux) is balanced by a wave *setdown* and *setup* in the nearshore. The setdown (Fig. 2) is a small lowering of

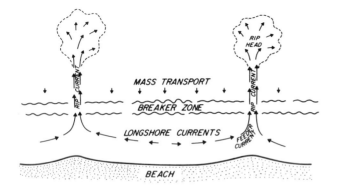

FIGURE 1. *The cell circulation of rip currents, associated longshore currents feeding the rips, and a mass transport shoreward to replace water lost from the surf through the rips. From Shepard and Inman (1950).*

the water level below the still water level (that level when no waves are present) occurring just outside the breaker zone, and the setup is a rise of the mean water level above the still water level within the surf zone. As can be seen in Fig. 2, the larger the size of the breaking waves the higher the setup within the surf zone. This mean water level rise is the important aspect with regards to the generation of a cell circulation, for if there is a longshore variation in the wave height, there will be a corresponding longshore change in the water level. Shoreward of the larger breaking waves the water level

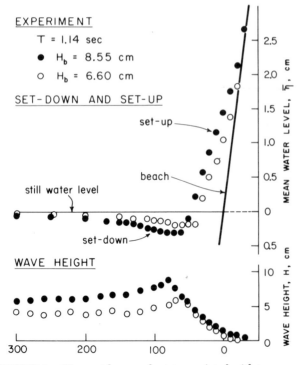

FIGURE 2. *Wave setdown and setup associated with two wave sizes. From Bowen (1969a).*

will be highest and so will cause the water to flow alongshore toward a position where the wave setup is lower. The longshore currents therefore converge toward positions of lowest wave breakers where the flow turns seaward as rip currents.

In addition to this longshore variation in wave setup, there is also a longshore component of the wave radiation stress that acts in a direction from large to small breakers. This adds a second term to the driving force for the cell circulation within the surf zone. It might be noted in Fig. 2 that the wave setdown outside the larger breakers is greater than that outside the smaller breakers. One might therefore expect a current beyond the breakers flowing parallel to shore from where the waves are lowest to where they are highest, in the opposite direction to the longshore current within the surf. It turns out, however, that the longshore variation in wave setdown is opposed and balanced by that longshore component of the radiation stress that supported the flow within the surf. Because of this balance outside the breaker zone, the cell circulation is driven entirely by forces acting within the surf zone.

The development of cell circulation outlined above due to longshore variations in wave setup and the component of the wave radiation stress is a summary of the formulation of Bowen (1969a). Bowen's paper may be examined for the mathematical development; a more complete summary is given by Komar (in press). Bowen and Inman (1969) go on to examine possible mechanisms for producing the initial longshore variations in wave height required for the generation of the cell circulation. The most obvious way is by wave refraction, the waves being high where the wave rays and energy converge and low where the rays diverge. A good example of this is the system of rip currents at Scripps Beach, La Jolla, California (Fig. 3). There the wave refraction is produced by a system of offshore submarine canyons. In the lee of the canyon the waves are smallest, and that is where the rip current occurs. Midway between the rips the wave heights are greatest (Fig. 3), and that is where the longshore currents diverge as they flow toward separate rips.

Bowen and Inman (1969) also explored the possible role of edge waves in producing longshore variations in wave heights and thus causing rip currents. Edge waves could account for the presence of rips on long straight beaches with a regular bottom topography where wave refraction clearly is not important. Bowen and Inman demonstrated both theoretically and experimentally within a wave basin that ordinary incident swell waves may generate standing edge waves on the beach that have the same period as the incoming waves. The interaction or summation of the swell and edge waves produces alternating high and low breakers along the

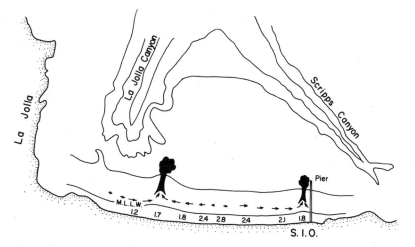

FIGURE 3. *Cell circulation at Scripps Beach, La Jolla, California, due to wave refraction over submarine canyons producing longshore variations in wave heights. Values shown along the shoreline are breaker heights in meters. From Shepard and Inman (1950).*

shoreline and therefore can produce a cell circulation. The edge waves are generally standing waves with their crests normal to the shoreline and their wavelengths from crest to crest parallel to the shoreline; they are perpendicular in orientation to the incoming swell waves. There are alternate positions in the longshore direction of nodes where there is no observable up-and-down motion of the water because of the edge waves, and antinodes where the full edge wave height is observed as an up-and-down motion. Several offshore variations in the edge wave height are possible, depending on which edge wave is present, but in general the height is a maximum at the shore, decreases rapidly offshore, and becomes small a short distance outside the breakers. The important factor in the interactions between edge waves and incoming swell is that they have the same period. Because of this the edge wave remains in phase with the breakers at one antinode position, and the incoming wave height and edge wave height add to give a large breaker. Conversely, at the next antinode position the edge wave height subtracts from the incoming wave height so that the breaking wave is smaller. This addition of incoming waves and edge waves to produce a longshore variation in wave heights is illustrated in Fig. 4. This will in turn produce a regular pattern of rip currents and cell circulation. The rips are again found in the positions of low breakers, that is, at every other antinode of the causative edge waves. If the edge waves have a period that differs from that of the swell waves, the positions where they add to form large breakers or subtract to form small waves will continuously shift about in the longshore direction and stationary rips cannot develop.

Bowen and Inman (1969, p. 5488) also made measurements of rip current spacings and accompanying wave parameters at El Moreno Beach on the northwest coast of the Gulf of California. The beach there has a steep face (tan $\beta = 0.148$), rising above a nearly horizontal low tide terrace, so that possible effects of wave refraction on rip current formation could be eliminated. In the four sets of field measurements obtained, there was very close agreement between the rip current spacing and the theoretical wavelength of the edge waves. This offers strong supporting evidence for an edge wave generation mechanism.

As we shall discuss later, the cell circulation can redistribute the beach sediment and therefore has a profound effect on the beach configuration. This also becomes a complicating factor with regard to the generation of the cell circulation in that the troughs scoured by the rips may act to stabilize the rip current positions. The cell circulation is then strongly affected by the beach topography and not completely free to respond to changing conditions of swell waves and edge waves. A number of studies have found, for example, shoreward moving currents over bars or shoals, a longshore current confined to a trough extending along the length of the beach, and narrow seaward-flowing rip currents passing through troughs that cut across the bar; see, for example, McKenzie (1958).

Sonu (1972a) has conducted an especially thorough study of the nearshore circulation in an area of irregular bottom topography at Seagrove, Florida. The main discovery of interest was that the breaker heights were uniform along the beach so that one would not at first expect a cell circulation. However, over the shoals the

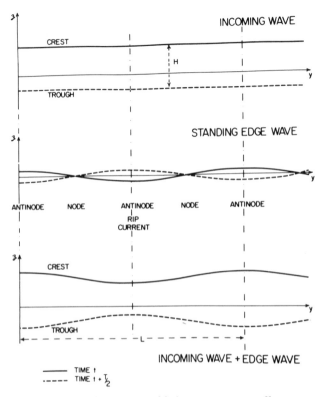

FIGURE 4. *Edge waves added to incoming swell waves to produce longshore variations in the wave breaker heights. From Bowen and Inman (1969).*

waves broke by spilling and tended to maintain the broken crest through the surf zone, while those entering the rip broke by plunging. Because of this difference in breaking, wave setup shoreward of the shoals was higher than shoreward of the plunging breakers at the rip current positions. Thus a gradient of the setup existed even though there was no variation in breaker height. Sonu (1972a) and Noda (1974) present theoretical analyses of nearshore circulation patterns produced by the interactions of incoming waves and local bottom topography.

Based on observations such as those of Sonu (1972a), a school of thought has formed that argues against any role played by edge waves in the control of cell circulation development, believing instead that the irregular topography came first and thereupon controls any water circulation. As pointed out by Noda (1974), there remains the basic question of how the original bottom deformation was formed if not by a cell circulation. In addition, rip currents can develop on plane smooth beaches with no topographic effects, both in the laboratory and on some natural beaches. Hino (1975) has developed an alternative explanation to edge waves for the origin of rip currents under such conditions. Hino's approach is a hydrodynamic instability theory where

there is a small initial disturbance to the wave setup which is otherwise constant in the longshore direction. Rip currents form with a preferred longshore spacing that is about four times the distance from the shore to the breaker zone. Whatever the cause of the cell circulation, the rip currents probably come first, causing sediment transport and bottom deformation. At a later stage the irregular bottom may be sufficient to provide the principal control over the nearshore circulation.

Currents Due to an Oblique Wave Approach

When waves break at an angle to the shoreline there is a longshore component to the wave radiation stress (momentum flux) that generates a longshore current that is continuous in the longshore direction rather than flowing out as rip currents. There have been many theories to account for this current but Bowen (1969b) was the first to attribute it to the longshore component of the radiation stress. This was also discussed by Longuet-Higgins (1970a,b) and Thornton (1971).

Longuet-Higgins (1970a) balanced the longshore thrust of the radiation stress directly against the frictional drag of the current to obtain the equation

$$\bar{v}_l = \frac{5\pi}{8} \frac{\tan \beta}{c_f} u_m \sin \alpha_b \cos \alpha_b \qquad (1)$$

where $\bar{v}_l$ is the longshore current at the midsurf position, α_b is the breaker angle, $\tan \beta$ is the beach slope, c_f is a frictional drag coefficient, and u_m is the maximum orbital velocity at the breaking wave position given by

$$u_m = \left[\frac{2E_b}{\rho h_b} \right]^{1/2}$$

E_b is the wave energy density at the breaker zone and h_b is the water depth at breaking (ρ is the water density). Equation 1 obtained by Longuet-Higgins on theoretical grounds is basically the same as the relationship

$$\bar{v}_l = 2.7 u_m \sin \alpha_b \cos \alpha_b \qquad (2)$$

obtained by Komar and Inman (1970) on the basis of the equivalence of two sand transport models. This is examined in the next section. The principal difference between (1) and (2) is the presence of the ratio $\tan \beta / c_f$. Equation 1 is tested in Fig. 5 against the available field data and (2) is tested in Fig. 6. It is apparent that the ratio $\tan \beta / c_f$ must be approximately constant, an increase in beach slope resulting in a corresponding increase in c_f. The near constancy of that ratio is further demonstrated in Komar (1975) by equating the midsurf velocity of (2) against the midsurf value of the solution for the complete velocity distribution across the surf.

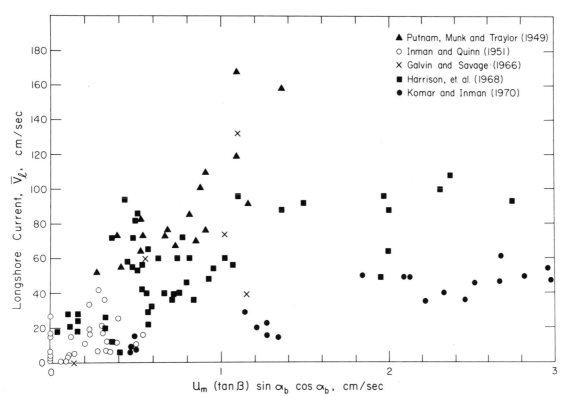

FIGURE 5. *Longshore current relationship determined by Longuet-Higgins (1970a), with the beach slope tan β included in the equation, tested with the avaiable field data. Compare with Fig. 6. From Komar and Inman (1970).*

Komar (1975) also utilizes the available laboratory data on longshore currents. The conclusion is that (2) fits all of the data, both laboratory and field, much better than any of the previous models. In addition, this formulation has the best theoretical basis as it is derived from radiation stress concepts by Longuet-Higgins (1970a).

Bowen (1969b), Longuet-Higgins (1970b), and Thornton (1971) obtained solutions for the complete velocity profile across the surf zone rather than just the midsurf position given by (1) and (2). In the complete solution they consider the horizontal transfer of momentum across the surf zone that redistributes the momentum and causes a smoothing of the velocity profile. This shifts the velocity maximum shoreward away from the breaker zone and also causes a coupling with waters beyond the breaker zone so that there may be some longshore current outside the breakers. Figure 7 gives the solutions of Longuet-Higgins (1970b) where P is the dimensionless parameter

$$P = \frac{\pi N \tan \beta}{\gamma c_f} \qquad (3)$$

and reflects the significance of the horizontal momentum transfer. The larger the value of P, the more important is the effect of horizontal transfer of momentum across

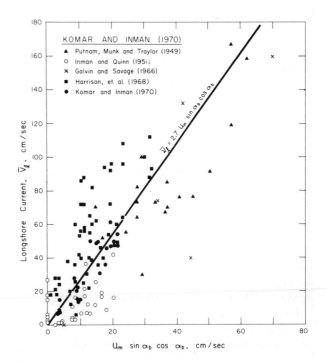

FIGURE 6. *Longshore current relationship obtained by Komar and Inman (1970) from the equivalence of two sand transport equations. A comparison with Fig. 5 indicates that tan β/c_f in (1) is approximately constant. From Komar and Inman (1970).*

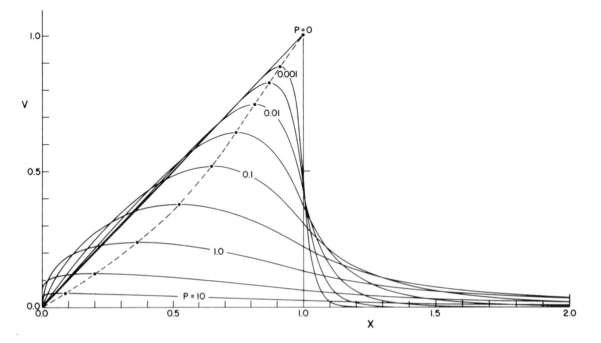

FIGURE 7. *Solutions of the longshore velocity distribution across the nearshore obtained by Longuet-Higgins (1970b).* $X = x/X_b$ *where* X_b *is the width of the surf zone,* *and* $V = \bar{v}/\bar{v}_1$ *where* $\bar{v}$ *is the local velocity and* $\bar{v}_1$ *is given by* (1). *From Longuet-Higgins (1970b).*

the surf zone. N is a numerical constant with limits $0 < N < 0.16$ in an expression for the horizontal eddy viscosity. The straight-line solution for $P = 0$ in Fig. 7 is where there is no horizontal eddy mixing, hence the discontinuity at the breaker zone with no current beyond the breakers. Comparison with the laboratory measurements of Galvin and Eagleson (1965) suggests a range for P from 0.1 up to 0.4. It is seen in Fig. 7 that for these values of P the current should be largely confined to the surf zone, decreasing rapidly outside the breakers. This conforms with observations.

LONGSHORE SAND TRANSPORT

The cell circulation can locally rearrange the sand to alter the shoreline configuration, but it is the longshore currents caused by an oblique wave approach that are primarily responsible for a longshore drift of beach sediments. This longshore movement of sand on beaches manifests itself most clearly when the transport is prevented through the construction of jetties, breakwaters, and groins.

Attempts at evaluation of the sand transport rate have relied most heavily on empirical correlations with

$$P_l = (ECn)_b \sin \alpha_b \cos \alpha_b \qquad (4)$$

where $(ECn)_b$ is the wave energy flux evaluated at the breaker zone, and α_b is again the breaker angle. P_l has

been variously called the longshore component of wave power or the longshore component of the wave energy flux. Objections have been raised to this terminology (Eaton, 1951; Longuet-Higgins, personal communication).

Watts (1953) obtained the first field measurements by which the sand transport rate could be related to the local wave characteristics, based on the rate at which a bypassing plant had to pump sand past the jetty at South Lake Worth Inlet, Florida. Caldwell (1956) obtained additional field data from California by measuring the rate at which sand placed on the beach was dispersed by the waves. They provided empirical equations that supplied approximate engineering "answers" with regard to the quantity of sand moving along the beach under a given set of wave conditions. However, besides providing little consideration of the physical processes that produce the transport, the empirical approach also gives rise in this instance to difficulties concerning the physical units in the equations.

Inman and Bagnold (1963) pointed out that the littoral transport rate should be expressed as an immersed weight transport rate, I_l, rather than as a volume transport rate, S_l. The two are related by

$$I_l = (\rho_s - \rho)ga'S_l \qquad (5)$$

where ρ_s and ρ are the sand and water densities, respectively, and a' is the correction factor for the pore space of the beach sand (approximately 0.6 for most beach sands). In the cgs system of units, S_l is given in cubic centimeters per second and I_l has units of dynes per second. The use of the immersed weight transport rate is based on a consideration by Bagnold (1966, 1963) of the problem of sediment transport in general from an energetics point of view. Under such considerations I_l is related to P_l through

$$I_l = K(ECn)_b \sin \alpha_b \cos \alpha_b = KP_l \qquad (6)$$

where K is a dimensionless proportionality coefficient. It is seen that one advantage of using I_l as a measure of the sand transport rate is that it has the same units as P_l so that (6) is dimensionally correct. A second advantage in using the immersed weight transport rate is that, as can be seen in (5), it takes into consideration the density of the sediment grains. Therefore a correlation between I_l and P_l is good for beaches composed of coral sand, coal, etc., as well as for the more usual quartz sand beaches.

Komar and Inman (1970) obtained simultaneous measurements of the wave and current parameters in the surf zone and the resulting longshore sand transport rate. Unlike the previous studies, the measurements

represent short periods of time, on the order of 2 to 4 hours. Quantitative measurements of littoral transport rate obtained from the time history of the center of mass of sand tracers (natural sand tagged with a thin coating of fluorescent dye). Two beaches were involved in the study: El Moreno Beach on the Gulf of California, a coarse sand beach with a steep beach face, and Silver Strand Beach, California, a fine-grained beach of low profile with an extensive surf zone. Our data are presented in Fig. 8 together with the other field and laboratory data. A straight line of unit slope fitted to the *field* data yields a coefficient $K = 0.77$ so that (6) becomes

$$I_l = 0.77(ECn)_b \sin \alpha_b \cos \alpha_b \qquad (7)$$

Note in the figure how nearly all the laboratory measurements fall below the straight line established from the field data; the straight line actually appears to form an upper limit to the plot. It has been shown for sand transport in rivers (Bagnold, 1966) that the immersed weight transport rate is proportional to the available power of the flowing water with a constant proportionality factor, only when the transport conditions are fully developed (sheet sand movement under river flood flow). Under lower flow regimes, those that are not fully

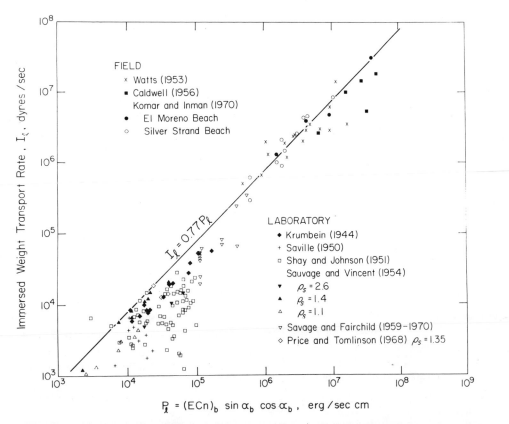

FIGURE 8. *The immersed weight sand transport rate* I_l *versus* P_l *of (4). From Komar and Inman (1970).*

developed and the sediment bottom is rippled, the available power to transport sand is less and the proportionality factor is lower and no longer constant. A similar effect can also be expected to occur in the transport of sand on beaches and could account for the observed distribution of plotting position of the laboratory data. The straight line of Fig. 8 represents the maximum transport conditions that can occur for a given P_l value, and therefore should exist as an upper boundary to the plotting of the laboratory data, just as it is seen to do. Any transport conditions that are not fully developed, whether in the laboratory or on natural gravel beaches, would fall below the line and K would be less than 0.77. If material is too coarse even to be moved by the wave action, then $K = 0$.

This correlation between the sand transport rate and P_l, although apparently successful, is still largely empirical and its formulation involved no real consideration of the mechanics of sand transport. A more fundamental examination of the processes of sand transport by combined waves and currents was undertaken by Bagnold (1963). This model was discussed in Chapter 8 and led to (13) of that chapter. Inman and Bagnold (1963) applied this model to the littoral zone where they assumed $\Omega \propto (ECn)_b \cos \alpha_b$ for the available power from the waves to place the sand in motion, and $\bar{u}_\zeta$ became the longshore current, so that the relationship is

$$I_l = K'(ECn)_b \cos \alpha_b \frac{\bar{v}_l}{u_m} \qquad (8)$$

where $u_m = (2E_b/\rho h_b)^{1/2}$ is again the maximum orbital velocity evaluated at the breaker zone. The data of Komar and Inman (1970) in Fig. 9 yield $K' = 0.28$ for the dimensionless coefficient.

Therefore (8) as well as (7) is successful in relating the sand transport rate to the wave and current parameters. Since we cannot have two independent equations relating the same parameters, there must be a relationship between the two equations. Komar and Inman (1970) concluded that the connection is through the mode of generation of the current $\bar{v}_l$ in (8). We simply solved (7) and (8) simultaneously to obtain

$$\bar{v}_l = 2.7 u_m \sin \alpha_b \qquad (9)$$

for the longshore current. This is seen to be the same as (2) above since $\cos \alpha_b \simeq 1$ for most field conditions. As already discussed, this relationship shows the best agreement with the available field and laboratory measurements of longshore currents even though it was originally based on the equivalence of two sand transport equations. It is also seen that a correlation between I_l and P_l depends primarily on the longshore current being due entirely to an oblique wave approach such

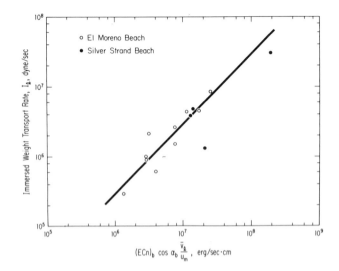

FIGURE 9. *Test of (8) for the longshore sand transport rate on beaches. From Komar and Inman (1970).*

that $\bar{v}_l$ is given by (2). Now if $\bar{v}_l$ were generated as a local wind-driven current, a tidal current, or the longshore currents of the cell circulation, then it would no longer be given by (2). Under such circumstances, to evaluate the resulting sand transport one must use (8), not (7), with either direct measurements of $\bar{v}_l$ or with some appropriate prediction of $\bar{v}_l$.

SHORELINE CONFIGURATION

The large-scale beach configuration tends to approximate an equilibrium in which the wave climate provides precisely the energy and mean wave approach angle required to transport and redistribute the sediments supplied to the beach. The beach configuration then is controlled mainly by the curvature and orientation of the refracted wave crests, the locations and importance of beach sediment sources and losses, and the position of headlands and other structures that block the littoral drift. Superimposed upon this general equilibrium curvature may be a hierarchy of rhythmic shoreline forms (Dolan and Ferm, 1968), ranging from rhythmic topography with cusp spacings up to 2000 m down to ordinary beach cusps.

This section first examines the large-scale equilibrium beach configuration, doing this by means of application of computer simulation models. Then the origin of the variety of rhythmic beach forms is considered.

Numerical Models of Shoreline Configuration

One approach to understanding the changing beach configuration under variable wave conditions is through

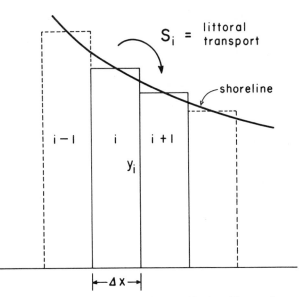

FIGURE 10. *Shoreline divided into cells of width* Δx. *Changes in the shoreline position are produced by the littoral drift* S_i, *from one cell to the next. From Komar* **(1973).**

application of computer simulation methods. Such methods applied to beaches are still in the infant stage but have a considerable potential for future application.

In a numerical simulation model on a computer the equations of sand transport along a beach and the continuity equation are solved together over a grid. As illustrated in Fig. 10, the shoreline is divided up into a series of cells of uniform width Δx and with individual lengths $y_1, \ldots, y_{i-1}, y_i, y_{i+1}, \ldots, y_n$ beyond some baseline. The narrower the cells (the smaller Δx), the more nearly the series of cells approximates the true shoreline. Changes in the shoreline configuration are brought

about by the littoral drift S_l (cubic meters per day) which shifts sand alongshore from one cell to the next. Considering one cell in the shoreline (Fig. 11), if S_{in} is the rate of sediment drift into this cell and S_{out} is the rate of drift out, then the net accumulation or erosion ΔV (cubic meters) in the compartment in the lapsed time Δt (days) is

$$\Delta V = (S_{in} - S_{out})\, \Delta t \qquad (10)$$

ΔV must be reflected in an advance or retreat of the shoreline. If Δy is the change in the shoreline position in the onshore-offshore direction in the increment of time Δt, then from the geometry of the wedge depicted in Fig. 11, we have

$$\Delta V = d\, \Delta y\, \Delta x \qquad (11)$$

where d is a linear dimension such that $d\, \Delta y$ gives the section area of beach eroded or deposited. Combining (10) and (11) yields

$$\Delta y = (S_{in} - S_{out})\, \frac{\Delta t}{d\Delta x} \qquad (12)$$

This continuity relationship may be solved for each of the cells, evaluating S_{in} and S_{out} with (7) or (8), so that shoreline changes can be determined over small increments of time Δt. If a particular cell has other sources or losses of sand besides littoral drift, then these can easily be included. For example, if the cell is at a river mouth supplying sand at a rate S_r (cubic meters per day), this term would be added to (12).

By using such techniques, Price et al. (1973) examined changes in a beach brought about by the construction of a long groin blocking the longshore drift of sand. The results of their computer simulation model compared

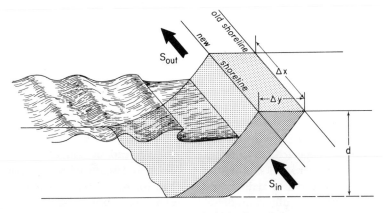

Volume, $\Delta V = d \cdot \Delta x \cdot \Delta y$

FIGURE 11. *One shoreline cell demonstrating how a change in sand volume* ΔV *within the cell is produced by the littoral drift in and out of the cell and how this is reflected in a change in the shoreline position* Δy. *From Komar* **(1973).**

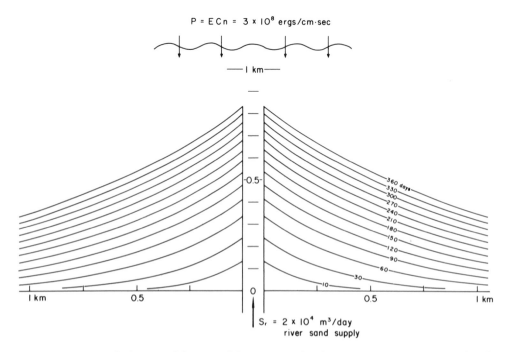

FIGURE 12. *Growth of a river delta to equilibrium as simulated on a computer. From Komar (1973).*

favorably with the actual shoreline changes experienced in a laboratory wave basin test.

Komar (1973) investigated the shoreline configuration where a river supplies sand to the beach and the waves act to redistribute the sand away from the river mouth. Of interest was the equilibrium shoreline shape wherein the waves would provide the necessary energy and wave breaker angles to redistribute the sand supplied by the river. Figure 12 shows the first 360 days of growth of a delta where the river supplies sand at the rate $S_r = 2 \times 10^4$ m³/day tending to build out the delta, whereas waves of energy flux 3×10^8 ergs/cm·sec work to flatten the shoreline. It is seen that after some 150 days there is a balance such that the delta continues to grow outward but at a steady rate, maintaining its overall shape in the process. This must be the equilibrium delta shape where the waves are just able to redistribute the sand along the shoreline. The further away from the river mouth, the smaller is the amount of sand remaining to be transported since there is deposition along the way. Because of this the breaker angles progressively decrease away from the river mouth. As expected, the greater the wave energy flux for a given river sand supply, the smaller are the breaker angles required to redistribute the sand and the flatter the resulting equilibrium delta.

Figure 13 shows a model of a pocket beach of 1 km length which reorients itself from its originally straight shoreline until it takes on the shape of the refracted wave crests. In an example like this where there are no sources or losses of sand the shoreline would take on exactly the

shape of the refracted wave crests until $\alpha_b = 0$ everywhere and no further sand transport occurs. A true equilibrium would be achieved. Under slightly varying wave conditions the pocket beach would wobble, attempting to keep up with the variable wave conditions. This can also be shown in computer simulation models (Komar, in press). This more closely fits what is found in nature, and under such conditions there is a "mean" shoreline position that represents a dynamic equilibrium. Figure 14 shows a stream supplying sand to a pocket beach. Because of the supply the beach builds outward within the bay but maintains a constant equilibrium configuration since the wave and river conditions are not varied.

Rhythmic Shoreline Features

Beaches commonly contain crescentic seaward projections of sediment which trend at right angles to the shoreline and are known as beach cusps, sand waves, shoreline rhythms, or giant cusps. A wide range of cusp spacings can be found on beaches. On the shores of ponds and small lakes the spacing may vary from less than 10 cm to 1 m. Similar small cusps can be generated in a laboratory wave basin. On ocean beaches with small waves the cusp spacing may be less than 2 m while those built by large storm waves may be 60 m apart. Dolan's (1971) measurements of "shoreline rhythms" from the North Carolina coast yielded spacings between successive cusps ranging from 150 to 1000 m,

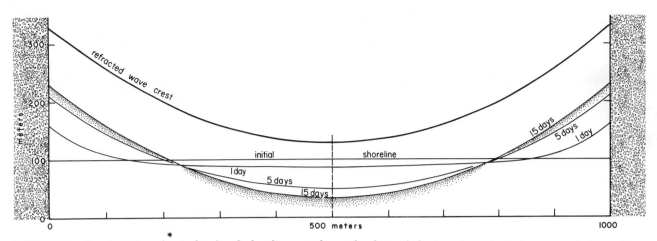

FIGURE 13. *Reorientation of a pocket beach shoreline to take on the shape of the incoming refracted wave crests. From Komar (in press).*

with most between 500 and 600 m. Shepard's (1952) "giant cusps" ranged up to 1500 m in spacing.

Attempts at classifying shoreline rhythmic features generally stress their spacings. Since it appears that cuspate shorelines with a wide range of spacings can be produced by a single mechanism of formation and that there may be more than one mechanism capable of producing a cuspate shoreline, only a genetic classification will be satisfactory. More important than the spacings of the shoreline cusps is their associated offshore morphology. On this basis we can distinguish between beach cusps and rhythmic topography (sand waves, giant cusps). Beach cusps commonly exist as simple ridges or mounds of coarse sediment stretching down the beach face, but where best developed on a steep beach, there exist deeper troughs offshore from the cusps and underwater deltas offshore from the embay-ments. In contrast, as distinguished by Hom-ma and Sonu (1963), rhythmic topography consists of a regular series of crescentic and inner bars with a regular spacing in the longshore direction (Fig. 15). In certain circumstances, the rhythmic bars give rise to a rhythmic series of cusps along the shoreline. At other times the rhythmic bars exist with an otherwise straight shoreline. In contrast with beach cusps, which are chiefly a subaerial feature, the underwater morphology is more important to the rhythmic topography and may only secondarily produce a series of cusps. In general, the spacing of the cusps associated with the rhythmic topography is larger than the spacings of beach cusps.

There is a considerable literature on beach cusps, their general features, and origin. Bowen (1973) has supported the opinion that edge waves are the most probable cause of the rhythmic spacing of beach cusps.

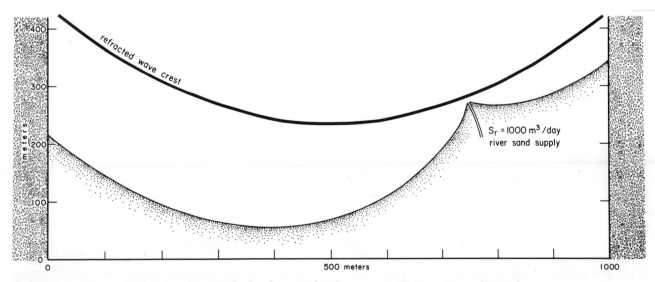

FIGURE 14. *Stream supplying sand to a pocket beach as simulated on a computer. From Komar (in press).*

A. Rhythmic Topography on Inner Bar

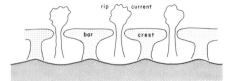

B. Crescentic Bars

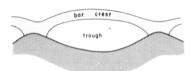

C. Combination

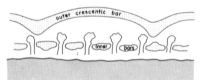

FIGURE 15. *Types of rhythmic topography. Crescentic bars and inner bars cut by rip currents provide the two distinctive types. Both may sometimes be found on the same beach. From Komar (in press).*

Bowen and Inman (1969, p. 5490) suggested that where cusps form on the inner beach face shoreward of a wide surf zone, the formation may be in response to an interaction of the re-formed wave bores following breaking with the edge waves which have a smaller modal number than that controlling the predominant cell circulation. At present only a poor correlation can be made between the cusp spacing and edge wavelengths. More investigation must be made of this proposed mechanism.

The evidence for the role of edge waves in the generation of crescentic bars, one form of rhythmic topography, is somewhat clearer. The crescentic bar (Figs. 15 and 16) is essentially a submerged sandbar, concave shoreward, which may or may not have an associated series of cusps along the shoreline. They appear to range in wavelength from about 100 to 2000 m with a predominance of 200 to 500 m. Bowen and Inman (1971) have convincingly argued that they are formed by the velocity field associated with edge waves on a sloping beach. They envision that sediments will drift about under the currents of the edge waves until they reach zones where the velocity is below the threshold of sediment motion, the sediment depositing in those zones. Their theoretical treatment demonstrated that such depositional zones have patterns that are basically the same as the observed crescentic bar morphology such as those of Fig. 16. Bowen and Inman also conducted a series of wave basin tests that confirmed the crescentic depositional pattern caused by edge waves. Sonu (1972b, p. 6629) has attacked the theory on the basis of requiring edge waves of period 40 to 50 seconds to explain the spacings of the crescentic bars, periods beyond the range of ordinary surface waves. Bowen and Inman (1972) argue that edge waves associated with crescentic bars are probably not generated from ordinary surface waves, but rather may be driven by surf-beat or possibly by local wind fields.

The rhythmic topography of the inner bar (Fig. 15) is in response to the interaction of the beach configuration with the cell circulation. This system is to a first approximation independent of the series of crescentic bars. Both may sometimes be found on the same beach. There is an extensive literature on this; Sonu (1973), Davis and Fox (1972), and Dolan (1971) are recent papers. If a nearshore cell circulation develops on an initially smooth sand beach, it soon shifts the sand about, usually into a series of cusps. Sometimes the cusps are found in the lee of the rips, at other times midway between the rips. As we have already discussed, once the shoreline configuration has been highly modified, it may largely control the nearshore circulation system. Komar (1971) has shown that cusps can develop because of rip currents but then the rip currents disappear because of a balance between the longshore variation in wave height and the oblique wave approach to the flanks of the cusps. Cusps may be produced by rip currents although the cell circulation is absent or in a much weakened condition at the time of observation.

CONCLUSION

Although considerable progress has been made in the past decade in understanding these aspects of nearshore processes, much work remains. It is apparent from this brief review that there is still considerable disagreement on the origin of rip currents and how they interact with an irregular beach topography. This will be a worthwhile line of research in the coming decade. Adequate numbers of careful field measurements of longshore currents and sand transport as related to the wave parameters are still lacking. There exist essentially no measurements of the distribution of longshore current velocities across the surf zone except where strongly controlled by beach topography. Thus we are unable to test the theoretical formulations of the longshore current velocity distribution. Finally, with one exception, the computer simulation approach to shoreline configuration changes has only been applied to simple geomorphic studies. This approach has much potential for applied problems of beach erosion and deposition and should be more widely utilized.

FIGURE 16. *Well-developed crescentic bars off an enclosed beach near Cape Kalaa, Algeria. From Clos-Arceduc (1962).*

SUMMARY

Two systems of nearshore currents are distinguished: (1) a cell circulation of rip currents, and (2) longshore currents caused by waves breaking at an angle to the shoreline. Rip currents can be generated by longshore variations in wave breaker heights, the longshore currents flowing from positions of largest breakers to positions of small breakers where the flow turns seaward as rips. In some cases the longshore variations in breaker heights are produced by offshore wave refraction. It has been suggested that in other cases edge waves interacting with the incoming swell waves can produce longshore variations in breaker heights and thus generate rip currents. A hydrodynamic instability theory has also been proposed for the origin of rip currents.

The cell circulation can redistribute the beach sediment and therefore have a profound effect on the beach configuration. Cuspate shoreline features can be generated in this way. The irregular bottom in turn can have a profound effect on the cell circulation, and in some cases control the patterns of circulation.

Waves breaking at an angle to the shoreline generate longshore currents that flow parallel to the shoreline. These currents are especially important in sand transport along the beach. The generation of these currents is best explained in terms of the momentum flux (radiation stress) of the waves, there being a longshore component under oblique waves.

When the sand transport is caused by an oblique wave approach to the shoreline, the sand transport rate is best predicted from the wave parameters with (7). If the sand transport results from longshore currents other than those from an oblique wave approach, then (8) must be utilized. The equations become identical when the longshore current is given by (9).

Longshore sand transport may result in changes in the shoreline configuration. These changes can be modeled by utilizing techniques of computer simulation. A continuity equation for the beach sand is applied to finite element cells approximating the shoreline shape. Sand transported from one cell to another brings about the shoreline changes.

There is a variety of rhythmic shoreline forms including beach cusps, sand waves, giant cusps, and crescentic bars. They range in spacing from a few centimeters to over 1000 m. There is still considerable debate on the origin of the features. In all probability, there is more than a single generation mechanism for otherwise similar appearing features.

SYMBOLS

a' correction factor for pore space of beach sand
Cn wave group velocity (energy flux velocity)
c_f frictional drag coefficient
E_b wave energy density at the breaker zone

g acceleration of gravity (981 cm/sec^2)

h water depth

h_b water depth at breaking

I_l total immersed weight sand transport rate

K dimensionless coefficient

K' dimensionless coefficient

L wavelength

N coefficient in expression for horizontal eddy viscosity

P dimensionless horizontal momentum transfer

P_l power relationship

S_l total volume sand transport rate

u_m maximum orbital velocity at the breaking wave position

u_0 average orbital velocity of waves

u_ξ near-bed unidirectional velocity component

$\bar{v}_l$ longshore current velocity at midsurf position

x coordinate axis in longshore direction

y coordinate axis in offshore direction

z height above bottom

α_b wave breaker angle

β beach slope angle

γ ratio of wave height to water depth

ρ density of water

ρ_s density of sediment grains

Ω power expended by waves

REFERENCES

Bagnold, R. A. (1963). Mechanics of marine sedimentation. In M. N. Hill, ed., *The Sea*, Vol. 3. New York: Wiley-Interscience, pp. 507–528.

Bagnold, R. A. (1966). An approach to the sediment transport problem from general physics. *U.S. Geol. Surv., Prof. Pap. 422-I*, 37 pp.

Bowen, A. J. (1969a). Rip currents. 1. Theoretical investigations. *J. Geophys. Res.*, 74: 5467–5478.

Bowen, A. J. (1969b). The generation of longshore currents on a plane beach. *J. Mar. Res.*, 37: 206–215.

Bowen, A. J. (1973). Edge waves and the littoral environment. *Proc. 13th Conf. Coastal Eng.*, 1313–1320.

Bowen, A. J. and D. L. Inman (1969). Rip currents. 2. Laboratory and field observations. *J. Geophys. Res.*, 74: 5479–5490.

Bowen, A. J. and D. L. Inman (1971). Edge waves and crescentic bars. *J. Geophys. Res.*, 76(36): 8662–8671.

Bowen, A. J. and D. L. Inman (1972). Reply. *J. Geophys. Res.*, 77(33): 6632–6633.

Caldwell, J. M. (1956). *Wave Action and Sand Movement near Anaheim Bay, California*. U.S. Army Beach Erosion Board, Tech. Memo. No. 68, 21 pp.

Clos-Arceduc, A. (1962). Etude sur les vues aériennes, des alluvions littorales d'allure périodique, cordons littoraux et festons. *Bull. Soc. Fr. Photogramm*, 4: 13–21.

Davis, R. A. and W. T. Fox (1972). Coastal processes and nearshore sand bars. *J. Sediment. Petrol.*, 42: 401–412.

Dolan, R. (1971). Coastal landforms: crescentic and rhythmic. *Geol. Soc. Am. Bull.*, 82: 177–180.

Dolan, R. and J. C. Ferm (1968). Concentric landforms along the Atlantic coast of the United States. *Science*, 159: 627–629.

Eaton, R. O. (1951). Littoral processes on sandy coasts. *Proc. 1st Conf. Coastal Eng.*, 140–154.

Galvin, C. J., Jr. and P. S. Eagleson (1965). *Experimental Study of Longshore Currents on a Plane Beach*. U.S. Army Coastal Eng. Res. Cent., Tech. Memo. No. 10, 80 pp.

Hino, M. (1975). Theory on formation of rip-current and cuspidal coast. *Proc. 14th Conf. Coastal Eng.*, pp. 901–919.

Hom-ma, M. and C. J. Sonu (1963). Rhythmic patterns of longshore bars related to sediment characteristics. *Proc. 8th Conf. Coastal Eng.*, pp. 248–278.

Inman, D. L. and R. A. Bagnold (1963). Littoral processes. In M. N. Hill, ed., *The Sea*, Vol. 3. New York: Wiley-Interscience, pp. 529–553.

Komar, P. D. (1971). Nearshore cell circulation and the formation of giant cusps. *Geol. Soc. Am. Bull.*, 82: 2643–2650.

Komar, P. D. (1973). Computer models of delta growth due to sediment input from rivers and longshore transport. *Geol. Soc. Amer. Bull.*, 84: 2217–2226.

Komar, P. D. (1975). Nearshore currents: Generation by obliquely incident waves and longshore variations in breaker height. In J. R. Hails and A. Carr, eds., *Proceedings, Symposium on Nearshore Sediment Dynamics*. New York: Wiley, pp. 17–45.

Komar, P. D. (in press). *Beach Processes and Sedimentation*. Englewood Cliffs, N.J.: Prentice-Hall.

Komar, P. D. and D. L. Inman (1970). Longshore sand transport on beaches. *J. Geophys. Res.*, 75(30): 5914–5927.

Longuet-Higgins, M. S. (1970a). Longshore currents generated by obliquely incident sea waves, 1. *J. Geophys. Res.*, 75(33): 6778–6789.

Longuet-Higgins, M. S. (1970b). Longshore currents generated by obliquely incident sea waves, 2. *J. Geophys. Res.*, 75(33): 6790–6801.

Longuet-Higgins, M. S. and R. W. Stewart (1964). Radiation stress in water waves, a physical discussion with applications. *Deep-Sea Res.*, 11(4): 529–563.

McKenzie, R. (1958). Rip current systems. *J. Geol.*, 66: 103–113.

Noda, E. K. (1974). Wave-induced nearshore circulation. *J. Geophys. Res.*, 79(27): 4097–4106.

Price, W. A., K. W. Tomlinson, and D. H. Willis (1973). Predicting changes in the plan shape of beaches. *Proc. 13th Conf. Coastal Eng.*, 1321–1329.

Shepard, F. P. (1952). Revised nomenclature for depositional coastal features. *Bull. Am. Assoc. Pet. Geol.*, 36: 1902–1912.

Shepard, F. P. and D. L. Inman (1950). Nearshore circulation related to bottom topography and wave refraction. *Trans. Am. Geophys. Union*, 31(4): 555–565.

Sonu, C. J. (1972a). Field observation of nearshore circulation and meandering currents. *J. Geophys. Res.*, 77(18): 3232–3247.

Sonu, C. J. (1972b). Comments on paper by A. J. Bowen and D. L. Inman, 'Edge wave and crescentic bars.' *J. Geophys. Res.*, 77(33): 6629–6631.

Sonu, C. J. (1973). Three-dimensional beach changes. *J. Geol.*, 81: 42–64.

Thornton, E. B. (1971). Variations of longshore current across the surf zone. *Proc. 12th Conf. Coastal Eng.*, 291–308.

Watts, G. M. (1953). *A Study of Sand Movement at South Lake Worth Inlet, Florida*. U.S. Army Beach Erosion Board, Tech. Memo. No. 42, 24 pp.

CHAPTER 14

Coastal Sedimentation

DONALD J. P. SWIFT

Atlantic Oceanographic and Meteorological Laboratories, Miami, Florida

The preceding chapters have discussed sedimentation in the intracoastal zone of lagoons and estuaries which lie seaward of the main shoreline, and on the open beach and associated surf zone. This chapter looks at sedimentation in the coastal zone as a whole, from the shoreline out to an indeterminate distance on the order of 5 km, where shelf flows are no longer affected by proximity to shore. From this perspective, the system of longshore sand transport beneath the zone of shoaling and breaking waves can be examined together with a deeper system of longshore sediment transport driven by intermittent wind or tidal flows. Time and space patterns of sediment input into this double system, the character of sediment transport, zones of temporary storage or permanent deposition, and the bypassing of sediment onto the shelf surface are analyzed. More complex patterns of sediment transport are also described, which result when coastal flows associated with straight coastal compartments interact with circulation in the erosional reentrants of rocky coasts or constructional inlets of lagoons and river mouths.

ONSHORE-OFFSHORE SEDIMENT TRANSPORT

In considering coastal sediment transport, it is convenient to divide the movement of sediment into an onshore-offshore component and a coast-parallel component, and to consider these separately before examining the coastal sediment budget as a whole. Coast-parallel transport is many times more intensive than onshore-offshore trans-

port, but it is the latter that determines morphologic changes at given coastal transects. Hence this chapter begins by examining the coast in profile.

Hydraulic Zones and Morphologic Provinces

When examined in cross section, the inner shelf is seen to consist of a regular succession of morphologic provinces, each associated with a distinctive zone of hydraulic activity (Fig. 1).

Subaerial environments of open coasts are most highly developed on barrier islands, where a zone of storm washover and eolian activity results in *washover flats* and *dune belts*, respectively. The intertidal swash zone builds the *beach foreshore*. The foreshore progrades seaward during fair weather by the addition of successive inclined sand strata to form the *beach prism*, a body of stored sand. The upper surface of the beach prism is the *beach backshore*. The zone of breaking waves may be divided into the breaker line, which tends to maintain a *breakpoint bar*, and a surf zone, in which a wave-driven littoral current flowing parallel to the beach is overridden by the bores of breaking waves. The littoral current tends to scour a *longshore trough*.

On unconsolidated coasts capable of relatively short-term response to the hydraulic regime, the inner shelf seaward of the breakpoint bar tends to exhibit two morphologic elements. A more steeply dipping *shoreface* extends to depths of 12 to 20 m. Its upper slope may be as steep as 1:10; its seaward extremity, at 2 to 20 km from shore, may slope as gently as 1:200. Beyond it lies

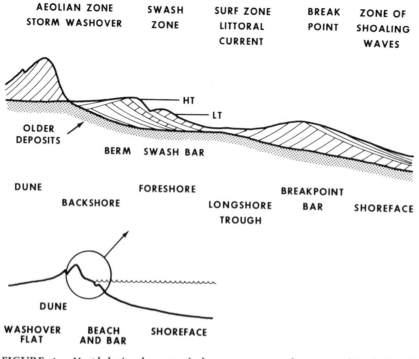

FIGURE 1. *Morphologic elements of the open coast and corresponding hydraulic provinces.*

the flatter *inner shelf floor* proper; the transition may be abrupt or very gentle. The upper shoreface, to a depth of perhaps 10 m, corresponds to the hydraulic zone of shoaling waves. The lower shoreface and inner shelf flow also experience the surge of shoaling waves, but their slopes, textures, and bed forms are equally a response to unidirectional shelf flows.

The Beach Profile

Circulation in the surf zone and the morphologic response of the substrate are described in Chapter 13. This section deals with the net effect of such hydraulic process and substrate response on the onshore-offshore sediment budget.

As a consequence of the enormous and nearly continuous expenditure of energy in the beach and surf zones, the topographic features of cohesionless sand found there may only exist as equilibrium or near-equilibrium responses to the circulation patterns described in the preceding chapter. The equilibrium is not a static one, however, as the characteristics of the wave regime that force the response are constantly changing, often more rapidly than the morphologic response can accommodate. As a consequence, the nearshore beach and surf zone topography is endlessly destroyed and rebuilt according to a complex cycle, as the nearshore wave regime and circulation pattern alternate between fair-

weather and storm configurations, and on a larger scale between the summer season of infrequent storms and the winter season of frequent storms (Davis and Fox, 1972); see Fig. 2.

FAIRWEATHER PHASE: BEACH AND BAR BUILDING. The cycle is controlled by two mechanisms: the wave regime and the net circulation pattern driven by it. During fair weather, waves tend to be far-traveled swells, of low amplitude and long period. The asymmetry of associated bottom wave surge is marked, with the landward stroke beneath the wave crest being significantly more prolonged and more intense than the seaward stroke beneath the trough (Chapter 8, Fig. 8). Peak orbital velocities may be separated by periods of 8 seconds or on windward coasts, markedly longer. These same fairweather swells tend to result in a relatively weak nearshore circulation pattern. Momentum flux, which is a function of wave height, is relatively low during fair weather both seaward and landward of the breaker, hence discharge through the littoral circulation cells is relatively low.

During fair weather, these two mechanisms, bottom wave surge and the littoral circulation pattern, cooperate to store sand in the beach prism. The wave regime appears to serve as a fractionating mill, dividing the available sand into a fraction undergoing mainly bed load transport, and a fraction undergoing mainly suspensive

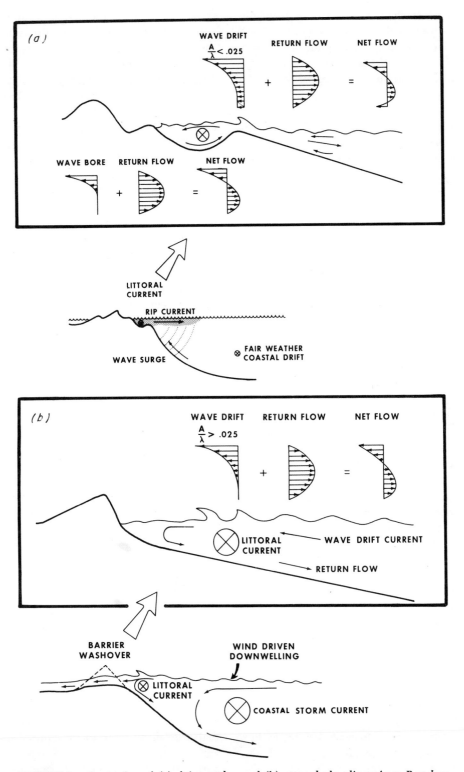

FIGURE 2. *Comparison of* (a) *fair-weather and* (b) *storm hydraulic regimes. Based on Longuet-Higgins (1953), Schiffman (1965), and Ingle (1966).*

transport. Sand coarser than a critical size threshold will be driven landward as bed load, by the landward asymmetry of bottom wave surge, toward the breakpoint. Longuet-Higgins (1953) and Russell and Osorio (1958) have undertaken calculations and experiments to determine the shore-normal components of flow averaged over a wave cycle, in the nearshore zone of shoaling waves. Their results indicate an increase in net landward flow with increasing height off the bottom, as a result of the failure of wave orbitals to close. Superimposed on this is a mid-depth return flow resulting in a three-layer flow system (Fig. 2). It is not entirely clear, however, if the latter component of flow would exist in nature as a response to wave setup, or if it was merely a wave tank artifact, induced by the continuity requirements of a closed system.

At the breakpoint, much of the energy heretofore available to drive sand landward over the bottom is lost to turbulence, and sand tends to accumulate as a breakpoint bar. Waves of oscillation turn into waves of translation (bores), in which water moves forward as a mass, and there is some evidence to indicate that landward of the breakpoint the velocity cross section averaged over the wave cycle changes to a two-layer system (Schiffman, 1965); see Fig. 2. An upper layer moves landward as a series of bores, and tends to be compensated by a basal return flow. This two-layer system is of course superimposed on the generally much stronger coast-parallel flow characteristic of the longshore trough. To the extent that the two-layer flow prevails, the bar crest is a zone of flow convergence, and its sand storage capability is readily understood. The bar builds upward until the rate of deposition of sand at the conclusion of wave breaking is equaled by resuspension during wave breaking. Depth of water over the bar at equilibrium is generally a third of the water depth prior to formation of the bar (Shepard, 1950).

Breakpoint bars tend to orient themselves normal to wave orthogonals. When deep-water orthogonals make a high angle to the shore, wave refraction does not fully eliminate this angle near the beach. Under these conditions, the bar tends to consist of series of *en echelon* segments, each aligned obliquely with respect to the beach, and alternating with rip current channels (Sonu et al., 1967).

Bar position is very sensitive to wave height, as this determines breakpoint position (Keulegan, 1948). If the tide range is appreciable, bar position will shift detectably through the tidal cycle. New bars tend to form during the peak or waning phases of a storm and to be slowly driven onshore as waves diminish during the ensuing fair-weather period, although an abrupt decrease in wave height may cause a second bar to form landward

of the first. During the period of landward migration of the bar, coarser bed load sand may bypass the bar and move onto the beach, if the waves are sufficiently long in period to re-form after breaking (King and Williams, 1949). Such bypassed sand will tend to accumulate as a swash bar (intertidal bar), or the plunge point bar itself will tend to migrate landward to the point where it is captured by intertidal processes, and becomes a swash bar (Fig. 3). As noted by King (1972), a swash bar may only form when the beach slope is lower than the maximum potential slope permitted by the grain size of the available sand; swash bars thus comprise attempts by the regime of wave swash and backwash to build to this ideal beach profile. Unlike plunge-point bars which are formed at a bottom current convergence, swash bars are formed by an abrupt bottom current deceleration. Their seaward slopes are swash current graded, but the landward slopes are lower than the angle of repose, and have the same net landward sense of sand transport.

Swash bars are the dominant bar on fine, flat beaches such as those of the central Gulf of Mexico, where the wave climate is mild and the supply of fine sand is abundant. They also tend to form on beaches with a high tidal range, where the bar is exposed to swash and backwash throughout much of the tidal cycle (ridge and runnel systems).

The landward movement of coarser fine sand during fair weather on open beaches may thus proceed as a sheet flow bypassing the bar, or migration of the bar up the beach, or more commonly as both. The result of this landward flux of sand is the formation of the beach prism of gently inclined sand strata, differentiated into the backshore beach (constructional upper surface subject to eolian action) and foreshore beach (swash-graded forward surface) separated by the berm (Fig. 1). If swash bar migration is the dominant mode of beach aggradation, then the berm will prograde seaward mainly by the welding to it of successive swash bars, and the internal structure of the beach prism will consist of seaward-dipping cross-strata sets, whose internal structures dip more steeply landward (Davis et al., 1972).

The ease with which breakpoint and swash bars can be constructed in wave tanks strongly suggests that these are indeed basic genetic types of bars. These two relatively simple types belong to a broad class of bed forms that arise in response to the mutual interaction of flow with the substrate. However, it has recently become apparent that much more elaborate patterns of bars may form more or less passively, in response to an innate pattern within the velocity field. Crescentic bars that form in response to standing edge wave patterns have been described in Chapter 13. On gently inclined shorefaces, shore-parallel bars may form in arrays of up to 30

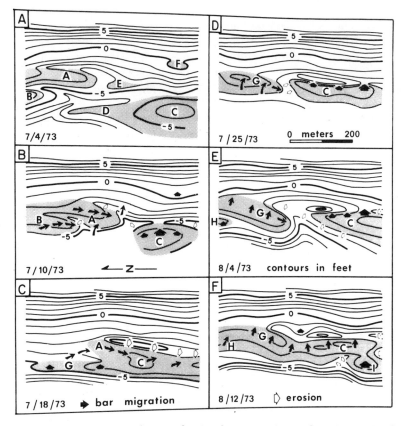

FIGURE 3. *Sequence of maps showing bar migration and erosion at South Beach, Oregon. Bars form below mean sea level and advance up beach at rate of 1 to 5 m/day. Under the influence of strong, southward-flowing currents they migrate southward at 10 to 15m/day. When they reach midtide level, they become stationary, or welded to the beach. From Fox and Davis (1974).*

units. Bowen (personal communication) has suggested that such multiple bar systems may form in response to standing waves generated by the partial reflection of low-amplitude, long-period (1–2 minutes) incident waves. Such complex bar patterns clearly amplify the fair-weather storage capacity of the surf zone.

As noted above, the fair-weather littoral hydraulic regime is a fractionating mill, which splits the available sand into bed load and suspended fractions. The behavior of the bed load fraction has been traced above. Sand thrown into suspension at the breakpoint and fine enough to stay in suspension in the turbulent surf zone will tend to be fluxed alongshore by the longshore flow in the surf zone, and out through a rip channel to rain out on the shoreface (Cook, 1969).

STORM PHASE: BEACH AND BAR DESTRUCTION. During a storm, the wave regime and the littoral circulation patterns cooperate to withdraw littoral sand stored during the preceding fair-weather period. Wave steepness (ratio of wave height to wavelength) increases beyond a critical value (Johnson, 1949), at which point bottom

wave surge asymmetry is no longer efficient in driving coarser sand landward as bed load. Waves during storms are locally generated, and they tend to be shorter in period and higher (more energetic) with higher maximum orbital velocities. More sand is thrown into suspension and the critical grain-size threshold between suspensive and tractive sand fractions is shifted to favor suspension. Suspension is more nearly continuous. At the same time, discharge through the littoral circulation cells is increased manyfold.

During the advent of a severe storm the sudden seaward shift in breaker position, plus the great intensification of seaward sand transport, may be sufficient to destroy the bar and beach prism altogether. Some sand is driven across the back beach and over the dunes in the form of a washover fan (if this area is low enough to be so flooded), but most is transported seaward through rip channels and in rip current plumes. Toward the end of the storm, fallout from rip currents accumulates as a series of coalescing aprons of sand on the shoreface. Lagoons that are flooded during the period of rising storm surge may cut new inlets and break out through their barrier islands. The associated sand-laden jets may greatly add to this shoreface fallout (Hayes, 1967). As the storm wanes, the bar re-forms, and the cycle begins anew.

The cyclic nature of sand storage on beaches has been quantitatively assessed by Sonu and Van Beek (1971) in a study of northern North Carolina beaches (Fig. 4).

They observed a sequence wherein a storm-degraded concave beach profile, representing minimum storage, passed by means of swash bar accretion to a convex profile of maximum storage, during a four-month period. They noted that the sense of sedimentation (erosion or accretion) was more strongly correlated with the direction of wave approach (and hence with wind direc-

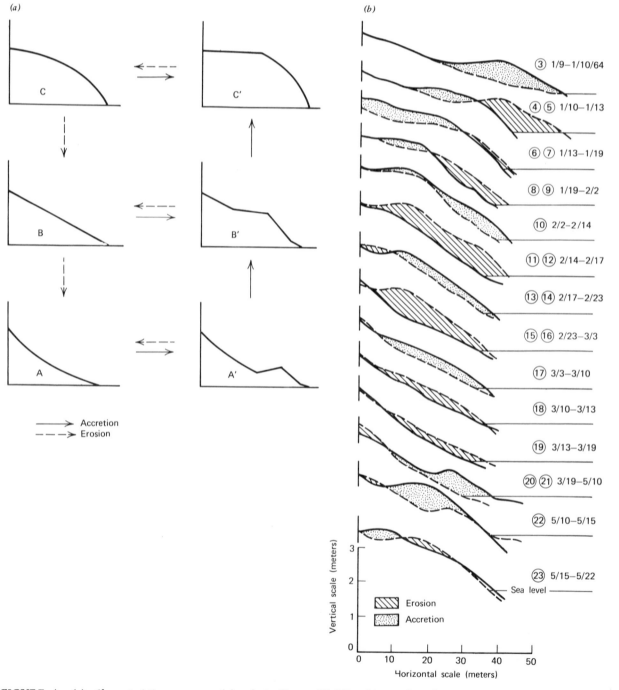

FIGURE 4. (a) *Characteristic sequences of beach profile change.* (b) *Observed sequences.* (c) *Observed sequences as a function of sediment storage* (Q) *and beach width* (S).

(d) *Time history of sand storage. From Sonu and Van Beek (1971).*

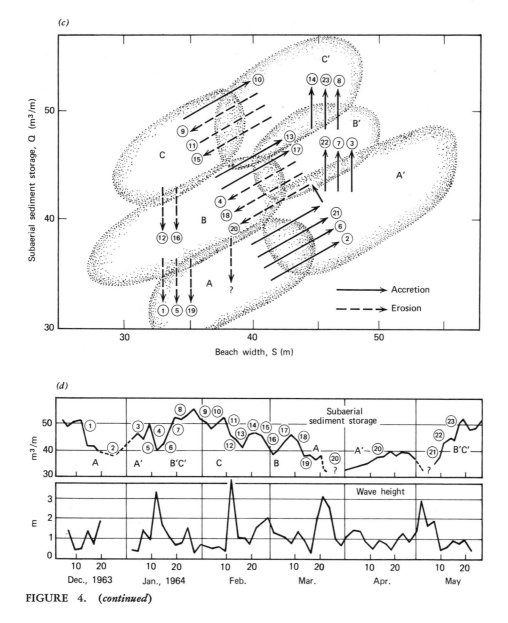

FIGURE 4. (*continued*)

tion) than with the wave steepness. Waves arriving from the northeast tended to cause erosion. These were associated with strong onshore winds and probably a wind-driven bottom return flow. It appears that during periods of strong alongshore or onshore winds, the system of littoral sand transport is no longer a closed system but discharges its sand into the wind-driven flow of the adjacent shelf floor. This deeper, intermittent system of transport is described in the following sections.

THE SHOREFACE PROFILE

Hydraulic Climate of the Shoreface

Far less is known about the circulation patterns of the shoreface and inner shelf than is known about the circulation patterns of the surf zone. Classical coastal workers, long preoccupied by the surf, have been indifferent to this topic, as have been physical oceanographers, whose habit has long been to hurry in their ships across the inner shelf, to the intellectual challenges of the large-scale planetary flows of the deep ocean basins. This situation is being rapidly reversed in view of rising public concern over the coastal environment (see Chapter 2), but old mental sets still linger.

The shoreface and inner shelf are a zone of transition, where the wave climate is still a major factor in shaping the seafloor, but where the shelf flow field is becoming of increasing significance in a seaward direction. There is some justice in the indifference of classical coastal workers to this hydraulic province. During periods of fair weather, the shelf flow on most coasts may be many times less intense than littoral drift (Fig. 2A). Its velocities, on the order of 1 to 10 cm/sec, are capable of moving whatever fines happen to be in suspension, but are not significant transporters of sand, although sand is repeatedly suspended by bottom wave surge at the crests of wave-generated ripples. Fair-weather flows, however, may be relatively complex in pattern, with nearshore reversals of the open shelf flow, induced by coastal promontories and by interaction with the tidal streams of inlets and estuary mouths.

Two kinds of inner shelf flows are quite significant in transporting sand and in molding coastal topography. On coasts with high tidal ranges, midtide current velocities associated with the passage of the coastal tidal wave may exceed 2 knots and locally attain 4 knots a few hundred meters seaward of the surf. Enormous volumes of sand are shifted on each tidal cycle, with significant net transport in the direction of the residual tidal current. Coastal tidal flows are poorly understood and tend to be rather complex because of strong interactions between tide-built topography and the tidal flow. Some examples are discussed in later sections (see pp. 294–295).

Intense coastal flows may also develop during storms (see Chapter 4). Such flows are far more infrequent than semidiurnal tidal currents, but unlike the latter, they occur on every coast, whether or not strong tidal cur

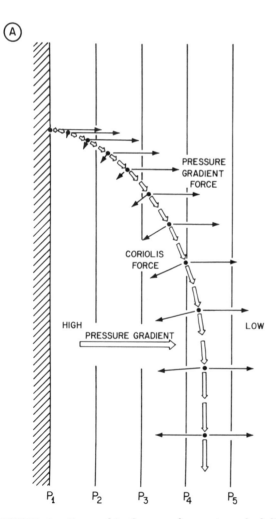

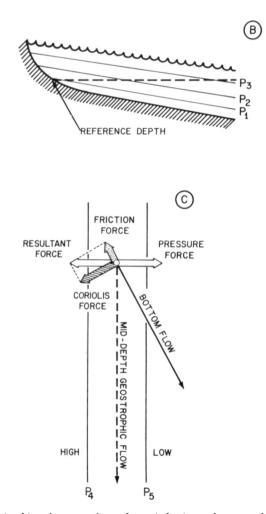

FIGURE 5. *Geostrophic flow on the continental shelf.* (A) *Parcel of water at a reference depth moves seaward in response to pressure gradient force. As it accelerates, it experiences a Coriolis force impelling it to the right of its trajectory. Eventually trajectory parallels isobars, and pressure force and Coriolis force balance.* (B) *Cross section of hypothetical shelf experiencing geostrophic flow: illustrating* relationship of sea surface slope, isobaric surfaces, and reference depth. (C) *Relationship between geostrophic flow and flow in bottom boundary layer. In latter case, a friction term enters the equation of motion, and the balance of forces occurs among a friction term, a pressure term, and a Coriolis term. Modified from Strahler (1963).*

rents occur. They, too, are significant transporters of sand. Without these storm-driven flows, the coasts of our planet would have a markedly different appearance.

Storms, whether of tropical or extratropical origin, are rapidly moving counterclockwise wind systems that may be a thousand or more kilometers in lateral extent. Winds intensify rapidly toward the storm center, and in hurricanes, by definition, exceed 74 mph. The extent to which the shelf water column will couple with storm winds depends on the trajectory of the storm with respect to the geometry of the shelf. Sustained regional coupling of water flow with wind flow appears to occur when the winds blow equatorward along the length of eastward-facing coasts (Beardsley and Butman, 1974) or blow poleward along the length of westward-facing coasts (Smith and Hopkins, 1972). Under such conditions, water in the surface layer will be transported landward as a consequence of the Coriolis effect. Coastal sea level will rise until the coastal pressure head balances bottom friction, and bottom water can flow seaward as rapidly as surface water flows landward. Beardsley and Butman report up to 100 cm of coastal setup under such conditions. Since the sea surface is inclined against the coast, there is a gradient of seaward-decreasing pressure at any reference depth. A parcel of water, accelerated by the pressure force, has its trajectory steadily deflected to the right by the Coriolis "force," until finally, it is flowing along the isobars and the pressure and Coriolis terms balance (Fig. 5).

INNER SHELF VELOCITY FIELD.   The complex velocity structure of the coastal zone is best approached in terms of the interaction of three major flow strata (Ekman, 1905; see Neumann and Pierson, 1966, p. 202). These are an upper boundary layer, a core flow, and a lower boundary layer (Fig. 6). The reader is advised to review Chapters 3 and 4 for a better understanding of this section.

The upper velocity boundary layer experiences strong wave orbital motion and, much of the time, a vertical velocity gradient imposed on it by wind stress. When the surface boundary layer is fully developed, surface water tends to move at 45° to the right of surface wind as a consequence of the Coriolis effect. Each successive lower layer moves at slower speed than the one above it, and is deviated successively further to the right (Ekman spiral). Net flow averaged over the depth of the layer trends 90° to the right of the surface wind. Above a critical Reynolds number this Ekman velocity structure becomes unstable, and is overprinted by a more complex structure, in which zones of upwelling and downwelling alternate, forming a pattern of horizontal helical vortices aligned parallel to or at a small angle to the

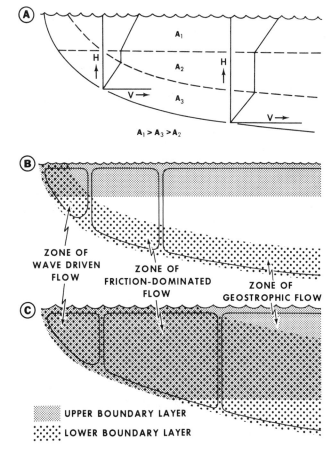

FIGURE 6.   *Velocity structure of the shoreface and inner shelf. (A) General form of velocity profiles through the upper boundary layer, core flow, and lower boundary layer, and relative values of eddy viscosity. (B) Velocity structure during a period of relatively mild flow. (C) Velocity structure during peak flow.*

mean flow direction (Langmuir circulation: Langmuir, 1925). The transition tends to occur at surface wind speeds of 10 km (Assaf et al., 1971). The coefficient of eddy diffusion A_v is relatively large in the surface layer as a consequence of wave-generated turbulence (Fig. 6); it must undergo an abrupt increase at the onset of Langmuir circulation.

Below the base of the layer, core flow extends, unmodified, down to the bottom boundary layer. In the core, water flows in slablike fashion, with little vertical shear. Core flows are generally geostrophic in the sense that in the equation of motion, the pressure term is primarily balanced by the Coriolis term (Fig. 5). However, a steady state geostrophic balance is rarely maintained for any length of time. The shelf pressure field is in a state of continual change, in response to the passage of the diurnal tidal wave and to the passage of weather systems. As the pressure field builds up and then decays, the flow must accelerate and decelerate in sympathy, constantly changing direction so that the pressure and Coriolis terms may balance. Such time-dependent flows are referred to as rotary tidal currents if mainly tide-forced, or inertial currents if mainly wind-forced.

The character of the bottom velocity boundary layer differs fundamentally from its surface analog. The surface boundary layer is externally forced, by the wind. Its velocity gradient, wave surge, and secondary flow patterns are overprinted on the core flow, and are carried along with it. The bottom velocity boundary layer is caused by frictional retardation of the core flow as it shears over the motionless substrate. Its lowermost meter exhibits a logarithmic velocity profile (Chapter 7), but the lower boundary layer as a whole is a thicker stratum, characterized by a velocity profile that is a reverse Ekman spiral. Frictional retardation of flow results in a deviation of boundary flow direction to the left of core flow, so that Coriolis and frictional terms may together balance the pressure term (Fig. 5C). The lowest layers, experiencing the greatest retardation, are deviated the furthest. Theoretical studies (Faller, 1963; Faller and Kaylor, 1966) suggest that this layer is also subject to helical flow structure above a critical Reynolds number. However, no field studies of this phenomenon have been undertaken. Such innate flow stabilities, and also turbulence induced by bottom roughness elements, would lead to an eddy coefficient larger than that of the core flow (Fig. 6A).

Three hydraulic provinces may be defined on the inner shelf on the basis of flow structure (Figs. 6B, C). Near the beach, the two boundary layers of the shelf flow field must completely overlap. In this zone the effects of the regional pressure gradient on water behavior are largely damped out as a consequence of frictional retardation. Oscillatory wave surge is the dominant water motion, giving rise to the complex nearshore circulation pattern described in the preceding chapter. A little farther seaward, the two boundary layers are more or less separate, but still occupy most of the water column. Flow is frictionally dominated; in the equation of motion the wind stress is largely balanced by friction. The effect of the Coriolis term is negligible in shallow water and there is little or no deviation of boundary flow with respect to core flow. The flow is Couette-like, in that there is a more or less linear velocity gradient from top to bottom. Still further seaward, the two boundary layers diverge significantly. The geostrophic core flow dominates the water column.

This pattern of coastal flow zonation must vary with the intensity of the regional and local wind fields. An intensified regional wind will accelerate core flow and increase the thickness of the bottom boundary layer. Intensification of the local wind field will cause the upper boundary layer to thicken, though not necessarily

at the same rate. The intensification of local wind may either lead or lag the intensification of wind on the adjacent shelf, depending on the trajectory of the weather system.

The net effect of a storm is to expand the width of the coastal flow zones and to displace the outer two zones seaward. There are few data available for such situations (see Chapter 4). From theoretical considerations, it appears that the upper and lower boundary layers may overlap far out on the shelf. Zonation becomes primarily a function of depth (Fig. 5C). In the zone of friction-dominated flow, the water accelerates in response to direct wind stress until the stress is balanced entirely by friction; the Coriolis term is not significant, and flow in this zone may take on the dimensions of a coastal jet (Csanady and Scott, 1974). The zone of friction-dominated flow will be a downwelling zone if local winds have an onshore component, or if regional coast-parallel winds result in onshore surface transport. It will be an upwelling zone if the reverse situation prevails (Cook and Gorsline, 1972).

The deeper, offshore flow may retain a primarily geostrophic balance of forces during a storm, although the friction term is necessarily more prominent. If overlap of the boundary layers extends through this zone, it is theoretically possible (Faller, 1971) that there be top to bottom overturn as a consequence of Ekman instability, with high-velocity, wind-driven surface water delivered to the seafloor in zones of downwelling.

The velocity structure of the shelf water mass follows a seasonal cycle that is coupled to the cycle of density stratification. During the summer, this upper velocity boundary layer is the same as the upper mixed layer. Wave turbulence and Langmuir circulation maintain the layer's mixed character, while the pycnocline tends to decouple upper boundary flow from core flow. During the fall, the thermal contrast is weakened by surface cooling. The increasing frequency and severity of storms cause steady erosion of the lower, stratified portion of the water column by Langmuir circulation (Faller, 1971) and the upper mixed layer thickens at the expense of the stratified water below. Meanwhile, a lower mixed layer may be induced by intensified turbulence in the bottom boundary layer, and may thicken until the density structure has simplified to a two-layer system (Charnell and Hansen, 1974). Further vigorous storm action will drive the weakening pycnocline down to the seafloor, so that there is no further impediment to top-to-bottom overturn by secondary flow components.

Sedimentation on the Upper Shoreface

The shoreface slope, with its gradient of seaward-decreasing grain size, occurs primarily in the zone of wave-

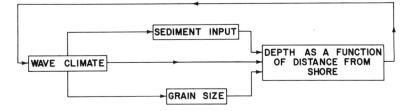

FIGURE 7. *Relationships of variables controlling slope of the shoreface.*

driven flow (Fig. 6) although its lower portion tends to extend into and be modified by the zone of friction-dominated flow. The slope and grain-size gradient of the shoreface have been generally considered to comprise a response to the regime of shoaling waves seaward of the breakpoint, in which depth as a function of distance from shore is itself a function of littoral wave power, sediment discharge, and grain size (Fenneman, 1902; Johnson, 1919, p. 211; Johnson and Eagleson, 1966; Price, 1954; Wright and Coleman, 1972; see Fig. 7. Johnson (1919, p. 211) has described this equilibrium relationship as follows:

The subaqueous profile is steepest near land where the debris is coarsest and most abundant; and progressively more gentle further seaward where the debris has been ground finer and reduced in volume by the removal of the part in suspension. At every point, the slope is precisely of the steepness required to enable the amount of wave energy there developed to dispose of the volume and size of debris there in transit.

The main line of inquiry into the forces maintaining the shoreface profile has led to the null-line hypothesis, evaluated in Chapter 8. The hypothesis has been expressed in its most complete form by Johnson and Eagleson (1966). It envisages shoreface dynamics in terms of a Newtonian balance of forces experienced by a sand particle on the shoreface, in which the downslope component of gravitation is opposed by the net fluid force averaged over a wave cycle. Since in shallow water, bottom orbital velocities are asymmetrical, with stronger landward surge (Chapter 8, Fig. 8), fluid forces are directed upslope. The gravitational force becomes more intense as the shoreline is approached and the slope increases. However, the fluid force increases yet more rapidly. As a consequence, for a given grain size there should be a null isobath, seaward of which particles of the critical size tend to move downslope, and landward of which they tend to move upslope. The equilibrium grain size should decrease with increasing depth. Hence, the shoreface sand sheet should tend to become finer downslope, as indeed it does. The shoreface slope at each point should be uniquely determined by the grain size of substrate and the intensity of bottom wave surge.

However, attempts to utilize null theory in the field have met with ambiguous or negative results (Miller

and Zeigler, 1958, 1964; Harrison and Alamo, 1964). Objections include: (1) slopes are not sufficiently steep over much of the shoreface (Zenkovitch, 1967, p. 120), and (2) slope sorting by waves tends to be overwhelmed by other processes, which as the authors of the theory admit, are not accounted for in null theory. No account, for instance, has been taken of the process of ripple sorting as described in Chapter 8 (p. 117). Wells (1967) has shown that divergence of onshore-offshore transport of a given grain size from its null isobath should occur as an innate response to higher order wave interactions, without regard to the gravitational force acting on the grains.

A perhaps more telling criticism of null-line theory is that a significant portion of shoreface sand travels not as bed load, but in suspension. Murray (1967) has performed tracer studies that indicate that on the upper shoreface, the dispersal of sand corresponds to the prediction of diffusion theory. Field observations by Cook and Gorsline (1972) have led them to conclude that the seaward-fining grain-size gradients of the shoreface are more likely to be caused by rip current fallout rather than by the null-line mechanism.

It may be more fruitful to approach the problem of shoreface maintenance from the point of view of energetics, rather than from the point of view of a balance of forces. Such an approach would view the depth at each point of a shoreface profile as a function of wave power at that point. The ideal wave-graded profile would be one that experiences at each point a maximum bottom orbital velocity equivalent to the threshold velocity of the size class of available sand. It should be possible to construct an algorithm for calculating water depth as a function of wave characteristics and bottom sediment grain size, based on the equations for bottom orbital velocity, for friction energy loss to the bottom, and for the shoaling transformations of waveform that have been presented in Chapter 6.

Lower Shoreface Sedimentation: Onshore-Offshore Sand Budget

It seems doubtful that such a model for maintenance of the shoreface profile by the wave regime would be sufficient to fully account for the distribution of slopes and

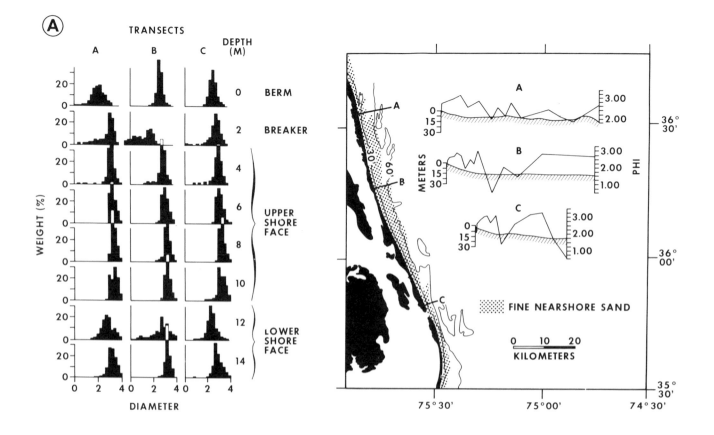

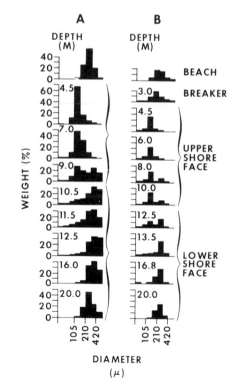

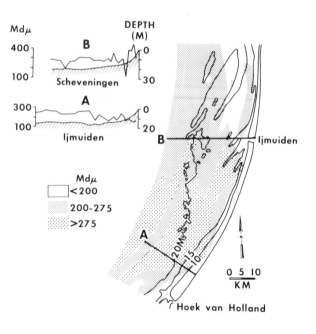

FIGURE 8. *Distribution of grain sizes on retrograding coasts.* (A) *The storm-dominated coast of Virginia–northern North* *Carolina. Data from Swift et al. (1971). (B) Dutch coast. Data* *from Van Straaten (1965).*

grain sizes associated with observed shoreface profiles. For instance, modern coasts whose historical records indicate that they are undergoing erosional retreat tend to consist of two distinctive grain provinces. From the breaker to a depth of about 10 m, the upper shoreface consists of fine, seaward-fining sand (Fig. 8). Seaward of 10 m, grain size on the lower shoreface and adjacent shelf floor is far more variable and generally markedly coarser.

We may account for the fine, upper shoreface sand province as a mantle of rip current fallout, whose slope is adjusted by the regime of shoaling waves (Cook, 1969). However, the lower shoreface province of coarse variable sand does not fit the model for wave maintenance of the shoreface. We may consider the hypothesis that it is instead a response to the deeper, intermittent high-intensity flows of the zone of friction-dominated flow (Figs. 2B and 6B, C).

Observations by Moody (1964, pp. 142–154) on the erosional retreat of the Delaware coast lend some support to this hypothesis (Fig. 9). In this area, the shoreface steepens over a period of years toward the ideal wave-graded profile, during which time the shoreline remains relatively stable. The steepening is both a depositional and erosional process. Moody notes that steepening was accelerated after 1934 because a groin system initiated then "presumably trapped sand, causing the upper part of the barrier between mean low water and −3 m to build seaward" (Moody, 1964, p. 142). However, erosion continued offshore at depths of 6 or 7 m below mean low water. The steepening process is not continuous, but varies with the frequency of storms and duration of intervening fair-weather periods. The slope of the barrier steepened from 1:40 to 1:25 between 1929 and 1954, but erosion on the upper barrier face between 1954 and 1961 regraded the slope to 1:40.

The steepening process is terminated by a major storm, during which time the gradient is reduced and a significant landward translation of the shoreline occurs. Moody (1964, p. 199) describes the Great Ash Wednesday Storm of 1962, bracketed within his time series, as having stalled for 72 hours off the central Atlantic coast. Its storm surge raised the surf into the dunes for six successive high tides. The shoreline receded 18 to 75 m during the storm. While much of the sand was transported over the barrier to build washover fans over 1 m thick, much more was swept back onto the seafloor by large rip currents and by the storm-driven seaward-trending bottom flow of the shoreface (Moody, 1964, p. 114); see Fig. 2B.

Moody's observations allow us to present a general model of shoreface maintenance, in terms of the onshore-offshore sediment budget. There seems to be little

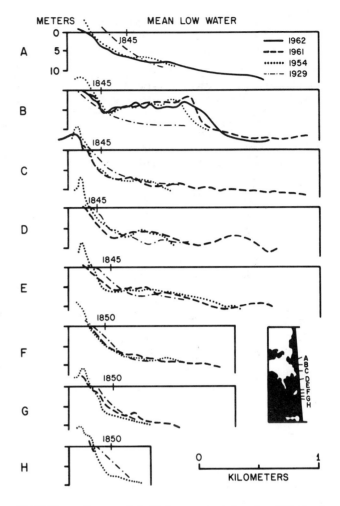

FIGURE 9. *Retreat of the Delaware coast, based on U.S. Coast and Geodetic Survey records and a survey by Moody. From Moody (1964).*

reason to doubt the applicability of the conventional wave-grading model to the upper shoreface, even if we cannot yet present this model in a quantitative manner. Upper shoreface textures and slopes are time-averaged responses to two opposing mechanisms, the seaward flux of suspended sand in rip currents on one hand, and the landward creep of bottom sand in response to the net landward sense of bottom wave surge on the other hand. The upper shoreface profile varies in cyclic fashion, with storage of sand mainly on the beach during the fair-weather summer season, and storage of sand mainly on the upper shoreface during the winter. For long periods of time, the upper shoreface profile may oscillate about the ideal wave-graded configuration.

During major storms, however, the upper shoreface system of wave-driven longshore sand flux interacts with coastal boundary of the storm flow field. Sand eroded from the beach and bar by storm waves passes seaward in intensified rip currents to the zone of friction-dominated flow (Fig. 2B), which during storms may take the form of a downwelling coastal jet. When this occurs, bottom flow on the lower shoreface will have a seaward component of flow, and the coastal sand transport system is transformed from a closed system of net sand storage to an open system of net sand loss. Sand raining out of rip currents will not come to rest, but will be transported obliquely seaward. If the storm is severe enough, the mantle of rip current fallout that accumulated during the preceding fair-weather period will be stripped off, and the underlying strata will be exposed to erosion.

This hypothetical scheme has not yet been adequately tested by field observations. However, as a hypothesis, it has a number of advantages. It provides a rationale for the Bruun model of erosional shoreface retreat (Fig. 10). Bruun (1962; see also Schwartz, 1965, 1967, 1968) noted the characteristic exponential curve of the inner shelf profile, and accepted the hypothesis that it constituted an equilibrium response to the hydraulic climate. With this premise adopted, it follows that a rise in sea level must result in a landward and upward translation of the profile, as long as coastwise imports of sand into the coastal sector under study are equaled by coastwise exports. The translation necessitates shoreface erosion and provides a sink for the debris thus generated beneath the rising seaward limb of the profile.

Moody's time series shows that over a 32 year period, shoreface erosion on the Delmarva coast was in fact nearly compensated by aggradation on the seafloor in accordance with the Bruun principle (Table 1). The small deficit is probably attributable to loss to washover fans, and through littoral drift to nearby Cape Henlopen spit.

Moody's studies provide us with insight into the processes governing the Bruun model. His observations indicate that the process of erosional retreat of the shoreface is not continuous. It is cyclic in a manner analogous to the annual cycle of the upper shoreface profile, but the period is related to the frequency of exceptional storms, and is on the order of years.

The model also provides a more detailed and satisfactory explanation for the origin of the surficial sands of shelves undergoing transgression than does the relict-Recent sediment model of Emery (1968). The surficial sand sheet of the shelf is a lag deposit created during the process of erosional shoreface retreat by the seaward transfer of sand during storms and its deposition on the adjacent shelf floor (Fig. 10A). The nearshore modern sands of the upper shoreface are a transient veneer of rip current fallout. Both textural provinces are "modern" in the sense of being adjusted to the prevailing hydraulic regime; both are "relict" in the sense of being derived

TABLE 1.   Sediment Budget from the Delmarva Coast

Sediment Source	Period	Average Volumetric Change* (m³/year)
Barrier (mean low water to toe of sand barrier)	1929–1961	−148,000
Sand dunes (mean low water to top of sand dunes)	1954–1961	−100,000 (estimated)
Offshore erosion (principally on northwest side of ridges)	1919–1961	−100,000
Erosion from bay inside Indian River Inlet	—	−69,000
	Total erosion	−417,000
Site of Deposition		
Tidal delta	1939–1961	+120,000
Barrier south of Indian River Inlet	1939–1961	+5,700
Offshore accretion	1919–1961	+256,000
	Total accretion	+381,700

Total erosion	−417,000	
Total accretion	+318,700	
Net erosion	−98,300 m³/year	

Source. From Moody (1964).

* "+" indicates accretion; "−" indicates erosion.

from the underlying substrate. The role of shoreface retreat in generating shelf sediments is explored further in Chapter 15.

Deposits of the Coastal Profile: Textures and Bed Forms

TEXTURES OF THE SHOREFACE.   The patterns of onshore and offshore sediment transport described in the preceding sections give rise to systematic distributions of sediment types and bed forms over the beach and shoreface. The beach and surf zones consist of alternating belts of finer and coarser sand, the absolute grain-size values depending on grain sizes available to the coast and on the hydraulic climate of the coast (Bascom, 1951). The coarsest grain sizes are found on the crest of the berm, in the axis of the longshore trough during the erosional phase of the beach cycle, and on the crest of the plunge point bar.

The distribution of grain sizes on retrograding shorefaces has already been described (Fig. 8). Upper shoreface sands tend to be fine grained to very fine grained, and become finer in a seaward direction. The grain-size

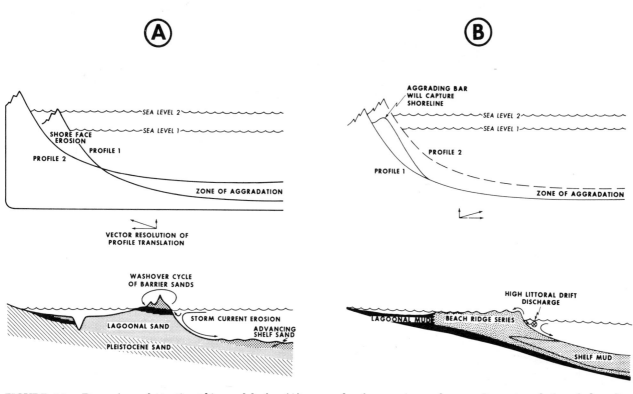

FIGURE 10. *Dynamic and stratigraphic models for* (A) *a retrograding and* (B) *a prograding coast during a rise in sea level. If coastwise sand imports are balanced by or are less than coastwise sand exports, the hydraulically maintained coastal profile must translate upward and landward by a process of* shoreface erosion and concomitant aggradation of the adjacent seafloor (Brunn, 1962). If coastwise sand imports exceed exports, as is the case for deltaic coasts, then the profile must translate seaward and upward. Based on Curray et al. (1969).

gradient is perhaps originally the result of progressive sorting (see p. 162) operating on the suspended sand load of rip current plumes; the underlying deposit becomes finer down the transport direction in the manner of loess or volcanic ash deposits. It is perhaps secondarily the result of adjustment to the landward increase in bottom orbital velocity, and to the mechanism of ripple sorting (p. 117).

On retrograding coasts such as the North Carolina and Dutch coasts (Fig. 8), the lower shoreface consists of variable but generally coarser sand. It is texturally adjusted to the coastal boundary currents associated with peak flow events. This material is of negligible thickness and constitutes a residuum mantling the eroding surface of the underlying older deposits. Its final resting place appears to be the adjacent seafloor, where it forms a discontinuous layer up to 10 m thick (Stahl et al., 1974).

Bimodal sands tend to occur at the contact between the two provinces, where the rip current fallout blanket thins to a feather edge. This contact advances downslope during fair-weather periods of upper shoreface aggradation, and retreats upslope during periods of storm erosion of the entire shoreface.

On prograding coasts, such as the western Gulf of Mexico (Bernard and Le Blanc, 1965) or the Costa de Nayarit (Curray et al., 1969), more sand is delivered by littoral drift during fair weather than can be removed by storms. The fine, seaward-fining sand of the upper shoreface extends down to the break in slope where it may become as fine as mud, and continues across the shelf floor (Fig. 10B).

Visher (1969) has observed size frequency distributions in the surf zone and on the shoreface (see Fig. 11). Moss' theory may be applied to his observations (see p. 162), but caution must be used, as Moss' theory relates to quasi-steady flows, while in the coastal marine environment a high-frequency flow oscillation due to wave surge tends to be superimposed on a steady flow component. Peak wave surge regularly induces the upper flow regime (Moss' rheologic regime), while the intervening flow may consist of one of the less intense stages.

In Fig. 12, the supercritical flows of swash and backwash in the intertidal zone have resulted in complex subpopulation assemblage (lower foreshore, 1.5 ft samples). The contact (C) population, consisting primarily of shell debris, comprises up to 10% of the total distri-

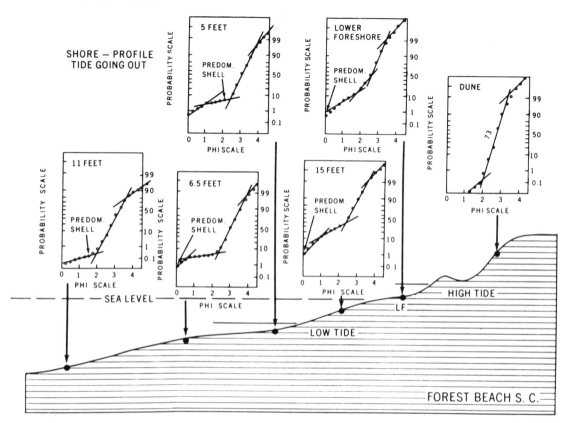

FIGURE 11. *Representative size frequency distributions of the shoreface. From Visher (1969).*

bution. A large C population is characteristic of Moss' rheologic regime; the B population, however, is less than 1%. While the hydraulic microclimate of rheologic flow is conducive to the incorporation of a large B population into the bed, such a response is presumably inhibited by the gross hydraulic structure of the surf zone; suspended fines are steadily flushed seaward through rip channels. Two framework (A) populations are locally present, reflecting perhaps discrete responses to swash and the slightly higher backwash velocities. Subtidal surf zone populations (5 and 6.5 ft) are similar, but the framework population is better sorted (corresponding segment of the cumulative curve is steeper). This sorting is further improved in the upper foreshore sample (11 ft sample). This sample has a reduced contact (C) population and an enriched interstitial (B) population. B population enrichment reflects the heavy rip current fallout of fine suspended sand experienced at this depth, and perhaps also the presence of Moss's fine ripple regime.

BED FORMS OF THE SHOREFACE. Clifton et al. (1971) have noted that the high-energy shoreface of southern Oregon is characterized by zones of primary structures that reflect the hydrodynamic subenvironment (Fig. 12). An "inner planar facies" occurs beneath the reversing supercritical flows of the swash zone; the associated

structure within the deposit consists of thin beds and laminae of gently inclined sand. The rhomboid ripple marks and antidunes that form in each backwash are rarely preserved.

Beneath the surf zone of gently sloping beaches lies an "inner rough facies" of shore-parallel ridges and troughs 1 to 2 m across and 10 to 50 cm deep. The flat-topped ridges tend to be steepest on the seaward side, and the ridges migrate seaward. During periods of strong littoral currents, troughs are more nearly perpendicular to land, and migrate downcurrent and offshore. The internal structure of this facies consists of medium-scale (units 4–100 cm thick) seaward-dipping trough cross-bedding.

Beneath the breakpoint lies an "outer planar facies." No bar existed here during the period of Clifton's study. Small ripples may form during the initiation of trough or crest surge, but the flow becomes supercritical during maximum surge, and the ripples are destroyed. The internal structure is horizontal lamination.

Clifton et al. describe the upper shoreface of the Oregon coast as the "outer rough facies." The characteristic bed forms are lunate megaripples, 30 to 100 cm high, and with spans (terminology of Allen, 1968a, pp. 60–62) between 1 and 4 m. Concave slopes face landward, and the ripples migrate landward at rates of 30 cm/hr.

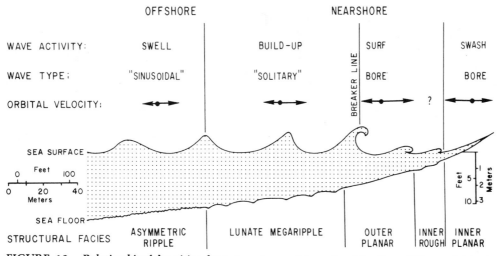

FIGURE 12.   *Relationship of depositional structures to wave type and activity. From Clifton et al.* (1971).

Crestal sands are notably coarser than trough sands. The resulting internal structure is a medium-scale cross-stratification with foresets dipping steeply landward.

The lower portion of the upper shoreface of the Oregon coast has an "asymmetrical ripple facies" of short-crested wave ripples 3 to 5 cm high with chord lengths (Allen, 1968a, pp. 60–62) of 10 to 20 cm. They may reverse asymmetry with each passing crest and trough, or reveal a persistent landward asymmetry. Crests become lower, longer, and straighter as the pattern is traced seaward. Interfering sets are common, weaker sets tending to occur as ladderlike rungs in the troughs of the stronger set. The angle between the two sets is bisected by the wave surge direction. The internal structure of this facies tends to consist of small-scale (less than 4 cm) shoreward-inclined ripple cross-lamination, interfingering with gently dipping, medium-scale scour and fill units.

A similar sequence of bed forms and textural provinces has been reported from the Georgia coast by Howard and Reineck (1972). The Georgia coast has a milder wave climate than those described above, although it is also characterized by strong tidal flows. The equilibrium configuration of the shoreface is rather different here (Fig. 13); the slope is much gentler, and the break between the fine sand of the upper shoreface and the coarse sand of the lower shoreface occurs as far seaward as 14 km from the beach.

An inner planar zone of laminated sand is equivalent to that of Clifton et al.'s (1971) but is markedly wider, extending from 0 to −1 m, 200 m from the beach. An inner rough facies (1–2 m depth) is equivalent to that of Clifton et al.'s, but is expressed as rippled, laminated sand, rather than megaripples. An outer planar facies (5–10 m depth) consists of laminae and thin beds with sharp, erosional lower contacts, grading upward into bioturbate texture. Howard and Reineck (1972) suggest deposition during storm intervals, alternating with periods of fair weather and bioturbation. An upper shoreface facies of fully bioturbated, muddy fine sand has no parallel in Clifton's study of the high-energy Oregon coast, and is a consequence of the high input of fine sand and reduced wave energy.

Seaward of 10 m, the muddy, gently sloping shoreface becomes markedly coarser, then gives way to a flatter seafloor of medium to coarse sand, characterized by heart urchin bioturbation and trough cross-stratification. Clifton and co-workers did not extend their study sufficiently far seaward to detect such a coarse lower shoreface and seafloor facies However, an equivalent facies does appear on retreating coasts of both North Carolina and Holland (Fig. 8).

Reineck and Singh (1971) have described shoreface and inner shelf deposits from the low wave energy, high mud input, prograding coast of the Gulf of Gaeta, Italy. The inner facies are rather similar to those of the Georgia coast. Ripple bedding is the main sedimentary structure out to 2 m. Below 2 m, laminated bedding becomes the main structure, and bioturbation becomes prominent, increasing seaward. Laminae are inferred to be deposited from graded suspensions after storms. At 6 m, sand gives way to silty mud, heavily bioturbated by *Echnocardium cordatum*. There is no equivalent of the coarse offshore facies of retreating coasts.

LONGSHORE SEDIMENT TRANSPORT

The seasonal cycle of onshore and offshore sand migration in the surf is superimposed on a much more intensive flux of sand parallel to the beach, under the impetus of the wave-driven littoral current. The mechanisms driving

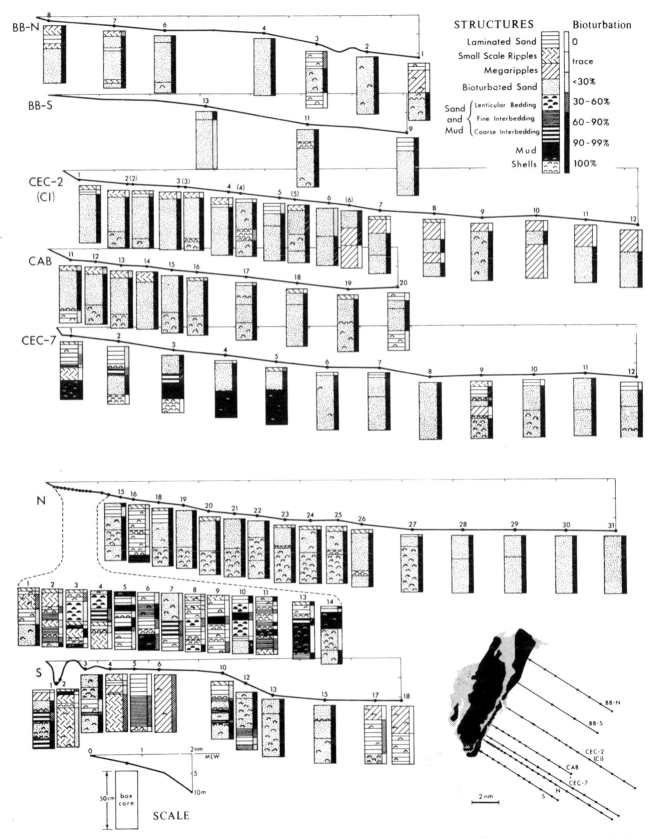

FIGURE 13. *Primary structures of the Georgia shoreface, as revealed by box cores. Complexity of lines N and S are due to their traversing the north flank of an estuary mouth shoal. From Howard and Reineck (1972).*

272

this littoral drift and equations for determining its transport rate are described in Chapter 13.

The propensity of this coastwise sand flux to aggrade or erode the shoreline can be understood by reference to a convenient graphical model presented by May and Tanner (1973). As a consequence of refraction of waves about a coastal headland, such a headland will tend to concentrate the wave rays on it and hence wave energy (see p. 76). As a consequence, the wave energy density (proportional to the spacing of wave rays) decreases steadily from point a on the headland to point e in the bay. These relationships are shown in highly schematic fashion in Fig. 14.

The longshore component of wave power P_L is a function of both the wave energy density and the breaker angle (see Equation 2, Chapter 13). It must therefore pass through a maximum between the point of greatest wave energy density a, and the point of greatest breaker

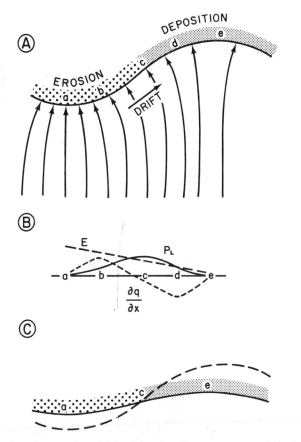

FIGURE 14. *Model for littoral sediment transport.* **(A)** *Wave refraction pattern, with wave approach normal to coast.* **(B)** *Resulting curves for energy density at the breaker* E *(dimensions* MT^{-2}*); longshore component of littoral wave power* P_L *(dimensions* MLT^{-3}*); and the littoral discharge gradient* $\partial q/\partial x$ *(dimensions* L^2T^{-1}*).* **(C)** *Advanced state of coastal evolution. After May and Tanner (1973).*

angle at c. The maximum is attained, however, closer to point c, as the gradient of wave energy density on this subdued model coast is relatively flat.

Both the sand transport rate I_L (see p. 247) and the discharge of sand (q) are proportional to P_L and vary with it. Therefore, the longshore discharge gradient $\partial q/\partial x$ varies as the derivative of P_L (Fig. 14). The sediment continuity equation (p. 190) states that the time rate of change of seafloor elevation along a streamline in the littoral current is proportional to the littoral discharge gradient under conditions of steady flow. In other words, if more sand is moving into a given section of shoreface than is moving out (negative $\partial q/\partial x$), then the seafloor of that section must aggrade. If, on the other hand, more sand is being exported than imported (positive $\partial q/\partial x$), the seafloor of that sector must erode (see the discussion on p. 190). In general, erosion occurs along a positive discharge gradient, and deposition occurs along a negative discharge gradient.

In the model of Fig. 14, this relationship means that the shoulder of the headland, from a to c, should erode, with the material being transported into the bay, to fill sections c through e. The same sort of process should occur on the other side of the bay (not shown) and the other side of the headland (not shown). If the direction of wave approach were held constant and normal to the regional trend of the coast, then a very peculiar coastline should eventually result. It would straighten out to a nearly east-west line running through c, but with a needlelike projection at a, the point of littoral drift divergence, and a similarly narrow indentation at e, the point of littoral drift convergence. On real coasts, however, the direction of wave approach is not constant, but fluctuates about the mean value, with changes occurring on a scale of hours to days. As the direction of wave approach fluctuates, so do the positions of points a and e, and the development of coastal reentrants and projections is suppressed.

If waves tend to approach at an angle instead of approaching normal to shore, and if coastal relief is more deeply embayed (Fig. 15), then a rather different distribution of longshore wave power will result. The locus of maximum deposition will be shifted from the bay head toward the tip of the adjacent headland where the gradients of wave energy density and breaker angle are the steepest. During a storm when the intensity of littoral drift discharge is the greatest, deposition at this point may be so intense that a discontinuity in the shoreface may occur, in the form of a spit that builds out across the bay as an extension of the headland shoreface. As the shoreline matures, headland retreat, spit extension, and bay head beach progradation occur simultaneously and in this model also, the final coastline is again straight.

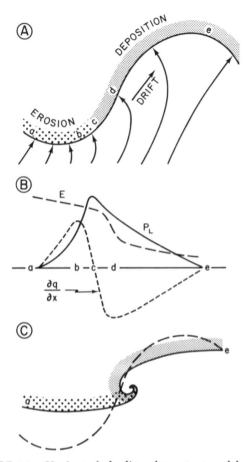

FIGURE 15. *Variant of the littoral transport model with a more deeply embayed coast and an oblique direction of wave approach. Conventions as in Fig. 14.*

Thus, as a consequence of the submarine refraction o waves about the shoals off headlands, the shoreface tends toward an equilibrium plan view as well as an equilibrium profile. Headlands tend to be suppressed and bays filled, because their existence leads to longshore wave power gradients that transfer sand from headland to bay head.

A similar smoothing process operates at deeper levels on the shoreface, where sand transport occurs in response to tide- and wind-driven currents. Such flows accelerate past the projecting headlands which impede them, and expand and decelerate off the bays. The result is a positive discharge gradient on the upcurrent sides of the headlands, and a negative discharge gradient on the downcurrent side. This pattern reverses when the currents themselves reverse, so that the lower shoreface off headlands experiences net erosion, while the lower shoreface off bays experiences aggradation. Thus the equilibrium plan view of a coast tends to be straight, to the extent that variations in the homogeneity of the sub-

strate and the rate of sand supply to the surf zone will permit.

A second characteristic of equilibrium coastal configuration is the adjustment of the trend of the coast to the angle of wave approach as mediated by the rates and locations that sand is put into and taken out of a littoral drift cell. A perfectly straight and infinitely long coast could ideally maintain any angle to wave approach, if the only source of sand were its own shoreface erosion. In fact, however, such an ideal straight coast is rarely attained. Sea level is rising or has been until very recently, and the straightening process must operate continuously as successive portions of the irregular subaerial surface are inundated. Depending on the degree of induration of the coast, an effective equilibrium is attained with less than the "climax" degree of coastal straightness; some irregularity usually persists, with at least subdued headlands serving as sand sources, and embayments serving as sand sinks. Locally, river mouths may serve as point sources of sand with high sand input rates. These result in deltas and an effective coastal equilibrium at less than climax straightness.

Komar (Chapter 13, Figs. 12, 13) has provided two examples of the adjustment of coastal trend to the angle of wave approach, and to the location of sources and sinks. His river mouth (Fig. 12) injects sand at a point in the littoral drift system at a rate greater than the system can initially accommodate, and the coastline progrades. As it does so, orientation of the coast at each point adjusts so that the river mouth protrudes as a delta. Eventually, an equilibrium configuration is attained so that each point along the shoreline maintains an incident wave angle sufficient to bypass the same amount of sand at every other point.

Komar's beach (Fig. 13) has no point source of sand. He starts with a straight beach and a landward-convex waveform, a form that might arise from offshore refraction over the rocky headlands enclosing a pocket beach. The center of the beach becomes a source and the ends become sinks; the shoreface adjusts to the wave refraction profile.

Komar's two examples correspond to two basic categories of coastlines. In the *swash alignment* (Davies, 1973, p. 123), each point on the coast tends to be oriented normal to the direction of wave approach, either as an initial condition or because the configuration of the shoreface and the wave refraction pattern have interacted until this is the case. Such an adjustment is only possible if there is a projecting headland in the downdrift direction as in the case of Komar's beach, or another reason to cause a sand sink and allow coastal progradation. *Drift alignments* (Davies, 1973, p. 123) are more nearly like Komar's delta model. These coasts are

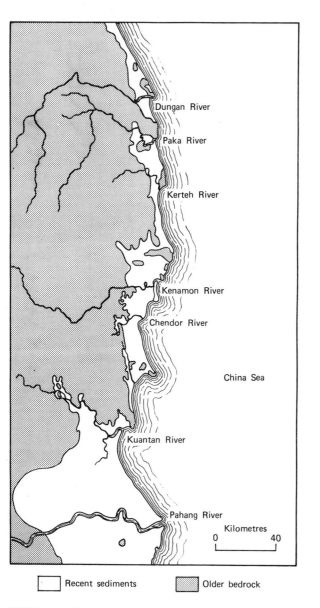

FIGURE 16. *Zetaform bays on the offset coast of eastern Malaya. Major wave approach direction is from the northeast. From Davies (1973).*

swash-aligned beach in the shadow zone immediately downdrift from the barrier, and a straight drift-aligned beach extending downdrift from the swash segment to the next headland (Fig. 16). If the variance of the direction of wave approach is high, then the shadow zone behind the barrier will be exposed to the direct approach of waves for a significant part of the time and will intermittently operate as a drift-aligned beach of reverse drift. The apex of the barrier will become a zone of net drift convergence, and hence a self-sustaining constructional feature.

The dynamics of zetaform bays have been discussed in detail, with summaries of earlier observations, by Silvester (1974, pp. 71–90).

TRANSLATION OF THE SHOREFACE

The preceding evaluation of the alongshore and on-shore-offshore components of sediment transport permits us to take a more general look at the coastal sediment budget and its effect on the shoreface stability.

It is clear from the preceding discussion that the shoreface profile will translate either landward or seaward (the coast will retreat or prograde) depending on whether the effects of the fair-weather regime, which tends to aggrade the shoreface, or the storm (or tidal) regime, which tends to erode it, are dominant. The sense of coastal profile translation further depends on coastwise gradient of sand discharge (whether sand imports and exports for the sector under consideration sum to a surplus or deficit). As a consequence of the onshore-offshore cycle of sand exchange, the nature of coastal translation will finally depend on which shoreface province, if any, is actually subjected to a sand surplus. The coastal transport system may be visualized as two coast-parallel pipes, corresponding to the wave-driven littoral drift near the beach and the intermittent storm- or tide-driven sand flux that occurs on the shoreface and inner shelf seaward of the breaker. These two pipes are connected by valves, corresponding to the onshore-offshore cycle of sand exchange. The factors listed above determine what valves are open, for how long, and the net sense of flow through the valves. We do not have the measurements of onshore-offshore sand transport that would allow us to document the manner in which this system actually works. Until we do, we must be satisfied with an exploration of possible limiting cases, by means of deductive reasoning (Fig. 17).

Four basic cases may be distinguished. On modern coasts, undergoing relatively rapid sea-level rise, the gradient of littoral drift discharge ($\partial q/\partial x$) is either positive or is so slightly negative that the resulting sand surplus is not sufficient to balance offshore transport

stabilized by competition between two opposing trends for control of the littoral transport system. As a coastal compartment becomes more nearly normal to wave orthogonals, littoral energy density increases; however, the longshore component of wave energy decreases. Maximum discharge tends to occur when the orthogonals of prevailing wave trains make an angle of 40 to 50° with the coast. A coast captured in this alignment will tend to be stable, as it is the alignment of maximum transport.

Coasts with closely spaced barriers to littoral drift, in the form of river mouths or projecting headlands, may form offset coasts consisting of successive zetaform bays (Halligan, 1906). These fishhook beaches have a curved

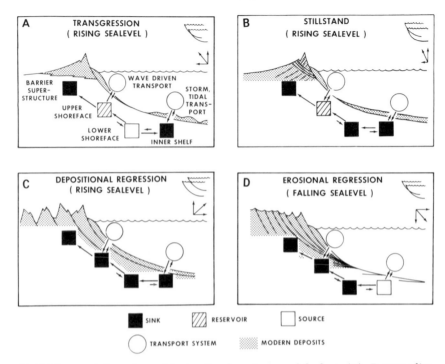

FIGURE 17. *Schematic models for the shore face sand budget. (A) Retrograding coastal sector with rising sea level and balance or deficit in coastwise sand flux. (B) Near-stillstand coastal sector with effect of rising sea level compensated by sand surplus associated with coastwise sand flux. (C) Prograding coastal sector with effect of rising sea level and reversed by sand surplus associated with coastwise sand flux. (D) Prograding coastal sector with falling sea level and balance in or sand deficit resulting from coastwise sand throughput.*

during storms. Under these conditions Bruun coastal retreat must prevail (Fig. 17*A*). Storm erosion must predominate over fair-weather aggradation on the shoreface as it translates landward and upward in response to sea-level rise. During fair weather, sand may be temporarily stored on the upper shoreface and beach (see Fig. 4*A*). The barrier superstructure becomes a long-term reservoir, receiving sand from the eroding shoreface by storm washover, storing it, and finally releasing it to the eroding shoreface. The inner shelf floor tends to become a sand sink, retaining the coarser sand transmitted to it by seaward bottom flow during storms, and releasing the finer fraction to the coastwise shelf flows. The coast of northern New Jersey appears to be undergoing such a retreat (Stahl et al., 1974).

Locally, however, the sand discharge gradient associated with the deeper storm-driven shelf flows may be steeply negative (Fig. 17*B*), resulting in a sand surplus. The surplus tends to be absorbed by inner shelf and lower shoreface aggradation and also by storm washover on the barrier. The upper shoreface, however, is relatively unaffected. The trajectory of the shoreface profile appears to be nearly parallel to the upper shoreface slope, resulting in stillstand of the shoreline. The

barrier system of coastal North Carolina immediately north of Cape Hatteras appears to be undergoing such a depositional stillstand, resulting in the opening of an anomalously wide lagoon behind it as sea level continues to rise (Swift, 1975).

If the littoral drift transport system has a strongly negative discharge gradient, as is the case downdrift from river mouths with a high sand discharge, the saturation of the upper shoreface with sand causes the flooding of all other inner shelf provinces as well, and the shoreface progrades by the successive capture of upper shoreface bars as beach ridges (Figs. 10*B* and 17*C*). The analysis of the Holocene history of the Costa de Nayarit by Curray et al. (1969) provides an excellent case history of such a coast.

Finally, we must consider the case of falling sea level. If, under such conditions, the net littoral drift input is negligible, not all portions of the shoreface will prograde. The shoreface must translate seaward down the gradient of the shelf in a reversal of the Bruun process. Under these conditions, successive beach–upper shoreface sand prisms undergo subaerial capture, and the lower shoreface and inner shelf undergo erosion as sea level drops (Fig. 17*D*). However, should the sand surplus due to littoral throughput increase, the sand budget must approach that of the subsiding prograding coast and a more rapid shoreline regression must result. Modern examples of such falling sea-level budgets are confined to regions of glacial rebound or tectonic uplift, but the surfaces of the world's coastal plains were molded by it during the withdrawal of the Sangamon (Riss-Würm) Sea (Oaks and Coch, 1963; Colquhoun, 1969).

A word on the factors controlling the steepness and curvature of the shoreface profile is in order at this point. Grain size is the most obvious control (Bascom, 1951); the coarser the sediment supplied to the coast, the steeper are the shoreface profiles. Shorefaces built of shingle may attain 30° slopes near the beach; shorefaces of sand are rarely more than 10° at their steepest, while shorefaces on muddy coasts are so flat as to be virtually indistinguishable from the inner shelf. Sediment input and the wave climate also affect the shape of the profile. In general, inner shelves experiencing a higher influx of sediment and a lower wave energy flux per unit area of the bottom are flatter, whereas inner shelves with a lower influx of sediment and a higher wave energy flux per unit area of the bottom are steeper (Wright and Coleman, 1972). Because of the complex interdependence of the process variables, cause and effect are difficult to ascertain; on a steeper shelf, for instance, grain size is coarser because the steeper slope results in more energy being released per unit area of the bottom; more energy is released because the coarser grain size results in a higher effective angle of repose. Or a reduced input of sand will allow the profile to attain the maximum steepness permissible under the prevailing wave climate, with a resultant higher rate of energy expenditure on the shoreface, and a consequent coarsening of its surface (Langford-Smith and Thom, 1969).

The relationship between the rate of sea-level displacement and the shape of the profile requires some thought. A number of workers have assumed that rapidly translating coasts are in a state of disequilibrium, and that equilibrium can only be realized on very slowly translating or stillstand coasts. This view results from an inadequate appreciation of the equilibrium concept and is tantamount to stating that only chemical reactions that have gone to completion are equilibrium reactions.

It is important to clearly distinguish between the concept of coastal maturity on one hand, and the concept of coastal equilibrium on the other. Davis (in Johnson, 1919) has assembled a spectrum of coastal types that suggest that the coastal profile passes through stages of "youth, maturity, and old age" in which the profile becomes increasingly flatter, until a final profile of static equilibrium is reached—ultimate wave base, in which

the continental platform has been shaved off to a level below which further marine erosion occurs so slowly as to be negligible. The scheme is unrealistic in that it fails to recognize the continuous nature and mutual dependence of the process variables of an equilibrium system. Some of these stages will occur as transient states after the sudden rejuvenation of a tectonic coast. But as the profile becomes increasingly mature, its rate of change decreases, until it attains the equilibrium configuration required by existing rates of such other process variables as sediment input and eustatic sea-level change. At this point the profile must continue to translate according to the Bruun (1962) model of parallel shoreface retreat, until the rate of one or another variable changes again. This equilibrium, of course, is only apparent if the coastal profile is examined over a sufficiently long period of time—on the order of decades. Shorter periods of observation will resolve the apparent "equilibrium" into a series of partial adjustments to periods of fair weather and periods of storms.

Only in cases of relatively rapid tectonism may hysteresis, or lagged response, occur, and strictly speaking, the term "disequilibrium" should be applied only to such cases. Slower changes in a process variable will allow continuous and compensating adjustment of profile, and while its shape changes, the profile is at all times in equilibrium. Coastal disequilibrium tends to be more apparent on rocky coasts, because of the greater response time of the indurated substrate, and because such coasts are more likely to be subject to tectonism.

Consequently, the effect of the rate of sea-level displacement in the equilibrium profile must depend on the initial slope of the substrate. On low coasts, where the initial slope is flatter than the maximum potential slope of the equilibrium profile, then the more rapid the sea-level displacement, the flatter is the resulting equilibrium profile (e.g., see Van Straaten, 1965). This relationship may be viewed as a function of work done on a substrate to build the optimum shoreface. As a coast advances more rapidly, successive shorelines experience the erosive effect of shoaling waves for shorter periods of time and the resulting profile is flat (immature). If, however, a coast undergoes stillstand, the climax or fully mature configuration can develop, which is the steepest profile possible for the available grain size of sand, rate of sediment influx, and hydraulic climate.

On high, rocky coasts, however, the initial slope of the substrate may be steeper than the mean, or even the maximum slope of the steepest profile permitted by these variables. Under such circumstances, the more rapidly transiting shorelines, since these have the least work done on them, have the least modified and hence steepest (most immature) profiles, while the most slowly moving shorelines are the most modified and hence flattest profiles.

As noted above, existing measurements of the coastal hydraulic climate and resulting sand transport are generally inadequate to define the coastal sand budget. It is possible, however, to extend the inferential models presented in Fig. 17 so as to take into account the effect of these variables on shoreface slope and curvature (Fig. 18).

COASTAL ENVIRONMENTS

The preceding sections of this chapter have described the onshore-offshore component of sediment transport, and also the flow of sediment parallel to the coast. Modes of shoreface displacement in response to rising and falling sea level have been considered. These insights are prerequisites to an examination of specific patterns of coastal sedimentation. But before we proceed to such an

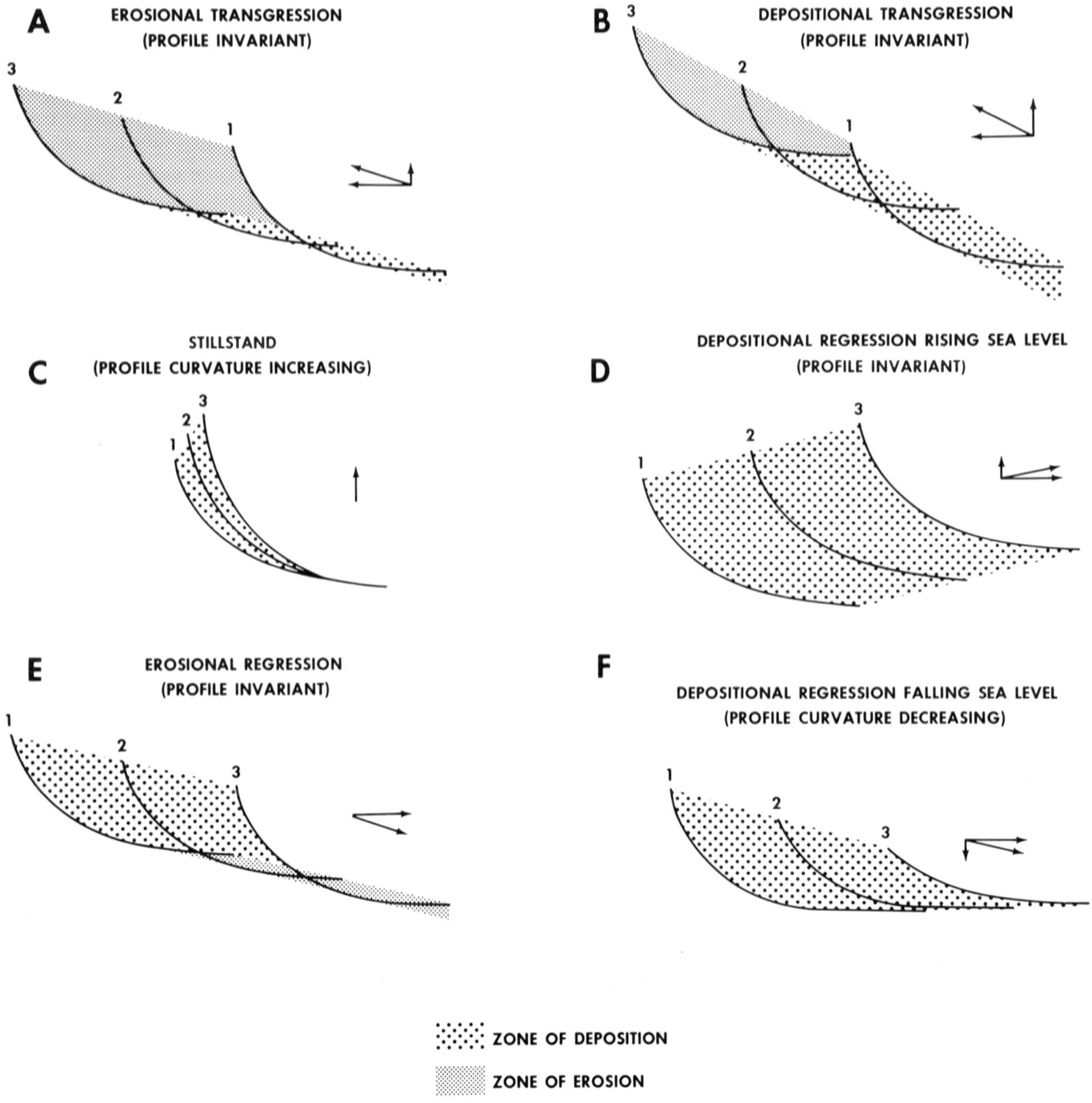

A   EROSIONAL TRANSGRESSION
(PROFILE INVARIANT)

B   DEPOSITIONAL TRANSGRESSION
(PROFILE INVARIANT)

C   STILLSTAND
(PROFILE CURVATURE INCREASING)

D   DEPOSITIONAL REGRESSION RISING SEA LEVEL
(PROFILE INVARIANT)

E   EROSIONAL REGRESSION
(PROFILE INVARIANT)

F   DEPOSITIONAL REGRESSION FALLING SEA LEVEL
(PROFILE CURVATURE DECREASING)

:::::: ZONE OF DEPOSITION

ZONE OF EROSION

FIGURE 18.   *Modes of shoreface translation as a function of (1) direction of profile translation and (2) change in profile curvature. Envelopes of erosion and aggradation are shown. Terms from Curray (1964).*

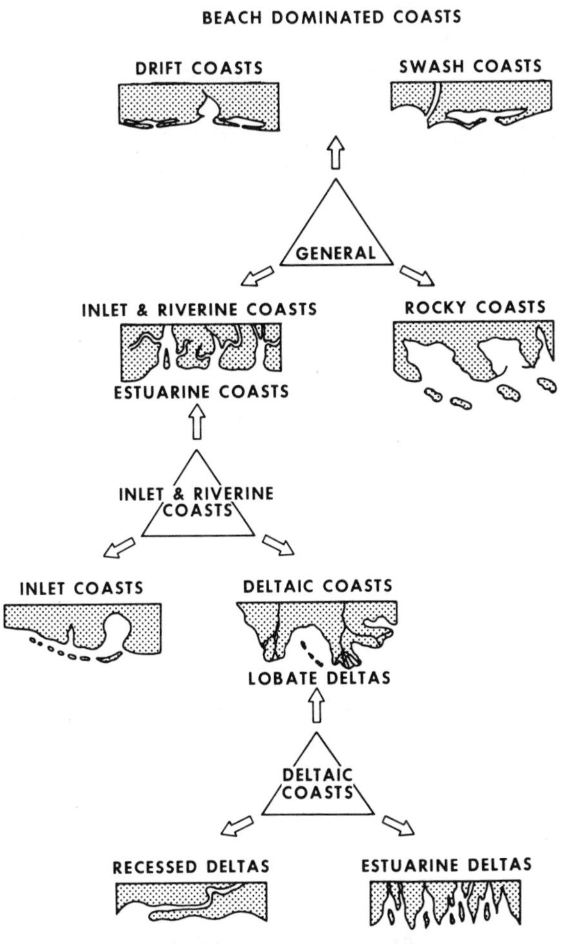

FIGURE 19. *Descriptive taxonomy of coasts.*

analysis, we should consider a scheme for classifying the coastal settings in which the transport patterns occur.

A study of maps of the world's coastlines suggests that the apparently unlimited variety of coastal configurations falls into a relatively small number of repeating patterns. Considerable thought has gone into coastal classification, and the reader is referred to the excellent summary of existing classifications presented by C. A. M. King (1972; also Chapter 15). The operational classification that is used in this text is presented in Table 2 and Fig. 19.

The most basic practicable division appears to be into coasts with substrates of crystalline or lithified sedimentary rock versus coasts bordering coastal plains, with substrates of unlithified sediment. Both lithified and unlithified coasts may adjust their configurations in response to the coastal wave climates but they do so at different rates, and in response to somewhat different mechanisms. Patterns of sedimentation may be relatively simple on straight rocky coasts, but coasts of structured rock may be so deeply embayed as to greatly complicate the pattern (crenulate rocky coasts).

Unconsolidated coasts are floored by easily eroded and flat-lying strata, and the surficial sediment tends to be both abundant in quantity and continuous in extent. On such coasts, wave-driven currents in the littoral zone and wind- or tide-driven currents farther offshore tend to build straight coastal segments of the sediment available to them. A basic second-order division in the morphology of unconsolidated coasts depends on the relative importance of straight coastal segments versus inlets that alternate with them.

TABLE 2.   A Coastal Taxonomy

Criterion	Coastal Type
Substrate indurated	Rocky coasts
Coast-parallel anisotropy	Straight rocky coasts
Coast-transverse anisotropy	Crenulate rocky coasts
Substrate unconsolidated	Unconsolidated coasts
Littoral drift dominant	Beach and barrier coasts
Low drift angle	Drift coasts (straight)
High drift angle	Swash coasts (cuspate)
Transverse drainage dominant	Inlet and riverine coasts
Fluvial drainage dominant	Deltaic coasts
Mild wave climate	Lobate deltas
Moderate wave climate	Arcuate deltas
Moderate-strong wave climate	Cuspate deltas
Strong wave climate	Recessed deltas
Tide modified	Estuarine deltas
Tidal drainage dominant	Estuarine coasts
Wave-modified tidal drainage	Inlet coasts

Inlets occur at river mouths, where they may be maintained by river and tidal flow or by purely tidal flow, where there is a tidal exchange between lagoons and the sea. A dense subaerial drainage net or a high tide range may cause inlets with their coast-normal flow to occupy over 50% of the shoreline, resulting in deltaic, estuarine, or tidal inlet coasts. Open coasts with rigorous wave climates and frequent strong wind-driven currents tend to have fewer inlets than coasts not so affected, resulting in mainland beach–barrier beach coasts.

This chapter has so far dealt mainly with the sedimentary regime of such simple, two-dimensional coasts. The succeeding sections examine in greater detail the modes of sand storage on beach–barrier island coasts, and also the modes of sediment storage on more complex coasts. The following chapter on shelf sedimentation stresses the role of varying coastal configurations in bypassing sediment to the continental shelf, and thus modulating the shelf sedimentary regime.

SAND STORAGE IN THE SHOREFACE

Storage in Low Retreating Shorefaces: Barrier Spits and Islands

BARRIER FORMATION. On most retreating coasts, the most important form of sand storage is within the shoreface itself, in the form of barrier spits and barrier islands. It would seem that along many coastal sectors, the coastal sedimentary regime rejects the primary shoreline formed by the intersection of the subaerial continental surface with the sea surface, and instead builds a secondary "barrier" shoreline seaward of the primary one. A characteristic of the equilibrium shoreface surface that as much as any mechanism is the basic "cause" of barrier islands and spits is its innate tendency toward two-dimensionality, its tendency to be defined by a series of nearly identical profiles in the downdrift direction. The equilibrium shoreface does not "want" a lateral boundary, since the wave and current field to which it responds does not generally have one. The initial conditions during a period of coastal sedimentation may, however, include such discontinuities, as in the case of a coast of appreciable relief (bay-headland coast) beginning transgression.

On such a coast shoreface surfaces will tend to be incised into the seaward margins of headlands exposed to oceanic waves, and will propagate by constructional means in the downdrift direction as long as material is available with which to build, and a foundation is available to build on. The basic mechanism is that described by May and Tanner (1973); see Fig. 14. Where the shoreline curves landward into a bay, the longshore component of littoral wave power decreases, and the alongshore gradient of sediment discharge $(\partial q/\partial x)$ is negative. The shoreface at that point must aggrade until the gradient approaches zero at that point, and the zone of negative gradient has moved downdrift. We give the lateral propagation of the shoreface into coastal voids the descriptive term "spit building by coastwise progradation" (Gilbert, 1890; Fisher, 1968).

However, the tendency of the shoreface to maintain lateral continuity also acts to prevent discontinuities as well as to seal them off after they have formed. In order to illustrate this, we may consider another set of initial conditions—a low coastal plain with wide, shallow valleys after a prolonged stillstand during which processes of coastal straightening by headland truncation and spit building have gone to completion. As this coastline submerges, the water, seeking its own level, will invade valleys more rapidly than headlands can be cut back. The oceanic shoreline, however, cannot follow, for if it should start to bulge into the flooding stream valleys, the bulge would become a zone of negative discharge gradient; hence the rate of sedimentation would increase to compensate for any incipient bulge. The shoreface would translate more nearly vertically than landward at this sector, until continuity along the coast was restored (Fig. 17E). Thus a straight or nearly straight oceanic shoreline must detach from an irregular inner shoreline, and be separated from it by a lagoon of varying width. This process of mainland beach detachment was first proposed by McGee (1890), and later described in detail by Hoyt (1967); see Fig. 20.

COASTWISE SPIT PROGRADATION VERSUS MAINLAND BEACH DETACHMENT. Much of the debate concerning origin of barriers deals with the relative importance of spit building versus mainland beach detachment (Fisher, 1968; Hoyt, 1967, 1970; Otvos, 1970a,b); see Chapter 12 (p. 223). The problem can be fully answered only by careful study of the field evidence, and as noted by several authors (Otvos, 1970a,b; Pierce and Colquhoun,

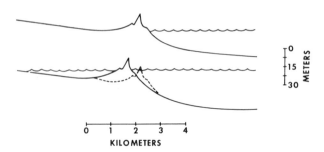

FIGURE 20. *Barrier island formation by mainland beach detachment. Modified from Hoyt (1967).*

1970) the evidence has frequently been destroyed by landward migration of the barriers. However, it is possible, in the time-honored deductive fashion of coastal morphologists, to consider the conditions most favorable to these two modes of barrier formation. Spits are certainly characteristic of coasts of high relief undergoing rapid transgression as described above [see the papers in Schwartz (1973)]. It seems probable that under such conditions mainly beach detachment would be severely inhibited. Even allowing for ideal initial conditions with a classic coast of old age (Fig. 21), where alluvial fans are flush with truncated headlands, detached mainland beaches would have a limited capability for survival. With significant relief, the submarine valley floors adjacent to retreating headlands must lie in increasingly deeper water after the onset of transgression. As the barrier grows into the bay, its submarine surface area must increase, and the capacity of littoral drift to nourish it may eventually be exceeded. As this point is approached, the combination of storm washover and shoreface erosion will cause the barrier to retreat until equilibrium is restored, a position which may be well inland from the tips of headlands. Both littoral wave power and sediment supply may be deficient in these inland positions, further jeopardizing the survival of

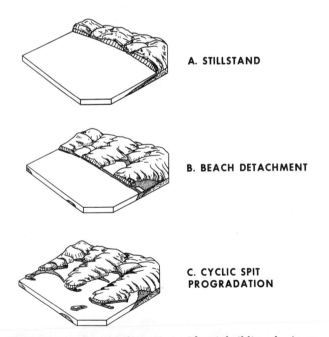

A. STILLSTAND

B. BEACH DETACHMENT

C. CYCLIC SPIT PROGRADATION

FIGURE 21. *Barrier formation with spit-building dominant. As a rugged coast passes from stillstand to transgression, a mature configuration is replaced by a transient state of mainland beach detachment, then by a quasi-steady state regime of cyclic spit building. This diagram also illustrates the relationship between the concepts of coastal equilibrium and coastal climax, since it consists of Johnson's (1919) stages of coastal maturity—portrayed in reverse sequence!*

the barrier. As the loop of the barrier into the bay becomes extreme, sediment supply from headlands is liable to capture by secondary spits formed during storms. These may prograde out toward the drowned valley thalweg until capacity is again exceeded and their tips are stabilized, further movement being limited to retreat coupled with that of the headland to which they are attached.

Finally the survival of primary barriers on such a coast would be limited by the tendency of submerging headlands to form islands. A spit tied to a promontory that becomes an island can retreat no further if a drowned tributary valley lies landward of it, but must instead be overstepped. The few unequivocal examples of transgressed barriers on the shelf floor appear to be overstepped, rock-tied spits (Nevesskii, 1969; McMaster and Garrison, 1967).

On the other hand, transgression of a coast of very subdued relief, such as is the case for most coastal plains, would tend to promote mainland beach detachment at the expense of spit formation, given initial conditions of a straight coast (Fig. 22). The depth of water in which detached bay mouth barriers would be built would be less, because the relief would be less. The upper, erosional zone of the shoreface (Fig. 10A) would be more likely to extend down into the pre-Recent substrate (Fig. 23A); hence erosion of the inner shelf floor would become as important a source of sand for the barrier as the erosion of adjacent headlands. With a rise in sea level, valley-front dune lines would grow upward. River mouths, initially deltaic, would flood as estuaries, while lagoons would creep behind the beaches toward the headlands on either side. Barriers would retreat in cyclic, tank-trend fashion by means of storm washover, burial, and reemergence of the buried sand at the shoreface (Fig. 10A). Coastal discontinuities sufficient to induce coastwise spit progradation would occur only locally. Thus, on a low, initially straight coast, barrier spits and barrier islands would preferentially form by mainland beach detachment rather than by coastwise progradation.

Storage in Prograding Shorefaces

The preceding discussion has identified barrier islands and spits as forms of sand storage on retreating coastlines. On prograding coastlines, sand storage occurs in beach ridges and cheniers; the two forms differ in that beach ridges are separated by sand flats, whereas cheniers are separated by, and rest on, mud deposits.

Sequences of beach ridges 15 to 200 m apart may form subaerial strand plains tens of kilometers wide. These are smaller scale features than the barriers, which

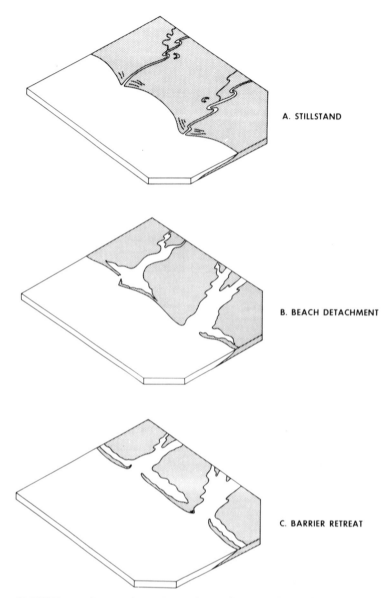

A. STILLSTAND

B. BEACH DETACHMENT

C. BARRIER RETREAT

FIGURE 22. *Barrier formation with mainland beach detachment as the dominant process. A mature low coast passes via main land beach detachment into a steady state regime of barrier retreat.*

characterize retreating and stillstand coasts; and barriers may, in fact, be locally comprised of beach ridge fields, as a consequence of minor frontal progradation or more extensive distal, coast-parallel migration (Hoyt and Henry, 1967). Curray et al. (1969) have presented a detailed study of what has been recognized as a classic strand plain coast, the Costa de Nayarit (Fig. 24). They postulate that each ridge forms as a plunge point bar, which in the presence of an oversupply of littoral sand, builds up close to mean low water. During a period of constant low swells, the bar may grow above this level as tides rise to the spring tide value (0.98–1.25 m);

the bar becomes a subaerial feature during the subsequent neap phase, and continues to grow by eolian activity (Fig. 10B).

Chenier plains form on coasts with a high suspended sediment input. In the classic chenier plain of the Louisiana coast west of the Mississippi Delta, the sand ridges support stands of live oaks (French, chêne), hence the name (Price, 1955). The formation of chenier plains has been ascribed to rapid progradation of mud flats during periods of high suspended sediment discharge from nearby rivers or delta distributaries. When distributaries crevasse and the subdeltas are abandoned, the

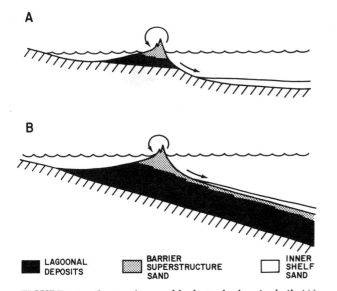

A

B

LAGOONAL DEPOSITS

BARRIER SUPERSTRUCTURE SAND

INNER SHELF SAND

FIGURE 23. *Contrasting sand budgets of a barrier built (A) on a gentle submarine gradient as in Fig. 22, and (B) on a steep submarine gradient as in Fig. 21. In (A), zone of shoreface erosion penetrates to Pre-Recent substrate, which becomes "income" for barrier nourishment. In (B) the barrier may only "borrow" from its own "capital" through shoreface erosion, and the heavy expenditure involved in paving the shelf with sand during barrier retreat may "bankrupt" the retreating barrier, which must either accelerate its retreat or be overstepped. In either case shoreface continuity is liable to be broken, resulting in cyclic spit building.*

downdrift coast becomes sediment starved. The mud flat erodes back and a beach ridge formed of the coarse sediment is thrown up by storm high tide. Todd (1968) has stressed the role of estuary mouths and inlets in the localization of chenier plains. He notes that littoral currents updrift of the inlet tend to decelerate during ebb tide because of a reduction of the coastwise pressure gradient by the ebb jet, resulting in sediment deposition. Littoral currents in the same locality are accelerated by proximity of the inlet during flood tide, but fine sediment deposited requires a greater velocity for erosion than for deposition, and in any case has already compacted. Hence chenier plains tend to be localized on the muddy, updrift sides of tidal inlets. Downdrift of the inlet, the coast may instead be starved for fines as a result of seaward transport or "dynamic diversion" of the littoral current by the ebb jet, and the littoral sand deposits have more nearly the character of a beach ridge sequence.

Otvos (1969) recognizes the role of inlets in localizing deposition, but notes that chenier deposition goes on for long distances beyond inlets. He cites new chronologic evidence from the Mississippi chenier plains to indicate that chenier ridge formation cannot be closely correlated with the abandonment of a subdelta mouth, and suggests that the intermittent shielding effect of nearby sub-

delta growth on the wave climate plays a greater role in cyclic chenier plain growth.

Beall (1968) has examined in detail sediment distribution and stratigraphy in the present shoreface of the western Louisiana shoreline (Fig. 25). He distinguishes between three main stratigraphic patterns. *Mud flats* are defined on their seaward margins by a break point bar zone of very fine sand. Midtidal and upper tidal flats are distinguished by progressively finer sand and increasing percentages of silt and clay. A thin sand storm beach may rest on eroded marsh sediments. The stratigraphy is complex. Apparently a period of increasing littoral sediment discharge results in progressive flattening of a shoreface, until the bar zone is triggered and becomes the maximum locus of sedimentation, prograding both landward and shoreward. A (submarine) mud flat zone is thereby initiated in the sheltered longshore trough, and prograbes toward the bar and landward.

Transitional beaches have largely erosional profiles, with thin bar, beach, and washover sands overlying the erosional surface near the high-water line. The sequence is typical of that of erosional transgression, where the thin sand cap is a transient fair-weather veneer. However, Beall interprets these transitional beaches as progradational, with rates of progradation intermediate between those of mud flats and those of "normal beaches."

Normal beaches consist of up to 1.7 m of seaward-fining fine sand, prograding seaward over an outer shoreface facies. Washover fans of normal beaches are thicker than those of transitional beaches. The three types of beaches described by Beall would appear to illustrate a temporal as well as a spatial sequence. Periods of rapid mud flat progradation are presumably followed by erosion, then the formation of transitional beaches, which prograde to become cheniers, then prograde more rapidly as mud flats.

SAND STORAGE OFF CAPES

South of the Middle Atlantic Bight of North America, the generally southwest-trending coastline has been molded into a series of large-scale cuspate forelands (Fig. 26). They are the response of the shoreface regime to a moderate to intense wave climate and a high variance in the direction of wave approach (Swift and Sears, 1974). Storm waves approach from the northeast, as is the case in the Middle Atlantic Bight, but the coast is also exposed to waves from more distant storms in the southeastern Atlantic. As a result, the cuspate forelands have been self-maintaining features throughout the postglacial period of sea-level rise and erosional shoreface retreat. Each foreland apex is a zone of littoral drift convergence.

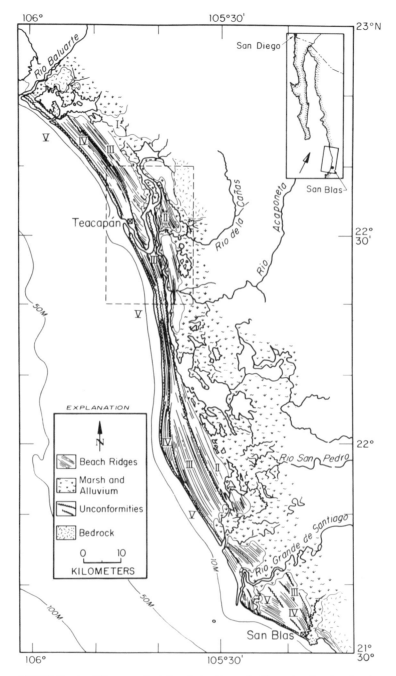

FIGURE 24. *The strand plain of the Costa de Nayarit, showing beach ridge sequences. Rio de la Cañas meanders through interlocking spits, indicating reversal of drift directions. From Curray et al. (1969).*

The resulting surplus of sand at the apex creates a coastal shoal. The shoal in turn maintains a pattern of wave refraction that drives littoral drift convergence (Swift and Sears, 1974); see Fig. 27.

The question arises as to how such a closely coupled feedback system begins. The answer is that in a sense, it does not matter. It is a truism that as process variables approach the instability threshold, any singularity in a water-substrate system will excite the feedback of the process and response that lends to the formation of bed forms. In the case of the cuspate Carolina coast, the initial conditions were probably the sequence of shelf-edge deltas during the Late Wisconsinan low stand, corresponding to the Peedee, Cape Fear, Neuse, and Pam-

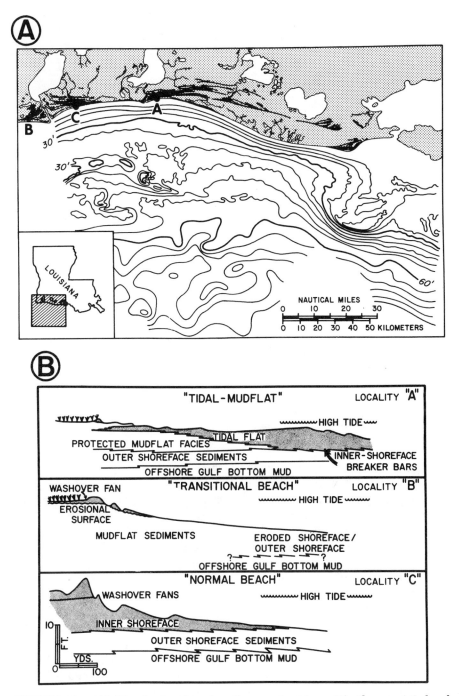

FIGURE 25. (A) *The chenier plan of southeastern Louisiana.* (B) *Characteristic beach configurations. See text for explanation. From Beall (1968).*

lico rivers (note river system of Fig. 26 and compare Fig. 27). The Appalachicola cuspate foreland of the Florida Panhandle is particularly suspect as having been formed by this mechanism (Swift, 1973). Other cape-associated shoals may occur as a consequence of the reduction in intensity of littoral drift around a rock-defended cape with consequent reduction in the compe-

tence of littoral drift (Tanner, 1961; Tanner et al., 1963). On offset coasts (forelands separated by zetaform bays; p. 275), the forelands may be triggered by cape extension shoals, either river mouths or rocky promontories (Davies, 1958). Forelands and cape-associated shoals also occur on swash-aligned coasts. These coasts tend to be inherently unstable, breaking into short "arcs

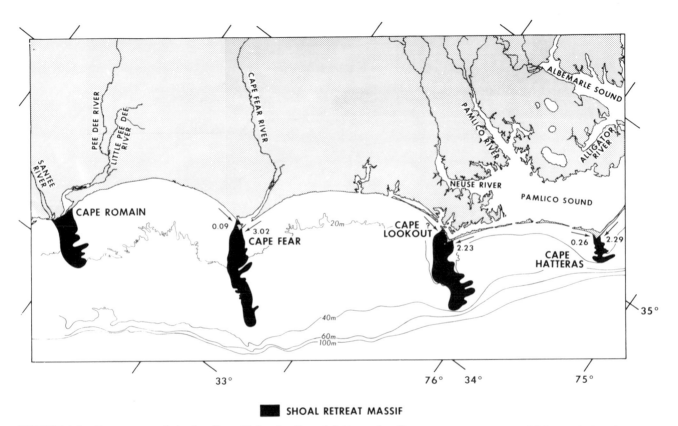

SHOAL RETREAT MASSIF

FIGURE 26. *Cuspate coast of the Carolinas. Values for littoral drift are in yd/year* $\times 10^{-3}$. *From Langfelder et al.* (1968).

of equilibrium," terminating in cuspate forelands with neither rivers nor outcrops required for cusp formation. Tidal inlets may evolve into cuspate forelands with associated shoals. Chincoteague Shoals, on the Delmarva coast, is an example of a barrier-overlap inlet that has become a cuspate spit with associated shoal (Fig. 36). Shoals developing over cuspate forelands may extend seaward the width of the shelf. Such cape extension shoals do not result from the seaward transport of sand, but rather from the landward translation of the cuspate foreland in response to rising postglacial sea level, together with the retreat of the associated littoral drift convergence. The seaward-trending shoal marks the retreat path of this convergence. Its response to the shelf regime is discussed in the next chapter.

SAND STORAGE IN INNER SHELF RIDGE FIELDS

Storm-Induced, Shoreface-Connected Ridges

A major category of inner shelf sand storage found on low retreating coasts is storage in shoreface-connected ridges (Fig. 28) and in associated inner shelf ridge fields. These features are up to 10 m high, 2 to 5 km apart, and

their crestlines may extend for tens of kilometers. Side slopes are rarely more than a degree. They typically converge with the shoreface at angles of 25 to 35° (Duane et al., 1972) and may merge with it at depths as shoal as 3 m. The best known development is on the coast of the Middle Atlantic Bight of North America, but they may be found on coastal charts as far south on the Atlantic coast as Florida, and around the Gulf coast littoral as far as Alabama. They also appear locally on the Texas coast. Allersma (1972) has reported them on the muddy coast of Venezuela, where they are dominantly composed of mud. They have been detected by ERTS satellite imagery on the Mozambique coast (John McHone, personal communication), and also appear on the southern littoral of the North Sea. With the exception of the Venezuelan coast, most settings are that of a low, unconsolidated coast undergoing Bruun erosional retreat (Fig. 104) in response to a moderate to strong wave climate and periodic intense storm or tidal flows.

Where best studied, on the Virginia–northern North Carolina coast (McHone, 1972), the ridges appear to have some of the response characteristics of wave-built bars at their inner ends where they merge with the shoreface. Like wave-built bars, their landward ends are asymmetrical, with steep landward flanks, although the

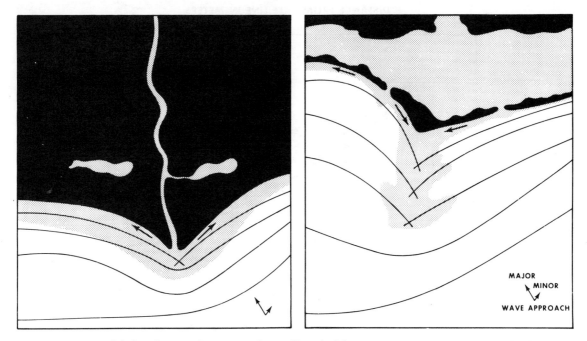

FIGURE 27. *Model for the transformation of a stillstand delta into a retreating cuspate foreland. From Swift and Sears (1974).*

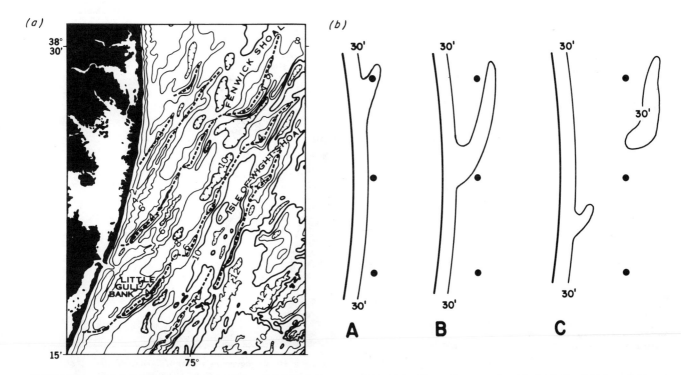

FIGURE 28. (a) *Shoreface-connected ridges of the Delmarva inner shelf, contoured at 2 fathom intervals. From ESSA bathymetric map 0807N-57. Ridges are in varying stages of detachment.* (b) *Schematic diagram of detachment sequence as inferred* from (a). *Dots represent hypothetical fixed points during a period of shoreface retreat and downcoast ridge migration. See text for explanation. From Swift et al. (1974).*

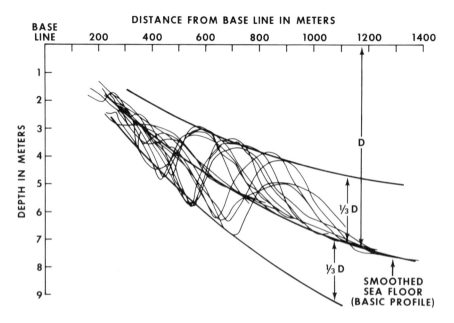

DISTANCE FROM BASE LINE IN METERS

FIGURE 29. *Superimposed profiles of the inner shoreface-connected ridge at False Cape, Virginia. From McHone (1972).*

reverse asymmetry tends to prevail further seaward. Envelopes of profiles indicate that, as in the case of their small-scale break-point counterparts, ridges built to a height of approximately one-third water depth, at which point wave agitation is sufficiently intense to preclude further growth. The troughs between the ridges and the shoreface are similarly excavated to one-third of water depth below the smoothed profile (McHone, 1972); see Fig. 29. At the False Cape Ridge Field, Virginia (McHone, 1972; Swift et al., in press), analysis of the wave climate suggests that waves are capable of breaking on some part of the inner ridge crest about 10% of the time. As a consequence of their oblique orientation and varying crestal depth, such ridges may utilize energy from a relatively broad spectrum of wavelengths.

As wave-built bars, however, the low-angle ridges are anomalous. They are much larger than surf zone bars and their oblique orientation is more nearly parallel to the direction of wave approach than normal to it. The ridges may be primarily a response to a downwelling coastal jet that comprises the coastal margin of the storm flow field (see p. 275), although storm wave action is clearly a complementary mechanism. At False Cape, Virginia, a 28 hour current-meter station revealed a steady southward and offshore flow on the order of 15 cm/sec at a distance of 8 cm off the bottom, subsequent to the passage of a cold front with winds in excess of 25 knots (Fig. 30). During this period, however, the anchored observation vessel maintained a wake trending southward and shoreward. The inferred structure of the coastal flow field during the observation period is pre-

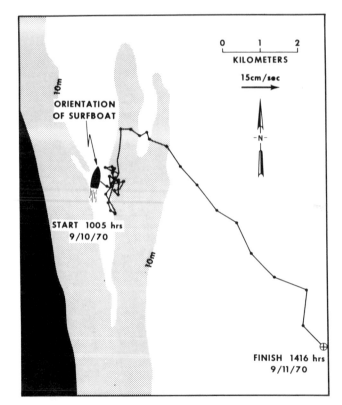

FIGURE 30. *Progressive vector diagram of storm bottom flow at the innermost ridge at False Cape, Virginia. Vectors represent velocities taken for 3 minute intervals every 30 minutes by two orthogonal Bendix Q-18 meters mounted in a plane parallel to the seafloor, 16 cm off the bottom. After passage of a cold front, bottom flow trended southeast obliquely seaward over ridge crest at velocities up to 18 cm/sec, while wake of anchored observation vessel streamed southeast, toward shore. Based on Holliday (1971).*

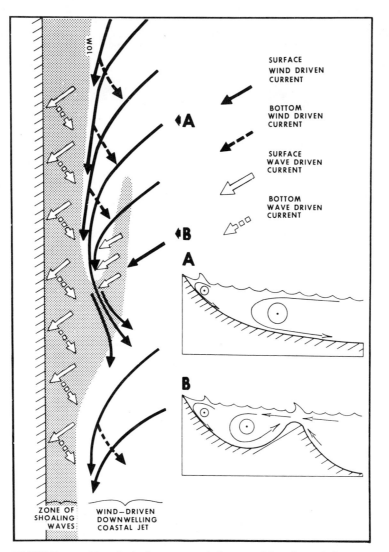

FIGURE 31. *Hypothetical structure of the coastal boundary of the storm flow field, based on Figs. 2 and 30.*

sented in Fig. 31. The observed pattern is interpreted as downwelling coastal flow intensified by constriction of the trough toward its southern end, and also by the setup of waves breaking on the inshore end of the trough. Mapping of the junction of this ridge with the shoreface on four successive occasions has revealed the presence of a shifting saddle, where storm flows presumably break out over the ridge base (McHone, 1972).

A grain-size profile over the ridge is extremely asymmetric (Fig. 32). Sands are coarsest in the landward trough and become steadily finer up the landward flank, are of relatively constant grain size across the crest, and become finer again down the seaward flank. Sorting is variable on the landward flank and crest but increases steadily down the seaward flank.

The profile is characteristic of a flow-transverse sand wave, and suggests that the ridges are responding as would a sand wave to the cross-shoal component of flow. As described in Chapter 10 (p. 166), bed shear stress increases up the upcurrent flank of a sand wave, attaining a maximum at the crest or just forward of it, then decreases down the downcurrent flank. Grain size would tend to decrease monotonically across such a shear stress maximum as a consequence of the progressive sorting mechanism; as sand is eroded out of the trough, the coarser grains are more likely to be trapped out in the initial portion of the transport path (Chapter 10, p. 162). On this particular ridge, however, size characteristics do undergo a reversal on the landward side of the crest, where maximum shear stress is to be expected.

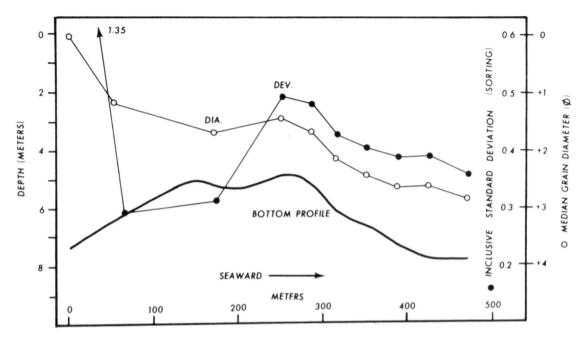

FIGURE 32. *Grain-size profile across the inner ridge at False Cape, Virginia (by Leonard Nero).*

Shoreface ridges are not simply giant sand waves, however, because flow crosses them at an oblique angle. The model presented in Fig. 31 suggests that they partake of the characteristics of both flow-transverse sand waves (Chapter 10, p. 166) and flow-parallel sand ridges (Chapter 10, p. 172).

Thus the ridges appear to represent the attempt of an intensified coastal flow to build an outer bank with materials scoured from the high-velocity axis of downwelling. The topography presented in Fig. 30 presents a clue to the historical development. The initial perturbation required to trigger these large-scale instabilities of the shoreface might be something on the order of the migrating "sand humps" of the surf zone described by Bruun (1954) and Dolan (1971). As the ridge and trough take form in response to the interaction of storm waves with the coastal storm flow, they would tend to be self-propagating. Enlargement of the trough by headward erosion and downward scour of the trough, and aggradation of the ridge crest by peak flow events would create a morphology that would amplify successive trough flows. Thus during a period of general shoreface retreat due to storm erosion, ridges would be carved out of the shoreface partly because the shoreface would retreat away from them, and partly because they themselves would tend to migrate obliquely offshore, extending their crestlines to maintain contact with the shoreface as they do so (Fig. 28). The sinuous pattern of crestlines on the inner shelf floor of the Delmarva peninsula suggests that ridge formation may be an episodic affair; troughs en-

large and trough flows amplify until the flow is intense enough to cut through the ridge base, whereupon the process is repeated farther down the shoreline. Shoreface-connected ridges are seen in all stages of detachment in Fig. 28. It is doubtful, however, if this is a full or adequate explanation of ridge genesis even in a qualitative sense. Any more comprehensive analysis must undertake to explain the relative orientations of peak bottom flow, the ridge crest, and the shoreline. As noted, neither the breakpoint bar nor the sand wave models meet this requirement.

The ridges are generally oriented along a trend that is intermediate between the dominant direction of storm wave approach and the coast-parallel trend of storm currents (narrow-angle ridges), perhaps as a consequence of the dual role of these elements in ridge genesis. Locally, however, they may be aligned along the direction of storm wave approach (wide-angle ridges: Duane et al., 1972); see Fig. 33. Such ridges would resemble the "finger bars" of Niedoroda and Tanner (1970), rather than break-point bars. While breaking waves tend to drive sand across break-point bars, the bottom surge of refracting waves tends to drive sand obliquely crestward and landward, up both sides of a finger bar. Wide-angle ridges exhibit the same textural and morphologic asymmetry as do narrow-angle ridges, indicating that they too are shaped by storm flow as well as by wave surge. But they presumably react to storm flow more nearly as a flow-transverse bed form than as a flow-parallel bed form, as in the case of narrow-angle

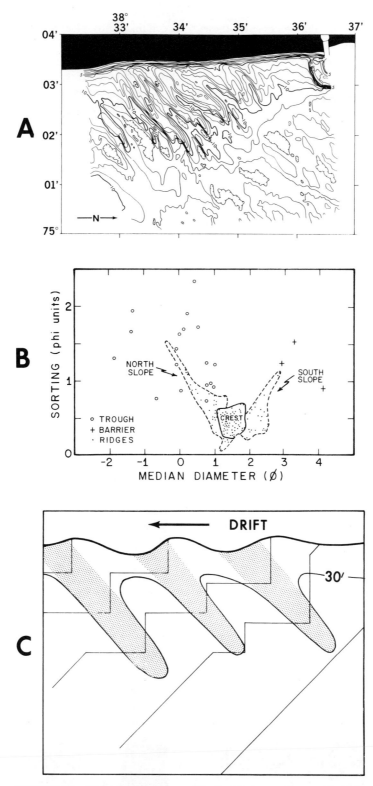

FIGURE 33. *Substrate response and hydraulic process for a wide-angle, shoreface-connected ridge system, Bethany Beach, Delaware. (A) Bathymetry. From Moody (1964). (B) Median diameter versus inclusive graphic standard deviation for sand samples. From Moody (1964). (C) Schematic model for the generation and maintenance of wide-angle ridges. Refracted waves converge toward ridge crests. Breaker angle is most intense at heads of trough. Storm flow is coast parallel and to the south.*

ridges. A full understanding of the genesis of these coastal sand bodies must await more detailed field measurements of the responsible flows.

Storm-Induced Inner Shelf Ridge Fields

Ridge fields on the inner shelf floor continue their morphologic identity and their characteristic pattern of grain-size distribution (Fig. 34). Relief continues at 10 m, and slopes for isolated inner shelf ridges are very similar to those of shoreface-connected ridges. Scour continues in troughs; the erosional surface cut by shoreface translation extends beneath the ridges and is locally exposed in trough axes (Swift et al., 1972a,b). Generally, however, it is veneered with a few decimeters of coarse, pebbly sand, overlaid by finer sand. The coarser sand is commonly exposed in elongate windows through the finer sand veneer. Sidescan sonar records suggest that the finer sand is moving as ribbonlike streamers over a coarser lag substrate. Ridge crests consist of medium to fine, well-sorted sand, with cross-stratified horizons (Swift et al., 1972a; Stubblefield et al., 1975). Flanks consist of fine to very fine sand and are distinctly asymmetrical in their textural pattern; seaward flanks are notably finer, and are locally steeper than landward flanks. Crestal sands, however, may be distinctly coarser than

the flank sands of either side, probably a response to winnowing by wave surge.

The inner shelf ridges themselves appear to be in a state of slow transit, wherever there is a bathymetric time series adequate to test this hypothesis (Figs. 35 and 36). The pattern of movement is a fairly consistent one, in which both shoreface-connected and isolated inner shelf ridges move along similar trajectories. Where the angle of convergence of the ridge crest with the shoreline is fairly large, the ridges are moving downcoast and offshore, extending their crestlines so as to maintain contact with the shoreface as they do so. Where the ridges are nearly coast-parallel, they are extending these crestlines downcoast, and may move either inshore or offshore, but more commonly offshore.

The considerations just discussed strongly suggest that inner shelf ridges continue to interact with the shelf flow field after detachment, in such a way as to maintain their morphologic and textural characteristics. In fact, ridged inner shelf topography occurs on sectors of the North American inner shelf where it cannot have

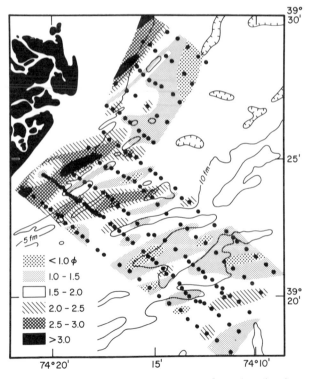

FIGURE 34. *Grain size distribution in the Brigantine inner shelf ridge field, New Jersey. Data of M. Dicken.*

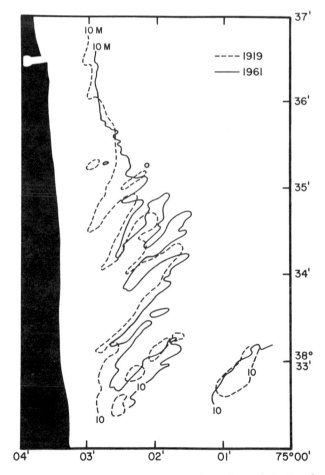

FIGURE 35. *Bathymetric time series from the Bethany Beach ridge field, Delaware, between 1919 and 1961. From Moody (1964).*

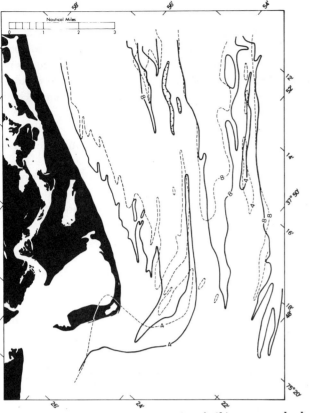

FIGURE 36. *Bathymetric time series of Chincoteague shoals, Delmarva (Delaware–Maryland–Virginia) coastal compartment. Ridges have migrated slightly offshore, and have extended their crestlines markedly to the south between the Coast and Geodetic surveys of 1881 (dashed line) and 1934 (solid line). From Duane et al. (1972).*

been formed by shoreface ridge detachment (Swift et al., 1974); see Chapter 15. It appears that the shelf hydraulic regime will adopt ridges from the retreating shoreline or mold them afresh in the substrate if the hydraulic regime is conducive to a ridged substrate. A possible mechanism for ridge maintenance involving helical flow structure in the storm flow field is presented in Chapter 10 (p. 173).

There is in addition a smaller scale bed form pattern on the inner shelf whose patterns of distribution are compatible with this hypothesis. These are the sand ribbons on the inner shelf of the Middle Atlantic Bight of North America, revealed by sidescan sonar. They tend to be 5 to 50 m wide, are of negligible relief, and tend to make angles of 10 to 45° with the shore. They are most commonly observed as dark streaks on sidescan sonar, which means that they are not true sand ribbons (streamers of finer sand over an immobile substrate of coarser sand or gravel) but are instead erosional windows in which a coarser substrate is locally exposed through a discontinuous sand sheet (Chapter 10, p. 170). Locally, however, the pattern anastomoses so as to create a true sand ribbon pattern (Fig. 37). The dark streaks in many areas are distinctly asymmetrical, with sharper landward boundaries (Chapter 10, Fig. 17). The streaky patterns occur on the smooth inner shelf or in ridge fields. In the latter case they are largely confined to troughs, where they tend either to parallel the trough axes or make a somewhat larger angle with the shoreline.

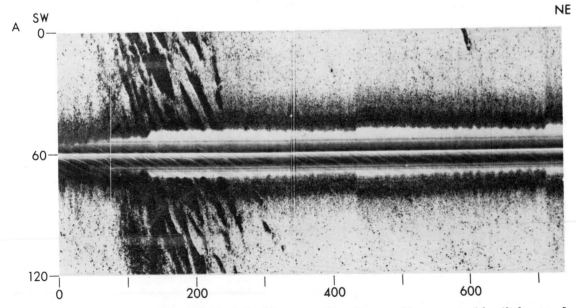

FIGURE 37. *Sidescan sonar record of sand ribbon pattern in the trough of a central shelf ridge. From McKinney et al. (1974). Dark band is a window of coarse trough sand traversed by streamers of fine (lighter toned) sand.*

As noted in Chapter 10, the larger ridges are probably responses to repeated flow events, whereas the smaller sand ribbons and erosional windows may be formed during a single flow event. The similar orientations and asymmetries of the sand ribbons and ridges suggest that both may be responses to a geostrophic flow regime in which secondary flow cells occur at several spatial scales. With fully developed secondary flow, the tendency for regional landward transport of surface water and regional seaward transport of bottom water (solid arrows, Fig. 38) would be suppressed in order to maintain continuity. Instead, cells rotating with the sense of regional shore-normal flow component would be enhanced, and cells rotating with the opposite sense would be suppressed. Detailed measurements of the velocity field in the vicinity of such offshore ridges during peak flow events would serve to test this model, and perhaps lead to alternative models.

Tide-Induced Inner Shelf Ridge Fields

Tide-dominated coasts such as the Anglian coast of England also tend to store sand in shoreface-connected and inner shelf ridges. The forcing mechanism for ridge formation must be in part the storm-augmented shelf flow field as in the case of such storm-dominated coasts as the Middle Atlantic Bight of North America, since the Anglian coast is also subjected to severe storms (Valentin, 1954). However, this coast experiences in addition the progressive tidal edge wave associated with the amphidromic tidal system of the North Sea (see Chapter 5, p. 60) on a twice daily basis; midtide coastal tidal velocities regularly exceed 2 knots.

As a consequence of the greater rate of energy expenditure in tidal flow than in wave- and wind-current-generated flow, storage of sand in the subaerial zone as barrier superstructure, or in the surf zone as a beach and surf prism, is greatly inhibited at the expense of submarine storage in tide-maintained sand bodies. The efficiency of this storage is greatly strengthened by the tendency of the coastal tidal wave to interact with a loose substrate by the formation of interdigitating ebb and flood channels separated by sand shoals which form effective sand traps (Robinson, 1966); see the discussion on page 177. As in the case of the Middle Atlantic

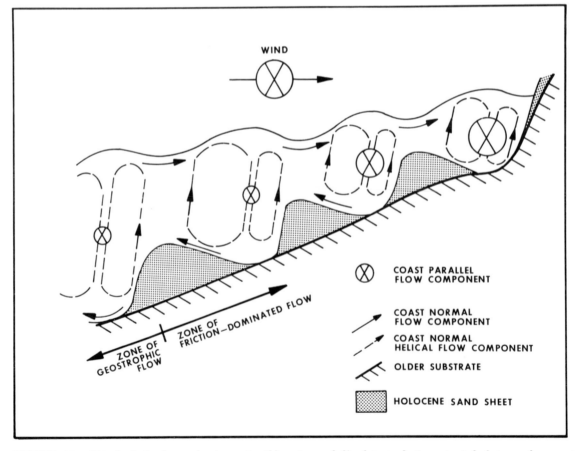

FIGURE 38. *Hypothetical scheme showing a possible mode of coupling between Ekman flow cells and a mobile inner shelf substrate during a period of strong downcoast winds. See text for explanation.*

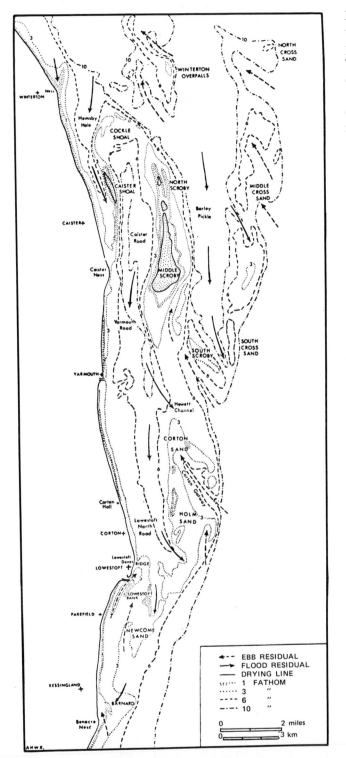

Bight, the nearshore zone of sand storage is not a stationary one but is translating landward in response to postglacial sea-level rise and erosional shoreface retreat. The primary element of storage is again a shoreface-connected sand ridge (Fig. 39). The angle between these ridges and the coast opens northward, into the direction of the advancing coastal tidal wave, and therefore the trough opens into the flood-dominated residual tidal flow. Downwelling may also occur during the flood tide, since the high velocity axis of trough flow will tend to converge with the rising trough axis. The outside of the ridge is shielded from the flood tide, and therefore experiences a greater ebb discharge. This diversion of flow might be expected to result in a sand circulation cell, with sand moving obliquely up the inner flank and over the crest during the flood tide, to be returned along the seaward flank during the ebb tide. As in the case of storm-maintained shoreface-connected ridges, probably any initial perturbation of the shoreface would result in such a self-maintaining system.

As in the case of the Middle Atlantic Bight, the ridges tend to migrate offshore and downcoast, in the direction of the residual tidal flow (Robinson, 1966), and tend to become detached and isolated on the inner shelf floor. However, unlike storm-maintained ridges, tidal ridges on the inner shelf floor tend to be unstable. Variations in the rate of offshore migration along the length of the ridges tend to result in self-propagating modifications of ridge morphology (Caston, 1972; Chapter 10, Fig. 21), whereby the ridge deforms into a sigmoidal pattern, because of the growth of secondary ebb- and flood-dominated channels, and may eventually split into three ridges.

As in the case of the Middle Atlantic Bight, isolated inner shelf ridges continue to be maintained, but the character of interaction between the flow field and substrate changes as the water column deepens. Ridge spacing increases as a function of flow depth (Allen, 1968b) as sand is partitioned between fewer, wider ridges. As channels widen, they cease to become wholly ebb- or flood-dominated, but are themselves partitioned into ebb-dominant and flood-dominant sides (Caston and Stride, 1970); see Chapter 10, Fig. 20). All channels have the same sense of shear. Thus the offshore ridges are sand circulation cells, but the sense of circulation is the same from ridge to ridge, instead of alternating between clockwise and counterclockwise as on the shoreface.

FIGURE 39. *Tide-maintained ridge topography on the inner Anglian shelf. Shoreface-connected ridges separate ebb- and flood-dominated channels. Ridges tend to migrate southward with time, and to detach from retreating shoreface. Ridges are nourished at the expense of shoreface, hence constitute cases of downdrift by-* *passing. Offshore ridges are probably being nourished at expense of nearshore ridges; if so, sand is moving seaward more rapidly than are the ridge forms. From Robinson (1966).*

SAND STORAGE AT COASTAL INLETS

Categories of Coastal Inlets

In addition to the coastal flow discontinuities found at capes and cuspate forelands and on ridged shorefaces, discontinuities also occur at coastal inlets, which likewise result in sand storage systems. The term coastal inlet is here used in its broadest sense, for a variety of coastal reentrants, defined by the ratio of salt- to freshwater discharge in their two-layer circulation systems, and in the extent to which their channel cross sections have equilibrated with the discharge (Table 3).

TABLE 3.   Categories of Coastal Inlet

	Constructional form (equilibrium channel) $\longrightarrow$	Erosional form (inherited river valley)
Saline/freshwater discharge ratio	Delta distributary	
	Tidal delta distributary or equilibrium estuary	
		Disequilibrium estuary
		Tidal channel–mud flat complex
	Tidal inlet	Bay

The main sequence of coastal inlet morphologies trends diagonally across Table 3, from delta distributaries entering tideless seas, through tidal distributary mouths and trumpet-shaped equilibrium estuaries to funnel-shaped disequilibrium estuaries, to tidal channel–tidal flat complexes and open bays. Tidal inlets are hybrid cases, in which an equilibrium channel has been fitted to a lagoon. Special effects such as mirror-image ebb, flood "tidal deltas," and offset barrier coasts result. Tidal inlet morphologies are continuous with equilibrium estuary morphologies and grade into them through intermediate cases in which a central channel meanders through a marsh-filled lagoon.

Equilibrium River Mouths

Delta distributaries (prograding river mouths) and equilibrium estuaries (retrograding river mouths) belong to a general class of river mouths. The cross-sectional area of river channels is a power function of river discharge (Leopold et al., 1964, p. 215). Where rivers enter the sea, a salt wedge intrudes beneath the fluvial jet, whose discharge is amplified by a two-layer (estuarine) circu-

lation (see p. 24). Most rivers enter tidal seas, and river mouth discharge is further amplified by the discharge associated with the semidiurnal tidal cycle, which propagates for some distance upstream. Thus a river mouth whose channel is in equilibrium with total discharge must expand rapidly through the tide-influenced zone toward the sea, resulting in a trumpet-shaped plan configuration.

At the river mouth proper, a variety of processes conspire to construct an arcuate, seaward-convex sand shoal (Fig. 40). The most fundamental factor is the hydrodynamic continuity relationship: Expansion of the fluvial jet results in rapid deceleration and loss of competence, and river sand is deposited in the form of a shoal. Estuarine circulation also plays a role; river mouth morphology and the circulation interact, so that the crest of the shoal becomes the leading edge of the salt wedge during flood stage (Fig. 41), or, if the tidal component of river mouth discharge is very large, the spring ebb tide terminus of the salt wedge, or both (Wright and Coleman, 1974; Moore, 1970; Farmer, 1971). The crest of the shoal thus becomes a bottom current convergence during periods of maximum sediment transport, and hence a reservoir for sand storage. A second major source of sand maintaining the river mouth shoal is littoral drift, which is diverted seaward along the shoal crest.

Sediment storage in tidal river mouths is mediated by the behavior of the tidal wave as it passes over the shoal crest. Here, as on open tidal coasts, tidal wave and substrate tend to interact to form interdigitating ebb- and flood-dominated channels separated by shoals that are efficient sand traps. The tide within the estuary is retarded by friction and is out of phase with the shelf tide; it continues to ebb after the shelf tide has already begun to flood. The two water masses tend to interpenetrate, with the main ebb jet passing out over the center of the shoal and the oceanic tide flooding on either side of it. The response of the shoal surface to this periodic flow pattern is an interdigitation of ebb- and flood-dominated channels, separated by a discontinuous, zigzag system of sandbanks (Ludwick, in press); see the discussion on page 177.

A further process modifying the surface of the shoal and enhancing its capacity for sand storage is the interaction between the tide-generated pattern of channels and sand ridges and incident wave patterns. The arcuate pattern of the shoal as a whole serves to focus wave energy on it. Sand ridges between ebb and flood channels tend to build toward the level of mean high tide. As their upper surfaces emerge into the intertidal zone they become swash platforms, on which intersecting patterns of wave trains tend to drive sand in the resultant, landward direction (Oertel, 1972); see Fig. 40.

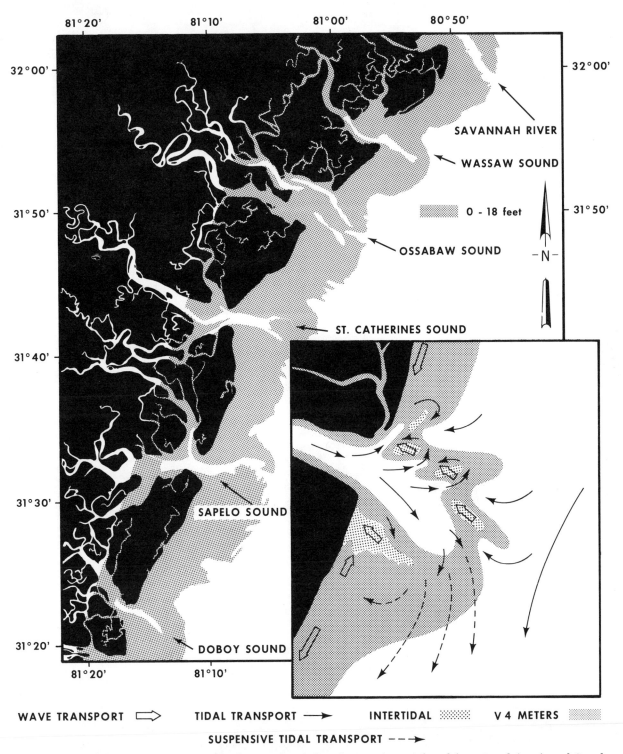

FIGURE 40. *Sedimentation patterns at the mouths of Georgia estuaries as inferred from Oertel (1972) and Oertel and Howard (1972).*

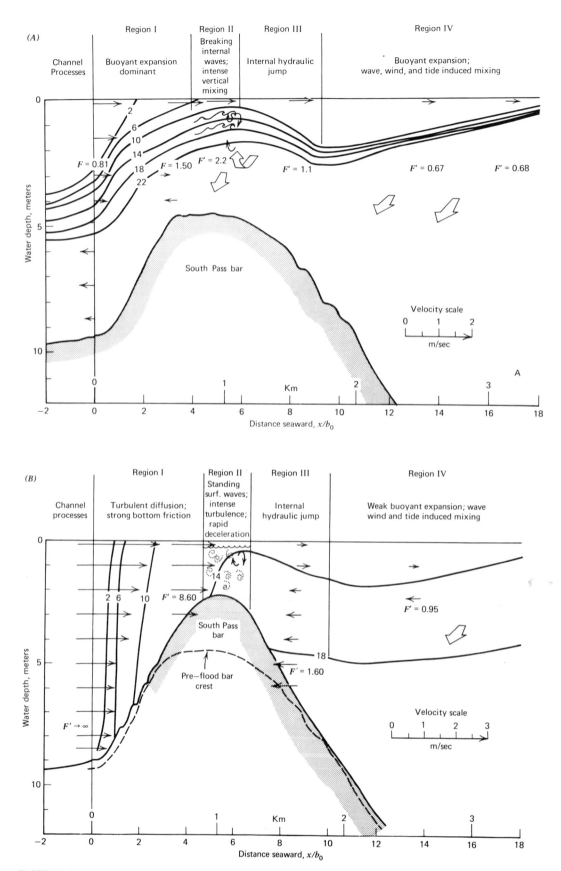

FIGURE 41. (A) *Cross section of density and flow structure of the South Pass of the Mississippi River during low river stage (October 25, 1969) and* (B) *during flood stage (April 5, 1973). Both sections taken during flood tide. From Wright and Coleman (1974).*

Despite the relatively rapid postglacial rise in sea level, some river mouths have been able to maintain equilibrium channels in which cross-sectional area is adjusted to discharge, as deltas (prograding river mouths) or as equilibrium estuaries (slowly retrograding river mouths; see Figs. 42*A,B*). Most, however, have not. Disequilibrium estuaries have resulted whenever aggradation of the estuary floor in millimeters per year has been less than the rate of sea-level rise, so that before any given segment of channel could close down to the required cross-sectional area, the main shoreline had passed it by.

Such "drowned" or disequilibrium estuaries are generally nearly funnel-shaped, rather than trumpet-shaped, as are the equilibrium forms. As a consequence of their higher ratio of saltwater to fluvial discharge, their river mouth shoals are retracted into the throat of the estuary and the interpenetration of ebb and flood channels becomes marked (Fig. 42*C*).

With a yet further increase of tidal over fluvial discharge such a coastal indentation may no longer be appropriately called an estuary, but simply a bay. Large bays experiencing high tidal ranges may build a tide

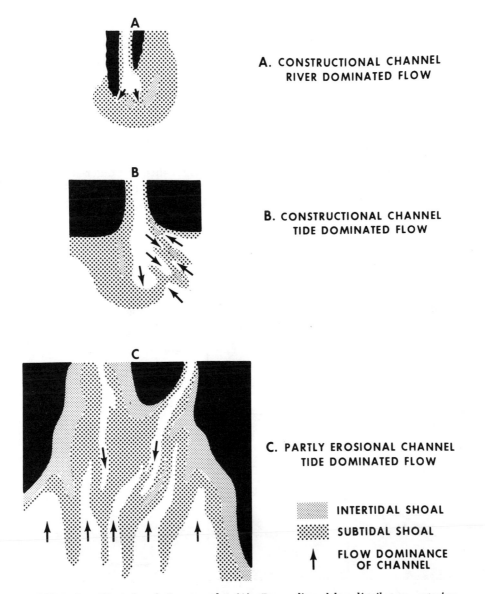

A. **CONSTRUCTIONAL CHANNEL RIVER DOMINATED FLOW**

B. **CONSTRUCTIONAL CHANNEL TIDE DOMINATED FLOW**

C. **PARTLY EROSIONAL CHANNEL TIDE DOMINATED FLOW**

INTERTIDAL SHOAL
SUBTIDAL SHOAL
↑ FLOW DOMINANCE OF CHANNEL

FIGURE 42. *Varieties of river mouths.* (A) *Prograding delta distributary entering tideless sea (based on Mississippi River Delta).* (B) *Equilibrium (discharge adjusted) tidal estuary mouth (based on Georgia coast estuaries).* (C) *Disequilibrium estuary mouth (based on Thames estuary). From Swift (1975b).*

flat–tidal channel complex at their heads as a consequence of net landward sediment transport by the shoaling tidal wave. These deposits are the functional equivalent of the tide-molded deposits of a disequilibrium estuary.

The patterns of sand storage in estuary mouths may be extremely elaborate. These dynamic topographies are of major concern to port authorities concerned with the maintenance of deep-water approaches. As in the case of the systems of open coasts, estuary mouth sand storage systems are in a state of continuous reorganization in response to the postglacial rise in sea level.

Kraft et al. (1974) have attempted to trace the transgressive history of the mouth of Delaware Bay by equating a series of transects across the modern bay with the time series of profiles that would be expected at a single point during transgression (Fig. 43). Here ridges first appear as subaqueous tidal levees on the edge of tidal flats marginal to tidal channels. Unlike the tidal sand ridges of open shelf seas, these ridges migrate away from their steep sides (Weil et al., 1974). As transgression proceeds, the channels service a larger and larger tidal prism and tend to widen. The effect on the levees is erosion on the steep, channel-facing side, and aggradation on the gently sloping side facing away from the channel. Weil et al. (1974) have attributed the submarine levees of Delaware Bay tidal channels to density-driven

secondary flow associated with the tidal cycle (Chapter 10, Fig. 26).

Inner estuary channels tend to be ebb-dominated perhaps because the upper estuary water mass tends to flood as a sheet, but tends to preferentially ebb through the channel system under the impetus of gravity discharge. Further down the estuary, as levees begin to build, the interfluves tend to become flood-dominated channels in their own right, although the dominance of channel and interfluves may locally be reversed. As previously noted, retardation of the tidal wave in the estuary results in a phase lag across the estuary mouth shoal, causing an interdigitation of ebb and flood channels, separated by partition ridges, across the crest of the shoal. Thus ridges initiated in the upper estuary may undergo a complex evolution as successive estuary environments and associated flow regimes pass over them. Individual ridges may maintain their integrity through this process or be replaced by related forms maintained by somewhat different mechanisms.

Modification of ridge morphology intensifies as the regional shoreline passes, and the lower estuarine regime is replaced by an open shelf regime. If the wave climate is intense, then the outer surface of the estuary mouth shoal is pushed back by erosional shoreface retreat in a fashion similar to that transpiring on the adjacent main-

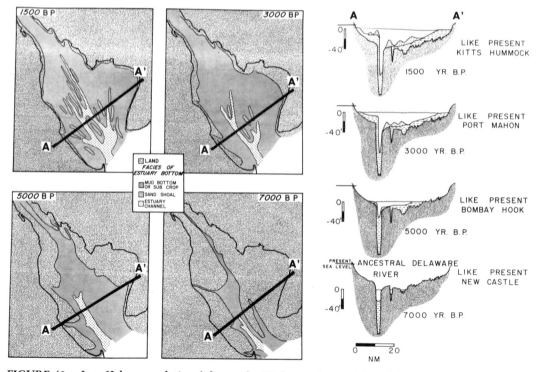

FIGURE 43.   *Late Holocene evolution of the mouth of Delaware Bay, as inferred from cross sections across the modern bay. Apron of sand extending into bay mouth is assumed to have prograded up the bay concurrently with the landward movements of the shoreline on either side. From Kraft et al. (1974).*

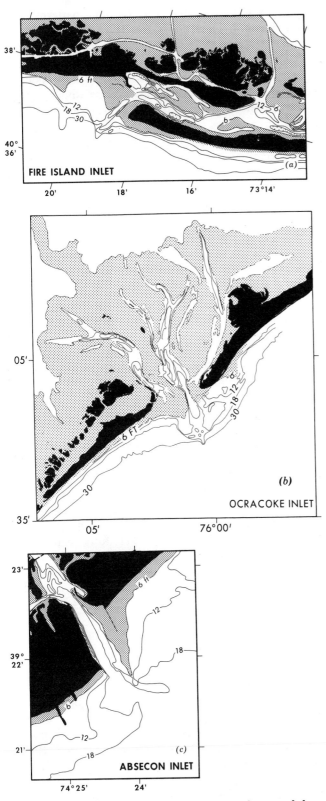

land coast. Frequently, however, the retreat path of the estuary is visible in the form of a cross-shelf channel and a ridge on the updrift side of the channel. On the Delaware inner shelf, such a ridge can be seen to mark the retreat path of the shoal on the north side of the estuary mouth, while the associated channel has formed by the retreat of the main flood channel of the estuary mouth (see Fig. 12, Chapter 15).

Coastal Inlets and Littoral Drift

The morphology of narrow estuary mouths and their analogs, coastal inlets, depends on the relative strengths of the river mouth or inlet jet, the wave-driven littoral current, and the tidal- and wind-driven components of the shelf flow field. Distributary mouths, subject to periodic flooding and entering relatively tideless, wave-sheltered seas, consist of subaerial levees capped by a lunate distributary mouth shoal (Fig. 42A). As a consequence of the Coriolis effect, flow is more intense on the right-hand side of the channel looking downstream, and as a consequence, the right-hand levee tends to extend itself farther seaward as in the case of Mississippi distributary mouths. If the inlet faces an open or tidal sea, then the wave- and tide-driven coastal flow is diverted seaward around the ebb tidal jet (Todd, 1968) and the shoal assumes a half-teardrop shape (Fig. 42B).

On barrier coasts, the pattern of sand storage at tidal inlets tends toward one of three basic patterns: overlap, symmetrical, or offset inlets (Fig. 44). While these patterns have long been recognized, the responsible transport systems and sand budgets are imperfectly understood (Hayes et al., 1970; Byrne et al., in press; Goldsmith et al., in press). As noted by Byrne (personal communication), the patterns appear to reflect the relative intensities of gross littoral drift (both up and down the coast) and net drift (the difference between mean annual upcoast discharge and mean annual downcoast discharge). If both the gross rates of drift and the net rate are high, a disproportionately high volume of sand storage may occur in the updrift barrier segment, and an overlap barrier may result (Fig. 44A). Where moderate gross rates of drift are associated with a strong net rate of drift, the situation favors a barrier offset inlet, in which the storage of sand on the downdrift side of the exterior shoal is favored (Fig. 44C). In one of the best studied barrier offset inlets, Wachapreague Inlet on the

FIGURE 44. *Representative examples of inlet morphology.* (a) *Fire Island Inlet, Long Island, a barrier-overlap inlet on a drift-aligned coast. Littoral drift dominates the ebb tidal jet.* (b) *Ocracoke Inlet, North Carolina. Nearly symmetrical inlet flow on a swash-aligned coast has resulted in sand storage in the wave-protected interior (flood delta) shoal. Ebb tidal jet dominates over littoral drift.* (c) *Absecon Inlet, New Jersey. Ebb-dominated flow has resulted in sand storage on the downdrift side of the inlet and an offset of the flanking barrier islands.*

Delmarva coast, the role of the lagoonal reservoir in modifying the hydraulic characteristics of the inlet is of paramount importance (Byrne et al., in press a). In lagoon-inlet systems where the ratio of the intertidal water prism to the subtidal volume is very large, a strong time-velocity asymmetry develops (see Postma, 1967). The strongest currents occur just before high tide, when the tidal channels have filled and the vast marsh surface is beginning to flood, and just after high tide, when the marshes are draining. Flows around low tide are weaker, as they are associated with the much slower discharge and recharge of the tidal creek system.

In addition, flood and ebb durations are dissimilar, with a greater ebb duration. This phenomenon is a consequence of the lagoonal basin's morphology and frictional characteristics (Byrne et al., in press a), and has been predicted by shallow water tidal theory for storage systems with sloping banks (Mota-Oliveira, 1970; King, 1974). In physical terms, the hydraulic head generated across the inlet by the flood tide is imposed on the deepest part of the lagoon relatively early in the tidal cycle. Here frictional retardation of flow is least efficient, and the resulting sea surface slope propagates rapidly across the lagoon, resulting in rapid water influx. The greatest potential drop across the lagoon surface during the ebb half-cycle occurs when the marsh surface is still uncovering. Frictional retardation of flow is more effective in the thin landward portions of the lagoonal water column, and the ebb is prolonged.

As a result of these modifications of the tidal cycle, the inlet operates in a bypassing mode. Sand is swept into the inlet from the updrift side, but does not penetrate very far before it is swept out again, and the prolonged ebb carries it into the storage area on the downdrift side of the external shoal. Here sand storage is enhanced by the refraction pattern of shoaling waves (Goldsmith et al., in press).

Symmetrical inlets are favored by swash-aligned coasts, where the ratio of the littoral component of wave power to tidal power is relatively low (Fig. 44B). Symmetrical inlets, particularly those backed by lagoons with relatively small intertidal prisms and relatively large subtidal volumes, tend to store sand primarily in the interior shoal within the lagoon.

SAND STORAGE ON ROCKY COASTS

Rocky coasts display the greatest complexity in three dimensions. Rocky hinterlands in a mature state of dissection result in embayed coasts with deep reentrants between rocky salients. If the substrate consists of folded metamorphic rocks, then it may have a well-defined

anisotropy of its own and truly baroque patterns may result (Fig. 45). The fields of wave refraction developed over the seaward extensions of headlands result in frequent reversals of the sense of littoral drift cells and closely spaced alteration of zones of littoral drift divergence and convergence. Because of the relative steepness of the regional seaward slope and the resistant nature of the substrate, wave energy is concentrated along a very narrow intertidal zone. Waves breaking against vertical surfaces can generate enormous instantaneous forces of tens of metric tons per square meter (Zenkovitch, 1967, p. 139). Rocky shores yield along planes of weakness to become mantled with boulders under this assault (Fig. 46) and the intertidal and subtidal talus slopes become grinding mills where attrition produces finer debris and continues to grind it finer until, at about the grade of medium sand, the immersed weight of grains is no longer adequate to result in significant chipping or cracking— as long as the particles are able to escape the proximity and nutcracker behavior of coarser particles. The interaction of intertidal and shallow subtidal wave forces with the three-dimensional complexity of rocky coasts results in such erosional forms as stacks, arches, and sea caves, and the constructional forms of looped, fringing, recurved, and cuspate spits, and tombolos that have long been the delight of coastal morphologists (Fig. 47). The constructional forms constitute localized depositional regressions and are usually comprised of sets and subsets of beach ridges reflecting stages in the feature's growth. If the net rate of sedimentation is sufficiently high relative to the rate of sea-level rise, these forms tend to grow and coalesce, and will ultimately form a continuous shoreface.

Rocky coasts are more nearly likely to be tectonically active than low, unindurated coasts, other things being equal, and the resistant character of their substrate may result in delays in the adjustment of the incised equilibrium profile to the crustal movement, if this adjustment is indeed attained. Comparison of rocky coasts from different parts of the world has revealed a continuum of adjustment from coasts as irregular as the margins of newly dammed reservoirs, to coasts whose adjustment has been complete, so that, by a combination of headland truncation and the filling in of bays, the coastline has been straightened in plan view and the shoreface has received the characteristic exponential curvature.

This continuum led Davis (1909) and Johnson (1919) to the concept of a cycle of coastal evolution in which, after an initial relative movement of sea level, the shoreline is straightened and the equilibrium profile passes through a cycle of youth, maturity, and old age. Zenkovitch (1967) has objected to the simplified assumptions of the model and suggests that three types of em-

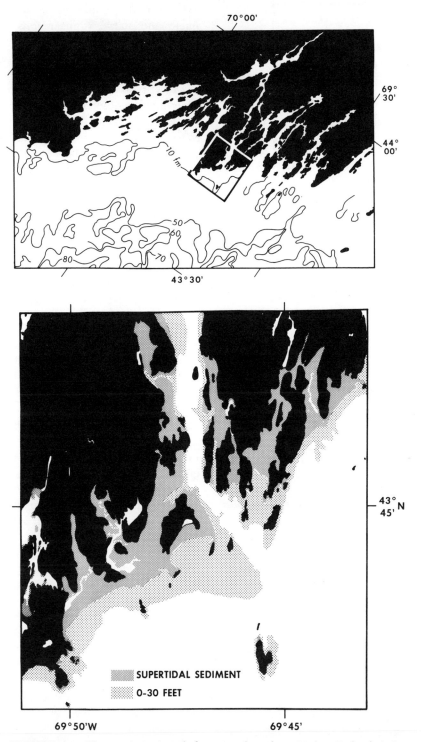

FIGURE 45. *Upper: A portion of the coast of southern Maine. Bedrock is iso-clinally folded schist and gneiss. Lower: Beginning of formation of constructional shoreface and estuary mouth shoal at mouth of Kennebec River; see the upper diagram for location.*

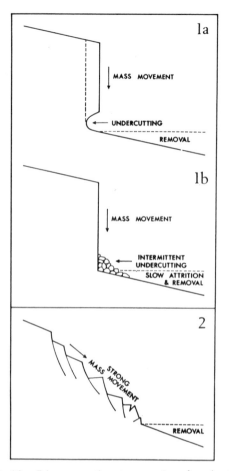

FIGURE 46. *Diagrammatic representation of major processes of cliff retreat and evolution. (1a) Undercutting and rapid removal of collapsed material. (1b) Undercutting and slow removal of collapsed material. (2) Mass movement and removal at various rates. From Davies (1973).*

bayed coasts may be distinguished on the basis of the relationship between the submarine slope and the equilibrium profile generated on it, as follows: (1) deepwater coasts where the submarine bottom passes immediately below the equilibrium profiles; (2) coasts with deep-water capes, where this is true only off capes, and (3) shallow-water coasts, where the submarine slope is everywhere above the equilibrium profile. The term "effective wave base" is probably best substituted for equilibrium profile here, for Zenkovitch concludes that sectors of coasts "above the profile of equilibrium" are those sectors that develop forms of accumulation (sandy beaches, barriers, spits, and tombolos; Fig. 47), and that shallow water coasts develop the most complex array of these features. Zenkovitch further traces subcycles of coastal evolution caused by feedback between evolving accumulation forms and the rocky substrate, or between two forms, whereby the growth of some spits into wave shadows behind headlands may distort their sub-

sequent pattern, and the growth of other spits may shield and starve younger spits, or induce yet others where none existed.

These subcycles are probably more common than the Davis-Johnson cycle, which requires an isostatic crustal movement or eustatic sea-level jump for rejuvenation. They may be observed on all stable rocky coasts undergoing transgression by postglacial sea-level rise. Such coasts probably do not evolve at all in the Davis-Johnson sense, but undergo steady state subsidence in a state of perpetual youth, maturity, or old age, depending on the degree of induration of the substrate and the amplitude of the inherited relief.

The relationship between the rate of sea-level rise and the relief and induration of the substrate also determines the geometry of sediment storage (Fig. 48). Cores off transgressed crystalline coasts of high relief might be expected to reveal a residual rubble overlain by fine-grained bay deposits. Overlying sand deposits of complex shape would reflect the passage of the outer shoreline with its array of accumulation forms. The upper surface of the sand horizon will have been beveled at least locally by shoreward profile translation, and offshore sands or muds may locally have accumulated over the surface of marine erosion. Off high crystalline coasts, the full sequence will rarely develop and will be completely missing off capes, where surf-rounded boulders may litter bare rock surfaces for kilometers offshore. Pocket beaches and spits may locally survive the transgressive process relatively intact; a rock-tied spit cannot retrograde with the ease of a low coast barrier.

On steep coasts transgressive deposits may be minimal. On steep coasts with very narrow shelves, submarine canyons may penetrate almost to the shoreface, to tap the littoral drift, through such gravity processes as sand creep. On steep deep-water coasts, prisms of beach shingle intermittently cascade to bathyal depths down steep rock slopes that may be erosion-modified fault scarps; sediment passes through the coastal zone by gravity bypassing (Fig. 49). As the coast is lower and softer, so will the sequence more nearly resemble the uniform sequence typical of the low coast transgression. Bay muds will more nearly resemble lagoonal muds, capped perhaps by nearly uniform sheets of backbarrier and shelf sands instead of lenticular remnants of spits and tombolos.

Regressive deposits occur on some rocky coasts, as a consequence of the Late Holocene reduction in the rate of sea-level rise, where sediment input is sufficient to reverse the sense of shoreline migration. In extreme cases, alluvial gravel cones may build out across the transgressive deposits. Bouldery topset beds may pass into foreset sands and then into bottomset muds within a few hun-

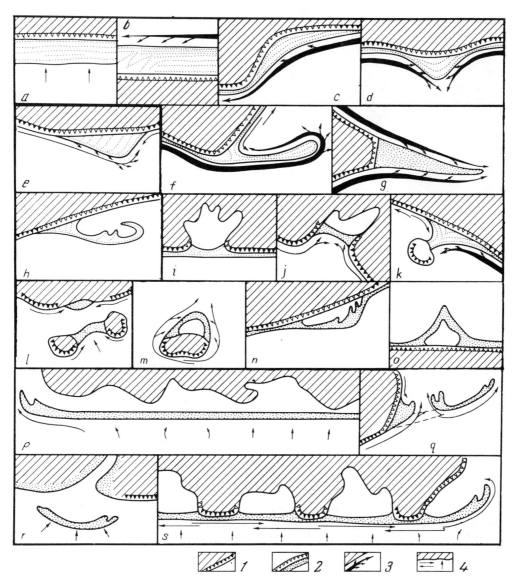

FIGURE 47. *Types of coastal accumulation forms, according to Zenkovitch (1967). Fringing: a. beach nourished from offshore; b, beach nourished from alongshore; c, beach filling an indentation; d, cuspate beach with bilateral nourishment; e, asymmetrical, cuspate beach with bilateral nourishment, attached at one end; f, spit with unilateral nourishment; g, arrow (spit with bilateral nourishment); h, spit on smooth coast; i, bay mouth barrier; j, midbay barrier; k, tombolo; l, interisland tombolo, doubly attached; m, looped spit with bilateral nourishment; n, looped spit with unilateral nourishment; o, cuspate spit, detached; p, barrier island; q, barrier island resulting from cutting of inlet; r, estuary mouth swash bar; s, barrier sequence. Symbols: (1) mainland and active cliff; (2) dead cliff and coast with beach; (3, 4) major and minor transport directions.*

dred meters. On steep, unstable coasts, such masses may periodically slump down the submarine slopes to the basin floor.

SUMMARY

In considering coastal sediment transport, it is convenient to divide the movement of sand into an onshore-offshore component and a coast-parallel component. Onshore-offshore transport occurs in two provinces. In the nearshore province of beach, longshore trough, plunge point bar, and upper shoreface, onshore-offshore transport is controlled by the regime of shoaling and breaking waves. Breakpoint bars are initiated during the waning phases of storms. During the ensuing fair-weather period they tend to migrate onshore, and weld

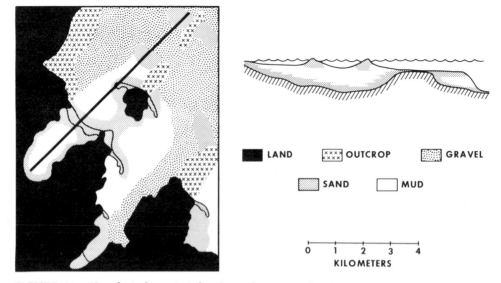

FIGURE 48. *Hypothetical stratigraphy of a rocky coast undergoing transgression.*

to the berm. The high, steep waves of storms tend to strip sand from the beach and transport it out to the surf zone, and the cycle begins anew. The cycle tends to be linked to the cycle of seasons in that offshore transport dominates during the period of winter storms, while onshore transport tends to dominate during the summer season of fair weather. The lower shoreface is a second province subject to onshore-offshore transport. The corresponding hydraulic regime is the zone of friction-dominated unidirectional flow that constitutes the coastal boundary of the shelf flow field. During storms (or peak tidal flows) velocity in this zone may be more intense than in the zone of quasi-geostrophic flow further offshore. Downwelling and a seaward component of bottom flow may occur in this zone during some storm flows, at the same time that sand is moving seaward in the surf zone, so that sand is transported off the shoreface altogether.

The interrelated behavior patterns of the zone of shoaling and breaking waves and zone of friction-dominated flows give rise on many coasts to a long-term cyclic pattern of advance or retreat of the coastal profile. The upper shoreface undergoes net aggradation and progradation over a period of years tending toward the ideal wave-graded profile. A major storm or period of severe storms will result in large-scale seaward transport of sand, causing flattening and significant landward translation of the profile. On coasts experiencing a net littoral drift surplus, fair-weather progradation is more effective than storm erosion, and the profile translates seaward and (in compensation for postglacial sea-level rise) upward. On coasts experiencing a net littoral drift deficit,

the storm regime controls the offshore-onshore sand budget, and the coastal profile undergoes landward and upward translation through a process of erosional shoreface retreat. The debris resulting from this process nourishes the leading edge of the surficial sand sheet that mantles the shelf.

The cycle of onshore-offshore transport is superimposed on a much more intensive flux of sand parallel to the beach, under the impetus of the wave-driven littoral flows, and wind- and tide-driven coastal currents. As a result, there is an innate tendency toward two-dimensionality of the shoreface, in that successive downcoast profiles tend to be very similar. Headlands experience a greater littoral wave energy density, greater breaker angles, and decreasing littoral sand discharge along the beach toward the adjoining bay. A pattern of transport away from headlands toward bays is superimposed on a regional direction of littoral sand transport determined by the prevailing direction of deep-water wave approach. The resulting alternation of littoral drift divergences and convergences may impose a three-dimensionality on an unconsolidated coast in the form of alternate cuspate forelands and zetaform bays. Three-dimensionality may also be inherited from the relief of a rocky surface undergoing transgression, or may be induced on an unconsolidated coast in the form of constructional river mouths and tidal inlets.

The beach and shoreface comprise major reservoirs of sand in the coastal sediment transport system. During a transgression, the superimposition of a straight, wave-maintained upper shoreface on an irregular surface results in the formation of two shorelines. An inner, la-

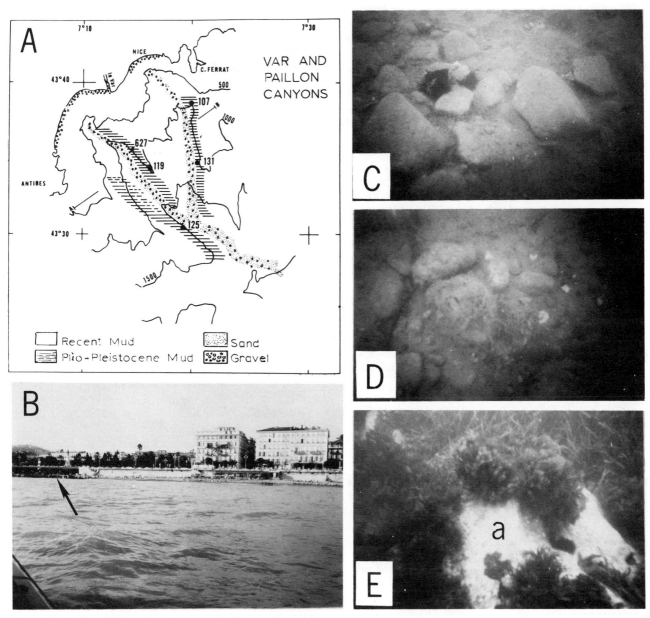

FIGURE 49. *Gravity bypassing on a recently formed continental margin, Provençal coast of France. (A) Axial facies of the Var and Paillon canyons. (B) Paillon River mouth (left) and pebble beach, Baie des Anges, Nice. (C, D) Boulder (up to 50 cm) mud admixtures at diving locality shown in (B); depth 25 m. (E) Large blocks of Jurassic limestone overgrown with* Poseidonia *near Cap Ferrat. From Stanley (1969).*

goonal shoreline approximates the intersection of still water level with the dissected subaerial surface undergoing erosion. An outer, oceanic shoreline of barrier spits and islands results from (1) the detachment of drift-nourished, wave-maintained beaches from the mainland as the rising sea floods the swales behind them, and (2) the lateral propagation of the shoreface from headlands across the mouths of adjoining bays. Sand is also stored in shoals that form at littoral drift convergences,

and in oblique-tending, shoreface-connected sand ridges that form at the foot of the shoreface in response to the storm wave regime and storm coastal currents. Where the littoral drift system intersects with the river- and tide-driven jets of river mouths and inlets, sand is stored in arcuate, seaward-convex shoals whose crests bear complex patterns of interdigitating ebb and flood channels, separated by sand ridges that build into the intertidal zone as swash platforms.

REFERENCES

Allen, J. R. L. (1968a). *Current Ripples*. Amsterdam: North-Holland, 433 pp.

Allen, J. R. L. (1968b). The nature and origin of bedform hierarchies. *Sedimentology*, **10**: 161–182.

Allersma, E. (1972). Mud on the oceanic shelf off Guiana. In *Symposium on Investigations and Resources of the Caribbean Sea and Adjacent Regions*. UNESCO, Paris (Unipub, N.Y.), pp. 193–203.

Assaf, G., R. Gerard, and A. L. Gordon (1971). Some mechanisms of oceanic mixing revealed in aerial photographs. *J. Geophys. Res.*, **76**(27): 6550–6572.

Bascom, W. N. (1951). The relationship between sand size and beach slope. *Trans. Am. Geophys. Union*, **32**: 866–874.

Beall, A. O. (1968). Sedimentary processes operative along the western Louisiana shoreline. *J. Sediment. Petrol.*, **38**: 869–877.

Beardsley, R. C. and B. Butman (1974). Circulation on the New England continental shelf: Response to strong winter storms. *Geol. Res. Lett.*, **1**: 181–184.

Bernard, H. A. and R. J. Leblanc (1965). Resume of the Quaternary geology of the Northwest Gulf of Mexico Province. In H. E. Wright, Jr. and D. J. Frey, eds., *The Quaternary of the United States*. Princton: Princeton Univ. Press, pp. 137–185.

Bruun, P. (1954). Migrating sand waves and sand humps, with special reference to investigations carried on in the Danish North Sea coast. *Conf. Coastal Eng.*, 5th Proc., 460–468.

Bruun, P. (1962). Sea level rise as a cause of shore erosion. *Proc. ASCE J. Waterw. Harbors Div.*, **88**: 117–130.

Byrne, R. J., P. Bullock, and D. G. Tyler (in press a). Response characteristics of a tidal inlet: A case study. In M. O. Hayes, ed., *Second International Estuarine Conference Proc., Myrtle Beach, S.C., Oct. 15–18, 1973:* New York, Academic Press.

Byrne, R. J., J. T. De Alteris, and P. A. Bullock (in press b). Channel stability in tidal inlets. *14th Conf. Coastal Eng. Proc., Copenhagen, 1974.*

Caston, V. N. D. (1972). Linear sand banks in the southern North Sea. *Sedimentology*, **18**: 63–78.

Caston, V. N. D. and A. H. Stride (1970). Tidal sand movement between some linear sand banks in the North Sea off northeast Norfolk. *Mar. Geol.*, **9**: M38–M42.

Charnell, R. L. and D. V. Hansen (1974). *Summary and Analysis of Physical Oceanography Data from the New York Bight Apex Collected During 1969–70*. MESA Rept. 74-3, National Oceanic and Atmospheric Administration. Washington, D.C.: U.S. Govt. Printing Office, 44 pp.

Clifton, H. E., R. E. Hunter, and R. L. Phillips (1971). Depositional structures and processes in the non-barred, high energy nearshore. *J. Sediment. Petrol.*, **41**: 651–670.

Colquhoun, D. J. (1969). *Geomorphology of the Lower Coastal Plain of South Carolina*. MS-15, South Carolina Development Board, Div. Geol., 36 pp.

Cook, D. O. (1969). Occurrence and geologic work of rip-currents in southern California. *J. Sediment. Petrol.*, **39**: 781–786.

Cook, D. O. and D. S. Gorsline (1972). Field observations of sand transport by shoaling waves. *Mar. Geol.*, **13**: 31–55.

Csanady, G. T. and J. T. Scott (1974). Baroclinic coastal jets in Lake Ontario during IFYGL. *J. Phys. Oceanogr.*, **4**: 524–541.

Curray, J. R. (1964). Transgressions and regressions. In R. L. Miller, ed., *Papers in Marine Geology—Shepard Commemorative Volume*. New York: Macmillan, pp. 175–203.

Curray, J. R., F. J. Emmel, and P. J. S. Crampton (1969). Lagunas costeras, un simposio. In *Mem. Simp. Int. Lagunas Costeras. UNAM-UNESCO, Nov. 28–30, 1967, Mexico, D.F.*, pp. 63–100.

Davies, J. L. (1958). Wave refraction and the evolution of curved shorelines. *Geogr. Stud.*, **5**: 1–14.

Davies, J. L. (1973). *Geographical Variation in Coastal Development*. New York: Hafner, 204 pp.

Davis, R. A. and W. T. Fox (1972). Coastal process and nearshore sand bars. *J. Sediment. Petrol.*, **42**: 401–412.

Davis, R. A., W. T. Fox, M. O. Hayes, and J. C. Boothroyd (1972). Comparison of ridge and runnel systems in tidal and non-tidal environments. *J. Sediment. Petrol.*, **32**: 413–421.

Davis, W. M. (1909). *Geographical Essays*, 1954 Dover edition. New York: Dover Publications, 777 pp.

Dolan, R. (1971). Coastal landforms: Crescentic and rhythmic. *Geol. Soc. Am. Bull.*, **82**: 177–180.

Duane, D. B., M. E. Field, E. P. Meisburger, D. J. P. Swift, and S. J. Williams (1972). Linear shoals on the Atlantic inner continental shelf, Florida to Long Island. In D. J. P. Swift, D. B. Duane, and O. H. Pilkey, eds., *Shelf Sediment Transport: Process and Pattern*. Stroudsburg, Pa.: Dowden, Hutchinson & Ross, pp. 447–499.

Ekman, V. W. (1905). On the influence of the earth's rotation on ocean currents. *Ark. J. Mat., Astron. Phys.*, **2**: 1–53.

Emery, K. O. (1968). Relict sediments on continental shelves of world. *Am. Assoc. Pet. Geol. Bull.*, **52**: 445–464.

Faller, A. J. (1963). An experimental study of the instability of the laminar Ekman boundary layer. *J. Fluid Mech.*, **15**: 560–576.

Faller, A. J. (1971). Oceanic turbulence and the Langmuir circulations. *Ann. Rev. Ecol. System.*, **2**: 201–233.

Faller, A. J. and R. B. Kaylor (1966). A numerical study of the instability of the laminar Ekman boundary layer. *J. Atmos. Sci.*, **23**: 466–480.

Farmer, D. G. (1971). A computer simulation model of sedimentation in a salt wedge estuary. *Mar. Geol.*, **10**: 133–143.

Fenneman, N. M. (1902). Development of the profile of equilibrium of the subaqueous shore terrace. *J. Geol.*, **10**: 1–32.

Fisher, J. J. (1968). Barrier island formation: Discussion. *Geol. Soc. Am. Bull.*, **79**: 1421–1426.

Fox, W. T. and R. A. Davis (1974). *Beach Processes on the Oregon Coast, July 1973*. Tech. Rep. 12, Office of Naval Research, Washington, D.C., 85 pp.

Gilbert, G. K. (1890). *Lake Bonneville*. U.S. Geol. Surv. Mono. 1, 438 pp.

Goldsmith, U., R. J. Byrne, A. H. Sallenger, and R. Driecker (in press). The influence of waves on the origin and development of the offset coastal inlets of the southern Delmarva Peninsula, Virginia. In M. O. Hayes, ed., *Second International Estuarine Conference Proceedings, Myrtle Beach, South Carolina, Oct. 15–18, 1973*. New York: Academic Press.

Halligan, G. H. (1906). Sand movement on the New South Wales coast. *Proc. Limnol. Soc. NSW*, **31**: 619–640.

Harrison, W. and R. M. Alamo (1964). *Dynamic Properties of Immersed Sand at Virginia Beach, Virginia*. Coastal Eng. Res. Cent., Tech. Memo. 9, pp. 1–52.

Hayes, M. O. (1967). *Hurricanes as Geological Agents: Case Studies of Hurricanes Carla, 1961, and Cindy, 1963*. Texas Bureau of Economic Geology, Report of Investigation 61, 54 pp.

Hayes, M. O., U. Goldsmith, and C. H. Hobbs (1970). Offset coastal inlets. In **12***th Conference Coastal Engineering Proceedings*, pp. 1187–1200.

Holliday, B. W. (1971). Observations on the hydraulic regime of the ridge and swale topography of the inner Virginia shelf. Masters Thesis, Old Dominion University, Inst. of Oceanography, Norfolk, Va.

Howard, J. D. and H. E. Reineck (1972). Physical and biogenic sedimentary structures of the nearshore shelf. *Senckenbergiana Marit.*, **4**: 81–124.

Hoyt, J. H. (1967). Barrier island formation. *Geol. Soc. Am. Bull.*, **78**: 1125–1136.

Hoyt, J. H. (1970). Development and migration of barrier islands, northern Gulf of Mexico, discussion. *Geol. Soc. Am. Bull.*, **81**: 3779–3782.

Hoyt, J. H. and V. J. Henry (1967). Influence of island migration on barrier island sedimentation. *Geol. Soc. Am. Bull.*, **78**: 77–86.

Ingle, J. C. (1966). *The Movement of Beach Sand*. New York: American Elsevier, 221 pp.

Johnson, D. W. (1919). *Shore Processes and Shoreline Development*. New York: Hafner (1965 facsimile), 584 pp.

Johnson, J. W. (1949). Scale effects in hydraulic models involving wave motion. *Trans. Am. Geophys. Union*, **30**: 517–527.

Johnson, J. W. and P. S. Eagleson (1966). In A. J. Ippen, ed., *Estuary and Coastline Hydrodynamics*. New York: McGraw-Hill, pp. 404–492.

Keulegan, G. H. (1948). *An Experimental Study of Submarine Sand Bars*. Beach Erosion Board Tech. Rep. 3, 42 pp.

King, C. A. M. (1972). *Beaches and Coasts*. New York: St. Martins Press, 570 pp.

King, C. A. M. and W. W. Williams (1949). The formation and movement of sand bars by wave action. *Geogr. J.*, **113**: 70–85.

King, D. B., Jr. (1974). *The Dynamics of Inlets and Bays*. Tech. Rep. 22, College of Engineering, University of Florida, Gainesville, 82 pp.

Kraft, J. C., R. E. Sheridan, R. D. Moose, R. N. Strom, and C. B. Weil (1974). Middle-Late Holocene evolution of the morphology of a drowned estuary system—the Delaware Bay. In *International Symposium on Interrelationships of Estuarine and Continental Shelf Sedimentation*. Inst. Géol. du Bassin d'Aquitaine, July 9–14, 1973, pp. 297–306.

Langfelder, J., D. Stafford, and M. Amein (1968). *A Reconnaissance of Coastal Erosion in North Carolina*. Dept. Civil Eng., North Carolina State University, Raleigh, N.C., 127 pp.

Langford-Smith, T. and B. G. Thom (1969). New South Wales coastal morphology. *J. Geol. Soc. Aust.*, **16**: 572–580.

Langmuir, I. (1925). Surface motion of water induced by the wind. *Science*, **87**: 119–123.

Leopold, L. B., M. G. Wolman, and J. P. Miller (1964). *Fluvial Processes in Geomorphology*. San Francisco: Freeman, 522 pp.

Longuet-Higgins, M. S. (1953). Mass transport in water waves. *Phil. Trans. Roy. Soc. London*, **245**: 535–581.

Ludwick, J. C. (in press). Tidal currents, sediment transport, and sandbanks in Chesapeake Bay entrance, Virginia. In M. O. Hayes, ed., *Second International Estuarine Conf. Proc. Myrtle Beach, S.C., Oct. 15–18, 1973*: New York, Academic Press.

McGee, W. D. (1890). Encroachments of the sea. *The Forum*, **9**: 437–449.

McHone, J. F., Jr. (1972). Morphologic time series from a submarine sand ridge on the south Virginia coast. Masters Thesis, Old Dominion University, Norfolk, Va., 59 pp.

McKinney, T. F., W. L. Stubblefield, and D. J. P. Swift (1974). Large scale current lineations on the Great Egg shoal retreat massif: Investigations by sidescan sonar. *Mar. Geol.*, **17**: 79–102.

McMaster, R. L. and L. E. Garrison (1967). A submerged Holocene shoreline near Block Island, Rhode Island. *Mar. Geol.*, **75**: 335–340.

May, J. P. and W. F. Tanner (1973). The littoral power gradient and shoreline changes. In D. R. Coates, ed., *Publications in Geomorphology*. Binghamton: State University of New York, 404 pp.

Miller, R. L. and J. M. Zeigler (1958). A model relating dynamics and sediment pattern in equilibrium in the region of shoaling waves, breaker zone, and foreshore. *J. Geol.*, **66**: 417–441.

Miller, R. L. and J. M. Zeigler (1964). A study of sediment distribution in the zone of shoaling waves. In R. L. Miller, ed., *Papers in Marine Geology—Shepard Commemorative Volume*. New York: MacMillan, pp. 133–153.

Moody, D. W. (1964). Coastal morphology and processes in relation to the development of submarine sand ridges off Bethany Beach, Delaware. Ph.D. Thesis, Johns Hopkins University, Baltimore, 167 pp.

Moore, G. T. (1970). Role of salt wedge in bar finger sand and delta development. *Am. Assoc. Pet. Geol. Bull.*, **54**: 326–333.

Mota-Oliveira, I. B. (1970). Natural flushing ability in tidal inlets. *Proc. 12th Coastal Eng. Conf., Washington, D.C.*, pp. 1827–1845.

Murray, S. P. (1967). Control of grain dispersion by particle size and wave state. *J. Geol.*, **75**: 612–634.

Murray, S. P. (1970). Bottom currents near the coast during Hurricane Camille. *J. Geophys. Res.*, **75**: 4579–4582.

Neumann, G. and W. J. Pierson, Jr. (1966). *Principles of Physical Oceanography*. Englewood Cliffs, N.J.: Prentice Hall, 545 pp.

Nevesskii, E. N. (1969). Some data on the postglacial evolution of Karkinit Bay and the accumulation of bottom sediments within it. In V. V. Longinov, ed., *Dynamics and Morphology of Seacoasts*. Akademiya Nauk SSSR Trudy Instituta Okeanologii, Vol. 48, 371 pp., translated from Russian by Israel Program for Scientific Translations; reproduced by the Clearinghouse for Federal Scientific and Technical Information, Springfield, Va., pp. 92–110.

Niedoroda, A. W. and W. F. Tanner (1970). Preliminary study of transverse bars. *Mar. Geol.*, **9**: 41–62.

Oaks, R. Q., Jr. and N. K. Coch (1963). Pleistocene sealevels, southeastern Virginia. *Science*, **140**: 979–983.

Oertel, G. F. (1972). Sediment transport of estuary entrance shoals and the formation of swash platforms. *J. Sediment. Petrol.*, **42**: 858–863.

Oertel, G. F. and J. D. Howard (1972). Water circulation and sedimentation at estuary entrances on the Georgia coast. In D. J. P. Swift, D. B. Duane, and O. H. Pilkey, eds., *Shelf Sediment Transport: Process and Pattern*. Stroudsburg, Pa.: Dowden, Hutchinson & Ross, pp. 411–427.

Otvos, E. G. (1969). A sub-recent beach ridge complex in southern Louisiana. *Geol. Soc. Am. Bull.*, **80**: 2353–2358.

Otvos, E. G. (1970a). Development and migration of barrier islands, northern Gulf of Mexico. *Geol. Soc. Am. Bull.*, **81**: 241–246.

Otvos, E. G. (1970b). Development and migration of barrier islands, northern Gulf of Mexico: Reply. *Geol. Soc. Am. Bull.*, **81**: 3783–3788.

Pierce, J. W. and D. J. Colquhoun (1970). Holocene evolution of a portion of the North Carolina coast. *Geol. Soc. Am. Bull.*, **81:** 3697–3714.

Postma, H. (1967). Sediment transport and sedimentation in the marine environment. In G. H. Lauff, ed., *Estuaries.* Washington, D.C.: Am. Assoc. Adv. Sci., pp. 158–180.

Price, W. A. (1954). Dynamic environments: Reconnaissance mapping, geologic and geomorphic, of continental shelf of Gulf of Mexico. *Trans. Gulf Coast Assoc. Geol. Soc.*, 4: 75–107.

Price, W. A. (1955). *Correlation of Shoreline Type with Offshore Bottom Conditions.* Dept. of Oceanography, Texas A&M University, Project 65: 75–107.

Reineck, H. E. and I. B. Singh (1971). Der Golf von Gaeta (Tyrrhenisches Meer). III. Die gefuge von vorstrand—und schelfsedimenten. *Senckenbergiana Marit.*, 3: 135–183.

Robinson, A. H. W. (1966). Residual currents in relation to shoreline evolution of the East Anglian coast. *Mar. Geol.*, **4:** 57–84.

Russell, R. C. H. and J. D. C. Osorio (1958). An experimental investigation of drift profiles in a closed channel. *Proc. 6th Conf. Coastal Eng.*, pp. 171–183.

Schiffman, A. (1965). Energy measurements in the swash-surf zone. *Limnol. Oceanogr.*, **10:** 255–260.

Schwartz, M. L. (1965). Laboratory study of sea level rise as a cause of shore erosion. *J. Geol.*, **73:** 528–534.

Schwartz, M. L. (1967). The Bruun theory of sea level rise as a cause of shore erosion. *J. Geol.*, **75:** 76–92.

Schwartz, M. L. (1968). The scale of shore erosion. *J. Geol.*, **76:** 508–517.

Schwartz, M. L. (1973). *Barrier Islands.* Stroudsburg, Pa.: Dowden, Hutchinson & Ross, 451 pp.

Shepard, F. P. (1950). *Longshore-Bars and Longshore-Troughs.* Beach Erosion Board Tech. Memo. 15, 32 pp.

Silvester, R. (1974). *Coastal Engineering II.* New York: American Elsevier, 338 pp.

Smith, J. D. and T. S. Hopkins (1972). Sediment transport on the continental shelf off of Washington and Oregon in the light of recent current meter measurements. In D. J. P. Swift, D. B. Duane, and O. H. Pilkey, eds., *Shelf Sediment Transport: Process and Pattern.* Stroudsburg, Pa.: Dowden, Hutchinson & Ross, pp. 143–180.

Sonu, C. J. and J. L. van Beek (1971). Systematic beach changes in the outer banks, North Carolina. *J. Geol.*, **74:** 416–425.

Sonu, C. J., J. M. McCloy, and D. S. McArthur (1967). Longshore currents and nearshore topographies. *Proc. 10th Conf. Coastal Eng.*, pp. 55–549.

Stahl, L., J. Koczan, and D. Swift (1974). Anatomy of a shoreface-connected ridge system on the New Jersey shelf: Implications for genesis of the shelf surficial sand sheet. *Geology*, **2:** 117–120.

Stanley, D. J. (1969). Submarine channel deposits and their fossil analogs (fluxoturbidites). In D. J. Stanley, ed., *The New Concepts of Continental Margin Sedimentation.* Washington, D.C.: Am. Geol. Inst., pp. DJS-9-1 to DJS-9-17.

Strahler, A. (1963). *The Earth Sciences.* New York: Harper and Row, 681 pp.

Stubblefield, W. L., J. W. Lavelle, T. F. McKinney, and D. J. P. Swift (1975). Sediment response to the hydraulic regime on the central New Jersey shelf. *J. Sediment Petrol.*, **45:** 337–358.

Swift, D. J. P. (1973). Delaware Shelf Valley: Estuary retreat path, not drowned river valley. *Geol. Soc. Am. Bull.*, **84:** 2743–2748.

Swift, D. J. P. (1975a). Barrier island genesis: Evidence from the Middle Atlantic Shelf of North America. *Sediment. Geol.*, **14:** 1–43.

Swift, D. J. P. (1975b). Tidal sand ridges and shoal retreat massifs. *Mar. Geol.*, **18:** 105–134.

Swift, D. J. P., D. B. Duane, and T. McKinney (1974). Ridge and swale topography of the Middle Atlantic Bight: Secular response to Holocene hydraulic regime. *Mar. Geol.*, **15:** 227–247.

Swift, D. J. P., B. W. Holliday, N. F. Avignone, and G. Shideler (1972a). Anatomy of a shoreface ridge system, False Cape, Virginia. *Mar. Geol.*, **12:** 59–84.

Swift, D. J. P., J. W. Kofoed, F. P. Saulsbury, and P. Sears (1972b). Holocene evolution of the shelf surface, central and southern Atlantic coast of North America. In D. J. P. Swift, D. B. Duane, and O. H. Pilkey, eds., *Shelf Sediment Transport: Process and Pattern.* Stroudsburg, Pa.: Dowden, Hutchinson & Ross, pp. 499–574.

Swift, D. J. P., R. B. Sanford, C. E. Dill, Jr., and N. F. Avignone (1971). Textural differentiation on the shoreface during erosional retreat of an unconsolidated coast, Cape Henry to Cape Hatteras, North Carolina. *Sedimentology*, **16:** 221–250.

Swift, D. J. P. and P. Sears (1974). Estuarine and littoral depositional patterns in the surficial sand sheet, central and southern Atlantic shelf of North America. In *International Symposium on Interrelationships of Estuarine and Continental Shelf Sedimentation.* Inst. Géol. du Bassin d'Aquitaine, Bordeaux, Mém. 7, pp. 171–189.

Tanner, W. F. (1961). Offshore shoals in an area of energy deficit. *J. Sediment. Petrol.*, **31:** 87–95.

Tanner, W. F., R. G. Evans, and C. W. Holmes (1963). Low energy coast near Cape Romano, Florida. *J. Petrol.*, **33:** 713–722.

Todd, T. W. (1968). Dynamic diversion: Influence of longshore current-tidal flow interaction on chenier and barrier island plains. *J. Sediment. Petrol.*, **38:** 734–746.

Valentin, H. (1954). Der landverlust in Holderness, Ostengland von 1852 bis 1952. *Die Erde*, **6:** 296–315. See also, Land loss at Holderness, 1852–1952. In J. D. Steers, ed., *Applied Coastal Morphology.* Cambridge, Mass.: MIT Press, 1971, pp. 116–137.

Van Straaten, L. M. J. U. (1965). Coastal barrier deposits in south and north Holland—in particular in the area around Scheveningen and Ijmuden. *Meded. Geol. Sticht.*, NS **17:** 41–75.

Visher, G. S. (1969). Grain size distribution and depositional processes. *J. Sediment. Petrol.*, **34:** 1074–1106.

Weil, C. B., R. D. Moose, and R. E. Sheridan (1974). A model for the evolution of linear tidal built sand ridges in Delaware Bay, U.S.A. In G. Allen, ed., *Estuary and Shelf Sedimentation: A Symposium. University of Bordeaux, July 1973.*

Wells, D. R. (1967). Beach equilibrium and second order wave theory. *J. Geophys. Res.*, **72:** 497–509.

Wright, L. D. and J. M. Coleman (1972). River delta morphology: Wave climate and the role of the subaqueous profile. *Science*, **176:** 282–284.

Wright, L. D. and J. M. Coleman (1974). Mississippi River mouth processes: Effluent dynamics and morphologic development. *J. Geol.*, **82:** 751–778.

Zenkovitch, V. P. (1967). *Processes of Coastal Development.* New York: Wiley, 738 pp.

CHAPTER 15

Continental Shelf Sedimentation

DONALD J. P. SWIFT

Atlantic Oceanographic and Meteorological Laboratories, Miami, Florida

The preceding chapter considered in detail the nature of hydraulic process and substrate response along the coast. This chapter examines patterns of sedimentation on the shelf as a whole. It reexamines the coastal boundary of the shelf as a source of sediment for the rest of the shelf, and as a zone which thus regulates the rate and character of sedimentation on the shelf surface. Chapter 14 described a "littoral energy fence" imposed upon coastal sedimentation by the landward-directed asymmetry of wave surge in shoaling water, which causes sediment to be retained on the shoreface. This chapter concerns itself with the mechanisms by which this dynamic barrier is penetrated, along the shoreface or at river mouths, and by which sediment is injected into the shelf dispersal system. The relative efficiencies of shoreface and river mouth bypassing during periods of transgression on one hand, and during periods of regression on the other are described. These varying efficiencies lead to two distinct shelf regimes: a passive regime in which the shelf sand sheet is generated by erosional shoreface retreat (autochthonous sedimentation) and a more active regime in which river mouth bypassing causes deposition across the shelf surface (allochthonous sedimentation). The chapter analyzes the transport patterns associated with these two regimes, and the resulting patterns of morphology, stratigraphy, and grain-size distribution. Portions of the material in this chapter have been presented elsewhere (Swift, 1974).

MODELS OF SHELF SEDIMENTATION

One of the first comprehensive models for the genesis of clastic sediments on continental shelves was a by-product of Douglas Johnson's (1919, p. 211) attempt to apply Davis' geomorphic cycle of youth, maturity, and old age to the continental shelf (see p. 277). Johnson saw the shelf water column and the shelf floor as a system in dynamic equilibrium, in which the slope and grain size of the sedimentary substrate at each point control, and are controlled by, the flux of wave energy into the bottom. He described the resulting surface as an exponential curve in profile, concave up, with the steeper segment being the shoreface. Grain size was considered to decrease as a function of increasing depth with distance from shore, as a consequence of the diminishing input of wave energy into the seafloor. The model derived its sediment from coastal erosion rather than from river input, a more broadly applicable interpretation than many subsequent textbooks have realized.

Despite its qualitative expression and limited applicability, the model was in advance of its time in its dynamical systems approach. However, this model could not withstand in its initial form the subsequent flood of data on the characteristic of shelf sediments. Shepard (1932) was the first to challenge it, noting that nautical charts of the world's shelves bore notations indicating that most shelves were veneered with a

311

complex mosaic of sediment types, rather than a simple seaward-fining sheet. He suggested that these patches were deposited during Pleistocene low stands of the sea, rather than during Recent time. Emery (1952, 1968) raised this concept to the status of a new conceptual model. He classified shelf sediments on a genetic basis, as *authigenic* (glauconite or phosphorite), *organic* (foraminifera, shells), *residual* (weathered from underlying rock), *relict* (remnant from a different earlier environment such as a now submerged beach or dune), and *detrital*, which includes material now being supplied by rivers, coastal erosion, and eolian or glacial activity. On most shelves, a thin nearshore band of modern detrital sediment is supposed to give way seaward to a relict sand sheet veneering the shelf surface.

A third, more generalized model for shelf sedimentation has been primarily concerned with the resulting stratigraphy. It incorporates elements of both the Johnson and Emery models. Like the Johnson model, it views the shelf surface as a dynamic system in a state of equilibrium with a set of process variables. The rate of sea-level change, however, is one of these variables; hence the effects of post-Pleistocene sea-level rise, as noted by Shepard and Emery, may be accounted for. The model may be referred to as the transgression-regression model, since it is generally expressed in these terms, or the coastal model, since it focuses on the behavior of this dynamic zone. It was first explicitly formulated by Grabau (1913), and more recently by Curray (1964) and Swift et al. (1972). In this model, the rate of sediment input to the continental shelf S, the character of the sediment G (grain size and mineralogy), the rate of energy input E, the sense and rate of relative sea-level change R, and slope L are seen as variables that govern the sense of shoreline movement (transgression or regression) and ultimately the character of shelf deposits.

The relationship may be expressed in quasi-quantitative form as

$$\frac{SG}{E} - \frac{R}{L} \propto T$$

The processes controlling shelf sedimentation are much too complex to be adequately described by this equation and there is no way to evaluate its variables adequately. The expression is useful, however, in helping to sort out relationships. The first term, SG/E, might be called the effective rate of coastal deposition. It increases with increasing S, the rate at which sediment is delivered to the shore. It increases with increasing grain size G, since coarser sediments are less easily bypassed across the shelf. It decreases with increasing E, the rate of wind and tidal energy input, since a more

rigorous hydraulic climate causes more sediment to be bypassed across the shelf.

The second term, R/L, might be called the effective rate of sea-level movement. It increases with increasing R, the absolute rate of sea-level movement (eustatic or tectonic), but decreases with increasing slope, L, of the coast. The steeper the slope, the greater the fall of sea level must be in order for the coast to advance a given distance. Also, with a greater slope, a greater volume of sediment must be delivered to the shoreline in order for the shoreline to prograde a given distance shoreward.

The equation tells us that the rate and sense of shoreline movement, T, whether landward (negative) or seaward (positive), depends on the relationship between these two terms. Basic elements of the relationship are presented graphically in Fig. 1, according to a scheme of Curray (1964). In Fig. 2, the history of the Nayarit coast of Mexico has been plotted.

The Coastal Boundary as a Filter: Shelf Sedimentary Regimes

The fundamental determinants of shelf sedimentation are the areal extent of the adjacent continent undergoing denudation, and its relief, climate, and drainage pattern. These factors control the quantity of sediment delivered to the shoreline, and its textural and mineralogical composition. However, the rate and sense of shoreline movement, as determined by the parameters described above, have a modulating effect on the shelf sedimentary regime.

It is helpful to think of the coastline as a "littoral energy fence" (Allen, 1970b, p. 169) in which the net landward flow associated with bottom wave surge tends

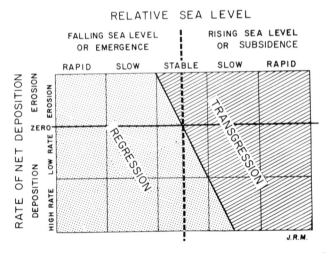

FIGURE 1. *Diagram of the effects of sea-level movement and the rate of coastal deposition on lateral migration of the shoreline. See text for explanation. From Curray (1964).*

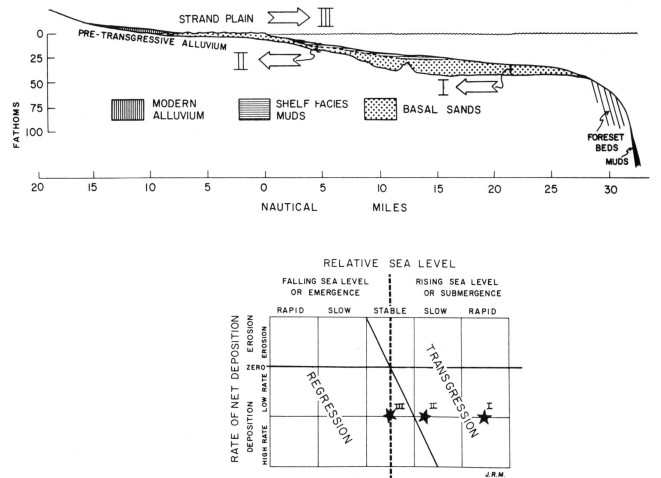

FIGURE 2. *Above: Diagrammatic section off the Costa de Nayarit, Mexico. See Fig. 10A, Chapter 14 for details of coastal stratigraphy. Below: Schematic representation of shoreline migration. From Curray (1964).*

to push sediment toward the shore. There are two basic categories of "valve" which regulate the passage of sediment through this dynamic coastal barrier into the transport system of the shelf surface. The shoreface may serve as a zone of sediment bypassing. The erosional retreat of the shoreface during a marine transgression bevels the subaerial surface being transgressed (Fig. 10A, Chapter 14), and spreads the resultant debris as a thin sheet over the shelf floor. The process by which the sediment is so transferred is described in the accompanying text. The process is a passive and indirect one; the sediment that is released has undergone long-term storage as flood plain, lagoonal, or estuarine deposits, or has been derived from an earlier cycle of sedimentation.

A second, more active route by which sediment may pass through the littoral energy fence is via the ebb tidal jet or flood stage jet of a river mouth. Patterns of river mouth bypassing are illustrated in Chapter 14 (Fig. 42). River mouth bypassing is more direct than shoreface by-

passing, but sediment must still undergo storage. Sand is stored in the throat of the river mouth, and fines are stored in marginal marshes and mud flats until the period of maximum river discharge, when the salt wedge moves to the shoal crest, and stored sediment is bypassed to the shoreface of the shoal front (Wright and Coleman, 1974); see Chapter 14, Fig. 41. It may undergo a second period of storage on the shoreface and inner shelf until the period of maximum storm energy (Wright and Coleman, 1973).

The mode of operation of these valves is dependent on basic parameters of coastal sedimentation. The spacing of river mouths is the fundamental determinant of the relative roles of shoreface versus river mouth bypassing. The character of the hydraulic climate is also important; an intense tidal regime increases the efficiency of river mouth bypassing, whereas an intense wave climate increases the efficiency of shoreface bypassing.

The rate and sense of coastal translation as described in the preceding section strongly affect the relative

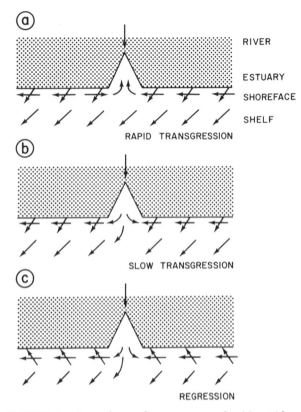

FIGURE 3. *Sense of net sediment transport for* (a) *rapid transgression,* (b) *slow transgression, and* (c) *regression. Offshore component of transport is exaggerated for continuity. See text for explanation.*

roles of river mouth and shoreface bypassing (Fig. 3). Rapid transgression results in disequilibrium estuaries which become sediment sinks (see Chapter 14, Fig. 42 and associated text), and shoreface bypassing must dominate (Fig. 3a). The resulting deposits consist of a transient veneer of surf fallout on the upper shoreface, and the residual sand sheet of the lower shoreface and adjacent shelf (see Chapter 14, Fig. 8 and discussion, p. 265. These two deposits correspond to the nearshore modern sand and shelf relict sand, respectively, of Emery (1968). Both deposits are relict in the sense that they have been eroded from a local, pre-Recent substrate, and both are modern in the sense that they have been redeposited under the present hydraulic regime. They are, in fact, palimpsest sediments (Swift et al., 1971) since they have petrographic attributes resulting from both the present and the earlier depositional environment. The term relict is best reserved for those specific textural attributes reflecting the earlier regime. Perhaps the most effective term for describing the relationship of these materials to the present depositional cycle is autochthonous (of local origin: Naumann, 1858), and a shelf sedimentary regime characterized by rapid transgression and by-

passing via shoreface erosion is described in this chapter as a regime of autochthonous shelf sedimentation.

With a slower rate of translation (Fig. 3b), estuaries can equilibrate to their tidal prisms (see Chapter 14, Fig. 42 and associated text). River mouth as well as shoreface bypassing becomes a significant source of sediment. More subtle, but equally important, is the effect of a slow transgression on the grain size of bypassed sediment. With a slower rate of shoreline translation the intracoastal zone of estuaries and lagoons can aggrade nearly to mean sea level. The resulting surface of salt marshes (or in low latitudes, mangrove swamps), threaded by high-energy channels, tends to serve as a low-pass, or bandpass filter, in the sense that the finer fraction of the sediment load is preferentially bypassed, while the coarsest fraction (and in the bandpass case, the finest fraction as well) is preferentially trapped out. In this process, migrating channels tend to select coarse materials for permanent burial in their axes. The surfaces of the tidal interfluves receive the finest material for prolonged storage or permanent burial. However, fine sands and silts are deposited as overbank levees and tend to be reentrained by the migrating channels; hence they have the highest probability of being bypassed to the shelf surface. This material is sufficiently fine to travel in suspension for long distances.

The estuaries of the Georgia coast have built a gently sloping shoreface of fine to very fine sand up to 20 km wide (Pilkey and Frankenberg, 1964; Henry and Hoyt, 1968); see Chapter 14, Fig. 13. This unusually wide and broad shoreface may be built by the combined contributions of shoreface and river mouth bypassing. Recent studies (Visher and Howard, 1974) suggest that the reversing tidal flows within the estuary constitute an efficient mechanism for the sorting of sands into size fractions, the spatial segregation of these fractions, and the bypassing of the finest sand out onto the shoreface.

There is clearly a contribution of sand from shoreface erosion; however, shoreface sands, like the adjacent shelf sands, contain trace amounts of phosphorite (Pilkey and Field, 1972), indicating erosion of the Miocene strata which underlie the shoreface between the closely spaced estuaries, and which floor the deep scour channels of the estuary mouths (Barby and Hoyt, 1964).

As the sense of coastal translation passes through stillstand to progradation (Fig. 3c), the shoreface becomes a sink rather than a mechanism for bypassing. Distributary mouths must further partition their prefiltered load between sand sufficiently coarse to be captured by the littoral drift and buried on the shoreface, and sand fine enough to escape in suspension in the ebb

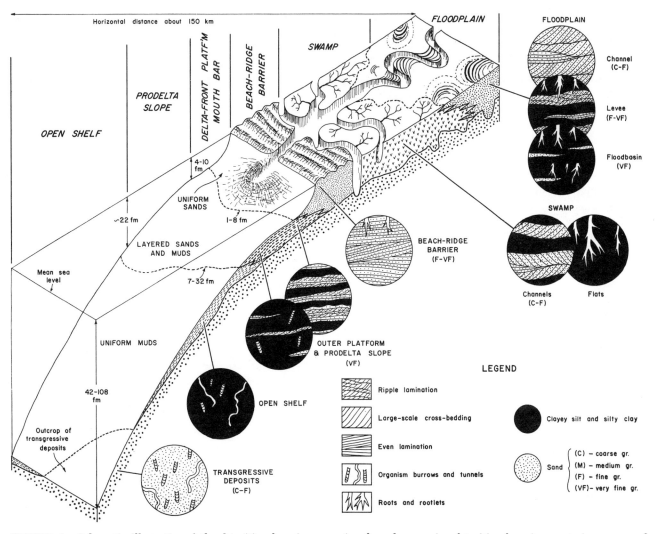

FIGURE 4. *Schematic illustration of the depositional environments and sedimentary facies of the Niger Delta and Niger shelf. Progressive size sorting of sediment results in a decrease in grain size through successive depositional environments in a seaward direction. From Allen (1970a).*

tidal jet, and be entrained into the shelf dispersal system. The shoreface behaves more nearly as a sediment trap, and bypassing occurs primarily through river mouths.

The Niger-Benue delta system is one of the best studied examples of differential sediment bypassing through a prograding, deltaic environment (Allen, 1964, 1970a). The Niger-Benue river system delivers about 0.9×10^6 m³ of bed load sediment and about 16×10^6 m³ of suspended sediment (Allen, 1964) to its delta each year. During peak discharge from September to May, average flow velocities range from 50 to 135 cm/sec, and gravel as well as sand is in violent transport. During low stages, flow velocities decrease to 37 to 82 cm/sec, enough to transport sand and silt. In the higher part of the flood plain, the Niger is braided; in the rest the Niger shows large meanders (Fig. 4). During high stages, levees are overtopped, crevasse develops, and bottom lands are flooded. Gravel and coarse sand are deposited as a substratum of braid bars and meander point bars, respectively, and are veneered with a top stratum of overbank clays. Silt undergoes temporary deposition in levees in the lower flood plain but these tend to be undermined, so that their deposits reenter the transport system.

Thus the flood plain environment serves as a skewed bandpass filter, with preferential bypassing of the medium and finer grades, preferential entrapment of the finest material over bank, and much coarse material being deposited in channel axes. This process continues through the tidal swamp environment, where the entrapment of fines dominates. Reversing tidal flows generate velocities of 40 to 180 cm/sec in tidal creeks, enough to move sand and gravel. Entrapment of fines overbank in the mangrove swamps is enhanced by the phenomena of slack high water and the prolonged period of reduced velocity associated with it. Fines then deposited begin to compact, and require greater velocities to erode them than served to permit their deposition.

Major channels, which pass through the intertidal environment to the sea, must store their coarser sediment during low water stages at the foot of the salt wedge, where the landward-inclined surface of zero net motion intersects the channel floor. During high water stages, stored bottom sediment must be rhythmically flushed out of the estuary mouth by the tidal cycle. Sand coarser than the effective suspension threshold of 230 μ (Bagnold, 1966) will be deposited on the arcuate estuary mouth shoals, where, after a prolonged period of residence in the sand circulation cells of the shoal (see Chapter 10, p. 177), it leaks into the downcoast littoral drift system. Finer sand is entrained into suspension by large-scale top-to-bottom turbulence in the high-velocity estuary throat (Wright and Coleman, 1974) and will be swept seaward with the ebb tidal jet, to rain out on the inner shelf (Todd, 1968) where it is accessible to distribution by the shelf hydraulic regime.

Shelves undergoing slow transgression or regression (Figs. 3b,c) thus experience a contrasting regime of allochthonous shelf sedimentation (Naumann, 1858) characterized by significant river mouth bypassing. In this regime there is a massive influx of river sediment whose grain size has been modified by passage through the coastal zone. Sheets of mobile fine sand and mud stretch from the coast toward the shelf edge. Shorefaces are broad and gentle and merge imperceptibly with a shallow inner shelf.

Sedimentation on tectonic continental margins is a special case of allochthonous shelf sedimentation so distinctive as to warrant designation as a third and equal category. Shelves subject to such a regime are narrow and steep, if developed at all. River mouth bypassing and fractionation of the sediment load occur here also. Rubble subaerial fans may pass over short distances into sandy marine deltas with bottomset mud beds. Gravity dispersal becomes a significant coastal bypassing mechanism. Submarine canyons may cut completely across narrow shelves to tap the littoral drift (Shepard, 1973, p. 140) and divert sand seaward by slow or rapid mass movements. Where shelves are altogether lacking, coarse littoral prisms cascade intermittently down slopes that are nearly tectonic surfaces, to bathyal depths (Stanley, 1969). Tectonic regimes on incipient shelves are beyond the scope of this chapter, partly because they are more appropriately discussed in

the chapter on slope sedimentation, and partly because of our ignorance, as this category is one of the last to be better known in the rock record (Stanley, 1969) than in modern environments.

AUTOCHTHONOUS PATTERNS OF SEDIMENTATION

Morphologic-Stratigraphic Patterns

Shelves undergoing autochthonous sedimentation characteristically have a varied and systematic pattern of relief. The pattern tends to be correlated with both the distribution of surficial sediment and the internal structure of the surficial sediment mantle, and hence is a morphologic-stratigraphic pattern. On shelves of high relief, the pre-Holocene surface is exposed at the surface over wide areas, and constitutes an additional control of the pattern.

Survival of Subaerial Patterns

On high-latitude shelves, relief may exceed 200 m. Much of this relief may be the consequence of pre-Holocene fluvial and glacial erosion of a crystalline substrate (Holtedahl, 1940, 1958), and of the dissection of flat-lying or gently inclined Cenozoic strata into cuestas and plateaulike remnants. On the North American Atlantic shelf, the Fall Line, where turbulent piedmont streams pass onto the coastal plain strata, intersects the shoreline at New Jersey (Fig. 5). To the north, the Fall Line cuesta, of gently inclined coastal plain strata, forms first islands (Long Island, Nantucket), then offshore banks (Georges Banks, the Nova Scotian Banks). Basins landward of the drowned Fall Line (Long Island Sound, the Gulf of Maine, the Nova Scotian basins) have inner margins of crystalline rock thinly veneered with coarse detritus. The basin centers have a lower stratum of glacial lake deposits overlain by Holocene marine mud.

Shelves of lower relief tend to be divided into broad, flat, plateaulike compartments by shelf valleys excavated during Quaternary low stands of the sea (Figs. 6 and 7). The outer margins of such shelves tend to consist of low-stand deltas, whose fronts are seaward-bulging shelf-edge scarps and whose landward margins may be marked by V-shaped, seaward-facing scarps that rise to the level of the inner shelf.

Subaerial morphologic elements smaller in scale than cuestas, basins, and shelf valleys seem in general to have been destroyed by erosional retreat of the shoreface, and the larger scale elements have often been subtly but pervasively modified by this process. This point can usually be demonstrated by a comparison of

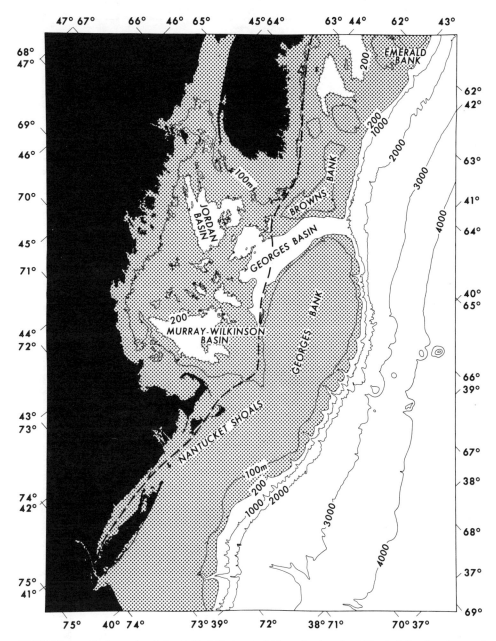

FIGURE 5.   *Bathymetry of the Gulf of Maine and Nova Scotian shelf. Dashed line is submerged extension of the Fall Line, separating gently dipping coastal strata from the crystalline substrate of the Appalachian orogenic belt. From Uchupi (1968).*

shelf morphology with the morphology of the associated subaerial surface. The coastal plains of the world bear a delicate fabric of high-stand scarps, separated by terraces overprinted with beach ridge fields, commonly dating from the last interglacial, or a high stand during the Würm-Wisconsin glacial epoch (see, e.g., Colquhoun, 1969; Oaks and Coch, 1963; Bernard and Le Blanc, 1965). However, most submarine shelves are relatively featureless (the Aquitaine shelf: Caralp et al., 1972) or bear complex patterns of sand ridges that are the conse-

quence of marine systems of sediment transport initiated after the passage of the shoreline (ridge and swale topography of Fig. 6).

Major exceptions to this rule are the littoral bed forms of carbonate coasts; fringing reefs, beach rock, and calcarenite dunes cement as they form, and are far more resistant to the destruction during the passage of the shoreline. Carbonate littoral and sublittoral features have been reported from many shelves (Kaye, 1959; Ginsburg and James, 1974; Van Andel and Veevers,

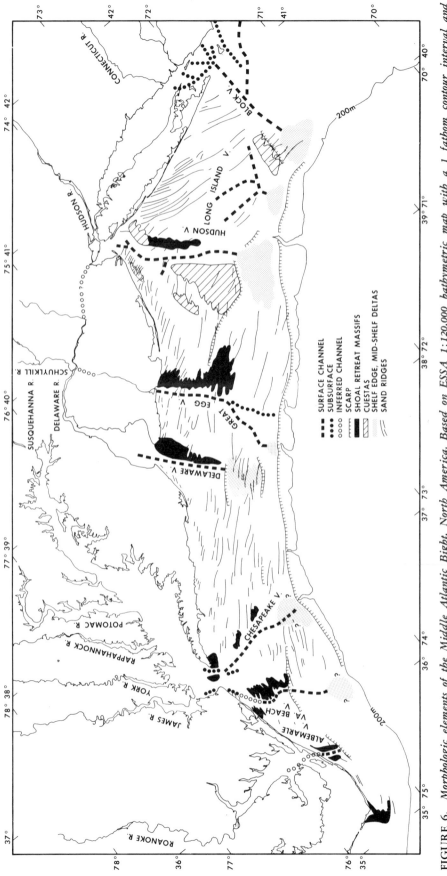

FIGURE 6. *Morphologic elements of the Middle Atlantic Bight, North America. Based on ESSA 1:120,000 bathymetric map with a 1 fathom contour interval and Uchupi (1970) endpaper map with a 4 m contour interval. From Swift (1975).*

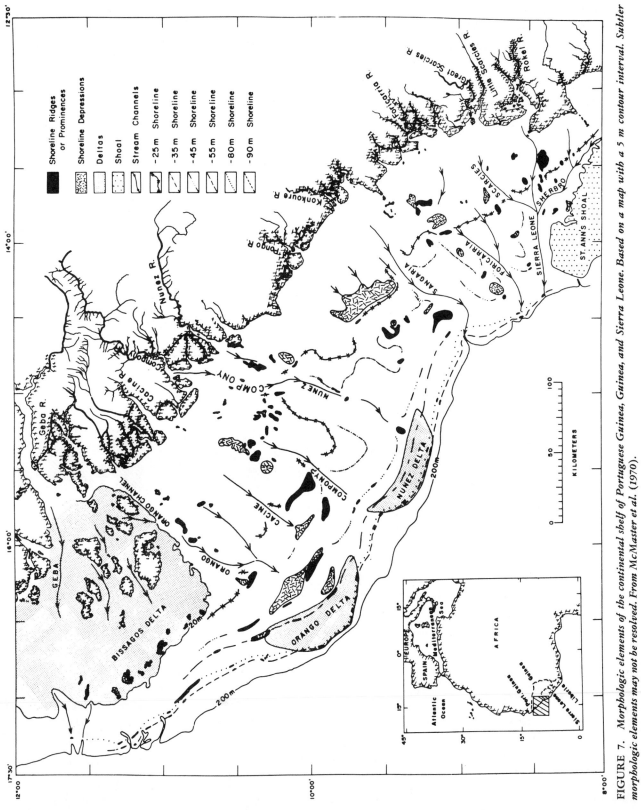

FIGURE 7. *Morphologic elements of the continental shelf of Portuguese Guinea, Guinea, and Sierra Leone. Based on a map with a 5 m contour interval. Subtler morphologic elements may not be resolved. From McMaster et al.* (1970).

319

1967; Stanley et al., 1968; Sarnthein, 1972). Other exceptions are the "perilittoral" deltas of terrigenous sand which seem to have survived transgression in the Gulf of Mexico (Curray, 1964, p. 299). The latter are large river-fed spits that grow in the direction of littoral drift, causing the river to flow parallel to the coast before it breaks out to the sea. End moraines have survived on the New England–Canadian shelf (King et al., 1972), but they were apparently emplaced seaward of the shoreline by grounded ice; King notes the vulnerability of glacial deposits on the present shoreline to glacial attack. The rinnentaler of the North Sea (subicestream channels) may likewise have been formed by an ice sheet seaward of the shoreline (Brouwer, 1964).

Survival of Nearshore Marine Patterns

THE SURFICIAL SAND SHEET. The most characteristic aspect of shelves undergoing autochthonous sedimentation is the discontinuous surficial sand sheet 0 to 10 m thick, deposited during the erosional retreat of the shoreface during the Holocene transgression (Fig. 8). On flat-lying constructional shelves such as the Middle Atlantic Bight of North America, relief elements on the surface of this sheet formed as the zones of nearshore sand storage (estuary mouth and cape extension shoals; shoreface-connected sand ridges: Swift et al., 1972), and both the surface morphology and the internal structure of the sand sheet bear little relation to the surface morphology and the internal structure of the older strata beneath (McClennen and McMaster, 1971). On shelves of greater relief, the surficial sediment blanket occurs as a thin drape over topographic highs, broken by substrate outcrops. In the adjoining basins, marginal sands, shed by highs, pass laterally into deposits of mud (Fig. 9).

The stratigraphy of the surficial deposits of the shelf is twofold. On shelves bordering low coasts a lower unit of fine sands and mud was deposited in the belt of lagoons and estuaries in advance of the main shoreline (Fig. 10). Its lower surface is ribbed with estuarine channel fillings that fill the buried drainage pattern of the pre-Recent substrate (Sheridan et al., 1974; Emery, 1968). Meandering of these channels in response to

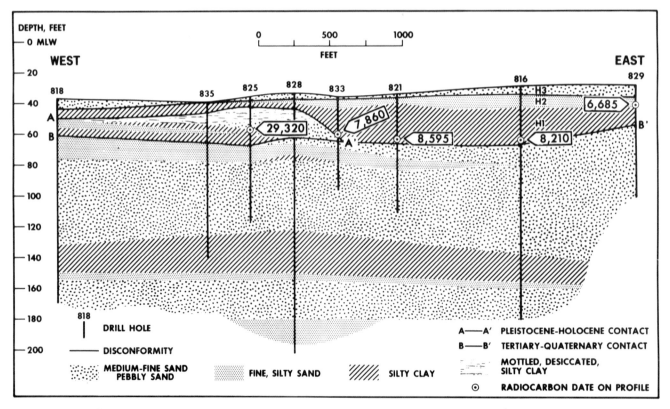

FIGURE 8.   *Surficial sand of the inner New Jersey shelf. Stratum H3 is a shelf sand. Stratum H2 is a backbarrier sand. Stratum H1 is a lagoonal mud. Thick zone in H1 is inferred to be a filled tidal scour channel whose axis is normal to the plane of the diagram. The sequence was produced by coastal retreat during the Holocene transgression. The barrier superstructure is represented in this sequence by an unconformity; its forward face underwent continuous erosional retreat (Chapter 14, Fig. 10A) and the resulting debris accumulated seaward of the shoreface as the leading edge of H2. From Stahl et al. (1974).*

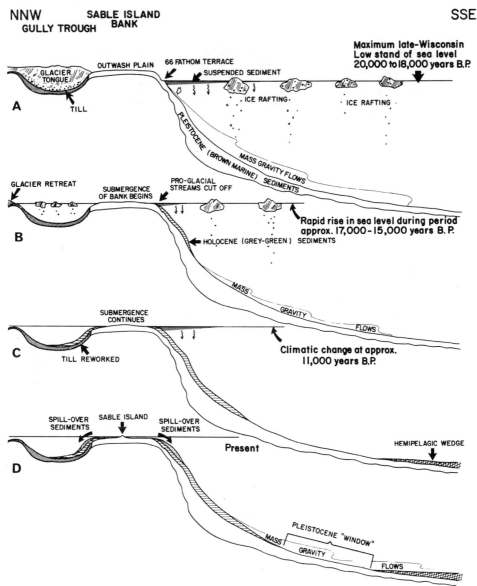

FIGURE 9. *Evolution of the surficial sand shelf on a glaciated shelf of appreciable relief (Nova Scotia shelf; compare with Fig. 5). From Stanley (1969).*

tidal flow has served to reduce the relief of the buried subaerial surface. Tidal inlets scour trenches into the lagoonal stratum, then backfill these trenches as they migrate downdrift (Hoyt and Henry, 1967; Kumar and Sanders, 1970). The lagoonal carpet is itself discontinuous; Pleistocene beach ridges and other topographic highs protrude through the lagoonal deposits as peninsulas or islands during their formation and are sheared off by shoreface retreat during passage of the main shoreline (Sheridan et al., 1974). On rocky coasts, lagoonal and estuarine deposits are confined to shelf valleys.

Passage of the main shoreline results in destruction of the barrier and the upper part of the lagoonal sequence,

and in the deposition of a second major stratum, a sheet of residual sand. This sand sheet overlies a surface of marine erosion whose areal geology is a patchwork of remnant lagoonal deposits and older substrate. On shelf sectors where the lagoonal carpet is well developed, this sand must travel from eroding headlands along the shoreface of retreating barriers, before being spread over the lagoonal carpet; or it is released as the retreating shoreface cuts into tidal inlet fills, or into estuarine channel sands scoured out of the pre-Recent substrate (Andrews et al., 1973). On shelves with poorly developed lagoonal strata, the retreating shoreface may be incised all the way through the lagoonal deposits and into

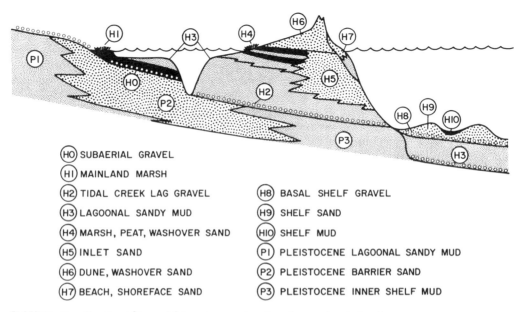

(HO) SUBAERIAL GRAVEL

(H1) MAINLAND MARSH

(H2) TIDAL CREEK LAG GRAVEL   (H8) BASAL SHELF GRAVEL

(H3) LAGOONAL SANDY MUD   (H9) SHELF SAND

(H4) MARSH, PEAT, WASHOVER SAND   (H10) SHELF MUD

(H5) INLET SAND   (P1) PLEISTOCENE LAGOONAL SANDY MUD

(H6) DUNE, WASHOVER SAND   (P2) PLEISTOCENE BARRIER SAND

(H7) BEACH, SHOREFACE SAND   (P3) PLEISTOCENE INNER SHELF MUD

FIGURE 10.   *Stratigraphic model for a low coast undergoing erosional shoreface retreat.*

Pleistocene sands, whose erosion provides material for the surficial sand sheet.

The basal layer of the surficial sand sheet is a thin, discontinuous gravel (Powers and Kinsman, 1953; Belderson and Stride, 1966; Veenstra, 1969; Norris, 1972) or shell hash rich in backbarrier and beach species (Fischer, 1961; Merrill et al., 1965; Milliman and Emery, 1968; Field, 1974). More exotic clasts are clay pebbles eroded from Early Holocene lagoonal deposits, elephant teeth (Whitmore et al., 1967), and concretions from Tertiary strata (Stanley et al., 1967). The basal gravel is rarely more than a meter thick. It is

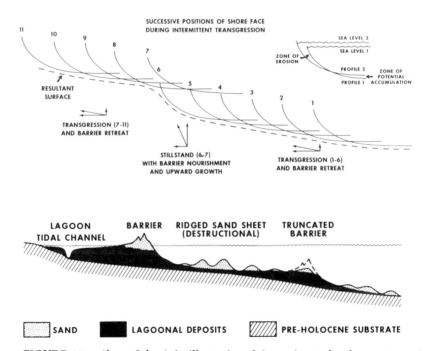

FIGURE 11.   *Above: Schematic illustration of intermittent shoreface retreat. As shoreface profile translates primarily landward in response to rising sea level, material eroded from shoreface accumulates on adjacent shelf as ridged sand sheet. Periods of primarily vertical translation of profile followed by periods of resumed landward translation result in truncated scarp. Below: Resulting stratigraphy. From Swift et al. (1973).*

commonly overlain by 1 to 10 m of sand, with a sublittoral molluscan fauna (Shideler et al., 1972).

TERRACES AND SCARPS. Most continental shelves are terraced, with terraces separated by scarps 10 m or more in height (Fig. 6). Their counterparts on the adjacent subaerial coastal plains mark Quaternary (or earlier) high stands of the sea. Shelf scarps appear to have resulted primarily from stillstands of the returning Holocene sea, although they may in some cases be reoccupied Pleistocene shorelines. On the Georgia coast, for example, the modern barrier island system is perched on the forward face of a Pleistocene shoreline, and the modern tidal inlets are reoccupied Pleistocene inlets (Hoyt and Hails, 1967).

Shelf scarps are drowned shorelines only in the broadest sense; more specifically, they are relict lower shorefaces. To form such a scarp would require a period of near stillstand during a general transgression. The shoreface profile will translate more nearly upward than landward (Fig. 11) during this period, by means of upper shoreface and barrier surface aggradation. At the resumption of rapid transgression, the superstructure of the stillstand barrier will resume its landward migration through the process of shoreface erosion and storm washover, leaving behind a truncated lower shoreface. If both the lower and upper shoreface undergo aggradation during the stillstand period, so that the ideal profile is realized at all times, then there will be no surface expression of the stillstand shoreline, although seismic profiles may reveal a buried scarp (Stanley et al., 1968).

SHELF VALLEY COMPLEXES AND SHOAL RETREAT MASSIFS. A second nearshore marine pattern of relief and sediment distribution that may survive from the nearshore environment during a marine transgression is a shelf valley complex. This term refers to the groups of morphologic elements that occur along the paths of retreat of estuary mouths on autochthonous shelves. Shelf valley complexes are composed of deltas, shelf valleys, and shoal retreat massifs (Figs. 6 and 12). A shoal retreat massif is a broad, shelf-transverse sand ridge of subdued relief that marks the retreat path of a zone of littoral drift convergence (Swift et al., 1972). It may be dissected by subsequent storm or tidal flows into a cross-shelf sequence of smaller coast-parallel ridges, and the term massif is used in the sense of a composite topographic high, itself consisting of smaller highs.

Such complexes are locally well developed on low-relief shelves such as the Middle Atlantic Bight of North America. Here they are largely constructional features molded into the Holocene sand sheet. The sand

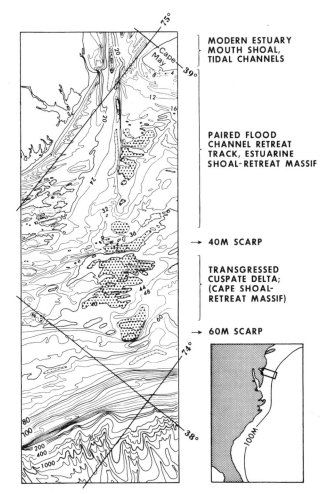

MODERN ESTUARY MOUTH SHOAL, TIDAL CHANNELS

PAIRED FLOOD CHANNEL RETREAT TRACK, ESTUARINE SHOAL-RETREAT MASSIF

→ 40M SCARP

TRANSGRESSED CUSPATE DELTA; (CAPE SHOAL-RETREAT MASSIF)

→ 60M SCARP

FIGURE 12. *The Delaware shelf valley complex, Delaware shelf of North America. Southward littoral drift of New Jersey coastal compartment is injected into reversing tidal stream of mouth of Chesapeake Bay. The resulting shoal is stabilized as a system of interdigitating ebb and flood channels, north of the main couplet of a mutually evasive ebb and flood channel. The shelf valley complex seaward of the bay mouth is the retreat path of the bay mouth sedimentary regime through Holocene time. Retreat of the main flood channel has excavated the Delaware shelf valley; retreat of the bay mouth shoal has left a seaward-trending shoal retreat massif on the shelf valley's north flank. From Swift (1973).*

sheet tends to completely fill the former subaerial valley cut by the river into the Pleistocene strata, and the shelf valley complex and the buried river channel may not everywhere coincide (Fig. 13).

Shelf valley complexes are built in serial fashion by the retreating shoreline. It is important to remember that the last high-energy depositional environment experienced by any given segment was the nearshore zone. As a consequence of remolding of preexisting deposits in this zone, elements of shelf valley complexes are not always what they seem. For instance, in Fig. 12, the topographic characteristics of the Delaware midshelf

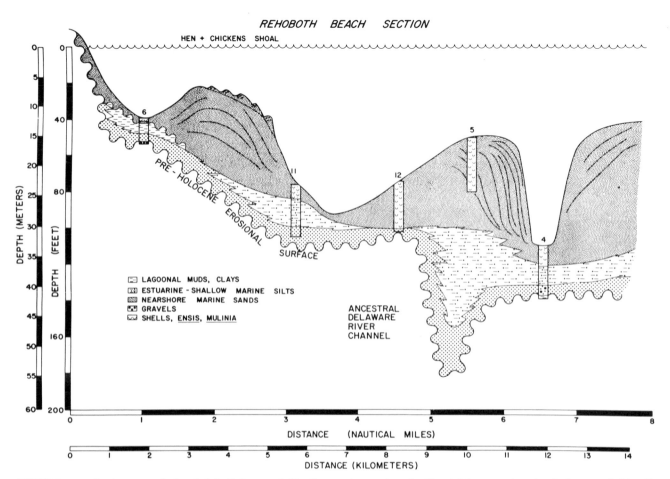

FIGURE 13. *Section across the head of the Delaware shelf valley complex based on vibracores and a 3.5 kHz seismic profile. Marine sand sheet with constructional tidal topography rests on Holocene lagoonal and older Pleistocene deposits. Delaware shelf valley occurs entirely within Pleistocene sands. Note offset between shelf valley and buried river channel. From Sheridan et al. (1974).*

delta suggest that the surface of this stillstand feature was successively remodeled at the resumption of transgression, first as a retreating cuspate foreland, then as cape shoal retreat massif, as illustrated in Chapter 14, Fig. 22. After a further period of stillstand indicated by a 60 m scarp, the coastal regime again changed, and the Delaware River mouth resumed retreat, this time as an estuary. The retreat path of this estuary mouth consists of a sharply defined submarine channel (shelf valley) flanked by a shoal retreat massif. The origins of these two features are easily deduced from uniformitarian reasoning. The shoal retreat massif may be traced into the modern north side shoal of the Delaware estuary mouth. This shoal is a sink for the littoral drift of the New Jersey coastal compartment, and is stabilized by a system of interdigitating ebb and flood channels. The shelf valley may be traced into the flood channel of a large ebb channel–flood channel couplet on the south side of the estuary mouth that accommodates most of the tidal discharge.

On the central and southern Atlantic shelf of North America, four basic morphologic provinces may be described on the basis of constructional morphologic elements inherited from the retreating shoreline (Fig. 14). In the Middle Atlantic Bight (Fig. 6), widely spaced master streams have resulted in widely spaced shelf valley complexes. The plateaulike interfluves between the shelf valley complexes bear ridge fields that were also generated by shoreface retreat (see Chapter 14, Fig. 28).

The more intense wave climate experienced by the Carolina salient has elicited a different response from the retreating river mouths. Capes Romain, Fear, Lookout, and Hatteras may have originally been cuspate deltas, associated with the Peedee, Cape Fear, Neuse, and Pamlico rivers (Chapter 14, Fig. 26). Retreat of these forelands has left large widely spaced shoal retreat massifs. South of Cape Romain the retreat of small, closely spaced cuspate forelands has generated a blanket of coalescing shoal retreat massifs on the adjacent shelf

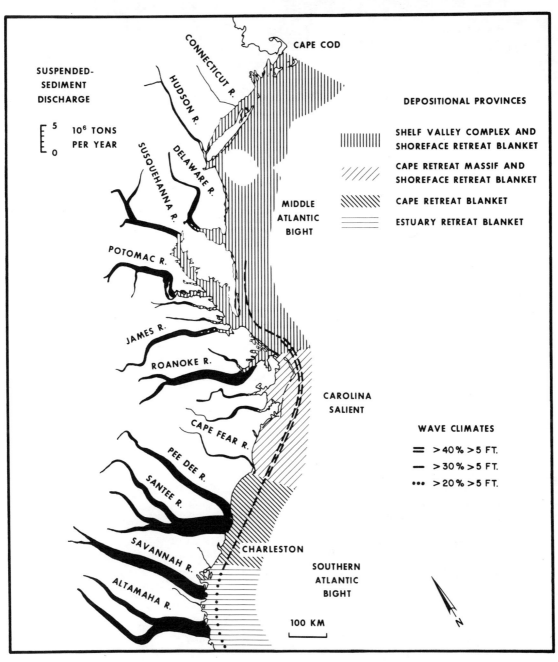

FIGURE 14. *Coastal sediment discharge (Meade, 1969) and wave climate of the Middle Atlantic Bight (Dolan et al., 1973) and resulting depositional provinces. From Swift and Sears (1974).*

(Fig. 15). Yet further south, the Georgia Bight experiences a high tide range, a milder wave climate, and the closely spaced river mouths are estuarine in configuration. Their retreat has generated a blanket of coalescing shelf valley complexes (Fig. 16).

Initiation of Modern Patterns

TEXTURAL AND MORPHOLOGIC PATTERNS ON A STORM-DOMINATED SHELF.   On two of the best studied autoch-

thonous shelves, the Middle Atlantic Bight of North America and the shelf around the British Isles, the hydraulic climate is sufficiently intense to overprint older subaerial and nearshore marine patterns of the surficial sand sheet with a modern textural and morphologic pattern.

In the Middle Atlantic Bight fair-weather flows are driven by the geostrophic response of the stratified shelf water column to freshwater runoff and to winds (McClennen, 1973; Bumpus, 1973); see Fig. 17. How-

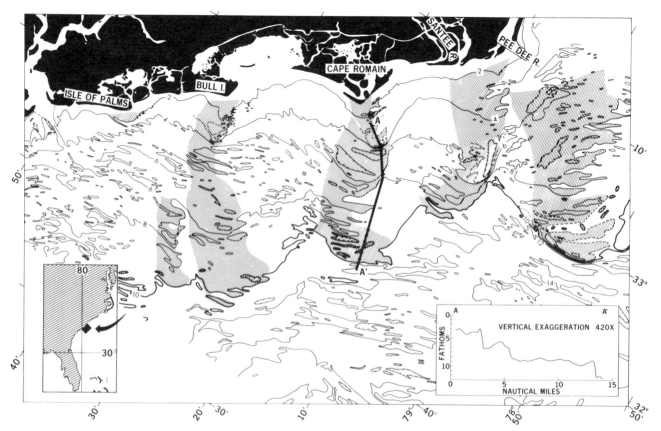

FIGURE 15. *Cuspate forelands and cape shoal-retreat massifs (stippled) of the South Carolina shelf. Note overprinting by ridge and swale topography. Contours in fathoms. From Swift et al. (1972).*

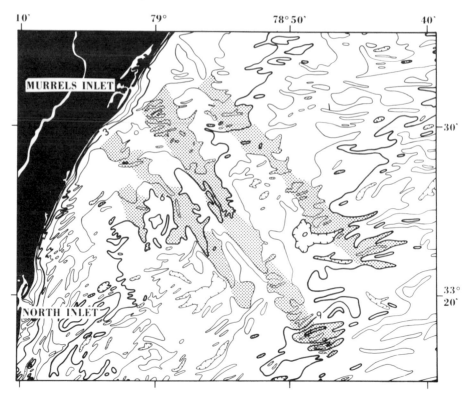

FIGURE 16. *Morphologic pattern of estuarine shoal retreat blanket, overprinted by ridge and swale topography, South Carolina coast. Highs are stippled. Contours in fathoms. From Swift and Sears (1974).*

326

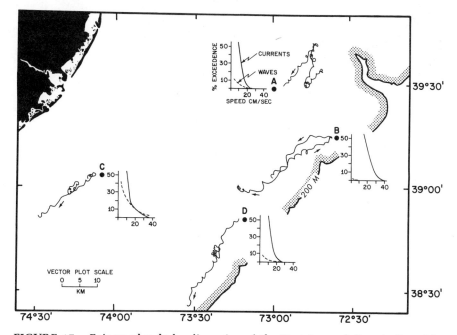

FIGURE 17. *Fair-weather hydraulic regime of the New Jersey shelf as indicated by Savonius rotor current meters mounted 1.5 to 2.0 m above the seafloor at four stations on the New Jersey shelf, for periods of 9 to 11 days in late spring. Progressive vector diagrams indicate a general southerly water drift, partly correlatable with wind directions (McClennen, 1973). Loops, spikes, and bulges on progressive vector diagrams are modulation by the semidiurnal tide. In percent exceedence diagrams, current velocities are compared with bottom wave surge calculated from wave climate data. All data from McClennen (1973).*

ever, neither these unidirectional flow components nor the superimposed wave oscillations (McClennen, 1973) and tidal oscillations (Redfield, 1956) are strong enough to result in significant bed load transport over broad areas. During the winter period of frequent storms, the water column is not stratified, and air-water coupling is more efficient (see discussion, pp. 263–264). The geometry of the Middle Atlantic Bight is especially conducive to strong flows during this period. When low-pressure systems pass over the bight, so that the isobars of atmospheric pressure parallel the isobaths of the shelf surface, the resulting winds blow southward down the length of the Middle Atlantic Bight, paralleling the curve of the shoreline, and induce a uniform setup of the shelf water mass against the coast of 40 to 60 cm. High-velocity "slablike" flows of remarkable longshore coherence result (Beardsley and Butman, 1974; Boicourt, personal communication).

The coastal boundary of these storm flows appears to initiate ridge topography at the foot of the shoreface (Duane et al., 1972); see Chapter 14, Figs. 28 and 31. However, it is clearly an oversimplification to describe the ridge topography of the Middle Atlantic Bight as a purely inherited topographic pattern. The ridges maintain their characteristic 10 m relief and textural patterns across the central shelf (Swift et al., 1974);

see Fig. 18. Troughs retain erosional windows in which Early Holocene lagoonal clays are thinly veneered with a lag deposit of pebbly sand. Calculations based on current-meter records suggest that the unidirectional components of storm flows are sufficient to mobilize the sandy bottom (Fig. 19), and to slowly level the ridge topography, if the topography were not in fact a continuing response of the seafloor to the modern hydraulic climate.

In several areas on the Middle Atlantic shelf, there is evidence to suggest that ridge topography may be initiated on the central shelf, if not already present as a survival from the nearshore environment. Off South Carolina, shoal retreat massifs are overprinted by a ridge topography even though the modern nearshore zone is not apparently forming ridges (Fig. 15). Elsewhere, the ridge pattern appears to have changed as the water column deepened and the shoreline receded during the course of the Holocene transgression. The estuary mouth shoal that is the landward end of the Delaware Massif (Fig. 12) has impressed into it a tide-maintained ridge pattern that trends normal to the shoreline and parallel to the sides of the estuary mouth. As the crest of the massif is traced seaward, the trend of the ridges and troughs superimposed on it shifts toward a shore-parallel orientation. The bay

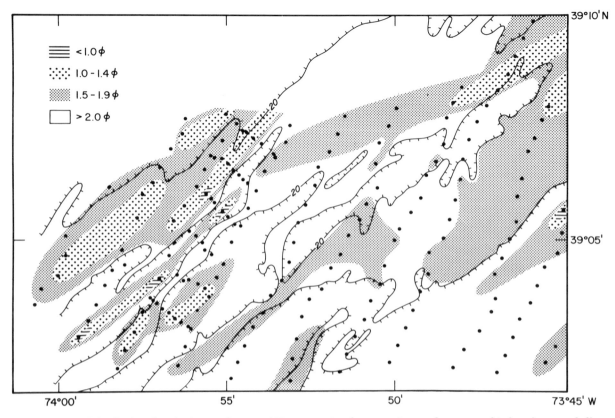

FIGURE 18. *Distribution of grain sizes on the central New Jersey shelf. Medium to fine sands occur on ridge crests. Fine to very fine sands occur on ridge flanks and in troughs. Locally,* *erosional contours in troughs expose a thin lag of coarse, shelly, pebbly sand over lagoonal clay. From Stubblefield et al. (in press).*

mouth ridges are oriented parallel to the reversing tidal flows of the bay mouth; the offshore ridges appear to parallel instead the geostrophic storm flows of the open shelf.

The Great Egg Massif, associated with the former course of the Schuylkill River across the shelf, has been heavily dissected into a transverse ridge pattern. Seaward of a scarp whose toe lies at 90 m, a second, small-scale ridge pattern with a somewhat different trend has been superimposed on the first (Fig. 20). Stubblefield and Swift (1975) have presented a model for the evaluation of the compound ridge pattern based on vibracores and 3.5 kHz seismic profiles collected in the area (Fig. 21). Radiocarbon dates indicate that the large-scale ridges appear to have formed immediately subsequent to the passage of the shoreline at approximately 11,000 BP (Fig. 21*A*). Internal stratification indicates that large-scale ridges grow by the accretion of conformable beds. Wide, large-scale troughs appear as zones of bare Pleistocene substrate, where the surficial sand sheet was never formed, or where its material was swept away to nourish the growth of adjacent ridges.

With continuing transgression and deepening of the water column, the ridges appear to have increased their spacing by means of lateral migration or the coalescence of adjacent ridges. Internal strata tend to dip more steeply than present ridge flanks, suggesting that toward the latter part of their history, ridge growth was mainly the consequence of lateral rather than vertical accretion

Small-scale troughs transect large-scale ridges, and tend to break large-scale ridges up into *en echelon* segments (Fig. 20). Where small-scale troughs cross large-scale troughs they are incised into the flat-lying Early Holocene and Pleistocene strata that floor the large-scale troughs. Small-scale troughs are commonly narrow features that do not penetrate through the Early Holocene lagoonal clay (Fig. 21*B*). Where this clay is in fact breached, so that the small-scale troughs penetrate the underlying sand, the troughs are noticeably wider, as though they had expanded by undercutting of the clay in a fashion analogous to the growth of a blowout on a grass-covered eolian flat (Fig. 21*C*).

The ridge topography of the Middle Atlantic Bight is accompanied by mesoscale bed form patterns, whose relationship to the ridge pattern is not clearly understood. The most ubiquitous mesoscale bed forms are the current lineations, which occur as sand ribbons or more

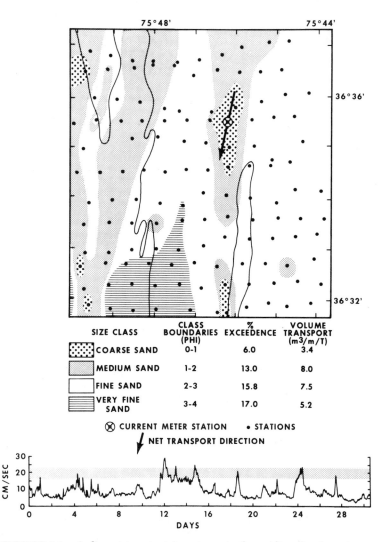

FIGURE 19. *Sediment transport in response to the unidirectional component of flow during the month of November 1972, in an inner shelf ridge field, False Cape, Virginia. Estimates based on Shield's threshold criterion, a drag coefficient of 3 × 10⁻³, and Laursen's (1958) total load equation. Values expressed as cubic meters of quartz per meter transverse to transport direction for time elapsed. Solid line is the 10 m isobath.*

commonly as linear erosional furrows. They may trend parallel to the trend of the ridge topography, or may cut across it, so as to make a larger acute angle with the shoreline (Fig. 22). Toward the southern end of the Middle Atlantic Bight, the shelf surface shoals, narrows and curves to the east. Sand wave fields appear, perhaps indicative of the acceleration of storm flows in response to the decreasing cross-sectional area of the shelf water column.

Ridges molded into the Albermarle shoal retreat massif bear sand waves on their crests (Fig. 23). Sand waves locally attain 2 m heights and angle of repose slopes. Sand wave crestlines are not quite normal to shore, suggesting that the ridge crests on which they are

found experience a seaward component of flow during storms. At Diamond Shoals, the southern extremity of the Middle Atlantic Bight, sand waves up to 7 m high occur between sand ridges, forming a reticulate pattern (see Fig. 27, Chapter 16).

Grain-size patterns in the Middle Atlantic Bight suggest that the storm flows that interact with the ridge topography and the mesoscale bed forms are capable of transporting at least the finer grades of sand for appreciable distances. The inner shelf sectors before the seaward-convex coastal compartments of the Middle Atlantic Bight exhibit a repeating pattern of grain-size distribution (Fig. 24). The northern half of each of these inner shelf sectors, where south-trending storm flows must

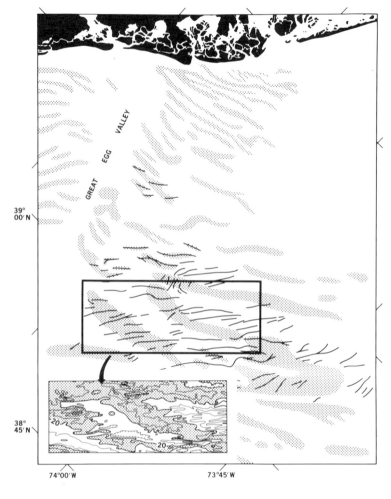

FIGURE 20. *Great Egg shelf valley and shoal retreat massif. Large-scale ridges in inset may date from a period when the ancestral Great Egg estuary was active. Nearshore large-scale ridges were probably formed by shoreface detachment during erosional retreat of the shoreface, after capture of the ancestral Schuylkill River by the Delaware River, and consequent reduction in discharge of the Great Egg estuary. See Fig. 6 for relationships of Schuylkill, Delaware, and Great Egg rivers. From Stubblefield and Swift (in press).*

presumably converge with the shoreline, tend to be floored with primarily medium- and coarse-grained sands, molded into a well-defined ridge topography. On the southern halves of the coastal compartment, where the shoreline tends to curve to the west, storm flows might be expected to expand and decelerate. Here the fine sand blanket of the shoreface extends across the inner shelf floor, as though nourished by material swept out of the ridge topography to the north. The schematic flow pattern in the lowest panel of Fig. 24 is not basic on detailed observations. It is intended to indicate that current flowing generally southwest parallel to the long dimension of the shelf will tend to converge with the northeastern portion of the shoreline and diverge from the southwestern portion of the shoreline.

A somewhat closer relationship appears to exist between flow geometry and sediment distribution in the vicinity of the shoal retreat massifs (Fig. 25). The ridge topography attains its maximum relief where it has been molded onto the crests of the massifs. The massifs do not exhibit bilateral symmetry; troughs are deepest and widest on the northern sides. As a trough axis is traced across the massif, erosional windows exposing the basal pebbly sand or the underlying clayey substrate become less frequent. The fine sands of the trough flanks tend to bridge across the trough floor.

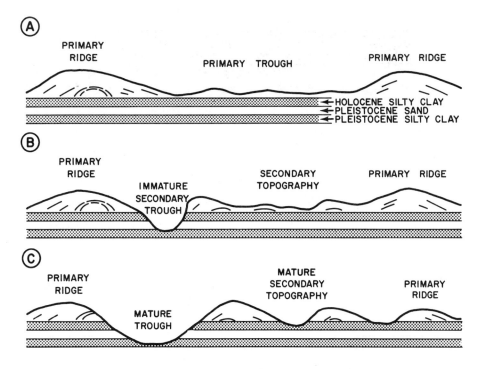

DEVELOPMENT OF RIDGE TOPOGRAPHY

FIGURE 21. *Ridge evolution on the central New Jersey shelf. (A) Ridge nuclei are formed during the process of ridge detachment and shoreface retreat, or by other means in the nearshore zone. Sand continues to be swept out of troughs onto ridges as water column deepens during the course of the Holocene transgression. (B) Seafloor scour during storms locally penetrates the Early Holocene lagoonal clay carpet, and a secondary trough forms, initially by downcutting. (C) Downcutting in the secondary trough decreases and lateral erosion increases as second silty clay layer is exposed. Secondary trough widens by undercutting of upper clay in "blowout" fashion. Sand from similar excavations upcurrent forms secondary ridges. From Stubblefield and Swift (in press).*

The trough axis tends to climb toward a low sill on the southern side of the massif; beyond this the seafloor drops off rapidly to the adjacent shelf valley. The valley floor commonly consists of fine to very fine featureless sand. The topography and grain-size pattern suggest that south-trending flows converge with the rising seafloor and accelerate up the northern flanks of the massifs. Fine sand swept out of the troughs is deposited in the zone of flow expansion and deceleration over the shelf valley south of the massif.

TEXTURAL AND MORPHOLOGIC PATTERNS ON A TIDE-DOMINATED SHELF. The tide-swept shelf around the British Isles (Stride, 1963) provides an interesting contrast with the storm-induced sedimentation of the Middle Atlantic Bight. Surges are at least as frequent here as in the Middle Atlantic Bight (Steers, 1971). However, much more work is done on the seafloor by the semidiurnal tidal currents associated with the amphidromic edge waves that sweep the margins of mar-

ginal shelf seas of western Europe (see Chapter 5, Fig. 4 and discussion, p. 60). The rotary tidal currents associated with these tidal waves are in fact analogous in some respects to the inertial wind-driven currents generated by storms. Midtide surface velocities in excess of 50 cm/sec (1 knot) are sustained over vast areas, and locally exceed 200 cm/sec. Ebb-flood discharge differentials result in currents residual to the tidal cycle, whose velocities may be as great as a tenth of the midtide value.

As a consequence of the higher rate of expenditure of energy on the seafloor, morphologic and textural patterns inherited from the retreating nearshore zone have been largely erased. Erosional shoreface retreat has resulted in a surficial sediment sheet that is comparable in many respects to that of the Middle Atlantic Bight (see Belderson and Stride, 1966). However, the poorly resolved sand transport patterns of the Middle Atlantic Bight are replaced by well-defined transport paths, with sand streams that diverge from beneath

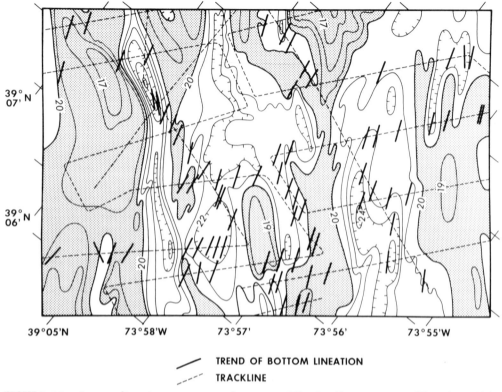

TREND OF BOTTOM LINEATION

TRACKLINE

FIGURE 22. *Current lineation patterns on the central New Jersey shelf. Bars indicating lineations are over 10 times as long as features that they rep-* *resent. They locally represent sets of lineations. High areas are stippled. Contours in fathoms. From McKinney et al. (1974).*

tide-induced "bed load partings" and flow down the gradient of maximum tidal current velocities until either the shelf edge or a zone of "bed load convergence" and sand accumulation is reached (Stride, 1963; Kenyon and Stride, 1970; Belderson et al., 1970); see Fig. 26.

Each stream tends to consist of a sequence of more or less well-defined zones of characteristic bottom morphology and sediment texture (Fig. 27). Streams may begin in high-velocity zones [midtide surface velocities in excess of 3 knots (150 cm/sec)]. Here rocky floors are locally veneered with thin (centimeters thick) lag deposits of gravel and shell. Where slightly thicker, the gravel may display "longitudinal furrows" parallel to the tidal current (Stride et al., 1972), a bed form related to sand ribbons (see Chapter 10, p. 170).

Between approximately 2.5 and 3.0 knots (125–150 cm/sec) sand ribbons are the dominant bed form (Kenyon, 1970). These features are up to 15 km long and 200 m wide, and usually less than a meter deep. Their materials are in transit over a lag deposit of shell and gravel. Kenyon has distinguished four basic patterns that seem to correlate with maximum tidal current velocity and with the availability of sand (Chapter 10, Fig. 15).

Further down the velocity gradient, where midtide surface velocities range from 1 to 2 knots (50–100 cm/sec), sand waves are the dominant bed form. Where the gradient of decreasing tidal velocity is steep or transport convergence occurs, this may be the sector of maximum deposition on the transport path. Over 20 m of sediment has accumulated at the shelf-edge convergence of the Celtic Sea, although it is not certain that this sediment pile is entirely a response to modern conditions.

The Hook of Holland sand wave field off the Dutch coast is one of the largest (15,000 km²) and the best known (McCave, 1971). The sand body is anomalous in that it sits astride a bed load parting; the sand patch as a whole may be a Pleistocene delta or other relict feature. Sand waves with megaripples on their backs grow to equilibrium heights of 7 m with wavelengths of 200 to 500 m in water deeper than 18 m; in shoaler water, wave surge inhibits or suppresses them. Elongate tidal ellipses favor transverse sand wave formation, and the sand waves tend to be destroyed by midtide cross

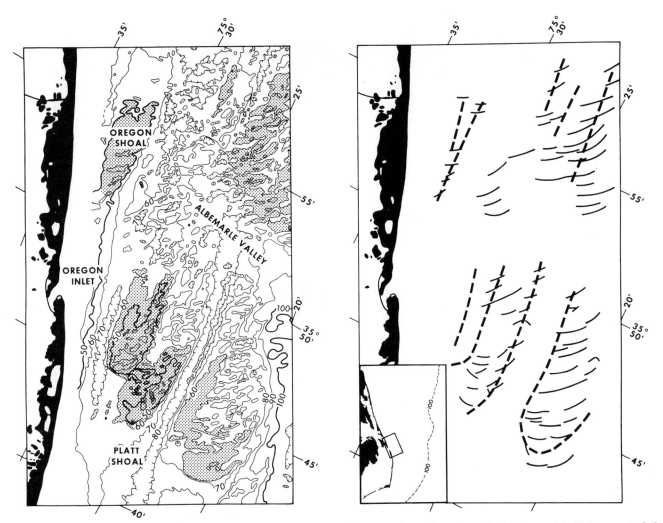

FIGURE 23. *Sand ridges with superimposed sand waves on the northern North Carolina inner shelf. Topographic highs are stippled. Contours in feet. From Swift et al.* (1973).

flow when the ellipse is less symmetrical. Under the latter condition, linear sand ridges may be the preferred bed form, as midtide cross flow would tend to nourish rather than degrade them (Smith, 1969). The triangular sand wave field is limited by a lack of sand on the northwest, by shoaling of the bottom and increasing wave surge on the coast to the south, and by fining of sand to the point that suspensive transport is dominant to the north (McCave, 1971).

Further down the velocity gradient, beyond the zones of obvious sand transport, there are sheets of fine sand and muddy fine sand and in local basins, mud. They lack bed forms other than ripples, and appear to be the product of primarily suspensive transport (McCave, 1971) of material that has outrun the bed load stream (see discussion, Chapter 10, p. 160). These deposits may be as thick as 10 m (Belderson and Stride, 1966), but

where they do not continue into mud, they break up into irregular, current-parallel or current-transverse patches of fine sand less than 2 m thick, resting on the gravelly substrate.

The complex pattern and mobile character of the shelf floor around the British Isles have led British workers to reject the relict model for the shelf sediments (Belderson et al., 1970). They note that it correctly draws attention to the autochthonous origin of the sediment, but that it fails to allow for its subsequent dynamic evolution. They propose instead a dynamic classification:

1. Lower sea-level and transgressive deposits, patchy in exposure, but probably more or less continuous beneath later material; largely the equivalent of a blanket (basal) conglomerate.

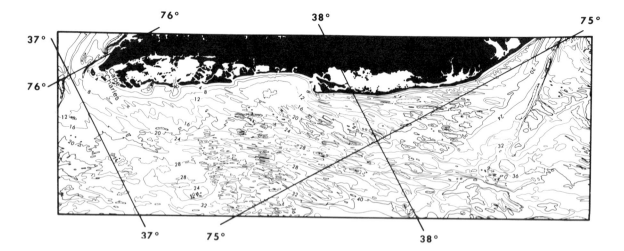

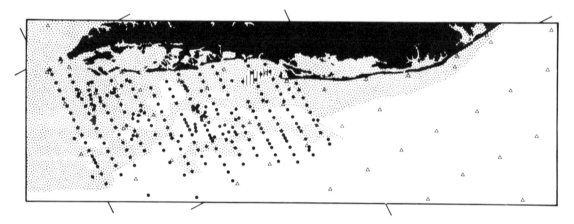

VERY COARSE TO MEDIUM SAND   FINE SAND   VERY FINE SAND, SILTY CLAY   △ WOODS HOLE DATA   ● VA. INST. MAR. SCI. DATA

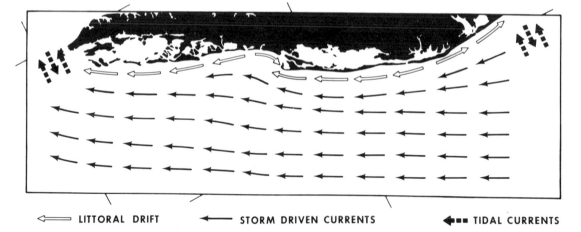

⇐ LITTORAL DRIFT   ← STORM DRIVEN CURRENTS   ◄■■ TIDAL CURRENTS

FIGURE 24. *Above: Bathymetry of the Delmarva inner shelf. From Uchupi (1970). Center: Distribution of sediment. From Hathaway (1971) and Nichols (1972). Below: Inferred* *direction of currents responsible for bed load sediment transport. Reproduced from Swift (1975).*

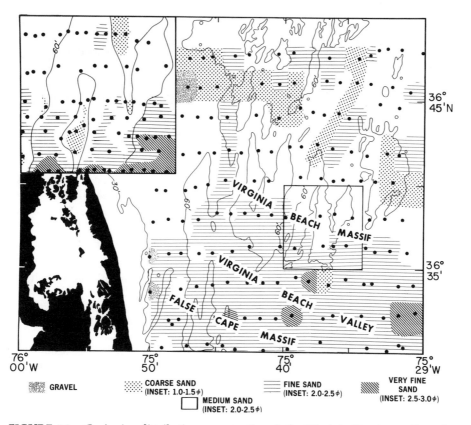

FIGURE 25. *Grain-size distribution on a portion of the Virginia Beach massif, and adjacent shelf valley.*

2. Material moving as bed load (over the coarser basal deposits) mainly well sorted sand and in places first-cycle calcareous sand.

3. Present sea-level deposits (category 2 sediment having come to permanent rest) consisting of large sheets to small patches, which range from gravel and shell gravel to sand and calcareous sands, muddy sands, and mud.

The implication is that of a shelf surface moving toward a state of equilibrium with its tidal regime. The degree of adjustment appears to be greater than in the case of the North American Atlantic Shelf, in that there is less preservation of nearshore depositional patterns. As a consequence of the intensity of the hydraulic climate, there is less on shelf storage (category 3) and more material in transit.

Locally, sand ridges similar to those of the Middle Atlantic Bight do occur. Like those of the Middle Atlantic Bight, they tend to be grouped in discrete fields. In some cases, it is possible to infer that these ridge fields are in fact shoal retreat massifs, generated by the retreat of a near shore depositional center during the course of the Holocene transgression (Swift, 1975). The clearest case may be made for the Norfolk Banks

(Houbolt, 1968; Caston and Stride, 1970; Caston, 1972); see Fig. 28. Here a series of offshore sand ridges may be traced into a modern nearshore generating zone (Robinson, 1966; see Chapter 14, Fig. 39) where sand is packaged by the specialized tidal regime of the shoreface into shapes hydrodynamically suited for survival on the open shelf (see discussion, p. 180). The Nantucket Shoals sector of the North American Atlantic shelf appears to constitute a similar evolutionary sequence of ridges (see Chapter 10, Fig. 30).

The Norfolk Banks are analogous to the cape shoal retreat massifs of the Carolina coast of North America, in that the generating zone is a coastal salient that serves as a sink for the nearshore sand flux. Other, more poorly defined ridge fields in the southern bight of the North Sea (Fig. 28) may be analogous to the estuarine shoal retreat massifs of the Middle Atlantic Bight in that they may have been generated by the retreat of the ancestral Rhine and Thames estuaries.

ALLOCHTHONOUS PATTERNS OF SEDIMENTATION

Shelves undergoing allochthonous sedimentation differ from autochthonous shelves in a variety of character-

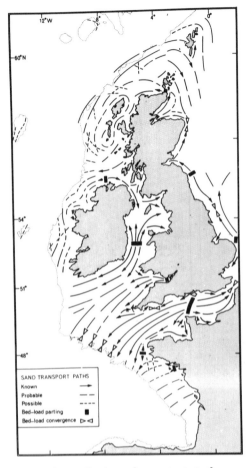

FIGURE 26. *Generalized sand transport paths around the British Isles and France, based on the velocity asymmetry of the tidal ellipse and the orientation and asymmetry of bed forms. From Kenyon and Stride (1970).*

istics. The most obvious is that allochthonous shelves tend to be floored by fine sands, fine muddy sands, or muds that have escaped from adjacent river mouths: autochthonous shelves in contrast are generally covered by coarser grained sand of local origin. Although surfaces of allochthonous shelves are constructional in nature, they tend to be smooth and featureless; their fine materials have traveled primarily in suspension, and the effective underwater angles of repose of the sediment may be too low to result in such large-scale bed forms as sand waves or sand ridges. However, such features are not totally unknown. Allersma (1972) has reported "mud waves" from the Venezuelan shelf that appear to be very similar to the shoreface-connected ridges of the Middle Atlantic Bight.

Transport on Allochthonous Shelves

Mechanisms of sediment transport on allochthonous shelves have been generally described by Drake in Chapter 9. Since this chapter stresses regional transport patterns, it seems worthwhile to summarize Drake et al.'s (1972) study of river-dominated sedimentation on the southern California shelf. This carefully documented, real time study of the dispersal of flood sediment is probably the most detailed report on the nature of allochthonous sediment dispersal available at the time of writing.

In January and February of 1969, southern California experienced two intense rainstorms which resulted in a record flood discharge. The freshly eroded sediment was a distinctive red-brown in contrast to the drab hue of

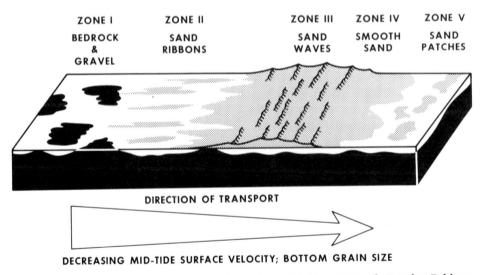

FIGURE 27. *Succession of morphologic provinces along a tidal transport path. Based on Belderson et al. (1970).*

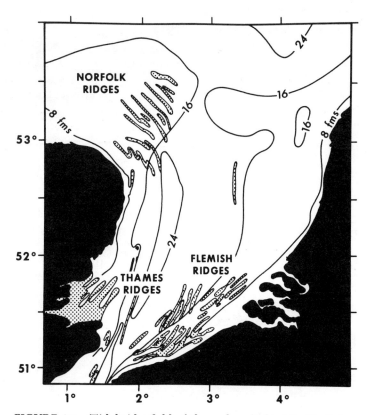

FIGURE 28. *Tidal ridge fields of the southern bight of the North Sea. Northernmost ridge field appears to constitute a shoal retreat massif, marking the retreat path of the nearshore tidal regime of the Norfolk coast. Ridge fields in the approach to the English Channel may have been initiated in an earlier, more nearly estuarine environment. From Houbolt (1968).*

the reduced shelf sediments. The flood deposit on the shelf could therefore be repeatedly cored and isopached, and its shifting center of mass traced seaward through time.

USGS stream records show that 33 to 45 $\times$ 10^6 metric tons of suspended silt and clay and 12 to 20 $\times$ 10^6 metric tons of suspended sand were introduced by the Santa Clara and Ventura rivers. By the end of April 1969, more than 70% of this material was still on the shelf in the form of a submarine sand shoal extending 7 km seaward, and a westward-thinning and -fining blanket of fine sand, silt, and clay existed seaward of that (Fig. 29A).

By the end of the summer of 1969, the layer extended further seaward, had thinned by 20%, and had developed a secondary lobe beneath the Anacapa current to the south (Fig. 29B). Eighteen months after the floods, the surface layer was still readily detectable. Considerable bioturbation, scour, and redistribution had occurred south of Ventura, but the deposit was more stable to the north (Fig. 29C).

A concurrent study of suspended sediment distribution in the water column revealed the pattern of sediment transport (Fig. 29D). Vertical transparency profiles, after four days of flooding, showed that most of the suspended matter was contained in the brackish surface layer, 10 to 20 m thick. Profiles in April and May revealed a layer 15 m thick, with concentrations in excess of 2 mg/l, and a total load of 10 to 20 $\times$ 10^4 metric tons. Since this load was equal to river discharge for the entire month of April, it must have represented lateral transport of sediment resuspended in the nearshore zone. Vertical profiles over the middle and outer shelf for the rest of the year were characterized by sharply bounded turbidity maxima, each marking a thermal discontinuity. These also were nourished by lateral transport from the nearshore sector where the discontinuities impinged on the sloping bottom. The near-bottom nepheloid layer was the most turbid zone in the inner shelf. This nepheloid layer was invariably the coolest, and was invariably isothermal, indicating that its turbidity was the result of turbulence generated

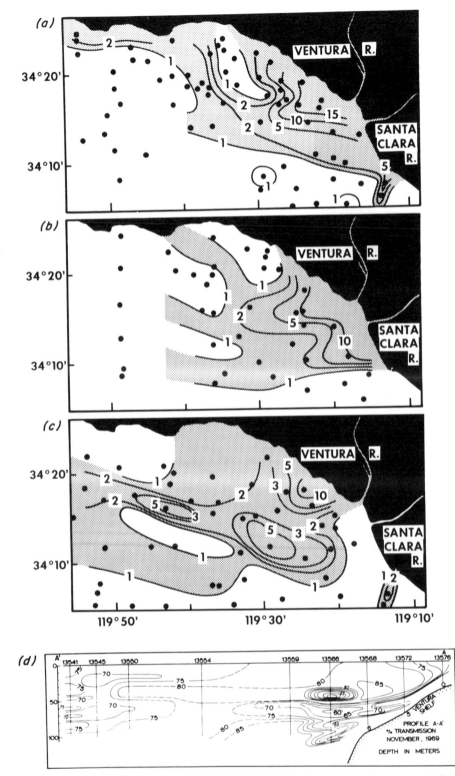

FIGURE 29. *Thickness of flood sediment (centimeters) on the Santa Barbara–Oxnard shelf in* (a) *March–April 1969;* (b) *May–August 1969; and* (c) *February–June 1970, based on cores.* (d) *East-west cross section showing vertical distribution of light-attenuating substances over Santa Barbara–Oxnard shelf. For clarity, the bottom 20 m of the water column is not contoured, but the percent transmission value at the bottom is noted. From Drake et al.* (1972).

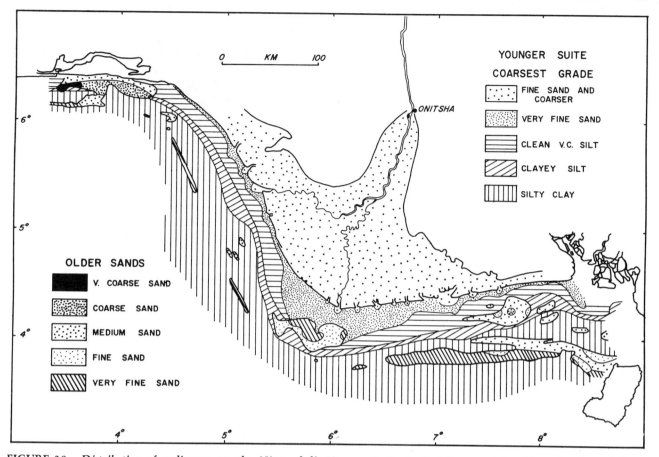

FIGURE 30. *Distribution of sediments on the Niger shelf. Young suite is of allochthonous origin; older suite of autochthonous sand is exposed in nondepositional windows. From Allen (1964).*

by bottom wave surge. Bottom turbidities ranged from 50 mg/l during the flood to 4 to 6 mg/l during the next winter, but were at no time dense enough to drive density currents.

Drake et al.'s study suggests that the transport of suspended sediment across shelves undergoing allochthonous sediment action starts with introduction by a river jet, and continues with deposition, resuspension, and intervals of diffusion and advection by coastal currents in a near-bottom nepheloid layer.

Depositional Patterns on Allochthonous Shelves

Fine sediments deposited on allochthonous shelves may occur as a seaward-thinning sheet (Fig. 30), or as a series of strips of fine sand or mud oriented generally parallel to the shoreline; see Figs. 31 and 32 (see also Venkatarathnam, 1968; McMaster and Lachance, 1969; and Niino and Emery, 1966). On shelves of equant or irregular dimensions, shelf sectors surfaced by far-traveled, fine-grained sediment may be more irregular in shape (Niino and Emery, 1966; McManus et al.,

1969; Knebel and Creager, 1973). Such allochthonous deposits tend to be separated by, or to enclose, nondepositional "windows" in which relatively coarse autochthonous sands are exposed. The disposition of these strips and sheets of allochthonous sediment is generally meaningful in terms of what is known of regional circulation patterns. Locally, the strips may underlie turbid, brackish water plumes that extend from river mouths under the impetus of buoyant expansion and inertial flow (Chapter 14, Fig. 41). Where such flows of high-turbidity water extend for long distances parallel to the coast or seaward across the shelf at promontories, they have been described by McCave (1972) as "advective mud streams" (Fig. 33). He cites Jerlov (1958) as describing such a mud stream running south from the Po Delta, over the mud bed shown in Fig. 34.

However, the presence of windows of older sand does not necessarily mean that the sediment pattern is a transient one, which must be eventually followed by a total masking of the old surface of transgression by fine sediment. Instead, the pattern may be a steady state one,

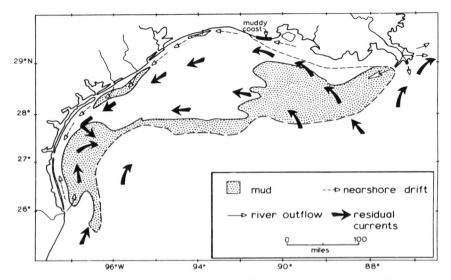

FIGURE 31. *Distribution of mud and generalized transport pattern on the western Gulf shelf of North America. From McCave (1972), after Van Andel and Curray (1960).*

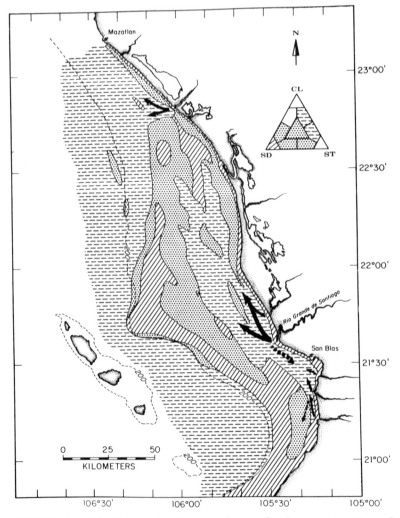

FIGURE 32. *Distribution of sediment and generalized transport pattern off the Nayarit coast, Pacific side of Mexico. Autochthonous sands occur in non-depositional windows. From Curray (1969).*

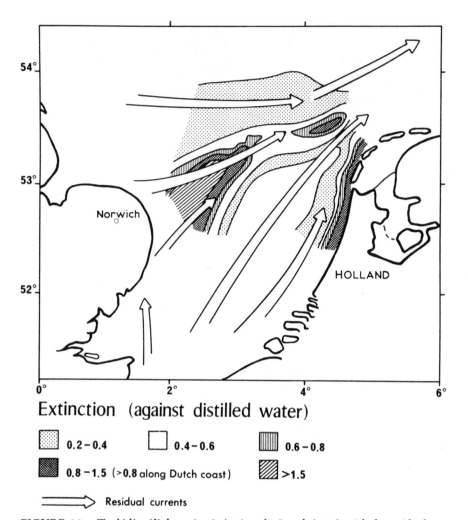

FIGURE 33. *Turbidity (light extinction) given by Joseph (1953) with the residual current pattern in the southern North Sea. Two advective mud streams are illustrated, one crossing from the English to the Dutch side of the area, the other running up the Dutch coast from the mouths of the Plime River. Actual sediment concentrations are higher in the latter. From McCave (1972).*

determined by the local relationship between the hydraulic activity (primarily wave surge) on the bottom and the near-bottom concentration of suspended sediment (Fig. 35), as well as by the regional transport pattern.

On autochthonous shelves, sand transport is primarily advective in nature, occurring during short, intense episodes of wind-driven or tidal flow, and textural gradients tend to reflect the direction of sand transport, with transport becoming finer down the transport path (see Figs. 25 and 27). The transport of fine sand and mud on shelves undergoing allochthonous sedimentation is also primarily advective in nature, in that the turbid water tends to flow as a mass in response to the regional circulation pattern. However, because of the greater role of reversing tidal currents and wave surge in dis-

tributing fine sediment it is convenient to think of fine sediment transport on allochthonous shelves as consisting of a dominant advective component, driven by the regional circulation pattern, and an important but subordinate diffusive component, driven by reversing tidal flows and wave surge.

The diffusive component of transport not only influences the regional pattern of fine sediment deposition as noted in Fig. 35 and Chapter 9, Fig. 15, but may also result in textural gradients within allochthonous sediment sheets that trend at an angle across the advective transport direction. On the Niger shelf, for instance, the dominant, advective transport direction is from east to west, under the impetus of the Guinea current (Allen, 1964). However, bottom sediments tend to become finer in a seaward (north to south) direction (Fig. 36). Sharma et al.

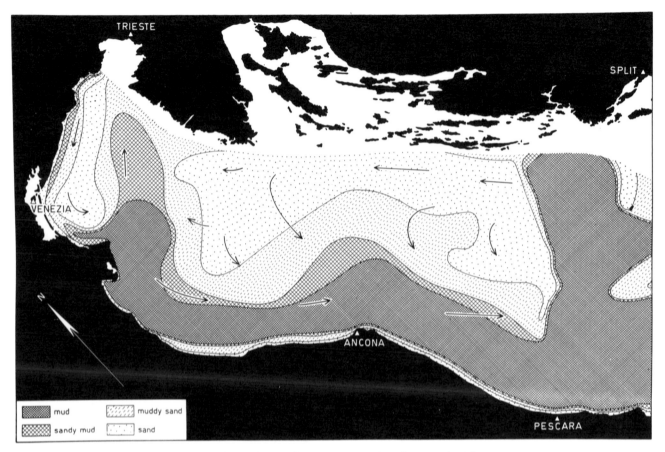

FIGURE 34. *Distribution of sediment in the northern Adriatic Sea. From Van Straaten (1965).*

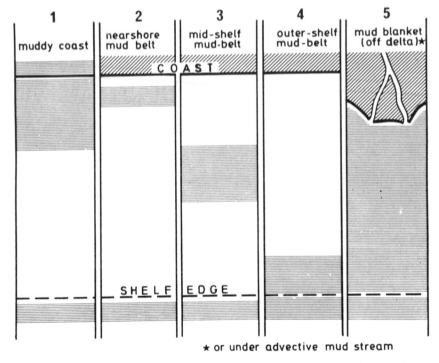

FIGURE 35. *Schematic representation of five cases of sites of shelf mud accumulation. Compare with Fig. 15 in Chapter 9. From McCave (1972).*

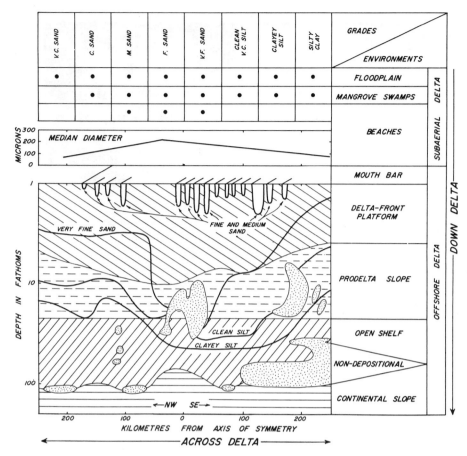

FIGURE 36. *Grain size in relation to sedimentary environments in Niger Delta area. In subaerial delta, all grades present are shown. In offshore part of delta, coarsest grade in near-surface layers is projected onto vertical plane perpendicular to axis of delta symmetry. From Allen (1964).*

(1972) have described grain-size gradients in Briston Bay of the Bering Sea that are more nearly related to the iso-baths than to the prevailing currents. They consider the textural gradients to be the consequence of wave surge diffusion (Fig. 37). The operative mechanism would be progressive sorting during the seaward diffusion of sediment. In this process, sediment that drifts seaward into deeper water during a transport event is likely to leave its coarsest fraction behind when reentrained, because of the weaker nature of deep-water wave surge (see discussion of progressive sorting, Chapter 10, p. 162.

Stratigraphy of Allochthonous Shelves

The tenfold reduction in the rate of eustatic sea-level rise experienced between 4000 and 7000 years ago (Milliman and Emery, 1968) has resulted in a shift from autochthonous to allochthonous regimes in a number of shelf sectors (Curray, 1964). River mouths servicing such shelves have equilibrated with their tidal prisms,

and have begun to bypass fine sediment in quantities sufficient to result in deposition on the shelf surface. Two characteristic stratigraphies have resulted, which may be correlated with the transport schemes illustrated in Figs. 3a and 3c. Where the shift in the balance between the rate of sedimentation and the rate of sea-level rise has not been adequate to cause coastal progradation, the coast has continued to undergo erosional shoreface retreat, or has approached stillstand conditions (see discussion of equation, p. 312). Patches and sheets of fine-grained sediments have accumulated more or less simultaneously over the sand sheet produced during the earlier period of erosional shoreface retreat (Fig. 38). Elsewhere, where the Late Holocene balance between sedimentation and sea-level rise has resulted in coastal progradation, the transgressive sand sheet passes landward beneath a veneer of mud some few meters thick into a thick littoral sand body deposited during stillstand, and a second, subaerial, sand sheet extends seaward over the inner portion of the mud veneer (see Chapter 14, Fig. 10B).

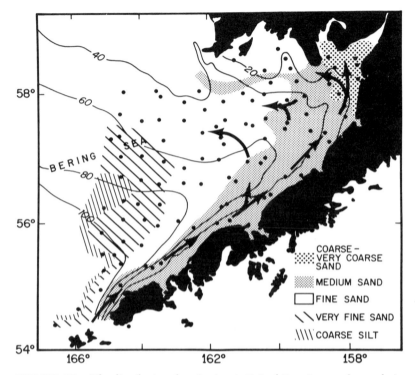

FIGURE 37. *The distribution of grain sizes in Bristol Bay. See text for analysis. From Sharma et al.* (1972).

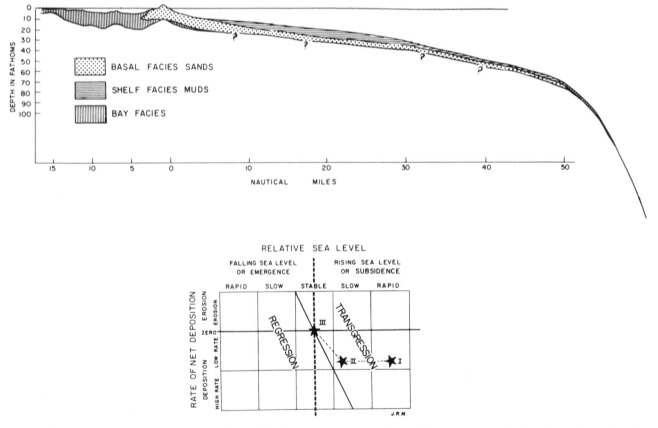

FIGURE 38. *Above: Generalized cross section of Late Quaternary sediments in a line perpendicular to the coast near Rockport,* *Texas. Below: Schematic representation of shoreline migration. From Curray* (1964).

The abrupt nature of the transition from Early Holocene autochthonous regimes and the recent nature of this transition have prevented us from observing on modern continental margins a third characteristic stratigraphy, which is widespread in the rock record, namely "marine onlap" (Grabau, 1913, Fig. 144). In this model, a prolonged period of relatively slow sea-level rise is accomplished by the transport scheme shown in Fig. 3b, where both shoreface and river mouth bypassing occur in a regime of transgressive allochthonous sedimentation. Subaerial and submarine depositional environments are linked by a unified pattern of sediment transport, and their landward displacement results in a threefold sequence of fluvial, marine marginal, and open shelf lithosomes beneath the shelf surface.

During the slow eustatic transgression of the Cretaceous, such a sequence was deposited on the North American shelf off North and South Carolina (Fig. 39). The present erosional surface approximates a time plane, and seaward decrease in grain size across this environment suggests progressive sorting through fluvial, estuarine, and marine environments (Swift and Heron, 1969). Shoreface erosion was an important source of sediment, as indicated by the internal unconformity that largely replaces the littoral sand facies. River mouth injection may also have been an important mechanism of coastal bypassing, since the fine-grained, open marine facies thickens seaward.

The Amazon shelf off Brazil, South America may represent a modern analog of such a transgressive autochthonous sequence (Milliman et al., in press). Milliman et al. describe the mud deposits of the Amazon shelf as a landward-thickening wedge, whose offshore portions were deposited during lower stands of sea level, by the predecessor of the coastal mud stream which presently trends northwest, from the mouth of the Amazon toward the Guyana coast (Fig. 40). Milliman and his associates suggest that the offshore surface of this mud deposit is at present experiencing an autochthonous sedimentary regime. They suggest that the net fine sediment budget of the offshore shelf is negative, with more fine material being lost to erosion than is replaced by diffusion from the coastal source, so that a silty lag is accumulating over its surface. More detailed investigations may indicate that this type of transgressive allochthonous regime is more common than now supposed.

SUMMARY

The rate and sense of shoreline movement have an important modulating effect on the shelf sedimentary regime. It is helpful to think of the coastline as a "littoral energy fence" in which the landward-oriented net surge of shoaling waves tends to push sediment back toward the beach. There are two basic categories of "valves" that serve to regulate the passage of sediment through this barrier into the shelf dispersal system: river mouths and the intervening expanses of shoreface.

During rapid transgressions, river mouths generally cannot adjust to their combined river and tidal discharges as fast as required by the rise of sea level, and they become sediment sinks. Sediment is bypassed through the coastal zone by the basically passive process of erosional shoreface retreat, which leaves the shelf surface veneered with a sandy residue.

During slow transgressions, estuarine channels are more likely to equilibrate to their discharge. Such channels are capable of bypassing sand as well as finer sediment, and sediment is supplied by both shoreface and river mouth bypassing.

During regressions, river mouth bypassing is dominant. Shorefaces become sand sinks, which advance seaward by means of the successive growth of beach ridges.

Rapid transgressions result in autochthonous shelf regimes, in which the surficial sediments are of in situ origin. Slow transgressions result in allochthonous shelf regimes. The sediment load is filtered during passage through a broad intracoastal zone of estuaries and lagoons, so that the fraction reaching the shelf is fine-grained and mobile, and may be dispersed for long distances across the shelf surface.

On autochthonous shelves, only such large-scale subaerial features as cuestas and river valleys seem able to survive transgression, and even these are strongly modified by passage of the shoreline. On shelves of low relief, most morphologic elements have formed at the foot of the retreating shoreface. Shelf valley complexes consist of shelf valleys, shoal retreat massifs, and deltas. In many cases shelf valleys are the retreat paths of estuary mouth scour channels, and do not always overlie the buried subaerial river channels. They tend to be paired with estuarine shoal-retreat massifs, the retreat paths of estuary mouth shoals. Littoral drift convergences at capes and headlands may also result in shoal retreat massifs. Scarps on autochthonous shelves do not seem to be drowned shorelines in the strict sense, but instead are truncated lower shorefaces formed during postglacial stillstands.

In the Middle Atlantic Bight of North America, both morphology and grain-size distribution patterns can be shown to be in part of post-transgressional origin, forming in response to storm flows. The shelf surface is characterized by a pervasive ridge and swale topography. It is locally forming at the foot of the retreating shoreface,

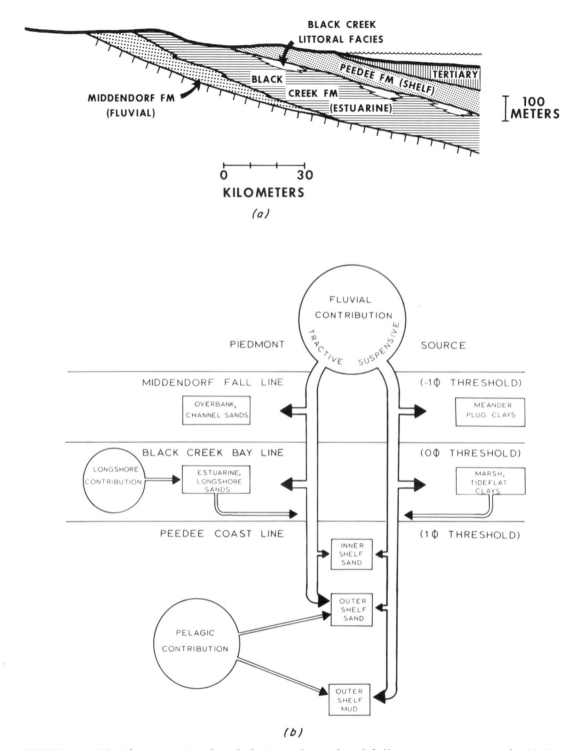

FIGURE 39. (a) *Schematic section through the Cretaceous Lumbee Group of North and South Carolina. The Middendorf, Black Creek, and Peedee Formations are deposits of landward-displacing fluvial, estuarine-* *lagoonal, and shelf environments, respectively.* (b) *Pattern of sediment transport as reconstructed from grain-size gradients and primary structures in outcrops. From Swift and Heron (1969).*

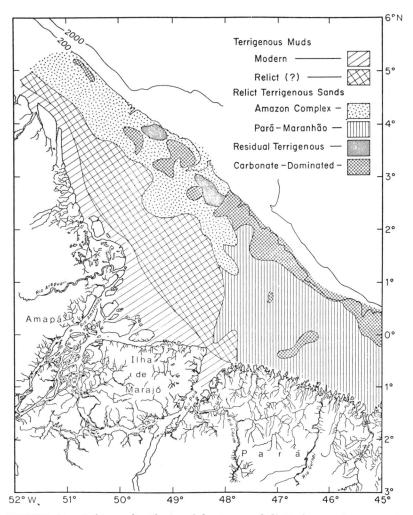

FIGURE 40. *Sediment distribution of the Amazon shelf. Modern terrigenous muds are deposited beneath a north-trending coastal mud stream. "Relict" muds are believed to have been deposited by the same mud stream during lower stands of sea level. From Milliman et al. (in press).*

in response to coastal boundary flow, but elsewhere appears to have developed more or less spontaneously further out on the shelf surface. Current lineations (sand ribbons and erosional furrows) are abundant. Coarse sand lags occur on highs, finer sands occur on their downcurrent slopes and in adjacent lows.

The shelf around the British Isles is an example of an autochthonous shelf that has reacted in a more vigorous fashion, in response to a high-intensity tidal regime. A well-organized pattern of sand dispersal consists of sand streams that extend from bed load partings to bed load convergences, or to the shelf edge. Nearshore morphologic elements have largely been obliterated. However, well-defined fields of tide-maintained sand ridges are probably analogous to the shoal retreat massifs of the Middle Atlantic Bight.

Allochthonous shelves occur adjacent to large rivers with sediment loads sufficiently large to locally slow or reverse the sense of the postglacial transgression. Transport is dominantly by water column advection (mud streams) but diffusion in response to bottom wave surge is important. Allochthonous deposits of fine sand and mud are commonly not continuous, but tend to leave "windows" in which autochthonous sands are exposed, Such windows are not necessarily transient phenomena. but may reflect areas in which the concentration of suspended sediment in the bottom nepheloid layer is counteracted by a relatively high level of "hydraulic activity." Textural gradients in autochthonous deposits may more nearly reflect the seaward, diffusive component of transport, rather than the coast-parallel advective component.

ACKNOWLEDGMENTS

I thank Paul E. Potter and Orrin H. Pilkey for their helpful criticisms of this chapter.

REFERENCES

Allen, J. R. L. (1964). Late Quaternary Niger Delta, and adjacent areas: Sedimentary environments and lithofacies. *Am. Assoc. Pet. Geol. Bull.*, **49**: 547–600.

Allen, J. R. L. (1970a). Sediments of the modern Niger delta: A summary and review. In J. P. Morgan, ed., *Deltaic Sedimentation, Modern and Ancient*. Soc. Econ. Paleontologists and Mineralogists Spec. Publ. 15, Tulsa, Okla., pp. 138–151.

Allen, J. R. L. (1970b). *Physical Processes of Sedimentation*. New York: American Elsevier, 248 pp.

Allersma, E. (1972). Mud on the oceanic shelf off Guiana. In *Symposium on Investigations and Resources of the Caribbean Sea and Adjacent Regions*. UNESCO, Paris (Unipub, N.Y.), pp. 193–203.

Andrews, E., D. G. Stephens, and D. J. Colquhoun (1973). Scouring of buried Pleistocene barrier complexes as a source of channel sand in tidal creeks, North Island Quadrangle, South Carolina. *Geol. Soc. Am. Bull.*, **84**: 3659–3662.

Bagnold, R. A. (1966). An approach to the sediment transport problem from general physics. *U.S. Geol. Surv. Prof. Pap. 422–1*, 37 pp.

Barby, D. G. and J. H. Hoyt (1964). An upper Miocene dredged fauna from tidal channels of coastal Georgia. *J. Paleontol.*, **38**: 67–73.

Beardsley, R. C. and B. Butman (1974). Circulation on the New England continental shelf: Response to strong winter storms. *Geol. Res. Lett.*, **1**: 181–184.

Belderson, R. H. and A. H. Stride (1966). Tidal current fashioning of a basal bed. *Mar. Geol.*, **4**: 237–257.

Belderson, R. H., N. H. Kenyon, and A. H. Stride (1970). *Holocene Sediments on the Continental Shelf West of the British Isles*. Inst. Geol. Sci. Rep. No. 70, pp. 157–170.

Bernard, H. A. and R. J. Le Blanc (1965). Resume of the Quaternary geology of the northwest Gulf of Mexico province. In H. E. Wright, Jr., and D. J. Frey, eds., *The Quaternary of the United States*. Princeton, N.J.: Princeton University Press, pp. 137–185.

Brouwer, J. E. (1964). The North Sea. In L. D. Stamp, ed., *The North Sea and the Law of the Continental Shelf*. Bude, Cornwall, England, Geographical Publications Ltd., World Land Use Survey Occ. Papers 5, pp. 3–14.

Bumpus, D. F. (1973). A description of circulation on the continental shelf of the east coast of the United States. *Prog. Oceanogr.*, **6**: 117–157.

Caralp, M., A. Klingebiel, C. Latouche, J. Moyes, and R. Prudhomme (1972). *Bilan Cartographique sur le Plateau Continental d'Aquitaine*. Bull. Inst Géol. Bassin d'Aquitaine, Numéro Special. 351 Cours de la Libération, 33 Talence, France, 25 pp.

Carter, L. (1973). Surficial sediments of Barkley Sound and the adjacent continental shelf, West Vancouver Island. *Can. J. Earth Sci.*, **10**: 441–459.

Caston, V. N. D. (1972). Linear sand banks in the southern North Sea. *Sedimentology*, **18**: 63–78.

Caston, V. N. D. and A. H. Stride (1970). Tidal sand movement between some linear sand banks in the North Sea off northeastern Norfolk. *Mar. Geol.*, **9**: M38–M42.

Colquhoun, D. J. (1969). *Geomorphology of the Lower Coastal Plain of South Carolina*. Geological Notes, Division of Geology, Columbia, South Carolina State Development Board, 36 pp.

Curray, J. R. (1964). Transgressions and regressions. In R. L. Mills, ed., *Papers in Marine Geology: Shepard Commemorative Volume*. New York: MacMillan, pp. 175–203.

Curray, J. R. (1969). History of continental shelves. In *The New Concepts of Continental Margin Sedimentation*. Washington, D.C.: Am. Geol. Inst., pp. JC-6-1 to JC-6-18.

Dolan, R., B. Hayden, G. Hoenenberg, J. Zeeman, and M. Vincent (1973). Classification of the coastal environments of the world, Part 1. *The Americas*. Tech. Rep. 1, Office of Naval Research, Washington, D.C., 163 pp.

Drake, D. E., R. L. Kolpack, and P. J. Fischer (1972). Sediment transport on the Santa Barbara–Oxnard shelf, Santa Barbara channel, California. In D. J. P. Swift, D. B. Duane, and O. H. Pilkey, eds., *Shelf Sediment Transport: Process and Pattern*. Stroudsburg, Pa.: Dowden, Hutchinson & Ross, pp. 301–332.

Duane, D. B., M. E. Field, E. P. Meisburger, D. J. P. Swift, and S. J. Williams (1972). Linear shoals on the Atlantic inner continental shelf, Florida to Long Island. In D. J. P. Swift, D. B. Duane, and O. H. Pilkey, eds., *Shelf Sediment Transport: Process and Pattern*. Stroudsburg, Pa.: Dowden, Hutchinson & Ross, pp. 447–499.

Emery, K. O. (1952). Continental shelf sediments of southern California. *Geol. Soc. Am. Bull.*, **63**: 1105–1108.

Emery, K. O. (1968). Relict sediments on continental shelves of the world. *Am. Assoc. Pet. Geol. Bull.*, **52**: 445–464.

Field, M. E. (1974). Buried strandline deposits on the central Florida inner shelf. *Geol. Soc. Am. Bull.*, **85**: 57–60.

Fischer, A. G. (1961). Stratigraphic record of transgressing seas in the light of sedimentation Atlantic coast of New Jersey. *Am. Assoc. Pet. Geol. Bull.*, **45**: 1656–1666.

Ginsburg, R. N. and N. P. James (1974). Holocene carbonate sediments of continental shelves. In C. A. Burk and C. C. Drake, eds., *The Geology of Continental Margins*. Berlin and New York: Springer-Verlag, pp. 137–156.

Grabau, A. W. (1913). *Principles of Stratigraphy*, 1960 facsimile edition of 1924 revision. New York: Dover, 1185 pp.

Hathaway, J. C. (1971). *Data File, Continental Margin Program, Atlantic Coast of the United States*, Vol. 2, *Samples Collection and Analytical Data*. Woods Hole Oceanographic Institution Ref. 71-15, U.S. Geol. Surv., Woods Hole, Mass., 446 pp.

Henry, V. J., Jr. and J. H. Hoyt (1968). Quaternary paralic shelf sediments of Georgia. *Southeast. Geol.*, **9**: 195–214.

Holtedahl, H. (1958). Some remarks on the geomorphology of the continental shelves off Norway, Labrador, and southeast Alaska. *J. Geol.*, **66**: 461–471.

Holtedahl, O. (1940). *The Submarine Relief off the Norwegian Coast*. Oslo: Norske Videnskaps-Akade, 43 pp.

Houbolt, J. J. H. C. (1968). Recent sediments in the southern bight of the North Sea. *Geol. Mijnbouw*, **47**: 245–273.

Hoyt, J. H. and J. R. Hails (1967). Pleistocene shoreline sediments in coastal Georgia: Deposition and modification. *Science*, **155**: 1541–1543.

Hoyt, J. H. and V. J. Henry, Jr. (1967). Influence of island migration on barrier island sedimentation. *Geol. Soc. Am. Bull.*, **78**: 77–86.

Jerlov, N. G. (1958). Distribution of suspended material in the Adriatic Sea. *Arch. Oceanogr. Limnol.*, **11:** 227–250.

Johnson, D. (1919). *Shore Processes and Shoreline Development.* New York: Wiley, 585 pp. (2nd ed., 1938).

Joseph, J. (1953). Die Trubangsverbaltnisse in der Sudwestlichon Nordsee Wahrend der "Gauss Fahrt" im Februar/Marz, 1952. *Ber. Dtsch. Wiss. Komm. f. Meeresforsch.* **13:** 93–103.

Kaye, C. A. (1959). Shoreline features and quaternary shoreline changes, Puerto Rico. *U.S. Geol. Surv. Prof. Pap.,* 317-B, 140 pp.

Kenyon, N. H. (1970). Sand ribbons of European tidal seas. *Mar. Geol.,* **9:** 25–39.

Kenyon, N. H. and A. H. Stride (1970). The tide-swept continental shelf sediments between the Shetland Isles and France. *Sedimentology,* 14: 159–175.

King, L. H., B. Maclean, and G. Drapeau (1972). The Scotian shelf submarine end moraine complex. In *24th Int. Geol. Congr., Montreal,* pp. 237–239.

Knebel, H. J. and J. S. Creager (1973). Sedimentary environments of the east-central Bering Sea continental shelf. *Mar. Geol.,* **15:** 25–47.

Kumar, N. and J. E. Sanders (1970). Are basal transgressive sands chiefly inlet filling sands? *Marit. Sediments,* 6: 12–14.

Laursen, E. M. (1958). The total sediment load of streams. *Proc. Am. Soc. Civil Eng.,* 84(HV1): 1530.

McCave, I. N. (1971). Sand waves in the North Sea off the coast of Holland. *Mar. Geol.,* 10: 149–227.

McCave, I. N. (1972). Transport and escape of fine-grained sediment from shelf areas. In D. J. P. Swift, D. B. Duane, and O. H. Pilkey, eds., *Shelf Sediment Transport: Process and Pattern.* Stroudsburg, Pa.: Dowden, Hutchinson & Ross, pp. 225–248.

McClennen, C. F. (1973). New Jersey continental shelf near bottom current meter records and recent sediment activity. *J. Sediment. Petrol.,* **43:** 371–380.

McClennen, C. E. and R. L. McMaster (1971). Probable Holocene transgressive effects on geomorphic features of the continental shelf off New Jersey, United States. *Marit. Sediments,* 7: 69–72.

McKinney, T. F. (1974). Large-scale current lineations on the Great Egg shoal retreat massif, New Jersey shelf. Investigation by side-scan sonar. *J. Sediment. Petrol.,* **17:** 79–102.

McKinney, T. F., W. L. Stubblefield, and D. J. P. Swift (1974). Large-scale current lineations on the central New Jersey shelf: Investigations by side scan sonar. *Mar. Geol.,* **17:** 79–102.

McManus, D. A., J. C. Kelley, and J. S. Creager (1969). Continental shelf sedimentation in an arctic environment. *Geol. Soc. Am. Bull.,* **80:** 1961–1984.

McMaster, R. L. and T. P. Lachance (1969). Northwestern African continental shelf sediments. *Mar. Geol.,* 7: 57–67.

McMaster, R. L., T. P. Lachance, and A. Ashraf (1970). Continental shelf geomorphic features off Portuguese Guinea, Guinea, and Sierra Leone (West Africa). *Mar. Geol.,* **9:** 203–213.

Meade, R. H. (1969). Landward transport of bottom sediments in the estuaries of the Atlantic coastal plain. *J. Sediment. Petrol.,* **39:** 229–234.

Merrill, A. S., K. O. Emery, and M. Rubin (1965). Oyster shells on the American continental shelf. *Science,* **147:** 395–400.

Milliman, J. D. and K. O. Emery (1968). Sea levels during the past 35,000 years. *Science,* **162:** 1121–1123.

Milliman, J. D., C. P. Summerhayes, and H. T. Barretto (in press). Quaternary sedimentation on the Amazon continental margin: A model. *Geol. Soc. Am. Bull.*

Naumann, C. T. (1858). *Lehrbuch der Geognosie,* Bd. 1, 2nd ed. Leipzig: Wilhelm Engelmann, 1000 pp.

Nichols, M. M. (1972). *Inner Shelf Sediments off Chesapeake Bay. I. General Lithology and Composition.* Spec. Sci. Rep. 64, Virginia Institute of Marine Science, Gloucester Point, Va., 20 pp.

Niino, H. and K. O. Emery (1966). Continental shelf off northeastern Asia. *J. Sediment. Petrol.,* **36:** 152–161.

Norris, R. M. (1972). Shell and gravel layers, western continental shelf, New Zealand. *N.Z. J. Geol. Geophys.,* **15:** 572–589.

Oaks, R. Q., Jr. and N. K. Coch (1963). Pleistocene sea levels, southeastern Virginia. *Science,* **140:** 979–983.

Pilkey, O. H. and M. E. Field (1972). Onshore transportation of continental shelf sediment: Atlantic southeastern United States. In *Shelf Sediment Transport: Process and Pattern.* Stroudsburg, Pa.: Dowden, Hutchinson & Ross, pp. 424–446.

Pilkey, O. H. and J. Frankenberg (1964). The relict-Recent sediment boundary on the Georgia continental shelf. *Georgia Acad. Sci. Bull.,* **22:** 1–4.

Powers, M. C. and B. Kinsman (1953). Shell accumulations in underwater sediments and the thickness of the traction zone. *J. Sediment. Petrol.,* **23:** 229–234.

Redfield, A. C. (1956). The influence of the continental shelf on the tides of the Atlantic coast of the United States. *J. Mar. Res.,* **17:** 432–448.

Robinson, A. H. W. (1966). Residual currents in relation to shoreline evolution of the East Anglian coast. *Mar. Geol.,* **4:** 57–84.

Sarnthein, M. (1972). Sediments and history of the postglacial transgression in the Persian Gulf and Northwest Gulf of Oman. *Mar. Geol.,* **12:** 245–266.

Sharma, G. D., A. S. Naidu, and D. W. Hood (1972). Bristol Bay: Model contemporary graded shelf. *Am. Assoc. Pet. Geol. Bull.,* **56:** 2000–2012.

Shepard, F. P. (1932). Sediments on continental shelves. *Geol. Soc. Am. Bull.,* **43:** 1017–1034.

Shepard, F. P. (1973). *Submarine Geology.* 3rd Edition. New York: Harper & Row, 517 pp.

Sheridan, R. E., C. E. Dill, Jr., and J. C. Kraft (1974). Holocene sedimentary environment of the Atlantic inner shelf off Delaware. *Geol. Soc. Am. Bull.,* **85:** 1319–1328.

Shideler, G. L., D. J. P. Swift, G. H. Johnson, and B. W. Holliday (1972). Late Quaternary stratigraphy of the inner Virginia continental shelf: A proposed standard section. *Geol. Soc. Am. Bull.,* **83:** 1787–1804.

Smith, J. D. (1969). Geomorphology of a sand ridge. *J. Geol.,* **77:** 39–55.

Stahl, L., J. Koczan, and D. Swift (1974). Anatomy of a shoreface-connected ridge system on the New Jersey shelf: Implications for genesis of the shelf surficial sand sheet. *Geology,* **2:** 117–120.

Stanley, D. J. (1969). Submarine channel deposits and their fossil analogs (fluxoturbidites). In D. J. Stanley, ed., *The New Concepts of Continental Margin Sedimentation.* Washington, D.C.: Am. Geol. Inst., pp. DJS-9-1 to DJS-9-17.

Stanley, D. J., G. Drapeau, and A. E. Cok (1968). Submerged terraces on the Nova Scotian shelf. *Z. Geomorphol., Suppl.,* **7:** 85–94.

Stanley, D. J., D. J. P. Swift, and H. G. Richards (1967). Fossiliferous concretions on Georges Bank. *J. Sediment. Petrol.*, **37:** 1070–1083.

Stanley, D. J., D. J. P. Swift, N. Silverberg, N. P. James, and R. G. Sutton (1972). *Late Quaternary progradation and sand spill-over on the outer continental margin off Nova Scotia, Southeast Canada.* Smithsonian Contr. Earth Sciences, **8:** 88 pp.

Steers, J. A. (1971). The East Coast floods, 31 January–1 February, 1953. In J. A. Steers, ed., *Applied Coastal Geomorphology.* Cambridge, Mass.: MIT Press, pp. 198–224, 227 pp. (reprinted).

Stride, A. H. (1963). Current-swept sea floors near the southern half of Great Britain. *Q. J. Geol. Soc. London*, **119:** 175–199.

Stride, A. H., R. H. Belderson, and N. H. Kenyon (1972). *Longitudinal Furrows and Depositional Sand Bodies of the English Channel.* Mém. Bur. Rech. Géol. Min. No. 79, pp. 233–244.

Stubblefield, W. L., J. W. Lavelle, T. F. McKinney, and D. J. P. Swift (1975). Sediment response to the present hydraulic regime on the Central New Jersey shelf. *J. sediment. Petrol.*, **45:** 337–358.

Stubblefield, W. L. and D. J. P. Swift (in press). Ridge development as revealed by sub-bottom profiles on the central New Jersey shelf. Submitted to *Mar. Geol.*

Swift, D. J. P. (1973). Delaware shelf valley: Estuary retreat path, not drowned river valley. *Geol. Soc. Am. Bull.*, **84:** 2743–2748.

Swift, D. J. P. (1974). Continental shelf sedimentation. In C. A. Burke and C. L. Drake, eds., *The Geology of Continental Margins.* Berlin and New York: Springer-Verlag, pp. 117–133.

Swift, D. J. P. (1975). Tidal sand ridges and shoal retreat massifs. *Mar. Geol.*, **18:** 105–134.

Swift, D. J. P. (in press). Continental shelf sedimentation. In R. Fairbridge, ed., *Encyclopedia of Sedimentology.* Princeton, N.J.: Van-Nostrand-Reinhold.

Swift, D. J. P., D. B. Duane, and T. F. McKinney (1973). Ridge and swale topography of the Middle Atlantic Bight, North America: Secular response to the Holocene hydraulic regime. *Mar. Geol.*, **15:** 227–247.

Swift, D. J. P. and S. D. Heron, Jr. (1969). Stratigraphy of the Carolina Cretaceous. *Southeast. Geol.*, **10:** 201–245.

Swift, D. J. P., J. W. Kofoed, F. P. Saulsbury, and P. Sears (1972). Holocene evolution of the shelf surface, central and southern Atlantic coast of North America. In D. J. P. Swift, D. B. Duane, and O. H. Pilkey, eds., *Shelf Sediment Transport: Process and Pattern.* Stroudsburg, Dowden, Hutchinson & Ross, pp. 499–574.

Swift, D. J. P. and P. Sears (1974). Estuarine and littoral depositional patterns in the surficial sand sheet, central and southern Atlantic shelf of North America. In *International Symposium on Interrelationship of Estuarine and Continental Shelf Sedimentation.* Inst. Géol. du Bassin d'Aquitaine, Mém. 7, pp. 171–189.

Swift, D. J. P., D. J. Stanley, and J. R. Curray (1971). Relict sediments, a reconsideration. *J. Geol.*, **79:** 322–346.

Todd, T. W. (1968). Dynamic diversion: influence of longshore current–tidal floor interaction on Chenier and barrier island plains. *J. Sediment. Petrol.*, **38:** 734–746.

Uchupi, E. (1968). Atlantic continental shelf and slope of the United States—Physiography. *U.S. Geol. Surv. Prof. Pap. 529–C*, 30 pp.

Uchupi, E. (1970). Atlantic continental shelf and slope of the United States: Shallow structure. *U.S. Geol. Surv. Prof. Pap. 524–I*, 44 pp.

Van Andel, Tj. H. and J. R. Curray (1960). Regional aspects of modern sedimentation in the northern Gulf of Mexico and similar basins, and Paleogeographic significance. In F. P. Shepard, F. B. Phegler and T. H. Van Andel, eds., *Recent sediments, northwest Gulf of Mexico.* Tulsa: Am. Assoc. Pet. Geol., pp. 345–364.

Van Andel, Tj. H. and J. J. Veevers (1967). *Morphology and Sediments of the Timor Sea.* Commonwealth of Australia, Bureau of Mineral Resources, Geology and Geophysics, Bull. 83, 172 pp.

Van Straaten, L. M. J. V. (1965). Sedimentation in the northwestern part of the Adriatic Sea. In W. F. Whitland and R. Bradshaw, eds., *Submarine Geology and Geophysics.* London: Butterworths, pp. 143–162.

Veenstra, H. J. (1969). Gravels of the southern North Sea. *Mar. Geol.*, **7:** 449–464.

Venkatarathnam, K. (1968). Studies on the sediments of the continental shelf off the regions Visakhapatnam–Pudimadaka and Pulicat Lake–Penner River confluence along the east coast of India. *Natl. Inst. Sci. India Bull.*, **38:** 472–482.

Visher, G. S. and J. D. Howard (1974). Dynamic relationship between hydraulic and sedimentation in the Altamaha estuary. *J. Sediment. Petrol.*, **44:** 502–521.

Whitmore, F. C., Jr., K. O. Emery, H. B. S. Cooke, and D. J. P. Swift (1967). Elephant teeth from the Atlantic continental shelf. *Science*, **156:** 1477–1481.

Wright, L. D. and J. M. Coleman (1973). Variations in morphology of major river deltas as functions of ocean wave and river discharge regimes. *Am. Assoc. Pet. Geol. Bull.*, **57:** 320–348.

Wright, L. D. and J. M. Coleman (1974). Mississippi River mouth processes: Effluent dynamics and morphological development. *J. Geol.*, **82:** 751–778.

CHAPTER 16

Shelf-Break Processes and Sedimentation

JOHN B. SOUTHARD

Department of Earth and Planetary Sciences, Massachusetts Institute of Technology, Cambridge, Massachusetts

DANIEL JEAN STANLEY

Division of Sedimentology, Smithsonian Institution, Washington, D.C.

The shelf break is defined as that sector of the continental margin where there is a marked change in gradient between the outermost continental shelf, or the outer edge of narrow platforms bounding emerged land and islands, and the much steeper continental slope. This zone has long been recognized as a topographic feature of the first order because it serves as a natural boundary between two major physiographic provinces, the continental shelf and the continental slope. The total length of the shelf break, serving as the boundary of the world's submerged outer continental margins, exceeds 300,000 km (Curray, 1966).

The stratigraphic and structural framework of the outer shelf and contiguous slope are regionally variable, as are the shape, depth, and surficial sediment cover of the shelf edge. The main emphasis here is on movement of fine to coarse materials within this sector and subsequent entrainment, or *spillover*, beyond the shelf. The break (as this feature is also called) is the zone through which most materials of terrestrial and in situ neritic origin (i.e., shelf carbonates, authigenic minerals, etc.) are transferred during their dispersal seaward to the deeper continental slope, base of slope, and abyssal plain environments. Yet surprisingly, less is known of processes that are active here at present than of those on the inner shelf and in the much deeper environments such as fans and rises.

Particular attention herein is paid to the plexus of processes which result in bed load movement of coarse fractions and resuspension of fine fractions. Transport of silt and clay fractions is discussed in Chapters 9 and 18. Mechanisms of sediment displacement are discussed in Chapter 11 and depositional patterns on the slope beyond the break are discussed in Chapter 17.

That the outer edge of the shelf and the uppermost slope in many areas are zones of active sedimentation has long been suspected. Empirical considerations by physical oceanographers (Fleming and Revelle, 1939; Part I, this volume) and early direct observations of water movement on the outer seafloor margin (cf. Stetson, 1937) support this concept. Recent technological advances applied to the shelf-break environment confirm that many processes, often operating concurrently, cause erosion of sand size and finer fractions, and that the importance of transport even at considerable depths is more substantial and more complex than previously imagined. We shall attempt to demonstrate that in terms of sediment transport the shelf break is almost as important a discontinuity as the shoreline.

VARIABILITY OF SHELF-BREAK MORPHOLOGY

The shelf break lies at an average distance of about 75 km (40 nautical miles) from shore at an average depth of 132 m (72 fathoms) according to Shepard (1963), although there is considerable deviation; breaks at depths of less than 20 to over 200 m and at distances from a few kilometers to more than 300 km from land are recorded. The average relief between the shelf break and base of the slope is almost 4 km. In a *grand tour du monde*, Shepard (1963, 1973) has illustrated the variability of sediment distribution on continental shelves as well as of the geographic configuration of the break. With the exception of narrow shelves, which are commonly shallower in regions of active coral growth, there does not appear to be a direct correlation between depth of break and width of shelf. And, interestingly enough, there does not appear to be a direct relation between depth of shelf margin and degree of wave exposure.

It is essential to define the break carefully if one is to measure accurately the depth of the break in a consistent manner. One practical definition is that the shelf edge is the point of the first major change in gradient at the outermost edge of the continental shelf (Wear et al., 1974). On shelf-to-slope bathymetric traverses the break is located on the landward point of a transition zone between the continental shelf and slope (method *C* in Fig. 1).

Fathometer profiles are usually highly distorted (vertical exaggeration magnified 10 to 40 times), and shelf-break configuration varies with ship speed. The

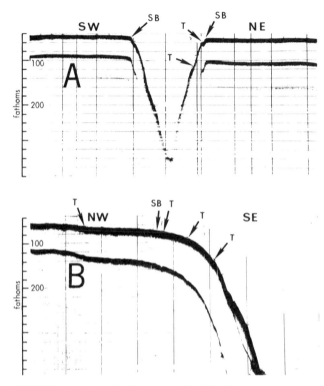

FIGURE 2. (A) *Profile across Norfolk Canyon; example of abrupt shelf break* (SB) *at 86 m* (47 *fathoms*). *Note terraces* (T) *at about 110 m* (60 *fathoms*) *and 216 m* (118 *fathoms*). (B) *Gradual gentle shelf-break type at 126 m* (69 *fathoms*) *on open shelf sector between Washington and Baltimore canyons. Arrows denote terraces. From Wear et al.* (1974).

most pronounced form, and certainly the most common, is termed the *sharp* break type; it often occurs between the outer edge of a low-relief, near-horizontal outer shelf surface [whose average slope is 0°07′; cf. Shepard (1963)] and a much steeper slope whose gradient exceeds 1:40, or 1°25′. This type, described by Dietz and Menard (1951), Heezen et al. (1959), and others, can be further subdivided into an *abrupt break* (angular and narrow; Fig. 2A) and a *gradual break* (arcuate and generally broader than the former; Fig. 2B), according to Wear et al. (1974). In contrast is the *gentle break*, defined by only a slight or subtle change in gradient. This latter type occurs off major deltas and off tectonically tilted or flexured continental margins such as the one off the eastern Pyrenees (Catalonian margin) in the northwestern Mediterranean (Fig. 3). Some gradational shape between the abrupt and gentle extremes is to be found along most of the world's outer shelves. More rarely, some margins, particularly in highly active, tectonically mobile regions, display no clearcut break at all.

Detailed regional surveys of shelf-break morphology between Hudson Canyon and Cape Hatteras off the

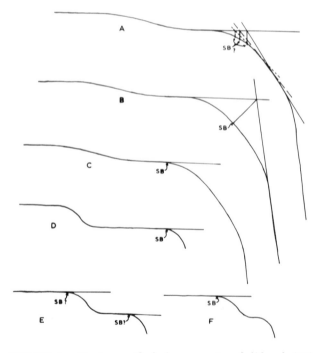

FIGURE 1. *Various methods for measuring shelf-break* (SB) *depth.* D–F *illustrate problems in selecting the depth when terraces are present at the break. From Wear et al.* (1974).

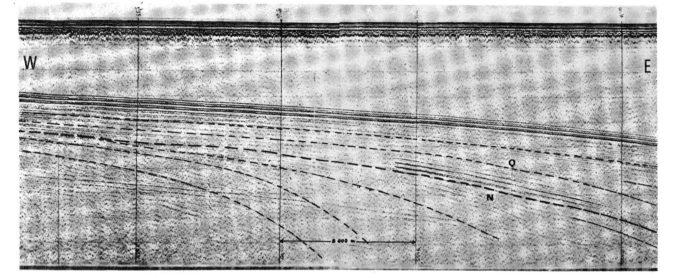

W   E

FIGURE 3. *Poorly defined break on the Catalonian margin east of the Pyrenees, Mediterranean Sea. Subbottom reflectors* *provide evidence of important subsidence during sedimentation. From Got (1973).*

Mid-Atlantic States (Wear et al., 1974) and off the southeastern United States (Macintyre and Milliman, 1970) reveal the variability of this feature even within one geographic province. The survey north of Cape Hatteras (Fig. 4) shows predominance of the sharp-break type, which occupies a relatively broad, arcuate transition zone (0.5–6.3 km) between the outer shelf and the uppermost slope (Wear et al., 1974). This survey also shows a notable variation in break depth along this 400 km stretch: Lack of uniformity in depth is apparent on histogram plots of open shelf, near-canyon (within 15 km of a canyon axis), and canyon-depth data (Fig. 5). Although average break depth approximates 110 m (60 km), depths ranging from 80 to 145 m (45–80 fathoms) are recorded along this relatively stable margin. Most open shelf traverses (those measured on traverses located more than 15 km from canyon heads) show a gradual break type, while the shelf edge near the canyon heads is of the abrupt type. Equally variable shelf-edge depth and configuration are noted off the southeastern United States between Cape Hatteras and Florida (Macintyre and Milliman, 1970) and along other margins where echo-sounding profiles have been collected systematically.

GEOLOGICAL CONSIDERATIONS AFFECTING THE SHELF BREAK

Knowledge of the outer continental margins in the various oceans of the world is uneven (cf. Fig. 4 of Chapter 25), but sufficient fathometer traverses are available to demonstrate that the break bounding the shelf-slope couple is not a uniform geomorphic feature. This is readily understandable when one considers the many factors that have shaped the shelf and contiguous slope (together, the two are termed *continental terrace*).

It is generally agreed that the structural and isostatic factors resulting in downwarping, flexuring, or vertical and lateral offset and displacement are of greatest significance. Superimposed on this primary structural control—and thus of second-order importance—is the strong imprint of Pliocene and Pleistocene geological events, which to a large degree molded the topography now observed. Most significant in this respect are large-scale eustatic oscillations in sea level related to the waxing and waning of the continental ice sheets, including lowering of sea level by about 125 m below the present mean during the period between 15,000 and 28,000 years ago (Curray, 1965). The coastline at times of major regressions was located at or near the break. When sea level was lowest, rivers, some of them very large, flowed seaward across the shelves and cut valleys on the subaerially exposed platform. Deltas formed at the mouths of some of these rivers, along the former coastlines near the shelf break, and also in deep embayments that now form the heads of submarine canyons.

A factor of third-order importance, but of major concern in this chapter, is the effect of subsequent physical, and to a lesser degree biological, shelf-break processes that continue to modify the surficial sedimentary cover of the shelf break by eroding and transporting sediment.

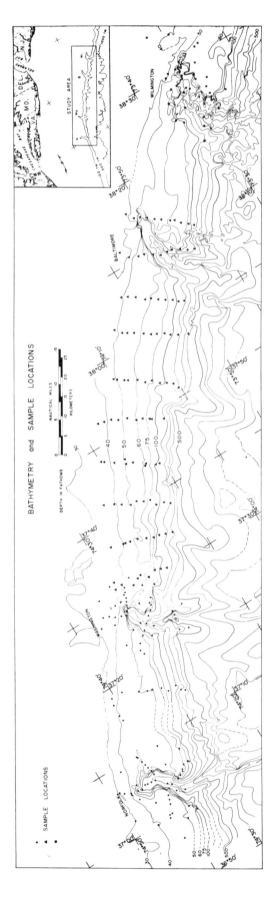

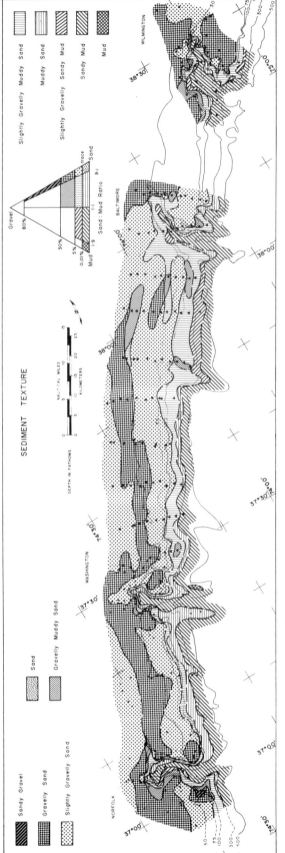

FIGURE 4. Outer continental margin off Mid-Atlantic States. Upper: Bathymetry showing Wilmington, Baltimore, Washington, and Norfolk canyons (cf. Fig. 5). Lower: Regional textural patterns of surficial sediment showing sand, gravelly muddy sand, and muddy sand draping across the break onto the slope. Offshelf spillover is evident on the open shelf sectors as well as in the canyons. From Wear (in preparation).

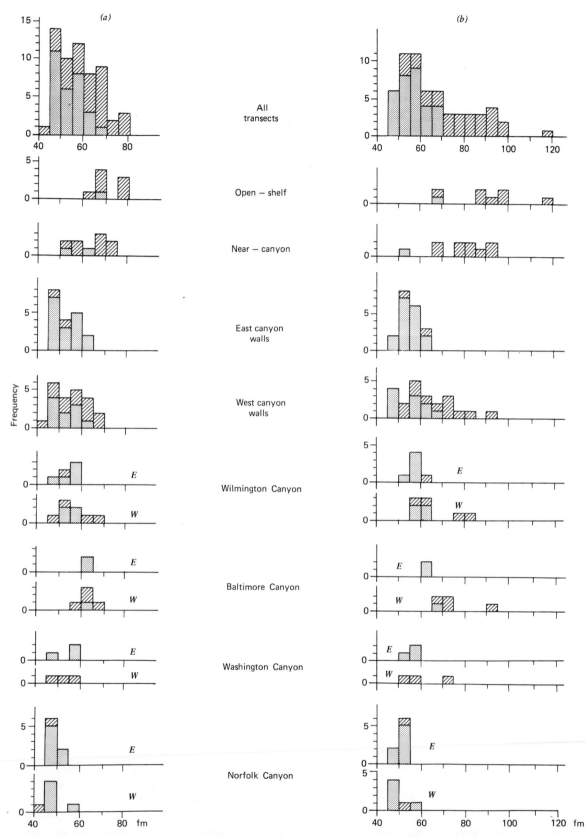

FIGURE 5. *Histograms depicting shelf-break depths (classes at 5 fathom, or 9 m, intervals) off the Mid-Atlantic States (see Fig. 4). Areas in black represent abrupt break* *types; hatched areas are gradual breaks. (a) Depths determined by method C in Fig. 1; (b) depths determined by method B in Fig. 1. From Wear et al. (1974).*

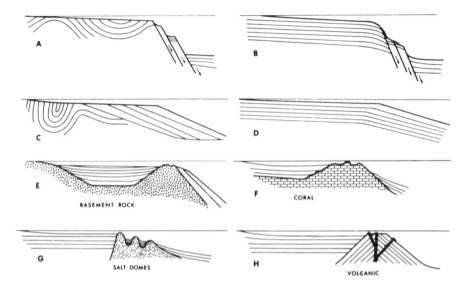

FIGURE 6. *Principal types of continental terraces showing different possible origins of the shelf-break and outer margin sector. From Shepard (1973).*

Geological Framework and Past Events Affecting the Shelf Break

The regional variability of continental margin shape cited above and the lack of direct correlation between break depth and present wave conditions together strongly suggest that the shelf break and immediately adjacent margin is largely a relict feature. Furthermore, marine geologists generally agree that in most instances the primary underlying origin of continental terraces is best explained in terms of structural origin, including modification by seafloor spreading, continental rifting, large-scale vertical displacement (subsidence, faulting), flexuring, or a combination thereof. Seismic interpretations show that shelf breaks often are localized at or near steep primary slopes (Worzel, 1968) and are not infrequently associated with tectonic, coral, salt, or volcanic dams (Emery, 1968b). Several examples of continental terrace types to which the shelf break may be related are illustrated in Fig. 6; additional interpretations are also presented by Emery (1969, Fig. 4), Inman and Nordstrom (1971), and Dietz and Holden (1974, Fig. 3). The reader interested in the relation between continental margin configuration and structural and stratigraphic framework is directed to recent reviews by Burk (1968), Emery (1968b, 1970), Curray (1969b), Dewey and Bird (1970), Maxwell (1970), Emery and Uchupi (1972), Burk and Drake (1974), Dietz and Holden (1974), and Lewis (1974).

More closely applicable to the discussion of shelf-break sedimentation are the effects of the recent geological events in the Pliocene and Quaternary which have

considerably modified the environment of the outer margin (Fig. 7). The significance of these events on the world's continental shelves has been summarized by Emery (1965), Curray (1965, 1967, 1969a), and Emery and Uchupi (1972). These and other workers call attention to the coincidence of the worldwide average shelf-break depth of approximately 130 m (Shepard, 1963) with the low stand of sea level during the Late Pleistocene (Wisconsin) glacial maximum. However, there are differences of opinion as to the exact position of earlier Quaternary eustatic low stands, inasmuch as variations in shelf-break depth may have been altered by Recent regional crustal movement and local "neotectonics." Identification of submarine terraces cut on the outermost shelf and upper slope sheds some light on this problem. Since it is probable that many of these notches and related rock exposures (Fig. 7) are the result of erosion in shallow water, including wave-cut benches near former shorelines (Bradley, 1958), terraces serve as valuable indicators of both eustatic sea level and neotectonic displacement. Recent mapping off the Mid-Atlantic States and other regions (review in Wear et al., 1974) indicates that earlier (pre-Wisconsin) low stands may have dropped to at least 200 m, and perhaps even deeper, below present mean sea level.

Further evidence of the major sea-level fluctuations that have modified the outermost shelf/slope sector is presented by high-resolution subbottom profiles that show well-developed unconformities (Fig. 8), truncated sequences (Fig. 9), and large cut-and-fill structures, some of them associated with terraces and former shorelines (e.g., profiles in Knott and Hoskins, 1968;

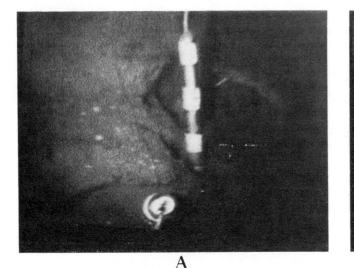

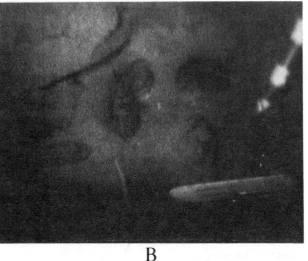

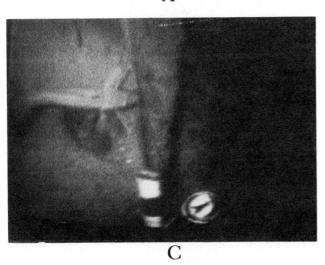

FIGURE 7. *Underwater television sequence showing cavernous rock outcrop just below the shelf break on the west wall of Wilmington Canyon (120–130 m). This unit forms low "cliffs" with a relief of 5 to 10 m, and is populated by numerous fish and occasional lobsters. The rock is poorly consolidated at the surface as indicated by breaking of outcrop by the penetrometer. From Stanley and Fenner (1973).*

Boillot and d'Ozouville, 1970; Rona, 1970; Shideler and Swift, 1972). Truncated subbottom reflectors and associated terraces and relict sediments on the outer shelf, and particularly those on the outer sectors of broad, tectonically stable margins, can be specifically related to the Quaternary oscillations. Shepard (1932) long ago recognized that the patchy and irregular distribution of coarse terrigenous fractions (sand, granules, gravel) on the outer shelves resulted from fluvial-delta, ice, and wind transport mechanisms that affected the subaerially exposed platforms during glacial stages (cf. Emery, 1968a; Shepard, 1973). These coarse sediments, which are associated with terraces and other shelf-edge topographic features, have been termed *relict* by Emery

(1952, 1968a). Thus, there is an implication that most surficial sediments and associated topographic features are unrelated to present conditions at these sites.

The importance of intense erosion of the outer shelf surface during early phases of rising sea level should be recalled in order to interpret properly the present distribution of texturally relatively "clean" sands near the shelf break. Stratigraphic models that relate the truncation of subbottom reflectors and position of coarse material near the break with alternating regressive/transgressive episodes (Fig. 10A) and with changing rates of subsidence versus sediment influx (Fig. 10B) have been proposed by Curray (1965, 1967). Also found associated with these relict shelf-edge sediments are vary-

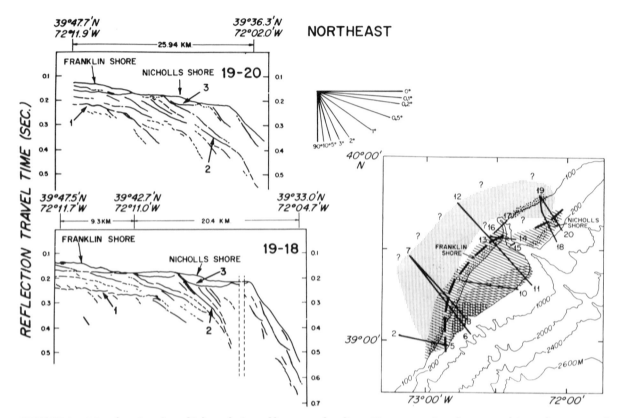

FIGURE 8.   *Line drawings from high-resolution subbottom records on the outer Atlantic margin near Hudson Canyon. Arrows denote major stratigraphic and erosional breaks, cut-and-fill structures, and terraces attributed to former*

shorelines. Note truncation of strata and irregular nature of progradational development at shelf edge. From Knott and Hoskins (1968).

ing amounts of weathered materials reworked from older units cropping out on the outer margins and, locally, authigenic minerals and nodules.

Although erosion and truncation have been a major result of the different phases of sea-level oscillation, there are also constructional topographic features resulting

from biogenic and physical processes. These include shelf-edge reefs (Zarudzki and Uchupi, 1968; Macintyre, 1972), linear ridges (Emery and Uchupi, 1972), and intercanyon deposits on the uppermost slope (Rona, 1970) that, at least in part, date from Late Pleistocene or Early Holocene time.

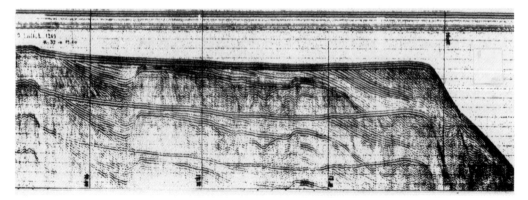

FIGURE 9.   *Sparker profile across shelf break on northern Spanish margin shows seaward-dipping strata covering irregular Tertiary bedrock surface.*

Note truncation of upper sediment sequences. From Boillot and d'Ozouville (1970).

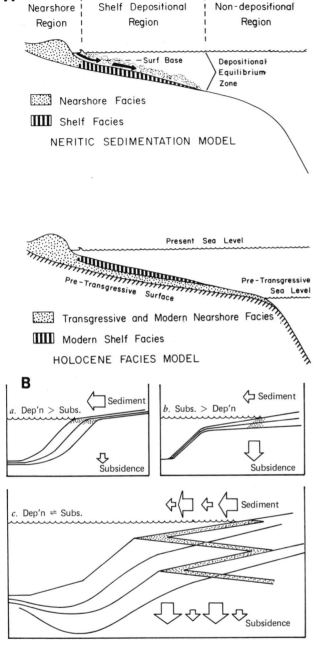

FIGURE 10. *Stratigraphic models depicting sediment facies distribution in terms of* (A) *eustatic sea-level oscillations and* (B) *rates of subsidence versus sediment influx. From Curray (1965, 1967).*

THE SHELF BREAK AS A ZONE OF ACTIVE SEDIMENTATION

Almost all detailed surveys of the outer margin provide evidence of recent sediment displacement at and near the shelf break. Although often qualitative and limited in time and scope, observations suggest that the present shelf edge is by no means the "Pleistocene museum" it was once believed to be (cf. Stanley, 1969). The degree to which the character of the shelf break reflects the modern hydraulic climate has been reexamined recently (Swift et al., 1971, 1972; Swift, Chapter 15, this volume). There is ample evidence showing that relict sediment has been reworked since the rise in sea level and continues to be modified at present, resulting in *palimpsest* sediment (p. 314).

It has been tacitly accepted that terrigenous sediment forming an important part of the deep-sea surficial cover once crossed the shelf and was transferred onto the slope, rise, and abyssal plains beyond. The currently popular concept is that most of this movement of silt and sand from shelf to deep-ocean basins took place at times of lower sea level in the Pleistocene by entrainment into canyon heads and subsequent downslope channelization in submarine valleys by gravity flow mechanisms (Middleton and Hampton, 1973; Chapter 11, this volume). Observations of the spillover of sand and gravel into canyon heads (Shepard and Dill, 1966; Got and Stanley, 1974; Stanley, 1974) and of frequent submarine cable breaks on the outer margin (Heezen and Hollister, 1971) are indications of post-Pleistocene to Recent displacement of sediment across the shelf edge and its loss seaward. While valid to a certain degree, this general dispersal model (Fig. 11) is in most cases an oversimplification of actual sedimentation patterns at the shelf break. There is no need to consider that movement of

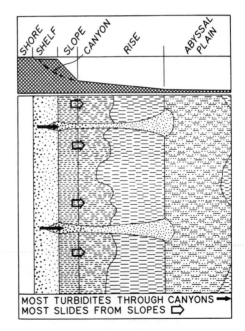

FIGURE 11. *Schematic of sediment facies and dispersal on the outer continental margin emphasizing offshelf transport of coarse material through canyons. From Emery (1969).*

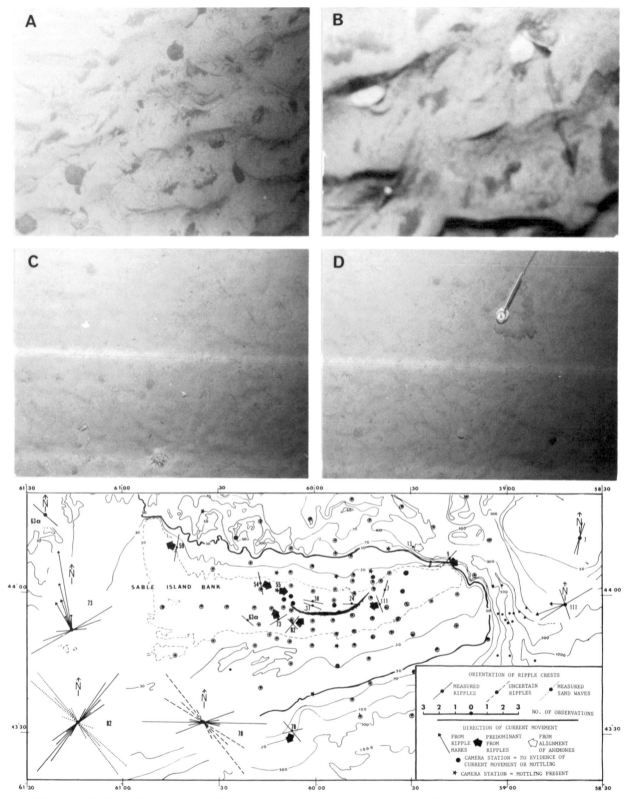

FIGURE 12. *Seafloor photographs provide evidence of recent movement of sand off Sable Island Bank. (A, B) Fresh ripple marks at station 9 near head of the Gully Canyon; sand dollars provide scale. (C, D) Older ripples at station 78. Depth on chart in fathoms. Modified from Stanley and James (1971).*

sand and mud is restricted to canyons. For instance, mapping of large sectors such as the one shown in Fig. 4 indicates the presence of sand drapes extending from the shelf edge onto the open-slope sectors away from canyons (Stanley et al., 1972). Furthermore, there is no reason to rule out bed load movement of sand and suspension transport of fine sediment landward away from the break.

Valuable in this respect are techniques that provide direct evidence of surficial sediment movement. Observations at the shelf edge made by research submersibles and by underwater camera (Stanley and ames, 1971) and underwater television (Stanley and

Fenner, 1973) show fresh ripple marks (Fig. 12) and, in some cases, actual movement of sand in response to bottom currents (Fig. 12). Detection of turbid water (Figs. 13*A*–*C*) resulting from high concentrations of suspended matter also provides some indication of shelf-edge erosion; this problem is discussed more fully by Pierce in Chapter 18.

Indirect evidence of recent transport is supplied by mapping of textural and mineralogical parameters of the surficial sediment along outer shelf and slope transects. Regional changes in grain-size parameters (Frank and Friedman, 1973) are particularly useful; see Fig. 4.

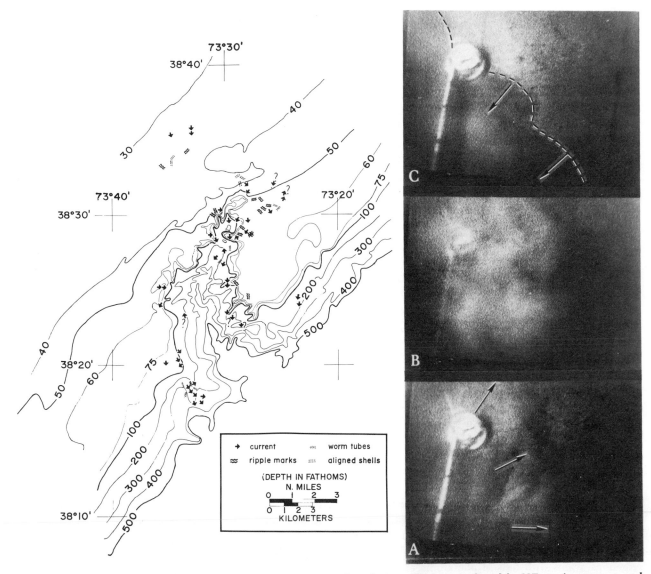

FIGURE 13.   *Left: Chart showing bottom current activity near shelf break adjacent to Wilmington Canyon, based on underwater television survey. Right: Sequence of photographs from video tapes illustrates turbidity initiated as compass strikes sandy mud bottom (A) at depth of 134 m. After a few seconds (B) mud brought into suspension is slowed by SSE moving current, and then (C) the cloud begins movement to south (at about 10 cm/sec) and out of view. Compass diameter, 7.5 cm. From Stanley and Fenner (1973).*

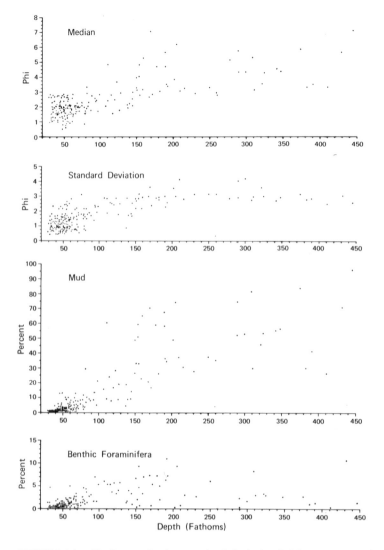

FIGURE 14. *Median grain size, standard deviation (phi), percentage of mud and benthic foraminifera versus depth on profiles off the Mid-Atlantic States (cf. Fig. 4). Break in trends occurs below break at depths exceeding 200 m. From Wear (in preparation).*

Plots of sediment size versus water depth generally show a progressive decrease of sand seaward of the shelf break (Fig. 14); the sharpest reduction of mean or median size and increase in percentage of mud occur well out on the slope (depths > 200 m). Mica content is usually low at the break (Doyle et al., 1968; Adegoke and Stanley, 1972) but increases markedly in finer sediment on the slope (Fig. 15). Furthermore, the presence on the upper slope of pelecypod valves and shells that must have originated at neritic depths and the mixing of sand-sized terrigenous grains with tests of planktonic organisms in the silt and clay fractions (Fig. 16) seaward of the break are additional evidence of offshelf displacement.

Large-scale bed forms are commonly present near the shelf edge. These may be of relict origin, as are the drowned shelf-edge reefs from the southeastern Atlantic shelf edge of North America (Pilkey et al., 1971; Macintyre and Milliman, 1970) or the lithified dunes on the Spanish Sahara shelf edge (Newton et al., 1973). In many places, however, shelf edge bed forms appear to be the result of recent currents, either tidal or storm-generated (see Swift, Chapter 15). Where the direction of sand wave migration has a component off the shelf, as off Cape Hatteras (Hunt et al., 1975), there must be loss of a significant volume of sand and also progradation of the slope (Rona, 1970; Shideler et al., 1972).

PHYSICAL PROCESSES AT THE SHELF BREAK

In considering the nature of sediment entrainment and transport at the shelf break, certain questions naturally

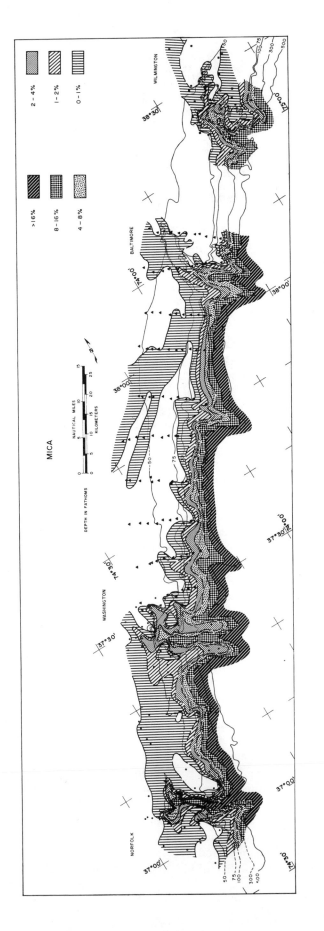

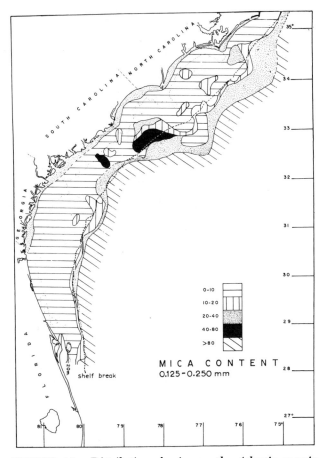

FIGURE 15. *Distribution of mica on the Atlantic margin. Left: In relative percent, off the Mid-Atlantic States. From Wear (in preparation). Right: In number of mica grains per 10,000 grains, off southeastern United States. From Doyle et al. (1968). Both areas show generally low values near the outer shelf and much higher content below the shelf break.*

come to mind: What is the nature of bottom water movements near the shelf break? In what ways do they differ from those acting elsewhere on the outer shelf? What special features of the resulting sediment transport might be expected? In this section an attempt is made to provide some answers to these questions, or at least to put them in perspective. Our aim is limited: Rather than surveying the nature of currents in the outer shelf and upper slope environments and the ways in which the erodible substrate responds to these currents (see Chapters 2–8, this volume) we will focus only on those kinds of bottom currents that seem to be distinctive of the shelf edge. This part of the chapter is necessarily more speculative than the preceding one, because not nearly as much is known about dynamics at the shelf edge as about morphology, sediment distribution, or structure.

The prevailing view is that the shelf edge is a locus of stronger turbulence or stronger currents, which keep

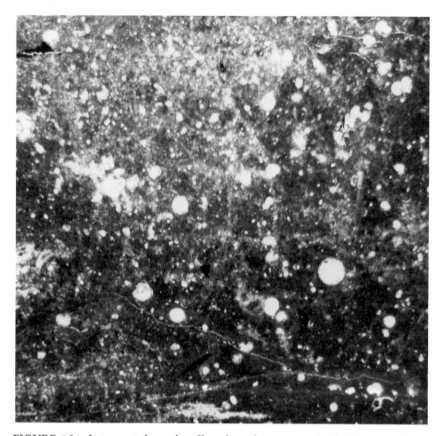

FIGURE 16. *Impregnated sample collected at about 1500 m in the Western Alboran Basin, western Mediterranean, showing sand-sized terrigenous material moved from the shelf margin deposited with planktonic foraminifera in fine-grained matrix.*

fine sediment in suspension or resuspend it frequently, so that fine material supplied from nearshore areas is transported to the continental slope or beyond without permanent deposition near the shelf break. This view is indeed in agreement with the common observation that bottom sediments on the outer shelf (except offshore of major fluvial sources of supply) are commonly free of fine fraction, despite an appreciable concentration of fine inorganic suspended sediment in the outer shelf water. There is a certain intuitive attractiveness in this picture, perhaps stemming from the analogy with the phenomenon of wake turbulence associated with separation attendant upon flow past an expanding boundary, or simply from the feeling that the shelf edge must in some way be more "exposed" than outer shelf or upper slope areas. But when enumeration of the physical processes that might be responsible for such heightened turbulence or currents is attempted, it is discovered that there are few processes to which we can appeal that are well established in either theory or observation. Despite recent direct observations of currents near the shelf break strong enough to produce substantial transport of coarser sediment (e.g., Lyall et al., 1971; Komar, this volume),

observations have not been extensive enough to establish definitely that bottom water movements, in fact, tend to be stronger at the break proper than elsewhere on the outer margin.

To illustrate both the possibilities and the difficulties involved in gaining an understanding of water movements at the shelf break, suppose that an ideally accurate long-term record of current speed and direction could be obtained at some point near the bottom. This record would be rather complicated: Velocities vary over time scales ranging from passage of tiny eddies in the near-bottom part of the boundary layer to slow changes in circulation caused by climatic changes, with variations on many intermediate scales caused by such factors as waves, tides, and various kinds of currents. At any given time one or another effect may be dominant, but typically there is a superposition of various effects, and it is usually difficult to sort these out. An actively expanding area of research in physical oceanography involves development of theoretical models for water movements on the shelf; since these commonly involve a degree of simplification in boundary conditions and water structure, for tractability of equations, they are not always specifically testable in terms of velocity records, and so only very general comparisons can be made.

The rest of this section deals with the various kinds of bottom currents that may be important in governing sediment movement at the shelf break. Until our observational knowledge of bottom currents and accompaying sediment transport is more extensive, it seems premature to attempt to evaluate the relative importance of any of these effects. It does not seem to be an exaggeration to say that we are at the stage of knowing something, but not enough, about a lot of possible motions and having some, but too little, field data to guide our thinking.

Distinctive Bottom Currents at the Shelf Break

Bottom currents in the vicinity of the break have a great variety of causes: passage of surface waves, which produce oscillatory motions at the shortest time scales, measured in seconds; barotropic tidal motions, which produce currents that vary over semidiurnal or diurnal periods; wind-driven currents produced by either storms or steadier seasonal wind systems; currents generated by differences in atmospheric pressure occasioned by passage of storms; and various thermohaline circulations, either specific to the shelf or areally more extensive but impinging on the shelf as well. Effects of various kinds of internal waves, which can be present provided that the shelf water is not so well mixed that no thermocline is developed, may be additionally important. Internal waves are known to be present in the oceans over a wide range of periods, and it is quite possible, even likely, that their interaction with sediments near the shelf edge could produce sediment movement largely unrelated to the currents noted above.

Near-bottom currents of these kinds govern both bed load transport of coarse sediment and resuspension of any fine sediment that finds a temporary resting place during periods of weak bottom currents. Their effect on the bottom is mediated by the frictional boundary layer that is developed within some meters of the bottom, whether purely oscillatory (as with the passage of surface waves) or varying slowly enough with time (as with the various currents of tidal periods or longer, or aperiodic currents) so that the flow structure in the near-bottom part of the boundary layer develops toward, or even attains, equilibrium. Since we have no theory and little experiment on which to rely in making the connection between boundary layer development and sediment entrainment in these usually unsteady flows, results must be qualitative at best, even if the nature of currents outside the boundary layer is known. The vertical structure of water movement throughout the entire water column is additionally important in determining the nature of both vertical mixing and lateral diffusion of suspended sediment near the shelf break (see Chapter 18).

SURFACE WAVES. Oscillatory horizontal motion of bottom water caused by passage of large surface waves is known to be one of the most important factors in bottom sediment transport on the inner shelf. This suggests that oscillatory bottom currents, even at shelf-edge water depths, should be great enough to produce some sediment transport (Fig. 17; Hadley, 1964b; Draper, 1967; Ewing, 1973), and this is supported by observations that symmetrical oscillation ripples are occasionally generated at depths as great as 200 m (Komar et al., 1972). Such sediment movement would not be a major factor in shaping the sediment bottom but could be important in resuspending any small amounts of fine sediment deposited on sand beds between major storms. There is no reason to expect that this sediment movement would be any stronger at the shelf break than further inshore. Ewing (1973), however, notes that refraction of large waves by irregularities in the trace of the shelf edge can produce a pattern of locally enhanced and diminished oscillatory bottom currents, so that at promontories of the shelf edge there would be much stronger bottom sediment transport.

TIDAL CURRENTS. Tidal currents are important on most shelves, and in many areas they are the dominant factor in sediment transport. Tidal currents are complicated in detail, depending on such things as nature of

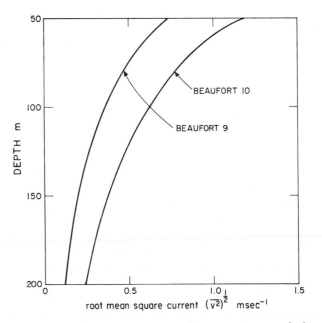

FIGURE 17. *Root-mean-square oscillatory current speeds for Beaufort-9 and Beaufort-10 wave conditions. From Ewing (1973).*

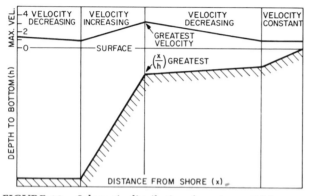

FIGURE 18. *Schematic distribution of maximum tidal current velocities over continental shelf where there is transport of water normal to the coast. From Fleming and Revelle (1939).*

the impinging deep-ocean tide wave, effect of the earth's rotation, topography of the shelf, and configuration of the coastline. But are tidal currents in well-mixed shelf waters stronger or weaker at the shelf edge than over the open shelf inshore? In many areas there should indeed be a noticeable maximum in tidal current velocities near the break (Fleming, 1938; Fleming and Revelle, 1939; Kuenen, 1939). This effect can easily be demonstrated by neglecting the earth's rotation and assuming that the tide wave, whether standing or progressive, is oriented parallel to the coastline, and then integrating the equation of motion to obtain velocity as a function of wave properties, local water depth, and distance from shore. Maximum tidal velocities are proportional to distance from shore divided by local water depth. Since this ratio is commonly greatest at the shelf edge, because seaward slope on the shelf is so small, the strongest tidal currents should in many cases be expected at the break (Fig. 18). The importance of this effect will vary with the nature of the tidal wave and with the importance of the earth's rotation; the greatest effect would be where the tide wave impinges approximately perpendicular to the coast on gently sloping, low-latitude shelves. Unfortunately, no systematic observations of this potentially important effect seem to have been made.

PRESSURE-INDUCED CURRENTS. Currents generated by wind stresses or by differences in atmospheric pressure occasioned by passage of major storms are important over most shelf areas.

Shelf currents generated by wind stress at the sea surface can be important or even dominant (e.g., Smith and Hopkins, 1972). If the continental shelf water is well mixed, currents can extend to the bottom, but there seems to be no reason to expect that bottom currents generated in this way would be any different at or near the shelf edge than elsewhere on the outer shelf. But Galt (1971) has found that migration of the pressure-

induced wave that is forced to travel beneath a moving storm can generate bottom currents that indeed are concentrated at the shelf break. Galt investigated the effect of such a pressure-induced wave generated by a storm moving in the offshore direction in a continental shelf/slope region, by numerical integration of appropriately simplified equations for a two-layer ocean. For a wide range of storm speeds and of layer thicknesses and density differences, the results indicate a current extending throughout the water column (the influence of the density stratification on the velocity field turns out to be small) parallel to the bottom contours, and reversing in direction as the storm passes. Speeds are of the order of 10 cm/sec in the vicinity of the shelf break but decrease substantially in both the offshore and onshore directions (Fig. 19). Physically these currents come about by virtue of the effect of the forced pressure-induced wave on extension and contraction of the vertical water column. Although there are no direct observations to support such a model, this sort of current could be an important additional factor, superimposed on tidal- and wave-generated water motions, in winnowing and resuspend-

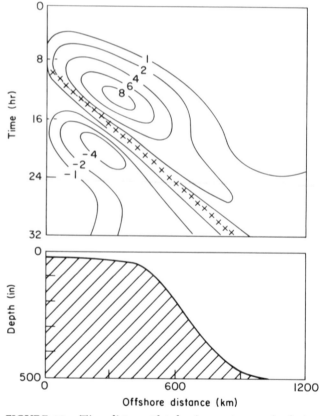

FIGURE 19. *Time-distance plot showing component of velocity parallel to shelf edge, in centimeters per second, produced by passage of a storm in the offshore direction. Storm track is marked by crosses. Lower diagram shows shelf-slope geometry assumed in the model. Schematized from Galt (1971).*

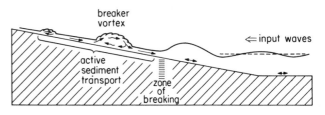

FIGURE 20. *Internal waves in a two-layer medium steepening and breaking on a slope. Length of arrows in the denser lower layer represents magnitude of flow (oscillatory or unidirectional). Dashed line at right marks mean position of interface. From Southard and Cacchione (1972).*

ing fine sediment, and in augmenting bed load transport of coarse sediment near the shelf break.

INTERNAL WAVES.   Waters over the continental shelf and upper slope can range from well mixed to strongly density-stratified, depending mainly on climate, season, and existence of currents that would tend to promote vertical mixing. If there is density stratification, internal waves with a wide range of amplitudes and periods, both progressive and standing, could be present on the interface between layers of differing density or within the zone of continuous vertical variation in density (Lafond, 1962). There have been many observations of internal waves over the shelf and slope; they are detected mainly from changes in vertical profiles at one or more stations in some area. These waves commonly, but not exclusively, are observed to have inertial or tidal periods; Gaul (1961) and Lee (1961) have observed shorter period waves propagating in continental shelf waters. But knowledge of the occurrence of such waves is still sketchy, and mechanisms for their generation are not very well understood (Wunsch, 1971).

There have been several suggestions, based largely on indirect evidence, that water movements produced by standing or progressive internal waves might move sediment on the continental shelf or slope (Emery, 1956; Cartwright, 1959; Lafond, 1961). It seems unlikely, however, that important sediment movement would be produced unless there is amplification or breaking of the waves caused by interaction with the bottom topography. Taking as a basis Wunsch's (1969) solutions for amplification of near-bottom velocities as first-mode internal waves propagate over a shoaling bottom, Cacchione and Southard (1974) have developed a criterion for incipient movement of bottom sediment by shoaling internal waves. They equate moments caused by fluid force and gravity force on exposed bed particles, as is commonly done for surface waves. Application of this criterion to sediment sizes present on the outer Atlantic continental shelf of North Atlantic indicates that shoreward propagation of internal waves, with frequencies and amplitudes known to be present in the oceans, would produce at least incipient movement of shelf bottom sediment.

It is not likely that near-bottom velocities produced simply by upslope amplification would be great enough for substantial sediment transport. However, under certain conditions, internal waves can break, thereby generating much stronger near-bottom velocities. Breaking has been studied experimentally in a continuously stratified medium by Cacchione and Wunsch (1974) and in a two-layer medium over low-density artificial sediment by Southard and Cacchione (1972). The onset of breaking is abrupt, and involves generation of near-

bottom velocities much greater than immediately downslope of breaking. A turbulent vortex or breaker is formed which travels upslope and is gradually dissipated (Fig. 20). A seemingly similar natural-scale phenomenon involving shoaling internal waves on the California shelf (Fig. 21) has been reported by Emery and Gunnerson (1973). In the experimental work, the bed upslope of breaking is swept by a steadier and thinner flow except during passage of a breaker; this results in net downslope transport of sediment in the form of asymmetrical ripples.

The experiments by no means represent a true dynamic scale model of the oceanic situation, but they suggest that there should be strong resuspension of fine sediment at the point of breaking and development of downslope-migrating ripples or even sand waves upslope of breaking. Although there is much circumstantial evidence for the occasional importance of this mode of sediment transport near the shelf break, nonexistence of concurrent field observations of bottom sediment transport and internal wave motions in a single area makes this only a speculative possibility.

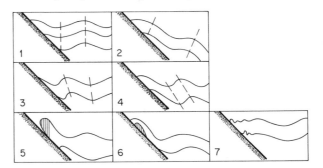

FIGURE 21. *Types of thermal sections over the continental shelf, Santa Monica Bay, California. (1) Internal waves with no effect by shoaling bottom; isotherms at different depths. (2) Internal waves with no effect by shoaling bottom: wave pattern superimposed on general temperature slope due to upwelling. (3) Internal swash: shallower isotherms precede deeper ones; marked steepening of wave front. (4) Strong internal swash: definite wave runup. (5) Internal surf: temperature inversion. (6) Internal surf: discrete bolus of cold water along bottom. (7) Internal surf: extreme irregularity in waveform near the bottom. From Emery and Gunnerson (1973).*

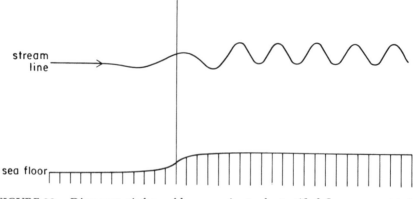

FIGURE 22. *Diagrammatic form of lee waves in steady stratified flow over a step in the seafloor. From Cartwright (1959).*

It is possible that standing lee waves in density-stratified waters over the outer continental shelf, produced by passage of the tide over the shelf edge, can cause spatial variations in bottom velocity great enough to construct large and relatively low-amplitude wavelike forms in a coarse sediment bottom (Fig. 22). Cartwright (1959) has developed a theory based on this lee wave concept to account for a series of regular and very widely spaced symmetrical sand waves off the coast of Britain, described by Cartwright and Stride (1958) and Carruthers (1963). This concept is analogous to the well-known phenomenon of standing waves in the atmosphere in the lee of a mountain range. The theory is attractive but the evidence is somewhat tenuous and largely circumstantial. The theory would be applicable only to shelves where tidal currents are fairly strong, and during that part of the year when there is marked density stratification.

CURRENTS IN SUBMARINE CANYONS.   It is widely recognized that currents on the outer shelf sweep sediments over the shelf edge both onto undissected slope sectors and into the heads of submarine canyons, which almost universally incise the outer shelves. The most widely accepted view is that this sediment is then transported down the canyon to and beyond the base of the continental slope by various mass-movement processes, including turbidity currents (see Chapters 11 and 17). Both textural evidence and recent direct observations show that relatively clear-water currents in the upper parts of the canyons are also commonly strong enough for continued winnowing and net downslope transport of the sediment. Lyall et al. (1971) and Stanley et al. (1972a) have measured currents of up to 20 cm/sec in Wilmington Canyon, southeast of the mouth of Delaware Bay. Shepard and Marshall (1973) and Shepard et al. (1974a,b) have made extensive observations of bottom currents in several submarine canyons off the California

coast: Velocities up to about 35 cm/sec, coherent with depth, alternate upcanyon and downcanyon with periods ranging from less than an hour to about semidiurnal tidal period. These coherent patterns of alternating currents propagate upcanyon at speeds of several tens of centimeters per second, suggesting that the currents are caused by focused internal tides or shorter period internal waves moving up the canyons. Downcanyon velocities are higher and of longer duration than upcanyon velocities, suggesting a net downcanyon transport of sediment in the form of migrating asymmetrical ripples or large sand waves. Cacchione (personal communication, 1973) has directly observed, from a submersible, active downcanyon migration of sand waves in the upper reaches of Hydrographer Canyon off the northeast coast of the United States.

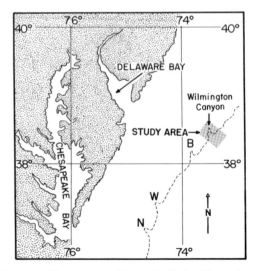

FIGURE 23. *Temperature (A) and salinity (B) sections at the shelf break near Wilmington Canyon. Shelf water extends to a depth of about 200 m on the uppermost slope. Modified from Lyall et al. (1971).*

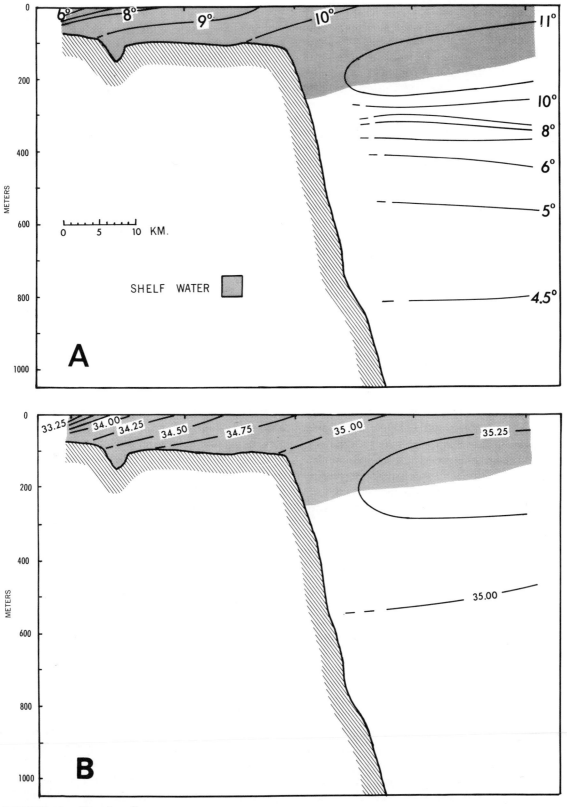

FIGURE 23 (Continued)

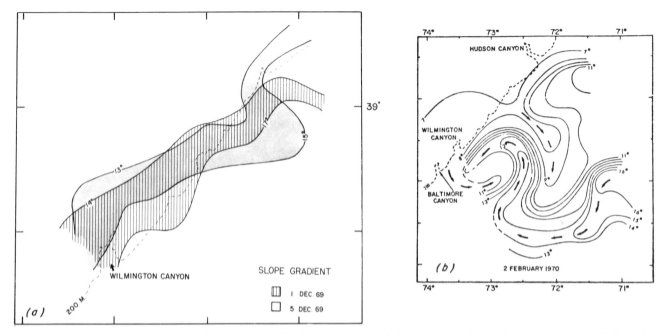

FIGURE 24. *Surface isotherms (in degrees centigrade, from airborne infrared radiation temperature measurements) showing (a) frontal water mass movement contiguous to the shelf break* *near Wilmington Canyon, December 1–5, 1969; and* (b) *intrusion of warm water of probable Gulf Stream origin near the shelf break on February 2, 1970. Modified from Lyall et al. (1971).*

SHEAR ZONES BETWEEN MAJOR CURRENTS.   Simply by virtue of its position, the area near the shelf edge tends to be a meeting ground between water masses of different origin (Fig. 23). Thus it occupies a zone of contact between shelf currents, either short term or long term, and larger scale currents in the open ocean. The resulting shear between such currents of different speeds and directions, with attendant production of large eddies (Fig. 24) with vertical scales of the same order as the thickness of the shelf water column, must subject the region at and somewhat inshore of the shelf edge to temporarily stronger bottom currents of variable velocity and duration. Effects of such interaction between water masses near the shelf edge have been observed off the east coast of the United States by Fisher and Gotthardt (1970) and Lyall et al. (1971).

ORGANISM-CONTROLLED EROSION AT THE SHELF BREAK

Not all of the factors now contributing to sediment entrainment are physical in origin. Benthic organisms alter surficial sediment surfaces and introduce bioclastic material, including fecal pellets, foraminiferal tests, pelecypod valves, etc. A typical shelf-edge silty shelly sand is illustrated in Fig. 12. Bottom photographs reveal the importance of biogenic components along the break;

these show moderate to large amounts of valves, broken shell hash, and organic traces on the rippled seafloor. Reworking by benthic organisms within the sediment and at the sediment-water interface disrupts stratification and fabric (Heezen and Hollister, 1971); see Fig. 13 of Chapter 17. The large populations of bottom dwellers such as polychaetes that expel materials into the water column and bottom-feeding fish that stir the seafloor (Stanley, 1971) contribute to the erosion in this sector and maintain a sizable amount of sediment in suspension (Fig. 25).

TRANSFER OF COARSE SEDIMENT FROM SHELF TO SLOPE

There is ample indirect evidence that silt and sand are transferred across the shelf edge onto the upper slope from temporary resting places on the outer shelf. This has been termed *offshelf spillover* or *shelf-edge bypassing* (Stanley et al., 1972b). This phenomenon is not necessarily dependent on the various processes discussed in the preceding section; even if these processes are unimportant, we still have to account for offshelf spillover. But we deal with this effect here because geologically it is one of the most characteristic aspects of the shelf break. Once past the shelf edge, sediment can be transported down the slope by various gravity-driven modes

FIGURE 25. *Benthic organisms maintain some sediment in suspension on the outer margin. Photograph shows flatfish eroding seafloor surface and producing tracks, on right. From Stanley (1971).*

of transport, such as turbidity currents, sand flows or debris flows (cf. Chapters 11 and 17; Middleton and Hampton, 1973).

Several modes of spillover can be distinguished. Among the most obvious are (1) entrapment of bed load moving in a sand stream by a submarine canyon incised into the shelf; (2) termination at the shelf edge of a shelf sand stream directed normally or obliquely offshelf, driven by either wind stress currents or tidal currents. In the following paragraphs we discuss some well-documented examples of these modes.

Entrapment of silt, sand, and gravel by canyon heads is perhaps the best-known mode of shelf-edge bypassing of coarse sediment. This spillover mechanism is prevalent both on narrow shelves (such as off southern California) and on wide outer margins (such as off the Mid-Atlantic States). Typically a sand stream controlled by some pattern of shelf currents is intercepted by a canyon without any effect on the current itself. Examples are described by Shepard and Dill (1966) and Stanley (1970).

Obliquely offshelf sediment transport may occur at the southern terminus of the Middle Atlantic Bight of North America (Hunt et al., 1975). Here the zone of shear between the north-flowing Gulf Stream and sporadic wind-driven southerly shelf currents shifts back and forth across the shelf edge (Fig. 26). There seems to be net migration of sand southward on the shelf toward this shelf-edge shear zone in the form of large

north-south sand ridges covered with southward-migrating sand waves (note crest orientation in Fig. 27). Seaward of the outermost sand ridge at Cape Hatteras a deeper terrace is mantled with a quartzose foraminiferal silt. This facies is the consequence of pelagic fallout of the Gulf Stream mixing with sand winnowed from the shelf surface, and is probably resulting in a seaward progradation of the shelf edge. The upper slope below the terrace is capped by foraminiferal silt which fills the upper tributaries of the Hatteras Canyon system (Rona, 1970).

An example of more normally directed movement of sand off an open shelf edge is recorded along the southern margin of the Celtic Sea (Hadley, 1964a). Here convergent bed load transport is probably the consequence of complex interaction between the oceanic tide seiche and the shelf edge (Fig. 28), but the specific mechanisms are obscure. Medium sand and coarser is periodically set in motion in water depths exceeding 100 m, and the offshelf component is substantial (Stride, 1963).

There are undoubtedly many intermediate cases between these cited examples. The important point is that, by whatever mode of transport, sediment is delivered from the shelf either to canyons or, in open shelf-break sectors, to the uppermost slope, and either progrades the shelf or is transported downslope and deposited beyond the base of the slope. Our understanding of these spillover processes will improve as we learn more about outer shelf bottom sediment transport.

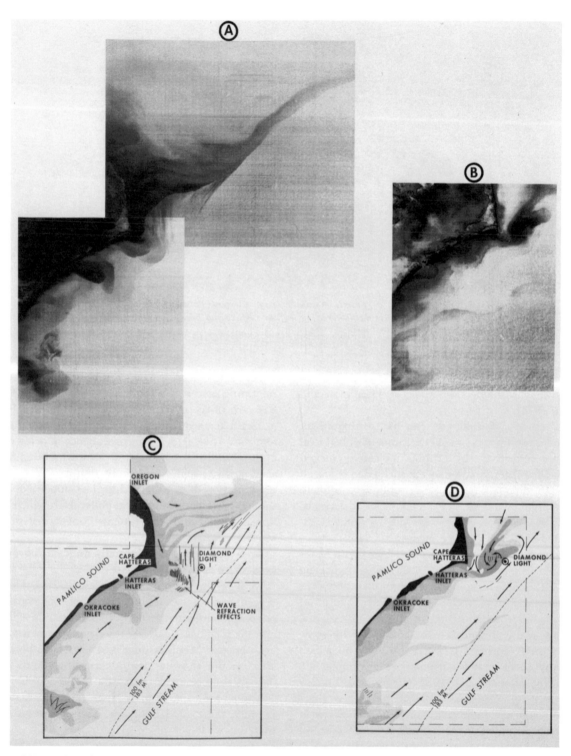

FIGURE 26. *The Cape Hatteras region showing average motion of the major water masses and the position of the Gulf Stream. Shear near the shelf break probably varies in part with the strength and direction of the wind. (A) Cape Hatteras on December 2, 1972 (ERTS-I satellite imagery). Suspended sediment plumes show motion of inshore water. Wind blowing 5 m per second from west.*

(B) Cape Hatteras on January 25, 1973. Wind from north at 5 m per second for several hours prior to photograph. Note intrusion of clearer Virginia Shelf water from the north and cyclonic eddy formed inshore of Diamond Light as the Virginia water encounters the Gulf Stream. (C, D) interpretations of (A) and (B). From Hunt et al. (1975).

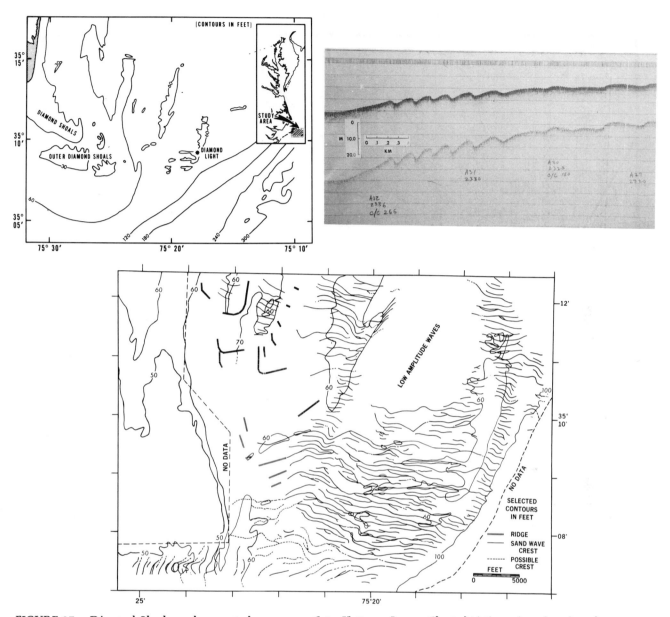

FIGURE 27. *Diamond Shoals sand wave study area near Cape Hatteras. Lower: Chart depicting orientation of sand wave crests. Upper: Sand wave train near shelf break. From Hunt et al. (1975).*

SUMMARY

Accelerated exploration of the seafloor in zones progressively deeper and more distant from shore tends to confirm the concept that has long been accepted by geologists, that the shelf-break surface serves only as a temporary resting place for sediment moving from terrigenous sources to ultimate depositional sites in deep marine environments. The distribution, composition, texture, and bed geometry of surficial sediment indicate that shelf-edge sectors where sediment is not frequently set into motion are the exception, not the rule.

Several kinds of water movements seem to be distinctive of the shelf edge, although theory and observation are not yet far enough advanced to supply firm estimates of relative importance. Tidal currents, especially on low-latitude, gently sloping shelves, should tend to be stronger than farther inshore on the outer shelf. Bottom currents produced by migration of the atmospheric pressure-induced wave that migrates beneath a major storm moving in the offshore direction are predicted by theory to result in a current on the order of 10 cm/sec concentrated at the shelf break. Both theory and experiment suggest that internal waves of tidal or shorter

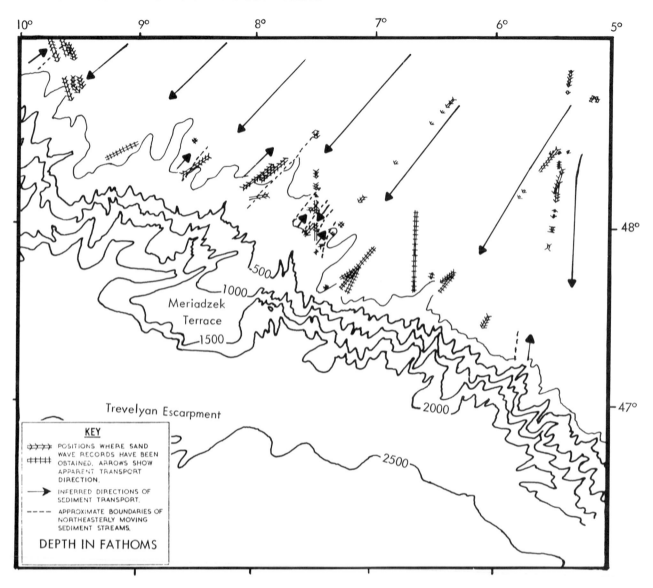

FIGURE 28. *Outer continental margin southwest of the English Channel showing position of sand waves and inferred directions of sediment transport. Offshelf spillover of sand occurs in this area in response to present hydraulic regime. Modified from Hadley (1964a).*

period propagating shoreward in thermoclines produce intensified near-bottom velocities near the shelf edge as they steepen and break. Large eddies generated at shear zones between major currents on the shelf and in the adjacent open ocean should locally increase bottom velocities in the shelf-break region. When superimposed on bottom currents caused by tides, waves, and winds that are known to be typical of the outer shelf environment, these distinctive currents act to produce the heightened bottom sediment movement that seems to be characteristic of the shelf edge.

One of the most characteristic aspects of the shelf edge is that a part of the bottom sediment that is con-tinually entrained and shifted on the outermost shelf is ultimately transferred across the shelf edge and onto the upper slope or into the deep ocean. The most important modes of such sediment spillover seem to be entrapment of bed load moving in a sand stream by a submarine canyon incised into the shelf, and termination at the shelf break of a shelf sand stream moving normally or obliquely offshelf.

ACKNOWLEDGMENTS

We thank C. M. Wear and R. E. Hunt, D. J. P. Swift, and H. D. Palmer for permission to use unpublished figures from their papers in press. Swift also provided useful suggestions for improving the text. Support to Southard was provided by the Office of Science and Technology, Office of Naval Research, Contract N00014-75-C-0291. Support to Stanley was provided by Smithsonian Research Foundation Grant 430035; ship time aboard the USCGC Rockaway was made available through the U.S. Coast Guard.

REFERENCES

Adegoke, O. S. and D. J. Stanley (1972). Mica and shell as indicators of energy level and depositional regime on the Nigerian shelf. *Mar. Geol.*, **13:** M61–M66.

Boillot, G. and L. d'Ozouville (1970). Etude structurale du plateau continental nord-espagnol entre Aviles et Llanes. *C.R. Acad. Sci. Paris*, **270:** 1865–1868.

Bradley, W. C. (1958). Submarine abrasion and wave-cut platforms. *Geol. Soc. Am. Bull.*, **69:** 967–974.

Burk, C. A. (1968). Buried ridges with continental margins. *N.Y. Acad. Sci. Trans.*, **30:** 397–409.

Burk, C. A. and C. L. Drake, eds. (1974). *The Geology of Continental Margins*. Berlin and New York: Springer-Verlag, 1009 pp.

Cacchione, D. A. and J. B. Southard (1974). Incipient sediment movement by shoaling internal gravity waves. *J. Geophys. Res.*, **79:** 2237–2242.

Cacchione, D. A. and C. I. Wunsch (1974). Experimental study of internal gravity waves over a sloping bottom. *J. Fluid Mech.*, **66:** 223–239.

Carruthers, J. N. (1963). History, sand waves and nearbed currents of La Chapelle Bank. *Nature*, **197:** 942–946.

Cartwright, D. E. (1959). On submarine sand-waves and tidal lee-waves. *Proc. Roy. Soc. London*, **A253:** 218–241.

Cartwright, D. E. and A. H. Stride (1958). Large sand waves near the edge of the continental shelf. *Nature*, **181:** 41.

Curray, J. R. (1965). Late Quaternary history, continental shelves of the United States. In H. E. Wright, Jr. and D. G. Frey, eds., *The Quaternary of the United States*. Princeton, N.J.: Princeton University Press, pp. 723–735.

Curray, J. R. (1966). Continental terrace. In R. W. Fairbridge, ed., *The Encyclopedia of Oceanography*. New York: Reinhold: pp. 207–214.

Curray, J. R. (1967). Morphology of pre-Quaternary continental terraces. *Proc. VIIth Int. Sediment. Congr.*, *Reading-Edinburgh*, 2 pp.

Curray, J. R. (1969a). History of continental shelves. In D. J. Stanley, ed., *The New Concepts of Continental Margin Sedimentation*. Washington, D.C.: Am. Geol. Inst., pp. JC-VI-1 to JC-VI-18.

Curray, J. R. (1969b). Shallow structure of the continental margin. In D. J. Stanley, ed., *The New Concepts of Continental Margin Sedimentation*. Washington, D.C.: Am. Geol. Inst., pp. JC-XII-1 to JC-XII-22.

Dewey, J. F. and J. M. Bird (1970). Mountain belts and the new global tectonics. *J. Geophys. Res.*, **75:** 2625–2647.

Dietz, R. S. and J. C. Holden (1974). Collapsing continental rises: Actualistic concept of geosynclines—A review. In R. H. Dott, Jr. and R. H. Shaver, eds., *Modern and Ancient Geosynclinal Sedimentation*. SEPM Spec. Publ. 19, pp. 14–25.

Dietz, R. S. and H. W. Menard (1951). Origin of abrupt change in slope at continental shelf margin. *Am. Assoc. Pet. Geol. Bull.*, **35:** 1994–2016.

Doyle, L. J., W. J. Cleary, and O. H. Pilkey (1968). Mica: Its use in determining shelf-depositional regimes. *Mar. Geol.*, **6:** 381–389.

Draper, L. (1967). Wave activity at the sea bed around northwestern Europe. *Mar. Geol.*, **5:** 133–140.

Emery, K. O. (1952). Continental shelf sediment of southern California. *Geol. Soc. Am. Bull.*, **63:** 1105–1108.

Emery, K. O. (1956). Deep standing waves in California basins. *Limnol. Oceanogr.*, **1:** 35–41.

Emery, K. O. (1965). Characteristics of continental shelves and slopes. *Am. Assoc. Pet. Geol. Bull.*, **49:** 1379–1384.

Emery, K. O. (1968a). Relict sediments on continental shelves of the world. *Am. Assoc. Pet. Geol. Bull.*, **52:** 445–464.

Emery, K. O. (1968b). Shallow structure of continental shelves and slopes. *Southeast. Geol.*, **9:** 173–194.

Emery, K. O. (1969). Continental rises and oil potential. *Oil Gas J.*, **67:** 231–243.

Emery, K. O. (1970). Continental margins of the world. In *The Geology of the East Atlantic Continental Margin*. **1**. ICSU-SCOR Working Party 31 Symposium, Cambridge, Rep. 70/13, pp. 7–29.

Emery, K. O. and C. G. Gunnerson (1973). Internal swash and surf. *Proc. Natl. Acad. Sci. U.S.A.*, **70:** 2379–2380.

Emery, K. O. and E. Uchupi (1972). *Western North Atlantic Ocean: Topography, Rocks, Structure, Water, Life, and Sediments*. Am. Assoc. Petrol. Geol. Mem. 17, 532 pp.

Ewing, J. A. (1973). Wave-induced bottom currents on the outer shelf. *Mar. Geol.*, **15:** M31–M35.

Fisher, A., Jr. and G. A. Gotthardt (1970). *Aerial Observation of Gulf Stream Phenomena, Virginia Cape Area, October 1968–May 1969*. U.S. Naval Oceanographic Office, Tech. Rep. 223, 23 pp.

Fleming, R. H. (1938). Tides and tidal currents in the Gulf of Panama. *J. Mar. Res.*, **1:** 192–206.

Fleming, R. H. and R. Revelle (1939). Physical processes in the oceans. In P. D. Trask, ed., *Recent Marine Sediments*. Am. Assoc. Pet. Geol., pp. 48–141.

Frank, W. M. and G. M. Friedman (1973). Continental shelf sediments off New Jersey. *J. Sediment. Petrol.*, **43:** 224–237.

Galt, J. A. (1971). A numerical investigation of pressure-induced storm surges over the continental shelf. *J. Phys. Oceanogr.*, **1:** 82–91.

Gaul, R. D. (1961). Observations on internal waves near Hudson Canyon. *J. Geophys. Res.*, **66:** 3821–3830.

Got, H. (1973). Etudes des correlations tectonique-sédimentation au cours de l'histoire Quaternaire du précontinent Pyrenéo-Catalan. Thesis, Perpignan, 294 pp.

Got, H. and D. J. Stanley (1974). Sedimentation in two Catalonian canyons, northwestern Mediterranean. *Mar. Geol.*, **2:** 164–167.

Hadley, M. L. (1964a). The continental margin southwest of the English Channel. *Deep-Sea Res.*, **11:** 767–779.

Hadley, L. M. (1964b). Wave-induced bottom currents in the Celtic Sea. *Mar. Geol.*, **2:** 164–167.

Heezen, B. C. and C. D. Hollister (1971). *The Face of the Deep*. London and New York: Oxford University Press, 659 pp.

Heezen, B. C., M. Tharp, and M. Ewing (1959). *The Floors of the Oceans*. Geol. Soc. Am. Spec. Pap. 65, 122 pp.

Hunt, R. E., D. J. P. Swift, and H. Palmer (1975). Constructional shelf topography—Diamond Shoals, North Carolina. *Geol. Soc. Am. Bull.*, **86** (in press).

Inman, D. L. and C. E. Nordstrom (1971). On the tectonic and morphologic classification of coasts. *J. Geol.*, **79**: 1–21.

Knott, S. T. and H. Hoskins (1968). Evidence of Pleistocene events in the structure of the continental shelf off the northeastern United States. *Mar. Geol.*, **6**: 5–43.

Komar, P. D., R. H. Neudeck, and L. D. Kulm (1972). Observations and significance of deep-water oscillatory ripple marks on the Oregon continental shelf. In D. J. P. Swift, D. B. Duane, and O. H. Pilkey, eds., *Shelf Sediment Transport: Process and Pattern*. Stroudsburg, Pa.: Dowden, Hutchinson & Ross, pp. 143–180.

Kuenen, Ph. H. (1939). The cause of coarse deposits at the outer edge of the shelf. *Geol. Mijnbouw*, **1**: 36–39.

Lafond, E. C. (1961). Internal wave motion and its geological significance. In *Mahadevan Volume*. Andrha Pradesh, India: Osmania University Press, pp. 61–77.

Lafond, E. C. (1962). Internal waves. 1. In M. N. Hill, ed., *The Sea*, Vol. I. New York: Wiley-Interscience, pp. 731–751.

Lee, O. S. (1961). Observations on internal waves in shallow water. *Limnol. Oceanogr.*, **6**: 312–331.

Lewis, K. B. (1974). The continental terrace. *Earth Sci. Rev.*, **10**: 37–71.

Lyall, A. K., D. J. Stanley, H. N. Giles, and A. Fisher (1971). Suspended sediment and transport at the shelfbreak and on the slope. *Mar. Technol. Soc. J.*, **5**(1): 15–26.

Macintyre, I. G. (1972). Submerged reefs off eastern Caribbean. *Am. Assoc. Pet. Geol. Bull.*, **56**: 720–738.

Macintyre, I. G. and J. D. Milliman (1970). Physiographic features on the outer shelf and upper slope, Atlantic continental margin, southeastern United States. *Geol. Soc. Am. Bull.*, **81**: 2577–2598.

Maxwell, A. E., ed. (1970). *The Sea. New Concepts of Sea Floor Evolution*. Vol. 4. New York: Wiley-Interscience, 664 pp.

Middleton, G. V. and M. A. Hampton (1973). Sediment gravity flows: Mechanics of flow and deposition. In G. V. Middleton and A. H. Bouma, eds., *Turbidites and Deep-Water Sedimentation*. SEPM Pacific Section Short Course, Anaheim, pp. 1–38.

Newton, R. S., E. Siebold, and F. Werner (1973). Facies distribution patterns on the Spanish Sahara continental shelf mapped with side-scan sonar. *Meteor Forsch.-Ergeb.*, **15**: 55–77.

Pilkey, O. H., I. A. Macintyre, and E. Uchupi (1971). Shallow structures: Shelf edge of continental margin between Cape Hatteras and Cape Fear, North Carolina. *Am. Assoc. Pet. Geol. Bull.*, **55**: 110–115.

Rona, P. A. (1970). Submarine canyon origin on upper continental slope off Cape Hatteras. *J. Geol.*, **78**: 141–152.

Shepard, F. P. (1932). Sediments on continental shelves. *Geol. Soc. Am. Bull.*, **43**: 1017–1034.

Shepard, F. P. (1963). *Submarine Geology*, 2nd ed. New York: Harper & Row, 557 pp.

Shepard, F. P. (1973). *Submarine Geology*, 3rd ed. New York: Harper & Row, 517 pp.

Shepard, F. P. and R. F. Dill (1966). *Submarine Canyons and Other Sea Valleys*. Chicago: Rand McNally, 381 pp.

Shepard, F. P. and N. F. Marshall (1973). Currents along floors of submarine canyons. *Am. Assoc. Pet. Geol. Bull.*, **57**: 244–264.

Shepard, F. P., N. F. Marshall, and P. A. McLoughlin (1974a). "Internal waves" advancing along submarine canyons. *Science*, **183**: 195–197.

Shepard, F. P., N. F. Marshall, and P. A. McLoughlin (1974b). Currents in submarine canyons. *Deep-Sea Res.*, **21**: 691–706.

Shideler, G. L. and D. J. P. Swift (1972). Seismic reconnaissance of post-Miocene deposits, Middle Atlantic continental shelf—Cape Henry, Virginia to Cape Hatteras, North Carolina. *Mar. Geol.*, **12**: 165–185.

Shideler, G. L., D. J. P. Swift, G. H. Johnson, and B. W. Holliday (1972). Late Quaternary stratigraphy of the inner Virginia continental shelf: A proposed standard section. *Geol. Soc. Am. Bull.*, **83**: 1787–1804.

Smith, J. D. and T. S. Hopkins (1972). Sediment transport on the continental shelf off of Washington and Oregon in light of recent current measurements. In D. J. P. Swift, D. B. Duane, and O. H. Pilkey, eds., *Shelf Sediment Transport: Process and Pattern*. Stroudsburg, Pa.: Dowden, Hutchinson & Ross, pp. 143–180.

Southard, J. B. and D. A. Cacchione (1972). Experiments on bottom sediment movement by breaking internal waves. In D. J. P. Swift, D. B. Duane, and O. H. Pilkey, eds., *Shelf Sediment Transport: Process and Pattern*. Stroudsburg, Pa.: Dowden, Hutchinson & Ross, pp. 83–97.

Stanley, D. J., ed. (1969). *The New Concepts of Continental Margin Sedimentation*. Short Course Lecture Notes, Am. Geol. Inst., Washington, D.C., 400 pp.

Stanley, D. J. (1970). *Flyschoid sedimentation on the Outer Atlantic Margin off Northeastern North America*. Geol. Assoc. Can., Spec. Pap. 7, pp. 179–210.

Stanley, D. J. (1971). Fish-produced markings on the outer continental margin east of the middle Atlantic States. *J. Sediment. Petrol.*, **41**: 159–170.

Stanley, D. J. (1974). Pebbly mud transport in the head of Wilmington Canyon. *Mar. Geol.*, **16**: M1–M8.

Stanley, D. J. and P. Fenner (1973). Underwater television survey of the Atlantic outer continental margin near Wilmington Canyon. *Smithsonian Contrib. Earth Sci.*, **11**: 54 pp.

Stanley, D. J., P. Fenner, and G. Kelling (1972a). Currents and sediment transport at the Wilmington Canyon shelfbreak, as observed by underwater television. In D. J. P. Swift, D. B. Duane, and O. H. Pilkey, eds., *Shelf Sediment Transport: Process and Pattern*. Stroudsburg, Pa.: Dowden, Hutchinson & Ross, pp. 621–644.

Stanley, D. J. and N. P. James (1971). Distribution of *Echinarachnius parma* (Lamarck) and associated fauna on Sable Island Bank, southeast Canada. *Smithsonian Contrib. Earth Sci.*, **6**: 1–24.

Stanley, D. J., D. J. P. Swift, N. Silverberg, N. P. James, and R. G. Sutton (1972b). Late Quaternary progradation and sand spillover on the outer continental margin off Nova Scotia, southeast Canada. *Smithsonian Contrib. Earth Sci.*, **8**: 88 pp.

Stetson, H. C. (1937). Current measurements in the Georges Bank canyons. *Trans. Geophys. Union*, *18th Ann. Meet.*, pp. 216–219.

Stride, A. H. (1963). Current-swept sea floors near the southern half of Great Britain. *Q. J. Geol. Soc. London*, **119**: 175–199.

Swift, D. J. P., D. B. Duane, and O. H. Pilkey, eds. (1972). *Shelf Sediment Transport: Process and Pattern*. Stroudsburg, Pa.: Dowden, Hutchinson & Ross, 656 pp.

Swift, D. J. P., D. J. Stanley, and J. R. Curray (1971). Relict sediment on continental shelves: A reconsideration. *J. Geol.*, **79:** 322–346.

Wear, C. M., D. J. Stanley, and J. E. Boula (1974). Shelfbreak physiography between Wilmington and Norfolk canyons. *Mar. Technol. Soc. J.*, **8:** 37–48.

Worzel, J. L. (1968). Survey of continental margins. In D. T. Donovan, ed., *Geology of Shelf Seas*. Edinburgh: Oliver & Boyd, pp. 117–152.

Wunsch, C. I. (1969). Progressive internal waves on slopes. *J. Fluid Mech.*, **35:** 131–144.

Wunsch, C. I. (1971). Internal waves. *EOS, Trans. Am. Geophys. Union*, **52:** 233–235.

Zarudzki, E. F. K. and E. Uchupi (1968). Organic reef alignments on the continental margin south of Cape Hatteras. *Geol. Soc. Am. Bull.*, **79:** 1867–1870.

CHAPTER 17

Sedimentation in Canyon, Slope, and Base-of-Slope Environments

GILBERT KELLING

Department of Geology and Oceanography, University of Wales at Swansea, Swansea, Great Britain

DANIEL JEAN STANLEY

Division of Sedimentology, Smithsonian Institution, Washington, D.C.

Submarine slopes and contiguous, deeper outer continental margin environments that border the major continental masses are marked by a varied, yet distinctive, morphology and molded by a unique plexus of processes. The following discussion is concerned with the salient physiographic features of continental slopes, rises, and aprons and with the forces that have shaped them. As sedimentologists, we are primarily interested in the nature and genesis of the materials that are deposited, either temporarily or permanently, in this milieu. The last section of this chapter considers the effects of human activities in shelf-break, submarine canyon, slope, and base-of-slope environments, in accord with the general tenor of this volume. The reader interested in the geotectonic status of modern slopes and the recognition of possible ancient analogs of marginal slopes and slope-related sequences is referred to the companion volume of this text (Stanley, 1969) and other recent syntheses (Donovan, 1968; Maxwell, 1970; Emery and Uchupi, 1972).

The area to be discussed extends from the shelf break to the abyssal or basin plain and thus includes both the slope proper (the region of maximum gradient) and the adjacent rise and deep-sea fan environments which frequently (but not invariably) intervene between mar-

ginal slopes and the essentially flat deep-sea plains. Aspects of slope sedimentation directly related to shelf-edge morphology and processes have been dealt with in Chapter 16 and accordingly are not considered here, except insofar as they relate directly to features and mechanisms characteristic of the slope and base-of-slope environments.

Technological advances in the past decade or so, together with the greatly expanded academic, governmental, and industrial efforts in oceanographic research, have produced a great volume of data concerning offshore environments, including the slope region. However, unlike areas such as the inner shelf, the evaluation of sedimentary processes in the slope and base-of-slope environments has remained a largely qualitative exercise. One of the intents here is to stress the need for quantification and modeling of the mechanisms responsible for molding these important regions of the marine realm and to indicate the practical value of the quantitative data already gleaned.

MORPHOLOGIC CONSIDERATIONS

Continental terrace is the generic term applied to the shelf and slope, which are closely related in a structural sense (Curray, 1966). The slopes bordering modern continental masses and linking shelf or platform environments to those of the deep sea or ocean floor present a physiographic diversity matching that of their terrestrial counterparts. However, in aggregate two major physiographic provinces may be recognized: the continental slope *sensu stricto*, and the continental rise or fan-apron complex. A similar division can be discerned, at least locally, in smaller, enclosed marine basins, giving rise to basin slope and basin apron provinces. The submarine escarpments around island arcs and along much of the Pacific periphery are steep and lack the rise or apron, passing directly into deep marginal trenches. The slopes associated with noncontinental features such as the midocean ridges are not considered here although they display many aspects of morphology and sedimentation that are comparable with those found on truly marginal slopes.

Together, the slope and rise regions constitute approximately 15% of the submerged surface of the earth (Menard and Smith, 1966; modified by Emery, 1969a, b), of which the rise and associated features account for more than half (Fig. 1). Gradients of the continental slope are generally low—3 to 6° are commonly quoted limits (Shepard, 1973; Stanley, 1969)—but may increase locally to values in excess of 15°, especially on the walls of submarine canyons. Maximum gradients commonly occur just below the shelf break, except on slopes defined by active faults (such as on the flanks of trenches),

where the angle may remain relatively constant or may even increase near the base of the slope (Shepard, 1973, pp. 279, 299). The nature and gradient of marginal slopes are greatly affected by the presence of modern or ancient deltas, such as those of the Mississippi, Niger, Nile, Rhone, or Ganges. Delta growth and rapid progradation may induce gradients as low as 0.5° (Mississippi) or 0.3° (Nile), and large arcuate deep-sea cones replace and overlap any continental rise that may be present (Shepard, 1955, 1960; Emery et al., 1966).

Gradients of the continental rise surface are generally less than 1:40 (Heezen et al., 1959) and decrease oceanward as the apron gradually merges into the virtually horizontal basin or abyssal plain. Off the east coast of North America, the broad rise is divisible, along much of its length, into upper and lower portions, the latter generally possessing a somewhat decreased gradient (Uchupi and Emery, 1967; Heezen and Tharp, 1968; Emery et al., 1970). The surface gradient of major deep-sea fans is comparable with that of the continental rise, although steepening is sometimes encountered, especially near the apex of the fan (Menard, 1964). The depth of the relatively pronounced slope-rise boundary is variable, averaging 4000 m on the Pacific periphery and around 3000 m on both margins of the Atlantic (Shepard, 1973, p. 279).

Although numerous classifications of coastlines and associated continental terrace topography have been proposed (see, *inter alia*, Johnson, 1919; Shepard, 1952;

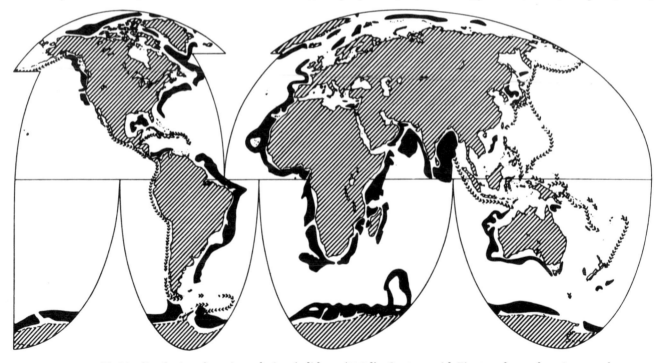

FIGURE 1. *Worldwide distribution of continental rises (solid black) and deep-sea trenches (chevron pattern). From Emery* *(1969b). Compare with Fig. 2 and note that rises are almost confined to trailing-edge margins.*

McGill, 1958; Bloom, 1965) it is only recently, with the benefit of continuous seismic reflection profiling, the deep-sea drilling program, and the unifying concept of plate tectonics and seafloor spreading, that the great variety of continental margin types have been marshalled into a relatively small number of basic categories (Curray, 1969; Dewey and Bird, 1970; Inman and Nordstrom, 1971; Reading, 1972; cf. Stanley, 1970). At the risk of oversimplifying the inherently complex natural situation, by modifying the coastline classification of Inman and Nordstrom (1971) to the total architecture of margins, it is possible to distinguish three major categories of margin (Fig. 2):

1. Tectonically *active* margins—mainly the "collision" type of Inman and Nordstrom [termed "subduction" type by Shepard (1973, p. 104)], but including margins located by transcurrent or transform faulting.

2. Tectonically *passive* margins—mainly the "trailing edge" type of Inman and Nordstrom (1971), which were further subdivided into "neo-," "Afro," and "Amero" subtypes.

3. Tectonically *buttressed* margins—mainly the "marginal sea" type of Inman and Nordstrom, protected behind island arcs.

Physiographically, type 1 margins are represented by mountainous hinterlands, narrow shelves, and relatively steep, fault-defined slopes, often displaying a step-and-platform (borderland) configuration, like that typical of the west coast of the United States and the Mediterranean Sea. Mature development of a wide continental rise is improbable in this situation but localized deep-sea fans are frequently found. A "pseudorise" may be present, as a result of marginal faulting, but will consist of foundered bedrock blocks with an irregular and usually thin veneer of sediment. The so-called Balearic Rise in the western Mediterranean (Stanley et al., 1974a) is an excellent example of this variety of marginal feature (Fig. 3). True subduction margins are flanked by deep trenches (Dewey and Bird, 1970) while these are normally lacking off margins defined by transcurrent (or tensional) faults.

Type 2 margins (except for the juvenile forms such as the Red Sea or the Gulf of California) border hinterlands of more subdued relief and are generally represented by wide shelves, prograding slopes of moderate gradient, and wide, laterally extensive continental rise wedges of sediment. The eastern margin of the United States (Fig. 7) and the continental edge of northwestern Europe are typical examples. Localized faulting of

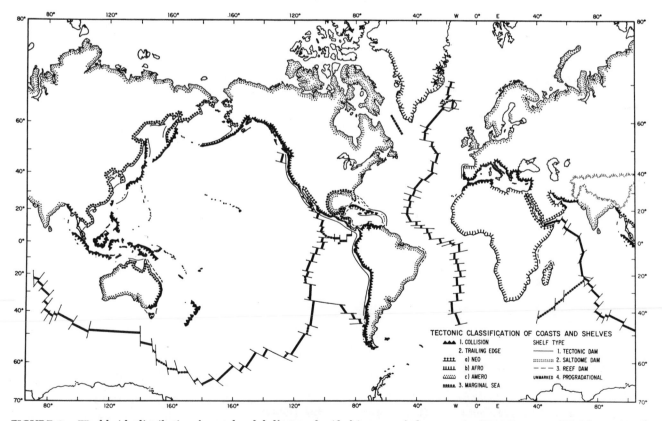

FIGURE 2. *Worldwide distribution of coastal and shelf types, classified in terms of plate tectonics. From Inman and Nordstrom (1971).*

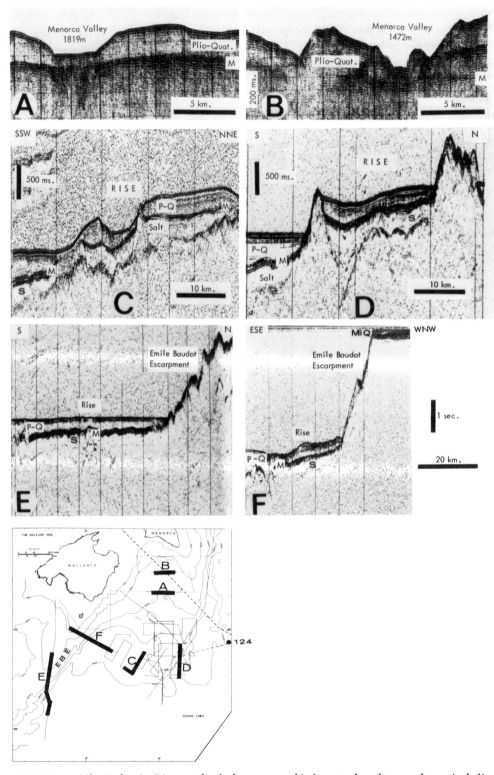

FIGURE 3. *The Balearic Rise, south of the Balearic Islands of Mallorca and Menorca (inset map), is a fan-shaped feature resulting from marginal faulting. Seismic profiles (30,000 J sparker)* *across this feature show large-scale vertical displacement of upper Miocene (M), Pliocene (P), and Quaternary (Q) sequences. From Stanley et al. (1974a).*

tensional or gravity-glide type may occur, giving rise to isolated blocks and marginal plateaus on the slope, as in the case of the Blake Plateau, south of Cape Hatteras (Fig. 4). It is important to appreciate that although type 2 margins are relatively stable in tectonic terms, they are not exempt from the broad flexuring of the continental terrace which appears to be an inevitable concomitant of the isostatic contrast between oceanic and continental crust (Lewis, 1974). Downwarping rates on stable margins, however, are generally much lower than those found on orogenic borders. Thus on the outer shelf off the northeastern United States, subsidence is about 10 cm/1000 years (Uchupi and Emery, 1967), whereas on the New Zealand margin rates on the order of 300 cm/1000 years have been measured (Lewis, 1971a).

Type 3 margins are less well defined in physiographic terms than the previous categories. They are generally characterized by very wide and shallow shelves with a thick sediment cover (South China Sea) but narrow shelves are known, especially where the flanking slope is fault- or fold-defined (Taiwan; Gulf coast south of the Rio Grande). The continental slope frequently has a very low gradient, sometimes modified by salt diapirs (as in the northern Gulf of Mexico).

The Continental Slope Proper

The slope proper generally possesses the most irregular topography of all the continental margin provinces. This irregularity may be ascribed to four principal

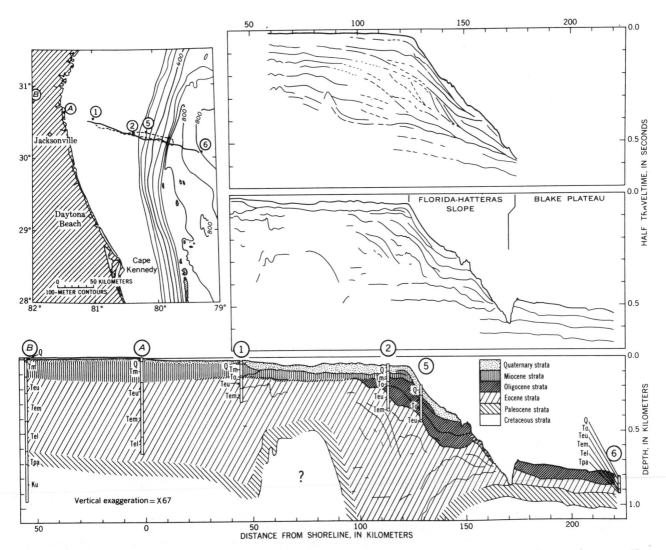

FIGURE 4. *Seismic profiles across the Blake Plateau region, southeastern United States, illustrating stratigraphic and tectonic relationships including displacement of Tertiary strata. Inset at top left indicates location of boreholes (A, B, 1–6) and inferred* stratigraphic section (solid line) shown below. The two profiles at upper right are line drawings of seismic reflection profiles located by dotted lines on the inset. From Emery and Zarudski (1969).

factors (erosion, deposition, diastrophism, and slumping) which may act independently or in concert to produce distinctive features. The most conspicuous features assigned to erosional processes are submarine canyons, which include some of the largest negative features on the surface of the earth and which transect marginal slopes around most of the major land masses (Shepard and Dill, 1966). Although seismic profiles generally indicate the eroded character of canyons (Fig. 5), the precise nature of the erosive processes is not so readily discerned. Moreover, the present form of many canyons appears to have originated from a combination of erosion and deposition (Rona, 1970a; Andrews et al., 1970; Shepard, 1973, p. 333).

In addition to the major canyon systems, which frequently traverse the entire marginal slope from shelf to basin plain or rise, the upper and middle portions of many slopes are dissected by smaller gully features. Most of these slope gullies are on the prograding front of modern or ancient deltas which are building out into deeper water, as off the active distributary mouths of the Mississippi Delta. Shepard (1955; 1973, p. 338) suggested that such gullies are due primarily to slumping and mass movement. Another mode of origin—erosion—should be considered in some regions such as the Congo, in view of the predominant trend of the gullies, which is down the slope. This downslope cutting may be associated with the development of highly dense underflows emanating from the distributaries during periods of river flood. Detailed investigations of gully systems on the Californian continental slope suggest that they were formed by enhanced erosion, possibly subaerial, during Pleistocene low stands of sea level (Inderbitzen and Simpson, 1971) although gullies on the midslope may have been enlarged and accentuated by differential deposition (Buffington and Moore, 1963) as well as slumping (Stanley and Silverberg, 1968).

In addition to the effects of differential deposition already referred to with respect to submarine canyons and gullies, slope irregularities ascribed to net deposition and sediment accumulation include the shallow ridges (wavelength approximately 100 m) encountered in the midslope region off New England on submersible dives by Emery and Ross (1968). These features have been assigned to the depositional activity of contour-following bottom currents, and direct measurements in this area indicated flow velocities as high as 70 cm/sec (Emery and Ross, 1968, p. 419).

Diastrophic effects on the slope include those attributable to large-scale foundering of the entire margin, such as the western Mediterranean example described by Bourcart (1960a) and to more localized faulting, folding, and salt tectonics. Except in favorable circumstances, proof of faulting on the slope is not readily

obtained. However, detailed seismic studies, especially those of the borderland topography off southern California, have revealed abundant evidence for the existence of both faulting and folding on the open slopes, as the study by D. G. Moore (1969) has demonstrated. Many of the faults (and folds) in this area affect near-surface sediments, indicating the geologic immediacy of the tectonism.

Comparable "neotectonic" modification of the upper sediment veneer is demonstrable along most of the tectonically active (collision type) margins defined by Inman and Nordstrom (1971). The inception of submarine canyon systems frequently may be linked to faulting (e.g., Yerkes et al., 1967). A remarkable series of close-spaced ridges trending along the slope off the east coast of Mexico represents the surface expression of an array of anticlinal structures revealed by seismic profiles (Bryant et al., 1968). Comparable, if smaller scale, phenomena have been observed off southern Portugal (Roberts and Stride, 1968) where they have been ascribed to slumping, but the upright style and basal coherence of the folds, together with their parallel trend to local landward structures, favor a tectonic origin.

A profusion of salt diapirs is now known to exist beneath the outer shelf and slope of the northwest Gulf of Mexico (Moore and Curray, 1963; Uchupi and Emery, 1968; Lehner, 1969) while similar features, associated with mud diapirs, also occur on the southern border of the Caribbean off the great delta of the Magdalena River (Uchupi, 1967a; Shepard, 1973, p. 229). Salt domes also are known from the continental margin of northwest Africa (Aymé, 1965; Rona, 1969, 1970b; Pautot el al., 1973; Solzansky, 1973). Where such features penetrate the seafloor they give rise to a highly irregular topography that may exert a considerable influence on indigenous sedimentary processes (Hoover and Bebout, 1974; Stanley et al., 1974b).

The effects of large-scale slumping and mass movement on marginal slopes are not always readily distinguishable from the results of diastrophism and especially gravitational tectonics (Rona and Clay, 1967). However, seismic profiling indicates that the irregular and outward-bowed or "lumpy" bathymetry that characterizes the lower slope in many regions is, in the main, caused by the emplacement and accumulation of large bodies of displaced rock and sediment (up to several scores of meters thick and several kilometers across), many of them contorted, which have slumped and slid downslope; see Fig. 6 (cf. Uchupi, 1967b, Fig. 4; Emery et al., 1970, Fig. 15; Stride et al., 1969, Figs. 6, 21). A contributory cause of upper and midslope irregularity is the scarps or hollows vacated by the slump masses (see Uchupi, 1968, p. C25). The strongly dissected slopes off

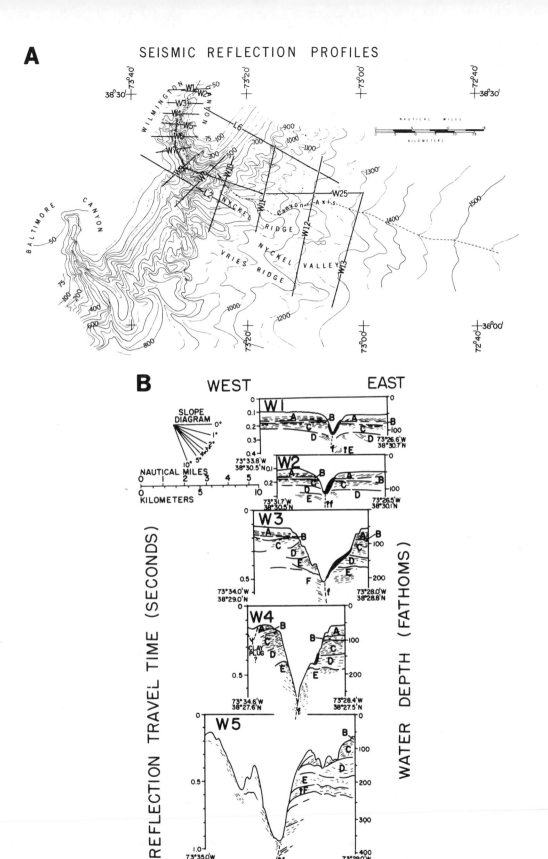

FIGURE 5. *Seismic profiles across the head of Wilmington Canyon, southeast of Delaware Bay, eastern United States. (A) Chart of Wilmington Canyon and associated features showing location of seismic reflection profiles illustrated in Figs. 5B and 6. (B) Interpreted sections illustrate the offset of layers on opposite canyon walls due to probable axial faulting (f) and also the thick asymmetric veneer of Quaternary sediment (shown in black). From Kelling and Stanley (1970).*

385

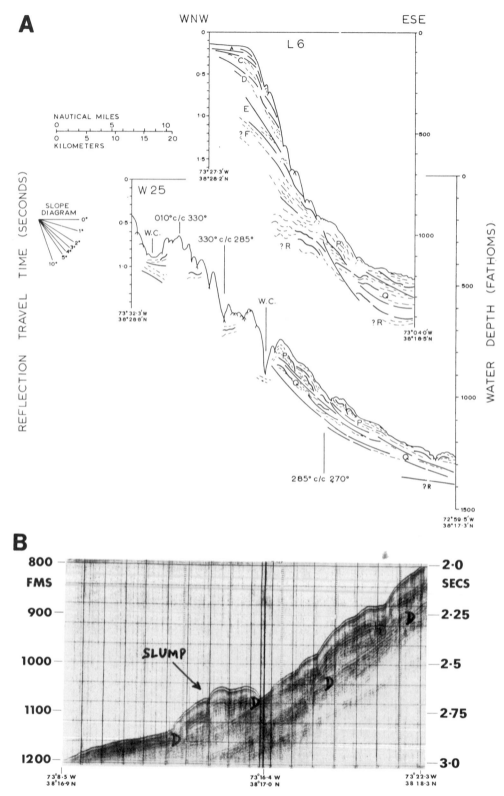

FIGURE 6. *Seismic profiles across outer shelf, slope, and upper continental rise in the vicinity of Wilmington Canyon (W.C.) off the eastern United States, illustrating probable slumping and gravitational displacement of strata. Profile location shown in Fig. 5A. (A) Interpreted line drawings of reflection profiles. Note truncation of some reflectors in midslope region (profile L6), curved slide planes and contorted strata within layers P and Q on the lower slope (all profiles). (B) Photograph of part of seismic reflection profile W25 across the lower slope and upper rise, illustrating slump deformation in near-surface reflectors of layer P, and internal surfaces of discordance (D). From Kelling and Stanley (1970).*

parts of the Nova Scotian shelf and in the region south of the Hudson Canyon contrast with laterally adjacent less scalloped slope topography and are probably attributable to slump scars only partially blanketed by younger sediment (Uchupi, 1968; Stanley and Silverberg, 1968; Kelling and Stanley, 1970).

Base-of-Slope Environments

The base-of-slope province is essentially a depositional zone characterized by a number of distinct subenvironments. Seismic profiles generally reveal an onlapping relationship between the gently sloping strata of the rise or apron and the more steeply inclined layers of the lower slope. Where onlapping (progradation) has proceeded farthest, the slope sediments pass gradually and laterally into those of the rise, surface declivities are low, and the break in gradient, which generally distinguishes the slope proper from the rise or apron, is scarcely discernible (Fig. 7). Such a situation is generally encountered where slope sedimentation rates have been enhanced by shelf bypassing, for example off delta mouths and in higher latitudes, where glacial low stands enabled vast quantities of reworked glaciofluvial detritus to be fed directly to the outer margin.

The most important subenvironments of the base-of-slope province are submarine fans, fan valleys, distributary channels, and suprafans (Fig. 8). Excellent descriptions of the morphology and sediments of these morphologic types have recently been published (Nelson and Kulm, 1973; Normark, 1974; Nelson and Nilson, 1974) and only a brief outline of the general physiography of deep-sea fans and their ancillary features is provided here.

Submarine fans (the term *cone* is frequently used to describe deep-sea fans associated with major active deltas, such as the Mississippi, Nile, or Ganges) are generally depicted as subarcuate wedges of sediment thickening to an apex at the point of emergence from the adjacent slope of the feeder canyon (Gorsline and Emery, 1959; Menard, 1960). The analogy with the alluvial fans of the subaerial realm has frequently been drawn (Shepard and Buffington, 1968). However, local circumstances and particularly the local tectonic situation can greatly modify this simple model. For example, Shepard et al. (1969) have shown that the well-developed La Jolla fan in fact is constructed from a thin veneer of sediment that covers block-faulted bedrock and attains a maximum thickness on the outer part of the fan.

Nevertheless, most submarine fans are characterized by flat-floored fan valleys leading directly from the submarine canyon mouths at the base of slope and generally bordered by levees, which are usually higher on the right-hand bank (looking downchannel) in the northern hemisphere (Menard, et al., 1965; Pratt, 1967; Normark, 1970). This part of the fan-channel system is essentially depositional and may be characterized by relatively straight or rapidly migrating and braided sectors, depending on the grade and amount of sediment being conveyed to this region. Levee height tends to decrease outward across the fan and the main channels simultaneously degenerate into a plexus of smaller, subradiating distributaries and isolated elongate depressions that are excavated into the middle fan surface. An upbulging zone termed the *suprafan* (Normark, 1970) characteristically occurs at the downslope end of the fan valley but this feature appears to be confined to fans composed of relatively coarse sediment (Normark, 1974,

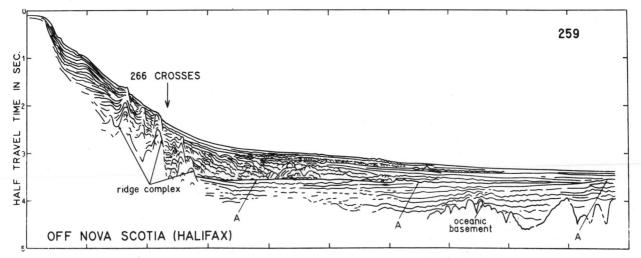

FIGURE 7. *Line drawing from seismic reflection profile across slope and rise off Halifax, Nova Scotia, illustrating a smooth, strongly progradational margin. Note presence of probable dis-* *placed (?slump) masses in subsurface layers above horizon A at slope-rise break just right of ridge complex. From Emery et al.* (1970).

MAP VIEW

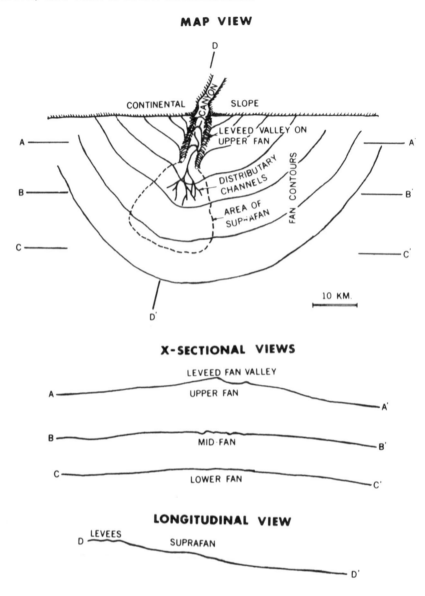

FIGURE 8. *Plan and cross sections of a deep-sea fan model to illustrate the principal features of growth. From Normark (1970).*

p. 65). In such fans, the ultimate low cone shape is achieved by lateral migration of the suprafan, combined with the depositional effects of the fine sediment spillover which is largely responsible for creating the outer fan segment. Fan valleys conveying only small amounts of relatively fine sediment may extend in leveed fashion much further out across the fan and the basin floor, as in the giant Bengal fan (Curray and Moore, 1971).

The coalescence and superimposition of adjacent fans are generally considered to have produced the continental rises (Menard, 1955; Shepard and Dill, 1966). However, the rise bordering the eastern margin of North America, which comprises fine-grained hemipelagic

deposits, turbidites, and large gravity slide and slump masses (Emery et al., 1970), has been ascribed to transportation and molding by deep, contour-following geostrophic currents (Heezen et al., 1966; Schneider et al., 1967). The persistence of the incised portions of several major fan valleys across the entire rise (Pratt, 1967) and the preservation within these valleys of well-defined thalwegs and lateral terraces (Stanley et al., 1971) suggest that the contour-following Western Boundary Undercurrent has not been able to modify significantly even the small-scale features of the rise surface (Fig. 9), most of which were probably formed in the periods of enhanced fan formation during Pleistocene low stands of sea level. Moreover, inspection of bathymetric charts of the Northwest Atlantic (e.g., those in Uchupi, 1968) indicates that the widest sectors of the rise occur opposite the major submarine canyons and deep-sea valleys (Laurentian–Sable Island system to the north; Hudson–Wilmington–Norfolk system further south). Bottom and subbottom profiles obtained parallel to the regional isobaths here also show that the rise consists of a series of coalescing megafan structures, some of which are more than 100 km in width (Stanley et al., 1971). Conolly and von der Borch (1967) have noted a similar relationship between rise width and canyon occurrence for the rise bordering the southern margin of Australia. These observations again indicate that, at least to date, redistribution of canyon-derived sediment by deep contour currents has been relatively ineffective in the formation of the rise.

An interesting feature of the lower continental rise and abyssal plains off the eastern seaboard of North America is the localized development of fields of linear waveforms (termed Lower Continental Rise Hills) with amplitudes of several tens of meters and wavelengths of a few kilometers (Fox et al., 1968; Rona, 1969); see Fig. 10. Similar features, termed "giant ripples," have been reported from the Argentine Basin (Ewing et al., 1971) and from the Bay of Cadiz off southwest Spain and Portugal (Kenyon and Belderson, 1973, p. 91).

Ballard (1966) suggested that these features are gravitational slide blocks, but their occurrence on relatively flat regions of the ocean floor, together with their persistence in depth—in some instances down to the late Cretaceous horizon A (Ewing et al., 1971, p. 75)—rules out a slide block origin for those giant wavelike bed forms. Since many of these waveforms possess asymmetrical profiles, which "face" consistently in one direction, they have been interpreted as giant bed forms constructed by geostrophic contour currents.

It is therefore accepted that large waveform structures may be depositional features, and Fox et al. (1968) have pointed out that in their examples the direction of migration of the Lower Rise Hills, inferred from their internal structure, is obliquely upcurrent and upslope with respect to the present geostrophic Western Boundary Undercurrent. These authors imply that the hills are, in effect, bed forms resembling giant antidunes in the conventional model of sand bed hydrodynamics. On the other hand, Ichiye (1968) and Ewing et al. (1971) state that giant ripples in the Argentine Basin migrate in the observed bottom current direction, like conventional ripples. Thus there appears to be some doubt concerning the external and internal geometry of these features and their orientation with respect to existing bottom currents.

Ichiye (1968) has ascribed formation of such giant ripples to fallout from stationary waves generated in near-bottom waters rich in suspended lutum (nepheloid layer). Direct fallout from the slowly oscillating internal waves would result in symmetrical bed forms, whereas asymmetry might be induced by superimposing the waves on a slow bottom current (geostrophic or otherwise). A thick nepheloid layer (Chapter 18), exists at present over much of the continental slope and deep basin off eastern North America (Ewing and Thorndike, 1965; Stanley, 1969) and in the Argentine Basin (Ewing et al., 1971) and it is probable that this layer was even more extensive and important as a medium of sedimentation in past periods of lowered sea level. The existence of stationary waves within this layer has yet to be demonstrated, but the theory outlined above appears to offer a satisfactory explanation for many of the features associated with giant ripples.

However, a recent survey in the Bahama outer ridge area with a 3.5 kHz profiler and sidescan sonar suggests that certain hyperbolic waveforms, previously interpreted as depositional features, are more probably erosional in origin, i.e., furrows excavated by bottom currents (Hollister et al., 1974; see Fig. 29 of Chapter 10.) These smaller erosional features occur on the flanks of the Lower Rise Hills and suggest either that the hydraulic regime in this region is of fluctuating character or that the hills are relict features (perhaps formed during the glacial epoch) which are now undergoing erosional modification.

GENERAL ASPECTS OF SEDIMENT DISTRIBUTION

Studies of sediment distribution on the outer continental margins (e.g., Uchupi, 1963; Milliman et al., 1972; Schlee, 1973; Uchupi and Emery, 1963; Connolly and von der Borch, 1967; Tooms et al., 1971; Bourcart, 1954; Leclaire, 1972; McMaster et al., 1971) indicate that the slope and base-of-slope regions generally possess a surficial covering of fine-grained sediment (clay and silt) which is usually ascribed to pelagic or hemipelagic deposition (Fig. 11A).

The term pelagic, as applied to sediments, is generally reserved for deep-sea sediment without coarse terrigenous material (examples are red clay and organic ooze), whereas the term hemipelagic is applied to sediments containing a mix of terrigenous material (clays, windblown sediment, etc.) and pelagic organisms. Hemipelagites form the predominant sediment type on most of the world's continental margins, whereas pelagites tend to be localized in more distal abyssal plains. Both thicknesses and rates of sedimentation tend to be highest on the continental margins (Fig. 12).

A

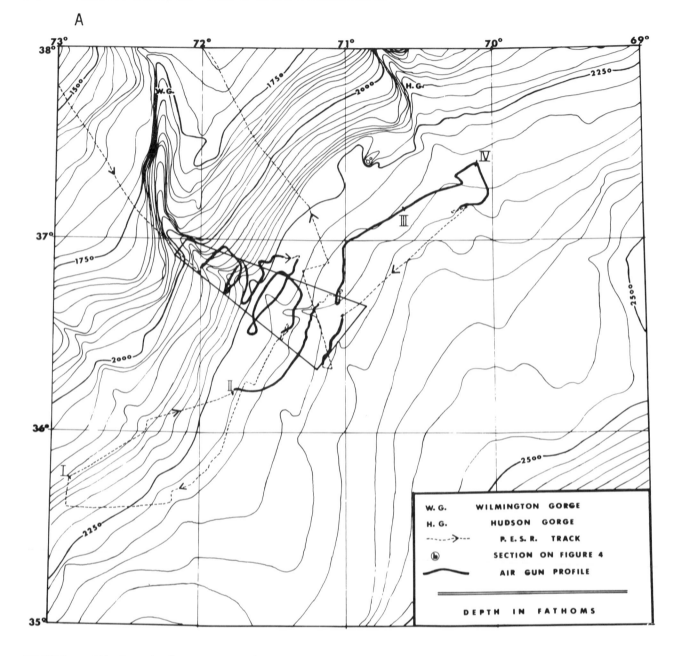

FIGURE 9.   (A) *Chart showing traverses on the upper portion of the lower rise east of the Middle Atlantic States.* (B) *Details of selected continuous bathymetric (PESR) profiles across the* *Washington, Hudson, and Wilmington valleys. Profiles made parallel to contours illustrate the megafan topography on the lower rise. From Stanley et al.* (1971).

Typically, lutite surfaces display an abundance of mounds, trails, and feeding traces of biogenic origin (Figs. 13A–E), indicative of the high infaunal productivity and low indigenous current activity in this region (cf. Hersey, 1967). However, bottom photographs reveal that certain areas of the continental slope and rise display abundant evidence of current scouring and erosion of clay and silt substrates (Figs. 13F,G).

Inferred directions of current movement may be downslope (Owen and Emery, 1967), but are more frequently subparallel to the isobaths and the strike of the slope (Emery and Ross, 1968; Heezen and Hollister, 1971; Hollister and Heezen, 1972; Ewing et al., 1971; Hollister and Elder, 1969; Kelling and Stanley, 1972a,b). Direct measurements of bottom currents in some of these regions indicate velocities in the general range of 10 to

B

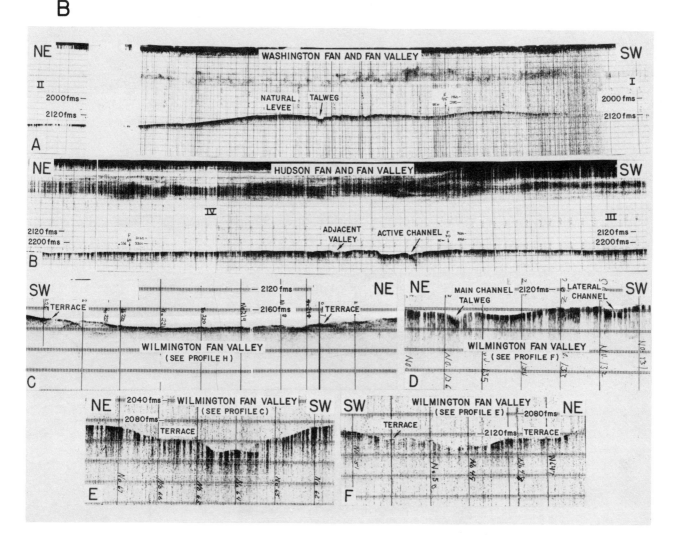

FIGURE 9.   (Continued)

30 cm/sec (Swallow and Worthington, 1961; Volke-mann, 1962; Heezen and Hollister, 1971), but locally velocities range to 70 cm/sec (Emery and Ross, 1968) or even higher. These features are attributable to oceanic bottom circulation systems of which the most widespread and important representatives appear to be the geo-strophic contour currents. These are preferentially developed on the western boundaries of major ocean basins as a result of contrasts in salinity, temperature, and tidal level accentuated by the Coriolis force. In the western Atlantic, the southward-flowing Western Boundary Undercurrent off North America and the northward-flowing Antarctic Bottom Current off South America (Fig. 14) affect marginal slopes, oceanic ridges, and elevations (Hollister and Heezen, 1972; Eittreim and Ewing, 1972).

The lutite cover of slope and rise regions, although prevalent, is by no means universally developed and localized areas of sand and gravel are commonly en-countered on slope and fan/rise surfaces. These coarse sediments can normally be assigned to one or more of the following factors:

1. *Fault-scarps and slump scars:* In the Gulf of California (Rusnak et al., 1964) and on the southern wall of the Puerto Rico Trench (Conolly and Ewing, 1967) rock outcrops near the base of steep, probably fault-defined slopes are associated with cobble and pebble talus gravels of local origin (see also Johnson et al., 1971, p. 64).

Smaller faults on the southwest flank of Nyckel Ridge, at the base of the continental slope near Wilmington Canyon, off the Mid-Atlantic States, have been detected

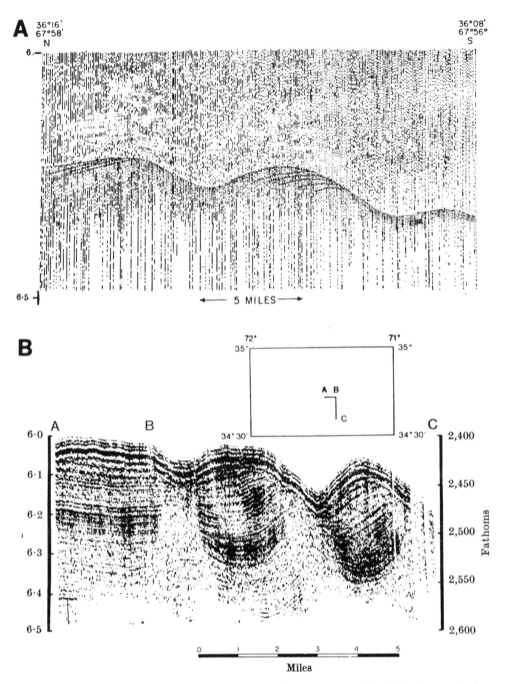

FIGURE 10. *Subbottom profiles across Lower Continental Rise Hills on the northwestern margin of the Hatteras abyssal plain, northwestern Atlantic. (A) Sounding record (12 kHz) shows shallow reflectors outcropping on the south flanks of these giant ripples. Figures on left margin are two-way travel times (in seconds). (B) Photograph of a seismic reflection profile reveals that the axes of the Rise Hills are displaced toward the south with increasing age (depth below seafloor). From Fox et al. (1968).*

A

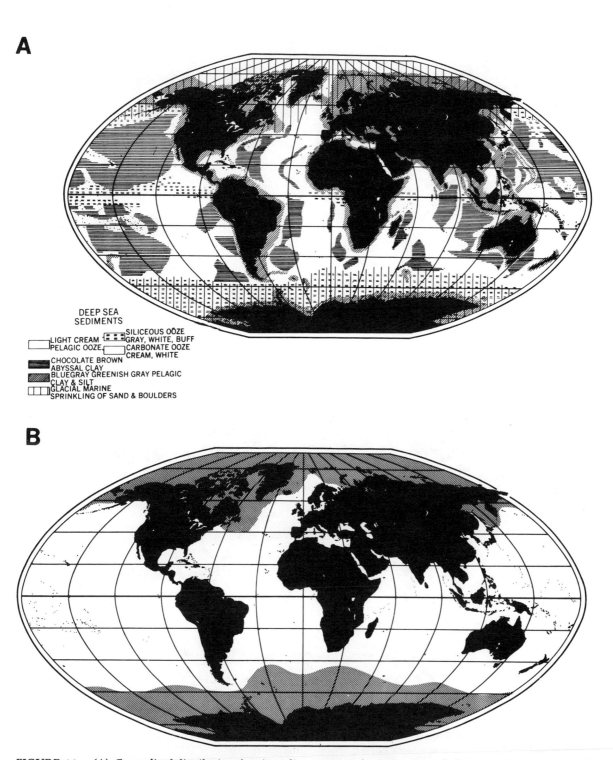

DEEP SEA
SEDIMENTS

LIGHT CREAM
PELAGIC OOZE

SILICEOUS OOZE
GRAY, WHITE, BUFF
CARBONATE OOZE
CREAM, WHITE

CHOCOLATE BROWN
ABYSSAL CLAY
BLUEGRAY GREENISH GRAY PELAGIC
CLAY & SILT
GLACIAL MARINE
SPRINKLING OF SAND & BOULDERS

B

FIGURE 11. (A) *Generalized distribution of major sediment types on the ocean floors.* (B) *Present-day extent of floating ice. Rafted sediment (boulders to clay grade) may occur on the seafloor beneath these areas. From Heezen and Hollister (1971).*

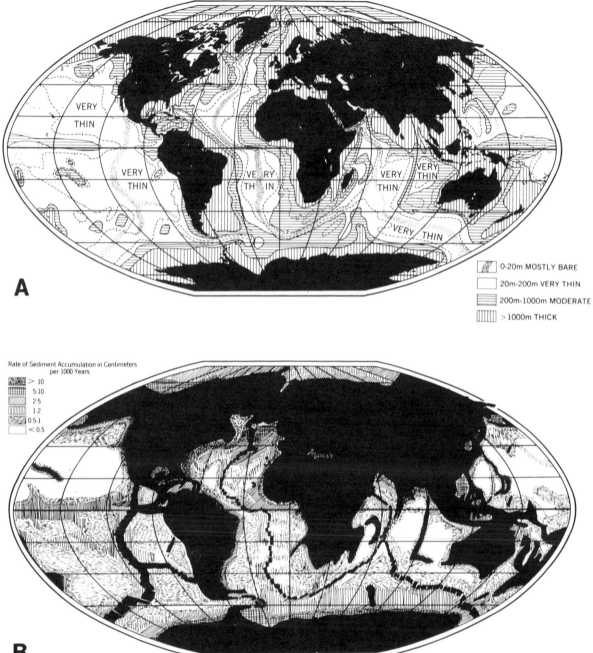

FIGURE 12. (A) *Distribution of ocean sediment thickness, based on seismic and echo-sounding profiles. Note that maximum thicknesses are generally found on the progradational margins of trailing-edge type, such as western Europe, Africa,* *and the eastern seaboard of North and South America (cf. Fig. 2). (B) Distribution of sediment accumulation rates for the world's oceans (in centimeters per thousand years). From Heezen and Hollister (1971).*

in seismic profiles (Kelling and Stanley, 1970, Fig. 13). Bottom photographs in this region (Stanley and Kelling, 1968, 1969) have revealed outcrops of well-jointed Tertiary sedimentary rocks, together with talus blocks, which appear to become abraded and size-sorted away from the outcrop (see Figs. 15*A–D*) and ultimately

generate a large discrete patch of rippled sand and gravel in an area which is otherwise floored by silty clay (Fig. 16).

2. *Powerful oceanic currents:* These may be either near-surface flows, such as the Gulf Stream, which appears to be responsible for the abundance of carbonate sand

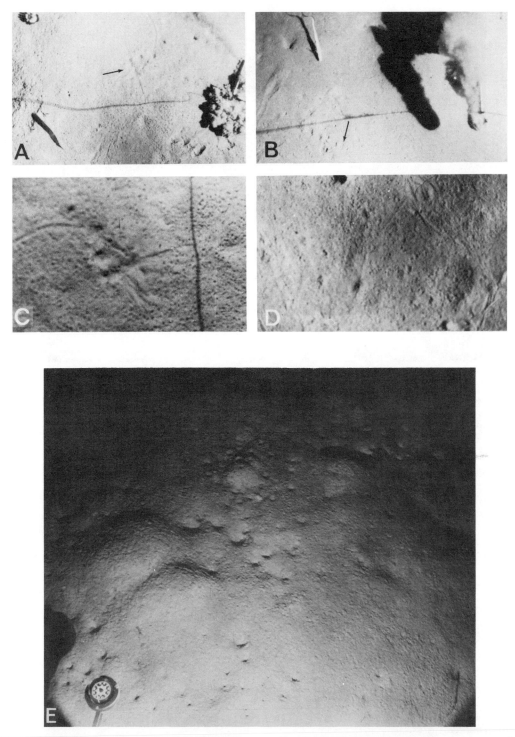

FIGURE 13. (A–E) *Lutite surfaces on the continental margin displaying traces of biogenic origin. (A–D) Fish-feeding marks on the lower slope in the vicinity of Wilmington Canyon. Note morid fish in (A) and halosaur fish in (B). From Stanley (1971a). (E) Highly bioturbate bottom displaying large and small mounds, and depressions of mud bottom in the Western Alboran Basin, western Mediterranean (depth 1500 m). The granular texture of the seafloor may represent fecal* *pellets. From Huang and Stanley (1972). (F) Mounded muddy sediment, slightly striated and lineated by erosive action of outflowing Mediterranean bottom waters (flowing from top right to lower left). East flank of Ceuta Canyon (Strait of Gibraltar region) at depth of 430 m. (G) Burrowed and mounded muddy sediment with marked erosional lineations (direction of flow shown by arrow). Axis of Ceuta Canyon at depth of 800 m.*

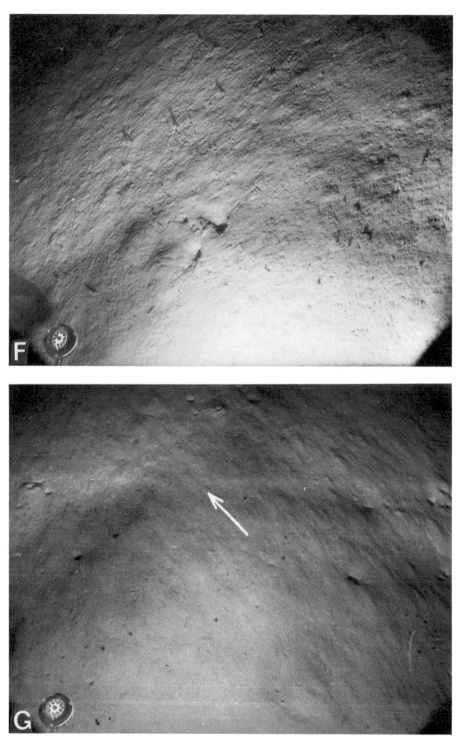

FIGURE 13 (Continued)

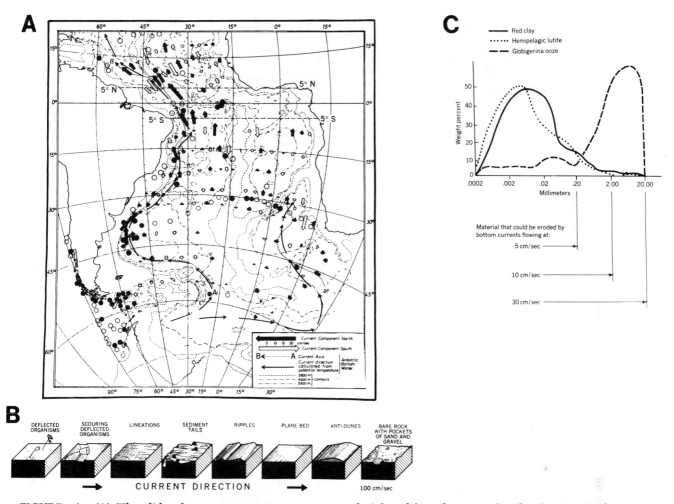

FIGURE 14. **(A)** *The solid and open arrows represent current-meter readings and illustrate how the deep geostrophic Antarctic Bottom Current skirts the continental margin of South America, leaving its imprint on the seafloor sediment in the form of ripples and lineations (dots). Relative current velocity is shown by the length of arrow.* **(B)** *Relative bottom current velocity may also be inferred from the nature of seafloor features. In this sequence, features indicating minimum velocities occur at the left.* **(C)** *Typical size distribution curves for common margin and deep-sea sediment types, related to the experimentally determined minimum current velocities required for their erosion. From Heezen and Hollister (1971).*

and rock outcrops on the upper slope between Florida and Cape Hatteras (Milliman et al., 1972), or near-bottom currents. The effects of the latter are most conspicuously displayed in the restricted straits linking larger oceanic basins, such as the Drake Passage (Heezen and Hollister, 1964, 1971), the Mallaca Strait (Keller and Richards, 1967), Rockall Bank–Feni Ridge area (Jones et al., 1970), the Gulf of Aden at the southern entrance to the Red Sea (Einsele and Werner, 1968, 1972), the Straits of Florida (Neumann and Ball, 1970), and the Gulf of Cadiz and Strait of Gibraltar (Heezen and Johnson, 1969; Kelling and Stanley, 1972a,b). In each of these areas, bottom samples and photographs from the slope and deep floor reveal evidence of rocky floors and walls covered by a relatively thin and mobile veneer of locally derived coarse sediment (Figs. 17 and

18). To the west of the Strait of Gibraltar, the outflowing Mediterranean undercurrent, which attains velocities in excess of 120 cm/sec within the strait (Heezen and Johnson, 1969), has spread a carpet of coarse, largely carbonate sand for nearly 200 km out into the Atlantic. Large ripples and dunes created by this outflow have been observed to depths of about 750 m (Kenyon and Belderson, 1973, p. 86).

3. *Glacial processes:* Especially common in higher latitudes are the isolated coarse clasts introduced by modern or ancient floating ice (Fig. 11B) and icebergs (Stanley and Cok, 1968; Stanley and Silverberg, 1968; Schlee and Pratt, 1970; Ewing et al., 1971; Schlee 1973; Heezen and Hollister, 1971, pp. 256–264). These erratic pebbles and cobbles are generally distinguishable from scarp talus deposits or the transported or residual gravels

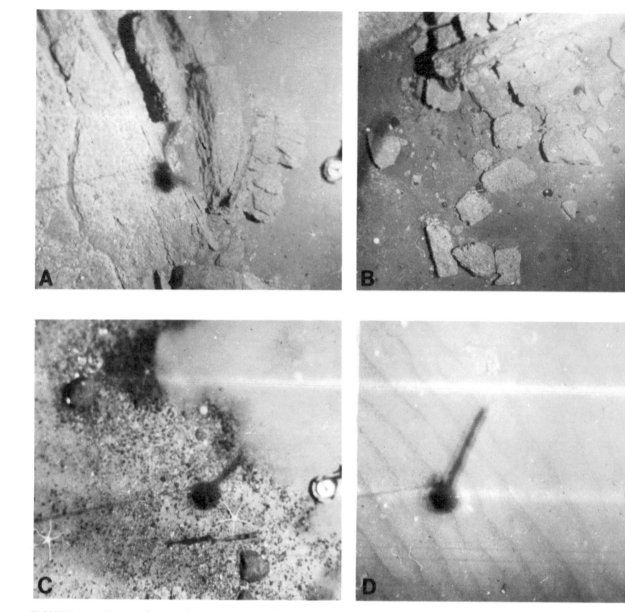

FIGURE 15.   *Bottom photographs from the base of the continental slope (2100 m) south of Wilmington Canyon, eastern seaboard of the United States, comprising a sequence moving progressively downslope from (A) to (D) over a distance of approximately 500 m. (A) Outcrop or ledge of consolidated rock (?sandstone), with prominent fractures (?joints) trending ENE–WSW. Rock is mantled by sandy sediment (at right). (B) Sand bottom downslope of rock ledge shown in (A), with numerous angular blocks* of well-jointed rock broken off the ledge and some darker (?manganese-coated), better-rounded pebbles. (C) Sandy gravel bottom, locally covered by silt (forming cloud stirred by compass striking seafloor). Dark pebbles and subrounded cobbles probably represent current-transported detritus downslope from rock ledge seen in (A) and (B). (D) Rippled sand (current toward NW) carrying occasional dark pebbles and larger, more angular blocks of rock (upper center) similar to that seen in (A) and (B).

associated with powerful tractional flows by means of (a) their exotic lithology, (b) their occurrence in much finer indigenous sediments, (c) their occurrence on topographic elevations such as seamounts (Pratt, 1968, Fig. 36) which "normal" mechanisms of transportation are incapable of reaching, and (d) the general absence of evidence (scouring and/or depositional bed forms) for powerful bottom currents in the vicinity of the clasts.

Such criteria are only appropriate where subsequent reworking by gravitative or hydrodynamic processes has not occurred. It is a common observation that icebergs and floating ice extend, at least seasonally, for considerable distances beyond the southern (or northern) landward limits of the polar ice caps. Thus it is probable that ice-rafted debris was conveyed into relatively low latitudes during periods of maximum glaciation in the

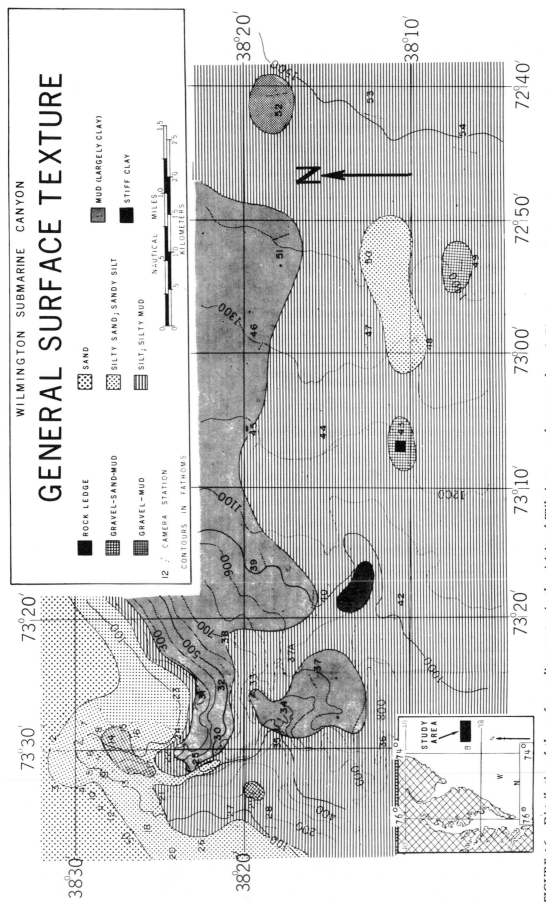

FIGURE 16. *Distribution of the surface sediment types in the vicinity of Wilmington Canyon, eastern margin of the United States. Surficial sand is almost confined to the shelf and the shallow canyon head although isolated patches of coarse sediment, associated with rock outcrops shown in Fig. 15, occur on the lower slope and upper rise. Depth in fathoms. From Stanley and Kelling (1968).*

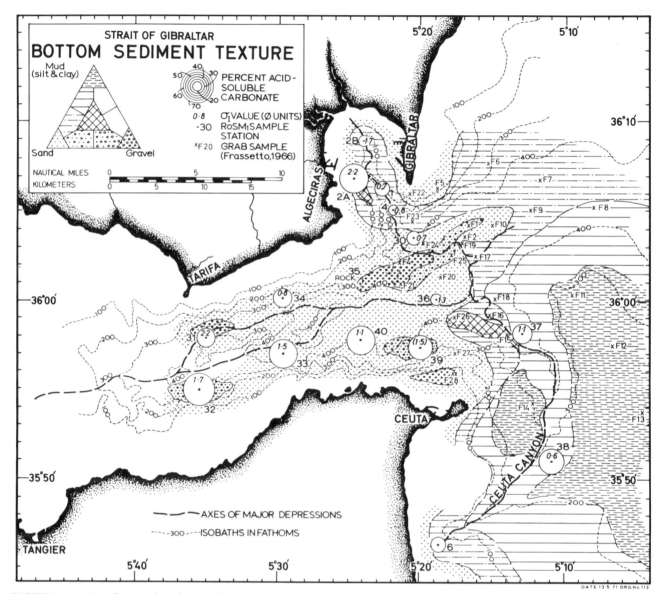

FIGURE 17. *Distribution of surficial sediment textures in the vicinity of the Strait of Gibraltar. Note the prevalence of sand and gravel (with some bare rock outcrops) in the Strait and the muddy nature of the floor in the shallow reaches of Gibraltar Canyon (between Algeciras and Gibraltar) and Ceuta Canyon. From Kelling and Stanley (1972b).*

Pleistocene and Late Tertiary (Hayes et al., 1973) epochs.

4. *Canyon-valley systems:* Many of the major depressions that transect the marginal slopes are acting (or have acted) as channels funneling sand and gravel from the shelf to the deep ocean floor. Consequently, they tend to be represented on sediment distribution maps as sublinear to winding shoestring units of coarse sediment. However, the surface manifestation of these coarse sediments is generally obscured by a blanket of modern mud (Fig. 19A), except in areas such as the California borderland province or the northwest Mediterranean where the shelf regime or the proximity of fluvial sources

ensures an almost continuous supply of sand and/or gravel through many of the canyon systems (see Hand and Emery, 1964; Shepard et al., 1969; Gennesseaux, 1966; Dangeard et al., 1968). The rapidity with which the passage of this coarse fraction from shelf to slope base may be achieved is attested by the preservation of kelp fragments (Piper, 1970, p. 219), tar-covered pebbles (Gennesseaux, 1966; Stanley, 1969), and even golfballs (Piper, 1970, p. 220) in the fan and base-of-slope sediments.

The apparent absence of coarse detritus from the surface deposits of less active canyons may be, in part, an artifact of the rather crude sampling techniques

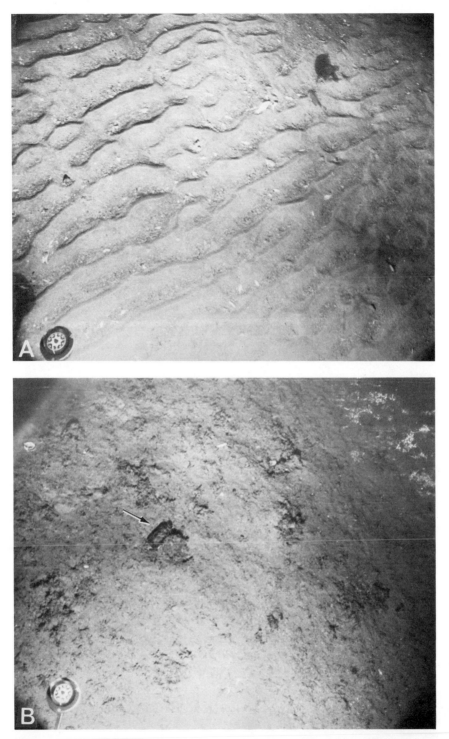

FIGURE 18. *Photographs of the seafloor in the Strait of Gibraltar region. (A) Shelly sand molded into ripples by strong bottom current flowing west. Note fish in top right corner. Station 39 (see Fig. 17), on southern side of the Strait Channel, depth 680 m. (B) Scoured rocky floor veneered by sand and encrusting calcareous organisms, mainly bryozoans and corals. Note bottle (arrow), free from sediment and trapped behind organic growth, indicating high current velocities (see Fig. 14B). Station 35 (see Fig. 17) on northern flank of the Strait Channel at a depth of 600 m. From Kelling and Stanley (1972b).*

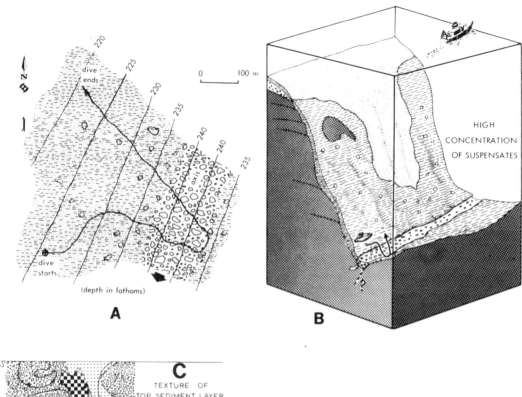

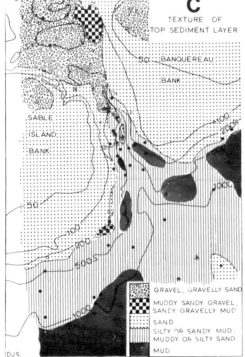

FIGURE 19. *Studies of canyons on the Atlantic margin of North America show the linear or tongue-shaped distribution of coarse-grained sediment in an otherwise fine-grained sediment setting. (A) Topography and surficial sediments in the head of Wilmington Canyon, off the Mid-Atlantic States as observed in a Deep Star 2000 dive. (B) Schematic showing (1) modern spillover of shelly, silty sands from the shelf margin into the canyon in response to current regime near the shelf break; and (2) slumps that rework canyon wall sediments down the canyon walls toward the axis. The pebbly mud axial fill observed may be a plug rafted along the top of debris in a state of flow. After Stanley (1974). (C) Similar phenomena in the Gully Canyon on the Nova Scotian margin. From Stanley (1967).*

employed. Submersible observations in such canyons have revealed in striking fashion the restricted width of the active axial channel within which most of the coarser sediment is being conveyed (Shepard, 1965; Andrews et al., 1970; Stanley, 1974); see Fig. 19B. Grab or dredge sampling and even photographic techniques frequently may fail to identify these narrow strips of mobile sediment in deeper zones of the slope or rise.

The provenance of the coarse elements within canyon sediments is variable. Some are derived directly from terrestrial sources through fluvial and littoral agencies but many clasts are contributed through erosion of the canyon walls (Trumbull and McCamis, 1967) and these may include resistant igneous and metamorphic rock types, as in several of the canyons off southern California (see review by Shepard, 1973, pp. 306–315). Slumping and reworking of pebbles from gravel beds and conglomerates exposed in canyon walls may also contribute spuriously rounded clasts to the axial sediments (Stanley, 1974, p. M3).

In practice, core data provide a more reliable key to the long-term sedimentary regimes existing in slope and base-of-slope environments. Cores obtained from open slopes at depths beyond the direct influence of shelf-sediment spillover (see Chapter 16) generally consist of lutite, frequently mottled by bioturbation (Stanley et al., 1972a, pp. 19–22) with subordinate, thin layers of silt and fine sand, usually parallel laminated but seldom graded. In some open slope areas the cored muds are structurally homogeneous but the texture is mixed, with appreciable proportions of fine sand and coarse silt (Connolly and von der Borch, 1967, pp. 200–202; Schlee, 1973, pp. L52–54). Although eolian transport, including ash falls, may have contributed some of these coarser grains (Fig. 20), intensive biological reworking of originally laminated sediment almost certainly accounts for most of this mixed texture (cf. Hesse et al., 1971; Stanley, 1971b).

There are surprisingly few published descriptions of cores from the mid- and lower slope reaches of submarine canyons. The Gully Canyon, off the Nova Scotian shelf, has yielded cores that include layers of poorly graded and ripple-laminated sand interbedded with bioturbated silty mud (Stanley, 1967; Stanley et al., 1972a). Some short cores (1 m) obtained from Wilmington Canyon, off the eastern seaboard of the United States, revealed only a slightly greater proportion of laminated sand than adjacent open slopes. On the other hand, Field and Pilkey (1971) report the presence of thick beds of well-graded sand from the axis of Hatteras Canyon. Cores from the somewhat atypical Great Bahama Canyon have yielded beds of graded carbonate sand and some gravel interbedded with gray pelagic ooze (Andrews et al., 1970, p. 1073). Beds of well-sorted oolite up to 18 cm thick which have also been sampled at depths of 1800 m and 3700 m in the same canyon system may have been introduced, or at least reworked, by indigenous current flow (Andrews et al., 1970, p. 1073).

The more active canyons, such as those cutting the steep slopes off southern California or off the Mediterranean coast of France, include layers of coarse to medium sand and gravel, several tens of centimeters thick, irregularly stratified and sometimes graded, with only minor interbeds of finer sediment (Bourcart, 1960a,b, 1964; Bouma and Shepard, 1964; Shepard and Dill, 1966; Shepard et al., 1969; Shepard, 1973). However, cores from Astoria Submarine Canyon on the Oregon coast (which appears to have been largely inactive in postglacial times) reveal unlaminated, poorly graded, and poorly sorted sands in relatively thin layers interbedded with mottled silty clays (Carlson and Nelson, 1969).

This brief review suggests that during more active phases the middle and lower canyon are characterized by transient accumulations of sandy sediment, probably conveyed in poorly segregated debris flows or immature turbidity currents. Periods of lower activity in the canyons are marked by a prevalence of hemipelagic and pelagic fines, and much biologic reworking, while the occasional increments of sand and silt supplied from the head or flanks generally show signs of tractional transport and reworking, perhaps by the indigenous bottom currents which appear to be a feature of many modern canyons (Shepard and Marshall, 1969; Reimnitz, 1971; Fenner et al., 1971; Keller et al., 1973; Shepard and Marshall, 1973a,b; Shepard et al., 1974).

Sediments cored from well-developed continental rise regions differ in several important respects from those encountered in discrete deep-sea fans. Cores and high-resolution seismic profiles from the continental rise off eastern North America show that this feature is constructed from a pile of hemipelagic sediments (Figs. 11 and 12), generally more than 1.5 km thick (Knott and Hoskins, 1968; Emery et al., 1970). The upper part of this pile contains thin but apparently widespread units of silt or sand, predominantly with well-developed parallel or cross-lamination [see Hollister and Heezen (1972) and Bouma and Hollister (1973) for reviews]. Graded sands are not uncommon but most are thinner and better sorted than in the adjacent canyons or basin plains. Laminae within the ungraded arenites are frequently marked by concentrations of heavy minerals, whereas the internal lamination of most graded beds is rich in lutum. Hollister and Heezen (1972) have concluded that these textural differences result from two fundamentally different processes of transportation and deposition and have termed the products "contourites" (thin, well-laminated, heavy mineral-placered sands and silts) and "turbidites" (thicker, well-graded, arenites with mud intercalations).

The contourites in this region are ascribed to tractional transportation along the rise by the Western Boundary Undercurrent, whereas the turbidites represent primary deposition from turbid flows introduced transversely, primarily via the submarine valleys. The majority of the turbidity currents are believed to have bypassed both

A

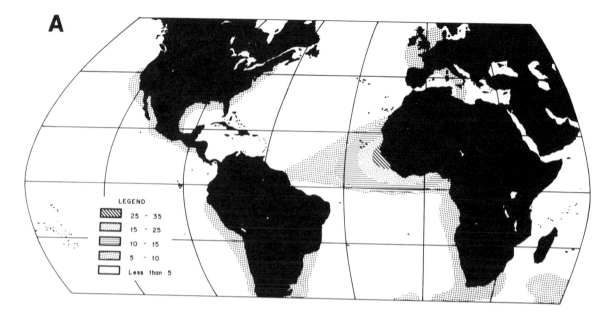

B

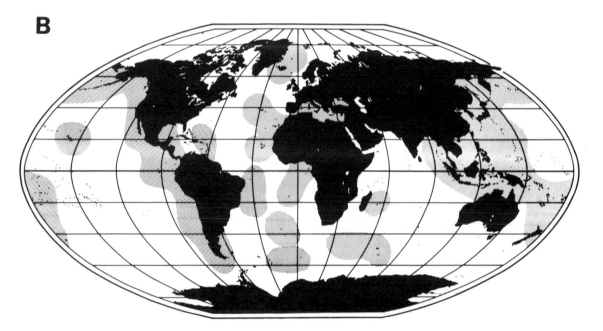

FIGURE 20. (A) *Eolian contribution to Atlantic sediments may be estimated from haziness of skies, due to transport of dust in the atmosphere. Numbers represent percentage of total number of observations of haze in* *northern winter.* (B) *Areas of the ocean where volcanic ash may form a significant element of the sediment blanket. From Heezen and Hollister (1971).*

slope and rise provinces to deposit their load on the Sohm and Hatteras abyssal plains, where thick, coarse-graded sands and silts form a major part of the unconsolidated sediment section (Hubert, 1964; Horn et al., 1971a); see Fig. 21.

Most of the thin-graded units encountered in the rise cores have a distal aspect (Walker, 1967). These (and

perhaps much of the associated lutite) may result from spillover of turbid flows overtopping the submarine levees bordering the canyon extensions and fan valleys which extend across the rise and onto the abyssal plains, where more proximal turbidites may have been deposited. Alternatively and additionally, the distal turbidites of the rise may be deposited from turbid flows

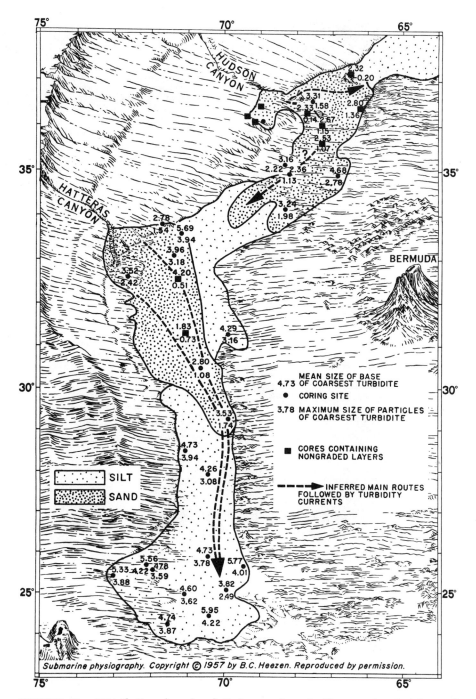

FIGURE 21. *Distribution of sand and silt in cores from the Hatteras abyssal plain, northwestern Atlantic. Grain size is indicated in phi units. From Horn et al. (1971a).*

of low density, originating from shelf-edge spillover (see Chapter 16), which mix with lutum suspended in deeper waters and in the nepheloid layer (Eittreim et al., 1972) and are deflected by the geostrophic flow of bottom waters parallel to the isobaths (Stanley, 1969, 1970). A further possibility is that dilute turbidity cur-

rents are generated from the slumps and slide bodies which form a significant element of the upper rise. Since these major slumps are derived mainly from the fine sediments of the open slopes the resulting suspensions will include a limited range of particle sizes and will probably possess the low density and high turbulence

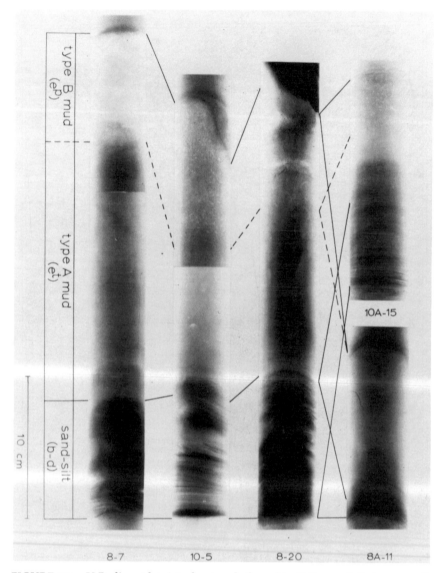

FIGURE 22. *X-Radiograph prints showing rhythms of turbidite sand-silt layers or silt laminae, followed by largely homogeneous type A mud (e^t), in turn followed by type B mud (e^p) exhibiting light speckling. The lower part of type A layers can be finely laminated (10–5, 8–20, 10A–15). One rhythm (8A–11) contains no type B layer. In the turbidite sand-silt layer in rhythm 10–5, note current ripple lamination between two cosets of lower and upper parallel lamination. Also note in type B mud layer two horizons of extra light speckled bands. The distinct horizontal lines in 8–7 and 10–5 are artifacts produced by the joining of two different photographic prints. From Rupke and Stanley (1974).*

(through incorporation of water into the irregular upper surface of the slump mass) appropriate to the formation of distal grading (Hampton, 1972); see Chapter 11.

It should be pointed out that many so-called hemipelagic muds actually are clayey silt turbidites (Stanley, 1969; Piper, 1972, 1973; Rupke and Stanley, 1974), on the basis of structures (as noted in x-radiographs, Fig. 22), composition, and texture. These graded mud units are petrologically different (lower amounts of pelagic

tests, higher amounts of terrigenous material, etc.) from the hemipelagic muds with which they are associated. Analysis of cores in the western Mediterranean basin shows that locally they account for almost half of the Late Quaternary section (Fig. 23), and are deposited at a rate of 2 to 3 units per 2000 years, while the hemipelagic units deposited from suspension accumulate much more slowly (10 cm/1000 years), according to Rupke et al. (1974).

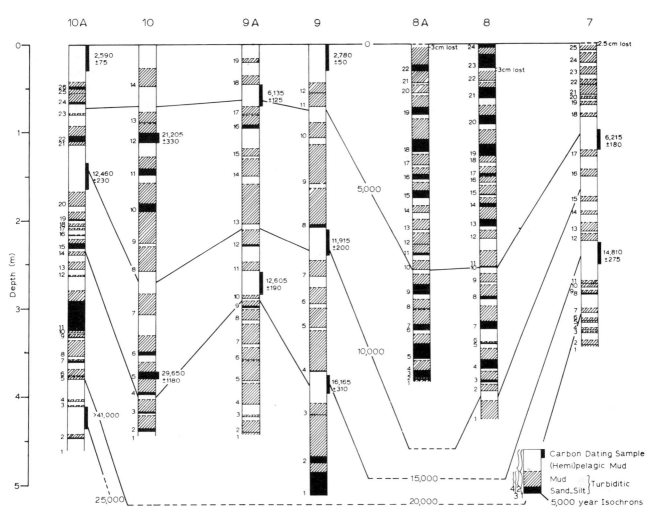

FIGURE 23. *Distribution of rhythms (numbered on left side of cores) consisting of turbidite sand-silt layers and two kinds of mud layers, type A (interpreted as turbiditic) and type B (interpreted as hemipelagic) in seven cores from the Balearic abyssal plain. Spacing between these cores corresponds to the distances between their respective locations. The position and extent of samples for radiocarbon dating and the corresponding ages are indicated.*

Isochrons are drawn between the cores, based on a calculation of the hemipelagic rate of sedimentation. Note that the radiocarbon dates in core 10 are determined not on type B muds (interpreted as hemipelagic), but on resedimented turbidite sands, causing a discrepancy between these dates and the ages indicated by the isochrons. From Rupke and Stanley (1974).

The discrete deep-sea fans of the California–Oregon margin present a different style of sediment distribution (Fig. 24). Here the occurrence, thickness, and nature of the coarse units are primarily a function of proximity to both the canyon mouth (as the immediate source of the sand and gravel) and the active channels (Normark, 1970; Haner, 1971; Nelson and Kulm, 1973). The typical distribution pattern of size or sedimentary structural sequence is elongate or elliptical rather than strictly concentric, since the change in flow character along the course of the channel is less abrupt than that encountered in transverse profile. This contrast is directly linked to the degree of flow confinement within the channel. Thus the thick, unstructured, and poorly sorted sands

and gravels of the inner fan valley, near the apex of the fan, are flanked by high levees of thin distal turbidites interbedded with hemipelagic muds, whereas the finer, well-graded (sequence Ta-e of Bouma, 1962) sands of the outer fan distributary channels may pass laterally across low levees into thinner and finer beds. These gradually develop "base cutout" Bouma grading sequences before merging into the hemipelagic muds of the interfan areas, which are often highly bioturbated (Piper and Marshall, 1969). Clearly, this type of pattern reflects the channelized flow of turbidity currents and other mass transport mechanisms into and through the fan-valley system (Normark, 1970; Nelson and Kulm, 1973).

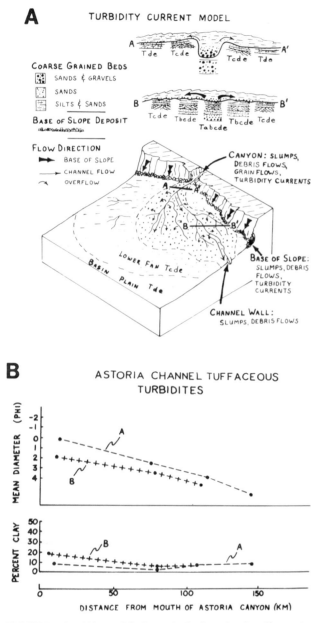

FIGURE 24. (A) *Model of a typical submarine fan illustrating distribution of grade and sedimentary structures in coarse-grained deposits from a single turbidity flow. Sequences* (T$_{cde}$, ...) *refer to the Bouma sequences of sedimentary structures within turbidite units.* (B) *Downchannel trends in thickness and texture of tuffaceous turbidites across a major deep-sea fan channel (Astoria Channel, northeastern Pacific). A and B refer to different turbidite units within the cored section. From Nelson and Kulm* (1973).

On most fans the sand layers in interchannel areas become both thinner and finer down the fan (Normark and Piper, 1972; Horn et al., 1971b) with an accompanying decrease in the sand-mud ratio. These changes are generally accompanied by a decrease in both the maximum and the mean clast diameter (Fig. 24B), again

compatible with a general decline in competency of the supplying flows. However, Piper (1970) has pointed out that on La Jolla Fan the coarsest sands actually occur on the outer edges of the feature and has explained this in terms of exceptionally powerful turbidity currents which erode the inner fan valley and which only start to deposit their load on the lower fan.

The striking leftward (looking downchannel) elongation of the west coast fans is generally ascribed to the Coriolis effect (Menard, 1955; Haner, 1971) which induces a tendency for breaching of the left-bank levees in meandering fan channels. This asymmetric development of fans is of particular significance in north-south aligned continental margins since it will promote coalescence and thus ultimately produce a true rise or continuous apron, unless structural events intervene.

There is a considerable body of evidence that most of the existing deep-sea fans off both coasts of North America (and probably off most other continental margins also) were developed mainly during the Pleistocene, primarily because of the greatly increased supply of sediment to the deep basins resulting from glacial lowering of sea level (Moore and Curray, 1974; Nelson and Kulm, 1973; Normark and Piper, 1972). The rise off eastern North America may then have arisen as a coalesced series of Pleistocene (and Pliocene?) fans. The DSDP borehole 106 on the upper continental rise southeast of New York City (Hollister et al., 1972) penetrated a 350 m Pleistocene section composed largely of sandy and silty turbidites. Sedimentation rates on the order of 20 cm/1000 years were established for this section, compared with 4 to 7 cm/1000 years for the Holocene (Emery et al., 1970). Such high rates are comparable with those for the Holocene section of the active west coast fans (Moore, 1969) and, taken in conjunction with the turbidite character of the coarse units, add further weight to the contention that the contour currents on the continental rise off eastern North America are responsible for modifying rather than constructing this wedge.

SEDIMENTARY PROCESSES ON THE OUTER MARGIN

The physiographic nature of the continental terrace and its outer margin is ultimately determined by major geotectonic, climatic, and meteorologic factors (Lewis, 1974, Dietz, 1963). The interplay of these factors creates a secondary set of sedimentary processes that are responsible for removing material both from the adjacent hinterland and the terrace itself and depositing it elsewhere, either on the terrace or in the deeper ocean

TABLE 1. The Principal Processes Affecting Sedimentation in Outer Margin Environments

Predominant Controlling Forces	Air/Water Interface	Water/Sediment Interface	Predominant Transport Mechanisms
Atmospheric and tidal	Wave agitation Wind-driven surface currents Eolian transport Glacial transport Tidal surface currents	Storm stirring on outer shelf Pelagic settling Tidal bottom currents Tidal canyon currents	Suspension, with subordinate traction
	Internal waves		
Thermohaline	Geostrophic surface currents	Geostrophic bottom currents	Suspension = traction
Gradient and gravity		Sliding and slumping Creep Debris flow Grain flow Turbidity flow	Mass movement, traction, and suspension all may be important
Biologic activity		Biogenic accumulation (tests, fecal pellets, etc.) Bioturbation and internal disturbance Surface erosion	Mainly in situ processes but some suspension; also traction or mass movement on steep slopes

beyond. The slope thus represents an interface in which erosion and deposition are in constant conflict and in delicate balance (see Stanley, 1969, Fig. DJS-5). Seismic profiles suggest that net erosion is a common condition on the upper parts of marginal slopes (see Southard and Stanley, Chapter 16) whereas sediment accumulation is associated with the deeper zones of the margin, although localized erosion may still be important.

The principal factors affecting sediment movement and deposition in the slope and base-of-slope environments can be grouped into four main genetic categories (Table 1). It is important to appreciate that no single one of these processes is diagnostic of, or confined to, the slope, fan, or rise/apron province. However, certain regions appear to be characterized by a particular combination of mechanisms which gives rise to a typical (but not necessarily unique) sedimentary sequence in that area.

Atmospheric/Tidal and Thermohaline Processes

Processes ascribed to these two intergrading groups of causes involve clear-water motion, both at the surface and on the seafloor, which under appropriate circumstances is capable of entraining and transporting sediment of varying grade either tractionally or in suspension. The volumetrically most important sedimentary agency on the outer margin is near-surface wave agitation which is primarily responsible for maintaining in

suspension the fine particulate matter ($<2 \mu$) which constitutes the bulk of the sediment deposited on the slope and at the base of the slope. The interaction of powerful surface currents such as the Gulf stream (which conveys 26 million m^3 water/sec at speeds of 2–3 m/sec) with the fine suspensate population is capable of introducing large quantities of lutum to the deep ocean (see Betzer et al., 1974, p. 24), and smaller, more ephemeral wind-driven currents may achieve the same end.

However, the precise manner in which this suspended matter is finally transferred to the seafloor is more debatable. The time-hallowed concept of a pelagic "snowfall of sediment" is not readily compatible with several aspects of the outer margin water mass and sediments. For example, well-developed thermal and salinity stratification is common, imposing barriers to simple settling (see Drake, 1971) and a subtle interlamination and intermixing of clay and fine silt is common in slope and rise lutites. The recognition and analysis, over the past few years, of "nepheloid layers" have thrown some light on the problems of lutite sedimentation. These layers, first recognized by means of their light-scattering properties (Ewing and Thorndike, 1965), are now known to exist over much of the northern Atlantic basin (Eittreim et al., 1969; Jones et al., 1970), the Argentine Basin (Ewing et al., 1971), the Arctic Ocean (Hunkins et al., 1969), and the circum-Antarctic region (Eittreim and Ewing, 1972).

In the northwestern Atlantic basin the nepheloid layer attains an average thickness of about 1 km and is best developed over the continental rise and adjacent abyssal plain (Eittreim and Ewing, 1972); see Fig. 25*A*. Here the layer appears to include modest concentrations (0.1 ppm, or 0.2 mg/l) of particulate material averaging about 12 μ in diameter, most of which is terrigenous clay (Eittreim and Ewing, 1972; cf. Jacobs et al., 1973). It is considered that the nepheloid layer represents a semi-permanent feature of deep ocean basins, being composed of a concentration of clay particles maintained in sus-pension by turbulence, indicated by values of A_z, the coefficient of eddy diffusion, which, at about 10^2 cm²/sec, is an order of magnitude greater than most deep clear-water measurements (Sverdrup et al., 1942). In the northwestern Atlantic this turbulent mixing is ascribed to the effects of the geostrophic Western Boundary Undercurrent and it is significant that well-developed nepheloid layers in other ocean basins appear to be associated with powerful bottom currents.

However, Eittreim and Ewing (1972) maintain that most of the suspended fines within the nepheloid layer

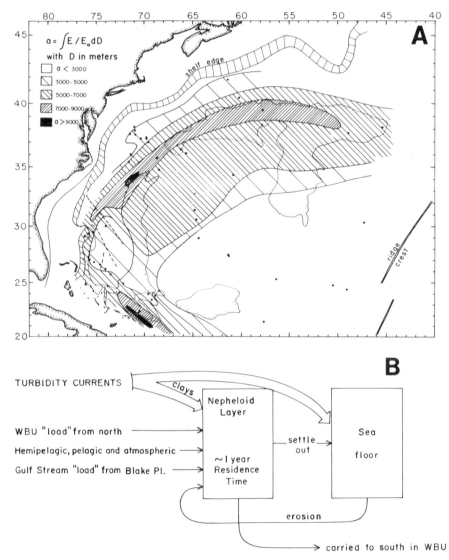

FIGURE 25. (A) *Concentration of suspended matter in bottom waters of the northwestern Atlantic, estimated from nephelometer readings. Values of $\int E/E_0 \, dD$ are proportional to volume of suspended matter comprising the nepheloid layer at each station. Note the zone of maximum concentration over the continental rise.* (B) *Hypothetical schematic for sediment input/output to and from the nepheloid layer. The five main sources of sediment on the left are shown in probable decreasing order of importance from top to bottom. WBU = Western Boundary Undercurrent. From Eittreim and Ewing (1972).*

ultimately are derived from terrigenous sources and are conveyed through submarine canyons to the deep basin by turbidity currents (see Fig. 25*B*). If this is the case, then it is likely that such turbidity currents are similar to the low-density turbid flows postulated in several studies (Moore, 1966; Gorsline et al., 1968; Piper, 1970), although certain studies have cast doubt on the stability of nearshore-derived turbid flows of this type (Drake et al., 1972). The relatively permanent character of the nepheloid layer indicates a periodicity considerably less than is usually considered reasonable for large turbidity currents. Suspension and entrainment of fines at the shelf edge and on the upper slope, induced by the turbulence associated with water mass movements during storms, may furnish a significant short-period alternative supply to the nepheloid layer (Stanley et al., 1971). Such effects are likely to be more pronounced in the vicinity of canyon heads (Ewing, 1973); see Chapter 18. Some incorporation of resuspended seafloor sediment is also possible (see Betzer and Pilson, 1971).

The existence, character, and possible sedimentary significance of deep geostrophic contour currents have already been noted. The observed velocities of these deep flows in the western and northern Atlantic and elsewhere are 10 to 30 cm/sec on average and are theoretically capable of transporting in traction particles of clay (Jones et al., 1970; Eittreim et al., 1972; Hunkins et al., 1969) to medium sand grade (Sundborg, 1956). However, much higher velocities are required in order to erode these materials and place them in motion. Various workers (Emery and Ross, 1968; Betzer et al., 1974) have commented on the apparent anomaly of high current velocities measured immediately above apparently undisturbed muddy seafloor; these observations provide additional testimony to the laboratory results which show that any degree of cohesion in a clay vastly increases its resistance to erosion.

One of the interesting aspects of this type of thermohaline flow and that associated with constricted or silled entrances such as the Malacca Strait, Florida Strait, or Strait of Gibraltar is that such currents are independent (in detail) of the bathymetry and may flow obliquely up, as well as down, slopes. Where sufficiently powerful, these bottom currents may convey not only silt or clay but also sand, and even gravel, for some distance up any gradient upon which the current impinges (Kelling and Stanley, 1972a,b). In these admittedly rare situations, determination of the provenance of the sand fraction is rendered hazardous since coarse sediment found on the slopes or basal aprons may have originated from lateral sources some distance upcurrent and hence may possess no compositional affinities with immediately adjacent

fluvial or shallow marine sediments (Stanley et al., 1975; cf. Kelling and Stanley, 1975).

Much evidence has come to light in recent years concerning the existence of bottom currents in submarine canyons; see Fenner et al. (1971), Shepard and Marshall (1973a), and Southard and Stanley (Chapter 16) for reviews. Indigenous flow in canyons generally appears to be oscillatory but is related only indirectly to tidal reversals, and maximum velocities are usually recorded in the downcanyon direction. This downcanyon component may be significantly reinforced by the effects of storms on the shelf adjacent to the canyon head (Reimnitz, 1971; Shepard and Marshall, 1973b). It is unknown to what depths in the canyon this type of flow may be effective but persisting water movements capable of molding ripples in the canyon floor sands have been observed in photographs and from submersibles to depths exceeding 1000 m (Shepard and Dill, 1966). Such currents may represent return underflows driven by internal waves within the canyon.

A further factor in deep margin sedimentation which is, as yet, poorly understood is the existence of internal waves. While the investigation of this phenomenon has a venerable history (Ekman, 1904) and has been the subject of considerable mathematical analysis (e.g., Wunsch, 1968), data concerning the generation and the sedimentological significance of internal waves are sparse (Lafond, 1961). The observations of Lafond, Wunsch (1968), Southard and Cacchione (1972), Cacchione and Southard (1974), and others (see Chapter 16, this volume) suggest that such waves are likely to attain their highest orbital velocities and their maximum bottom shear near the base of steeply inclined marginal slopes and therefore may be of particular importance in basins such as the western Mediterranean which are bordered by steep slopes. From measurements of vertical motion in the New England slope Voorhis (1968) and others concluded that internal waves were the principal carriers of energy in this environment and might be capable of entraining sediment.

Eolian transport (Fig. 20*A*) of particulate material has long been recognized as a significant factor in the supply of detritus to the deep oceans (Sverdrup et al., 1942) and especially to those regions isolated from aqueous supply of terrigenes by distance or physical barriers (Eriksson, 1961, 1965). The enhanced sedimentation rates of regions bordering continental masses generally obscure any eolian input, hence this process is not considered to be important in most slope or base-of-slope environments. However, Rona (1971) has postulated that eolian supply from adjacent arid regions has been a major process in the construction of the Cape Verde

Plateau and its adjacent slope and rise, off northwest Africa.

Gravity-Controlled Processes

The common occurrence in cores from the deep sea of layers of sand and silt, often graded and sometimes displaying a sequence of internal structures attributed to waning energy (Bouma, 1962; Walker, 1967), has long been adduced as evidence for the importance of turbidity currents as sedimentary agencies in this milieu. As with most scientific theories, however, a period of disenchanted reaction and qualification followed the decade or so of enthusiastic acceptance (Walker, 1973). It is now recognized that the classic turbidity current represents one end member of a spectrum of intergrading processes, all of which are sediment-fluid mixtures capable of flowing downslope and which are thus dependent on gravity (Middleton and Hampton,1973; and Chapter11).

Sliding and slumping are gravity-controlled processes in which cohesive materials yield and move under gravity in a more or less coherent manner. True slides display little internal deformation as a result of their translation downslope, whereas slumps normally undergo some plastic deformation. In both types the failure normally occurs along discrete shear planes. Factors influencing the rate of failure include (a) the degree of consolidation of the sediment, and (b) its drained or undrained nature.

Consolidation in large measure controls the shear strength, but internal pore pressures induced by high rates of sedimentation can effectively reduce the undrained strength to the point where failure may occur in clay on slopes as low as 3° (Morgenstern, 1967). In fact, failure is known to have occurred on even lower gradients (Lewis, 1971b) and one possible explanation may be earthquake vibration of underconsolidated sediment (Lewis, 1971b, p. 108).

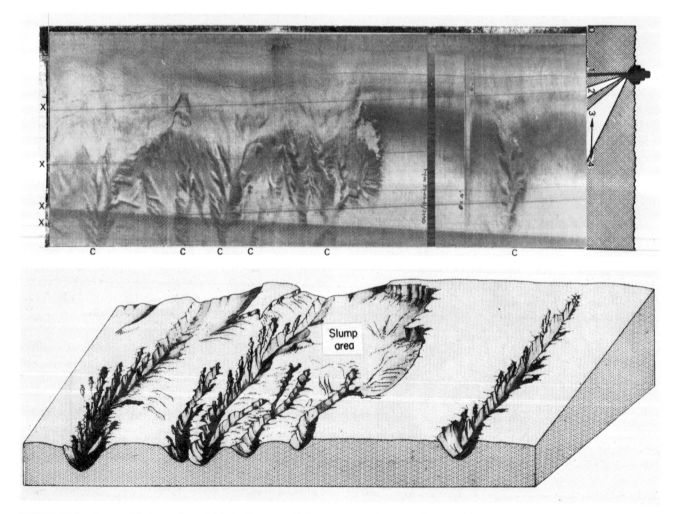

FIGURE 26. *Sonograph (at top) and block diagram (below) of an area near the top of the continental slope at the western edge of the Celtic Sea (northeastern Atlantic), showing canyon head axes (c on top photograph), subsidiary gulleys, and large slumped areas. From Belderson and Stride (1969).*

An associated phenomenon, appropriate to the eroded portions of canyon walls and the upper slope, is that of undrained loading (Hutchinson and Bhandari, 1971) where the headward part of the potential slide mass is subjected to additional loading by debris discharged from steeper slopes to the rear. Measured values of shear strength on slope sediments frequently indicate a highly stable condition even in areas where recent slumps and slides are in evidence (see Moore, 1961; Morelock, 1969) and it is clear that factors other than inherent cohesion must be involved to promote failure. As has been pointed out already, the volumetric significance of this type of mass motion on outer margins is considerable. Historical submarine slides such as that off the Grand Banks, Newfoundland, have displaced many scores of cubic kilometers of material (Heezen and Drake, 1964). Moreover, although earthquakes appear to be responsible for the most spectacular displacements, it is probable that smaller slides, due primarily to overloading through excessive accumulation of sediment on unstable slopes, are a prime cause of the irregular, scalloped aspect of the upper continental slopes in many areas, as shown in the sidescan sonographs (Figs. 26 and 27) of Belderson and Stride (1969) and Belderson et al. (1972).

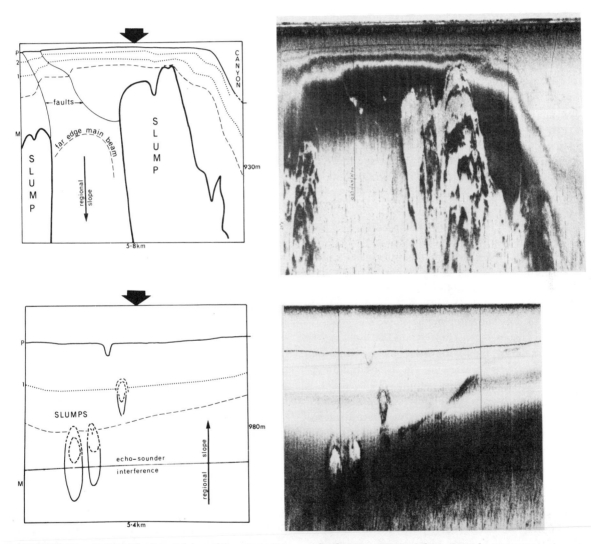

FIGURE 27. *Sonographs (on right) and interpretations (on left) of areas of slumping near the top of the continental slope. From Belderson et al. (1972). Upper: Rough bands on the sonograph are attributed to ground which has been deformed into transverse ruckles during downslope movement. The smooth floor is crossed by two faint, curved lines, possibly young faults associated with the slumping. Mediterranean Sea, off Algeria; minimum water depth, 88 m. Lower: The sonograph represents a view up toward the top of the slope, whose smoothness is broken by slump scars which appear as small, almost circular relief features, facing down the slope. Mediterranean Sea, off Mallorca; minimum water depth, about 200 m. Recent work suggests that these subcircular features may be related to the escape of groundwater (Belderson, personal communication, 1974).*

A variant of the slump or slide is the slow *creep* of cohesive (or cohesionless) material which is observed in the heads of some submarine canyons (Shepard and Dill, 1966). In this case, if the analogy with subaerial situations is valid, failure occurs along a large number of subparallel shear planes.

Incorporation of more fluid into the slump or creep mass, or a further increase in the external or internal shear stresses, may lead to the production of a *mass flow*, where internal coherence essentially has been lost. The precise nature of the ensuing process will then be dependent on the size distribution of the displaced material, the gradient and character of the slope on which it is moving, and the nature and strength of any external forces (such as an aqueous current) affecting the flow. The characteristics of the various types of gravity-controlled mass sediment flow process are detailed elsewhere in this volume (Chapter 11). Here we merely list the distinctive aspects of the more important types of mass gravity flow and indicate their role in outer margin environments.

Grain flows represent a category of mass flow in which the energy sustaining the sediment in motion derives from grain-to-grain interactions. This mechanism is confined to cohesionless sediments such as gravels, sands, and some silts. In an aqueous medium, floor gradients in excess of about 18° are required to induce forward motion of a pure grain flow, although this angle may be reduced by introducing clay into the interstitial fluid, thus promoting the buoyancy of the system (Middleton, 1970). Cessation of forward motion in the grain flow may result from a decrease in slope or from loss of interstitial fluid, leading to consolidation of the flow. Very rapid deposition then ensues, virtually freezing the flow.

Fluidized flows derive their supportive energy from the excess pore pressures generated in the interstitial fluid in a cohesionless sediment-fluid system. The "quick" behavior of some wet sands is probably the most familiar example of this mechanism. In nature, fluidization normally results from some reorganization of the sediment (such as change in packing style of component particles) which induces the increased pore pressures sufficient to overcome interparticle frictional forces and thus permit motion of the system.

Although differing fundamentally in mode of origin from grain flows, the transporting potential of fluidized flows appears to be governed by physical constraints qualitatively comparable to those controlling the behavior of grain flows. Consequently, both these categories of mass flow display similar, but not identical, depositional histories.

Debris flows are an important type of mass flow, created by remolding of slide and slump materials, in which there is generally a wide range of grain sizes, the larger particles being supported by a clay-water mixture that possesses finite cohesion or strength. Forward motion is normally sluggish but can take place on very low slopes (Curry, 1966). However, cessation of forward motion is generally abrupt, resulting in very rapid deposition. Thus deposits of debris flow origin are very ill sorted and almost ungraded, with few internal structures, except for some clast alignment (Lindsay, 1968).

Hampton (1972) has demonstrated experimentally that debris flows may generate turbidity flows, primarily through mixing with surrounding water at the snout of the flows and the creation of a turbulent layer overlying the debris flow. Cessation of motion in the latter does not necessarily result in similar behavior in this superjacent layer, which will normally possess sufficient kinetic energy to proceed downslope as a turbidity flow. Thus it is possible for the deposits of a debris flow to be followed directly upward by a turbidite of proximal aspect. Such "bipartite" units are encountered in the fills of some ancient submarine canyons (Fig. 28).

Turbidity flows (currents) are the mass flow mechanisms that have been investigated most thoroughly in the laboratory and (inferentially) in ancient sediments. These flows differ from the other phenomena considered here in that the solid material is essentially supported by turbulence within the flow which is generated by the downslope motion of the flow. This motion is induced initially by the density contrast between the flow (with its suspended sediment) and the surrounding clear water.

Most natural turbidity currents are probably surge-type flows, triggered by a quasi-instantaneous event such as a slump or earthquake, and they are evolving systems, with discrete portions, each characterized by somewhat different dynamic conditions (Middleton, 1970). In the context of the present discussion, the differing behavior of the head and body of turbidity flows appears pertinent on two counts.

First, Komar (1972) maintains that the differing velocities of head and body lead to the formation of an intervening "neck" region in a channelized turbidity current (Fig. 29). Since the ratio V/U is related to the Froude number, a reduction in the latter (for example, by a decrease in the bottom slope) when V and U are equal, must lead to a relative increase in thickness of the body over the head and vice versa for an increase in gradient. The implication for deep-sea channels is that the head (in which coarse sediment is concentrated) is likely to spill such material out of the steeply inclined canyon or upper fan valley, whereas on the gently sloping outer fan finer grained sediment carried in the body of the turbidity flow may escape the channel by overspill.

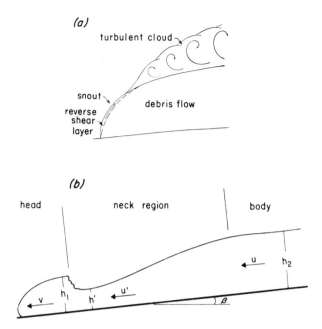

FIGURE 29. *Diagrammatic profiles of* (a) *the front of an experimental subaqueous debris flow* (*from Hampton, 1972*), *and of* (b) *a turbidity flow traveling down a slope of angle β showing development of the head, neck, and body regions of the flow* (*from Komar, 1972*).

FIGURE 28. *Typical bipartite unit in submarine channel-fill sequence of Lower Silurian age from Rhayader, central Wales. Dark hemipelagic mudstone at base is overlain by coarse sediment of probable gravel flow origin abruptly succeeded by graywacke-arenite of proximal turbidite aspect. The scoured base of the overlying gravel flow deposit is seen at the top of the photograph. Units of this type represent 40% of this particular channel-fill succession. Hammer provides scale.*

Role of Sediment Gravity Flows in Outer Margin Sedimentation

Since it is not usually possible to observe these short-lived processes directly in the deeper ocean beyond canyon heads, any estimate of the relative importance of individual processes at present must be inferential. The problem is compounded by the complex interrelationships that exist between the various types of mass flow (Middleton and Hampton, 1973; and Chapter 11). Thus, in the course of being transported from shelf to basin floor a given body of sediment may pass through several phases involving different support mechanisms (for example, grain flow to debris flow to turbidity flow). However, some notion of the nature of the final transportation phase can be gained from a consideration of the mode of deposition and Fig. 30 indicates the general order of depositional mechanisms predicted to operate in each of the main types of sediment gravity flow outlined above. Such mechanisms are reflected in the structures and textures preserved in the resulting sediment. Unfortunately, apart from a few small areas studied in detail, our knowledge of the geographic distribution of vertical structural/textural sequences in the coarser units found in the outer margin environment is as yet inadequate to document empirically the relative efficacy of the individual mass flow processes.

Second, there is the question of the "hydraulic jump" which occurs as a flow passes from a supercritical (Froude number greater than unity) to a subcritical condition. Turbidity flows possessing the generally assigned range of densities and velocities will probably undergo such a hydraulic jump at the break in slope from submarine canyons (average gradient about 0.050: Shepard and Dill, 1966, Appendix) to the fan channels, with much lower axial slopes (Middleton, 1970; Komar, 1971). Such a jump results in the incorporation within the flow of a substantial volume of water, with a concomitant reduction in the effective density and a probable increase in turbulence, leading to rapid deposition of coarse sediment just downstream of the jump point and more efficient autosuspension of the remaining finer material.

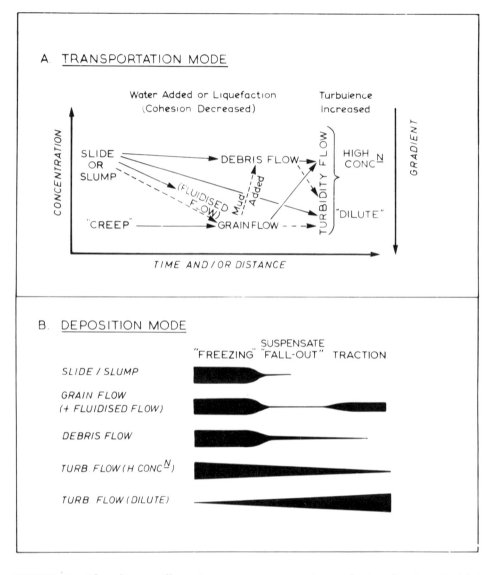

FIGURE 30. *Flow diagram illustrating possible interrelationships of the main types of sediment-transporting gravity-controlled mechanisms in the depositional mode. The relative width of the black "fingers" for each of the transportation mechanisms listed on the left is indicative of the relative importance of each of the three main modes of deposition listed at the top. See text for discussion.*

Nevertheless, a crude qualitative approach can be made by examining the limiting conditions (of slope, flow density, etc.) that govern the motion of each type of flow. Thus, while observations by divers and from submersibles indicate that grain flows [creating the "rivers of sand" quoted by Shepard and Dill (1966) and others] are a significant feature of the shallow canyon heads of the California margin, the limiting slope of 18° for inertial (high-velocity) flows must confine the occurrence of such deposits to exceptionally steep sections of the canyon axes, even where the overall gradient of the marginal slope approaches 10°.

Grain flows might also originate on the steep walls of canyon or fan channels but would be very limited in their occurrence. Unusual conditions, such as the growth of reefs on the margins of steep trench-defining slopes, may also enable debris to convey talus to considerable depths (Byrne, 1974). However, admixture of mud into the grain flow, conferring strength to the "matrix" and transforming the process to debris flow, may enable motion to continue on much lower slopes and will also permit large clasts to be carried downcanyon. It is probable that most of the exotic pebbles and cobbles encountered on deep-sea fans have been transported in this manner rather than in a fully turbulent suspension.

The great volumetric importance of turbidity flows is attested by the abundance of graded sands in base-of-slope and submarine fan sequences, as well as the association of graded beds found on abyssal plains with known episodic events such as the 1929 Grand Banks earthquake and slump (Heezen and Drake, 1964). The relative rarity of sandy graded units in the near-surface sequences cored on well-developed continental rises suggests that at the present time most turbidity currents conveying sand bypass the rise and reach the abyssal plain via a few deeply incised fan valleys. However, it is contended that the quantitative importance of *mud turbidites* forming the rises remains underestimated. Where detailed analyses of sediment textures and structures are available they reveal that a substantial proportion of the hemipelagic sequences is, in fact, composed of graded silt and clay units (Piper, 1973; Rupke and Stanley, 1974). In terms of total sediment input to the rise, these mud turbidites are probably more important than the sandy turbidites.

It is interesting to speculate on the possible evolutionary process of prograding margins (Fig. 31). At the inception of deposition, marginal slopes are high and the break in slope is pronounced. Under these conditions, debris flows and grain flows may proceed downcanyon and these deposits accumulate near the break at the base of the slope. True turbidity flow also may be generated on the upper slopes and undergo a hydraulic jump at the break of slope (fan apex), leading to further deposition on the upper fan. Progressive reduction of the break in slope angle ensues, probably accentuated by the accumulation of slides and slumps and leading to a decreasing incidence of debris and grain flows. The effects of the hydraulic jump on turbidity flows also become less pronounced, enabling these flows to proceed further across the fan surface. This combination of effects ultimately gives rise to a gently inclined, relatively smooth outward gradient of rise aspect in place of the earlier complex of more steeply inclined individual fans superimposed on a flat basin plain.

Biologic Processes

While the intervention of organisms in the deeper marginal environments does not produce the obvious effects on sedimentation that are created by the activities of shelf biota such as the reef builders or shelly benthos, the subtle working of several types of organisms may lead to important long-term results in the deeper zones of the ocean. The quantitative importance of the group of

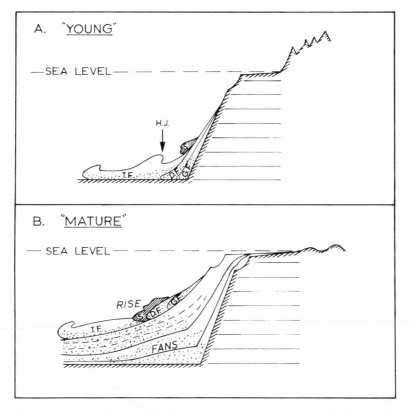

FIGURE 31. *Speculative diagrams illustrating the relative importance of the principal types of gravity-controlled processes in determining the configuration of outer continental margins. GF = grain flow; DF = debris flow; S = slump; TF = turbidity flow; HJ = hydraulic jump. See text for explanation.*

biologically controlled processes in the slope and base-of-slope provinces as yet cannot be defined with any accuracy but organisms appear to play some part in the initiation and maintenance of critical sediment paths such as submarine canyons, and may, therefore, have a role of considerable importance in outer margin sedimentation. The constructive and accumulative aspect (reefs, biostromes, nannoplankton oozes), which is the dominant feature of organically engendered sediments on the shelf and in the deeper reaches of the ocean, is less readily apparent in the canyons and slope regions where the role of organisms appears to be primarily disruptive. Two main modes of organic modification of outer margin sediments are distinguished. The first mode is concerned with the effects of biologic reworking of soft sediments, while the second details the erosive capabilities of organisms in these deep environments.

Bioturbation and internal disturbance derive from the various forms of biogenic activity which cause some degree of reorganization within the inorganic deposits of the canyon, slope, or base-of-slope regions. The plethora of infaunal and benthonic forms in these regions is attested by the virtually ubiquitous occurrence of mounds, burrows, trails, and tracks in photographs and television pictures of the seabed (Fig. 32*A*). Furthermore, these features are merely the superficial manifestations of the prevalent reworking of sediment which cores reveal as persisting to considerable depths below the surface; see Fig. 32*B* (cf. Owen et al., 1967). Such activity may affect sedimentation in four main ways.

First, many animals, both surface-grazing forms such as holothurians, and the ingesting infauna, such as annelid and polychaete worms, process sediment and eject it as fecal pellets which occur in profusion on many muddy lower slope surfaces. Incorporation of these pellets not only modifies original textures and physical structures, but may also alter significantly the geotechnical properties (see below).

Second, the activities of burrowing and benthic organisms in muddy sediments may provide substantial increments of suspended sediment to the near-bottom waters, at least on a local basis. Expulsion of muddy waters by burrowers is frequently observed and is accompanied by the bottom-stirring activity of crustaceans, some coelenterates, and, especially, benthic fish (Stanley, 1971a). While the total quantity of suspensate released by individuals is probably small, in some critical regions such as canyon walls or axes, the coalescing small turbid flows formed in this way [and analogous flows due to instrumental impact have been observed on videotape; see Stanley et al. (1972a)] may reinforce existing processes of downslope transfer and perhaps

provide a triggering mechanism for large-scale sediment movement.

The third aspect of biogenic disturbance is primarily a constructive process of local significance. It involves the construction of mounds of sand and silt by various animals. Where these occur on preexisting steep gradients, such as an offreef slope (Byrne, 1974), the critical angle for initiation of sand flow movement may be exceeded on the flanks of the mounds, especially if some mud is biologically admixed with the sand. A gradual downslope movement may ensue which may then persist beyond the lower limit of mounding.

Finally, the burrowing propensities of various types of animals may significantly alter the original structure, packing, texture, and porosity/permeability of the sediment. As a consequence, the internal shear strength and load capacity of the deposit may be changed (Stanley, 1971b). In some cases, cementation of grains attributable to biogenic burrowing may lead to an increase in sediment strength (Einsele, 1967), but in most instances the combined effect of sediment homogenization by prolific and continuous burrowing and the introduction of water through the open orifices is to increase the excess pore pressures in clayey or silty sediments and to enhance underconsolidation, both of which are means of promoting shear failure and slumping (Richards, 1965; Morgenstern, 1967) in sediments prone to failure.

Such a process is likely to be most important in the vicinity of submarine canyons, where the sedimentary fill is generally regarded as being in an unstable condition (Shepard and Dill, 1966, pp. 296–311). In this connection it may be significant that submarine canyons frequently appear to possess a distinctive benthic biota (Stanley and Kelling, 1969; Rowe, 1971), perhaps because of the nature of the substrate.

Biologic erosion again appears to be most prevalent, or at least most effective, in submarine canyons. Submersible and aqualung observations in some west coast canyons have revealed the importance of boring organisms as erosive agencies even in the solid rock of canyon walls (Dill, 1964; Warme et al., 1971; Shepard, 1973, p. 308). Similar observations have been made in some of the canyons off Georges Bank, New England (Dillon and Zimmerman, 1970), where the steepness of canyon walls appears to be maintained by periodic collapse of surficial sediment through burrowing, rather than boring. Another interesting example of localized biologic erosion was observed in an underwater television survey of Wilmington Canyon, off the Middle Atlantic States (Stanley et al., 1972a). Here, benthic fish of the *Urophycis* type were observed in hollows excavated not only out of soft, silty sediment but also out of poorly indurated

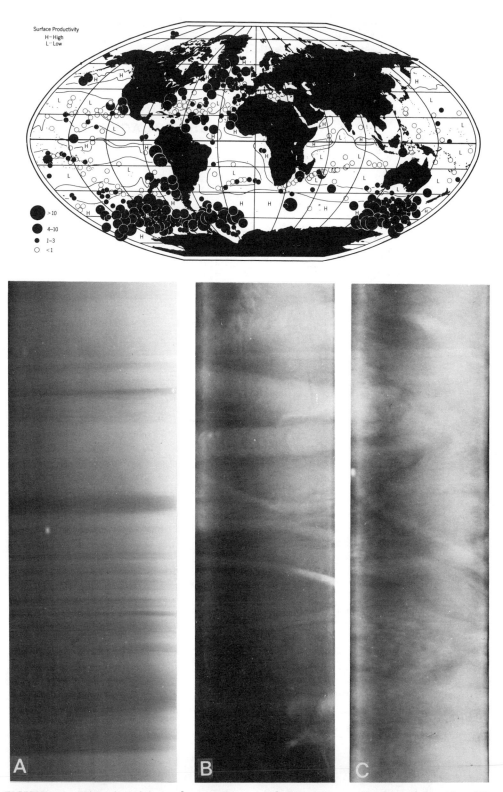

FIGURE 32. *Biogenic activity produces various degrees of reorganization in seafloor deposits, particularly in continental margin environments. Upper: Map showing density of large visible bottom-living organisms based on seafloor photographs. From Heezen and Hollister (1971). Lower: X-Radiographs (courtesy D. Lambert, AOML-NOAA) of selected core sections collected on the* *Atlantic outer margin off North America. (A) Undisturbed laminated silt and clay on upper rise (2721 m) off the Hudson Submarine Valley. (B) Burrows near base of slope (1908 m) east of Hydrographer Canyon. (C) Intensely mottled and bioturbate mud on lower slope (1628 m) east of Hydrographer Canyon.*

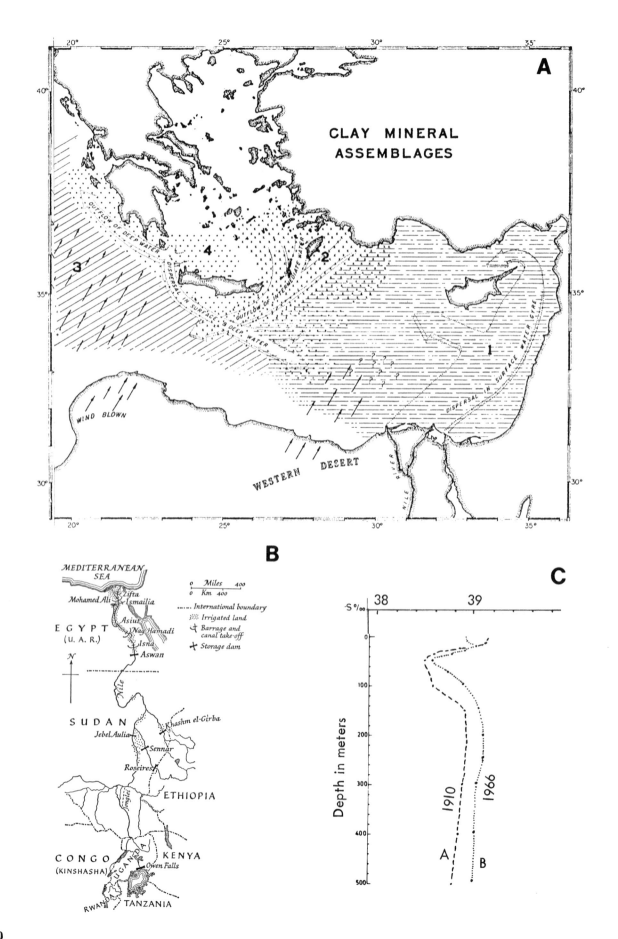

A

CLAY MINERAL ASSEMBLAGES

3

4

2

1

OUTFLOW OF DEEP WATER

DISPERSAL IN SURFACE & DEEP WATER

OUTFLOW OF INTERMEDIATE

DISPERSAL IN SURFACE WATER GYRE

WIND BLOWN

WESTERN DESERT

NILE RIVER

B

MEDITERRANEAN SEA

Miles 400
Km 400

International boundary
Irrigated land
Barrage and canal take-off
Storage dam

Zifta
Mohamed Ali
Ismailia

EGYPT
(U. A. R.)

Asiut
Nag Hamadi
Isna
Aswan

N

Nile

SUDAN
Khashm el-Girba
Jebel Aulia
Sennar
Roseires

ETHIOPIA

CONGO
(KINSHASHA)

KENYA
Owen Falls

UGANDA

RWANDA
TANZANIA

C

S ‰   38   39

Depth in meters

0

100

200

300

400

500

1910
A

1966
B

sandstones forming ledges around the outer canyon head, at depths (to over 200 m) which preclude an origin through subaerial erosion, even during glacial low stands of sea level. If fish are indeed responsible for creating the oval, steep-sided depressions in which they have been observed, then not only is biologic activity responsible for some aspects of canyon erosion, but it has provided a novel (internal) source of sediment to be reworked through the canyon.

HUMAN ACTIVITY AND ALTERED SEDIMENTATION PATTERNS IN THE OUTER MARGIN

Man's activities in the outer continental margin are increasing, and there is a realization that accelerated modification of even the distant deep marine sectors is likely to affect, in one way or another, human well-being. In shallow and more proximal shelf regions the need of effective means of assessing and controlling man's influence is recognized (if not always acted upon!). Not enough attention has been paid to problems, actual and potential, created as man invades the outermost shelf and deeper regions. The following discussion briefly outlines some of the more obvious human activities and needs that are likely to interfere with sedimentary processes on the shelf-break, slope, and base-of-slope environments.

Four main areas of interaction where sedimentologists will probably be required to apply their professional skills in an advisory capacity are distinguished: (1) artificial modification of nearshore environments, which affects offshore sedimentation patterns; (2) problems and side effects related to ever increasing waste disposal; (3) alteration of sedimentation patterns related to the exploitation of natural resources; and (4) sedimentation and engineering problems related to the emplacement of man-made structures and equipment on the seafloor. For a more comprehensive discussion of each of these aspects, the reader is referred to Chapters 20 to 24 in Part IV of this volume.

Problems Created by Artificial Modification of Margin Environments

One of the most direct effects of human intervention is the reinforcement or diminution of sediment supply to che shelf edge and heads of submarine canyons running close to shore in regions where shelves are either narrow or locally absent. The increasing tendency to dam, thannelize, or otherwise direct large rivers may result in interference with natural cycles involving outer margin sedimentation. Examples include the Colorado in southern California and the Ebro and Nile rivers in the Mediterranean, where the impoundment of waters has resulted in a markedly diminished supply of sediment to the river mouths and deltas. This sediment deficit is most conspicuous in the adjacent shelf areas but ultimately must find its parallel effects on the frontal slopes and flanking deep-sea cones and fans. The effects of dredging and marked alteration of the channel and bed load configuration of rivers bordering narrow continental margins (the ongoing modification of the Var River on the French Riviera is a good example) are noted almost immediately seaward on the adjacent submarine slope. This type of activity and the construction of groynes, piers, sea walls, or harbor-dredging works may leave their marks on both shallow and deep sediment budgets and processes (see, e.g., Shepard, 1973, p. 237).

The construction of the Aswan High Dam on the Nile River is a good example of the type of artificial modification that can induce sedimentation changes in more distant outer margin environments (Fig. 33). The immediate effect of the damming of this, the major river in the eastern Mediterranean, has resulted in a sudden reduction of the annual discharge, mainly through increased evaporation from reservoirs and irrigated ground. Since 1968 the annual discharge has averaged only one-tenth of the average amount for the period 1956 to 1964 (Gerges, 1973). A marked reduction in the total sediment volume carried into the Levantine Basin has been measured, as has the marked alteration of the salt/freshwater interface and of the delta-front salinity patterns of the southeastern Mediterranean (Fig. 33C). Minimum changes in salinity now occur in winter instead of late summer/autumn as previously observed, and the circulation patterns in the southern Levant have also been significantly altered. Thus, emplacement of the dam is responsible for the sudden change in the position of the turbidity maximum and, thus, rates of deposition and the distribution of fine-grained sediment. These changes and the altered delta-front nutrient

FIGURE 33. *The Nile River has been one of the major factors influencing sedimentation in the eastern Mediterranean, as shown in A. The Nile clay mineral assemblage (1) was previously transported as far north as Cyprus and Turkey. From Venkatarathnam and Ryan (1971). (B) Emplacement of the Aswan High Dam has considerably altered this dispersal pattern and produced* *in only a few years' time deleterious effects on delta development, fishing, and coastal erosion. From Worthington (1972). (C) Summer salinity changes between 1910 (Nielson station 156, 32°26'N, 26°51'E, July 29, 1910) and 1966 (Ichtiolog Station 30, August 9, 1966). Data courtesy of S. Morcos (UNESCO).*

patterns have had consequences on the biota of the region (Worthington, 1972), involving particularly serious repercussions on the fishing industry on the outer Egyptian margin (George, 1972). Furthermore, reduction in the bed load carried to the coast has locally accelerated the erosion of the delta margin (Kassas, 1972). It is not inconceivable that some long-term and far-reaching hydrologic effects could also ensue since some of the major water masses of the Mediterranean originate in the Levantine Basin—for example, formation of the Intermediate Water mass, as discussed by Wüst (1961) and Morcos (1972).

But what about more specific changes in sedimentation on the slope? The diminution of suspended fines and reduction in suspension-rich underflows of the type that in the past may have moved downslope as well as across the Nile Cone will undoubtedly decrease the rate of progradation of the submarine delta slope. In the one instance, the decreased rates of sedimentation on the slopes could be accompanied by a reduction in the amount of trapped pore water, and the resulting increase in sediment cohesion might well reduce rates of mass gravity failure such as slumping. On the other hand, changes in salinity and depositional rates on the slope might well alter the composition of the benthic com-

munities on the slope. The resulting change in bioturbation rates of the seafloor sediment could alter mass physical properties so that there would be a reduction in sediment strength and, thus, an increased possibility of failure. Some limited assessment of the physical properties of the Nile Cone sediment in cores (Einsele, 1967; Keller and Lambert, 1972) was undertaken prior to construction of the dam, and further work planned in this area will undoubtedly aid in the assessment of post-Aswan Dam slope sedimentation.

Problems Related to Waste Disposal Activities

Although the deep sea has long been considered as the ultimate trash bin for man's refuse, the expenses involved have precluded, in most cases, a direct transfer and disposal of wastes in outer margin sites far from shore. But with the decreasing availability of land fill and nearshore dumpsites in heavily populated and industrialized regions, municipal authorities must consider seriously the transfer offshore, by barge and by pumping, of sewage, industrial wastes, and dredge spoil, which finally may be dumped onto the outer continental margins.

Generally it is assumed that materials dumped into the heads of canyons or directly on the upper slope are

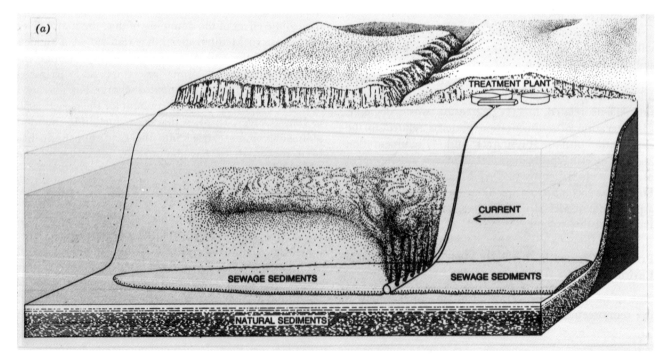

FIGURE 34.   (A) *Oceanic disposal of waste from a sewage-treatment plant is portrayed. The system is for a plant with a capacity of about 100 million gallons a day. Effluent from the plant flows a distance of from 2 to 5 miles through an outfall pipe that is from 6 to 12 ft in diameter. For about a quarter of a mile at its end the pipe has dozens of 6 in. discharge ports. The mostly liquid material it discharges rises to the thermocline, which is the bound-ary between deep, cool water and the warmer surface layer. Prevailing current mixes material and moves it to one side or the other, depending on wind and tide. Some solid particles settle on the bottom. (B, C) Effect of outfall on a community of polychaete worms is depicted in the lower diagrams. The worm population is reduced for about 3 miles, and thereafter is about the same as in uninfluenced seabed. From Bascom (1974).*

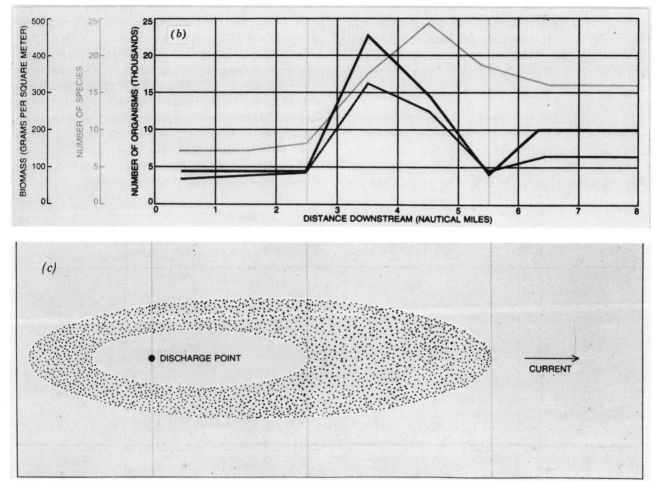

FIGURE 34   (Continued)

likely to find their way to deeper base-of-slope environments by mass gravity and suspension processes. Canyons that head close to shore are considered particularly obvious dumping grounds as well as some sites distant from shore, such as off the east coast of the United States. Two examples of the latter case are the head of the Hudson Canyon 175 km southeast of New York City, and the shelf edge off the wide New Jersey shelf.

Baseline surveys prior to dumping should be made in order to carefully evaluate natural dispersal of sediment, particularly sediment in suspension, in response to seasonal and even diurnal variation. Sediment entrainment is by no means always downslope as is demonstrated by the upslope and onshelf movement of slope water masses in response to storms (Lyall et al., 1971), breaking of internal waves on the upper slope, tidal effects, surges, and other processes active on the outer shelf margin (see Chapter 16). Evaluation of these factors would help minimize the landward return of material, particularly less dense pollutants of silt and clay size, and would provide a basis for the determination of

volume, rate, composition, and form of waste discharge (containerized, slurry, etc.). Furthermore, monitoring is essential in order to minimize effects of lateral displacement of toxic wastes by water mass movement, particularly in the case of small closed seas. Although some workers believe that long-term deleterious effects can be kept to a minimum (Bascom, 1974), the uncontrolled introduction of pollutants into semiclosed systems such as the Caspian (damage to the sturgeon fishing industry), Black, and Mediterranean seas (Brambati, 1972) already has clearly demonstrated that the out-of-sight, out-of-mind tack is by no means always advisable. A cautiously optimistic view is only warranted if, as Bascom (1974) suggests, controls are maintained so as to absorb the effects of man's intrusion of the marine realm.

Sedimentological and geochemical interest should be sparked by the fact that artificial matter in suspension or solution can serve as an excellent tracer in defining regional water mass patterns and related sediment dispersal paths (Fig. 34). The sinking of near-surface masses in the northern Mediterranean regions (Stommel et al., 1971, Fig. 8), where sources and introduction of municipal and industrialized wastes are greatest, can entrain and displace pollutants far from their original site of introduction. Trace element and suspended sediment studies would shed light on the movement of the deep Mediterranean water mass of which as yet little is known (Miller, 1972).

Other experiments pertinent to outer margin sedimentation include the tracing of man-made products on slopes and in canyons. The recovery of tar-coated pebbles in some Riviera canyons, for example (Gennesseaux, 1966), provides a means to determine realistically the

rates of downslope motion from the coast to the basin plain. Larger objects specifically placed in canyon heads (Shepard and Dill, 1966) serve to measure the rates of downslope movement by creep and mass gravity processes such as slumping and debris flow. The monitoring of mineral slurries including coal ash (Foster and Stone, in Middleton, 1966), bauxite residues (Fig. 35) in the northern Mediterranean (Bourcier and Zibrowius, 1973), and copper tailings (as proposed off southern Puerto Rico) pumped directly on the slope undoubtedly can shed important light on the mechanics of processes such as grain flow and turbidity current motion.

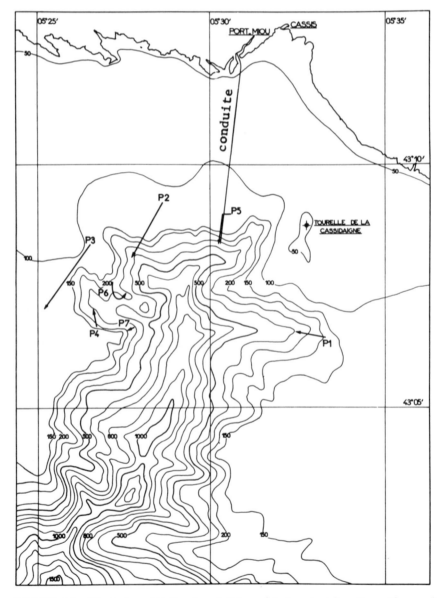

FIGURE 35. *Chart shows pipeline (conduite) used to transport bauxite residues and red mud slurries across a narrow shelf directly into the head of the Cassidaigne Canyon south of Cassis on the French Mediterranean margin. Monitoring of the effects is conducted by means of submersible dives (shown by arrows P3, P2, P5). The artificially introduced red muds periodically shift downslope. From Bourcier and Zibrowius (1973).*

Sedimentological Problems Related to Exploiting Natural Resources

Economic growth continues to be a high-priority goal of the United States and other industrial countries, and this growth goes hand-in-hand with a marked increase in energy consumption. A National Production Council (1972) projection for the 15 year period from 1971 to 1985 indicates an annual rate of growth in energy consumption by the United States of between 3.4 and 4.4%. Unless alternatives such as geothermal power, solar energy, energy from refuse, and conventional nuclear reactors are rapidly developed, there is every indication that exploration and exploitation will move from the coastal and shelf regions to the outer margin sectors. Technological advancements fostered by active exploitation of the North Sea, deep-sea drilling (JOIDES), and offshore mining now make it feasible to exploit the outermost shelf with potential sedimentary consequences on the slope and deeper sectors. Permits for exploitation of oil and gas on the outermost shelf are now under consideration (Fig. 36), and it will not be long before active drilling of the potentially valuable thick sedimentary flank and rise sequences (Fig. 1) at the base of continental slopes is initiated. However, it is in these outer margin environments that technology is least advanced to counter possible accidents such as blowouts. The reader interested in the technology assessment and present national policy-making for energy and environment in the outer continental shelf (OCS) sector and beyond is referred to a recently published National Science Foundation-supported summary (Kash et al., 1973).

Evaluation of the sediment regimes at the shelf break is now more than purely an academic exercise. Breaking internal waves and other shelf-edge processes cited earlier usually result in sediment "unmixing" (Swift et al., 1971), i.e., a bypassing of fine-grained fractions and concentration of coarser or denser lag material. This reworking of relict sediments in high-energy conditions can concentrate commercially valuable minerals such as diamonds and heavy mineral suites. The mining for these and for manganese nodules at greater depths by various types of airlift pumping systems (Fig. 37A) involves some disruption of the seabed by lifting sediment to the surface along with large volumes of ocean bottom water. The disposal of large quantities of dense sediment-water mix produces a situation that has sedimentological ramifications. The water-sediment slurry is either pumped over the side of a ship (Fig. 37B) or returned to the seafloor by pipeline. A preliminary evaluation of the artificially raised deep water on the Blake Plateau showed that vertical displacement of sedimentary

material enhances phytoplankton growth and productivity (Amos et al., 1972). One could envision methods to dispose of these dense wastes by artificially produced turbidity currents downslope. The eventual effects of reintroduction of rejected sediment-water mixes in the water mass as well as on the seafloor must be considered.

Marine Engineering Problems and Sediment Dynamics

The placement of structures and equipment on the seafloor is a risky business. We have shown earlier that slopes, zones of low to poor stability, are not uncommonly prone to failure. The possibility of damage or loss increases on slopes subject to rapid sedimentation such as off deltas (weakly cohesive sediments of low strength and high pore water pressure), in zones of relatively steep slopes that are affected by the sudden introduction of sediment (the irregular seasonal input of suspension-rich water on margins bounding the Mediterranean is a good example), and in tectonically active regions such as the circum-Pacific belt, Caribbean, and other mobile belts (see Fig. 2). Marine geotechnics (Richards, 1967) and the analysis of physical reactions of sediment subjected to the placement of structures and other artificial modification (Inderbitzen, 1974) serve to minimize damage or loss.

Where and how should a structure be placed on the seafloor? Generalizations as to seafloor stability are to be avoided, for in situ studies have shown that properties such as shear strength vary rapidly within short distances due to bioturbation changes and differences in depositional history (Hagerty, 1974). However, the coupling of laboratory testing with in situ measurements enables reasonable predictions to be made, i.e., in some cases within 30% of the results expected by conventional soil mechanics theory (Simpson et al., 1974).

Bottom current activity and spillover phenomena limit stability at the shelf break. Scour of footings of fixed structures such as towers or of mobile platforms is notably accentuated by these processes during storms and hurricanes. The rupture of deep-sea cables (Heezen and Ewing, 1955; Heezen, 1959; Ryan and Heezen, 1965; Krause et al., 1970; Heezen and Hollister, 1971), pipelines, and similar features on slopes has long been of interest to sedimentologists. Turbidity currents (Fig. 38), slumps, and creep, particularly in and close to canyons, have been cited as the most common origin of breaks. The effects of flooding by major rivers (Fig. 38A) and submarine erosion by sediment moved by strong bottom currents, generally in constricted areas such as shallow banks and straits (cf. Gibraltar area: Heezen and Johnson, 1969, Fig. 31), are also recognized as important causes of failure.

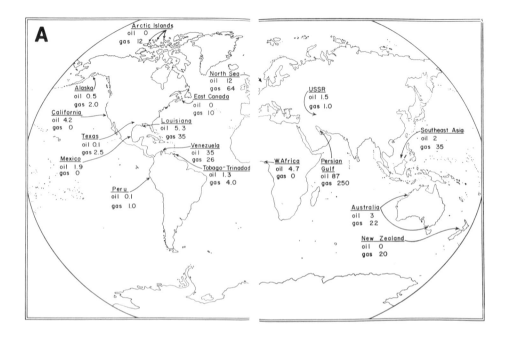

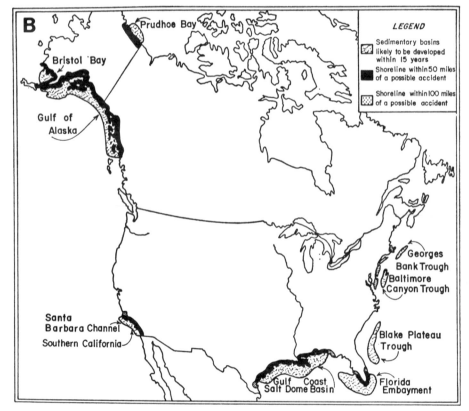

FIGURE 36. *Maps showing* (A) *worldwide offshore oil and gas reserves as of March 1973* (*oil in billion barrels; gas in trillion cubic feet*), *and* (B) *sectors of the North American outer continental shelf* (OCS) *likely to be developed in the next 15 years. Margins that may be affected in case of oil spills are also shown. From Kash et al.* (1973).

Sudden displacement of sand and of large cement blocks placed in canyon heads off Southern California has been cited by Shepard and Dill (1966). Interesting in this respect is the recent failure of large structures recorded on the outer margin off Gabon, Central West Africa. Large petroleum reservoirs placed on a linear sand ridge at the narrow shelf break near Cape Lopez (north of the Ogooue River) were suddenly shifted and downdropped to deeper water. Preliminary investigations of the physical properties of the sediment (M. Le Fournier, personal communication) suggest that fluidi-

zation, probably of the type that results in the "quick" flow of sand, produced the sudden failure of the ridge and resulting shift of the reservoirs. This and the river-of-sand phenomenon recorded in some canyons off California are some of the few examples of the mass flow process actually observed in a deep-water environment.

Analysis of such failures is extremely valuable, for it serves to increase our understanding of the outer margin transport processes, of which so little is yet known. We can expect that some remarkable discoveries on downslope sediment entrainment will be made in the next

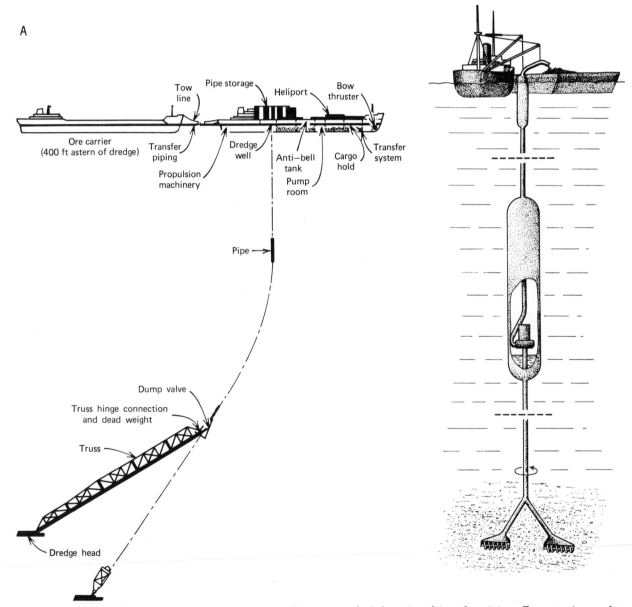

FIGURE 37. **(A)** *Two of the hydraulic systems proposed for mining seafloor nodules. System on the left, proposed by Deepsea Ventures, is powered by an airlift pump. System on the right is powered by a centrifugal dredge pump submerged to about 15% of the depth of dredging. From Mero (1972).* **(B)** *Aerial photo-*

graph of dye injected into the mining effluent 8 minutes after discharge from Deepsea Miner. The normal outboard discharge can be seen amidships on the port side. From Amos et al. (1972).

FIGURE 37   (Continued)

few years as a direct result of the progressive intrusion into what has been, up to now, a no-man's-land.

mantling most slope and base-of-slope regions is fine-grained (lutite) and strongly bioturbated. Localized areas of surficial sand and gravel are associated with axial portions of submarine canyons and fan channels, active fault scarps and slump scars, and regions subject to glacial processes or to the activity of powerful oceanic currents. Cores obtained from open slopes generally display laminated or biogenically homogenized lutite. Cores from well-developed continental rises contain layers of ungraded laminated fine sand, attributed to contour-current deposition, and graded sands and silts of distal turbidite character, interbedded with predominant muds, some of turbidite origin.

The principal features of the sedimentary processes operating in this outer margin province are outlined. These processes encompass atmospheric and tidal forces (including wave and storm stirring, pelagic settling, and semipermanent bottom currents, within and outside submarine canyons), thermohaline systems (including internal waves and geostrophic bottom currents), gravity-controlled mechanisms (slumping, creep, grain flow, fluidized flow, debris flow, and turbidity flow), and biologic processes (burrowing, internal disturbance, and biologic surficial erosion).

The influence of human activities on sedimentation patterns in the outer margin is discussed briefly in terms of four main areas of interaction: (a) artificial interference with nearshore environments resulting in offshore modifications, (b) problems related to waste disposal at sea, (c) exploitation of natural resources, and (d) emplacement of man-made structures on the seafloor.

SUMMARY

Environments intervening between the shelf break and the abyssal or basin plain are assigned to two major physiographic provinces: the continental slope and the continental rise or fan-apron complex. The slope proper may be modified by diastrophic folding and faulting while salt diapirs are locally important. Lower slope topography frequently is determined by large-scale slumping while submarine gullies and larger canyons are excavated into many slopes. Base-of-slope provinces include submarine fans, with channel, levee, and interchannel subenvironments and the continental rises, which may have originated through lateral coalescence of fans, although molding by geostrophic contour-following bottom currents also may be important. Sediment

ACKNOWLEDGMENTS

We are much indebted to many colleagues, too numerous to name individually, whose ideas and encouragement over the years have contributed much to the understanding of outer margin processes. More specifically, we are grateful to Douglas Hamilton, Gerard V. Middleton, Jack W. Pierce, and Donald J. P. Swift who read various drafts of this chapter and offered many penetrating and constructive comments. Douglas Lambert and Selim Morcos are thanked for permission to use unpublished illustrative material. Most of the original results reported here derive from studies funded for Gilbert Kelling by the United Kingdom Natural Environment Research Council (Research Grant GR/3/899) and for Daniel J. Stanley by the Smithsonian Research Foundation (Research Grant 450137) and the National Geographic Society (Research Grant 606).

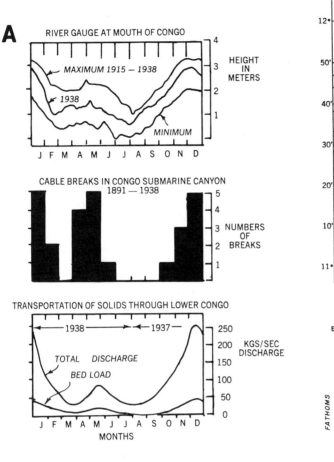

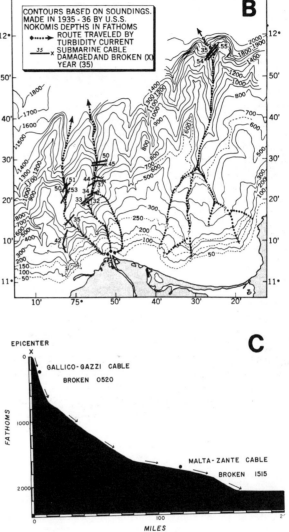

FIGURE 38. *Submarine cable breaks are frequently related to river discharge and turbidity currents. (A) Diagrams showing cables broken by turbidity currents which occur most frequently when total Congo River discharge and bottom-trans-ported sediment (bed load) are at a maximum. Cables broken by turbidity currents (B) off the Magdalena River, Colombia and (C) in the Messina Canyon (Mediterranean). From Heezen and Hollister (1971) and Ryan and Heezen (1965).*

REFERENCES

Amos, A. F., C. Garside, K. C. Haines, and O. A. Roels (1972). Effects of surface-discharged deep-sea mining effluent. In D. H. Horn, ed., *Ferromanganese Deposits on the Ocean Floor.* Washington, D.C.: Natl. Sci. Found., pp. 271–281.

Andrews, J. E., F. P. Shepard, and R. J. Hurley (1970). Great Bahama Canyon. *Geol. Soc. Am. Bull.,* **81:** 1061–1078.

Aymé, J. M. (1965). The Senegal salt basin. In *Salt Basins Around Africa,* New York: American Elsevier, pp. 83–90.

Ballard, J. A. (1966). Structure of the Lower Continental Rise Hills of the western North Atlantic. *Geophysics,* **31:** 506–523.

Bascom, W. (1974). The disposal of waste in the ocean. *Sci. Am.,* **231:** 16–25.

Belderson, R. H., N. H. Kenyon, A. H. Stride, and A. R. Stubbs (1972). *Sonographs of the Sea Floor, a Pictorial Atlas.* Amsterdam: Elsevier, 185 pp.

Belderson, R. H. and A. H. Stride (1969). The shape of submarine canyon heads revealed by Asdic. *Deep-Sea Res.,* **16:** 103–104.

Benjamin, T. B. (1968). Gravity currents and related phenomena. *J. Fluid Mech.,* **31:** 209–248.

Betzer, P. R. and M. E. Q. Pilson (1971). Particulate iron and the nepheloid layer in the western North Atlantic, Caribbean, and Gulf of Mexico. *Deep-Sea Res.,* **18:** 753–761.

Betzer, P. R., P. L. Richardson, and H. B. Zimmermann (1974). Bottom currents, nepheloid layers and sedimentary features under the Gulf Stream near Cape Hatteras. *Mar. Geol.*, **16:** 21–30.

Bloom, A. L. (1965). The explanatory description of coasts. *Z. Geomorphol.*, **9:** 422–436.

Bouma, A. H. (1962). *Sedimentology of Some Flysch Deposits.* Amsterdam: Elsevier, 168 pp.

Bouma, A. H. and C. D. Hollister (1973). Deep ocean basin sedimentation. In G. V. Middleton and A. H. Bouma, eds. *Short Course, Turbidites and Deep Water Sedimentation.* Soc. Econ. Paleontologists and Mineralogists, Pacific Section, Anaheim, pp. 79–118.

Bouma, A. H. and F. P. Shepard (1964). Large rectangular cores from submarine canyons and fan-valleys. *Am. Assoc. Pet. Geol. Bull.*, **48:** 225–231.

Bourcart, J. (1954). *Le Fond des Océans.* Paris: Presses Universitaires de France, 108 pp.

Bourcart, J. (1960a). Carte topographique du fond de la Méditerranée occidentale. *Bull. Inst. Océanogr. Monaco,* **1163:** 20 pp.

Bourcart, J. (1960b). Sur la répartition des sédiments observés en Mediterranée occidentale. *Int. Geol. Congr., 21st Copenhagen, 1960, Rep.,* Session, **23:** 7–18.

Bourcart, J. (1964). Les sables profonds de la Mediterranée occidentale. In A. H. Bouma and A. Brouwer, eds., *Turbidites.* Amsterdam: Elsevier, pp. 148–155.

Bourcier, M. and H. Zibrowius (1973). Les "boues rouges" déversées dans le canyon de la Cassidaigne (région de Marseille). *Tethys,* **4:** 811–842.

Brambati, A. (1972). Sedimentology and pollution in the Mediterranean: A discussion. In D. J. Stanley, ed., *The Mediterranean Sea: A Natural Sedimentation Laboratory.* Stroudsburg, Pa.: Dowden, Hutchinson & Ross, pp. 711–721.

Bryant, W. R., J. Antoine, M. Ewing, and B. Jones (1968). Structure of Mexican continental shelf and slope. *Am. Assoc. Pet. Geol. Bull.,* **52:** 1204–1228.

Buffington, E. C. and D. G. Moore (1963). Geophysical evidence on the origin of gullied submarine slopes, San Clemente, California. *J. Geol.,* **71:** 356–370.

Byrne, R. V. (1974). The deposition of reef-derived sediment upon a bathyal slope: The deep off-reef environment, north of Discovery Bay, Jamaica. *Mar. Geol.,* **16:** 1–20.

Cacchione, D. A. and J. B. Southard (1974). Incipient sediment movement by shoaling internal gravity waves. *J. Geophys. Res.,* **79:** 2237–2241.

Carlson, P. R. and C. H. Nelson (1969). Sediments and sedimentary structures of Astoria canyon–fan system. *J. Sediment. Petrol.,* **39:** 1269–1282.

Conolly, J. R. and M. Ewing (1967). Sedimentation in the Puerto Rico Trench. *J. Sediment. Petrol.,* **37:** 44–59.

Conolly, J. R. and C. C. von der Borch (1967). Sedimentation and physiography of the sea floor south of Australia. *Sediment. Geol.,* **1:** 181–220.

Curray, J. R. (1960). Sediments and history of Holocene transgression, continental shelf, northwest Gulf of Mexico. In F. P. Shepard, F. B. Phleger, and Tj. H. van Andel, eds., *Recent Sediments, Northwest Gulf of Mexico.* Tulsa, Okla.: Am. Assoc. Pet. Geol., pp. 221–266.

Curray, J. R. (1966). Continental terrace. In R. W. Fairbridge, ed., *The Encyclopedia of Oceanography.* New York: Reinhold, pp. 207–214.

Curray, J. R. (1969). Shallow structure of the continental margin. Lecture 12. In D. J. Stanley, ed., *The New Concepts of Continental Margin Sedimentation.* AGI Short Course Notes, Am. Geol. Inst., Washington, D.C., pp. JC-XII-1 to JC-XII-22.

Curray, J. R. and D. G. Moore (1971). Growth of the Bengal deep-sea fan and denudation in the Himalayas. *Geol. Soc. Am. Bull.,* **82:** 563–572.

Curry, R. R. (1966). Observations of alpine mudflows in the Ten Mile Range, central Colorado. *Geol. Soc. Am. Bull.,* **77:** 771–776.

Dangeard, L., M. Rioult, J. J. Blanc, and L. Blanc-Vernet (1968). Resultats de la plongée en soucoupe No. 421 dans la vallée sous-marine de Planier, au large de Marseille. *Bull. Inst. Océanogr. Monaco,* **67:** 1–22.

Dewey, J. F. and J. M. Bird (1970). Mountain belts and the new global tectonics. *J. Geophys. Res.,* **75:** 2625–2647.

Dietz, R. S. (1963). Wave-base, marine profile of equilibrium, and wave-built terraces: A critical appraisal. *Geol. Soc. Am. Bull.,* **74:** 971–990.

Dill, R. F. (1964). Contemporary submarine erosion in Scripps Submarine Canyon. Unpublished Ph.D. Thesis, Scripps Inst. of Oceanography, University of California, San Diego, 269 pp.

Dill, R. F. (1967). Effects of explosive loading on the strength of seafloor sands. In A. F. Richards, ed., *Marine Geotechnique.* Urbana: University of Illinois Press, pp. 291–299.

Dillon, W. P. and H. B. Zimmerman (1970). Erosion by biological activity in two New England submarine canyons. *J. Sediment. Petrol.,* **40:** 542–547.

Donovan, D. T., ed. (1968). *Geology of Shelf Seas.* Edinburgh: Oliver & Boyd, 160 pp.

Drake, D. E. (1971). Suspended sediment and thermal stratification in Santa Barbara channel, California. *Deep-Sea Res.,* **18:** 763–769.

Drake, D. E., R. L. Kolpack, and P. J. Fischer (1972). Comments on the dispersal of suspended sediment across the continental shelves. In D. J. P. Swift, D. B. Duane, and O. H. Pilkey, eds., *Shelf Sediment Transport: Process and Pattern.* Stroudsburg, Pa.: Dowden, Hutchinson & Ross, pp. 307–332.

Einsele, G. (1967). Sedimentary processes and physical properties of cores from the Red Sea, Gulf of Aden, and off the Nile Delta. In A. F. Richards, ed., *Marine Geotechnique.* Urbana: University of Illinois Press, pp. 154–169.

Einsele, G. and F. Werner (1968). Zusammensetzung, Gefuge und mechanische Eigenschaften rezenter Sediment vom Nildelta, Roten Meer und Golf von Aden. *Meteor Forsch.-Ergeb.,* Reihe C, **1:** 21–42.

Einsele, G. and F. Werner (1972). Sedimentary processes at the entrance Gulf of Aden/Red Sea. *Meteor Forsch.-Ergeb.,* Reihe C, **10:** 39–62.

Eittreim, S. and M. Ewing (1972). Suspended particulate matter in the deep waters of the North American Basin. In A. L. Gordon, ed., *Studies in Physical Oceanography,* Vol. 2, New York: Gordon & Breach, pp. 123–168.

Eittreim, S., M. Ewing, and E. M. Thorndike (1969). Suspended matter along the continental margin of the North American Basin. *Deep-Sea Res.,* **16:** 613–624.

Eittreim, S., A. L. Gordon, M. Ewing, E. M. Thorndike, and P. Bruchhausen (1972). The nepheloid layer and observed bottom currents in the Indian–Pacific Antarctic Sea. In A. L. Gordon, ed., *Studies in Physical Oceanography,* Vol. 2. New York: Gordon & Breach, pp. 19–35.

Ekman, V. W. (1904). On dead water. In *Norwegian North Polar Expedition 1893–1896, Scientific Results*, **5:** No. 15.

Emery, K. O. (1969a). The continental shelves. In *The Oceans, A Scientific American Book*. San Francisco: Freeman, pp. 41–52.

Emery, K. O. (1969b). Continental rises and oil potential. *Oil Gas J.*, **67:** 231–243.

Emery, K. O., B. C. Heezen, and T. D. Allan (1966). Bathymetry of the eastern Mediterranean Sea. *Deep-Sea Res.*, **13:** 173–192.

Emery, K. O. and D. A. Ross (1968). The topography and sediments of a small area of the continental slope south of Martha's Vineyard. *Deep-Sea Res.*, **15:** 415–422.

Emery, K. O. and E. Uchupi (1972). *Western North Atlantic Ocean: Topography, Rocks, Structure, Water, Life and Sediments*. Am. Assoc. Pet. Geol. Mem. 17, p. 532.

Emery, K. O., E. Uchupi, J. D. Phillips, C. O. Bowin, E. T. Bunce, and S. T. Knott (1970). Continental rise off eastern North America. *Am. Assoc. Pet. Geol. Bull.*, **54:** 44–108.

Emery, K. O. and E. F. K. Zarudski (1969). Seismic reflection profiles along the drill holes on the continental margin off Florida. *U.S. Geol. Surv. Prof. Pap. 581–A*, 1–8.

Eriksson, K. G. (1961). Granulométrie des sédiments de l'ile d'Alboran, Méditerranée occidentale. *Bull. Geol. Inst. Univ. Uppsala*, **15:** 269–284.

Eriksson, K. G. (1965). The sediment core No. 210 from the Western Mediterranean Sea. *Reports of the Swedish Deep-Sea Expedition 1947–1948*, **8:** 397–594.

Ewing, J. A. (1973). Wave-induced bottom-currents on the outer shelf. *Mar. Geol.*, **15:** M31–M35.

Ewing, M., G. Carpenter, C. Windisch, and M. Ewing (1973). Sediment distribution in the oceans: The Atlantic. *Geol. Soc. Am. Bull.*, **84:** 71–88.

Ewing, M., S. Eittreim, J. Ewing, and X. Le Pichon (1971). Sediment transport and distribution in the Argentine basin: Part 3, Nepheloid layer and processes of sedimentation. *Phys. Chem. Earth*, **8:** 49–77.

Ewing, M. and E. M. Thorndike (1965). Suspended matter in deep ocean water. *Science*, **147:** 1291–1294.

Fenner, P., G. Kelling, and D. J. Stanley (1971). Bottom currents in Wilmington submarine canyon. *Nature Phys. Sci.*, **229:** 52–54.

Field, M. E. and O. H. Pilkey (1971). Deposition of deep-sea sands: Comparison of two areas of the Carolina continental rise. *J. Sediment. Petrol.*, **41:** 526–536.

Fox, P. J., B. C. Heezen, and A. M. Harian (1968). Abyssal antidunes. *Nature*, **220:** 470–472.

Gennesseaux, M. (1966). Prospection photographique des canyons sous-marins du Var et du Paillon (Alpes-Maritimes au Moyen de la Troïka). *Rev. Géogr. Phys. Géol. Dyn.*, **8:** 3–38.

George, C. J. (1972). The role of the Aswam High Dam in changing the fisheries of the southeastern Mediterranean. In M. T. Farvar and J. P. Milton, eds., *The Careless Technology*. Garden City, N.J.: The Natural History Press, pp. 159–178.

Gerges, M. A. (1973). The damming of the Nile River and its effects on the hydrographic conditions and circulation pattern in the southeastern Mediterranean and the Suez Canal. In N. C. Hulings, ed., *Symposium on the Eastern Mediterranean*, 2 pp.

Gorsline, D. S., D. E. Drake, and P. W. Barnes (1968). Holocene sedimentation in Tanner Basin, California continental borderland. *Geol. Soc. Am. Bull.*, **79:** 659–674.

Gorsline, D. S. and K. O. Emery (1959). Turbidity-current deposits in San Pedro and Santa Monica basins off Southern California. *Geol. Soc. Am. Bull.*, **70:** 279–290.

Hagerty, R. (1974). Usefulness of spade cores for geotechnical studies and some results from the northeast Pacific. In A. L. Inderbitzen, ed., *Deep Sea Sediments*. New York: Plenum Press, pp. 169–186.

Hampton, M. A. (1972). The role of subaqueous debris flow in generating turbidity currents. *J. Sediment. Petrol.*, **42:** 775–793.

Hand, B. M. and K. O. Emery (1964). Turbidites and topography of north end of San Diego trough, California. *J. Geol.*, **72:** 526–542.

Haner, B. E. (1971). Morphology and sediments of Redondo submarine fan, Southern California. *Geol. Soc. Am. Bull.*, **82:** 2413–2432.

Hayes, D. E., L. A. Frakes, et al. (1973). Leg 28 deep-sea drilling in the southern ocean. *Geotimes*, **18** (6): 19–24.

Heezen, B. C. (1959). Natural processes contributing to mechanical failure of deep-sea cables. *Geophys. J.*, **2:** 142–163.

Heezen, B. C. and C. L. Drake (1964). Grand Banks slump. *Am. Assoc. Pet. Geol. Bull.*, **48:** 221–225.

Heezen, B. C. and M. Ewing (1955). Orleansville earthquake and turbidity currents. *Am. Assoc. Pet. Geol. Bull.*, **39:** 2505–2514.

Heezen, B. C. and C. D. Hollister (1964). Deep-sea current evidence from abyssal sediments. *Mar. Geol.*, **1:** 141–174.

Heezen, B. C. and C. D. Hollister (1971). *The Face of the Deep*. London and New York: Oxford University Press, 659 pp.

Heezen, B. C., C. D. Hollister, and W. F. Ruddiman (1966). Shaping of the continental rise by deep geostrophic contour currents. *Science*, **152:** 502–508.

Heezen, B. C. and G. L. Johnson (1969). Mediterranean undercurrent and microphysiography west of Gibraltar. *Bull. Inst. Océanogr. Monaco*, **69:** 1–95.

Heezen, B. C. and M. Tharp (1968). *Physiographic Diagram of the North Atlantic Ocean (Revised)*. Geol. Soc. Am. Chart.

Heezen, B. C., M. Tharp, and M. Ewing (1959). *The Floors of the Oceans*. Geol. Soc. Am. Spec. Pap. 65, 122 pp.

Hersey, J. B., ed. (1967). *Deep Sea Photography*. Baltimore: Johns Hopkins Press, pp. 55–67.

Hesse, R., U. von Rad, and F. Fabricius (1971). Holocene sedimentation in the Strait of Otranto between the Adriatic and Ionian seas (Mediterranean). *Mar. Geol.*, **10:** 293–355.

Hollister, C. D. and R. B. Elder (1969). Contour currents in the Weddell Sea. *Deep-Sea Res.*, **16:** 99–101.

Hollister, C. D., J. I. Ewing, et al. (1972). *Initial Reports of the Deep Sea Drilling Project*, XI. Washington, D.C.: U.S. Government Printing Office, 1077 pp.

Hollister, C. D., R. D. Flood, D. A. Johnson, P. Lonsdale, and J. B. Southard (1974). Abyssal furrows and hyperbolic echo traces in the Bahama Outer Ridge. *Geology*, **2:** 395–400.

Hollister, C. D. and B. C. Heezen (1972). Geologic effects of ocean 'bottom currents': Western North Atlantic. In A. L. Gordon, ed., *Studies in Physical Oceanography*, Vol. 2. New York: Gordon & Breach, pp. 37–66.

Hoover, R. A. and D. G. Bebout (1974). Structural and topographic control of sediment deposition, Magdalena slope, offshore northern Colombia. *Abstr. Ann. Meet., Am. Assoc. Pet. Geol., San Antonio, April 1974*, p. 46.

Horn, D. R., M. Ewing, B. M. Horn, and M. N. Delach (1971a). Turbidites of the Hatteras and Sohm abyssal plains, western North Atlantic. *Mar. Geol.*, **11:** 287–323.

Horn, D. R., M. Ewing, M. N. Delach, and B. M. Horn (1971b). Turbidites of the north east Pacific. *Sedimentology*, **16:** 55–70.

Huang, T.-C. and D. J. Stanley (1972). Western Alboran Sea: Sediment dispersal, ponding and reversal of currents. In D. J. Stanley, ed., *The Mediterranean Sea—A Natural Sedimentation Laboratory.* Stroudsburg, Pa.: Dowden, Hutchinson & Ross, pp. 521–559.

Hubert, J. F. (1964). Textural evidence for deposition of many western north Atlantic deep-sea sands by ocean bottom currents rather than turbidity currents. *J. Geol.,* **72:** 757–785.

Hunkins, K., E. M. Thorndike, and G. Mathieu (1969). Nepheloid layers and bottom currents in the Arctic Ocean. *J. Geophys. Res.,* **74:** 6995–7008.

Hutchinson, J. N. and R. K. Bhandari (1971). Undrained loading, a fundamental mechanism of mudflows and other mass movements. *Geotechnique,* **21:** 353–358.

Ichiye, T. (1968). Marine geological research and exploration. *Under Sea Technology Handbook/Directory.* Arlington, Va.: Compass Publ., 10 pp.

Inderbitzen, A. L., ed. (1974). *Deep-Sea Sediments.* New York: Plenum Press, 497 pp.

Inderbitzen, A. L. and F. Simpson (1971). Relationship between bottom topography and marine sediment properties in an area of submarine gullies. *J. Sediment. Petrol.,* **41:** 1126–1133.

Inman, D. L. and C. E. Nordstrom (1971). On the tectonic and morphologic classification of coasts. *J. Geol.,* **79:** 1–21.

Jacobs, W. B., E. M. Thorndike, and M. Ewing (1973). A comparison of suspended particulate matter from nepheloid and clear water. *Mar. Geol.,* **14:** 117–128.

Johnson, D. W. (1919). *Shore Processes and Shoreline Development.* New York: Wiley, 584 pp.

Johnson, G. L., P. Giresses, L. Dangeard, and W. H. Jahn (1971). Photographies de fonds bathyaux et abyssaux de l'Océan Atlantique entre 30°N et l'Equateur. *Bull. Bur. Res. Géol. Min., 2e Ser.,* Sect. IV, No. 1: 59–95.

Jones, E. J. W., M. Ewing, J. I. Ewing, and S. L. Eittreim (1970). Influences of Norwegian Sea overflow water on sedimentation in the northern North Atlantic and Labrador Sea. *J. Geophys. Res.,* **75:** 1655–1680.

Kash, D. E., I. L. White, K. H. Bergey, M. A. Charlock, M. D. Devine, R. L. Leonard, S. N. Salomon, and H. W. Young (1973). *Energy Under the Oceans.* Norman: University of Oklahoma Press, 378 pp.

Kassas, M. (1972). Impact of river control schemes on the shoreline of the Nile Delta. In M. T. Farvar and J. P. Milton, eds., *The Careless Technology.* Garden City, N. J.: The Natural History Press, pp. 179–188.

Keller, G. H. and D. N. Lambert (1972). Geotechnical properties of submarine sediments, Mediterranean Sea. In D. J. Stanley, ed., *The Mediterranean Sea: A Natural Sedimentation Laboratory.* Stroudsburg, Pa.: Dowden, Hutchinson & Ross, pp. 401–415.

Keller, G. H., D. Lambert, G. Rower, and N. Staresinic (1973). Bottom currents in the Hudson Canyon. *Science,* **180:** 181–183.

Keller, G. H. and A. F. Richards (1967). Sediments of the Malacca Strait, south east Asia. *J. Sediment. Petrol.,* **37:** 102–127.

Kelling, G., H. Sheng, and D. J. Stanley (1975). Mineralogic composition of sand-sized sediment on the outer margin off the Mid-Atlantic States: Assessment of the influence of the ancestral Hudson and other fluvial systems. *Geol. Soc. Am. Bull.,* **86:** 853–862.

Kelling, G. and D. J. Stanley (1970). Morphology and structure of Wilmington and Baltimore submarine canyons, eastern U.S.A. *J. Geol.,* **78:** 637–660.

Kelling, G. and D. J. Stanley (1972a). Sedimentary evidence of bottom current activity, Strait of Gibraltar region. *Mar. Geol.,* **13:** M51–M60.

Kelling, G. and D. J. Stanley (1972b). Sedimentation in the vicinity of the Strait of Gibraltar. In D. J. Stanley, ed., *The Mediterranean Sea: A Natural Sedimentation Laboratory.* Stroudsburg, Pa.: Dowden, Hutchinson & Ross, pp. 489–519.

Kelling, G. and D. J. Stanley (1975). A model for longitudinal transport within a modern multi-source basin. *Proc. IXth Inter. Congr. Sedim., Nice, 1975.* Theme V: 237–242.

Kenyon, N. and R. H. Belderson (1973). Bed forms of the Mediterranean undercurrent observed with side-scan sonar. *Sediment. Geol.,* **9:** 77–99.

Knott, S. T. and H. Hoskins (1968). Evidence of Pleistocene events in the structure of the continental shelf off the northeastern United States. *Mar. Geol.,* **6:** 5–43.

Komar, P. D. (1971). Hydraulic jumps in turbidity currents. *Geol. Soc. Am. Bull.,* **82:** 1477–1488.

Komar, P. D. (1972). Relative significance of head and body spill from a channelized turbidity current. *Geol. Soc. Am. Bull.,* **83:** 1151–1156.

Krause, D. C., W. C. White, D. J. W. Piper, and B. C. Heezen (1970). Turbidity currents and cable breaks in the western New Britain Trench. *Geol. Soc. Am. Bull.,* **81:** 2153–2160.

Lafond, E. C. (1961). The isotherm follower. *J. Mar. Res.,* **19:** 33–39.

Leclaire, L. (1972). Aspects of Late Quaternary sedimentation on the Algeria precontinent and in the adjacent Algiers-Balearic Basin. In D. J. Stanley, ed., *The Mediterranean Sea: A Natural Sedimentation Laboratory.* Stroudsburg, Pa.: Dowden, Hutchinson & Ross, pp. 561–582.

Lehner, P. (1969). Salt tectonics and Pleistocene stratigraphy of northern Gulf of Mexico. *Am. Assoc. Pet. Geol. Bull.,* **53:** 2431–2479.

Lewis, K. B. (1971a). Growth rates of folds using tilted, wave-planed surfaces: Coast and continental shelf, Hawkes Bay, New Zealand. In B. W. Collins and R. Fraser, eds., *Recent Crustal Movements.* Roy. Soc. N.Z. Bull., **9:** 225–231.

Lewis, K. B. (1971b). Slumping on a continental slope at 1°–4°. *Sedimentology,* **16:** 97–110.

Lewis, K. B. (1974). The continental terrace. *Earth Sci. Rev.,* **10:** 37–71.

Lindsay, J. F. (1968). The development of clast fabric in mud flows. *J. Sediment. Petrol.,* **38:** 1242–1253.

Lyall, A. K., D. J. Stanley, H. N. Giles, and A. Fisher, Jr. (1971). Suspended sediment and transport at the shelf-break and on the slope, Wilmington Canyon area, eastern U.S.A. *Mar. Technol. Soc. J.,* **5:** 15–27.

McGill, J. T. (1958). Map of coastal landforms of the world. *Geogr. Rev.,* **49:** 402–405.

McMaster, R. L., J. D. Milliman, and A. Ashraf (1971). Continental shelf and upper slope sediments off Portuguese Guinea, Guinea and Sierra Leone (West Africa). *J. Sediment. Petrol.,* **41:** 150–158.

Maxwell, J., ed. (1970). *The Sea,* Vol. 4 (in two parts). New York: Wiley.

Menard, H. W. (1955). Deep-sea channels, topography and sedimentation. *Am. Assoc. Pet. Geol. Bull.,* **39:** 236–255.

Menard, H. W. (1960). Possible pre-Pleistocene deep-sea fans off central California. *Geol. Soc. Am. Bull.,* **71:** 1271–1278.

Menard, H. W. (1964). *Marine Geology of the Pacific*. New York: McGraw-Hill, 271 pp.

Menard, H. W. and S. M. Smith (1966). Hypsometry of ocean basin provinces. *J. Geophys. Res.*, **71**: 4305–4325.

Menard, H. W., S. M. Smith, and R. M. Pratt (1965). The Rhône deep-sea fan. In W. F. Whittard and R. Bradshaw, eds., *Submarine Geology and Geophysics*. London: Butterworths, pp. 271–285.

Mero, J. L. (1972). Potential economic value of ocean-floor manganese nodule deposits. In D. R. Horn, ed. *Ferromanganese Deposits on the Ocean Floor*. Washington, D.C.: Natl. Sci. Found., pp. 191–203.

Middleton, G. V. (1966). Experiments on density and turbidity currents. *Can. J. Earth Sci.*, **3**: 523–546.

Middleton, G. V. (1970). Experimental studies related to problems of flysch sedimentation. In J. Lajoie, ed., *Flysch Sedimentology in North America*. Geol. Assoc. Can. Spec. Paper 7, pp. 253–272.

Middleton, G. V. and M. A. Hampton (1973). Sediment gravity flows: Mechanics of flow and deposition. In: G. V. Middleton and A. H. Bouma, eds., *Short Course, Turbidites and Deep Water Sedimentation*. Soc. Econ. Paleontologists and Mineralogists, Pacific Section, Anaheim, pp. 1–38.

Miller, A. R. (1972). Speculations concerning bottom circulation in the Mediterranean Sea. In: D. J. Stanley, ed., *The Mediterranean Sea: A Natural Sedimentation Laboratory*. Stroudsburg, Pa.: Dowden, Hutchinson & Ross, pp. 37–42.

Milliman, J. D., O. H. Pilkey, and D. A. Ross (1972). Sediments of the continental margin off the eastern United States. *Geol. Soc. Am. Bull.*, **83**: 1315–1334.

Moore, D. G. (1961). Submarine slumps. *J. Sediment. Petrol.*, **31**: 343–357.

Moore, D. G. (1966). Structure, litho-orogenic units and postorogenic basin fill by reflection profiling: California continental borderland. Doctoral Dissertation, U.S. Navy Electronics Lab., San Diego, 151 pp.

Moore, D. G. (1969). *Reflection Profiling Studies of the California Continental Borderland: Structure and Quaternary Turbidite Basins*. Geol. Soc. Am. Spec. Pap. 107, 142 pp.

Moore, D. G. and J. R. Curray (1963). Sedimentary framework of continental terrace off Norfolk, Virginia and Newport, Rhode Island. *Am. Assoc. Pet. Geol. Bull.*, **47**: 2051–2054.

Moore, D. G. and J. R. Curray (1974). Midplate continental margin geosynclines: Growth processes and Quaternary modifications. In R. H. Dott, Jr., ed., *Modern and Ancient Geosynclinal Sedimentation*. Soc. Econ. Paleontologists and Mineralogists Spec. Publ. No. 9, pp. 26–35.

Morcos, S. A. (1972). Sources of Mediterranean Intermediate Water in the Levantine Sea. In A. L. Gordon, ed., *Studies in Physical Oceanography*, Vol. 2. New York: Gordon & Breach, pp. 185–206.

Morelock, J. (1969). Shear strength and stability of continental slope deposits, western Gulf of Mexico. *J. Geophys. Res.*, **74**: 465–482.

Morgenstern, N. (1967). Submarine slumping and the initiation of turbidity currents. In A. F. Richards, ed., *Marine Geotechnique*. Urbana: University of Illinois Press, pp. 189–220.

Nelson, C. H. and L. D. Kulm (1973). Submarine fans and deep-sea channels. In G. V. Middleton and A. H. Bouma, eds., *Short Course, Turbidites and Deep Water Sedimentation*. Soc. Econ. Paleontologists and Mineralogists, Pacific Section, Anaheim, pp. 38–78.

Nelson, C. H. and T. H. Nilson (1974). Depositional trends of modern and ancient deep-sea fans. In R. H. Dott, Jr. and R. H. Shaver, eds., *Modern and Ancient Geosynclinal Sedimentation*. Soc. Econ. Paleontologists and Mineralogists Spec. Publ. No. 9, pp. 69–91.

Neumann, A. C. and M. M. Ball (1970). Submersible observations in the Straits of Florida: Geology and bottom currents. *Geol. Soc. Am. Bull.*, **81**: 2861–2874.

Normark, W. R. (1970). Growth patterns of deep-sea fans. *Am. Assoc. Pet. Geol. Bull.*, **54**: 2170–2195.

Normark, W. R. (1974). Submarine canyons and fan valleys: Factors affecting growth patterns of deep-sea fans. In R. H. Dott, Jr., ed., *Modern and Ancient Geosynclinal Sedimentation*. Soc. Econ. Paleontologists and Mineralogists Spec. Publ. No. 9, pp. 56–68.

Normark, W. R. and D. J. W. Piper (1972). Sediments and growth pattern of Navy deep-sea fan, San Clemente Basin, California Borderland. *J. Geol.*, **80**: 198–223.

Owen, D. M. and K. O. Emery (1967). Current markings on the continental slope. In J. B. Hersey, ed., *Deep-Sea Photography*. Baltimore: Johns Hopkins University Press, pp. 167–172.

Owen, D. M., H. L. Sanders, and R. R. Ressler (1967). Bottom photography as a tool for estimating benthic populations. In J. B. Hersey, ed., *Deep-Sea Photography*. Baltimore: Johns Hopkins University Press, pp. 229–234.

Pautot, G., V. Renard, J. Daniel, and J. Dupont (1973). Morphology, limits, origin and age of salt layers along the South Atlantic African margin. *Am. Assoc. Pet. Geol. Bull.*, **57**: 1658–1671.

Piper, D. J. W. (1970). Transport and deposition of Holocene sediment on La Jolla deep-sea fan, California. *Mar. Geol.*, **8**: 211–227.

Piper, D. J. W. (1972). Turbidite origin of some laminated mudstones. *Geol. Mag.*, **109**: 115–126.

Piper, D. J. W. (1973). The sedimentology of silt turbidites from the Gulf of Alaska. In L. D. Kulm et al., eds., *Initial Reports of the Deep Sea Drilling Project*, *18*. Washington, D.C.: U.S. Govt. Printing Office, pp. 847–867.

Piper, D. J. W. and N. F. Marshall (1969). Bioturbation of Holocene sediments on La Jolla deep-sea fan. *J. Sediment. Petrol.*, **39**: 601–606.

Pratt, R. M. (1967). The seaward extension of submarine canyons off the northeast coast of the United Coast. *Deep-Sea Res.*, **14**: 409–420.

Pratt, R. M. (1968). Atlantic continental shelf and slope of the United States: Physiography and sediments of the deep sea basin. *U.S. Geol. Surv. Prof. Pap. 529–B*.

Reading, H. G. (1972). Global tectonics and the genesis of flysch successions. *Proc. Int. Geol. Congr.*, *24th Session, Montreal, 1972*, Sect. 6, pp. 59–66.

Reimnitz, E. (1971). Surf-beat origin for pulsating bottom currents in the Rio Balsas submarine canyon, Mexico. *Geol. Soc. Am. Bull.*, **82**: 81–89.

Richards, A. F. (1965). Geotechnical aspects of recent marine sediments, Oslofjord, Norway (Abstract). *Am. Assoc. Pet. Geol. Bull.*, **49**: 356.

Richards, A. F. (1967). *Marine Geotechnique*. Urbana: University of Illinois Press, 327 pp.

Roberts, D. G. and A. H. Stride (1968). Late Tertiary slumping on the continental slope of southern Portugal. *Nature*, **217**: 48–50.

Rona, P. A. (1969). Linear "Lower Continental Rise Hills" off Cape Hatteras. *J. Sediment. Petrol.*, **39:** 1132–1141.

Rona, P. A. (1970a). Submarine canyon origin on the upper continental slope off Cape Hatteras. *J. Geol.*, **78:** 141–152.

Rona, P. A. (1970b). Comparison of continental margins of eastern North America at Cape Hatteras and north-western Africa at Cap Blanc. *Am. Assoc. Pet. Geol. Bull.*, **54:** 129–157.

Rona, P. A. (1971). Bathymetry off central northwest Africa. *Deep-Sea Res.*, **18:** 321–327.

Rona, P. A. and C. S. Clay (1967). Stratigraphy and structure along a continuous seismic reflection profile from Cape Hatteras, North Carolina, to the Bermuda Rise. *J. Geophys. Res.*, **72:** 2107–2130.

Rowe, G. T. (1971). Observations on bottom currents and epibenthic population in Hatteras submarine canyon. *Deep-Sea Res.*, **18:** 569–582.

Rupke, N. and D. J. Stanley (1974). Distinctive properties of turbiditic and hemipelagic mud layers in the Algéro-Balearic Basin, western Mediterranean Sea. *Smithsonian Contrib. Earth Sci.*, **13:** 40 pp.

Rupke, N., D. J. Stanley, and R. Stuckenrath (1974). Late Quaternary rates of abyssal mud deposition in the Western Mediterranean Sea. *Mar. Geol.*, **17:** M9–M16.

Rusnak, G. A., R. L. Fisher, and F. P. Shepard (1964). Bathymetry and faults of the Gulf of California. In Tj. H. van Andel and G. G. Shor, Jr., eds., *Marine Geology of the Gulf of California*. Tulsa, Okla.: Am. Assoc. Pet. Geol., pp. 59–75.

Ryan, W. B. F. and B. C. Heezen (1965). Ionian Sea submarine canyons and the 1908 Messina turbidity current. *Geol. Soc. Am. Bull.*, **76:** 915–932.

Schlee, J. (1973). Atlantic continental shelf and slope of the United States—Sediment texture of the north eastern part. *U.S. Geol. Surv. Prof. Pap. 529–L*, 64 pp.

Schlee, J. and R. M. Pratt (1970). Atlantic continental shelf and slope of the United States—Gravels of the north-eastern part. *U.S. Geol. Surv. Prof. Pap.*, *529–H*, 39 pp.

Schneider, E. D., P. J. Fox, C. D. Hollister, H. D. Needham, and B. C. Heezen (1967). Further evidence of contour currents in the western North Atlantic. *Earth Planet. Sci. Lett.*, **2:** 351–359.

Shepard, F. P. (1952). Revised nomenclature for depositional coastal features. *Am. Assoc. Pet. Geol. Bull.*, **36:** 1902–1912.

Shepard, F. P. (1955). Delta-front valleys bordering the Mississippi distributaries. *Geol. Soc. Am. Bull.*, **66:** 1489–1498.

Shepard, F. P. (1960). Mississippi Delta: Marginal environments, sediments and growth. In F. P. Shepard, F. B. Phleger, and Tj. H. van Andel, eds., *Recent Sediments, Northwest Gulf of Mexico*. Tulsa, Okla.: Am. Assoc. Pet. Geol., pp. 56–81.

Shepard, F. P. (1965). Submarine canyons explored by Cousteau's Diving Saucer. In W. F. Whittard and R. Bradshaw, eds., *Submarine Geology and Geophysics*. London: Butterworths, pp. 303–311.

Shepard, F. P. (1973). *Submarine Geology*, 3rd ed. New York: Harper & Row, 517 pp.

Shepard, F. P. and E. C. Buffington (1968). La Jolla submarine fan-valley. *Mar. Geol.*, **6:** 107–143.

Shepard, F. P. and R. F. Dill (1966). *Submarine Canyons and Other Sea Valleys*. Chicago: Rand McNally, 381 pp.

Shepard, F. P., R. F. Dill, and U. von Rad (1969). Physiography and sedimentary processes of La Jolla submarine fan and fan-valley, California. *Am. Assoc. Pet. Geol. Bull.*, **53:** 390–420.

Shepard, F. P. and N. F. Marshall (1969). Currents in La Jolla and Scripps submarine canyons. *Science*, **165:** 177–178.

Shepard, F. P. and N. F. Marshall (1973a). Storm-generated current in La Jolla submarine canyon, California. *Mar. Geol.*, **15:** M19–M24.

Shepard, F. P. and N. F. Marshall (1973b). Currents along floors of submarine canyons. *Am. Assoc. Pet. Geol. Bull.*, **57:** 244–264.

Shepard, F. P., N. F. Marshall, and P. A. McLoughlin (1974). "Internal waves" advancing along submarine canyons. *Science*, **183:** 195–198.

Simpson, F., A. L. Inderbitzen, and A. Singh (1974). Initial penetration and settlement of concrete blocks into deep-ocean sediments. In A. L. Inderbitzen, ed., *Deep-Sea Sediments*. New York: Plenum Press, pp. 303–326.

Solzansky, V. (1973). Origin of salt deposits in deep-water basins of Atlantic Ocean. *Am. Assoc. Pet. Geol. Bull.*, **57:** 589–590.

Southard, J. B. and D. A. Cacchione (1972). Experiments on bottom sediment movement by breaking internal waves. In D. J. P. Swift, D. B. Duane, and O. H. Pilkey, eds., *Shelf Sediment Transport: Process and Patterns*. Stroudsburg, Pa.: Dowden, Hutchinson & Ross, pp. 83–97.

Stanley, D. J. (1967). Comparing patterns of sedimentation in some modern and ancient submarine canyons. *Earth Planet. Sci. Lett.*, **3:** 371–380.

Stanley, D. J. (1969). Sedimentation in slope and base-of-slope environments (Lecture 8). In D. J. Stanley, ed., *The New Concepts of Continental Margin Sedimentation*. AGI Short Course Notes, Am. Geol. Inst., Washington, D.C., pp. DJS-8-1 to DJS-8-25.

Stanley, D. J. (1970). Flyschoid sedimentation on the outer Atlantic margin off northeast North America. In J. Lajoie, ed., *Flysch Sedimentology in North America*. Geol. Assoc. Can. Spec. Pap. 7, pp. 179–210.

Stanley, D. J. (1971a). Fish-produced markings on the Atlantic continental margin off north-central United States. *J. Sediment. Petrol.*, **42:** 159–170.

Stanley, D. J. (1971b). Bioturbation and sediment failure in some submarine canyons. *Vie et Milieu, Suppl.*, **22:** 541–555.

Stanley, D. J. (1974). Pebbly mud transport in the head of Wilmington Canyon. *Mar. Geol.*, **16:** M1–M8.

Stanley, D. J. and A. E. Cok (1968). Sediment transport by ice on the Nova Scotian shelf. In *Ocean Sciences and Engineering of the Atlantic Shelf*. Philadelphia: Mar. Technol. Soc., pp. 109–125.

Stanley, D. J., P. Fenner, and G. Kelling (1972a). Currents and sediment transport at the Wilmington Canyon shelfbreak, as observed by underwater television. In D. J. P. Swift, D. B. Duane, and O. H. Pilkey, eds., *Shelf Sediment Transport: Process and Pattern*. Stroudsburg, Pa.: Dowden, Hutchinson & Ross, pp. 621–644.

Stanley, D. J., H. Got, O. Leenhardt, and Y. Weiler (1974a). Subsidence of the western Mediterranean Basin in Pliocene-Quaternary time: Further evidence. *Geology*, **2:** 345–350.

Stanley, D. J. and G. Kelling (1968). Sedimentation patterns in the Wilmington Submarine Canyon area. In *Ocean Sciences and Engineering of the Atlantic Shelf*. Philadelphia: Mar. Technol. Soc., pp. 127–142.

Stanley, D. J. and G. Kelling (1969). Photographic investigation of sediment texture, bottom current activity, and benthonic organisms in the Wilmington Submarine Canyon. *U.S. Coast Guard Oceanogr. Rep.*, **22:** 1–95.

Stanley, D. J. and G. Kelling (1970). Interpretation of a levee-like ridge and associated features, Wilmington submarine canyon, eastern United States. *Geol. Soc. Am. Bull.*, **81:** 3747–3752.

Stanley, D. J., G. Kelling, J.-A. Vera, and H. Sheng (1975). Sands in the Alboran Sea: A model of input in a deep marine basin. *Smithsonian Contrib. Earth Sci.*, **15:** 1–50.

Stanley, D. J., F. W. McCoy, and L. Diester-Haass (1974b). Balearic abyssal plain: An example of modern basin deformation by salt tectonism. *Mar. Geol.*, **17:** 183–200.

Stanley, D. J., H. Sheng, and C. P. Pedraza (1971). Lower continental rise east of Middle Atlantic States: Predominant dispersal perpendicular to isobaths. *Geol. Soc. Am. Bull.*, **82:** 1831–1840.

Stanley, D. J. and N. Silverberg (1968). Recent slumping on the continental slope off Sable Island Bank, southeast Canada. *Earth Planet. Sci. Lett.*, **6:** 123–133.

Stanley, D. J., D. J. P. Swift, N. Silverberg, N. P. James, and R. G. Sutton (1972b). Late Quaternary progradation and sand spillover on the outer continental margin off Nova Scotia, southeast Canada. *Smithsonian Contrib. Earth Sci.*, **8:** 1–88.

Stommel, H., A. Voorhis, and D. Webb (1971). Submarine clouds in the deep ocean. *Am. Sci.*, **59:** 716–722.

Stride, A. H., J. R. Curray, D. G. Moore, and R. H. Belderson (1969). Marine geology of the Atlantic continental margin of Europe. *Phil. Trans. Roy. Soc. London*, **A264:** 31–75.

Sundborg, A. (1956). The river Kläralven—A study of fluvial processes. *Geogr. Ann.*, **38:** 127–316.

Sverdrup, H. U., M. W. Johnson, and R. H. Fleming (1942). *The Oceans, Their Physics, Chemistry and General Biology.* Englewood Cliffs, N.J.: Prentice-Hall, 1087 pp.

Swallow, J. C. and L. V. Worthington (1961). An observation of a deep countercurrent in the western North Atlantic. *Deep-Sea Res.*, **8:** 1–9.

Swift, D. J. P., D. J. Stanley, and J. R. Curray (1971). Relict sediments on continental shelves: A reconsideration. *J. Geol.*, **79:** 322–346.

Tooms, J. S., C. P. Summerhayes, and R. L. McMaster (1971). Marine geological studies on the north-west African margin. In F. M. Delany, ed., *The Geology of the East Atlantic Margin*, Part 4. London: Inst. Geol. Sci., pp. 9–26.

Trumbull, J. V. A. and M. J. McCamis (1967). Geological exploration in an East Coast submarine canyon from a research submersible. *Science*, **158:** 370–372.

Uchupi, E. (1963). Sediments on the continental margin off the eastern United States. *U.S. Geol. Surv. Prof. Pap. 475–C*, 132–137.

Uchupi, E. (1967a). Bathymetry of the Gulf of Mexico. *Trans. Gulf Coast Assoc. Geol. Soc., 17th Ann. Meet.*, pp. 161–172.

Uchupi, E. (1967b). Slumping on the continental margin southeast of Long Island. *Deep-Sea Res.*, **14:** 635–639.

Uchupi, E. (1968). Atlantic continental shelf and slope of the United States—Physiography. *U.S. Geol. Surv. Prof. Pap. 529–C*, 1–30.

Uchupi, E. and K. O. Emery (1963). The continental slope between San Francisco, California, and Cedros Island, Mexico. *Deep-Sea Res.*, **10:** 397–447.

Uchupi, E. and K. O. Emery (1967). Structure of the continental margin off Atlantic Coast of the United States. *Am. Assoc. Pet. Geol. Bull.*, **51:** 223–234.

Uchupi, E. and K. O. Emery (1968). Structure of continental margin off Gulf Coast of United States. *Am. Assoc. Pet. Geol. Bull.*, **52:** 1162–1193.

Venkatarathnam, K. and W. B. F. Ryan (1971). Dispersal patterns of clay minerals in the sediments of the eastern Mediterranean Sea. *Mar. Geol.*, **11:** 261–282.

Volkmann, G. (1962). Deep current observations in the western North Atlantic. *Deep-Sea Res.*, **9:** 493–500.

Voorhis, A. D. (1968). Measurements of vertical motion and the partition of energy in the New England slope water. *Deep-Sea Res.*, **15:** 599–608.

Walker, R. G. (1967). Turbidite sedimentary structures and their relationship to proximal and distal depositional environments. *J. Sediment. Petrol.*, **37:** 25–43.

Walker, R. G. (1973). Mopping up the turbidite mess. In R. N. Ginsburg, ed., *Evolving Concepts in Sedimentology.* Baltimore: Johns Hopkins University Press, pp. 1–37.

Warme, J. E., T. B. Scanland, and N. F. Marshall (1971). Submarine canyon erosion: Contribution of marine rock burrowers. *Science*, **173:** 1127–1129.

Worthington, E. B. (1972). The Nile catchment-technological change and aquatic biology. In M. T. Farvar and J. P. Milton, eds., *The Careless Technology.* Garden City: The Natural History Press, pp. 189–205.

Wunsch, C. (1968). On the propagation of internal waves up a slope. *Deep-Sea Res.*, **15:** 251–258.

Wüst, G. (1961). On the vertical circulation of the Mediterranean Sea. *J. Geophys. Res.*, **66:** 3261–3271.

Yerkes, R. F., D. S. Gorsline, and G. A. Rusnak (1967). Origin of Redondo submarine canyon, Southern California. *U.S. Geol. Surv. Prof. Pap. 575–C*, C97–C105.

CHAPTER 18

Suspended Sediment Transport at the Shelf Break and Over the Outer Margin

J. W. PIERCE

Division of Sedimentology, Smithsonian Institution, Washington, D.C.

The two preceding chapters have defined the physical processes active at the shelf break which result in sediment entrainment and transport, including offshelf spillover, of sand-sized sediment onto the slope and beyond. The same processes result in resuspending silt- and clay-sized sediment or preventing its deposition. These finer particles tend not to accumulate permanently at the shelf break and are ultimately moved from terrestrial sources into deeper water environments by various suspension-transport processes. This chapter deals with the movement of finer (cohesive) sediments over the outer continental margin.

The boundary between predominantly sand and silty clay on the sea floor surface, called the "mud line" by several authors, lies parallel or subparallel and generally slightly seaward of the shelf break so that the continental slope and rise are areas of predominantly mud deposits. The processes giving rise to these fine-grained accumulations have been interpreted from the sediment record, often with conflicting results (Stanley et al., 1971, 1972b; Field and Pilkey, 1971; Pilkey and Field, 1972). The role and relative importance, in shaping the outer continental margin, of downcanyon and downslope transport processes (cf. discussion in Kelling and Stanley, Chapter 17) versus deposition from dilute (or not so dilute) suspensions, excepting turbidity currents, are still the subject of debate. An understanding of the relative importance of these processes is funda-mental to development of theories on formation of continental rise and base-of-slope sequences.

An evaluation of the relative roles of these processes is hindered by several problems. At present, data are lacking in quantity, are usually the result of one-time sampling, and are in several forms which may not be correlatable. Suspensates are a nonconservative (or, at best, only semiconservative) property of water masses. Because of this property, concentrations change in time and it will be years, if ever, before the lateral distribution of suspended materials can be mapped in the manner now done for bottom sediments. Time series and synoptic data gathering would greatly assist in providing some of the much needed information on the persistence of currents, resuspension, gravitational settling, and depositional rates. All in all, more could be done over the outer margins with present technology, but the logistics are infinitely more complex than obtaining data in estuaries or on continental shelves. Suspended sediment transport in the latter environments is discussed in Chapter 9.

NATURE AND DISTRIBUTION OF DATA

Data can be reported as mass per unit volume, resulting from weighed samples, in analog form as the result of the effect of suspensates on a light beam, as counts of particles per unit volume, or as transformations of the above mentioned data.

Published data on the distribution of suspended particulates indicate that the four environments (outer shelf, slope, rise, abyssal) have been studied about equally (Table 1). The number of papers counted is far from all inclusive: It does not include the very large number of Russian papers and is limited primarily to major Western publications, with a few additions from Japanese and Russian papers, either published in English or with an English summary. It is doubtful, though, that the total number of published reports is more than four times the number reported here.

TABLE 1.   Number of Published Papers on Suspended Sediment for Major Environments*

	Number of Papers	Analog Data Only	Surface Samples Only	Concentration Data Reported for Subsurface Waters
Abyssal	25	9	6	12
Rise	18	6	4	8
Slope	22	3	7	13
Outer shelf	24	0	6	18

* Columns give the number of papers and type of data format.

Papers in which concentration values are given for subsurface waters make up a much higher percentage of the total for the outer shelf and rise environments than for waters over the abyssal environment. Conversely, analog representation of concentrations alone constitutes a higher percentage for the abyssal environment. Thus, for deep water, there are very few published data on the quantitative distribution of suspensates. This paucity of data from deep water is due in no small part to the logistics of obtaining sufficiently large water samples which would permit determination of the mass of the suspended particles.

Information from the deeper water masses is often in the form of analog data: the effect of the suspensates on transmitted or scattered light. Considerable difficulty is encountered in transforming light-scattering or transmission data to mass concentration (Jerlov, 1953; Drake, 1971; Ewing et al., 1971). Neither light transmission (attenuation) nor light scattering has a direct relation to the mass of material suspended in the water.

Scattering is strongly affected by surface area. Hence small particles, which are present in the greatest number, contribute immensely to surface area but little to the mass, and scatter relatively more light than larger particles. Transmission of light is affected by absorption, scattering, and the size spectrum of the suspended particles. Dissolved substances are strong absorbers in certain wavelengths; light scattered from the direct path is lost; and long wavelengths of light may not be affected by small particles. Some of these problems can be alleviated, or at least reduced, by using monochromatic sources, collimated beams, and/or polarization.

SUSPENSIONS AT THE SHELF BREAK

In whatever manner fine material escapes from the nearshore zone and is transported across the shelf, the distribution of suspensates in time and space over the slope and rise is far from uniform. The distribution is largely dependent on currents, waves (surface or internal), input at the coastline, width of the shelf, and general circulation patterns of water masses as well as the nonconservative nature of the suspensions.

Material can be carried to the shelf break by advective transport, especially where the shelf is relatively narrow. This is especially pronounced off major rivers that have built deltas across the shelf, e.g., Irrawaddy (Rodolfo, 1969), Mississippi (Scruton and Moore, 1953), Nile (Jerlov, 1953; Emelyanov and Shimkus, 1972), and Orinoco (Nota, 1958). Even in these areas, rapid settling occurs. In relatively short distances (a few tens of kilometers), the surface waters are relatively free of river particles (Jerlov, 1953) whereas nearbottom waters retain relatively high concentrations of suspensates past the shelf break (Emelyanov and Shimkus, 1972). Photographs from space or remote-sensing imagery of turbid plumes crossing the shelf, such as the famous Apollo IX photograph of Cape Hatteras, North Carolina (Emery, 1969), indicate advective transport to the shelf break (see also Chapter 16, Fig. 26). Shelf circulation can move material to the outer edge of the shelf (Fig. 1). The path of particles that cross the shelf by diffusive processes is less clear. Both Schubel and Okubo (1972) and McCave (1972) argue for diffusion to account for some bypassing of the shelf and transport of suspended material to the shelf edge.

Resuspension of fine-grained sediment, deposited on the inner shelf as a result of floods, and seaward transport by geostrophic currents are well documented off southern California (Drake et al., 1972); see page 147. Reworking of bottom sediments and consequent resuspension of the fines by waves and currents can produce concentrations observed in near-bottom waters at the shelf break, even in areas where very little fine material is present in the bottom sediments. For example, there are 125 grams of silt and clay in the upper 1 cm of a m²

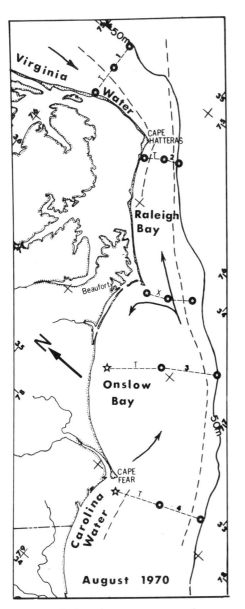

FIGURE 1. *Circulation of water masses and transport of suspended material on shelf off southeastern United States, March 1971. Transport occurs to the shelf edge at several places. Dashed lines indicate boundaries between water masses. From Pierce et al. (1972).*

Surface Water

The amount of suspended material in surface waters at the shelf break is highly variable and dependent, in part, on biologic productivity. Up to 90% of the suspensates in the surface waters at the shelf break may consist of organic compounds that are either combustible or oxidizable. A large portion of the remaining material consists of the hard parts of various organisms (pteropods, coccolithophorids, diatoms, radiolaria, and foraminifera), either whole or comminuted. Concentrations range from less than 0.125 mg/l to nearly 1.45 mg/l off the Atlantic coast of the United States (Manheim et al., 1970; Buss and Rodolfo, 1972; Pierce et al., 1972). Similar values, less than 0.8 mg/l, have been reported at the edge of the Santa Barbara, California shelf (Drake et al., 1972) and in the eastern Gulf of Mexico (Manheim et al., 1972; Harris, 1972).

Someone unfamiliar with such concentrations may feel lost with such numbers. Two examples may be cited to place these in perspective. A concentration of 1 mg/l could also be reported as 1 ppm. Ocean salinities are reported as parts per thousand. Thus, where a concentration of suspended material is reported as 0.5 mg/l in normal seawater, there will be about 35 grams of dissolved salts and 0.0005 gram of suspensates in 1 liter of water.

Pipet analysis is a common procedure for analyses of the size distribution of fine sediments. Folk (1961) recommends about 15 grams of silt and clay be used in 1 liter of water. This is a concentration of 15,000 mg/l. Obviously, the amount of material suspended in a liter of marine waters is a rather small mass. This meager amount has given rise to the many problems of adequately sampling the material.

Concentrations in the waters over the outer shelf are higher off major rivers and may undergo seasonal changes. In the Gulf of Mexico, concentration contours are shifted seaward off the Mississippi River, where more than 1 mg of material is suspended in a liter of outer shelf water (Manheim et al., 1972; Harris, 1972). Lisitsyn (1966) found 3 to 5 mg/l of suspensates in the western Bering Sea off the Anadyr River in the spring and 2 to 3 mg/l in the fall. The continental shelves off the Mississippi and in the western Bering Sea are relatively narrow.

Material can also be transported in suspension across wide shelves, depending on river discharge, sedimentation rates, and currents. Emery et al. (1969) report over 5 mg/l of total suspended material at the shelf edge of the East China Sea, approximately 550 km off the Yellow and Yangtze rivers. Half of this (2.5 mg/l) is believed to be of terrigenous origin. Concentrations at

of the seafloor which has 1% silt and clay, 50% void space, and an average grain density of 2.5 g/cm³. If all of the silt and clay particles were resuspended into the lower 5 m of the water column, a concentration of 25 mg/l would result. This value is far higher than that found over the outer shelf.

Diffusion, currents (Pierce et al., 1972), cascading (McCave, 1972, Fig. 96; Nelson et al., 1973), or movement of thermal fronts (Lyall et al., 1971) can then carry this suspended material out over the slope.

the shelf break approximately 350 km off the Mekong River are between 0.25 and 0.5 mg/l (Parke et al., 1971), over an order of magnitude less than off the Yellow and Yangtze. These marked differences may be due to the amount of material contributed by the rivers (40,600 mg/l by Yellow–Yangtze; 487 mg/l, Mekong); to control by currents as the effluent from the Mekong appears to be swept into the Gulf of Thailand (Parke et al., 1971) and that from the Yellow–Yangtze toward the shelf (Emery et al., 1969); or to different depositional rates in the two areas.

The concentrations reported are generally total suspended load, which includes material of both terrigenous and biologic origin. Rivers, in addition to the suspended terrigenous sediment load, discharge large amounts of dissolved nutrients, which stimulate phytoplankton production and give rise to a large and varied plankton community. Thus, the suspended particulate material off the mouths of rivers include large amounts of biogenic material, both skeletal and soft.

Increases in concentrations in the surface waters at the shelf edge may also be due in part to wave conditions (erosion and/or resuspension). Drake et al. (1972) found increased turbidity in the waters off California due to resuspension. Waves associated with storms stir the bottom sediments to considerable depths of water and cause resuspension of large amounts of material. Off North Carolina, it was found that concentrations of suspensates were more than doubled over prestorm values by the passage of a hurricane (Rodolfo et al., 1971). Within a week, concentration values had returned to prehurricane values. Unfortunately, ship time was not available to trace the diffusion of the material out of the area. The very basic question remains: Was the fine material redeposited or was it removed from the study area? It is extremely difficult to schedule hurricanes and ships to coincide in time; such circumstances arise only by happenstance. If the material off North Carolina followed the patterns observed by Drake et al. (1972) off California, the fines in the surface layers would follow the current patterns, and so in this case would be swept northeastward by the Gulf Stream. Diffusion of the material suspended lower in the water column is speculative although some probably was redeposited, some may have spilled over the edge of the shelf, and some may have been entrained in the Gulf Stream.

Subsurface Water

Distribution of suspended material below the surface is highly variable. Turbid layers may be interposed between layers of clearer water (Drake et al., 1972). High concentrations may also be present near the seafloor (Jerlov, 1953; Drake et al., 1972; Emelyanov and Shimkus, 1972).

Concentrations near the bottom at the shelf edge off California ranged from 2 mg/l to nearly 10 mg/l (Drake et al., 1972). Off the middle and southern U.S. Atlantic seaboards, near-bottom concentrations range from 0.06 to 4.37 mg/l (Smithsonian Institution, unpublished data). These changes are highly weather dependent, generally higher in the winter during the larger wave regime, and lower in the summer and fall. Mineral matter makes up from 33 to 100% of this material, averaging 79%.

Turbid layers within the water column have been reported by Jerlov (1953), Bouma et al. (1969), and Drake et al. (1972). These turbid layers are located at pycnoclines (density interfaces or zones of vertical variation in density). It appears that the turbid layers off California are formed by detachment of a part of a near-bottom turbid layer at density interfaces and lateral spreading (see Fig. 29 of Chapter 15). It has been suggested that these midwater turbid layers may be the result of material settling through the water column and detention at a pycnocline; it so, all such interfaces should be associated with increases in turbidity. Such is not the case as pycnoclines exist that have no attendant increase in turbidity. Movement of the material seaward and downward is not a simple case of diffusion and gravitational settling.

SEAWARD DISPERSION

It should be stressed that lateral dispersion of suspended material is wholly dependent on movement of the water mass in which the material is entrained. Vertical movements are controlled by oceanic circulation and eddy diffusion with the added complication of gravitational settling of the particles. Transport of suspended material into deeper water is controlled largely by the salinity and/or temperature differences within the water column. Seldom, if ever, are concentrations of suspended material high enough to affect the density of the water to any great extent (Drake, 1971; Eittreim and Ewing, 1972), although such has been suggested (Bouma et al., 1969). Concentrations approaching 100 mg/l would be necessary to overcome density differences imposed by observed temperature changes off California (Drake et al., 1972). Density differences in deep casts off the east coast of the United States, brought about by both salinity and temperature changes, range from 0.00039 to 0.00172 g/cm^3. These density differences require concentrations of 390 to 1720 mg/l in order for the added mass of suspended material to overcome the changes imposed by the salinity and temperature. Houk and Green (1973), in laboratory experiments with distilled water, used concentrations from 20 to 650 mg/l to determine formation and descent rate of suspension fingers; these concentrations are far in excess of what is found in nature. In the presence of a strong halocline (change in salt concentration) or thermocline, it is doubtful if the mass of suspended material is ever sufficient to overcome the density differences imposed by temperature and salinity changes between water masses. In a weakly stratified system, the concentrations necessary to affect the density significantly would be much lower. On the other hand, in a weakly stratified system would the material be held at the interface long enough to accumulate sufficient material?

Any model for predicting the routes of suspended particulates from the shelf to the deep sea must be developed with the guidelines given above. Four models have been suggested (Fig. 2). These are not mutually exclusive and show considerable overlap. Other combinations of processes are possible.

Moore (1969) suggested that excess density imparted by the suspended material was sufficient to maintain gravity flow near the bottom (Fig. 2A). It does not seem possible, as pointed out by Drake et al. (1972), that such a process is possible under conditions generally prevailing in the ocean.

Drake et al. (1972) developed the idea of transport of a near-bottom turbid zone by currents on the shelf with detachment of the turbid zone at density changes in the water column (Fig. 2D). This detachment would occur whenever the water in which the particles are suspended encounters denser water. Detached turbid layers are present in canyons off Southern California (see Fig. 10 of Chapter 9).

Cascading (Nelson et al., 1973) does not differ greatly from the previous model except that only one zone of increased turbidity exists within the water mass (Fig. 2C). This mechanism could move nearshore waters and the associated suspensates seaward, given the proper conditions of rapid cooling. No resuspension is necessary during cascading although some may occur. Pak (1974) shows one turbid layer off the west coast of South America.

Advection and diffusion may cause bypassing of the shelf (Schubel and Okubo, 1972). The particles that bypass the shelf settle through the water column to accumulate at the uppermost pycnocline (Fig. 2D). Renewed settling occurs only to have the particles collect at the next lower density discontinuity. This

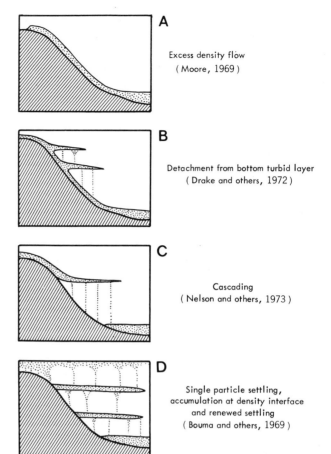

A

Excess density flow
(Moore, 1969)

B

Detachment from bottom turbid layer
(Drake and others, 1972)

C

Cascading
(Nelson and others, 1973)

D

Single particle settling,
accumulation at density interface
and renewed settling
(Bouma and others, 1969)

FIGURE 2. *Schematic models suggested for transport of suspended material from the continental shelf into the deep sea.*

interrupted settling continues until the particles reach the bottom. Alternatively, in the absence of density stratification, settling will continue uninterrupted to the bottom.

Settling of the particles through the pycnoclines presumably occurs and the particles either form another turbid layer at the next lower pycnocline, become incorporated into a near-bottom turbid layer (nepheloid layer), or are deposited. Renewed settling may occur after a finite time at the density interface and after aggregation of small particles into larger ones (see the section on deposition from suspension). Deposition does not imply a permanent site inasmuch as resuspension may occur.

Eventually, the material must approach the seafloor, where it is either deposited or becomes incorporated into the nepheloid layer, a term applied to the near-bottom turbid layer in the deep ocean (Ewing and Thorndike, 1965).

If settling through the water occurs and the material is incorporated into the near-bottom turbid layer, what happens to it? The transport of the suspensates will be dictated by the general water circulation. If the near-bottom circulation is directed offshore or down submarine canyons, then the material will be removed from the shelf system. On the other hand, the shelf circulation system may be such that there is an onshore drift of near-bottom water and sediment would be transported back toward the coast (Meade, 1969), where some of it could be reentrained in the upper waters by turbulence.

Submarine canyons can act as conduits for material from the shelf to the continental rise. Some of this material may move as a true suspension rather than in turbidity currents. Channelized currents, biologic reworking, or minor slumping can produce large volumes of suspended material. General downcanyon currents (Shepard and Marshall, 1969; 1973; Stanley et al., 1972a, Stanley, 1974) would eventually move this material onto the rise and into the nepheloid layer. Although upcanyon currents may prevail over relatively long periods, the net movement is believed to be downcanyon (Keller et al., 1973). Replenishment of the reservoir of fine sediment in the canyon heads would occur by reworking of shelf deposits (Lyall et al., 1971) and/or by diffusion across shelf into the canyon head, which may act as a sump for suspended material (Beer and Gorsline, 1971).

Some material may be returned to the outer shelf by upslope or upcanyon currents or by upwelling. Although the processes involved in such movement by water masses are well established, it is doubtful that the mass of sediment so transported is very large. Eventually, the forces of gravity will prevail and move the material downslope.

Upwelling of deep waters is most common off the western coasts of land masses (eastern edge of ocean basins). Few samples of sediment have been acquired in areas of major upwelling, most sampling having been accomplished for the purpose of measuring nutrients and productivity or biomass. Ewing and Connary (1970) show that near-bottom turbid layers rise to within 1000 m of the surface over the continental slope off northern California. They attribute the shallow depth of this layer to input from the Columbia River. If the shallow depth to the nepheloid layer occurred in areas of intense upwelling, it would not be unreasonable to expect some of the suspended sediment to be moved onto the shelf.

Upwelling is not as common along the western margins of the ocean basins but is known to occur. Slope waters encroach onto the shelf during periods when a mass deficiency occurs on the shelf, as with strong offshore winds (Blanton, 1971; Stefansson et al., 1971). Our data (unpublished) indicate that when these deeper waters encroach upon the shelf, sediments suspended in the water are carried onto the shelf and may be deposited, albeit only temporarily.

NEPHELOID LAYER

Jerlov (1953) was probably one of the first to discuss extensively the suspended material in the deep layers of the oceans. He suggested that particle content was, in many cases, a unique parameter of many water masses and might be useful as a tracer. Over long distances, the nonconservative nature of the suspensates creates problems because of settling of material to the bottom or into lower water masses and introduction of new material from above.

Equipment developed over the past few years permits in situ "analog" measurement of concentrations of suspended material in a column of water. Nephelometers measure the scattering of light; transmissometers measure attenuation of a light beam over a measured path length. These devices (primarily the nephelometer) have shown the existence of turbid layers near the seafloor over extensive areas of the ocean at depths greater than the shelf break. This near-bottom turbid water in deep water has become known as the nepheloid layer, although other turbid zones may exist within the water column in certain areas and could be rightly termed nepheloid layers.

The largest amount of information available on the nepheloid layer has been obtained by scientists at Lamont-Doherty Geological Observatory (Ewing and Thorndike, 1965; Eittreim et al., 1969, 1972; Ewing and Connary, 1970; Ewing et al., 1971; Connary and Ewing, 1972; Eittreim and Ewing, 1972; Jacobs et al., 1973).

Shaping of some continental rises has been attributed to western boundary bottom currents (e.g., Heezen et al., 1966; Ewing et al., 1971; Field and Pilkey, 1971; Hollister and Heezen, 1972). Deposition of material from the nepheloid layer is believed to account for a large part of the sediment on the lower continetal slope and in the continental rise (e.g., Eittreim and Ewing, 1972; Hollister and Heezen, 1972). Although the nepheloid layer is, for the most part, over the abyssal plains, its importance, along with the bottom currents, in sedimentation on the lower continental margin warrants treatment here.

Deep Ocean Circulation

Localization of the nepheloid layer in the ocean is dependent on the general circulation of deep ocean waters. A nepheloid layer over the continental rise has been reported from the eastern margins of the Americas, associated with the western boundary bottom currents; off the western coast of the Americas, associated with river effluent or resuspension of bottom sediments; and along the west coast of central Africa. It is also present in much of the abyssal areas of the North Pacific and Indian oceans. Although content and concentration change from locale to locale, the nepheloid layer appears to be nearly worldwide in extent.

The nepheloid layers, being near-bottom features, are associated with the dense, cold water masses found in the deeper parts of the ocean basins. These waters originate at the surface in high latitudes. Moving away from the polar areas, they sink below the lighter waters that arise in the lower latitudes. These cold, dense water masses can be traced for great distances, using parameters such as salinity and temperature.

In the southern hemisphere, cold water from the environs of Antarctica sinks below the surface at the Antarctic convergence. This cold dense water flows northward as an identifiable water mass, the Antarctic Bottom Water (AABW). The Coriolis effect on the moving water controls the path to a great extent with the flow patterns being further modified by continents, island groups, and submarine topographic features.

This water has been traced northward to near southern Australia in the Indian–Pacific Antarctic Sea (Eittreim et al., 1972); to the Bermuda area in the North Atlantic (Hollister and Heezen, 1972); in the Guinea and Angola basins of the eastern equatorial Atlantic; and north of Hawaii in the North Pacific (Gordon and Gerard, 1970). Ewing and Connary (1970) suggest that the AABW may also reach the Aleutian Trench.

Cold water in the North Atlantic originates in the Norwegian Sea or Labrador Sea (Jones et al., 1970). As it flows south, it plunges beneath the warmer surface water. The Western Boundary Undercurrent is part of this cold water mass. This bottom current has been reported as far south as the Antilles (McCoy, 1969), although the water mass itself (North Atlantic Deep Water) has been traced, at intermediate depths, to the Antarctic region and into the Pacific and Indian oceans.

Thickness and Distribution of the Nepheloid Layer

A near-bottom turbid layer is persistent over much of the North Pacific and is associated with the AABW. In the Central Pacific, the top of the nepheloid layer is at approximately 4150 m and light-scattering intensity increases to the bottom (Ewing and Connary, 1970). Although the thickness of the layer is over 1700 m in places, the scattering intensity is not strong in the

Central Pacific. Plank et al. (1972) failed to detect a nepheloid layer in the North Central Pacific, although they did in the Aleutian Trench area. Nephelometer readings at the same station, taken three years apart, indicate a remarkable consistency in depth to the top of the layer and in light-scattering ability.

In the northeastern Pacific, the top of the nepheloid layer is at about 4150 m in the abyssal area and rises to less than 2500 m off the tip of Baja California (Ewing and Connary, 1970). Off the northwestern United States, the top of the layer rises to a depth of less than 1700 m.

The turbid zone between Antarctica and Australia is at depths of 2 to 4 km, becoming deeper to the north (Eittreim et al., 1972). Thickness averages about 1 km.

The nepheloid layer in the Argentine Basin is between 720 and 2200 m thick (Ewing et al., 1971). It covers most of the Argentine Basin and laps onto the continental rise of Argentina. The layer has a consistent light-scattering ability from north to south throughout the basin. Some of the material associated with the nepheloid layer is undoubtedly transported north out of the Argentine Basin as the AABW passes through the Vema Gap in the Rio Grande Rise. The AABW in the North Atlantic is relatively clear and does not carry as great a suspended load as in the South Atlantic. This is believed due to lack of access to source areas which would replace material that is deposited. The Amazon River may be the northernmost source of material to the AABW.

The nepheloid layer in the North Atlantic is associated with the Western Boundary Undercurrent, part of the North Atlantic Deep Water. The water mass is displaced to the western edge of the basin as it flows south; the eastern part of the basin is occupied by AABW, moving north. The thickness varies from a few hundred meters to more than a kilometer (Fig. 3). This layer and its effect on the rise are discussed in detail later.

Concentrations

Concentrations of material in the near-bottom nepheloid layer are much higher than those of overlying waters but still considerably less than those found in estuarine and nearshore waters. Concentrations are reported on weight/volume, the conventional manner; on counts/ volume, the easiest and with the least error in very dilute concentrations; or area/volume (Eittreim and Ewing, 1972).

In the North Atlantic, concentrations are on the order of 0.01 to 0.1 mg/l when volumes are converted to weight, assuming a density of 2 g/cm^3 (Eittreim and

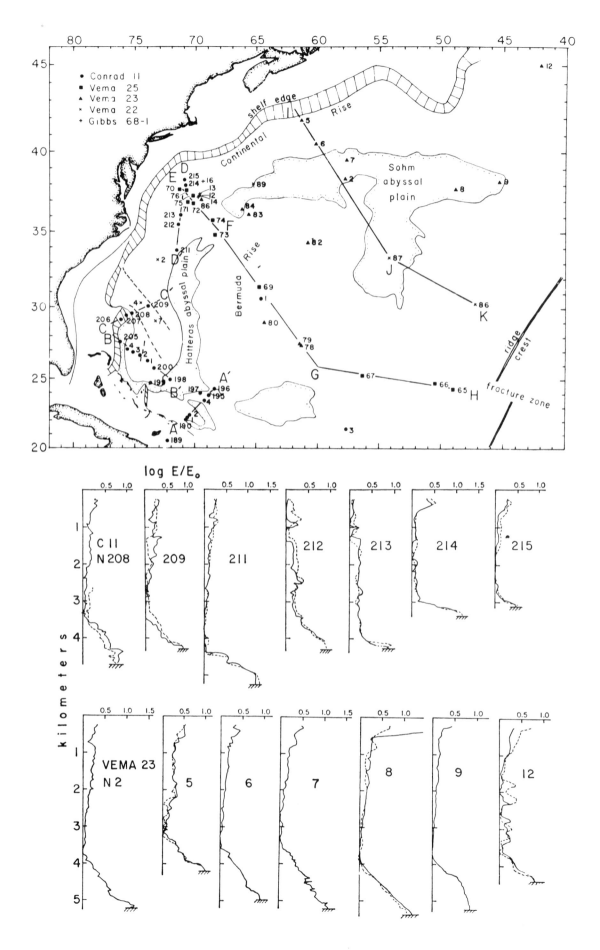

Ewing, 1972). Recent samples (August 1974) off the coast of Maine had 0.24 mg/l total material of which 0.23 mg/l was nonoxidizable (i.e., mineral). Betzer et al. (1974b) report 0.003 mg/l for clear ocean water off Cape Hatteras, with the nepheloid layer having concentrations of 0.025 to 0.030 mg/l. This suggests concentrations about an order of magnitude greater in the nepheloid layers than in the clear water, similar to the findings of Jacobs et al. (1973). Betzer et al. (1974a) report more than 0.016 mg/l over the continental rise off Venezuela and more than 0.20 mg/l over the abyssal plains to the east. The former concentration would be associated with North Atlantic Deep Water; the latter, with Antarctic Bottom Water. Concentrations in the Argentine Basin have been estimated, in the absence of samples, to lie between 0.03 and 0.3 (Ewing et al., 1971). Light-scattering values in the Argentine Basin are as high or higher than those of the North Atlantic (Eittreim and Ewing, 1974). Thus, concentrations along the Argentine continental rise must be 0.1 mg/l or more.

Transport and Mass in the Nepheloid Layer

The dearth of data on concentrations within the nepheloid layer makes calculation of transport difficult, to say the least. Calculations of the amount of sediment contained within the nepheloid layers or the amount of material transported through a cross section are interesting exercises, although, at the present state of knowledge, they tend to be highly speculative. Even with this limitation, it should be done as a first attempt to quantify these important processes.

Eittreim and Ewing (1972) calculate that the Western Boundary Undercurrent transports 9×10^{12} cm^3/year into the North American Basin at the northern end while 3×10^{12} cm^3/year leaves at the south end. This is considerably less than the 12.3×10^{13} cm^3/year that is estimated to be deposited on the basin floor. Betzer et al. (1974b) calculate a flux past a section off Cape Hatteras as 2.8×10^{12} g/year, considerably less than that of Eittreim and Ewing (1972). If a density of 2.0 g/cm^3 is assumed, the mass of material reported by Eittreim and Ewing is between 6 and 18×10^{12} g/year. Based on the higher estimated concentrations of Ewing et al. (1971) for the Argentine Basin, the sediment transport into the basin in the nepheloid layer is estimated to be between 4.8×10^{13} and 18.9×10^{13} g/year.

Transport in the nepheloid layer is considerable despite the relatively low concentrations and the slow flow of the currents in many places. The overriding control is, of course, the very immense thickness of the nepheloid layer, thicknesses of a kilometer over thousands of square kilometers.

In the Argentine Basin, Ewing et al. (1971) estimate the volume of the nepheloid layer to be 10^{18} liters. If concentrations in the layer vary between 0.03 and 0.3 mg/l, there will be between 30 and 300 million metric tons of material suspended in the nepheloid layer.

What sort of sediment thickness would result over the continental rise, if by some catastrophic event, all the material in the nepheloid layer was deposited? Using the minimum and maximum values of Ewing et al. (1971), between 3 and 30 mg would be deposited over each square centimeter. Assuming a density of newly deposited sediment of 1 g/cm^3, as did Ewing et al., a layer between 0.03 and 0.3 mm would result. These thicknesses are equivalent to a grain of silt and sand, respectively. Compacted to a density of 2 g/cm^3, the layer would be between 0.015 and 0.15 mm (15–150 μm). Thus, deposition from the nepheloid layer will be insignificant over a short time span. Such deposition during geologic time will give rise to thick homogeneous deposits.

DEPOSITION FROM SUSPENSION

Many of the comments by Drake (Chapter 9) are applicable to deposition from suspension in the open ocean. Our understanding of these processes is very limited and it seems that many of the currently held concepts will have to be discarded or greatly modified. Among these are concepts of material settling without interruption through the entire water column and of single-particle deposition. The clay-mineral distribution in seafloor sediments strongly suggests rapid deposition after introduction into the sea (Biscaye, 1965) as does the marked latitudinal zonation of clay minerals on the seafloor (Griffin et al., 1968; Lisitzin, 1972). Gross (1967) concluded that particle transit time through the water column is rapid, far in excess of the settling velocities calculated for single particles.

Recent data suggest that much of the suspended sediment occurs in aggregates that contain large amounts of

---

FIGURE 3. *Upper: Location of nephelometer stations in the North Atlantic. Lower: Profiles of light scattering in the North Atlantic. Deflection to the right indicates increased light scatter, hence greater turbidity. Clearest water shows least deflection and is generally used as a baseline. Solid lines are values obtained during lowering of the instrument; dashed, during retrieval. Location of stations on 3A. From Eittreim and Ewing (1972).*

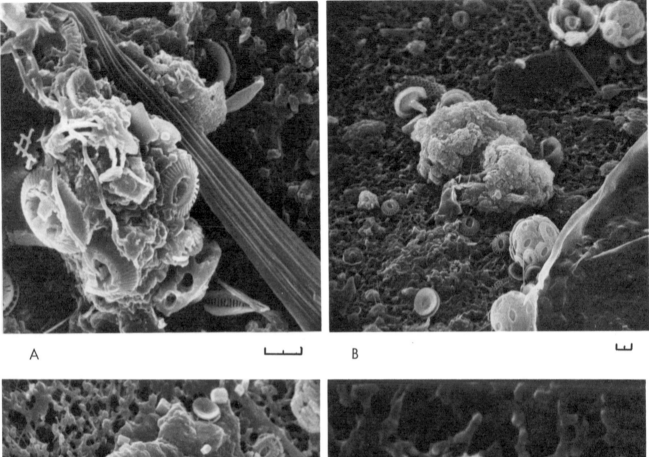

A

B

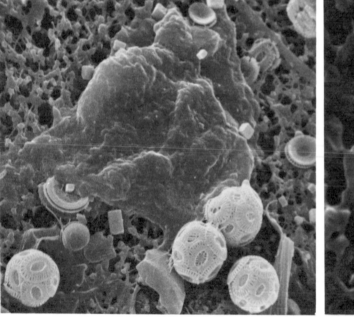

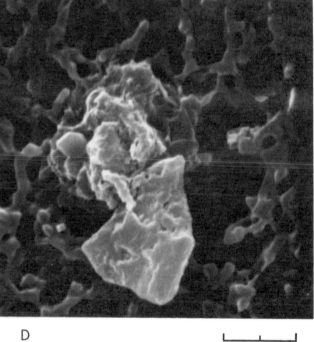

C

D

FIGURE 4. *Scanning electron micrographs of aggregates suspended in waters off the southeastern United States. Binding material appears to be organic in most cases. Coccospheres appear* *to be bound to sediment particles by definite strands. Scale by each illustration is 2 μm.*

organic material, both soft and mineral (Folger, 1970; Eittreim and Ewing, 1972; Jacobs et al., 1973; Manheim et al., 1972; Bornhold et al., 1973). These composite particles contain large numbers of individual mineral grains as well as organic matter (Fig. 4). Scanning electron micrographs have shown the existence of well-formed fecal pellets, aggregates held together by "mucus," and other particles in which the binding mechanism seems to be by organisms secreting calcium carbonate(?).

Forces holding the aggregates together range from weak to strong. It is becoming apparent that processes other than flocculation (i.e., physiochemical forces) play an important role in the formation of composite particles (Meade, 1972). The role of organisms has been underestimated and that of flocculation overestimated.

Jacobs et al. (1973) determined that aggregates make up between 0.7 and 10%, by count, of the material in the nepheloid layers and from 10 to nearly 20% in clear water. Our data (unpublished) show that from 38 to 100%, by count, of the suspended material is in aggregates. The discrepancy between these two sets of values can be explained by the difference in resolving power of the binocular microscope (Jacobs) and the scanning electron microscope. Even some extremely small particles ($<2\mu$m) consist of many smaller grains, resolvable only under high magnification ($>5000\times$).

Aggregates are much larger than the single particles with which they are associated in the water column. Being larger, the aggregates contain a much greater mass of material, as a part of the total, than count statistics would indicate.

One of the major questions is the rate of settling of these composite particles. Measurements can be made of the diameters of the aggregates and, from this, reasonable estimates can be made of the volumes associated with the diameters. This does not give the mass of material within the aggregate, since this also depends on density. There is a growing body of evidence that the density of composite particles is considerably less than that of single grains ($\sim$2.65 g/cm³).

The decrease in density is due to inclusion of organic matter (Hobson, 1967) and water (Krone, 1963; 1972; Sakamoto, 1972; Eittreim and Ewing, 1972). The density of aggregates that contain large amounts of organic matter is considerably lower than that of single mineral particles since organic matter has a density of about 1.1 g/cm³ (Hobson, 1967; Bassin et al., 1972).

Krone (1963, 1972) calculated that densities of aggregates varied from 1.06 to 1.25 g/cm³ for estuarine sediments while Sakamoto (1972) calculated densities down to 1.006 g/cm³. This latter value was for floccules

formed across a transition between fresh water and 5‰ salinity. Eittreim and Ewing (1972) used a density of 1.8 g/cm³ in conversion of diameter to mass. McCave (1970) estimated densities of 1.7 g/cm³, although he suggested 1.4 g/cm³ was possible (McCave, 1972) and used 1.59 g/cm³ (McCave, 1975).

We have taken several samples off the southeastern United States by means of Niskin bottles, being careful not to disaggregate the material, and subjected the suspensates to examination under a scanning electron microscope. Both individual particles and aggregates were present in all samples. In all cases, aggregates were larger than the associated single particles in the same sample (Table 2). A crude estimate can be made of the density of the aggregates by assuming that single particles and associated aggregates have the same settling velocity. Our data give an average density of the aggregates as 1.43 g/cm³, ranging from 1.07 to 1.80 g/cm³ (Table 2).

This value is not unreasonable according to the logic of Krone (1963). Aggregates consist of single particles and large aggregates are often the result of aggregate-aggregate contact. The density of the aggregates decreases because of the incorporation of water in the interparticle and interaggregate voids. Bottom sediment is one order of aggregation higher than the order just being deposited. Thus, the densities of suspended composite particles should be approximately the same as or slightly higher than that of newly deposited bottom sediments. Bulk densities of the upper 5 to 8 cm of deep-sea sediments are around 1.3 g/cm³ in the Gulf of Mexico (Cernock, 1970) and 1.56 g/cm³ for Whiting Basin (Bennett et al., 1970). These values must be considered maximum for the uppermost layers of sediment on the seafloor. They were obtained as average over the upper 5 to 8 cm, a large section considering the slow depositional rates. The upper few millimeters may, in fact, be much lower.

Calculated settling rates, using the densities and diameters of Table 2 and Stokes' law, range from 0.56 to 3 m/day. For the size ranges in our samples, these settling rates agree with those estimated by McCave (1975). It should be stressed that the settling rates are *lower* than those of single particles of the same size but far higher than the settling velocities of the individual units within the composite particles. Thus, any estimate of the residence time in the water column of a single particle derived from size analysis of bottom sediment, after dispersion, is likely to be off by an order of magnitude or more if the particle was originally part of an aggregate.

Our samples had very few aggregates with diameters larger than 40 μm, less than 2% of the total number.

TABLE 2. **Size of Aggregates and Particles Suspended in Water over the Outer Shelf, Slope, and Rise off the Southeastern United States***

| Sample | Location | Depth (m) | | Diameter (µm) | | Density |
		Bottom	Sample	Particles	Aggregates	Aggregates
E19171	34°10.8'N 75°40.4'W	2200	2000	3.92	6.38	1.58
E19174	33°45.5'N 76°43.0'W	45	42	5.26	28.42	1.07
E19179	33°17.5'N 77°22.7'W	36	3	3.98	15.38	1.12
E19180	33°10.0'N 77°45.5'W	43	37	8.6	14.98	1.51
E19182	32°25.6'N 79°11.1'W	41	37	7.66	10.88	1.76
E19212	28°50.0'N 80°06.5'W	73	3	3.32	4.58	1.80
E19205	30°21.8'N 79°34.8'W	745	515	3.68	7.49	1.38
E19221	27°49.8'N 79°40.0'W	550	500	6.42	17.48	1.22
			Average	5.36	13.24	1.43

* Density was estimated by assuming that aggregates and particles in a sample have the same settling velocity.

This is in line with the ideas of Sheldon et al. (1972) and McCave (1975). Smaller particles make up the bulk of the suspended material while large particles are rare. The mass of the larger composite particles dictates high settling rates and rapid removal from the water column. McCave (1975) suggests that large particles contribute a substantial amount of material to bottom sediments but their rapid removal and rarity in the water column cause samples of suspended material to be biased.

Theory and experimental data on deposition of cohesive sediment from flowing water have not settled the question of the velocities at which erosion starts nor when transport ceases and deposition begins. Advances have been dramatic in the past few years. Erosion-deposition curves derived from various studies are presented by Hollister and Heezen (1972).

Partheniades (et al., 1969; 1972) has examined transport, deposition, and resuspension of cohesive sediments in the laboratory using fresh or distilled water as the fluid medium. The concentrations of suspended material used were far higher than normally found in any natural system and several orders of magnitude greater than concentrations found in oceanic waters. Direct

extrapolation to the question of deposition from suspension over the outer continental margin is dangerous.

Partheniades found that a given flow can maintain a constant portion of an initial sediment load in suspension, irrespective of the initial load. That is, the ratio between the equilibrium load and the initial load is nearly a constant. Thus, for very high initial concentrations, a greater amount of material will remain in suspension than in more dilute suspensions. The equilibrium concentration is that portion of the suspended load that never reaches the bed, whereas the deposited part is that which settles out and is not resuspended. Thus, from his experiments, deposition and resuspension of coehsive sediments cannot occur concurrently.

One of the overriding controls in settling is the formation of "flocs," which are much larger than the individual particles. This agrees with the findings of Krone (1963, 1972). Those flocs that settle to the bottom are strongly bonded and, upon reaching the bottom, are strongly bound to the preexisting substrate. Flocs not strongly bound are disrupted by internal shear in the fluid and reentrained.

Lonsdale and Southard (1974) determined the threshold velocities necessary to erode red clay from the

North Pacific. For a plane bed, the threshold velocity was dependent on water content, decreasing with increasing water content. The lowest velocity, necessary to erode clay with about 82% water, was 18.7 cm/sec (at the bed surface). This surface velocity would result from a current at 30.4 cm/sec at a height of 1 m above the bottom. Other runs of the experiment showed velocities (at 1 m) ranging up to 89 cm/sec before mass failure occurred. Addition of coarse material lowered the minimum value.

Betzer et al. (1974b) found that average current velocities on the rise off Cape Hatteras, as measured by four current meters, were from 6.8 to 12.6 cm/sec toward the southwest. Two other meters recorded average velocities of 0.3 and 0.8 cm/sec toward the northeast and southeast. They also suggest that fluctuations can occur in position, direction, and velocity of the Gulf Stream and the deep currents. Zimmerman (1971) found velocities averaging about 20 cm/sec along the contours of the continental rise off New England. Thus, the currents are sufficient to transport fine silt and clay.

Rabinowitz and Eittreim (1974) measured near-bottom current velocities greater than 20 cm/sec in the Labrador Sea. These currents are part of the circulation pattern associated with the North Atlantic Western Boundary Undercurrent. The velocities, although only instantaneous measurements, are of sufficient magnitude to possibly erode the bottom and prevent deposition.

It is also interesting that Schneider et al. (1967) make a very strong case for erosion and deposition by the Western Boundary Undercurrent, citing bottom photographs, echogram character, and the presence of red lutites off North Carolina, which presumably originated off Nova Scotia. Clay-mineral ratios on the rise suggest clays with northern affinities (Pierce, 1970). On the other hand, Betzer et al. (1974b) found little evidence of the sediment surface being affected by bottom currents although they actually measured currents up to 47 cm/sec in the area.

It has been suggested that bottom currents such as the Western Boundary Undercurrent are far from a steady state condition (Betzer et al., 1974b). Fluctuations probably exist in the velocities, direction, and location of the current. Deposition in any one area could alternate with transportation or erosion, depending on the magnitude of the fluctuations. Once deposition has taken place, velocities necessary to resuspend cohesive sediments may seldom be exceeded except in the core of the current.

Thus, the processes that dominate the origin of the rise are still debatable and will probably remain a controversial issue for some time.

SOURCE OF SEDIMENT IN THE NEPHELOID LAYER

That the terrigenous material in the nepheloid layer was originally derived from the continents is obvious. Terrestrial erosion is supplying, or has supplied in the past, all this material, except for infinitesimal amounts of cosmogenic material and possibly some slight amount removed by submarine erosion of volcanic rocks. The amount of volcanic dust in the nepheloid layer is relatively low (Ewing and Connary, 1970), although this does not negate the fact that such material locally may be an important part of pelagic deposits.

Several sources (Fig. 5) are intuitively evident as potential contributors of sediment to the nepheloid layer; outward diffusion of fluvial discharge, resuspension of bottom sediments, introduction of glacial or glacio-fluvial sediments at high latitudes, and material that has settled through the entire water column. The latter would be introduced into the surface layers by outward diffusion from the continents or by eolian transport to the ocean (Folger, 1970).

OUTWARD DIFFUSION OF FLUVIAL DISCHARGE.   The increase in light-scattering ability off some major rivers has been noted. Ewing and Connary (1970) note an increase in scattering and an increased thickness of the nepheloid layer in the northeastern Pacific off the Columbia River. Plank et al. (1973) suggest rivers of Central America as a source for some of the material in the Panama Basin. Whether this is true diffusion from the river mouth or whether it occurs in stages with periods of deposition interspersed with resuspension is not known, although the work of Drake et al. (1972; see p. 150) would suggest transport in stages.

LOW-DENSITY FLOWS DOWN CANYONS.   Currents in canyons, biological reworking of bottom sediments, and minor slumping can produce suspended material in the waters in submarine canyons (Stanley et al., 1972a). With general downcanyon movement (Bennett et al., 1970), this material could move into the nepheloid layer.

TURBIDITY CURRENTS.   Downcanyon movement of turbidity currents will carry a mixed sediment out onto the rise and plain. Upon encountering the currents associated with the nepheloid layer, the finer fraction of the sediment could be winnowed out and incorporated into this turbid zone. Ewing et al. (1971) believe this to be a very important mechanism in the Argentine Basin; Stanley et al. (1971, 1972b) and Eittreim and Ewing (1972) suggest it for the North Atlantic.

RESUSPENSION OF BOTTOM SEDIMENTS.   The currents with which the nepheloid layer is associated in certain

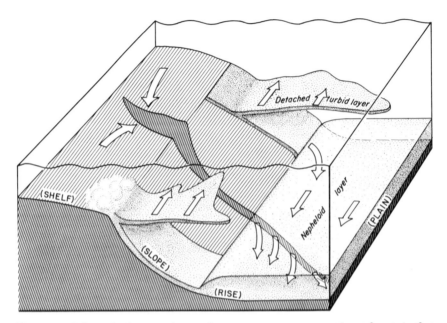

Figure 5. *Schematic of source of some of suspensates over outer continental margin that may contribute to nepheloid layer on rise.*

areas attain high velocities. The many current lineations on the seafloor indicate reworking or re-forming of the bottom sediment and certainly suggest resuspension (Hollister and Heezen, 1972). An increase in suspended sediment concentrations in the Panama Basin has been attributed to resuspension (Plank et al., 1973). Betzer et al. (1974b) also indicate resuspension by the Western Boundary Undercurrent in the North Atlantic. Suspension by waves at the shelf break and downslope movement also can contribute material, probably a significant amount at various times. Currents impinging upon the bottom may also sweep material out over the slope and rise (Betzer et al., 1974b). The ability of organisms to disturb and resuspend bottom sediments should not be overlooked.

INTRODUCTION AT HIGH LATITUDES.   The water masses that have near-bottom nepheloid layers all originate in high latitudes. There is a certain security, then, in ascribing at least some of the material in the nepheloid layer to high-latitude processes. Undoubtedly some of the fine material does originate in high latitudes. Eittreim et al. (1972) believe reworking of shelf sediments around Antarctica may contribute a large portion of the initial material in the Antarctic Bottom Water, although this has been questioned (Ewing et al., 1971; Siegel, 1973). Schneider et al. (1967) suggest the seafloor off the Canadian Maritimes as a source of material on the rise off eastern North America. The current velocities reported in this area appear to be sufficient to resuspend material (Rabinowitz and Eittreim, 1974).

The process would, for these two examples, not be greatly different from resuspension; only the location would differ.

PELAGIC CONTRIBUTIONS.   This term is used here to denote all those materials that have settled through the entire water column. This would include particles that have diffused outward from land in the surface layers as well as those introduced into the surface layers by eolian transport, generally considered hemipelagic. It would also include the hard parts of planktonic organisms that can survive solution at such depths. Eolian transport from land must contribute a significant fraction of the material found in the surface layers of the open ocean (Folger, 1970; Prospero and Carlson, 1972; Chester et al., 1972). In the northeast trade-wind belt, Chester et al. (1972) suggest a very large percentage. For bottom sediments of the continental margin, the amount is probably not very significant, except locally, because of the proximity of the margin to terrigenous sources. Ewing and Connary (1970) argue strongly that eolian-transported material is insignificant in the mass of the nepheloid layer in the Pacific. Their argument is that this coarser material would settle quickly through the nepheloid layer and hence, because of a small residence time, contribute very little to the mass suspended in this near-bottom turbid layer. The presence of very clear water above the nepheloid layer would also negate a continuous "rain" of material from the surface. Eittreim and Ewing (1972) do not believe that eolian dust contributes much to the mass in the nepheloid layer of the North Atlantic unless the residence time is much longer than currently assumed. Plank et al. (1972), based on light-scattering data, believe that the material in the clear water above and the nepheloid layer below are of the same type. They suggest that the nepheloid layer is the result of particle settling through the water column with the increased concentration near the bottom the result of turbulence from the bottom, which maintains the particles in suspension. This would suggest settling of particles to the bottom plus resuspension and lateral movement after deposition. Turbulence alone will not maintain particles in suspension. However, when coupled with a boundary that can act as a source and sink, turbulence will result in turbidity.

When considering the pelagic contribution to the nepheloid layer, some comments should be directed toward the biogenic contributions. For the most part, the hard parts of diatoms, radiolarians, foraminiferids, and other organisms are relatively large compared to the individual suspended particles. They are often associated with aggregates or fecal pellets and may settle at accelerated rates. Their rapid removal from the water column would act against their presence in samples. Our data indicate that biogenic carbonate and silica contribute very little to the number of particles but may account for a large part of the mass. Mostly, we have found coccolithophorids, radiolarians, and diatoms associated with suspended sediment. We have yet to find foraminifera or pteropod tests in the suspended fraction.

Organic material contributes very little to the total mass of suspended particulates in deeper layers. Available data (unpublished) show that between 10 and 22% of the total is decomposable organic matter. This agrees with conclusions of Menzel (1967) and Hobson (1967) that the distribution of organic carbon is stable in deeper depths and little influenced by productivity at the surface.

In summary, it would appear that all processes, with the exception of settling through the entire water column, can contribute to the mass of material suspended in the nepheloid layer. The relative amount from each would depend on the geographic area and the circumstances.

SUSPENDED SEDIMENTS AND SHAPING OF THE CONTINENTAL RISE

As mentioned earlier, the interpretation of sediment data from the continental rise has led to divergent opinions as to the relative importance of contour currents (boundary currents) versus downcanyon transport and coalescing fans. The shape of the outer continental margin has been attributed entirely to contour currents (Heezen et al., 1966; Schneider et al., 1967). Opposed to this are those who propose a series of large coalescing fans at the ends of canyons and youthful-appearing fan valleys, unmodified by a mud cover (Stanley et al., 1971). The two opposing viewpoints can be resolved by combining the two processes. Downcanyon transport can be a source of some of the fine sediment for the currents, maintaining the turbidity levels or increasing them (Fig. 5). Coarser material transported to the rise remains as a lag deposit, resulting in coalescing fans off active canyons, being unimportant where canyons are few in number or where those that are present are inactive.

It should be pointed out, also, that the measured current velocities often are sufficient to resuspend sediment. Reworking of Pleistocene deposits at the mouths on canyons, which were undoubtedly more active conduits during lower stands of sea level, can provide material for the nepheloid zone. The processes now active may be modifying preexisting geomorphic features.

Cold water is generated in high latitudes (Figs. 6 and 7): the Norwegian Sea for the North Atlantic (Jones et al., 1970) and the Antarctic Convergence for the South Atlantic (Ewing et al., 1971). These cold water masses have an existing suspended load that is the result of resuspension of bottom sediments in the polar regions and/or glacial input.

The cold water flows into the ocean basins through gaps in the sills at relatively high velocities, over 100 cm/sec in places. Because of the high velocities in the channeled areas, no deposition and possibly erosion occurs (Ewing et al., 1971). The inflowing current is not steady state but occurs in pulses that are, in part, seasonal (Jones et al., 1970).

The currents are affected by the Coriolis force (deflected to the right in the northern hemisphere). Thus, two cold water masses may be present in one basin. In the North Atlantic, North Atlantic Deep Water is present on the western side of the North American Basin, Antarctic Bottom Water on the east (Figs. 7 and 8).

After entering the more open basins, current velocities drop, resulting in formation of sedimentary deposits that appear homogeneous on seismic records. With deposition, concentration of material suspended in the nepheloid layer decreases (Ewing et al., 1971). In order to counteract deposition and maintain the nepheloid layer, additional input is necessary.

Periodic inputs from submarine canyons are size segregated by current winnowing, giving rise to coarser

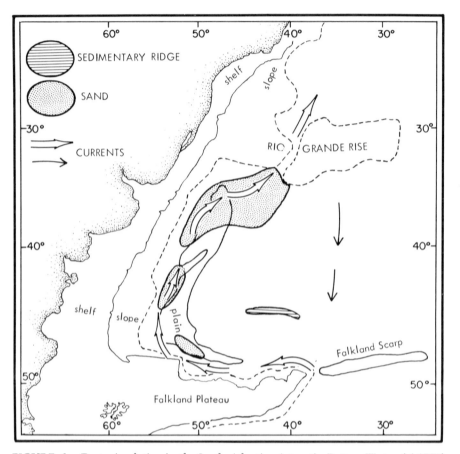

FIGURE 6. *Deep circulation in the South Atlantic. Antarctic Bottom Water (AABW) enters Argentine Basin through a gap in Falkland Scarp. The high velocities associated with the water here preclude deposition and may cause erosion. Drop in velocity to the north plus circulation has given rise to a sedimentary ridge near the geographic center of the basin. Most of the AABW flows along the Argentine Rise, winnowing material contributed down the canyons. The finer sediment is removed, leaving sand deposits on the rise. Much of the AABW leaves the basin through the Vema Gap on the Rio Grande Rise.*

deposits at the canyon mouths with finer material carried farther (Fig. 6; see Ewing et al., 1971). Currents in the core of the water mass are sufficiently strong to transport sands some distance which form thin cross-bedded units on the rise (Hollister and Heezen, 1972). Each cross-bedded unit, of which the greatest number occur on the rise, can represent a strong pulse in the current or another downcanyon movement. The fines are winnowed out and transported from the area, being deposited near the margins of the current or in areas where the current slows sufficiently.

Erosion of bottom materials can give rise to depressions (Fig. 6; see Ewing et al., 1971) or put tracers into suspension, such as distinctly colored material (Heezen et al., 1966). With sufficient data, source of this material could be isolated.

After leaving areas of active or previously active canyons, the nepheloid layer loses material so that concentration of suspended material decreases and vertical eddy diffusion may increase the thickness of the layer (Eittreim and Ewing, 1972). Unless material is added, the water mass becomes much less turbid, such as the Antarctic Bottom Water in the North Atlantic (Fig. 8). Material can be added by resuspension in areas of channeling or by settling from upper layers after being resuspended by currents (Fig. 5). It is thought that the Gulf Stream, in sweeping across the Blake Plateau, resuspends material which settles into the nepheloid layer (Betzer et al., 1974b; Eittreim et al., 1969).

The combination of currents and downcanyon movement, induced by internal waves, surges, tidal currents, or other processes, should give rise to a sediment that is a composite of the two processes. At the mouths of canyons, coalescing fans should be formed of material that is relatively coarser than the sediments making up the outer margin away from the canyon influence. In

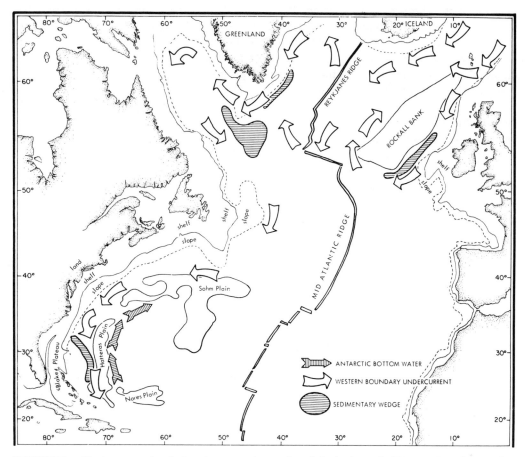

FIGURE 7. *Deep-water circulation in part of the North Atlantic. Velocity decreases in currents have given rise to several sedimentary wedges. North Atlantic Deep Water flows along the western* *edge of the basin as the Western Boundary Undercurrent. Antarctic Bottom Water moves northward on the east side of the Hatteras plain.*

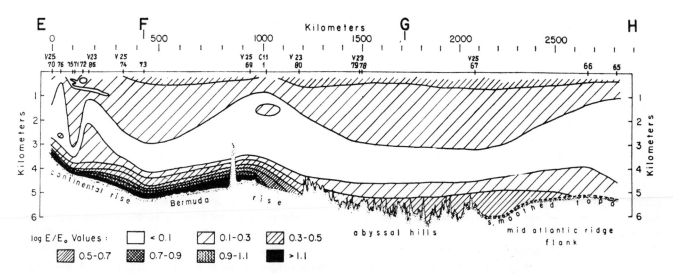

FIGURE 8. *Cross section of nephelometer values extending from the continental rise off New Jersey to near the Mid-Atlantic Ridge. Location of the section shown in the upper diagram of Fig. 3. North Atlantic Deep Water is shown by high suspended sediment* *values on the western side of the basin, lapping onto the rise. Antarctic Bottom Water is shown on the eastern side of the basin, also with high values of suspended sediment. From Eittreim and Ewing (1972).*

453

general, that portion of the rise downcurrent from the canyon outlet should be influenced more by material that has moved downcanyon than the side that is up-current; hence, there are coarser deposits. If the current is strong enough, there may be little bathymetric expression of the canyon contribution, having been modified by the persistent current.

It would be worthwhile to compare the sediments and bathymetric form of the deposits from outer margins which are influenced by the boundary currents and suspended sediment deposition in contrast to those that are not so influenced. Thus, a comparison would be appropriate between the eastern and western sides of the Atlantic Ocean to see if there is a fundamental difference between the two areas.

APPLICATIONS

Studies of suspended particulate material over the outer continental margin have little direct application in providing solutions to the problems besetting mankind and the environment. Such studies do point out the worldwide nature of some of these problems.

Man-made particles or those derived from man's activities are becoming a significant part of the material suspended in surface waters. Processed cellulose fibers are common in the shipping lanes off the east coast of the United States and closely resemble the fibers of disaggregated toilet paper (Manheim et al., 1970). The occurrence of talc in the same area (Pierce et al., 1971) may be from the same source, as talc is the filler in most white toilet papers. Soot and fly ash are also common in the shipping lanes and off major cities.

Crude oil lumps are quite common on the sea surface and have been noted for more than half a century (Moss, 1971). The universal use of oil for fuel in ships and the increase in amount of oil transported have led to an increase in oil spills and discharge into the sea. It has been suggested that natural oil seeps may contribute large amounts of oil to marine waters (Wilson et al., 1974) but it is doubtful that much, if any of this, reaches the surface to contribute to the number of floating lumps (Blumer, 1972). Although the fate of oil from seeps is not well understood, that issuing at depths greater than 60 ft appears to be deposited on the sea-floor or transported into deeper basins (Fischer and Stevenson, 1973).

Plastic particles are also common in surface waters, but more common in shipping lanes and off major ports (Colton et al., 1974). The densities of most of these particles are less than 1.0 g/cm³; hence, they remain suspended indefinitely in the water unless incorporated into denser aggregates. At the present levels of abun-dance, these particles seem to present little to no danger to the environment. They can act as substrates for some of the biota and may absorb polychlorinated biphenyls from seawater.

The effects of marine mining have not been fully assessed at present. There is little doubt that some change will accrue but the effects of such activity in the deep sea have not been evaluated, inasmuch as most of the mining will occur in shelf areas. Turbidity un-doubtedly will be increased in the area of active opera-tions (see Chapter 17, Fig. 37B). If occurring on the shelf areas, some of this resuspended material could be transported into the deeper waters or, if occurring in the deep sea as a result of manganese mining, the resus-pended material may be introduced directly into the near-bottom nepheloid layer. Resuspension of the bottom sediments will change their physiochemical environment, giving rise to potential for release or adsorption of metals, increased oxygen demand from reduced sediments, and scavenging of organic material (Padan, 1971).

There have been suggestions that canyons may be used as conduits to transport the wastes of civilization into the deep oceans. Mankind would then have an un-fillable disposal site for wastes generated by modern technology. The treatment of such wastes before disposal will determine, to a large extent, the effect of such practices on the deep-sea environment. In some cases, introduction of some wastes, such as sulfuric acid wastes, appear to be beneficial, or at least not detri-mental (Ketchum and Ford, 1952).

Of more concern are the effects of nondegradable materials, metals, and organic matter. Concentration levels and residuals in the area of discharge are im-portant variables because residuals prevent repopulation of the area in the event of an initial kill. Transfer of minerals to seawater tends to release some absorbed metals to the water (Turekian, 1971), which could lead to high levels of dissolved metals in the area of discharge. Some chlorinated hydrocarbons have a tendency to adsorb onto sediment particles; if these hydrocarbons become attached to minerals in the nepheloid layer, dispersal might occur over wide areas. Introduction of organic material could increase the chemical and biological oxygen demand. In deeper waters, isolated from the surface, the increase in oxygen demand could drastically reduce oxygen availability, and limited vertical mixing would preclude replenishment of the oxygen. In time, some of the deeper water masses might approach anoxic conditions.

As pointed out earlier, little is known of the effect of introduction of wastes into the deep sea. Most of the conclusions are speculative, at least in part, or are based on laboratory experiments or on results of research in shallow water. Consequences of introduction of wastes may be drastic; on the other hand, little change may occur.

SUMMARY

There is a paucity of information on suspensates in the water column over the outer continental margin. Concentrations of total material, a large part of which is biogenic, in surface waters range from 0.125 to 5 mg/l, generally higher off major rivers. Concentrations below the water surface are highly variable, both vertically and laterally. Temporal variations are due to changing oceanographic conditions. Increased concentrations are often present at density interfaces within the water column.

Lateral dispersion of suspended material is wholly dependent on water mass movements. Offshelf transport can occur by movement of a near-bottom layer of turbid water, cascading of cooled waters, and advection and diffusion across the shelf with subsequent settling through the water column.

The nepheloid layer, a near-bottom turbid layer, exists over extensive areas of the deep ocean basins; lap onto the rise and lower slope; and are associated with dense, cold water masses. The distribution of the nepheloid layer is controlled by movement of these cold, dense waters, primarily the Antarctic Bottom Water and the North Atlantic Deep Water. Thickness of this layer is over 2 km in places.

Concentrations in the nepheloid layer are estimated to range from 0.01 to 0.24 mg/l, nearly an order of magnitude larger than that of the overlying "clear" water. Between 30 and 300 million metric tons of sediment is in the nepheloid layer of the Argentine Basin which, if deposited, would result in a lamina between 0.03 and 0.3 mm thick.

A very large part of the material suspended in seawater occurs in composite particles or aggregates not as single grains. Aggregates, with a greater settling velocity than the single particles in the aggregates, effectively remove fine suspended material from the water column. The density of the composite particles is lowered by incorporation of water and organic matter to around 1.43 g/cm^3.

Deposition and resuspension of cohesive sediment do not occur concurrently from the same current, as opposed to noncohesive transport where both may occur simultaneously. Resuspension starts around 30 cm/sec. Below this velocity, transport and deposition occur. Near-bottom current measurements in the boundary currents are often near or greater than this velocity so that transport, deposition, and resuspension occur as a result of fluctuating current velocities.

The continental rise is probably the result of a combination of coalescing fans and shaping by boundary currents. Downcanyon transport and formation of coalescing fans will dominate where boundary currents are weak or absent or where canyons are extremely active. Where there are few canyons or where they are not active conduits, boundary currents and deposition from the nepheloid layer will dominate.

Studies of suspended material per se over the continental margin have little direct application to solutions of environmental problems. The studies do point out the potential worldwide nature of some of the problems.

ACKNOWLEDGMENTS

Many authors contributed to this chapter by publishing the results of their research, without which this chapter could not have been assembled. Special thanks are due to S. L. Ettreim and David Drake for permission to use their original figures.

L. B. Isham did the art work for the original figures presented here. Assistance of the staff of the Scanning Electron Microscope Laboratory, National Museum of Natural History, is acknowledged.

Original research involved in this section was supported partially by the Smithsonian Research Foundation Awards 430027 and 430070. Support in the form of ship time aboard the Duke University R/V EASTWARD, supported by NSF grants GB 17545 and GA 27725, is acknowledged.

REFERENCES

Bassin, N. J., J. E. Harris, and A. H. Bouma (1972). Suspended matter in the Caribbean Sea: A gravimetric analysis. *Mar. Geol.*, **12:** M1–M5.

Beer, R. M. and D. S. Gorsline (1971). Distribution, composition and transport of suspended sediment in Redondo Submarine Canyon and vicinity (California). *Mar. Geol.*, **10:** 153–175.

Bennett, R. H., G. H. Keller, and R. F. Busby (1970). Mass property variability in three closely spaced deep-sea cores. *J. Sediment. Petrol.*, **40:** 1038–1043.

Betzer, P. R., K. L. Carder, and D. W. Eggimann (1974a). Light scattering and suspended particulate matter on a transect of the Atlantic Ocean at 11°N. In R. J. Gibbs, ed., *Suspended Solids in Sea Water*. New York: Plenum Press, pp. 295–314.

Betzer, P. R., P. L. Richardson, and H. B. Zimmerman (1974b). Bottom currents, nepheloid layer, and sedimentary features under the Gulf Stream near Cape Hatteras. *Mar. Geol.*, **16:** 21–29.

Biscaye, P. E. (1965). Mineralogy and sedimentation of Recent deep-sea clay in the Atlantic Ocean and adjacent seas and oceans. *Bull. Geol. Soc. Am.*, **76:** 803–832.

Blanton, J. (1971). Exchange of Gulf Stream Water with North Carolina Shelf Water in Onslow Bay during stratified conditions. *Deep-Sea Res.*, **18**: 167–178.

Blumer, M. (1972). Submarine seeps: Are they a major source of open oil pollution? *Science*, **176**: 1257–1258.

Bornhold, B. D., J. R. Mascle, and K. Harada (1973). Suspended matter in the surface waters of the eastern Gulf of Guinea. *Mar. Geol.*, **14**: M21–M31.

Bouma, A. H., R. Rezak, and R. F. Chmelik (1969). Sediment transport along oceanic density interfaces. *Geol. Soc. Am. Abstr.*, **7**: 259–260.

Buss, B. A. and K. S. Rodolfo (1972). Suspended sediment in continental shelf waters off Cape Hatteras, North Carolina. In D. J. P. Swift, D. B. Duane, and O. H. Pilkey, eds., *Shelf Sediment Transport*, Stroudsburg, Pa.: Dowden, Hutchinson & Ross, pp. 263–279.

Cernock, P. J. (1970). *Sound Velocities in Gulf of Mexico Sediments as Related to Physical Properties and Simulated Overburden Pressures.* Tech. Rep. 70-5-T, Texas A.&M. University, College Station, 114 pp.

Chester, R., H. Elderfield, H. H. Griffin, L. R. Johnson, and R. C. Padgham (1972). Eolian dust along the eastern margins of the Atlantic Ocean. *Mar. Geol.*, **13**: 91–105.

Colton, J. B., F. D. Knapp, and B. R. Burns (1974). Plastic particles in surface waters of the northwestern Atlantic. *Science*, **185**: 491–497.

Connary, S. D. and M. Ewing (1972). The nepheloid layer and bottom circulation in the Guinea and Angola basins. In A. L. Gordon, ed., *Studies in Physical Oceanography*, Georg Wust Tribute, 2. New York: Gordon & Breach, pp. 169–184.

Drake, D. E. (1971). Suspended sediment and thermal stratification in Santa Barbara Channel, California. *Deep-Sea Res.*, **18**: 763–769.

Drake, D. E., R. L. Kolpak, and P. J. Fischer (1972). Sediment transport on the Santa Barbara–Oxnard Shelf, Santa Barbara Channel, California. In D. J. P. Swift, D. B. Duane, and O. H. Pilkey, eds., *Shelf Sediment Transport*, Stroudsburg, Pa.: Dowden, Hutchinson & Ross, pp. 307–331.

Eittreim, S. and M. Ewing (1972). Suspended particulate matter in the deep waters of the North American Basin. In A. L. Gordon, ed., *Studies in Physical Oceanography*, Georg Wust Tribute, 2. New York: Gordon & Breach, pp. 123–167.

Eittreim, S. and M. Ewing (1974). Turbidity distribution in the deep waters of the Western Atlantic trough. In R. J. Gibbs, ed., *Suspended Solids in Sea Water*. New York: Plenum Press, pp. 213–225.

Eittreim, S., M. Ewing, and E. M. Thorndike (1969). Suspended matter along the continental margin of the North American Basin. *Deep-Sea Res.*, **16**: 613–624.

Eittreim, S., A. L. Gordon, H. Ewing, E. M. Thorndike, and P. Bruchhausen (1972). The nepheloid layer and observed bottom currents in the Indian–Pacific Antarctic Sea. In A. L. Gordon, ed., *Studies in Physical Oceanography*, Georg Wust Tribute, 2. New York: Gordon & Breach, pp. 19–35.

Emelyanov, E. M. and K. M. Shimkus (1972). Suspended matter in the Mediterranean Sea. In D. J. Stanley, ed., *The Mediterranean Sea: A Natural Sedimentation Laboratory*. Stroudsburg, Pa.: Dowden, Hutchison & Ross, pp. 417–439.

Emery, K. O. (1969). The continental shelves. *Sci. Am.*, **221**: 106–122.

Emery, K. O., Y. Hayashi, T. W. C. Hilde, K. Kobayaski, J. H. Koo, C. Y. Meng, H. Niino, J. H. Osterhagen, L. M. Reynolds, J. M. Wageman, C. S. Wang, and S. J. Wand (1969). Geological structure and some water characteristics of the East China Sea and the Yellow Sea. *Tech. Bull. ECAFE*, **2**: 3–43.

Ewing, M. and S. B. Connary (1970). Nepheloid layer in the North Pacific. *Geol. Soc. Am. Mem.*, **126**: 41–82.

Ewing, M. and E. M. Thorndike (1965). Suspended matter in deep ocean water. *Science*, **147**: 1291–1294.

Ewing, M., S. L. Eittreim, J. I. Ewing, and X. LePichon (1971). Sediment transport and distribution in the Argentine Basin. 3. Nepheloid layer and processes of sedimentation. In L. H. Ahrens, F. Press, S. K. Runcorn, and H. C. Urey, eds., *Physics and Chemistry of the Earth* 8. New York: Pergamon, pp. 49–77.

Field, M. E. and O. H. Pilkey (1971). Deposition of deep-sea sands: Comparison of two areas of the Carolina continental rise. *J. Sediment. Petrol.*, **41**: 526–536.

Fischer, P. J. and A. J. Stevenson (1973). Natural hydrocarbon seeps along the northern shelf of the Santa Barbara Basin, California. *Offshore Tech. Conf.*, **1**: 159–164.

Folger, D. C. (1970). Wind transport of land-derived mineral, biogenic, and industrial matter over the North Atlantic. *Deep-Sea Res.*, **17**: 337–352.

Folk, R. L. (1961). *Petrology of Sedimentary Rocks.* Austin, Texas: Hemphills, 154 pp.

Gordon, A. L. and R. D. Gerard (1970). Bottom potential temperature in the North Pacific. *Geol. Soc. Am. Mem.*, **126**: 23–29.

Griffin, J. L., H. L. Windom, and E. D. Goldberg (1968). The distribution of clay minerals in the World Ocean. *Deep-Sea Res.*, **15**: 433–459.

Gross, M. G. (1967). Sinking rates of radioactive fallout particles in the northeast Pacific Ocean. *Nature*, **216**: 670–672.

Harris, J. E. (1972). Characterization of suspended matter in the Gulf of Mexico-I. Spatial distribution of suspended matter. *Deep-Sea Res.*, **19**: 719–726.

Heezen, B. C., C. D. Hollister, and W. F. Ruddiman (1966). Shaping of the continental rise by deep geostrophic currents. *Science*, **152**: 502–508.

Hobson, L. A. (1967). The seasonal and vertical distribution of suspended particulate matter in an area of the northeast Pacific Ocean. *Limnol. Oceanogr.*, **12**: 642–649.

Hollister, C. D. and B. C. Heezen (1972). Geologic effects of ocean bottom currents, western North Atlantic. In A. L. Gordon, ed., *Studies in Physical Oceanography*, Georg Wust Tribute, 2. New York: Gordon & Breach, pp. 19–35.

Houk, D. and T. Green (1973). Descent rates of suspension fingers. *Deep-Sea Res.*, **20**: 757–761.

Jacobs, M. B., E. M. Thorndike, and M. Ewing (1973). A comparison of suspended particulate matter from nepheloid and clear water. *Mar. Geol.*, **14**: 117–128.

Jerlov, N. G. (1953). Particle distribution in the ocean. In H. Petterson, ed., *Reports of the Swedish Deep-Sea Expedition*, Vol. III. Goteborg: Elanders, pp. 73–125.

Jones, E. J. W., M. Ewing, J. I. Ewing, and S. L. Eittreim (1970). Influences of Norwegian Sea overflow water on sedimentation in the northern North Atlantic and Labrador Sea. *J. Geophys. Res.*, **75**: 1655–1680.

Keller, G. H., D. Lambert, G. Rowe, N. Staresinio (1973). Bottom currents in the Hudson Canyon. *Science*, **180:** 181–183.

Ketchum, B. H. and W. H. Ford (1952). Rate of dispersion in the wake of a barge at sea. *Trans. Am. Geophys. Union*, **33:** 680–683.

Krone, R. B. (1963). *A Study of Rheological Properties of Estuarial Sediments*. Tech. Bull. 7, Comm. on Tidal Hydraulics, U.S. Army Corps of Engineers, 91 pp.

Krone, R. B. (1972). *A Field Study of Flocculation as a Factor in Estuarial Shoaling Processes*. Tech. Bull. 19, Comm. on Tidal Hydraulics, U.S. Army Corps of Engineers, 62 pp.

Lisitsyn, A. P. (1966). *Recent Sedimentation in the Bering Sea*. Akad. Nauk. Inst. Okeanol., 614 pp.

Lisitzin, A. P. (1972). *Sedimentation in the World Ocean*. Soc. Econ. Paleontologists and Mineralogists Spec. Publ. 17, 218 pp.

Lonsdale, P. and J. B. Southard (1974). Experimental erosion of North Pacific red clay. *Mar. Geol.*, **17:** M51–M60.

Lyall, A. I., D. J. Stanley, H. N. Giles, and A. Fisher, Jr. (1971). Suspended sediment transport at the shelf-break and on the slope. *Mar. Technol. Soc. J.*, **5:** 15–26.

McCave, I. N. (1970). Deposition of fine-grained suspended sediment from tidal currents. *J. Geophys. Res.*, **75:** 4151–4159.

McCave, I. N. (1972). Transport and escape of fine grained sediment from shelf areas. In D. J. P. Swift, D. B. Duane, and O. H. Pilkey, eds., *Shelf Sediment Transport*, Stroudsburg, Pa.: Dowden, Hutchinson & Ross, pp. 225–248.

McCave, I. N. (1975). Vertical flux of particles in the ocean. *Deed-Sea Res.*, **22:** 491–502.

McCoy, F. W. (1969). Bottom currents in the western Atlantic Ocean between Lesser Antilles and the Mid-Atlantic Ridge. *Deep-Sea Res.*, **16:** 179–184.

Manheim, F. T., J. C. Hathaway, and E. Uchupi (1972). Suspended matter in the surface waters of the northern Gulf of Mexico. *Limnol. Oceanogr.*, **17:** 17–27.

Manheim, F. T., R. H. Meade, and G. C. Bond (1970). Suspended matter in surface waters of the Atlantic continental margin from Cape Cod to Florida Keys. *Science*, **167:** 371–376.

Meade, R. H. (1969). Landward transport of bottom sediments in estuaries of the Atlantic Coastal Plain. *J. Sediment. Petrol.*, **39:** 222–234.

Meade, R. H. (1972). Transport and deposition of sediments in estuaries. In B. W. Nelson, ed., *Environmental Framework of Coastal Plain Estuaries*. Geol. Soc. Am. Mem. 133, pp. 91–120.

Menzel, D. W. (1967). Particulate organic carbon in the deep sea. *Deep-Sea Res.*, **14:** 229–238.

Moore, D. G. (1969). *Reflection Profiling Studies of the California Continental Borderland: Structure and Quaternary Turbidite Basins*. Geol. Soc. Am. Spec. Pap. 107, 142 pp.

Moss, J. E. (1971). Petroleum—The problem. In D. W. Hood, ed., *Impingement of Man on the Oceans*. New York: Wiley-Interscience, pp. 381–419.

Nelson, D. D., J. W. Pierce, and D. D. Colquhoun (1973). Sediment dispersal by cascading coastal water. *Geol. Soc. Am. Abstr. Programs*, **8:** 423–424.

Nota, D. J. G. (1958). Sediments of the western Guinea shelf. *Meded. Land. Wageningen*, **58:** 98 pp.

Padan, J. W. (1971). Marine mining and the environment. In D. W. Hood, ed., *Impingement of Man on the Oceans*. New York: Wiley-Interscience, pp. 553–561.

Pak, H. (1974). Distribution of suspended particles in the Equatorial Pacific Ocean. In R. J. Gibbs, ed., *Suspended Solids in Sea Water*. New York: Plenum Press, pp. 261–270.

Parke, M. L., K. O. Emery, R. Szymankiewicz, and L. M. Reynolds (1971). Structural framework of continental margin in South China Sea. *Am. Assoc. Pet. Geol. Bull.*, **55:** 723–751.

Partheniades, A. (1972). Results of recent investigations on erosion and deposition of cohesive sediments. In H. W. Shen, ed., *Sedimentation*, Einstein Tribute. Ft. Collins, Colo., pp. 20-1–20-39.

Partheniades, E., R. W. Cross, III, and A. Ayora (1969). Further results on the deposition of cohesive sediments. *Proc. 11th Conf. Coastal Eng.*, ASCE, New York, pp. 723–742.

Pierce, J. W. (1970). Clay mineralogy of cores from the continental margin of North Carolina. *Southeast. Geol.*, **12:** 33–51.

Pierce, J. W., D. D. Nelson, and D. J. Colquhoun (1971). Pyrophyllite and talc in waters off the southeastern United States. *Mar. Geol.*, **11:** M9–M15.

Pierce, J. W., D. D. Nelson, and D. J. Colquhoun (1972). Mineralogy of suspended sediment off the southeastern United States. In D. J. P. Swift, D. B. Duane, and O. H. Pilkey, eds., *Shelf Sediment Transport*, Stroudsburg, Pa.: Dowden, Hutchinson & Ross, pp. 281–305.

Pilkey, O. H. and M. E. Field (1972). Lower continental rise east of the Middle Atlantic States: Predominant sediment dispersal perpendicular to isobaths: Discussion. *Geol. Soc. Am. Bull.*, **83:** 3537–3538.

Plank, W. S., H. Pak, and J. R. V. Zaneveld (1972). Light scattering and suspended matter in nepheloid layers. *J. Geophys. Res.*, **77:** 1689–1694.

Plank, W. S., J. R. V. Zaneveld, and H. Pak (1973). Distribution of suspended matter in the Panama Basin, *J. Geophys. Res.*, **78:** 7113–7121.

Prospero, J. M. and T. N. Carlson (1972). Vertical and aereal distribution of Saharan dust over the western equatorial North Atlantic Ocean. *J. Geophys. Res.*, **77:** 5255–5265.

Rabinowitz, P. D. and S. L. Eittreim (1974). Bottom current measurements in the Labrador Sea. *J. Geophys. Res*, **79:** 4085–4090.

Rodolfo, K. S. (1969). Suspended sediments in surface Andaman Sea water off the Irrawaddy Delta, northeastern Indian Ocean. *Geol. Soc. Am. Abstr.*, **7:** 190–191.

Rodolfo, K. S., B. A. Buss, and O. H. Pilkey (1971). Suspended sediment increase due to Hurricane Gerda in continental shelf waters off Cape Lookout, North Carolina. *J. Sediment. Petrol.*, **41:** 1121–1125.

Sakamoto, W. (1972). Study on the process of river suspension from flocculation to accumulation in estuary. *Bull. Ocean Res. Inst., Tokyo Univ.*, **5:** 46 pp.

Schneider, E. D., P. J. Fox, D. C. Hollister, H. D. Needham, and B. C. Heezen (1967). Further evidence of contour currents in the western North Atlantic. *Earth Planet. Sci. Lett.*, **2:** 351–359.

Schubel, J. R. and A. Okubo (1972). Comments on the dispersal of suspended sediment across the continental shelves. In D. J. P. Swift, D. B. Duane, and O. H. Pilkey, eds., *Shelf Sediment Transport*. Stroudsburg, Pa.: Dowden, Hutchinson & Ross, pp. 333–346.

Scruton, P. C. and D. G. Moore (1953). Distribution of surface turbidity off Mississippi Delta. *Am. Assoc. Pet. Geol. Bull.*, **37:** 1067–1074.

Sheldon, R. W., A. Prakash, and W. H. Sutcliffe, Jr. (1972). The size distribution of particles in the ocean. *Limnol. Oceanogr.*, **17**: 327–340.

Shepard, F. P. and N. F. Marshall (1969). Currents in La Jolla and Scripps submarine canyons. *Science*, **165**: 177–178.

Shepard, F. P. and N. F. Marshall (1973). Currents along the floors of submarine canyons. *Am. Assoc. Pet. Geol. Bull.*, **54**: 244–264.

Siegel, F. R. (1973). Possible important contributors to Argentine Basin lutites: Argentine Rivers. *Mod. Geol.*, **4**: 201–207.

Stanley, D. J. (1974). Pebbly mud transport in the head of Wilmington Canyon. *Mar. Geol.*, **16**: M1–M8.

Stanley, D. J., P. Fenner, and G. Kelling (1972a). Currents and sediment transport at the Wilmington Canyon shelfbreak, as observed by underwater television. In D. J. P. Swift, D. B. Duane, and O. H. Pilkey, eds., *Shelf Sediment Transport*. Stroudsburg, Pa.: Dowden, Hutchinson & Ross, pp. 621–644.

Stanley, D. J., H. Sheng, and C. P. Pedraza (1972b). Lower con-

tinental rise east of the Middle Atlantic States: Predominant sediment dispersal perpendicular to isobaths: Reply. *Geol. Soc. Am. Bull.*, **83**: 3539–3540.

Stanley, D. J., H. Sheng, and C. P. Pedraza (1971). Lower continental rise east of the Middle Atlantic States: Predominant sediment dispersal perpendicular to isobaths. *Geol. Soc. Am. Bull.*, **82**: 1831–1840.

Stefansson, U., L. P. Atkinson, and D. F. Bumpus (1971). Hydrographic properties and circulation of North Carolina shelf and slope waters. *Deep-Sea Res.*, **18**: 383–420.

Turekian, K. K. (1971). Rivers, tributaries, and estuaries. In D. W. Hood, ed., *Impingement of Man on the Ocean*. New York: Wiley-Interscience, pp. 9–73.

Wilson, R. B., P. H. Monaghan, A. Osanik, L. C. Price, and M. A. Rogers (1974). Natural marine oil seepage. *Science*, **184**: 857–865.

Zimmerman, H. B. (1971). Bottom currents on the New England continental rise. *J. Geophys. Res.*, **76**: 5865–5876.

Sedimentation and Environmental Management

Part IV consists of a series of essays which apply the principles outlined in the preceding chapters to problems of environmental management. Effective management of the continental margin environment requires an awareness of the chemical and biological systems as well as of the system of sediment transport. Chapter 19 by P. G. Hatcher and D. A. Segar, and Chapter 20 by S. Saila survey these broad topics. Chapters 21 through 23 deal primarily with problems in applied physical sedimentation. D. B. Duane describes techniques of beach and harbor defense in Chapter 21, while H. D. Palmer (Chapter 22) considers problems of seafloor stability with respect to the emplacement of structures. In Chapter 23, D. B. Duane assesses the impact of sand and gravel mining on the shelf floor. C. Hard and H. D. Palmer examine the problems associated with ocean dumping in Chapter 24; these problems are notably interdisciplinary in nature, and the reader is advised to review the chapters dealing with chemical and biological oceanography.

CHAPTER 19

Chemistry and Continental Margin Sedimentation

PATRICK G. HATCHER and DOUGLAS A. SEGAR

Atlantic Oceanographic and Meteorological Laboratories, Miami, Florida

The coastal zone is an area that has been defined as the junction between two major biomes where the land meets the ocean (Ketchum, 1972). Since many processes in both environments are intensified at their boundaries, the continental shelves, although limited in size, are areas of dynamic change. Therefore, the coastal zone has been known as an area of intense hydraulic, depositional, chemical, and biological activity.

The chemical transformations occurring during sedimentation on the continental shelf are of fundamental importance in environmental processes such as the corrosion of materials, removal of contaminants from the water column, provision of food for benthic organisms, renewal of nutrient supplies in the water column, and transport of sediment.

Whenever man has used the ocean, chemical changes have been brought about in that environment. These effects are at their greatest on the continental margins where man's presence is pervasive and his impact increasingly intense. Shipping, ocean dumping, river discharge, offshore drilling, dredging, and bottom trawling all affect the chemistry of the ocean ecosystem, often in subtle ways. The change presenting probably the greatest potential for environmental degradation is the addition of toxic contaminants to the ecosystem. Solid phases are good scavengers for many contaminants so that perhaps the most important feature of continental shelf sedimentation from an environmental standpoint is the detoxification of contaminants and their removal to the sediments. Such removal may be essentially permanent or may be only temporary, with the toxic chemical species being released during diagenesis and bioturbation of the sediments.

THE SIGNIFICANCE OF THE SHELF ENVIRONMENT TO MAN'S ACTIVITIES

Aside from being the most productive areas in the world oceans, the continental shelves are the nursing grounds for a majority of the commercially important marine organisms. The nations of the world annually harvest over 10 billion pounds of commercial fish from the coastal zone (Pruter, 1972). Therefore, the natural balance of the shelf environment is important to the many fisheries that have established themselves along the coast. Although the bounty is plentiful, the natural processes that control the populations are delicately balanced. Biological systems of the shelf are subjected to extreme chemical and physical changes in their environment, and many organisms subsist at or near their tolerance limits for such changes. Any imbalance in these environmental conditions (especially chemical) may alter the biological structure.

The shelf environment provides an abundance of mineral resources. Recently, the supply of petroleum products has become critically short. The continental shelves house a tremendous supply of untapped oil. In the United States alone the amount exceeds five times the present known land reserve of oil (American Gas Association, 1971). It is expected that extensive drilling for oil will occur on our continental shelves in the near future. The shelf environment will feel the impact, with its delicate chemical balance threatened by petroleum products, some of which are highly toxic.

The large sand and gravel deposits located near metropolitan areas such as New York and Boston are a significant resource (Schlee et al., 1971). The demand for large quantities of sand and gravel in the building industry has generated interest in these deposits as a valuable source for the expanding needs of the future. Mining of these deposits could alter the biological and chemical balance of the shelf by releasing excess nutrients, organic materials, and toxic substances which had previously been removed from the natural chemical cycle.

Various recreational activities such as boating, sport fishing, and swimming constitute important uses of the continental shelf. Loss of aesthetic quality of water and sediment would directly impact these recreational facilities. Inversely, the ever increasing use of the continental shelf for recreation has led to a steady decline in water quality over the years, again endangering the delicate chemical and biological balance.

Finally, because of its proximity to land and its accessibility, the coastal zone has been, for centuries, the area where major population centers and industries have evolved. The continental shelf has felt this effect through the centuries. Now, as populations increase, the shelf is becoming endangered by the many chemical and biological wastes associated with urban growth. In metropolitan New York alone, 4.6 million metric tons of wastes are being dumped onto the shelf annually. These wastes include dredged material, construction wastes, acid-iron wastes, and sewage sludge (Gross, 1972). Obviously, the chemical constituents present in these wastes have a significant effect on the chemical and biological balance in the New York Bight.

CHEMICAL CHARACTERISTICS OF NEARSHORE SEDIMENTS

Bottom sediments are extremely complex mixtures of many different solid phases. Deep-sea sediments are predominantly fine-grained in texture. In contrast, nearshore sediments comprise a large range of grain sizes. Because of the many physical forces tending to redistribute these sediments, the spatial heterogeneity in composition is very large.

The principal phases of marine sediments include biologically and abiologically produced carbonates, alumino-silicate minerals, quartz, iron and manganese hydroxide phases, biogenic silica, and biologically produced organic matter. The relative proportions of these components are determined by a multitude of factors.

However, one factor is predominant in the coastal zone—the presence of river runoff. Despite the relatively high productivity of many coastal areas, calcium carbonate contents of continental margin sediments are low particularly adjacent to the major river systems. The major components of shelf sediments are quartz, amorphous silica, various feldspars, clay minerals, and a wide variety of heavy minerals all of which are characteristic of river drainage.

Coastal sediments in most instances differ from deep ocean sediments only in the wider range of concentrations of organic matter and trace components. Coastal sediments can range from 0.2 to 10% organic carbon (Gross, 1972) or in anoxic basins up to as much as 7% organic carbon (Richards, 1965). Generally, nearshore sediments have relatively high silica contents derived from river runoff. Exceptions are found where this runoff is quantitatively small or where carbonate precipitation occurs, as in the Bahamas. The trace element concentrations of nearshore sediments generally are lower than in deep ocean sediments and tend to approximate those of igneous rocks (Table 1). This is simply a reflection of the dominance of erosional debris from the continents in most nearshore sediments.

TABLE 1.   The Distribution of Trace Elements in Marine Sediments and Igneous Rocks

Trace Element	Igneous Rocks* (ppm)	Nearshore Sediments† (ppm)	Deep-Sea Clays‡ (ppm)
Cr	65	100	77
V	130	130	330
Cu	42	48	570
Pb	13	20	160
Ni	50	55	290
Co	19	13	120
Sn	2	21	20
Ba	530	750	2200
Sr	340	<250	590
Zr	150	160	150
Ga	17	19	20

* Average trace element content of basalts, granites, and granodiorites. Adapted from Turekian and Wedepohl (1961).

† Adapted from Wedepohl (1960) with the exception of Sn and Sr which are taken from El Wakeel and Riley (1961).

‡ Adapted from El Wakeel and Riley (1961), Goldberg and Arrhenius (1958), and Young (1954).

Organic matter in nearshore sediments is generally a complex mixture arising from a multiplicity of sources. Inputs derived from the leaching of terrigenous organic matter and from phytoplankton in shelf waters all combine to form this complex mixture. Although the major portion of organic matter is uncharacterized, some of the more significant contributors are known. These include humic acids which are primarily derived from terrigenous soils (Kononova, 1966), carbohydrates, amino acids, and proteins. Carbohydrates generally constitute approximately 5% of the organic matter (Rittenberg et al., 1963; Degens, 1967). However, in some cases, up to 35% of the organic matter is composed of carbohydrates (Segar et al., 1975; Hatcher and Keister, in preparation). Amino acids generally constitute between 10 and 90% of the organic matter in Recent shelf sediments (Degens, 1967). Other components such as fatty acids, hydrocarbons, phenolic compounds, and aromatic compounds constitute less than 1% of the organic matter.

SOURCES OF CHEMICAL CONSTITUENTS

The river transport of sediment to the oceans has been estimated at approximately 3.6×10^{15} g/year (Turekian, 1971). However, only a small fraction reaches the deep sea with the major portion being deposited on the continental shelf and slope. The primary constituents of these sediments are quartz, various feldspars, clay minerals, and others which are primarily dependent on the igneous rock compositions in the watersheds. The organic matter so derived is characteristic of soil organic matter from the watershed and therefore is high in humic materials.

Hydrogenous components constitute a major fraction of continental shelf sediments. These include materials resulting from the formation of solids from dissolved components by inorganic reactions in the water column (Goldberg, 1954). Carbonates, phosphates, some silicates, sulfates, oxides, and hydroxides are the major components of this class.

Biogenous material is another significant component of sedimentary material on the continental shelf. This includes carbonate shell debris, siliceous skeletons, and indigenous organic matter.

A small but often significant fraction of marine sediments is eolian-transported material. Volcanic and wind-blown debris and, more recently, volatile synthetic organic materials and industrial trace metal aerosols are all introduced to marine sediments via the atmosphere.

EFFECTS OF COASTAL PROCESSES ON THE DISTRIBUTIONS OF CHEMICAL SPECIES

Importance of the Estuarine Interface

Most of the chemical species, including contaminants, reach the ocean via the rivers of the world. They enter the oceans both in solution and as suspended particles. As river water mixes with seawater, significant changes take place in the physical and chemical environment in the water column. Ionic strength is increased and the pH usually rises significantly. These changes produce dramatic shifts in the equilibria between the solid and dissolved phases. In this context, the estuarine interface should be defined as including not only the estuary itself but also the area of open ocean within which the bulk of marine sediments are initially deposited.

The changes in equilibria occurring between dissolved and particulate chemical species in passing the estuarine interface are poorly known. The general events observed for trace metals (Turekian, 1971) and many organic compounds are as follows:

1. The increased concentrations of major ions in solution compete effectively for adsorption and ion and ligand exchange sites on particle surfaces, releasing the sorbed contaminants to solution (Kharkar et al., 1968).

2. The increased ionic strength and conductivity of the solution induces a "salting out" effect that causes finely divided colloidal and particulate matter to aggregate and settle out.

3. The increased pH and ionic strength together interact to reduce the solubility of a number of elements. Iron and manganese precipitate as hydroxides, and aluminum, silicon, and several other elements may precipitate as complex hydrated oxides. Dissolved organic materials such as fulvic and humic acids may also precipitate at this interface.

The three processes known to occur in the estuarine interface take place at different rates and at different geographical locations in the mixing zone determined by a complex of environmental factors not yet understood. Turekian (1971) has described the general features of trace metal cycles in Long Island Sound (Fig. 1). The Sound is essentially the seaward end of an estuary, fresh water being added at the southern end and from numerous rivers and streams along both its shorelines. The chlorosity distribution (Fig. 2) indicates that progressive mixing with seawater takes place along the length of the Sound. At the eastern end of the Sound, seawater enters as a bottom layer and lower salinity water leaves at the surface, predominantly in the upper 20 ft of the water column. The distributions of dissolved cobalt, nickel, and silver in and just outside the Sound are all similar (Figs. 3–5). Despite the considerable river-borne input of industrial contaminants and natural metals, the dissolved trace element concentrations are relatively uniform along the length of Long Island Sound. In fact, the water leaving the Sound appears to have somewhat lower trace metal concentrations that the sea-

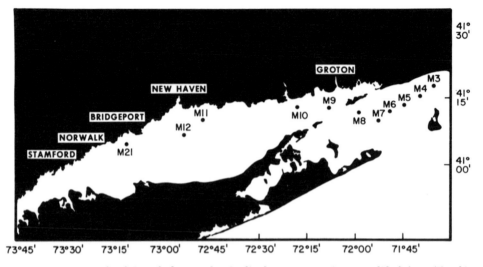

FIGURE 1. *Long Island Sound showing longitudinal transect stations. Modified from Turekian* (*1971*).

water entering. This illustrates the efficiency with which sedimentation processes in the estuarine interface remove contaminants from the water column.

Mechanism of Scavenging from Solution

In order to be able to predict and understand the fate of dissolved solids added to the oceans in the coastal zone, we must establish the mechanisms by which these elements or compounds are removed from seawater and added to the sediments. The scavenging mechanisms of elements, particularly trace metals, have been the subject of extensive study during the history of oceanography. Despite this effort, we still have only a very incomplete knowledge of the factors that control the concentrations of most elements in solution.

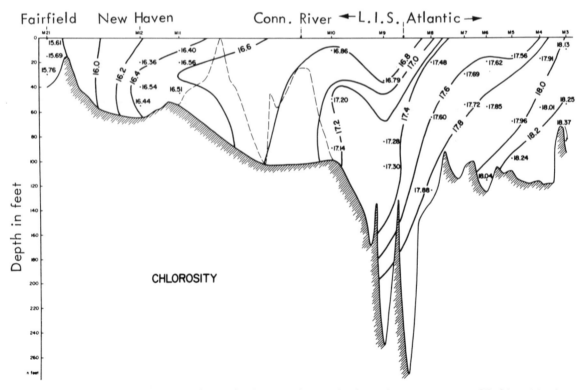

FIGURE 2. *Chlorosity distribution on longitudinal section of Long Island Sound, summer 1964. Modified from Turekian* (*1971*).

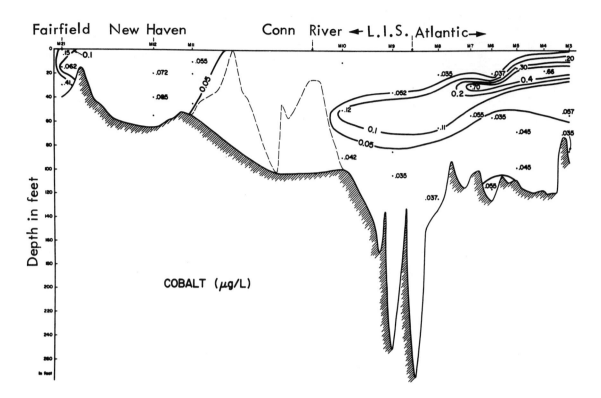

FIGURE 3. *Cobalt distribution on a longitudinal section of Long Island Sound, summer 1964. Modified from Turekian (1971).*

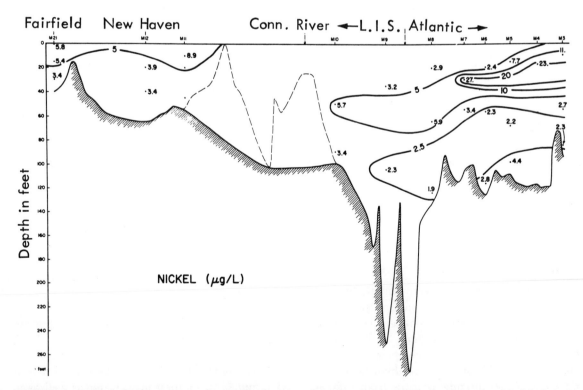

FIGURE 4. *Nickel distribution on a longitudinal section of Long Island Sound, summer 1964. Modified from Turekian (1971).*

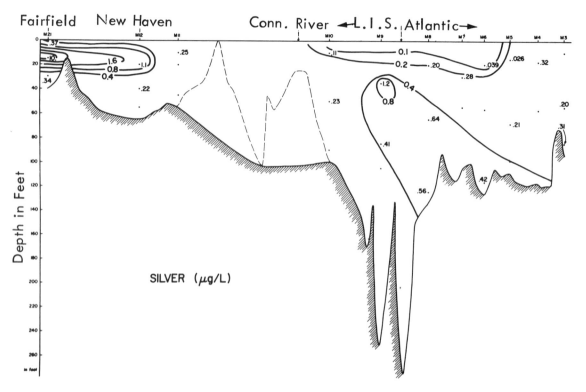

FIGURE 5. *Silver distribution on a longitudinal section of Long Island Sound, summer 1964. Modified from Turekian (1971).*

Krauskopf (1956) conducted a theoretical and experimental evaluation of direct precipitation, adsorption, and biological scavenging processes in controlling the concentration of various elements in seawater. His conclusions have remained little changed despite the considerable amount of subsequent research. The concentrations of all but a very few elements (Ca, Sr, Ba) in open ocean water are many times below their solubilities. Therefore, direct precipitation of their insoluble salts cannot occur. An exception to this rule occurs in anoxic waters (e.g., Norwegian fjords, Black Sea, etc.) where precipitation of many elements with extremely insoluble sulfides may occur. Another exception occurs at the estuarine interface. River waters, having a lower ionic strength and lower pH than ocean water, contain high concentrations of elements such as silicon, aluminum, iron, and manganese. These elements all have very insoluble hydrated oxides which precipitate at the interface.

The chemical interactions between solid surfaces and solutes are extremely complex. Mechanisms such as adsorption, ion exchange, ligand exchange, coprecipitation, and mixed crystal formation can all lead to the scavenging of elements or organic compounds from solution. Krauskopf (1956) showed that hydrous manganese and ferric oxides, organic detritus from different sources, and certain clay minerals are all effective in removing metals from seawater, whereas quartz and calcium carbonate are not. Lowman et al. (1966) have observed the precipitation or scavenging of several elements in experiments where filtered river and seawater were mixed.

The ocean waters of the continental margins generally contain much larger quantities of suspended particles than are found in the deeper oceans (Manheim et al., 1970; Buss and Rodolfo, 1972). In addition, a large fraction of the particles suspended in the coastal zone have been recently introduced to the marine environment and are out of equilibrium with the dissolved phase. The surface activity of such particles is therefore great, and they are potentially efficient scavengers of dissolved trace components. Hence, the coastal zone is an area where efficient scavenging of allochthonous constituents occurs.

Composition of Seawater and the Concept of Residence Time

Knowledge of sediment chemistry is important for its application to studies of the transport of elements and organic compounds among the solid, dissolved, and biological phases. Over 75 elements have been detected or determined in solution in seawater as well as many organic compounds (Riley and Chester, 1971). These dissolved components are not in equilibrium with solid phases in the oceans. Therefore, transfers take place between the dissolved phase and both nonliving and living solid particles in the ocean. As there is a constant supply of dissolved components to the oceans from the land, the net effect of these transports for most elements or compounds is a loss from the water to the sediments. Marine sediments act as a sink for material added to the ocean, although the removal and deposition may be either temporary or permanent.

The upper few centimeters of sediments generally contain trace elements in excess of those found in all of the overlying water column including suspended particulates and biota. This is illustrated by the distributions of Co, Fe, Mn, and Zn in a Texas Bay ecosystem (Parker et al., 1963); see Table 2. Similar distributions have been observed in coastal ecosystems in many parts of the world. As the upper layer of sediments in the continental shelf undergoes extensive bioturbation and reworking by wave, tide, and current energy, transport of chemical compounds between the overlying water column and the sediments down to depths of 10 cm or more will take place readily and will be more rapid than similar processes in the deep ocean where diffusion is the dominant transport mechanism.

TABLE 2. Distribution of Total Mass of Cobalt, Iron, Manganese, and Zinc in Water, Sediments, and Biota of a Texas Salt Marsh

Element	Mass (mg/m²)			
	Bay Water (1 m depth)	Sediments (upper 3 cm)	Plants	Animals
Co	0.5	12.0	3.8	0.0012
Fe	30	30,000	1,496	0.414
Mn	5	2,740	700	0.075
Zn	8	900	350	0.173

Source Adapted from Parker et al. (1963).

The composition of river water is very different from that of the ocean. Some elements, notably sodium, potassium, chlorine, calcium, magnesium, and strontium, are found in very much lower concentrations in river water than in seawater. Others (trace transition elements, aluminum, and silicon) are found in higher concentrations in river discharge than in the ocean. As the composition of river water has not changed drastically over geological time, the composition of the oceans does not necessarily reflect the accumulation of river water in the ocean basins. Water itself is removed from the

oceans by evaporation and recycled by rainfall. About 0.002% of the water in the oceans is recycled by this mechanism each year. Although some dissolved salts are entrained in aerosols and recycled by rivers, another mechanism (removal to the sediments) must be operative in order for the present concentrations of elements in the ocean to have been achieved. The concentrations of elements in seawater therefore represent a balance between their input from rivers and atmospheric aerosols and their removal to the sediments.

From a knowledge of the total quantity of an element dissolved in the ocean and either the rate of input from land or the sedimentation rate, the mean residence time of an element in seawater can be calculated (Goldberg, 1965). The mean residence time is simply the time it would take new inputs to double the existing concentration in the absence of removal mechanisms. The residence time is also the length of time required for sedimentation to remove all of an element from the oceans if no new inputs were made and if the sedimentation rate for the element did not depend on its concentration. The observed residence times for various elements range from as little as 100 up to hundreds of millions of years (Riley and Chester, 1971). The alkali and alkaline earth elements have residence times in excess of 10^6 years, as the processes of sedimentation do not efficiently scavenge them from solution. In contrast, elements such as aluminum, iron, chromium, titanium, beryllium, and thorium have residence times of less than 1000 years. No calculated residence times are available for organic compounds because of the limited knowledge of their concentration, distribution, and degradation rates in rivers, the ocean, and marine sediments.

Mean residence time defines the average length of time that any atom will spend in the ocean before being removed to the sediments. However, the residence time of some compounds may be much shorter in the coastal zone where sedimentation is more intense. In addition, the sedimentation of elements may occur much more rapidly than calculated from sediment accumulation rates, as much of the biologically sedimented matter will be readily decomposed and elements returned to the water column. Such internal cycling of elements will not affect the ocean mean residence time but may be critically important in determining the short-term fate of contaminants introduced to the ocean.

Influence of Seasonal Changes

Although different for the various geographical areas, the seasonal variations in water column physics and chemistry greatly influence the distribution of chemical species in shelf sediments.

In the spring, the water column is relatively uniform in chemical composition except at the mouth of major rivers. There, a halocline often exists because of the increased river flow which also supplies dissolved and particulate chemical species to the water column. Increased sunlight causes an increase of photosynthetic activity with its associated production of large quantities of organic matter and other biogenous materials.

Organisms extract many chemical species (both natural and contaminant) from the dissolved state during growth. These species are either returned to the water column or retained in the sediments when the organisms die. Bordovskiiy (1965) points out that biological production is concentrated just shoreward of the continental slope and just seaward of the shore zone, and that biogenic organic matter in association with fine suspended sediments settles in depressions on the shelf. The Hudson Shelf Valley is one such depression where deposition of fine sediments has increased the total organic carbon concentrations compared to the sandy shelf sediments (Fig. 6).

In the summer months, solar heating of the water column creates a thermal stratification in most temperate regions. During this time, photosynthetic activity is reduced because of depletion of nutrients in the surface layers. The sedimentation of biogenic material therefore decreases.

Reduced wind and wave activity are characteristic of the summer months. The various sedimented materials accumulate and postdepositional changes begin to occur in sediments little disturbed except by burrowing and filter-feeding organisms.

In the fall and winter months, wind and wave activity increase, thereby breaking up the thermocline. The water column becomes mixed, replenishing nutrients to the surface layer, and photosynthetic activity may increase if light is available. Because of the increased hydraulic energy which ultimately affects most areas of the shelf, sediments now are resuspended and transported away from the original deposition site. This affects the distribution of chemical constituents in the sediments as precipitation and dissolution processes occur. Chemical species formed in the sediments by postdepositional processes may be dissolved and reprecipitated elsewhere.

Influence of Sediment Transport Processes

Erosion of nearshore or shoreface sediments can supply tremendous amounts of sedimentary materials to the shelf (see Chapter 14). These and other shelf sediments

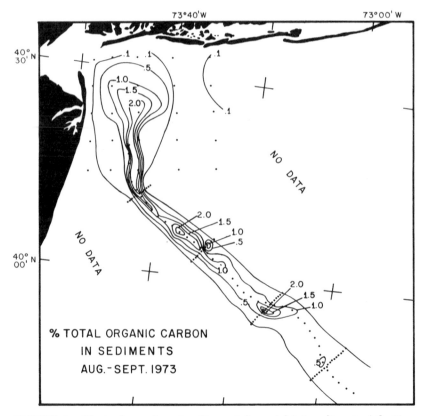

FIGURE 6. *The total organic carbon (percent dry weight) in sediments of the New York Bight (the dots represent the sampling stations).*

are distributed by various transport processes. The chemical compositions of shelf sediments at different locations are determined by both physical sorting and chemical changes occurring during transport. Dissolution and precipitation of various chemical phases occur. Deposited organic materials, for example, are subjected to microbial decomposition processes leading to dissolution of more labile materials and retention of the refractory material. For this reason, organic matter associated with highly mobile sediments takes on a more refractory character than organic matter associated with sediments accumulating in low-energy environments. As organic matter is usually associated with fines (Hunt, 1961; Bordovskiiy, 1965; Froelich et al., 1971) the high organic carbon concentration in mud bottoms on the shelf is not unexpected. Studies in the New York Bight (Fig. 6) and the Hudson Canyon (Fig. 7; Keller, 1973) indicate that organic carbon contents are highest in the depressions. Similar considerations may also be important in controlling the trace element distributions (Fig. 8).

Effects of Tidal Excursions

In the open ocean, tidal variations are small. However, as sea meets land, tidally induced fluctuations in the water level have a profound effect on erosional processes, and thereby on the distributions of chemical compounds.

Along the eastern coast of the United States, the tidally drained coastal wetlands are primarily characterized by grassy marshes and mangrove swamps. These wetlands are highly productive areas (Teal and Teal, 1969) with the detritus feeders forming the base of the food chain. They are also notable as sinks for sediments and as nutrient traps (Grant and Patrick, 1969). Although they are important scavengers of certain chemical species, the coastal wetlands release other compounds such as organic matter in the form of humic materials. With each ebb and flood tide, chemical exchange takes place between the tidal flats and the shelf waters. Many of the outgoing chemical constituents eventually precipitate on the shelf.

Influences of the Benthic Communities

As light penetrates to the bottom in many parts of the shelf and as production of detritus in the overlying water column is high, the shelf sediments support diverse and often abundant benthic biota. Attached macroalgae, various filter-feeding organisms, numerous classes of worms, and an abundance of microbial species are constantly altering the chemical balance at or below the sediment-water interface. These organisms utilize the various sedimented components as food sources, with their biomass and excreta becoming important components of shelf sediments. It is estimated that 75% of the benthic fauna on the shelf depends on detritus and associated organisms for a food source (Riley and Chester, 1971). A large proportion of organic matter produced by photosynthetic organisms is hydrolyzed by bacteria before reaching the sediments (Menzel and Goering, 1966). However, in shelf sediments, as much as 30 to 60% of

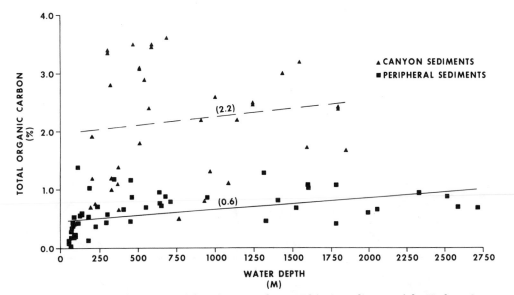

FIGURE 7. *The total organic carbon (percent dry weight) in sediments of the Hudson Canyon and adjacent shelf. From Keller (1973).*

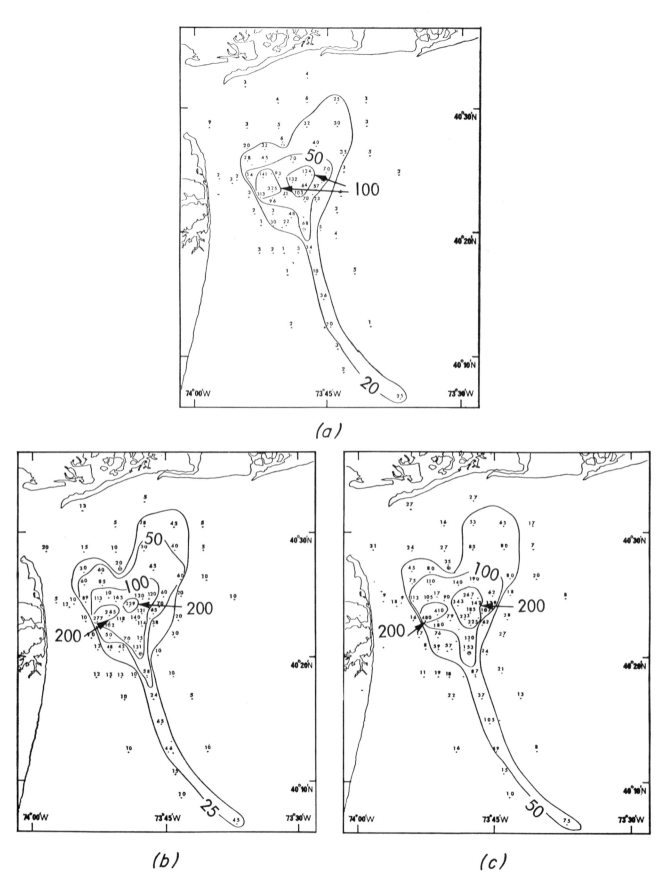

FIGURE 8. *Distribution of copper* (a), *lead* (b), *and zinc* (c) *in surface sediments from the New York Bight* (μg/g). *From Carmody et al.* (1973).

the organic matter is hydrolyzable and readily available to benthic organisms as a food source (Bordovskiiy 1965); see Table 3. A large portion of this readily hydrolyzable organic matter forms the cellular substance of bacterial biomass. As the organic matter is hydrolyzed or decomposed, a large fraction is returned to the water column as inorganic nutrients and other organic compounds (Riley and Chester, 1971).

TABLE 3. Distribution of Readily Hydrolyzable Organic Matter in Recent Bering Sea Sediments (Mean Data)

Type of Sediment	Total Organic Carbon (TOC) (% of dry sediment)	Content of Readily Hydrolyzable Carbon	
		% of Sediment	% of TOC
Medium-grained sands	0.32	0.19	59.4
Coarse silts	0.76	0.21	27.6
Fine silt muds	0.97	0.26	26.7
Silt-clay muds	1.38	0.30	21.8
Clay muds	0.54	0.14	25.7

Source. After Bordovskiiy (1965).

Chemical Changes in Sediments After Deposition

After deposition, sediments undergo diagenetic chemical changes that tend to bring them more nearly into equilibrium with their aqueous environment. This environment itself changes significantly. Oxygen in the sedimentary pore waters is consumed as organic matter introduced to the sediments is bacterially degraded. Unless the oxygen is renewed, sulfide formation takes place as bacteria utilize sulfate for oxidation of the organic matter. Within undisturbed sediments, diffusion is the only mechanism leading to renewal of oxygen in the pore waters (Fig. 9). Diffusion also permits solutes released during bacterial decomposition and chemical equilibrium to be released from the sediments. This is an extremely slow process. Transport of significant quantities of solutes through even 1 cm of sediment depth requires many years or even hundreds of years. Nevertheless, the distribution of elements in the sediments can be affected by postdepositional diffusion as illustrated by the distribution of manganese in certain deep-sea sediments and their pore waters (Fig. 2). Manganese is released to solution as its relatively soluble sulfide and diffuses upward, precipitating again in the oxidizing zone at the sediment surface. In this instance, manganese is not released into the water column, but elements or compounds released by sediments and not resorbed or precipitated may well diffuse out into the water column.

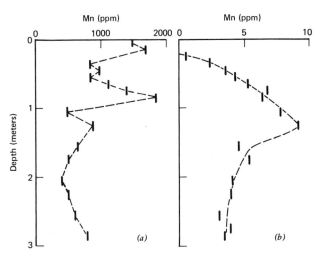

FIGURE 9. *The distribution of manganese in the sediment and interstitial water of an Arctic deep-sea core.* (a) *Distribution of manganese in the sediment.* (b) *Distribution of manganese in the interstitial water. From Riley and Chester (1971).*

In the coastal zone, sediments are continually resuspended and reworked both by biological action and by wave, tide, and current energy. Unless the sedimentation rate of organic matter is very rapid, the sedimentary pore waters do not maintain permanent concentrations of sulfide but are intermittently flushed with oxygenated seawater. Two important conditions must be fulfilled for metals and other constituents to be released to solution during diagenesis and transported back into the overlying water column. First, the decomposition of most of the detrital organic matter reaching the sediments and the equilibration of inorganic phases with seawater must take place before permanent burial is achieved. Migration of solutes into the overlying water column may thus take place by advective processes which are considerably faster than diffusion. Second, the periodic renewal of oxygen in the pore waters is necessary to either prevent metal sulfide precipitation or permit reoxidation and dissolution of such sulfides if they are formed. It is probable that significant release of metals and organic compounds occurs from coastal sediments particularly from organic matter and solid material dumped into the ocean directly. Unfortunately very little is known about these processes, although increased concentrations of dissolved trace metals have been observed in the water column close to the sediments in estuaries (J. H. Carpenter, personal communication), in the coastal zone, and even in the deep oceans (Segar, unpublished data). More information is required before it can be established with any degree of confidence whether or not the rapid removal of contaminants to the sediments at the estuarine interface and at outfalls and dumpsites is permanent or only temporary. Clearly this is important, as contaminants permanently held in the sediments can have little interaction with the biosphere.

Organic components undergo extensive diagenesis in the top layers of sediment. Proteins are immediately hydrolyzed by bacteria to free amino acids (Lindblom and Lupton, 1961; Hare, 1969) which dissolve in the pore waters and, later, may diffuse out. In fact, it has been suggested that extensive migration of amino acids occurs in certain coastal sediments (Rittenberg et al., 1963). Carbohydrates also undergo extensive hydrolysis to free sugars with increasing burial time (Degens et al., 1964; Hatcher, 1974). Since free sugars represent an important nutritional substance for many organisms, this process may be important in determining the ability of a particular sediment to support various benthic biota. In addition to carbohydrates and proteins, other relatively labile organic constituents undergo early diagenetic changes. These result in a predominance of resistant humiclike materials below the first few meters of shelf sediments.

Over geological time, organic constituents of coastal sediments undergo more extensive alteration and some may eventually be transformed to petroleum. Fatty acids may be converted to petroleum alkanes (Cooper and Bray, 1963; Eisma and Jurg, 1969). Chlorophyll derivatives may be converted to aromatic compounds and to the branched-chain components of crude oil (Orr et al., 1958). Various other changes in organic matter composition of sediments with time have also been delineated (Eglinton, 1969).

CHEMICAL FACTORS AFFECTING PHYSICAL PROPERTIES AND ERODIBILITY OF SEDIMENTS ON THE SHELF

Physical processes affect the distribution of chemical components in shelf sediments. Conversely, the chemical properties also exert a significant influence on the physical properties of sediments. No studies appear to have been reported concerning the influence of chemical properties on the susceptibility of sediment on the seafloor to erosion and resuspension. Variations in sediment physical properties due to chemical factors may be expected to affect sediment stability and ultimately sediment transport. A few studies have reported the relationships among chemical species and the mass physical properties of sediments. Chemical cementation and cohesion appear to be two properties largely controlled by sediment chemistry.

The effects of silica, carbonate minerals, and amorphous oxides on sediment cementation have been reported by Nacci et al. (1974). Organic compounds such

as barnacle cements are also important cementing agents. Cementation of mineral grains affects the transport of materials on the shelf by controlling both erodibility and effective particle size of the eroded material.

The cohesive property of a sediment is strongly affected by its organic components (Soderblom, 1966). Studies on various terrestrial soils have indicated that polysaccharides exert tremendous binding forces between mineral grains (Martin, 1971). Hulbert and Given (1975) suggest that organic matter may be important in determining the cohesive property (shear strength) of marine sediments. Bacterial mucus is composed mostly of mucopolysaccharide secretions and may also have an effect on the cohesion of finer particles on the shelf (see Chapter 18).

The ionic strength of pore waters has been shown to affect the cohesion of sediments (Lambe, 1958). The ionic chemical species control the electrostatic attractions of the clays within the bulk sediments.

Cementation and cohesion of a sediment occur because of chemical interactions between sediment grains. The amount of hydraulic activity needed to resuspend or transport the sediment increases with increased cohesion and cementation. For example, muds in the axis of the Hudson Shelf Valley are not swept away by currents which are known to be sufficient for transport of fine sands (Lavelle et al., 1975). Cohesion obviously plays a major role in maintaining the integrity of the substrate.

IMPACT OF HUMAN ACTIVITIES ON SEDIMENT CHEMISTRY

The oceans and their underlying sediments contain quantities of trace metals and natural organic materials that are vast when compared to the quantities that man releases. However, contaminant release by man takes place primarily at point sources in the coastal zone such as rivers, outfalls, and ocean dumpsites. Contaminant materials are mixed only slowly into the ocean, while they are often rapidly scavenged by the coastal sedimentation process. Therefore, anomalies have been observed in the chemistry of sediments close to shore which may be ascribed with reasonable certainty to human activities (Gross, 1972; NOAA, 1975; Segar and Pellenbarg, 1973).

Introduction of Contaminants

Anomalous sediment chemistry may be observed in areas of the shelf zone where man's activity is intense, although historical data are often missing and it is difficult to establish that these anomalies are not natural.

The sediment close to a power plant cooling water outfall at Turkey Point in Florida has anomalously high concentrations of a number of trace elements including V, Cu, and Ni (Segar and Pellenbarg, 1973); see Fig. 10. These anomalously high concentrations could be caused by emission of particulate metals from the power plant, emission of dissolved metals that are subsequently scavenged by entrained particles, or by concentration of naturally occurring metals due to geochemical or ecological changes other than direct metal release by the power plant.

In the New York Bight, dumpsites for sewage sludge, cellar dirt, dredge spoil, and acid waste have been used for many years (Gross, 1972). Anomalously high sedimentary concentrations of metals (Fig. 4) are found in the New York Bight but only in the topographic depressions (Carmody et al., 1973). Anomalous concentrations are not found immediately at the dumpsites themselves. It is probable that the dumped material, which is predominantly fine-grained, is simply redistributed by the active sediment transport processes of the continental shelf. Clearly in the absence of historical sediment chemistry data, the sediment transport processes must be understood before it can be concluded with certainty that the anomalous concentrations in the topographic lows are caused by contaminant material.

In addition to the metal contaminants, many synthetic organic compounds not found naturally are added to the oceans. Many of these compounds are not easily biodegraded (e.g., polychlorobiphenyls) and are concentrated in the sediments. Detectable concentrations of these types of compound have been found in many marine sediments, particularly those from the continental shelf (e.g., Murray and Riley, 1973). Recent studies of DDT and PCB compounds in sediments off the coast of California (McDermott and Heesen, 1974; Young, 1974) have indicated that increased levels of these contaminants are directly traceable to municipal waste effluents.

In the past decade, increasing concern has arisen over pollution of the shelf environment by petroleum products. More than 1 billion tons of crude oil and oil products is transported across the continental shelf each year. Oil spillage by accidental disasters such as the Torrey Canyon, by bilge pumping, and by tanker deballasting, contribute possibly between 5 and 10 million tons of petroleum per year to the marine environment (Ehrhardt and Blumer, 1972). Crude oil entering the marine environment is first subjected to extensive evaporation and dissolution, with the remaining fractions coagulating into floating tar lumps that may eventually settle to the sediments (Morris and Butler, 1973). Petroleum products also sorb to particulate matter and even-

tually become incorporated into sediments. In the process, they become available to the marine food chain and pose a long-term toxicity problem (Tissier and Oudin, 1973). Aromatic hydrocarbons are naturally present in very low concentrations in unpolluted sediments (Brown et al., 1972). However, their rate of incorporation into the sediments of heavily polluted areas is significant enough to achieve toxic levels (Tissier and Oudin, 1973).

Introduction of Radioactive Wastes

Recently the ocean dumping of radioactive wastes has aroused some concern for the shelf sediment environment. Peaceful uses of nuclear reactors, fuel fabrication plants, and reprocessing plants contribute on the order of 10^5 Ci/year (National Academy of Sciences, 1971). Iodine-151, strontium-90, cesium-137, tritium, argon-39, and krypton-88 are radionuclides that present major problems in the environment. Nuclear testing is also a potential hazard as atmospheric fallout is expected to affect shelf sediments.

Dredging and Mining Operations

Dredging and mining operations have increased recently as a result of accelerating urbanization and land development. Any sediment disturbance imposes an effect on the shelf environment as the sediment/water equilibrium is destroyed. Release of contaminants that have previously been removed from the system poses a significant stress on areas affected by mining operations. Additionally, entrainment of fine, silty material into the water column causes oversiltation which destroys the delicate food-gathering mechanisms of many filter-feeding organisms. The release of large quantities of sulfides and organic matter causes the water column to be stripped of essential oxygen for respiration. The decrease in water clarity affects light penetration, thereby decreasing photosynthetic activity.

Ocean Dumping

Large quantities of waste material including dredge spoil, solid wastes, incinerator residue, garbage, sewage sludge, chemical wastes, discarded military equipment and munitions, and construction debris are disposed of annually by dumping in the coastal zone (NOAA, 1974). For example, in the New York metropolitan area alone, 10^7 tons of sewage sludge, 2×10^7 tons of dredge spoil, and 6×10^6 tons of chemical wastes are annually released to the New York shelf environment (Environmental Protection Agency, 1975). This amount of contamination is bound to have an effect on the chemistry of the

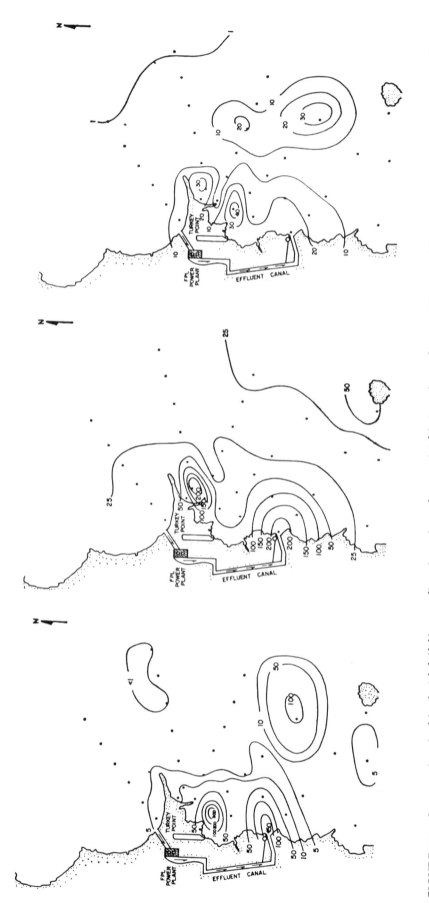

FIGURE 10. Concentrations (μg/g) of nickel (left), vanadium (center), and copper (right) in surface sediments of the area of Biscayne Bay, Florida, adjacent to the Turkey Point power plant cooling water discharge. Modified from Segar and Pellenbarg (1973).

sediments. The most notable changes that can be attributed to the sewage are the increase in organic carbon concentration in the sediments near the dumpsites, the increase in levels of trace metals (Carmody et al., 1973), and the reduction of the dissolved oxygen concentrations through BOD (Sandy Hook Laboratory, 1972). In fact, sediments in the dumpsite area have been labeled the "dead sea" because of the general absence of benthic organisms (Pearce, 1972).

SUMMARY

Chemical interactions involving sediments of the continental margins are extremely complex. River drainage with its associated chemical constituents, biological intensification, wind and wave activity, depositional and erosional processes, and last but not least the presence of man with his waste products interact to produce an extremely dynamic chemical ecosystem on the continental shelf. This complexity and the resulting heterogeneity have made it extremely difficult to describe the chemical characteristics of the coastal zone and the chemical interactions therein.

Nearshore sediments are complex mixtures of allochthonous and autochthonous constituents whose grain size range varies considerably between samples collected within several meters of each other. The relative proportions of allochthonous and autochthonous constituents varies as a function of distance from shore and proximity to major rivers. Aside from certain subtropical areas, the major components of shelf sediments are silica, various feldspars, clay minerals, a variety of heavy minerals, and a complex mixture of organic materials. The sources of the various constituents have been only partially and poorly identified. River transport supplies not only the shelf but also the deep oceans with considerable quantities of many of the various chemical compounds encountered in sediments and the water column. Both hydrogenous and biogenous components also comprise a significant fraction of shelf sediments. Eolian-transported constituents are thought to be only minor fractions in all marine sediments.

As chemical constituents are supplied to the shelf environment, various processes tend to redistribute these components among solid and dissolved phases. The most important of these processes occur when river water encounters ocean water at the estuarine interface. The increase in ionic strength causes various soluble and colloidal particles to precipitate. These precipitates scavenge other chemical constituents from solution. This process is an important regulating factor in the geochemical balance of many trace components in the world's oceans.

Several variations are extremely important in shelf waters. Seasonal changes in biological activity and river flow coupled to changes in the hydraulic regime constantly modify the nature and amounts of chemical compounds reaching shelf sediments. In addition, the redistribution of materials by sediment transport processes is seasonally dependent, and the sediments of the shelf have a continuously altering seafloor distribution. Different chemical constituents are associated with different particle sizes of material. Therefore physical sorting of shelf sediments affects the distribution of these constituents on the shelf.

Tidal fluctuations aid in movement of sediments and cause a corresponding effect on the distribution of chemical species. Although bioturbation is not considered to be a sediment transport process by some, it substantially affects the distribution of chemical constituents on the seafloor.

Once deposited, organic and inorganic chemical compounds undergo diagenesis whereby they are solubilized or precipitated. Diffusion, advection, and bioturbation all serve to transport the products of diagenesis within the sedimentary column and into the water column.

Just as physical and biological processes alter the distribution of chemical constituents on the shelf, variations of chemical constituents influence the physical processes of sedimentation. Cohesion and cementation of mineral grains may be controlled by organic and inorganic chemical agents, and these parameters affect the sediment transport processes.

The most important influence on the distribution of chemical constituents on the shelf in many areas is the presence of man and his many waste disposal and industrial practices. Adjacent to large urban centers, the shelf is constantly being supplied contaminants resulting from man's inability to recycle waste materials effectively. The shelf is used as a convenient repository for such materials as dredge spoils, cellar dirt, sewage and sewage sludge, chemical wastes, radioactive wastes, petroleum products, discarded military equipment, and various other chemical contaminants. Although the oceans are thought to be capable of assimilating vast quantities of our waste products, it is becoming clear that disposal of even limited quantities of contaminants in the shelf zone may cause critical problems. The chemistry of the shelf ecosystem is in a state of dynamic equilibrium, which may be destroyed if present practices are continued. Recent studies have indicated that the present contaminant loadings have caused significant changes in the chemistry of coastal sediments.

ACKNOWLEDGMENTS

We thank the NOAA Marine Ecosystem Analysis (MESA) program for their support of the New York Bight investigation.

REFERENCES

American Gas Association (1971). Reserves of crude oil, natural gas liquids, and natural gas in the United States and Canada and the U.S. productive capacity as of December 31, 1970. *Am. Gas Assoc., Am. Pet. Inst., Can. Petrol. Assoc.*, **25**: 1–256.

Bordovskiiy, O. K. (1965). Accumulation of organic matter in bottom sediments. *Mar. Geol.*, **3**: 33–82.

Broecker, W. S. (1974). *Chemical Oceanography*. New York: Harcourt Brace & Jovanovich, 214 pp.

Brown, F. S., H. J. Baedecker, A. Nissenbaum, and I. R. Kaplan (1972). Early diagenesis in a reducing fjord, Saanlich Inlet, British Columbia—III. Changes in organic constituents of sediment. *Geochim. Cosmochim. Acta*, **36**: 1185–1203.

Buss, B. A. and T. S. Rodolfo (1972). Suspended sediments in continental shelf waters off Cape Hatteras, North Carolina. In D. J. P. Swift, D. B. Duane, and O. H. Pilkey, eds., *Shelf Sediment Transport: Process and Pattern*. Stroudsburg, Pa.: Dowden, Hutchinson & Ross, pp. 263–279.

Carmody, D. J., J. B. Pearce, and W. E. Yasso (1973). Trace metals in sediments of New York Bight. *Mar. Pollut. Bull.*, **4**: 132–135.

Cooper, J. E. and E. E. Bray (1963). A postulated role of fatty acids in petroleum formation. *Geochim. Cosmochim. Acta*, **27**: 1113–1127.

Degens, E. T. (1967). Diagenesis of organic matter. In G. Larsen and G. V. Chilingar, eds., *Diagenesis in Sediments*. Amsterdam: Elsevier, pp. 343–390.

Degens, E. T., J. H. Reuter, and K. N. F. Shaw (1964). Biochemical compounds in offshore California sediments and sea waters. *Geochim. Cosmochim. Acta*, **28**: 45–63.

Eglinton, G. (1969). Organic geochemistry. The organic chemist's approach. In G. Eglinton and M. T. J. Murphy, eds., *Organic Geochemistry*. Berlin and New York: Springer-Verlag, pp. 20–73.

Eglinton, G. and M. T. J. Murphy, eds. (1969). *Organic Geochemistry*. Berlin and New York: Springer-Verlag, 828 pp.

Eisma, E. and J. W. Jurg (1969). Fundamental aspects of the generation of petroleum. In G. Eglinton and M. T. J. Murphy, eds., *Organic Geochemistry*. Berlin and New York: Springer-Verlag, pp. 676–698.

El Wakeel, S. K. and J. P. Riley (1961). Chemical and mineralogical studies of deep-sea sediments. *Geochim. Cosmochim. Acta*, **25**: 110–146.

Environmental Protection Agency (1975). *Ocean Disposal in the New York Bight*. Technical Briefing Report #2, USEPA, Region II, 86 pp.

Ehrhardt, M. and M. Blumer (1972). The source indentification of marine hydrocarbons by gas chromatography. *Environ. Pollut.*, **3**: 179–194.

Froelich, P., B. Golden, and O. H. Pilkey (1971). Organic carbon in sediments of the North Carolina continental rise. *Southeast. Geol.*, **13**: 91–97.

Goldberg, E. D. (1954). Marine geochemistry. *J. Geol.*, **62**: 249–265.

Goldberg, E. D. (1965). Minor elements in sea water. In J. P. Riley and G. Skirrow, eds., *Chemical Oceanography*, Vol. 1. New York: Academic Press, pp. 163–196.

Goldberg, E. D. and G. O. S. Arrhenius (1958). Chemistry of the Pacific pelagic sediments. *Geochim. Cosmochim. Acta*, **13**: 153–212.

Grant, R. R., Jr. and R. Patrick (1969). Tinnisum Marsh as a water purifier. In *Two Studies of Tinnisum Marsh, Delaware and Philadelphia Counties*. Washington, D.C.: Pa. Conserv. Found., pp. 80–118.

Gross, M. G. (1972). Geologic aspects of waste solids and marine waste deposits, New York metropolitan region. *Geol. Soc. Am. Bull.*, **83**: 3163–3176.

Hare, P. E. (1969). Geochemistry of proteins, peptides, and amino acids. In G. Eglinton and M. T. J. Murphy, eds., *Organic Geochemistry*. Berlin and New York: Springer-Verlag, pp. 438–463.

Hatcher, P. G. (1974). A study of the organic geochemistry of Mangrove Lake, Bermuda. M.S. Thesis, University of Miami, Florida. 231 pp.

Hood, D. W., ed. (1971). *Impingement of Man on the Oceans*. New York: Wiley-Interscience, 738 pp.

Hulbert, M. H. and D. N. Given (1975). Geotechnical and chemical property relationships for Wilkinson Basin, Gulf of Maine, sediments. *J. Sediment. Petrol.* (in press).

Hunt, J. M. (1961). Distribution of hydrocarbons in sedimentary rocks. *Geochim. Cosmochim. Acta*, **22**: 37–49.

Keller, G. K. (1973). Sedimentary dynamics within the Hudson submarine canyon. *Proc. of Symposium Relations sédimentaires entre estuaires et plateaux continentaux, Bordeaux, France, July 1973*, p. 49.

Ketchum, B. H. (1972). *The Water's Edge: Critical Problems of the Coastal Zone*. Cambridge, Mass.: MIT Press, 393 pp.

Kharkar, D. P., K. K. Turekian, and K. K. Bertine (1968). Stream supply of dissolved silver, molybdenum, antimony, selenium, chromium, cobalt, rubidium, and cesium to the oceans. *Geochim. Cosmochim. Acta*, **32**: 283–298.

Kononova, M. M. (1966). *Soil Organic Matter*, 2nd ed. New York: Pergamon, 544 pp.

Krauskopf, K. B. (1956). Factors controlling the concentrations of 13 rare metals in sea water. *Geochim. Cosmochim. Acta*, **9**: 1–32B.

Lambe, W. T. (1958). The structure of compacted clay. *J. Soil Mech., Found. Div.*, Pap. **1654**: 1–34.

Lavelle, J. W., G. H. Keller, and T. L. Clarke (1975). Possible bottom current response to surface winds in the Hudson shelf channel. *J. Geophys. Res.*, **80**: 1953–1956.

Lindblom, G. and M. D. Lupton (1961). Microbial aspects of organic geochemistry. *Dev. Ind. Microbiol.*, **2**: 9–22.

Lowman, F. G., K. K. Phelps, R. McClin, V. Roman De Vega, I. Oliver de Padovani, and R. J. Garcia (1966). Interactions of the environmental and biological factors on the distribution of trace elements in the marine environment. In *Disposal of Radioactive Wastes into Seas, Oceans and Surface Waters*. Vienna: IAEA, pp. 249–266.

McDermott, D. J. and T. C. Heesen (1974). Inventory of DDT in sediments. In *Coastal Water Research Project, Annual Report, 1974*, pp. 123–129.

Manheim, F. T., R. H. Meade, and G. C. Bond (1970). Suspended matter in surface waters of the Atlantic Continental margin from Cape Cod to the Florida Keys. *Science*, **167**: 371–376.

Martin, J. P. (1971). Decomposition and binding action of polysaccharides in soil. *Soil Biol. Biochem.*, **3**: 33–41.

Menzel, D. W. and J. J. Goering (1966). The distribution of organic detritus in the ocean. *Limnol. Oceanogr.*, **11**: 333–337.

Morris, B. F. and J. N. Butler (1973). Petroleum residues in the

Sargasso Sea and on Bermuda beaches. *Proc. Conf. Prevention and Control Oil Spills, 1973*, pp. 521–529.

Murray, A. J. and J. P. Riley (1973). The determination of chlorinated aliphatic hydrocarbons in air, natural waters, marine organisms and sediments. *Anal. Chim. Acta*, **65**: 261–270.

Nacci, V. A., W. E. Kelly, H. C. Wang, and K. R. Demars (1974). Strength and strain characteristics of cemented deep-sea sediments. In A. L. Inderbitzen, ed., *Deep-Sea Sediments, Physical and Mechanical Properties*. New York: Plenum Press, pp. 129–150.

National Academy of Sciences (1971). *Radioactivity in the Marine Environment*. Washington, D.C.: Natl. Acad. Sci., 272 pp.

NOAA (1974). *Report to the Congress on Ocean Dumping and Other Man-Induced Changes to Ocean Ecosystems, October 1972 through December 1973*, 84 pp.

NOAA (1975). *Technical Background Relating to Offshore Dumping Assessment—An Interim Report*. In R. L. Charnell, ed., NOAA Tech. Memo., ERL, MESA-1, 83 pp.

Orr, W. L., K. O. Emery, and J. R. Grady (1958). Preservation of chlorophyll derivatives in sediments off Southern California. *Bull. Am. Assoc. Pet. Geol.*, **42**: 925–962.

Parker, P. L., A. Gibbs, and R. Lawler (1963). Cobalt, iron and manganese in a Texas bay. *Publ. Inst. Mar. Sci., Univ. Texas*, **9**: 28–32.

Pearce, J. B. (1972). The effects of solid waste disposal on benthic communities in the New York Bight. In M. Ruivo, ed., *Marine Pollution and Sea Life*. London: FAO Publ., pp. 404–411.

Pruter, A. T. (1972). Foreign and domestic fisheries for groundfish, herring, and shellfish in continental shelf waters off Oregon, Washington, and Alaska. Woods Hole, Mass. Workshop on Critical Problems of the Coastal Zone, unpublished paper, 16 pp.

Richards, F. A. (1965). Anoxic basins and fjords. In J. P. Riley and G. Skirrow, eds., *Chemical Oceanography*, Vol. I. New York: Academic Press, pp. 611–646.

Riley, J. P. and R. Chester (1971). *Introduction to Marine Chemistry*. New York: Academic Press, 465 pp.

Riley, J. P. and G. Skirrow (1965). *Chemical Oceanography*. New York: Academic Press, Vol. 1, 712 pp.; Vol. 2, 508 pp.

Rittenberg, S. C., K. O. Emery, J. Hulsemann, E. T. Degens, R. C. Fay, J. H. Reuter, J. R. Grady, S. H. Richardson, and E. E. Bray (1963). Biogeochemistry of sediments in experimental Mohole. *J. Sediment. Petrol.*, **33**: 140–172.

Sandy Hook Laboratory (1972). *The Effects of Waste Disposal in New York Bight*. Final report submitted to U.S. Army Corps of Engineers, Coastal Eng. Res. Cent, 762 pp.

Schlee, J., D. Folger, and C. O'Mara (1971). *Bottom Sediments of the N.E. U.S.: Cape Cod to Cape Ann, Mass.* USGC Open File Report and Misc. Geol. Invest. Rep. No. 1-746.

Segar, D. A., P. G. Hatcher, G. A. Berberian, L. E. Keister, and M. A. Weiselberg (1975). The chemical and geochemical oceanography of the New York Bight apex region as it pertains to the problem of sewage sludge disposal. Part III. In R. L. Charnell, ed., *Technical Background Relating to Offshore Dumping Assessment—An Interim Report*. NOAA Tech. Memo., ERL, MESA-1, pp. 62–83.

Segar, D. A. and R. E. Pellenbarg (1973). Trace metals in carbonate and organic rich sediments. *Mar. Pollut. Bull.*, **4**: 138–142.

Soderblom, R. (1966). Chemical aspects of quick clay formation. *Eng. Geol.*, **1**: 415–431.

Stumm, W. and J. J. Morgan (1970). *Aquatic Chemistry, An Introduction Emphasizing Chemical Equilibria in Natural Waters*. New York: Wiley-Interscience, 583 pp.

Teal, J. M. and M. Teal (1969). *Life and Death of the Salt Marsh*. New York: Audubon Ballantine.

Tissier, M. and J. L. Oudin (1973). Characteristics of naturally occurring and pollutant hydrocarbons in marine sediments. *Proc. Conf. Prevention and Control Oil Spills, 1973*, pp. 205–214.

Turekian, K. K. (1971). Rivers, tributaries and estuaries. In D. W. Hood, ed., *Impingement of Man on the Oceans*. New York: Wiley-Interscience, pp. 9–74.

Turekian, K. K. and K. H. Wedepohl (1961). Distribution of the elements in some major units of the earth's crust. *Geol. Soc. Am. Bull.*, **72**: 175–191.

Wedepohl, K. H. (1960). Suprenanalytische Untersuchungen an Tiefseetonen aus dem Atlantik, Ein Beitrag zue Deutung der Geochemischen Sonderstellung von pelagischen tonen. *Geochim. Cosmochim. Acta*, **18**: 200–231.

Young, D. R. (1974). DDT and PCB in dated sediments. In *Coastal Water Research Project, Annual Report, 1974*, pp. 135–137.

Young, E. J. (1954). Trace elements in recent marine sediments. *Geol. Soc. Am. Bull.*, **65**: 13–29.

CHAPTER 20

Sedimentation and Food Resources: Animal-Sediment Relationships

SAUL B. SAILA

Graduate School of Oceanography, University of Rhode Island, Kingston, Rhode Island

Benthic communities and various classification schemes for them are briefly reviewed. It is shown that the unimodal negative binomial model describes the observed frequency distribution of benthic invertebrates from the New York Bight, and suggestions for sampling based on this information are made. A reanalysis of some benthic sample data is utilized to demonstrate a relationship between benthic biomass and sediment particle size on the Atlantic continental shelf.

A review is made of some of the impacts of man's activities, such as dredging and ocean waste disposal, on animal-sediment relationships. An important problem area is considered to be predicting the rate of recolonization of benthic organisms from disturbed sediments and estimating the nature of the new equilibrium number of species on these disturbed areas. A benthic recolonization model is adapted from previous work on island biogeography. Empirical data from dredge spoil studies in Rhode Island Sound are applied to this model. Results suggest that the benthic recolonization process in this area is relatively slow. An extension of the model is suggested which incorporates delay terms to provide a more realistic portrayal of the recolonization process.

BRIEF HISTORICAL BACKGROUND OF BOTTOM COMMUNITY STUDIES

The history of bottom faunal investigations of the seafloor has been documented by many writers. Examples of re-

views which cover background material well include Allee et al. (1949), Hedgpeth (1957), and Holme (1964). A thorough discussion of deep-sea benthos with some consideration of the significance of bottom sediments has recently been provided by Menzies et al. (1973). Descriptions of lagoon and reef sediments and their fauna are given by Wiens (1962). Coral reef ecology and morphology have been carefully documented by Stoddart (1969). However, carbonate sediments, their formation, and fauna are not considered in any detail in this review.

Some of the early quantitative studies in benthic ecology were initiated by C. G. J. Petersen near the beginning of the twentieth century. Petersen's work was first confined to Danish fjords. From his investigations the concept of animal communities among members of the bottom fauna was developed. Petersen did not, however, imply that benthic communities were structurally similar to the better studied plant communities; rather he suggested gross similarities. A review of benthic communities is given by Jones (1950) and a synthesis of the community concept is given by Thorson (1957), in which similar macrobenthic communities found throughout the world are described. A recent and detailed review of the benthic communities of the European North Atlantic continental shelf has been provided by Glémarec (1973).

The founding of marine and fisheries laboratories in many countries from about the turn of this century focused interest on benthic ecology with special reference to the benthos as fish food. This resulted in numerous investigations of the food, feeding, and availability of foods for many important species of fish. However, some areas received little attention. For example, the benthic fauna of Georges Bank, which is one of the more productive fishing grounds of the world, received little early attention by benthic ecologists. This paucity of information may be explained in part by the fact that the interests of the early benthic biologists apparently turned quickly to problems in physiology and taxonomy in this region rather than benthic ecology. This is especially regrettable since Rowe (1971) has demonstrated that surface productivity ranks next to depth in controlling benthic biomass, and the Georges Bank region is one of high primary productivity.

BOTTOM DEPOSITS AND ORGANISMS

On a global basis very large areas of the seafloor are covered by unconsolidated sediments consisting of muds, sands, and gravel. The proportion of the seabed occupied by rocky outcrops is very small, and these areas are usually limited to parts of the continental slope, submarine canyons, ridges, and certain areas on the continental shelf.

Virtually all types of marine sediments are inhabited by assemblages of organisms. Those benthic animals that live on the surface are usually termed epifauna and those that burrow or lie buried beneath the surface are called infauna. The latter usually maintain some connection with the surface of the sediment by means of tubes or siphons.

The depth to which infauna may burrow into the sediment is still a subject of some controversy. Molander (1928) has investigated the vertical distribution of benthic animals using a device that permitted separation of the sample into strata of different depths. He concluded from this study that most individuals were found in less than 10 cm of substrate, but with some polychaetes occurring to 15 cm. However, MacGintie (1939) has reported certain benthic animals at depths of about 61 cm on intertidal flats in California. It appears that most benthic infauna are restricted to depths of less than 15 cm in the sediments, and only relatively large forms, especially pelecypod mollusks and certain polychaetes, are occasionally found at greater depths.

On level bottoms where the sediment consists of unconsolidated sands or muds the infauna is considered to be more important than epifauna, in both numbers and weight per unit area. However, the epifauna includes the majority of species of benthic animals. The epifauna reaches its greatest development in shallow tropical areas. Coral reefs are a good example of a habitat with high epifaunal development.

According to Thorson (1971) the benthos consists of about 157,000 species, with the great majority of these living in depths of less than 200 m. The epifauna includes about four-fifths of all benthic animals by species. Motile epifaunal species tend to be active predators, and they are protected from attack by powers of locomotion as well as by having protective appendages or hard ectoskeletons. The infauna are frequently soft-bodied, and this group relies on burrowing for protection. The infauna are a major component of the food of demersal fish, especially on the continental shelf areas of major continents.

COMMUNITY DESCRIPTIONS AND CLASSIFICATION SCHEMES

The populations of all species of organisms in a defined area are frequently termed a community, and the community of populations with its physical environment is called an ecosystem. Although benthic animals are frequently described as communities, they are not independent entities. Benthic animals are a part of the food web involving the whole water column. A persistent belief in ecology is that the community is an organized, distinct entity. This belief is reflected in the current interest in measuring species diversity, and in numerous references to energy flow in communities. It should be recognized that a community appears to be a subjective entity and its species composition must be strictly defined in order to be useful for descriptive purposes.

Thorson (1957) has grouped benthic communities by the generic name of the dominant form—such as *Macoma*, *Tellina*, amphipod, and foraminifera communities. Sanders (1958) has followed the example founded by Petersen (1913) by combining the names of characteristic species to provide a similar form of community classification.

These Petersen-type bottom communities are characterized by constant species which are dominant in terms of numbers and weight. The existence of such communities has been questioned by some workers on the grounds that quantitative data on relatively few species have been analyzed adequately. A recent series of computer analyses of Petersen's original data were made by Stephenson and his co-workers (1972). Results of computer classification involving binary data compared favorably with Petersen's original classification, and the existence of Petersen-type communities was demonstrated in the majority of cases. It was also demonstrated from these analyses that a classification using both numbers and weight was not possible. Separate number and weight classifications, however, revealed Petersen-type communities, but the results differed from Petersen's interpretation. The presentation of results from this form of classification is difficult because of the complexity of the data. In addition, it should be appreciated that there are further problems of analysis and interpretation if spatial distributions and temporal changes are taken into account.

Some extremely convincing evidence of an association between bottom type and bottom fauna has been provided from the classification by Thorson (1957) of parallel, sublittoral, bottom communities in different geographical regions. This parallelism is characterized by the same or similar genera inhabiting the same type of bottom at similar depths at widely different latitudes. The concept appears to apply mainly to cold water fauna, and it breaks down when applied to tropical and some warm temperate regions. The diversity of species seems to increase disproportionately in these areas.

Studies of estuarine communities have been well summarized by Carriker (1967) and Green (1968). Two generalizations seem to emerge from this work. The first is that the fauna of estuaries is derived primarily from the sea, and the second is that there is a reduction in taxa from high numbers of species at the mouth of the estuary to low numbers of species where the salinity change is greatest near the head of the estuary.

The character and distribution of sediments in atolls depend on the character of the atoll. Such factors as its depth and the number and depth of passes affect the nature of the sediment. However, when atoll sediments are considered as a group it appears that Halimeda debris is the most important constituent followed by fine debris, coral, foraminifera, mollusk shells, and miscellaneous elements. The organism community of lagoons will vary according to habitat, such as lagoon mud flats, sand flats, foreshore, etc. Quantitative studies on carbonate sediments and their benthic fauna are not abundant. At present it is possible to describe these relationships only in a qualitative sense.

Methods of analysis of community structure include a variety of measures of species diversity and its components. Information-theory diversity measures have been widely applied in benthic ecology. A synthesis of some much used measures of diversity and evenness, and a unified concept of diversity have recently been presented by Hill (1973). He has shown that there is a continuum of diversity measures that estimate the effective number of species present. The measures differ only in their tendency to ignore or include the relatively rare species.

SAMPLING PROBLEMS

In order to understand animal-sediment relationships and to permit rational environmental management decisions, quantitative data on the distribution of benthic organisms are necessary. The spatial distribution of benthic organisms seldom appears to be random or regular. Instead it is frequently patchy or contagious. That is, the sample variance is significantly greater than the sample mean. The frequent occurrence of contagious distributions in benthic organisms is not surprising since many environmental factors are also unevenly distributed. In addition many species of organisms have some tendency to aggregate and this produces patches and gaps even without the influence of abiotic environmental factors. These behavioral and environmental factors acting alone or in combination, produce diverse patterns of contagious distributions and it is difficult to find a single mathematical distribution that fits these patterns. However, it has been demonstrated (Saila and Gaucher, 1966) that the frequency distributions for many marine invertebrate species do not deviate significantly from the unimodal negative binomial model.

Menzies et al. (1973) considered patchiness in some members of the abyssal benthic community by estimating the ratio of the sample variance to the mean, which was termed a "coefficient of dispersion." On the basis of this index they partitioned samples into categories labeled contagious, random, or regular. They suggested that the ophiuroid *Ophiomusium lymani* was more often randomly distributed than contagiously. The same seemed to be true of a polychaete, *Hyalinoecia artifex*, and a hermit crab, *Parapagurus pilosimanus*. It should be emphasized that these inferences were made from an analysis of bottom photographs. Recent work by the author and some colleagues in analyzing benthic data from the New York Bight area clearly supports the previous assertions concerning the fit of the negative binomial model to benthic invertebrate data. Table I illustrates the results of goodness-of-fit tests of the negative binomial distribution for 22 species of marine invertebrates. Only 2 of the 22 species deviated significantly from the model.

Briefly the negative binomial is a two-parameter (μ and k) distribution with the probability density function:

$$f(X) = \frac{(k + X - 1)!}{X!(k - 1)!} \left(\frac{\mu}{k + \mu}\right)^X \frac{(\mu + k)^{-k}}{k} \quad (1)$$

where the mean is μ, and the variance is $\mu(1 + \mu/k)$. The mean μ is estimated by $\bar{X} = (\sum_{i=1}^{n}(X_i/n)$ and k is estimated numerically by its maximum likelihood estimator $\hat{K}$. Bliss and Fisher (1953) describe several methods of estimating the parameters of this distribution.

Although it seems clear from the data of Table I that the chi-square goodness-of-fit tests for the negative binomial distribution were rejected in only 2 of the 22 cases examined, the values for the parameter k were

TABLE 1. Results of Chi-Square Goodness-of-Fit Tests of the Negative Binomial Distribution to Sample Frequency Distributions for 22 Species of Marine Invertebrates from the New York Bight*

Species	X	k	χ^2	D.F.	
Ampharaete acutifrons	0.5645	0.07486	4.83	2	Accept
Arctica islandica	0.2849	0.11800	1.34	2	Accept
Cancer irroratus	0.7151	0.09299	2.50	3	Accept
Capitella capitata	4.2204	0.04418	4.88	3	Accept
Cerianthus americanus	8.8118	0.25161	16.63	11	Accept
Drilonereis longa	0.4624	0.06728	0.27	2	Accept
Edotea triloba	0.3495	0.14026	3.96	2	Accept
Eteone longa	0.7688	0.07583	0.11	3	Accept
Glycera dibranchiata	1.0215	0.18971	3.52	4	Accept
Leptocheirus pinquis	4.7634	0.02161	14.35	1	Reject
Mulinia lateralis	0.3763	0.02334	1.67	1	Accept
Nassarius trivittatus	0.3011	0.30100	0.50	1	Accept
Nepthys incisa	5.8710	0.17031	19.89	9	Reject
Ninoe nigripes	5.0269	0.14334	14.57	8	Accept
Paraonis gracilis	9.5430	0.05004	2.53	4	Accept
Pherusa affinis	4.2581	0.14458	5.00	7	Accept
Pitar morrhuana	1.2043	0.06796	7.54	3	Accept
Prionospio malmgreni	82.6236	0.08190	5.00	5	Accept
Tellina agilis	2.6183	0.16231	12.14	7	Accept
Yoldia limatula	0.8280	0.10876	0.48	4	Accept
Cossura longocirrata	19.4946	0.04685	5.97	3	Accept
Diastylis sculpta	1.0968	0.07193	0.50	1	Accept

* The sample frequency distributions are derived from the preliminary sample data provided by the Sandy Hook Laboratory of National Marine Fisheries Service.

somewhat variable. It should be appreciated that problems can arise in estimating the parameter k of the negative binomial distribution for such a large number of species so widely dispersed both spatially and temporally. To calculate a value of k somewhat more representative of these species the regression method of Bliss (1958) has been utilized. Two sample statistics, x and y, are calculated for each species in the region as

$$x = \mu^2 - \frac{s^2}{N} \qquad (2)$$

and

$$y = s^2 - \mu \qquad (3)$$

where s^2 is the sample variance for each species, and μ is the sample mean by species.

The regression equation for Fig. 1 is

$$Y = -5.70X + 1.04 \qquad (4)$$

for $n = 18$, with a correlation coefficient r of 0.9715.

The calculated value for $1/b$ is -0.1754 in this case. Therefore, the common value of k, which is termed k_c, is -0.1754. A check to determine whether the value of k is relatively constant by species is obtained from Fig. 2. If the values of k by species are relatively constant for the New York Bight area, then there should be relatively little clustering or trend in the points of this figure. No trend seems evident although there is some indication of clustering. The advantage of a common value of k is to facilitate sample size determination in sequential sampling. Obviously only one expected value is required to estimate the maximum number of samples required to detect differences between two or more sets of conditions in this case. Saila et al. (1965) have described in detail the application of sequential sampling techniques for marine resource surveys, and the technique for calculating expected sample size for any observed mean is illustrated in the above-mentioned report.

In benthic studies the sampling procedure often involves repeatedly removing a sample quadrat (sampling unit) at random from a relatively continuous habitat, which is the sediment. If the spatial distribution of benthic organisms is contagious and if the size of the sampling unit or quadrat is much larger or smaller than the average size of clumps of individuals, and these changes are regularly or randomly distributed, then the dispersion of the population is apparently random. Nonrandomness is not detected in these instances. Thus a problem arises about how to take samples and what quadrat size to use, because of the creation of artificial sampling units. In general most samplers used in benthic ecology, such as the Smith-McIntyre and Shipek samplers, take relatively small sampling units, 0.1 m² or less. These small sampling units will detect nonrandomness in animals found in the sediment in all instances except in the case where there are only a few individuals in a clump.

Because of the possible effect of quadrat size on the measurement of aggregation it may be desirable to have a measure of dispersion independent of the size of the sampling unit. One method for measuring dispersion is by means of an index derived by Morisita (1959). If one assumes the population consists of patches of individuals of different densities, and that within a clump the individuals are randomly spaced, then Morisita's index I_δ can be used as follows:

$$I_\delta = \frac{\sum_{i=1}^{n} n_i(n_i - 1)}{n(n - 1)} N \qquad (5)$$

where N is the number of samples, n_i is the number of individuals in the ith sample, and n is the total number of individuals in all samples.

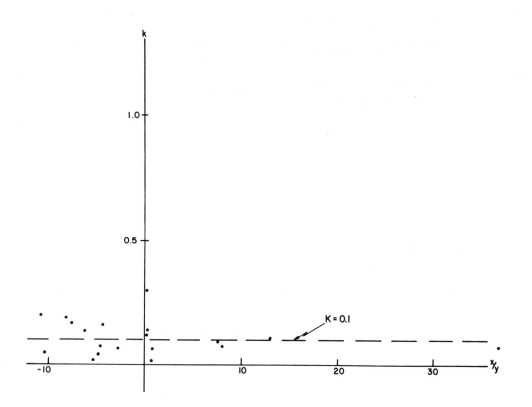

FIGURE 1. *Example of the method of calculating a common k value for the New York Bight invertebrates. The common value of k, k_c is* *equal to 1/b where b is the slope of the regression line.*

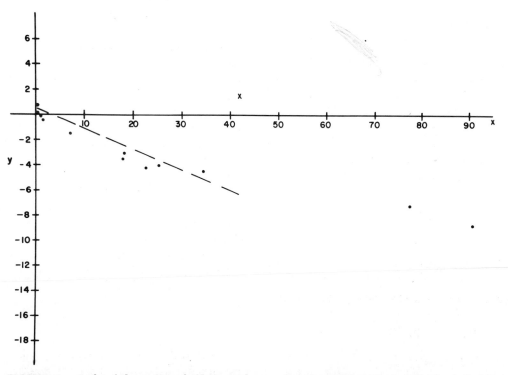

FIGURE 2. *A plot of the reciprocal of k for each invertebrate species from the New York Bight* *against x/y as defined in the text. The dashed line is the regression line.*

RELATIONSHIP OF FAUNAL COMMUNITIES TO CHARACTER OF SEDIMENTS AND RATES OF SEDIMENTATION

Biological Activity Related to Environmental Variates

Many morphological and physiological adaptations of benthic animals appear to be related to the mass properties of sediments. These animal-sediment associations have been described by Sanders (1958), McNulty et al. (1962), Rhoads and Young (1970), and Johnson (1971), among others. It seems clear that these animal-sediment relations can be utilized in predicting to some extent at least the impacts of man's activities on the marine environment. The major change which occurs is the alteration of a given substrate and its replacement by another. This results in the replacement of one community with another some time after the environmental perturbation. Changes in erosional and depositional patterns and in sediment composition are the consequences of dredging, dredge spoil deposition, erection of structures, and improper management of the watershed or coastal zone.

Burrowing activities of deposit-feeding organisms in the upper 10 cm of the bottom contribute significantly to the mass properties of bottom muds Rhoads (1967, 1972) has clearly demonstrated that intensive burrowing of subtidal muds by deposit-feeding organisms produces a granular surface layer 5 to 10 mm thick. This uncompacted zone contains 50 to 60% water by weight and exhibits thixotropic properties. Such muds are easily resuspended by weak tidal currents. Work by Winston and Anderson (1971) in a New Hampshire estuary showed that the amount of bioturbation (burrowing activity) decreased from the mouth to the head of the estuary as the salinities became lower and the bottom fauna changed in composition. Four levels of bioturbation were observed—ranging from high values in sandy bottoms near the mouth to no activity on a silty bottom near the head. The highest amounts of activity were associated with the polychaete genus *Nereis*.

Distribution of Organisms in Relation to Sediment Properties

An inventory of benthic fauna and bottom sediments on Georges Bank, a relatively shallow sill extending between Massachusetts and Nova Scotia at the mouth of the Gulf of Maine, has been made by Wigley (1961, 1962). According to his studies the benthic fauna on this highly productive fishing ground were composed by weight of the following groups: Mollusks 41%, Echinodermata 31%, miscellaneous groups 17%, Annelida 6%, and Crustacea 5%. The average weight of benthic fauna was $156.6/g$ m^2 and the average number of specimens was 1690 m^{-2}. Wigley's investigation demonstrated marked differences in species composition and relative abundance associated with bottom sediments. The greatest biomass of organisms occurred in coarse-grained sediments, and the finest grained sediments had the lowest values. A reanalysis of Wigley's data showing the relationship between sediment particle size and biomass of benthic organisms is shown in Fig. 3. The product moment correlation coefficient (r) between the logarithms of the two variables was found to be 0.84, a value that is highly significant. Wigley's data are somewhat at variance with inferences drawn by Purdy (1964) who

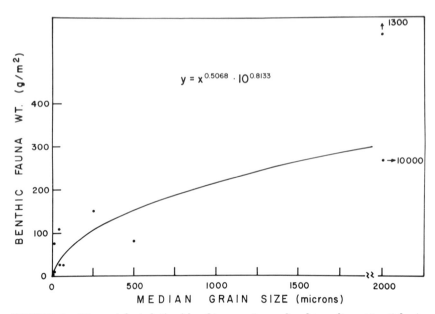

FIGURE 3. *Wet weight (g/m^2) of benthic organisms related to sediment particle size in the Georges Bank region. Reanalysis of data from Wigley (1962).*

suggested that both ends of the sand-to-mud texture continuum represent stress habitats for most invertebrates. It should be recognized that the high biomass values for Wigley's coarse sediments were caused by a relatively few large suspension-feeding mollusks.

In general, it would appear that in rapidly shifting sand deposits both benthic microfauna and macrofauna tend to be characterized by a relatively low population density and relatively few species. Exceptions to this statement occur in the case of certain large suspension feeders, such as some species of bivalve mollusks. Sand deposits under less current-agitated conditions have greater substrate stability with an increasing number of species and a greater population density. The suspension-feeding fauna are the dominant type here. As sediment texture becomes finer, the proportion of suspension feeders decreases, and the deposit feeders increase until they dominate. The above-mentioned relations result from the following: (a) weak bottom currents associated with accumulations of finer sand deposits transport less suspended food over a given area per unit time, and (b) the weak currents allow deposition of organic detritus which increases the density of deposit feeders. In extremely fine-grained sediments the fauna is generally characterized by low population densities and low taxonomic diversity, because of poor interstitial circulation and/or oxygen depletion.

Some preliminary observations have been made on invertebrates of economic importance in relation to sediment properties in estuaries and on the continental shelf. Pratt (1953) has found that the numerical abundance of *Mercenaria mercenaria*, the hard clam in a Rhode Island estuary, was positively correlated with the presence of large particles as minor constituents of a fine sediment. Wells (1957) found a similar correlation in Chincoteague Bay. Pratt (1953) also demonstrated that the growth rate of this organism depended on the nature of the substratum. Populations of the hard clam raised in sand grew about 24% faster than those living in sandy mud. Saila et al. (1967) considered several sediment properties which included particle size, cation exchange capacity, organic carbon, total nitrogen, available phosphorus, and water depth in an effort to discriminate between high- and low-abundance samples of *M. mercenaria* in an estuary. Using all of the above-mentioned variables in a linear discriminant function it was found that only the organic carbon content of the sediment and sediment particle size greater than 2 mm diameter contributed significantly to a separation of the two sample groups. The final discriminant function was found to be

$$Z = -X_2 + 7.07X_5 \qquad (6)$$

where X_2 refers to particle size greater than 2 mm and X_5 refers to the organic carbon content of the sediment.

It is evident from this equation that organic carbon seems to have the greater discriminating power. Bader (1954) has described the role of organic matter in determining the distribution of pelecypods in marine sediments, and the above-mentioned study reinforces his statements regarding the correlation between sediment organic matter and pelecypod abundance.

Postulated associations between motile organisms and sediment characteristics are obviously more difficult to make than for nonmotile forms. However, a limited amount of work has been done in attempting to relate shrimp abundance and distribution to the nature of the substrate. Springer and Bullis (1954) suggested that in the Gulf of Mexico pink shrimp were more abundant on calcareous mud and shell sands whereas white and brown shrimp were more abundant on terrigenous mud. These studies were further confirmed by Williams (1958). More recently Grady (1971) has also shown a positive association between the catch of shrimp of the genus *Peneaus* in the Gulf of Mexico with the organic content of the sediment. That is, high catches occurred on the grounds with the highest organic matter content.

A recent review by Menzies et al. (1973) adequately summarizes available information concerning animal-sediment relations in the outer continental shelf and abyssal regions. The relationship demonstrated for continental shelf areas—namely that the percentage of deposit-feeding animals increases with the increasing clay content—apparently does not apply to abyssal regions. There appears to be an increase in sestonophages in the red clays of the abyss, and red clay is one of the finest sediments known.

Deep-sea benthos are surprisingly diverse considering the apparent limitations in diversity of food resources and microhabitats and the low density of food. The basis for this diversity is currently under discussion (Dayton and Hessler, 1972; Grassle and Sanders, 1973).

Menzies et al. (1973) have summarized the variations in benthic biomass as a function of depth as follows. In the intertidal zone biomass can be estimated in thousands of grams per square meter. The continental shelf from 60 to 200 m depth has biomass values between 150 and 500 g/m² except in the Antarctic where exceptionally high values have been found. Between 400 and 1000 m, biomass is usually in tens of grams per square meter. At depths between 1000 and 4000 only about 1 g/m² is present. Measurements of abyssal biomass are rare and there may be considerable variation according to specific regions. In addition anomalies caused by oxygen deficiencies and the presence of toxic substances must be recognized. However, there appear to be certain regularities in the distribution of biomass at abyssal depths according to recent investigations. In general biomass decreases with increasing depth from land, and it seems to be greatest in high latitudes and lowest in low latitudes. The biomass of red clay deposits always appears to be low, less than 0.1 g/m². On the other hand, there seem to be higher biomass values in some of the deep trenches.

IMPACTS OF DREDGING, DUMPING, OIL SPILLS, AND CONSTRUCTION ON ANIMAL-SEDIMENT RELATIONSHIPS

The apparent needs to develop and exploit new sources of raw materials continuously and to dispose of increasing quantities of waste combine to ever increase the impingement of man on the oceans. Every engineering and mining operation in the marine environment may be expected to alter in some manner the quantity, distribution, and deposition of sediments (Sherk and Cronin, 1970). Consequently, these operations may impose temporary or enduring changes in suspended loads and surface sediments either in localized areas or over considerable distances. Sites selected for mining operations, ocean dumping, or oil exploration are often inhabited by communities of organisms used by man directly or indirectly as food for organisms consumed by man. These sites frequently become centers of conflict between fisheries and conservation interests versus other interests. Thus, it is important that the biological consequences of localized or widespread resuspension, redistribution, and resorting of bottom deposits be understood before larger scale intensification of these activities takes place.

The effects of mining, dredging, spoil, and waste disposal as well as other sediment-resuspending activities have generated considerable research effort. Some examples of relevant studies that deal with benthic ecology include Cronin et al. (1970), Filice (1959), First (1972), Jenkinson (1972), Maurer et al. (1974), Postma (1967), Pratt et al. (1973), Rounsefell (1972), Saila et al. (1972), Sherk and Cronin (1970), Sherk (1971), and Sykes and Hale (1970). No effort is made to provide a comprehensive literature review in the material that follows. Instead, an effort is made to summarize some of the relevant observations.

Effects of Dredging and Spoil Disposal

It seems obvious that there will be a significant reduction in the density of benthic organisms at dredging and disposal sites immediately following operations because of the mechanical agitation of the habitat. However, there are differing opinions and reports concerning the recolonization rate of dredged areas and spoil dumps. In general, it appears that few, if any, truly estuarine species of benthic invertebrates are expected to survive

transfer to a truly marine environment. Thus, the major question concerns the nature of the recolonization process.

The direct effects of spoil dumping on marine animals may be a result of changes in grain-size distribution, turbidity, anoxia, toxic hydrocarbons, and heavy metals. In general terms, the smallest organisms are expected to have the highest probability of being destroyed. A few benthic animals seem to be able to reach the surface successfully even after burial to depths greater than 20 cm.

It also appears that most marine organisms are able to tolerate rather high concentrations of suspended sediments for short periods of time. The motile forms, such as fish, are often able to detect and avoid undesirable areas. Laboratory experiments have recently been conducted by Schubel and Wang (1973) in which eggs of four species of fish were incubated in suspensions of variable concentrations of natural, fine-grained sediment. Results of this study showed that concentrations up to 500 mg/l had no statistically significant effect on the hatching success of the four species, one of which was the striped bass. Some delay in hatching time was experienced at sediment concentrations in excess of 100 mg/l. Davis (1960) found that in natural silt concentrations below 750 mg/l the percentage of clam eggs developing normally was not significantly different from the control cultures but decreased progressively in higher silt concentrations. The data from Davis have been reanalyzed and are presented in Fig. 4. It seems that the results of studies with developing invertebrate eggs and fish eggs are reasonably similar. Davis also demonstrated that growth of clam larvae was adversely affected by silt concentrations above 750 mg/l.

The disposal of sewage sludge is another source of sediments as well as being a major source of environmental perturbation. Gross (1970) has documented the contribution of the New York metropolitan area to New York Bight sediments, and Pearce (1970) has demonstrated some of the effects of these deposits on benthic communities of this area. Discharged sewage sludge now covers a portion of the bottom in the New York Bight, and it has had a significant adverse effect on the numerical abundance and diversity of certain benthic organisms in the affected areas. The results of an ongoing marine ecosystem analysis program in the New York Bight are expected to increase substantially our understanding of sewage sludge and other waste disposal effects on the marine environment and its biota.

Oil Spills

Hydrocarbons in the marine environment are derived from natural sources, organisms, and submarine seeps,

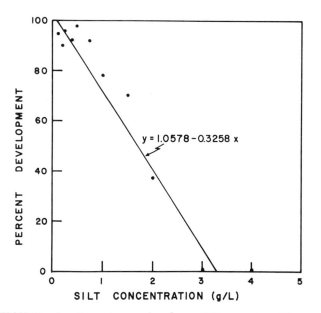

FIGURE 4. *Percentage of clam* (Mercenaria, Venus mercenaria) *eggs developing to the straight hinge larval stage related to the silt concentration of the medium. Reanalysis of data from Davis* (1960).

and are introduced through man's activity. The toxicity of various oils and oil products is highly variable, depending on a number of factors. Crude oil and other oil products spilled on the sea or on the shore are altered by evaporation, dissolution, and bacterial or chemical attack.

A coordinated biological and chemical study of the long-term effects and fate of a coastal oil spill in West Falmouth, Massachusetts has been documented. The results of those studies have been summarized by Blumer (1973). This spill involved about 650 tons of No. 2 fuel oil in a small coastal area. In addition to immediate effects of the spill on organisms, there apparently were some long-term effects that were associated with the incorporation of oil into the sediments and the spread of oil-laden sediments. Oil was also reported to be incorporated into certain bivalve mollusks. Additional studies are currently under way to determine how severe and long term these reported effects have been. A brief review of the ecological implications of oil spills has been provided by Hay (1974).

Nadeau and Roush (1972) have compiled a recent bibliography on the effects of oil pollution. From the bibliography it is evident that relatively little is yet known concerning the effects of oil in sediments upon marine biota. Blumer et al. (1970) observed contamination of shellfish by oil pollution. On the other hand, Fauchald (1971) did not observe such results with benthic fauna in the Santa Barbara channel. In general, it appears that both fish and shellfish may at times become tainted by certain products associated with oil spills. The nature of the substances and the conditions that maximize the removal of these taints are not yet resolved.

Recolonization Model Development

In order to predict and understand the effects of various types of activities that result in changing the physical and/or chemical properties of marine sediments, it is essential to understand something about the nature of recolonization of the disturbed areas. Some empirical information concerning sediment-animal relations is already available.

Johnson (1971) has suggested that mud species of benthic invertebrates are better able to invade clean sands than clean sand species are able to invade muds. This ability is presumably related to feeding and reproduction. Most clean sand species probably cannot tolerate large amounts of silt and clay, whereas mud species have special adaptations to do so. On the other hand, mud species have mechanisms for discarding larger grains from the gill or food tracts.

These observations suggest a means for predicting the sequence of faunal changes following alteration of the substrate, such as by dredging and spoil disposal in the marine environment. For example, introduction of silt- or clay-size particles into a clean sand area would be expected to reduce the clean sand assemblage. The more ubiquitous mud species from nearby muddy areas would probably first appear, and the most diversely occurring sand species would also tend to appear soon. Later the more ubiquitous of the sand species would be found among the more diverse muddy sand community.

The above is one qualitative aspect of the recolonization problem. Another aspect involves attempting to predict the rate of recolonization and the new equilibrium number of species to be expected on the disturbed area. In order to answer these questions a dredge spoil species equilibrium model has been applied. The model is based on the idea that as an area is recolonized the recolonization or immigration rate I decreases and the extinction rate E increases until the two are equal—producing a species equilibrium S. The recolonization rate is a rising curve that ascends at a continually decreasing rate. Because recolonization and extinction tend toward convergence from the beginning, their difference $dS/dt = I - E$ is always decreasing. The number of species present is therefore the integral through time of this difference.

The regularity of some observed area-species relations and correlations between the slope of the area-species curve and the degree of isolation of islands prompted

MacArthur and Wilson (1967) to develop a basic species equilibrium model. This model and some extensions of it have been applied to dredge spoil recolonization. The following definitions and notation are taken primarily from the above-mentioned linear model:

λ_s = the total immigration rate (the number of new species arriving into a given area per unit time)

P = the total number of species in the "pool" (the number of species found in surrounding source areas)

μ_s = the total extinction rate (the rate at which species already in the given area become extinct)

S = the number of species already present in a given area

$\hat{S}$ = the number of species present at equilibrium (this occurs when $\lambda_s = \mu_s$)

λ_A = the average immigration rate of new species, per species, into a given area where S species are present (this is assumed to be constant)

μ_A = the average extinction rate per species (assumed constant)

Using this notation, the time rate of increase in the number of species is also the total immigration rate minus the total extinction rate. That is,

$$\frac{dS}{dt} = \lambda_s - \mu_s = \lambda_A(P - S) - \mu_A S \qquad (7)$$

At equilibrium the number of species is denoted by S and at equilibrium, $dS/dt = 0$ by definition. Thus,

$$\left.\frac{dS}{dt}\right|_{S=\hat{S}} = \lambda_A(P - \hat{S}) - \mu_A \hat{S} = 0 \qquad (8)$$

A slight rearrangement of (8) yields

$$S = \frac{\lambda_A P}{\lambda_A + \mu_A} \qquad (9)$$

Equation 9 can be utilized to predict the equilibrium number of species from the linear model.

For the case of a dredge spoil site and surrounding area described by Saila et al. (1972) some preliminary estimates of model coefficients have been empirically determined. Available information from sample surveys of the surrounding area indicate that a reasonable approximation of P, the number of invertebrate species in surrounding areas, is 150. During the first sampling period in 1968 it was found that the dredge spoil area contained 32 species of benthic organisms. A second sampling of the same area two years later revealed that there were 55 species in the spoil area. However, 9 of

the original 32 species had disappeared from the area and 32 new species had immigrated to the area during this time. From these data we find that the average immigration rate λ_A is the total immigration rate divided by the number of species not yet in the area. In this case it is $(32/2)/(150 - 32) = 0.14 = \lambda_A$. The average extinction rate μ_A is the total extinction rate divided by the number of species already in the spoil area, or in our case it is $(9/2)/(32) = 0.14 = \mu_A$. If these coefficients are applied to (9), the following result is obtained:

$$\hat{S} = \frac{\lambda_A P}{\lambda_A + \mu_A} = \frac{0.14(150)}{0.14 + 0.14} \sim 75 \qquad (10)$$

From this preliminary information on estimated immigration and extinction rates we predict that the ultimate equilibrium species number for the dredge spoil will be about 75 species. Obviously, this prediction is based on the assumptions underlying this simple equilibrium model.

It can be shown that the solution to the differential equation (7) is

$$S = \frac{\lambda_A P}{\lambda_A + \mu_A} \, 1 - \exp[-(\lambda_A + \mu_A)t] \qquad (11)$$

As $t \to \infty$, $\exp[-(\lambda_A + \mu_A)t]$ approaches zero, and S approaches $\hat{S}$, which has already been defined as equal to $\lambda_A P/(\lambda_A + \mu_A)$. The rate of approach to equilibrium can be utilized to derive a rate of turnover at equilibrium.

For the purposes at hand we select an arbitrary function of S which will be 95% of $\hat{S}$, or $0.95S$. We multiply both sides of (9) by this quantity to obtain

$$0.95\hat{S} = \frac{\lambda_A P}{\lambda_A + \mu_A} \times 0.95 \qquad (12)$$

We next apply (11) and note that

$$S = 0.95\hat{S} = \frac{\lambda_A P}{\lambda_A + \mu_A} \, 1 - \exp[-(\lambda_A + \mu_A)t_{0.95}] \qquad (13)$$

Combining (12) and (13) we find that

$$1 - \exp[-(\lambda_A + \mu_A)t_{0.95}] = 0.95 \qquad (14)$$

where $t_{0.95}$ is the time required to fill the spoil area to 95% of its equilibrium population. By rearrangement and taking natural logarithms of (14) the following expression is developed:

$$t_{0.95} = \frac{-\ln(0.05)}{\lambda_A + \mu_A} = \frac{2.996}{\lambda_A + \mu_A} \qquad (15)$$

Substitution of the previously obtained values of 0.14 for λ_A and 0.14 for μ_A into (15) gives a value of 10.7 or approximately 11 years as the time required for the dredge spoil benthic invertebrate population to reach 95% of its final equilibrium population. This value is significantly greater than an empirical value estimated by Reish (1961) as two years in a dredged boat harbor in southern California.

Two important quantities have been derived. The first relates to the estimated number of invertebrate species that are expected on the spoil area when equilibrium is reached, and the second is related to the time that will be required for the establishment of near-equilibrium conditions. The new equilibrium is estimated at about one-half the number of species found in the area previously. The amount of time required (less than 15 years) is considered to be reasonable for a temperate latitude.

One important consideration that has not yet been examined in detail is the relative biomass of the spoil area versus the surrounding region. Estimates of this biomass will probably change until the new equilibrium is reached. However, further data are highly desirable even during the transition period toward equilibrium.

In the case of the transition zone, it was found that 58 species were present on the transition area during the first sampling period and 68 species were found two years later. Of these 68 species, it was found that 18 were new to the area between the first and second sampling periods. Also it was observed that eight species had apparently disappeared from the area.

Utilizing (10) and (15) in a manner similar to that described in detail above, it was estimated that the transition zone might reach equilibrium with about 88 species and that 95% of the equilibrium species would be found in the area within a period of approximately 18 years.

These data would suggest that some of the diversity of the transition zone may have been lost. However, no data on biomass values are yet available to permit inferences about the standing crops or relative productivity of the transition zone in contrast to surrounding undisturbed areas.

It is obvious that the model described above is but a first approximation to conditions in the real world. In addition to questions one may have regarding the actual shapes of the immigration and extinction curves, it seems evident that in the real world a newly reached equilibrium is not constant nor is it approached monotonically. Instead, the equilibrium would be expected to fluctuate over time and, instead of being approached monotonically, overshoots and undershoots would also be observed. The simple equilibrium model described above can be modified to allow for oscillatory solutions by introducing time delays.

The mathematical description of the simple equilibrium system requires use of time derivatives of the de-

pendent variables—thus leading to a differential equation. If one adds constant time delays into the system, one gets delay-differential equations. These equations include terms with derivatives, and various terms of a delay-differential equation are not all evaluated at the same instant in time. Solutions to delay-differential equations are somewhat more complicated than ordinary differential equations, and numerical integration with a digital computer is utilized.

One of the early delay-differential equations used for the Rhode Island Sound dredge spoil recolonization study is

$$\frac{dS(t)}{dt} + 0.14S(t) + 0.14S(t-1) = 0.14 \times 150 \quad (16)$$

with an initial function: $\exp[3.46574(T+1)] - 1$. It is clear that the model coefficients are the same as those used previously. That is, $P = 150$, $\lambda_A = 0.14$, and $\mu_A = 0.14$. In this case the initial function involving no species at time $t - 1$ and 32 species at time $t = 0$ was used. This model predicts 56 species for year 2 as was observed and equilibrium at about 11.0 years. Oscillations in the equilibrium have been introduced into this model. However, they may be too small to reflect real conditions in the actual environment.

In any event, the big shortcoming of the first model described is the assumption of instantaneous response in the model coefficients. That is, the various terms of the simple equilibrium model are equated at the same time. In the new model an instantaneous response in the extinction rate is assumed but the immigration rate is affected by what happened one year previously. Thus, the latter model seems considerably more realistic.

In general, it appears that our knowledge of animal-sediment relationships, albeit limited, is improving to the point where simple models for prediction and environmental management decisions are being developed and applied. Some of the limitations in empirical data have been pointed out. These, hopefully, will be resolved in the relatively near future.

SUMMARY

The history of bottom faunal investigations traces the introduction and utilization of the concept of animal communities. In spite of its high biological productivity and significance to commercial fisheries the Georges Bank region of the Northwest Atlantic Ocean has not received much attention by benthic ecologists until fairly recently.

The benthos consists of about 157,000 species of organisms, the majority of which are found in depths of less than 200 m. Benthic organisms are conveniently divided into epifauna, the animals that live on the surface of sediments, and infauna, those that lie buried beneath the surface. The depth of burial by benthic organisms is still a matter of some controversy, but 15 cm appears to be a reasonable maximum depth for most species.

The concept of benthic communities, although much used, has recently been questioned by some workers. Computer classification techniques applied to certain published descriptions of benthic communities have demonstrated that classification by both numerical abundance and weight is not possible. Classification by number or weight was successful, but the results did not necessarily coincide with more arbitrary techniques. The concept of parallel, sublittoral bottom communities as developed by Thorson seems to apply mostly to cold water organisms and breaks down in tropical regions. Estuarine benthos are derived primarily from the sea, and the number of estuarine species increase from the head to the mouth of the estuary.

Although a variety of methods have been applied for measuring benthic species diversity, they appear to be highly correlated and differ only in their tendency to ignore or include the relatively rare forms.

The spatial distribution of benthic organisms appears to be patchy or contagious. It has been found that the frequency distributions of many marine invertebrate organisms do not deviate significantly from the unimodal negative binomial model. Only for 2 of 22 species of organisms examined was the chi-square goodness-of-fit test rejected. A regression technique for calculating a common value of k, one of the parameters of the negative binomial distribution, was illustrated.

Community succession can be utilized to some extent to predict the effects of man's activities on the marine environment. Burrowing activities of deposit-feeding organisms contribute significantly to the mass properties of certain bottom sediments.

A reanalysis of some data from the Georges Bank area demonstrated that the greatest biomass of benthic organisms occurred in the coarse-grained sediments. The high biomass in coarse sediments may have been due to a few large suspension-feeding mollusks. In general, it seems that both extremely fine and extremely coarse sediments may represent stress habitats for most marine invertebrates. With reference to economically important bivalve mollusks, it appears that their numerical abundance is positively correlated with coarse sediment particle size and the organic carbon content of the sediment. The catches of motile invertebrates, such as shrimp, also appear to be correlated with the organic content of the sediment.

The relationship between benthic biomass and water depth indicates a reduction of biomass with increasing depth, and seems to be highest in the high latitudes.

The effects of environmental perturbations on animal-sediment relationships have received considerable attention, and significant reductions in the diversity and density of benthic invertebrates have been reported in some instances. The early life history stages of most marine organisms are the most vulnerable to the effects of suspended sediments. The effects of various oils and oil products on animal-sediment relations are less well known.

A basic species equilibrium model developed for island biography has been modified to apply to the recolonization of dredge spoil which has been moved from one environment into another. The model is considered applicable to other types of benthic recolonization. It is based on constant coefficients relating to the rate of immigration and extinction of organisms, and it predicts the equilibrium of species and the time required for the equilibrium to be reached. Empirical data from a temperate region (Rhode Island) suggest that new species equilibria would be established but that the time involved in this process is approximately a decade. An extension of the original model using time delays in one of the model coefficients is suggested.

ACKNOWLEDGMENTS

The author is indebted to his colleague, Sheldon D. Pratt, for helpful suggestions and criticisms. Dr. John D. Pearce of the National Marine Fisheries Service, Sandy Hook Marine Laboratory, Highlands, New Jersey, has been a source of inspiration and guidance as well as being instrumental in obtaining financial support for portions of this material.

SYMBOLS

I_δ	Morisita's index of dispersion
$\hat{K}$	maximum likelihood estimator of the parameter k
k	parameter of the negative binomial distribution
P	the total number of species in the "pool" (the number of species found in surrounding source areas)
S	the number of species already present in a given area
$\hat{S}$	the number of species present at equilibrium (this occurs when $\lambda_s = \mu_s$)
$\bar{X}$	sample mean

x,y sample statistics used for calculations of k_c, a common value of the parameter k

λ_A the average immigration rate of new species, per species, into a given area where S species are present (this is assumed to be constant)

λ_s the total immigration rate (the number of new species arriving into a given area per unit time)

μ parameter of the negative binomial distribution

μ_A the average extinction rate per species (assumed constant). Using the notation above, the time rate of increase in the number of species is also the total immigration rate minus the total extinction rate

μ_s the total extinction rate (the rate at which species already in the given area become extinct)

χ^2 chi square

REFERENCES

Allee, W. C., O. Park, A. Emerson, T. Park, and K. Schmidt (1949). *Principles of Animal Ecology.* Philadelphia: Saunders, 837 pp.

Bader, R. G. (1954). The role of organic matter in determining the distribution of pelecypods in marine sediments. *J. Mar. Res.*, **13**: 32–47.

Bliss, C. I. (1958). The analysis of insect counts as negative binomial distributions. *Proc. X Int. Congr. Entomol.*, **2**: 1015–1032.

Bliss, C. I. and R. A. Fisher (1953). Fitting the negative binomial distribution to biological data and a note on the efficient fitting of the negative binomial. *Biometrics*, **9**: 176–200.

Blumer, M. (1973). Scientific aspects of the oil spill problem. In R. G. Pirie, ed., *Oceanography.* London and New York: Oxford University Press, pp. 453–464.

Blumer, M., G. Souza, and J. Sass (1970). Hydrocarbon pollution of edible shellfish by an oil spill. *Mar. Biol.*, **5**: 195–202.

Carriker, M. (1967). Ecology of benthic invertebrates: A perspective. In G. Lauff, ed., *Estuaries.* Washington, D.C.: Am. Assoc. Adv. Sci., pp. 442–487.

Cronin, L. E., R. B. Biggs, D. A. Flemer, H. T. Pfitzenmeyer, F. Goodwyn, W. L. Dovel, and D. E. Ritchie (1970). *Gross Physical and Biological Effects of Overboard Spoil Disposal in Upper Chesapeake Bay.* Natl. Res. Inst. Spec. Rep. 3, University of Maryland, 66 pp.

Davis, H. C. (1960). Effects of turbidity-producing materials in sea water on eggs and larvae of the clam *Venus* (*Mercenaria*) *mercenaria. Biol. Bull.*, **118**: 48–54.

Dayton, P. K. and R. R. Hessler (1972). The role of disturbance in the maintenance of deep-sea diversity. *Deep-Sea Res.*, **20**: 643–659.

Fauchald, K. (1971). The benthic fauna in the Santa Barbara channel following the January 1969 oil spill. In *Biological and Oceanographical Survey of the Santa Barbara Oil Spill, 1969–1970*, Vol. 1. Allen Hancock Found. Sea Grant Publ. No. 2, p. 61.

Filice, F. P. (1959). The effects of wastes on the distribution of bottom invertebrates in the San Francisco Bay estuary. *Wasmann J. Biol.*, **17**: 7–12.

Glémarec, M. (1973). The benthic communities of the European North Atlantic continental shelf. *Oceanogr. Mar. Biol. Ann. Rev.*, **11**: 263–289.

Grady, J. R. (1971). The distribution of sediment properties and shrimp catch on two shrimping grounds on the continental shelf of the Gulf of Mexico. *Proc. Gulf and Caribbean Fisheries Inst. 23rd Ann. Sessions*, pp. 139–148.

Grassle, J. F. and H. L. Sanders (1973). Life histories and the role of disturbance. *Deep-Sea Res.*, **20**: 643–659.

Green, J. (1968). *The Biology of Estuarine Animals.* Seattle: University of Washington Press, 401 pp.

Gross, M. G. (1970). New York metropolitan region—A major sediment source. *Water Resour. Res.*, **6**(3): 927–931.

Hay, K. G. (1974). Oil and the sea—The ecological implications of a controversial invasion. *Mar. Technol. Soc. J.*, **8**: 19–20.

Hedgpeth, J. W. (1957). Estuaries and lagoons. II. Biological aspects. *Geol. Soc. Am. Mem.*, **67**(1): 693–729.

Hill, M. O. (1973). Diversity and evenness: A unifying notation and its consequences. *Ecology*, **59**: 427–432.

Holme, N. A. (1964). Methods of sampling the benthos. *Adv. Mar. Biol.*, **2**: 171–260.

Jenkinson, I. R. (1972). Dredge dumping and benthic communities. *Mar. Pollut. Bull.*, **3**: 102–105.

Johnson, R. G. (1971). Animal-sediment relations in shallow water benthic communities. *Mar. Geol.*, **11**: 93–104.

Jones, N. S. (1950). Marine bottom communities. *Biol. Rev.*, **25**: 283–313.

MacArthur, R. H. and E. O. Wilson (1967). *The Theory of Island Biogeography.* Princeton, N.J.: Princeton University Press, 203 pp.

MacGintie, G. E. (1939). Littoral marine communities. *Am. Midl. Nat.*, **21**: 28–55.

McNulty, J. K., R. C. Work, and H. B. Moore (1962). Some relationships between the infauna of the level bottom and the sediment in South Florida. *Bull. Mar. Sci. Gulf Caribbean*, **12**: 322–332.

Maurer, D., R. Biggs, W. Leathem, P. Kinner, W. Treasure, M. Otley, L. Watling, and V. Klemas (1974). *Effect of Spoil Disposal on Benthic Communities near the Mouth of Delaware Bay.* DEL-SG-4-74, 231 pp. (processed).

Menzies, R. T., R. Y. George, and G. T. Rowe (1973). *Abyssal Environment and Ecology of the World Oceans.* New York: Wiley, 488 pp.

Molander, A. R. (1928). Investigations into the vertical distribution of the fauna of the bottom deposits in the Gullmar Fjord. *Svenska Hydrogr.-Biol. Komm. Skr., N. S. Hydr.*, **6**(6): 1–5.

Morisita, M. (1959). Measuring the dispersion of individuals and analysis of the distributional patterns. *Mem. Fac. Sci. Kyushu Univ. Ser. E (Biol.)*, **2**: 215–235.

Nadeau, R. J. and T. H. Roush (1972). Biological effects of oil pollution. *Selected Bibliography II.* EPA-R-2-72-055, 62 pp.

Pearce, J. B. (1970). The effects of solid waste disposal on benthic communities in the New York Bight. *FAO Tech. Conf. Mar. Pollut. (Rome).* Pap. F/K: MP/70/E-99, 12 pp.

Petersen, G. G. J. (1913). Valvation of the sea II. The animal communities of the sea-bottom and their importance for marine zeogeography. *Rep. Dan. Biol. Stn.*, **21**: 1–44, 68.

Postma, H. (1967). Sediment transport and sedimentation in the estuarine environment. In G. H. Lauff, ed., *Estuaries*. Washington, D.C.: Am. Assoc. Adv. Sci., Publ. No. 83, pp. 158–184.

Pratt, D. M. (1953). Abundance and growth of *Venus mercenaria* and *Callocardia morrhuana* in relation to the current of bottom sediments. *J. Mar. Res.*, *XII*: 60–74.

Pratt, S. D., S. B. Saila, A. G. Gaines, and J. E. Krout (1973). *Biological Effects of Ocean Disposal of Solid Waste*. Mar. Tech. Rep. Series No. 9, University of Rhode Island, 53 pp.

Purdy, E. G. (1964). Sediments as substrates. In J. Imbrie and N. D. Newell, eds., *Approaches to Paleoecology*. New York: Wiley, pp. 238–271.

Reish, D. J. (1961). A study of benthic fauna in a recently constructed harbor in southern California. *Ecology*, **42**: 84–91.

Rhoads, D. C. (1967). Biogenic reworking of intertidal and subtidal sediments in Barnstable Harbor and Buzzards Bay, Massachusetts. *J. Geol.*, **75**: 461–476.

Rhoads, D. C. (1972). Mass properties, stability, and ecology of marine muds related to burrowing activity. *Mar. Geol.*, **13**: 391–406.

Rhoads, D. C. and D. K. Young (1970). The influence of deposit-feeding organisms on sediment stability and community trophic structure. *J. Mar. Res.*, **28**: 150–177.

Rounsefell, G. A. (1972). Ecological effects of offshore construction. *J. Mar. Sci.*, **2**(1): 1–208.

Rowe, G. T. (1971). Benthic biomass and surface productivity. In J. D. Costlow, ed., *Fertility of the Sea*. New York: Gordon & Breach, pp. 441–454.

Saila, S. B., J. M. Flowers, and R. Campbell (1965). Applications of sequential sampling to marine resource surveys. *Ocean Sci. Eng.*, **2**: 782–802.

Saila, S. B., J. M. Flowers, and M. T. Cannario (1967). Factors affecting the relative abundance of *Mercenaria mercenaria* in the Providence River, Rhode Island. *Proc. Natl. Shellfish Assoc.*, **57**: 83–89.

Saila, S. B. and T. A. Gaucher (1966). Estimation of the sampling distribution and numerical abundance of some mollusks in a Rhode Island salt pond. *Proc. Natl. Shellfish Assoc.*, **56**: 73–80.

Saila, S. B., S. D. Pratt, and T. T. Polgar (1972). *Dredge Spoil Disposal in Rhode Island Sound*. Mar. Tech. Rep. No. 2, University of Rhode Island, 48 pp.

Sanders, H. L. (1958). Benthic studies in Buzzards Bay. I. Animal-sediment relationships. *Limnol. Oceanogr.*, **3**: 245–258.

Schubel, J. R. and J. C. S. Wang (1973). *The Effects of Suspended Sediment on the Hatching Success of Perca flavescens (yellow perch), Morone americana (white perch), Morone saxatilis (striped bass), and Alosa pseudoharengus (alewife) eggs*. Chesapeake Bay Inst. Spec. Rep. 30, Ref. 73-3, 77 pp.

Sherk, J. A. (1971). *The Effects of Suspended and Deposited Sediments on Estuarine Organisms, Literature Summary and Research Needs*. Chesapeake Biol. Lab. Contrib. 443, 73 pp.

Sherk, J. A., Jr. and L. Cronin (1970). *The Effects of Suspended and Deposited Sediments on Estuarine Organisms*. Natl. Res. Inst., University of Maryland, Ref. 70-10, 61 pp.

Springer, S. and H. R. Bullis, Jr. (1954). Exploratory shrimp fishing in the Gulf of Mexico. Summary report for 1952–54. *Comm. Fish. Rev.*, **16**: 1–16.

Stephenson, W., W. T. Williams, and S. D. Cook (1972). Computer analysis of Petersen's original data on bottom communities. *Ecol. Monogr.*, **42**: 387–465.

Stoddart, D. R. (1969). Ecology and morphology of recent coral reefs. *Biol. Res.*, **44**: 433–498.

Sykes, J. E. and J. Hale (1970). Comparative distribution of mollusks in dredged and undredged portions of an estuary with a systematic list of species. *Fish. Bull.*, **68**(2): 299–306.

Thorson, G. (1957). Bottom communities (sublittoral or shallow shelf). *Geol. Soc. Am. Mem.*, **67**: 461–534.

Thorson, G. (1971). *Life in the Sea*, translated by M. C. Meilgaard and A. Laurie. New York: McGraw-Hill, 256 pp.

Wells, H. W. (1957). Abundance of the hard clam *Mercenaria mercenaria* in relation to environmental factors. *Ecology*, **38**: 123–128.

Wiens, H. J. (1962). *Atoll Environment and Ecology*. New Haven, Conn.: Yale University Press, 532 pp.

Wigley, R. L. (1961). Bottom sediments of Georges Bank. *J. Sediment. Petrol.*, **31**: 165–188.

Wigley, R. L. (1962). Benthic fauna of Georges Bank. *Trans. 26th North American Wildlife and Natural Resources Conf.*, pp. 310–317.

Williams, H. B. (1958). Substrates as a factor in shrimp distribution. *Limnol. Oceanogr.*, **3**: 283–290.

Winston, J. E. and F. E. Anderson (1971). Bioturbation of sediments in a northern temperate estuary. *Mar. Geol.*, **10**: 39–49.

CHAPTER 21

Sedimentation and Coastal Engineering: Beaches and Harbors

DAVID B. DUANE*

Coastal Engineering Research Center, Fort Belvoir, Virginia

Waves and currents vary, among other things, with geography, water level (stage of tide), season, and offshore slopes. The net effect of wave and current forces impinging upon a shoreline is to change the morphology of that coastline through processes broadly grouped as sedimentation. That man early recognized the effect of moving water on the land and its possible detrimental effects upon the fruits of his labors is clearly noted in the New Testament writings of Matthew:

So then, everyone who hears these words of mine and obeys them will be like a wise man who built his house on the rock. The rain poured down, the rivers flooded over, and the winds blew hard against that house. But it did not fall, because it had been built on the rock. (Matt. 7:24, 25)

In a very general way the model of sediment transport can be thought of as movement of a sand grain from some source such as a headland, to a barrier beach, to a dune, into an inlet, or to an offshore sink. This process of sediment transport has been active since the beginning of geologic time and the record of such processes is preserved in the stratigraphic column. This natural process becomes of concern to man when it interferes with or affects what man wants to do. There is no complaint of coastal erosion until structures or valued property are threatened with destruction; no complaint

of beach erosion until the sand beach a family enjoys is disappearing; and there is no harbor or inlet "siltation" problem until ships cannot enter a harbor or pass through an inlet.

Design of engineering works in the coastal zone considers not only the materials to be employed and the forces they must withstand, but also considers the modifications the works will introduce in the natural sedimentation processes operating in the coast and nearshore zone. The basic concepts involve interactions of the atmosphere, hydrosphere, and lithosphere, such as generation of water motion and directions of flow, characteristics of the flow, water levels and their periodicity, bed form generation and movement, and sediment entrainment and transport.

A proper engineering design, like a good research experiment design, should provide a clear statement or definition of the problem, its causes, methods of solution, and design requirements for solution. Factors for which data are obtained and studies made relate to geology, hydrology, ecology, and economics. Considerable detailed information on engineering is found in several excellent publications. Two very well-known publications are Weigel's (1964) *Oceanographical Engineering* and the U.S. Army Corps of Engineers' (1973) *Shore Protection Manual*. For those unfamiliar with some of the terms used in the coastal zone, the *Shore Protection Manual* contains a glossary of terms concerning engineering, structures, and processes.

* Present address: National Sea Grant Program, National Oceanic and Atmospheric Administration, Rockville, Maryland 20852.

TIME SCALES

Knowledge of the processes of sedimentation, i.e., erosion, transportation (or transfer), and deposition in the coastal zone, is of value to the engineer and the geologist. However, the time scale of interest in these processes often differs; the scale utilized by the typical coastal engineer is short (a few seconds to a few decades), whereas that of the typical marine geologist is longer (a few hundred to a billion years). In both disciplines, however, the knowledge satisfies an economic need as well as an intellectual need. Problems, analytical techniques, and ranges of precision differ because of the differences in time scale.

The time scale under consideration here is that of several months to at most about 100 years. Such time spans are not the magnitude with which most geologists are familiar. These require that he use tools and techniques that will permit measurement, analysis, and interpretations of higher precision and accuracy than is usually necessary for stratigraphic studies and that he develop a good working relation with engineers, architects, and the general public—all users of science.

Economic Considerations

With today's knowledge and technology any engineering need in the coastal zone can be met, given enough time. However, the ability is not enough, nor is the desire. There must be an economic payout such that tangible—dollar—benefits exceed by more than unity, the costs of construction.

As long as offshore conditions, such as waves, can only be expressed in terms of statistics any structure will suffer damage sooner or later. Optimum design is reached with that structure (or engineering activity, such as beach nourishment) that meets the requirements at minimum total cost. The concept is shown schematically in Fig. 1. To establish the relations shown schematically in Fig. 1 requires knowledge of the cost of construction and behavior of the structure, anticipated damage, and characteristics of the sea and sea-floor. Data on one or more of these items are usually not known with great accuracy, but with reasonable assumptions good estimates can be made with respect to optimum design characteristics and concomitant costs. However, with today's changing laws and mores, intangible benefits are considered which may act to override a cost-benefit ratio in one direction or another.

Usually the duration of time considered in economic analysis of a design is 50 years, that time over which initial costs and maintenance are considered. Other aspects of economics include consideration of

1. the type of material employed, i.e., the useful life expectancy of those materials and therefore an anticipated maintenance/replacement rate;
2. anticipated damage and economic loss due to failure, i.e., when a harbor/entrance channel fails to function properly and shoals, thereby restricting the draft of vessels that may use the site;
3. rate of rise or fall of sea/lake level, i.e., will the work be left high and dry, drowned, or in some other way rendered useless or can it be designed for modification to make it functional within some range of change in water level; and
4. sediment transport (flux), i.e., what is the rate of downdrift shoreline retreat or the rate of harbor shoaling requiring mitigating efforts?

Materials are not germane to this discussion and are not mentioned again, but the other tangible factors used in an economic analysis are considered in the sections that follow.

Sea and Lake Level Changes

Sea level fluctuates with varying periodicity (frequency) and magnitude in response to geologic, atmospheric, and astronomic factors. The mechanism, range, and frequency of astronomic tides are well known, discussed, and predictable. They are always considered in design. Wave height, period, and frequency of occurrence are considered in design and also can be predicted with considerable confidence. Wind setup (the vertical rise in the still water level on the leeward side of a body of water caused by wind stress on the surface) or storm surge can also be predicted with considerable confidence

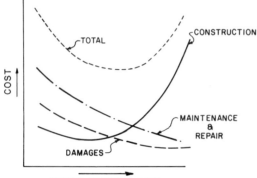

FIGURE 1. *Schematic illustration depicting how various cost-design factors affect total cost of a project. If the design factor (load) is small, construction cost will be low but damage and maintenance costs will be commensurately higher. As the design factor is increased, construction costs rise but damage and maintenance costs decrease.*

(Bodine, 1971; U.S. Army Corps of Engineers, 1973). Nevertheless, these sea-level changes which develop and persist for relatively short periods of time can be extremely destructive to developed or populated sectors of coastal zones and can produce marked changes in the morphology of the coast (Hayes and Boothroyd, 1969; U.S. Army Corps of Engineers, 1963).

Sea-level fluctuations throughout geologic time, in response to tectonic, isostatic, and eustatic factors, are, with the exception of tsunamis, slow but inexorable. When considering the investment of energy, resources, and dollars invested in engineering, knowledge of the rate of long-term secular trends is important to the coastal zone. Because of the multiple causative factors, a long sequence of observations is required in order to discern trends in the rate and direction of sea-level fluctuations. Hicks (1972) and Hicks and Shofnos (1965) have studied the array of tide gages on U.S. coastlines. They discuss various trends and validity of trends developed from 31 year data series for 43 stations. Highlights of data from their studies, which show rising trends in sea level for most U.S. coastlines and a marked increase in rate for the northeast during the past eight years, are presented in Table 1 and Fig. 2.

Large lakes, such as Lakes Superior, Michigan, Huron, Erie, and Ontario, are subject to water-level fluctuations much as are the oceans. Long-term changes in lake levels result form both tilting of the basin due to differential uplift or downwarping of the earth's crust, and more importantly in modern times, from actual changes in the water volume of the basin. Isostatic rebound, due to the recovery of the crust from the load of Pleistocene glaciers, has caused the northern areas of the Great Lakes to rise relative to the southern areas. Estimates of modern uplift vary among different investigators but all are rather small, 0.1 to 1 ft/100 miles/100 years (MacLean, 1963). Long-term lake-level fluctuations are usually thought of as volumetric changes, predominately climatic in origin, which can be summarized as

$$S = (P - E) + (I - O)$$

where
S = change in storage volume
P = precipitation in drainage area
E = evaporation
I = inflow
O = outflow

These processes have been discussed by Bajorunas (1963), DeCooke (1968), and Rowe (1969). Deviations in any of these processes from their average balanced rates lead to changes in water storage volumes and therefore in long-term lake levels.

Mean lake levels are lower now than they were when data collection was initiated in 1860 (Fig. 2). However, a quadratic polynomial fit to mean annual lake levels

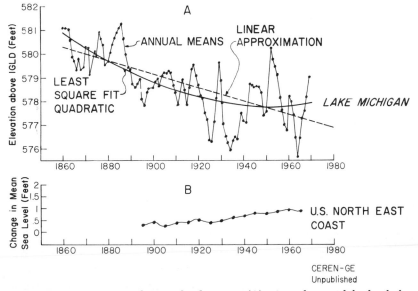

CEREN-GE
Unpublished

FIGURE 2. *Mean annual water-level curves:* **(A)** *Annual mean lake levels for Lake Michigan–Huron for the period of record. Superimposed are a linear and polynomial approximation to the data; the polynomial provides a superior fit.* **(B)** *Annual mean sea-level curve for the U.S. northeast coast; same scale as* **(A)***. Data sources:* **(A)** *NOAA-Lake Survey Center, Ann Arbor, Michigan;* **(B)** *Hicks (1972). From unpublished report, U.S. Army Coastal Engineering Research Center.*

TABLE 1. Apparent Secular Trends (cm/decade) in Sea Level for the United States*

Northeast Coast		Southeast and Gulf Coast		West Coast	
Portland, Maine	1.62	Charleston, S.C.	1.80	Juneau, Alaska	−13.05
Portsmouth, N.H.	1.65	Ft. Pulaski, Ga.	1.98	Sitka, Alaska	−2.04
Boston, Mass.	1.07	Fernandina, Fla.	1.25	Ketchikan, Alaska	0.30
Woods Hole, Mass.	2.68	Mayport, Fla.	1.55	Seattle, Wash	2.59
New London, Conn.	2.29	Miami Beach, Fla.	1.92	Astoria, Oreg.	−0.91
New York City	2.87	Key West, Fla.	0.73	Crescent City, Calif.	−1.34
Sandy Hook, N.J.	4.57	Pensacola, Fla.	0.40	San Francisco, Calif.	1.92
Atlantic City, N J.	2.83	Eugene Is., La.	9.05	Los Angeles, Calif.	0.43
Annapolis, Md.	2.87	Galveston, Tex.	4.30	LaJolla, Calif.	1.92
Hampton Roads, Va.	3.20			San Diego, Calif.	1.43

Source. From Hicks (1972).

* Positive (rising) unless indicated by −.

for Michigan–Huron, which is statistically superior to a linear fit, shows that since 1955 the general trend of annual average lake level has been rising very slightly (Fig. 2).

Property loss and damage in the Great Lakes, and in particular Lake Michigan–Huron in recent years, attests to the practical significance of these annual and long-term lake-level changes to the shore property owner. Although temporal and spatial cycles and scales of lake-level changes are different than sea-level changes, the analogy is clear.

While the consequence of these trends over 100 years might not greatly affect a breakwater designed for a 100 year storm, far-reaching effects could be placed on communities developed near tidelands with a shallow offshore (bay) slope, for the amount of encroachment (horizontal displacement of the shoreline), b, is equal to the amount of water-level rise, a, divided by the tangent of the slope. The bay coast of Sandy Hook, New Jersey, part of the new Gateway National Recreation Area, has an average slope of 0.98°. If the rate of sea-level rise at Sandy Hook remained the same over the next 100 years, the sea level would rise approximately 46 cm and the still water level on the bayshore would advance approximately 30 m, and in some locales nearly 50 m. However, projecting the recent eight year rate over the next 100 years, the still water level on the bayshore would advance approximately 55 m, and in some locales over 100 m.

PROCESSES

In large measure, the basic processes of sediment transport, the result of many complex processes, have been discussed in the preceding chapters. An oversimplified

but perhaps useful conceptual model is to consider sediment transport as a two-step process involving (a) entrainment of a particle by wave action, and (b) displacement of that particle by wind-, tide-, or wave-generated currents.

Bathymetric and shoreline configurations reflect a response of the lithosphere to fluid forces; consolidated materials such as granitic rock respond slowly whereas unconsolidated materials such as sand respond quickly. The manifestation of this process is spit growth, inlet migration, and shoreline retreat or advance. To a developed coast this means beach erosion, harbor shoaling, and inlet modification. Aspects of littoral transport considered for engineering and management of the coastal zone are depicted in Fig. 3.

Beach Erosion

Simply stated, beach erosion is loss of land. A shoreline may be considered in equilibrium if the amount of sediment leaving a reach of coast is equivalent to the amount of sediment entering that same reach from the upcoast direction:

$$V_{in} - V_{out} = 0: \quad \text{stable beach}$$
$$> 0: \quad \text{accreting beach}$$
$$< 0: \quad \text{eroding beach}$$

A change in sediment can be attributed to natural causes, such as a change (usually a rise) in sea or lake level, causing an adjustment of the profile; or interruption in the supply of sediment by storage in inlets; or diversion offshore. Man-induced interruptions can occur at the source by damming of streams (Watts, 1963) or the placement of a barrier such as harbor jetties across the littoral "stream" (Figs. 4A,B). Beach

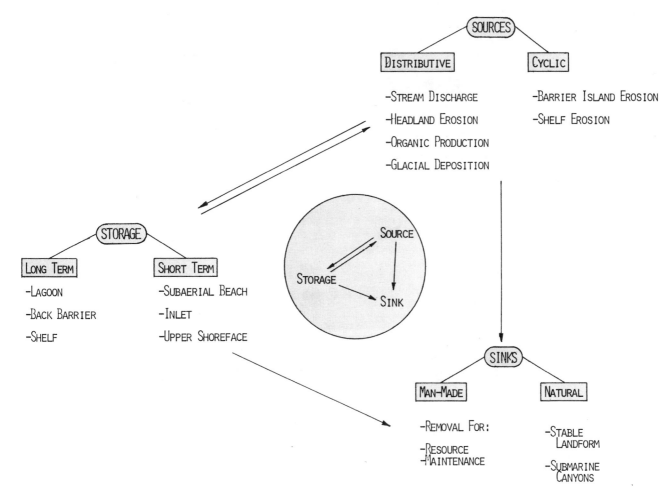

FIGURE 3. *Natural cycle of sediments in the coastal zone. Sources are viewed as either distributive (i.e., they only contribute material to the coast) or cyclic (i.e., recycling of material originally received from a distributive source). Eroding barrier islands are by their nature a cyclic source. Storage areas are the subenvironments of the coastal zone where sediments are temporarily impounded and later returned to the system. Short term is arbitrarily defined as storage for years and decades; long-term storage occurs for centuries and millenia. Sink, as used here, is a final depositional site for sediments coming from either a source or a temporary storage area. Removal of sediment from the coastal zone for either resource use or maintenance dredging is a man-made sink. Loss of material to the outer shelf or the deep sea and to subaerial landforms through coastal progradation, are examples of natural sinks. From Field and Duane (in press).*

recession decreases as distance downdrift from a barrier to littoral transport increases because more of the locally eroded material is replaced by material eroded farther upstream.

The rate of erosion or progradation of the coastline may be determined (estimated) by measuring the time rate of change of the shoreline contours (Kraft and Caulk, 1973; Seibel, 1972; Stafford, 1971). The precision of such derivative data depends on the accuracy and scale of the original land, sea, or airborne surveys.

A means of estimating the volumes of sediment in transport on an average annual basis is achieved through the use of shoreline change maps, especially at locales of accumulation such as spits and capes (Kraft, 1971; Kraft and Caulk, 1973; Caldwell, 1966; Carr, 1965). While this technique is often referred to as providing an estimate of the volume of littoral drift, at best it is a measure of the rate at which material has been trapped, i.e., that portion of littoral transport which bypassed the trap cannot be determined; this technique therefore underestimates the volume transport rate. Another factor to be considered in such determinations is the limiting offshore depth to which computations are carried. Such a depth (termed profile close-out depth) may be estimated on the basis of repetitive bathymetric profiles obtained in a region over a period of years and by noting the depth beyond which

FIGURE 4.   *Aerial views illustrative of the effect of harbor jetties on coastlines.* (A) *Pentwater Harbor, Michigan, on Lake Michigan, June 1974 (photo courtesy of U.S. Army Coastal Engineering Research Center). Predominant littoral drift direction is south.* (B) *Indian River Inlet, Delaware, on the Atlantic Ocean, June 1974 (photo courtesy of M. E. Field). Predominant littoral drift direction is north (top).*

little or no change in depth occurs. Unpublished studies by the U.S. Army Corps of Engineers Coastal Engineering Research Center indicate such depths on open coastlines to be approximately 15 ft for the Great Lakes, approximately 20 ft for the U.S. Atlantic coast, and 25 ft for the U.S. Pacific Coast of southern California.

The geometric shape developed for computation of volume changes should have a base at least equal to that of profile close-out depth. If a littoral drift depocenter such as a spit or cape is being studied for an estimation of average annual littoral transport volume, selection of the close-out depth should be made to coincide with the change in slope marking the convergence of the shoreface and the inner shelf. A tabulation of volume changes associated with different depths near Chincoteague Inlet, Virginia is provided in Figs. 5A,B.

A relatively simple but precise method of determining the volume of irregular shaped masses is by means of the prismoidal formula

$$V = \frac{H}{6}(S_T + 4S_M + S_B)$$

where V is volume, H is height of the section, S_T and S_B are cross-sectional areas of the top and bottom bases, respectively, and S_M is the cross-sectional area of the midsection.

Estimates of volume transport rates based on use of shoreline change maps, associated either with open coasts or spits and capes, are satisfactory in that short-term fluctuations in the profile changes are averaged out. Short-term cycles in beach and offshore profiles which occur in response to tides, storms, and seasons, have

FIGURE 4 (Continued)

been documented and reported upon by numerous workers (Abele, in press; Bajorunas and Duane, 1967; Hayes, 1971; Krumbein and James, 1974; Orme, in press; Shepard, 1950; Sonu, 1968; and Sonu and VanBeek, 1971).

It is often assumed that sand removed from a beach during storm wave conditions (ratio of deep-water wave height H_0 to deep-water wavelength $L_0 = 0.025$) is moved offshore where it is temporarily stored until after the storm when fair-weather swell conditions ($H_0/L_0 = 0.020$) move the sand back to the beach. Such cycles have been observed in laboratory studies (Rector, 1954; Scott, 1954) and inferred to occur on the basis of field studies (Hayes, 1971; Shepard, 1950; Sonu and VanBeek, 1971). However, most of these studies, both laboratory and field, have been two-dimensional in form. Such transport is predicted by theory (Ippen, 1966), and has been observed (Yasso, 1965; Ingle, 1966; Duane, in preparation; James, in preparation). However, no studies are known that

address the relative volume difference in offshort transport versus longshore transport during a storm. Theories of longshore current (Galvin, 1967; Longuet-Higgins, 1970) and verification of these theories (Bruno and Hiipakka, 1973; Balsillie, personal communication) subjectively argue for a conclusion that the volume of sediment transported alongshore is vastly greater than the volume transported offshore along coastal sectors.

Knowledge of the rate of shoreline adjustment over the long term is applied to (1) decisions such as whether or not to develop a segment of coast; (2) the design and location of that portion of a coastal protection project structure which is the last line of defense; (3) legal considerations leading to condemnation proceedings or determination of setback limits; and (4) estimates of the frequency, nature, and cost of maintenance over the life of the project. Knowledge of the loss (and recovery) of beach or coast due to single short-period phenomena is similarly utilized. It is important therefore to make observations over a period long enough to

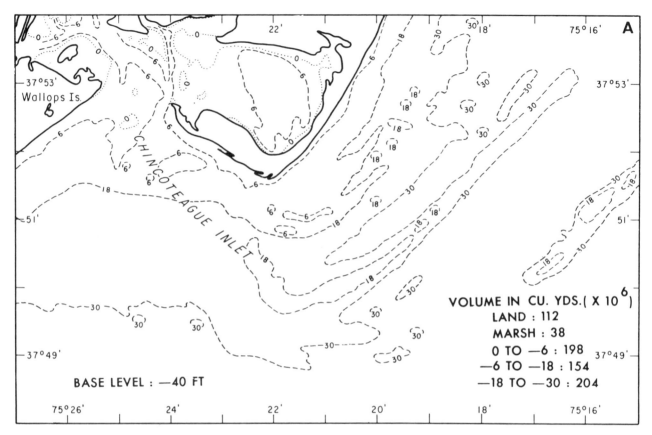

FIGURE 5. *The prograding spit at the south end of Assateague Island, Chincoteague Inlet. (A) Bathymetry surrounding Fisherman's Point. Using the prismoidal formula, various volumes of sand have been computed for different limiting (close-out) depths* as noted. (B) *Growth of Fishing Point (i.e., change in low-water datum contour) for a 31 year period. From unpublished report, U.S. Army Coastal Engineering Research Center.*

determine if annual or seasonal cycles are present, and what the scale and nature of the changes are.

Harbor Shoaling

A harbor shoals or an inlet is modified as a result of the entrapment of sediment transported as bed load or suspended load to its point of accumulation. The source of sediment involved in shoaling varies for different harbors, depending on many factors. Some harbors, such as New York City or Charleston, South Carolina, shoal largely as a result of river- or ocean-derived fine-grained sediment. Harbors at the coastline that require entrance training structures (jetties) usually shoal as the result of bed or nearbed transport in the littoral zone. Examples of these types are Santa Barbara, California; Barnegat, New Jersey; or Wrightsville Beach (Masonboro Inlet), North Carolina.

Because the material shoaling harbors or inlets is derived from material in transport in the littoral stream, harbor shoaling is intimately related to beach processes, volume transport, and seasonal changes in volume and direction of sediment transport. A dredged channel is usually required to provide navigable depths for vessels using the harbor. Training structures, which can be a variety of sizes and shapes (Fig. 6), not only channel flow through the inlet and protect the vessels from waves but also serve to keep the navigation channel free from sediment moving along the coast.

Knowledge of the rate of sediment transport is therefore extremely important to harbor design and shoaling. The size of the vessel using the harbor dictates certain parts of the harbor design such as dredged channel depth (which together with offshore geomorphology dictates the length of the channel), width of harbor opening, and length and configuration of jetties. The volume of sediment in transport past the harbor entrance influences the rate of shoaling, frequency of harbor dredging, and the effects on downdrift coastlines as a result of interruption of littoral transport.

With respect to harbor construction and maintenance then, it is very important to have knowledge of the volume of sediment in transit past the harbor site, and also to know the rates upcoast and downcoast as well.

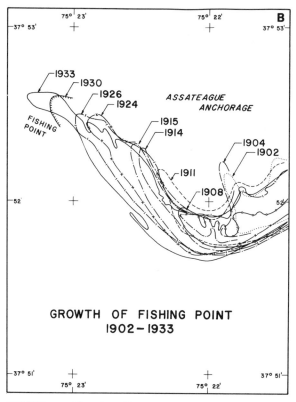

GROWTH OF FISHING POINT 1902-1933

FIGURE 5 (Continued)

Some techniques for estimating volume transport were explained in the preceding section. Those methods are not too satisfactory, however, because they tend to underestimate the volume of sediment in transport. On coastlines where harbors already exist, an estimate of littoral transport can be obtained by compiling the amount of sediment trapped in harbor entrances over a known time period. Such data can be obtained by comparison of results of periodic hydrographic surveys or estimates of the amount or material dredged and the time between dredging operations.

Attempts have been made to relate transport rates thus determined to incident wave characteristics, in particular wave height (Das, 1972; U.S. Army Corps of Engineers, 1973; R. O. Bruno, personal communication); see Fig. 7. Where shoaling volumes are used to provide the basis for transport curves, quantities predicted from the derived curves are likely to be minimums as there is no way to account for the quantity of material in transit which is not trapped to form the shoal.

Dredged channels and jetties disrupt the flow of sand downcoast past a site. The jetties form the base of a holding basin, or trap, which will accumulate sand until its holding capacity is exceeded, at which time sediment will begin to bypass the jetties and shoal the harbor. Once the harbor is shoaled, sediment will, as in

the case of a natural inlet, bypass the entrance and continue its transit downcoast. How long a period is involved before the holding capacity is exceeded is a function of the littoral transport, size and configuration of the jetties, and offshore slope. The holding capacity and wave protection criteria influencing the jetty design must consider variations in the oceanographic season, where it exists (Figs. 8A,B).

Rate of shoaling in the harbor entrance influenced by littoral transport rate and oceanographic seasons is also affected by the geometry of the dredged entrance channel, size (volume) of the harbor, and tide (water) level fluctuation.

Insofar as a harbor interrupts littoral transport, some effect on the downdrift coastal areas will be felt (Fig. 4). As a natural inlet will bypass sand downdrift, so too would a man-made inlet, i.e., a harbor. However, at or before the time in which a man-made harbor will begin to bypass sand naturally, it ceases to function in the manner designed and the shoaling material is physically removed and placed elsewhere. In the past, and in some instances today, the dredged material was removed from the littoral stream. At present it is common practice to use the dredged material for beach nourishment, or in some other fashion to return the dredged material to the littoral zone. Such a procedure is referred to as sand bypassing.

ENGINEERING

A need for engineering in the coastal zone is predicated upon socioeconomic factors. At the request of the U.S. Congress, the Corps of Engineers undertook a study of the national shoreline and published a report of findings (U.S. Army Corps of Engineers, 1971). This National Shoreline Study showed that about 2700 miles of the U.S. coastline is in a critical erosion condition (critical erosion is defined as that condition along the coast where actions to halt erosion may be justified to protect man's interest). Estimated cost to implement remedial works for the 2700 miles of shore is $1.8 billion. Periodic beach nourishment (i.e., placement of suitable sand on the beach) would be utilized in many cases for the remedial works; the estimated annual cost for performing the periodic nourishment is $23 million. Other interesting information developed in the National Shoreline Study shows that 17 shore protection projects, protecting 171 miles of shoreline, are currently under way at an estimated cost of $423 million for which the federal government's share of the cost is about 65% ($279 million). If it is assumed that results of research contributed 5% savings in the projects currently under way, this would

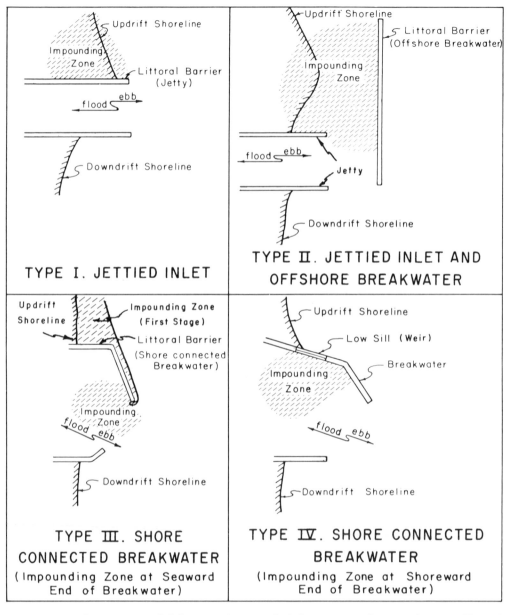

TYPE I. JETTIED INLET

TYPE II. JETTIED INLET AND OFFSHORE BREAKWATER

TYPE III. SHORE CONNECTED BREAKWATER
(Impounding Zone at Seaward End of Breakwater)

TYPE IV. SHORE CONNECTED BREAKWATER
(Impounding Zone at Shoreward End of Breakwater)

FIGURE 6. *Schematic views of different configurations of structures used to stabilize and protect navigable inlets from waves and or littoral drift. Predicted location of zone of sediment accumulation is shaded assuming predominant direction of littoral drift is from top of each view toward bottom. From Watts (1965) in U.S. Army Corps of Engineers (1973).*

represent about $21 million savings; even a 1% savings on total costs would be $4.2 million. Following initial coastal construction and with particular regard to single projects, it is common practice to extend or modify existing structures based on the history of sedimentation in and around the structure. (Usually the same effects can be projected to construction of new works in the same geographic area.) These single projects vary in

total first costs, but probably average about $3 million each. Therefore, if results produced by research would provide modification leading to savings of 5%, the dollar saving would be $150,000 each. With regard to annual savings on maintenance, if one considered a typical coastal structure as above, where sand is the shoaling medium, annual maintenance could be about 10% of the first cost, or approximately $300,000.

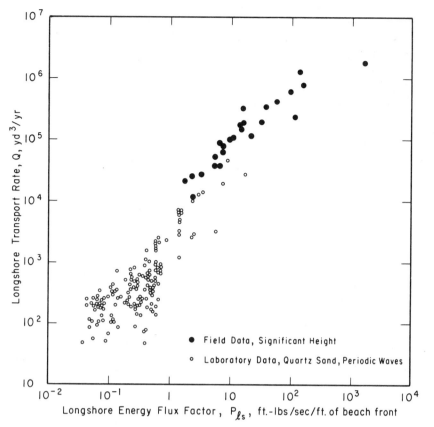

FIGURE 7. *Graph of predicted longshore transport versus longshore energy. Note scatter of data on log-log plot. From U.S. Army Corps of Engineers (1973).*

Using the assumption of a research study leading to a savings of 5%, the annual savings in maintenance would be $15,000.

In the coastal zone (including the Great Lakes) annual quantities dredged average about 300,000,000 yd³ for maintenance, and about 80,000,000 yd³ in new work, with total annual cost exceeding $150,000,000 (Boyd et al., 1972). Considering outer bar and entrance channel, bay channel, and harbor, approximately 165,000,000 yd³ is dredged annually. The cost of handling varies ($0.20–$3.00), but the nationwide average approximates $0.40 yd⁻³ (Boyd et al., 1972). In the coastal zone, approximately 75,000,000 yd³ of this material is classed as mixed sand and silt while 22,000,000 yd³ is gravel, sand, and shell. At $0.40 yd⁻³, moving this material amounts to $38.8 million, and if research resulted in 5% savings, more than $1.5 million would be involved.

There is a real need, therefore, to conduct research that can lead to increased knowledge of processes in the coastal zone so that improved engineering can be applied to activities in the coastal zone. Good research requires good data; the quality of data from a statistical sense

improves with an increase in the number of observations and the duration of time over which the observations have been made (Fig. 8). Waves and currents can be observed visually from the shore with good results (Berg, 1968; Bruno and Hiipakka, 1973; Szuwalski, 1970) or from instruments or dye placed in the littoral or nearshore zone (Harris, 1972; Teleki et al., 1973; Williams, 1969). Shoreline change maps, as discussed previously in this chapter, and determinaiton of time series of beach changes (Davis and Fox, 1972; Everts, 1973; Krumbein and James, 1974; Hayes and Boothroyd, 1969; Saylor and Hands, 1970; Sonu and VanBeek, 1971) are means of furthering knowledge of coastal process and response elements. Tracer studies provide information on variation of transportability of different sand particle sizes (Yasso, 1965; James, in preparation; Duane, in preparation); they also indicate there are differences in the rate of transport in the shore-normal direction (Duane, 1970).

Tracers also can be used to evaluate the effect of structures or engineering works on coastal processes (Duane, 1970; Duane, in preparation; Hart, 1969). Of similar value are the more conventional field data col-

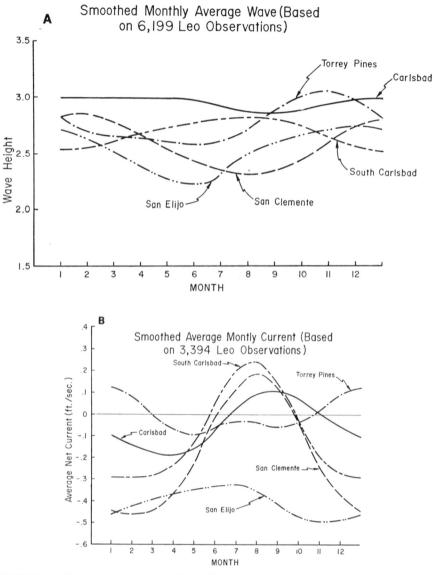

FIGURE 8. *Variations in wave height and longshore current characterizing two seasons at five sites in Southern California; sites are between Los Angeles and San Diego. Leo (littoral environment observations) measurements are simple, visual observations usually made at least once per day at reporting stations (Berg, 1968). (A) Average wave height, in feet, observed over a period of approximately two years (1969–1971). (B) Longshore current, landward of breakers, observed over same period as in (A). Negative sign refers to southerly directed flow. The dual season is more clearly expressed by current than by wave height. From Coastal Engineering Research Center, unpublished report on Oceanside Harbor.*

lection techniques used for evaluation of structures (Keith and Skjei, 1974; Orme, in press; Vallianos, 1970), of future disposal operations (Pararas-Carayannis, 1973) and offshore dredging (Courtenay et al., 1974). Research and evaluation of existing works contribute both to increased knowledge of the coastal zone leading to improved engineering and to a more effective and equitable use of the coastal zone.

Protective Beaches

Beaches are very effective in absorbing and dissipating wave energy, and therefore afford protection to the adjoining backshore areas. In a sense beaches are similar to structures such as sea walls, bulkheads, and revetments in that all serve to separate a land area from a water area. Sea walls, bulkheads, and revetments are used where it is necessary to maintain the shore in a forward position relative to adjacent shores, where there is little littoral material supply, little or no protective beach, or where it is desired to maintain a depth of water along the shoreline. Furthermore, these structures afford protection only to the land immediately behind and none to adjacent areas upcoast or downcoast. In contrast, a beach, through littoral processes, provides sand for natural nourishment and protection to downdrift areas.

Selection of a protective beach as the shore protection method, and particularly the periodic replenishment of a beach (termed artificial beach nourishment), means that long reaches of shore may be protected at relatively low cost as compared with costs of other shore protection methods. Also, such a treatment directly remedies the basic cause of most erosion problems—a deficiency of natural sand supply—and therefore provides benefits to the shore downdrift from the immediate problem area. A widened beach will also probably have considerable value as a recreation feature.

Planning for the design of a protective beach involves the following: (1) geometry of the fill (i.e., beach berm elevation and width, determination of adjusted foreshore slope); (2) determination of direction and volume of littoral transport (i.e., deficiency of supply to the problem area); (3) determination of composite grain-size characteristics of native material in the active littoral zone (considering both time and space); (4) specification of borrow (fill) material for initial placement and subsequent nourishment; (5) availability of specified sand (location, quantity, and cost).

Beach berms are formed by the deposit of material by wave action. Height of the berm is related to cyclic changes in water level, normal beach and nearshore slopes, and wave climate. A storm berm, higher in elevation than a normal berm, is formed, as the name implies, by storm wave action. Several alternative techniques are available to estimate the berm design height, i.e., measurement of natural berm height; comparison with other sites with similar coastal characteristics; and/or application of suitable wave data to estimation of wave runup (Savage, 1958). Design of berm width will depend on the purpose of the restoration or nourishment and the condition of the coast. Regardless, records of determination of the seasonal- or storm-related changes in beach width need to be incorporated in the determination of the design width.

If the protective beach is to have a recreational function also, estimates of the numbers of persons to use the beach on a daily basis are also considered. The present federal standard is 75 ft² of dry beach per bather (U.S. Army Corps of Engineers, 1973). Fill should be planned so the toe extends to that water depth where alongshore transport would reasonably still be expected to occur. As mentioned previously, that depth should be approximately 15 ft for the Great Lakes and 25 ft for the ocean coasts. Based on previous experience, design slopes are usually computed (for determining fill volume) in the range of 1:20 to 1:30 (flatter than natural beach slopes) from low-water datum to the intersection with the natural bottom. It is impractical to attempt to grade submarine slopes to such a value for they will adjust quickly to a natural slope commensurate with local wave and current conditions and grain size of the material placed (Fig. 9).

Direction of predominant littoral transport can be estimated by analysis of wave data, longshore current studies, and examination of coastline offset at engineering structures. The deficiency of material supply to the beach is the rate at which material supply must be increased to balance transport so that no net loss occurs. Studies of shoreline change maps or aerial photos provide information applicable to volume changes if no other survey data are available. A relationship where 1 ft² of change in beach surface area equals 1 yd³ of beach material is considered to provide usable values for estimating a deficiency of material supply (U.S. Army Corps of Engineers, 1973).

Determination of composite grain-size characteristics of native material in the active littoral zone is important to the specification of fill material. In a spatial and temporal sense, sufficient samples must be obtained to describe adequately the size distribution of beach (and potential borrow) material. Description of the "native" sand is the basis for subsequent determination of the suitability of potential borrow sites for fill as well as for determining the quantity of borrow material placed in order to have 1 yd³ remain. This description is usually accomplished by collecting sediment samples across a beach profile between the berm and the seaward limit of profile change, and averaging the grain-size distributions, thus providing a "composite" grain-size distribution from which the composite phi mean and phi sorting are estimated. The analogous composite can be made for the borrow area such that

$$Dc_n = \frac{\sum_0^n fi_n}{n}$$

where Dc = composite distribution

n = number of size classes

fi = frequency of material in each class of interval i

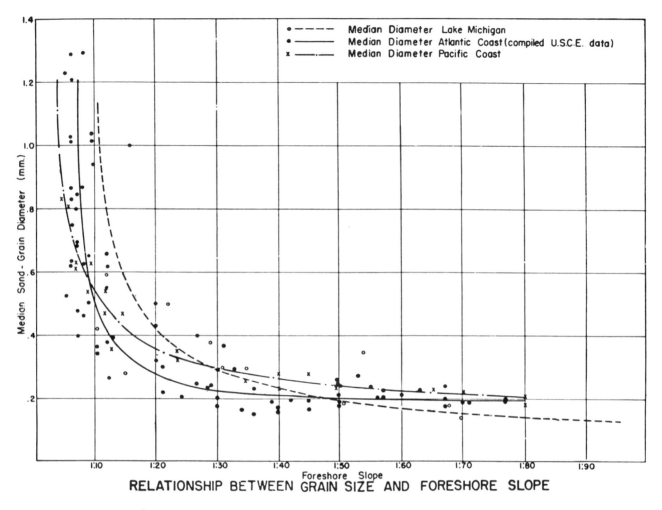

RELATIONSHIP BETWEEN GRAIN SIZE AND FORESHORE SLOPE

FIGURE 9. *Relationship between median grain diameter and foreshore slope. From U.S. Army Corps of Engineers (1966).*

When using historical data it is not uncommon to find only a tabulation of phi means and sorting. In such circumstances it is still possible to compute the composite mean and sorting using phi notation (Krumbein, 1957) as follows:

$$M_C = \frac{M_A + M_B}{2}$$

where M_C = composite mean for all profiles in the section

M_A = largest mean

M_B = smallest mean

$$\sigma^2 = \bar{\sigma}^2 + \frac{(M_B - M_A)^2}{12} + \frac{(M_B - M_A)^2}{6(N-1)}$$

where σ^2 = variance of computed composite curve

$\bar{\sigma}^2$ = mean variance of individual distributions

M_A, M_B = extreme phi means

N = number of samples in the set

Consideration of the suitability of sediment for beach fill material is based on previous experience with fill and the concepts expressed by Krumbein and James (1965) as later modified by the U.S. Army Corps of Engineers (1973) and James (in press). The underlying assumption is that the sediment on the beach to be filled (native material) is used to predict a grain-size distribution that will occur by modification, through natural processes, to the borrow material placed on the beach. The quantity defined, $R_{\phi crit}$, is a ratio that represents the minimum volume of borrow material required to produce a unit volume of material having the same grain-size distribution as the native material. The ratio is based on a simple mathematical model which estimates grain-size effects by direct comparison between native and borrow material grain-size distribution characteristics (Krumbein and James, 1965). This early work has been modified and upgraded to provide guidelines as published in the *Shore Protection Manual* (U.S. Army Corps of Engineers, 1973). Figure 10, reproduced

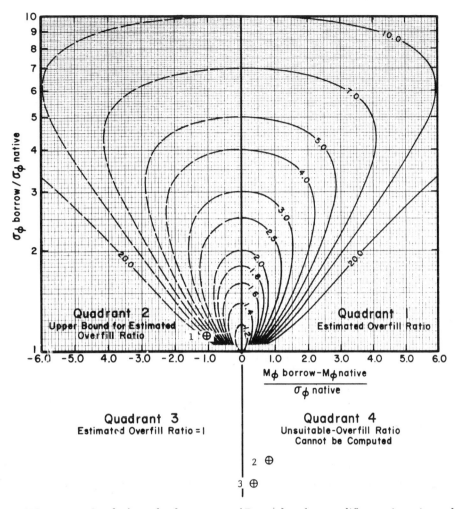

FIGURE 10. *Graph of equal value contours of* $R_{\phi crit}$ *plotted versus difference in native and borrow material textures. Note the log scale and that the abscissa is unity, not zero.* $R_{\phi crit}$ *values for the 1965 coarse fill at Presque Isle, Pennsylvania is plotted as point 1, the fine fill as point 2. Similarly the 1954 fill placed on Revere Beach, Massachusetts is plotted at point 3.*

from the *Shore Protection Manual*, provides a graphical means of determining $R_{\phi crit}$. Note that no curves are plotted below the horizontal axis; this is because it is mathematically impossible to produce a log-normal size distribution with poorer sorting than that of the native material by selective removal of material. Points plotting in quadrant 1 occur when the borrow material is finer $(M_B > M_N)$ and more poorly sorted than native material $(\sigma_B > \sigma_N)$; conversely, points plotting in quadrant 2 occur when the borrow material is coarser $(M_B < M_N)$ and more poorly sorted than native material.

[*Note.* The following two paragraphs of this chapter are derived from the U.S. Army Corps of Engineers (1973) and unpublished reports and correspondence of the Coastal Engineering Research Center.]

Conceptually it is easy to visualize the processes acting to modify a beach filled with borrow material

having finer mean grain size, but which is more poorly sorted than native sand. When sand is mechanically deposited on the beach, waves immediately start a sorting and winnowing action on the surface layer of the fill, moving finer particles seaward, and leaving coarser material shoreward of the plunge point. This sorting and winnowing process continues until a "layer," composed of a distribution of the coarser particles compatible with the predominant wave spectrum, armors the beach and renders the slope relatively stable. However, if the "armor" is disturbed due to a storm, the underlying material is again subjected to the sorting and winnowing processes. Material finer than that previously exposed on the natural beach face will, if exposed on the surface during a storm, move seaward to an equilibrium depth compatible with its size, thus tending to form inshore and foreshore slopes that might be flatter than those considered normal prior to placement of fill.

The processes acting to modify a beach filled with borrow material having a larger mean grain size, but more poorly sorted than native sand, are the same as those active in the situation described above. However, finer grain sizes of beach sand are more readily transported than coarser sizes. Therefore the processes that act to sort, or modify, the grain-size distribution of borrow material placed upon a beach will be most active on the finer grain sizes and it is expected that much of the excess (cubic yards) coarse material, computed by $R_{\phi crit}$ and placed as fill, will remain on the beach as a stabilized profile. In this case the grain-size distribution of the stabilized fill should be coarser and should not completely match the native sand grain-size distribution. The coarser material may also be expected to produce a beach profile steeper than the native beach profile. If this concept is representative of the actual occurrence, the amount of borrow material lost from the fill will be less than that calculated from $R_{\phi crit}$; the computed value of $R_{\phi crit}$ should therefore be considered to represent a maximum volume. Probably the best value of use is between 1.0 and 1.5, with the value approaching 1.0 as the difference in means increases.

Some empirical information pertinent in this consideration may be found in the results of the coarse sand fill placed on Presque Isle Peninsula, Erie, Pennsylvania in 1965. That operation and follow-up surveys are discussed by Berg and Duane (1968). Coarse fill used in this instance was added on a one-to-one basis, i.e., coarse sand was placed in the quantity necessary to meet a desired project dimension; $R_{\phi crit}$ was computed (after the fact) as 11.5 (Fig. 10). The Presque Isle coarse fill ($M_B = 0.4$; $\sigma_B = 1.5$) was considerably more stable than fine fill ($M_B = 2.2$; $\sigma_B = 0.66$) placed in an adjacent cell. Another case providing support to this model is a 1954 beach fill operation at Revere Beach, Massachusetts. Mean grain diameter of the 1954 borrow material was finer than the composite Revere Beach sand and was also better sorted (poorly graded). In the context of the above, the 1954 borrow plots in quadrant 4 on Fig. 10. Following placement, there was rapid loss of fill material. Noting where this borrow material plots on Fig. 10 (quadrant 4) adds credence to the model which predicts instability of the fill. General statements of the applicability of $R_{\phi crit}$ to various characteristics of native and fill material are given in Table 2.

In a recent study, James (in press) proposes a model for determining a relative retreat ratio as a means of evaluating the relative suitability of alternate borrow sources in a manner similar to that for determining the overfill ratio. This model predicts that the relative retreat rate increases with an increasing difference between the native and borrow means; the finer the borrow material, the higher is the predicted retreat rate. Additionally, more poorly sorted borrow material results in lower retreat rates, perhaps because the more

TABLE 2. Applicability of $R_{\phi crit}$ Calculations for Various Combinations of the Graphic Phi Moments of Borrow and Native Material Grain-Size Distribution

Quadrant in Fig. 10	Relationship of Phi Means	Relationship of Phi Standard Deviations	Response to Sorting Action
1	$M_{\phi B} > M_{\phi N}$ Borrow material is finer than native material	$\sigma_{\phi B} > \sigma_{\phi N}$	Best estimate of required overfill ratio is given by $R_{\phi crit}$
2	$M_{\phi B} < M_{\phi N}$ Borrow material is coarser than native material	Borrow material is more poorly sorted than native material	Required overfill ratio is probably less than that computed for $R_{\phi crit}$
3	$M_{\phi B} < M_{\phi N}$ Borrow material is coarser than native material	$\sigma_{\phi B} < \sigma_{\phi N}$	The distributions cannot be matched but the fill material should all be stable, may induce scour of native material fronting toe of fill
4	$M_{\phi B} > M_{\phi N}$ Borrow material is finer than native material	Borrow material is better sorted than native material	The distributions cannot be matched. Fill loss cannot be predicted but will probably be large

Source. From U.S. Army Corps of Engineers (1973).

poorly sorted material contains a larger fraction of coarser material which provides a more stable "armor."

Once selection of material has been completed for use in beach restoration or nourishment, there are a number of means of borrowing it and placing it on the beach. These are discussed in the *Shore Protection Manual* (U.S. Army Corps of Engineers, 1973) and Chapter 23 and are not mentioned further here. In the case of beach restoration, it is common practice to place material to project dimensions essentially synchronously along the length of the project. In the case of periodic nourishment, it is sometimes economically best to develop a "feeder beach" at one or more updrift locations and let natural processes transport the nourishment material to beach sectors downdrift of the feeder beach.

An important natural land form of the coastal zone is the sand dune. Comprised of a variety of sizes, shapes, and locations in the coastal area, they play different roles in shore protection and littoral transport. Foredunes, which form landward of the beach, are created by the combined action of wind, sand, and vegetation, and are the most important to the shore; they act as a sink/reservoir of sand for the littoral stream and as a levee to prevent waves and high waters from flooding low-lying areas behind the dunes.

Siting structures seaward of foredunes is not a sound act, nor is the leveling and/or destroying of them to provide a level lot or an unobstructed view. Dunes may be destroyed by a variety of natural processes: waves and tides of severe storms, drought, disease in the stabilizing vegetation, or overgrazing. Naturally occurring dunes can be maintained and stabilized by various means. Dunes can be created, where none before existed, by construction of mechanical structures such as fences of various types, or by use of vegetation (Jagschitz and Bell, 1966; Savage and Woodhouse, 1969; Woodard et al., 1971; Woodhouse et al., 1974).

Groins

A groin is a shore-protective structure designed to either build or maintain a protective beach by trapping littoral drift or to retard erosion of an existing beach. Groins are usually perpendicular to the shore and extend from a point landward of possible shoreline recession into the water a sufficient distance to stabilize the shoreline. They are relatively narrow in width and may vary in length from tens of feet to several hundred feet.

Groins differ from jetties structurally and functionally in that jetties generally are larger with more massive components, and are used primarily to direct and/or confine stream or tidal flow at the mouth of a river or inlet to a bay and prevent littoral drift from shoaling the channel. In some sections of the country, groins are mistakenly referred to as jetties.

Groins may be classified as permeable or impermeable, high or low, and fixed or adjustable. They may be constructed of timber, steel, stone, concrete, or other materials, or combinations thereof. Impermeable groins have a solid or nearly solid structure that prevents littoral drift from passing through the structure. Permeable groins have openings through the structure of sufficient size to permit passage of appreciable quantities of littoral drift. Some permeable stone groins may become impermeable with heavy marine growth. A series of groins acting together to protect a long section of shoreline is commonly called a *groin field*, although *groin system* is a preferable term. Two excellent reference works on groins are papers by Berg and Watts (1967), who discuss variations in groin design, and Balsillie and Bruno (1972), who have compiled an annotated bibliography of groins.

A groin creates a total or partial barrier to littoral drift moving in that part of the littoral zone between the seaward limit of breaking waves and the limit of wave uprush. The extent to which the littoral transport is so modified by a groin depends on the height, length, and permeability of the groin. The manner in which a groin modifies the rate of littoral transport is about the same whether it operates singly or as one of a system, provided spacing between adjacent groins is adequate. A schematized plan and section of a groin and a photograph of a rubble groin are pictured in Fig. 11. A typical groin system is illustrated in Fig. 12.

Groins may be used to stabilize a beach (subject to intermittent periods of advance and recession) and build or widen it by trapping littoral material, and reduce the rate of littoral transport out of an area (hence stabilize that shore sector) by reorienting a section of the shoreline to an alignment more nearly perpendicular to the major direction of wave approach.

When considering use of groins as shore protection methods, it is critical to assess several factors. These are (a) the extent to which the downdrift beach will be damaged if groins are used; (b) economic justification for groins in comparison with stabilization by nourishment only; (c) adequacy of natural sand supply to ensure that groins will function as desired; and (d) adequacy of shore anchorage of the groins to prevent "flanking" as a result of downdrift erosion. Where the supply of littoral drift is insufficient to permit the withdrawal from the littoral stream of enough material to fill the groin or groin system without damage to downdrift areas, artificial placement of fill may be required to fill the groin and thus minimize the reduction of littoral drift to downcoast areas.

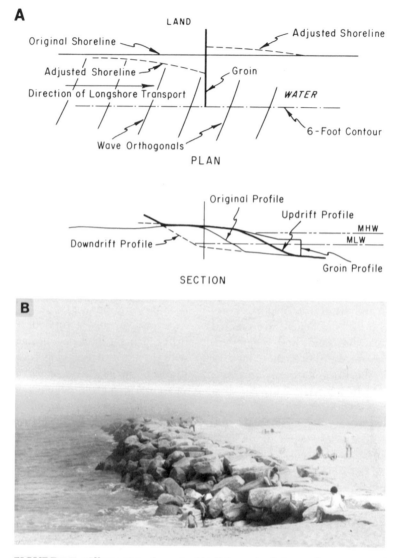

FIGURE 11. *Illustrations of a groin.* (A) *Schematized plan and cross-sectional view of a groin and its effect on the shoreline. From U.S. Army Corps of Engineers (1973).* (B) *Photo of a rubble groin at Plum Island, Massachusetts, 1968. Littoral drift at this location is from right (south) to left (north); tide range approximately 3 m (photo by the author).*

Restoration and widening of beaches, with and without groins, have become more widely employed in recent years. Several examples from various coasts are enumerated as follows: *Pacific Coast*; Redondo, Newport, and Oceanside, California; *Gulf Coast*, Harrison County, Mississippi; *Atlantic Coast*: Ocean City, New Jersey; Virginia Beach, Virginia; and Wrightsville Beach, North Carolina.

Harbor Shoaling

As stated earlier, inlets, either natural or improved to meet navigation requirements, tend to interrupt littoral transport along the shore. In the case of natural inlets

that have a well-defined bar formation on the seaward side of the inlet, a portion of the littoral drift ordinarily moves across the inlet by way of the outer bar, but the supply reaching the downdrift shore is usually intermittent rather than regular, with the result that the shore downdrift from the inlet is normally unstable for a considerable distance. If the strength and duration of tidal flow from open water through the inlet into an interior body of water are greater than those of the return flow, sediment will accumulate in the form of an inner bar, middle ground shoal, or tidal delta, thereby reducing the supply of littoral material available to nourish downdrift shores. In the case of migrating inlets the outer bar normally migrates with the inlet, but the

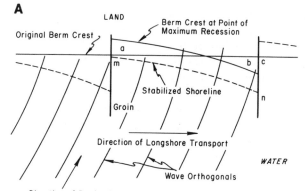

FIGURE 12.   *Illustrations of a groin field. (A) Schematized plan view and resulting changes in the shoreline. From U.S. Army Corps of Engineers (1973). (B) Oblique aerial photo of a groin field, Wallops Island, Virginia, June 1974. Predominant littoral drift is from bottom (south) to top (north) of photo; site is just south of Chincoteague Inlet, Fig. 5 (photo courtesy of M. E. Field).*

inner bar does not; the inner bar increases in length as the inlet migrates, thus increasing the volume of material inside the inlet.

When in the course of conducting commerce (or in the pursuit of recreation) the water depth over the inlet shoal is too shallow for safe navigation, dredging is usually required. When the natural depth of an inlet is increased by dredging, either through the outer or inner bars of the channel, additional storage area is created to trap the available littoral drift, thereby reducing the quantity that would naturally pass the inlet. If the material dredged (either initially or for channel maintenance) is deposited beyond the limits of the littoral zone, as in the case of disposal in deep water at sea, the supply to the downdrift shore may be virtually eliminated with consequent erosion at a rate equivalent to the reduction in supply.

The quantity of sediment available to shoal a harbor or its entrance is a function of the gross sediment transport and the hydraulics of flow in the inlet. The hydraulics of flow in the inlet are governed by the length and shape of the inlet channel(s), historical changes in the inlet, the harbor/bay area, tide range, and wind- and tide-induced currents (see Chapter 5). Accurate predictions of velocities are critical to proper design of navigational aids and prediction of inlet stability. Baines (1957) and Van de Kreeke (1967) discuss the mathematical analyses of flow through tidal inlets, and Keulegan

(1967) studied the hydraulics of a simplified inlet-bay system and presents a filling or repletion coefficient that expresses an inlet's capability to fill its harbor or bay. Determination of the repletion coefficient governs the prediction of harbor/bay water surface elevations as well as estimates of velocities in the inlet. These factors of course affect the type and volume of sediment transported and deposited in the inlet and inlet-associated shoals.

For harbors in the coastal zone, the primary source of sediment is littoral drift. On open coasts the net littoral drift is the factor critical to most engineering works, while for inlets it is both net and gross transport. Methods of estimating the volume of littoral drift were discussed previously. Sediment passing an inlet, either as suspended or bed load, is available for transport into or past the opening, depending on the hydraulics of the inlet. In some instances, where harbor breakwaters parallel the shoreline or approaching wave crests for considerable distance, another possible source of sediment *within* the harbor is that carried over the breakwater when waves or wave uprush overtop the structure. Some aspects of sediment transport in the vicinity of a harbor are provided in Fig. 13, aerial views of Oceanside, California. Patterns of sediment accumulation at an entrance can be deduced from Lagrangian studies of currents in the harbor vicinity (Saylor, 1966) and by study of the time history of bathymetric changes (Bajorunas and Duane, 1967).

The normal method of inlet improvement has been to provide jetties extending into the offshore flanking the inlet channel. Jetties may have any or all of the following functions: (1) to block the entry of littoral drift into the channel; (2) to serve as training walls to increase the velocity of tidal currents and thereby flush sediments from the channel; (3) to serve as breakwaters to reduce wave action in the channel; and (4) in the case of inlets in barrier chains, to serve to prevent further inlet migration. In cases where there is no predominant direction of littoral transport, jetties also serve to stabilize the adjoining coastal shores. In the more common cases where littoral drift in one direction predominates, jetties cause accretion of the updrift shore and erosion of the downdrift shore (Figs. 4 and 6).

Proper siting and spacing of jetties for the improvement of an inlet are important, and may require model studies (Ahrens, 1967). Determination of jetty length and configuration involves consideration of navigational, hydraulic, and sedimentation factors. It is almost a certainty that dredging will be required, for even the channel through the outer bar to San Francisco Bay, several miles seaward of the Golden Gate, requires periodic dredging to permit passage of normal-size ocean-going vessels.

In addition to affecting sedimentation in harbor inlets, jetties also affect erosion and accretion of adjacent shores (Fig. 4). A jetty (other than the weir type) interposes a total littoral barrier in that part of the littoral zone between the seaward end of the structure and the limit of wave uprush on the beach. Accretion takes place updrift from the structures at a rate proportional to the longshore transport rate, and erosion occurs downdrift at about the same rate. The quantity of the accumulation depends on the length of the structure and the angle at which the resultant of the natural forces strikes the shore. Structures perpendicular to the shore have greater impounding capacity for a given length, and thus are usually more economical than those at an angle, because perpendicular jetties can be shorter and still reach the same depth. If the angle is acute, channel maintenance will be required sooner because of littoral drift passing around the end of the structure. Planning for jetties at an entrance should include some method of bypassing the littoral drift to eliminate or reduce channel shoaling and erosion of the downdrift shore.

Sedimentation effects are, in most instances, the source of greatest concern in design or maintenance. Elimination of such problems is highly desirable, and can be accomplished by extending jetties into water depths beyond that where significant transport occurs. However, this practice can be very costly. For example, to extend a jetty, with a crest elevation of $+15$ ft, 2:1 side slopes, and 20 ft crest width from 15 to 20 ft water depth, would require on the order of 70,000 tons of stone. At a cost of \$20 ton^{-1} in place, it would cost nearly \$1.5 million to extend that jetty from 15 to 20 ft water depth. If maintenance dredging at that same site cost \$0.40 yd^3, approximately 3.75 million yd^3 of shoaled material could be removed. At 300,000 yd^3/year shoaling rate, 12 years' dredging could be carried out for the same cost of extending one jetty into 5 ft deeper water.

Because the jetty would continue to block downcoast movement of sand, the impounded material would still have to be bypassed to the downdrift beach. As a consequence of these factors, it is often best to accept the fact of some shoaling, attempt to minimize it, and try to make it accumulate where it is least troublesome to navigation and most easily removed and bypassed downdrift. Figure 6 illustrates the impoundment areas of several configurations of jettied harbor entrances.

Simplified, the volume of sediment in transit is the product of water discharge through a unit cross section and sediment concentration,

$$Q_{sed} = Q_{water} \times C_{sed}$$

and water discharge is the product of water velocity and unit cross section,

$$Q_{water} = V_{water} \times L^2$$

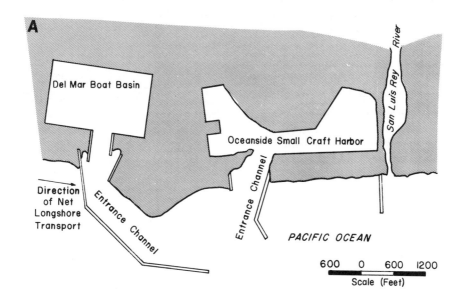

FIGURE 13. *Views of Oceanside Harbor, California. Seasonal oceanographic conditions for short distances upcoast and downcoast are documented in Fig. 8. (A) Schematic plan of the harbor complex. From U.S. Army Corps of Engineers (1973). (B) Oblique aerial photo looking upcoast, 1345 hours, January 26, 1973, tide +2 ft MLLW and rising; waves from west, H = 7 to 9 ft, T = 9 to 11 seconds. Note turbidity (suspended sediment plumes) contrast inside and outside harbor and the vortex "half-streets" shed by the outer breakwater. Dye streaks outside of harbor show downcoast movement of water mass past harbor entrance, while dye in the harbor entrance diffused. Bed material transport traced by radionuclide-tagged sand on January 27, 1973 showed movement downcoast parallel to shore and breakwaters past harbor entrance. From Coastal Engineering Research Center, unpublished report.*

Because it is not possible to directly affect sediment concentration in nature, it is necessary to change the hydraulic characteristics of the inlet channel to decrease velocity of flow and therefore to decrease sediment transport past a point. By overdeepening a section of the channel, the cross section is increased, thereby decreasing velocity, and a depositional basin is created. Sediment can then theoretically accumulate in the depositional basin in a volume equivalent to that of the basin before altering the project channel dimensions. If littoral drift is known to any degree of accuracy, it is possible to plan schedules and budget for periodic maintenance dredging and bypassing through placement of the dredged material on a downdrift feeder beach.

CLOSING REMARKS

Examples of the day-to-day effects of the principles discussed herein are found on the coastlines of the world. Details and examples are documented in Weigel's (1964) *Oceanographical Engineering*, the U.S. Army Corps of Engineers (1973) *Shore Protection Manual*, proceedings of the biennial Coastal Engineering Conferences (ASCE), and the monthly journals of the American Society of Civil Engineers, which the reader is encouraged to consult.

If *research* in the physical sciences, applied to problems, can be considered *early engineering*, then what is or is not done to our coastal zones depends on good science. Effective engineering (which should include non-construction as well as construction) depends on clear understanding of the multiple interactive processes operative in the coastal zone. Understanding requires knowledge gained from high-quality research—data collection, reduction, analysis, and interpretation. It is clear that enough knowledge of coastal and nearshore processes now exists to describe, with varying accuracy and precision, the processes active in this environment. Therefore it is possible to provide qualitative predictions of the effect of engineering works on those processes and subsequent consequences; in some cases it is possible to provide quantitative predictions.

Decisions regarding utilization of the coastal and near-shore zone must be weighed against competing demands and their effect upon one another. Objective information provided by good science and engineering is a prime requirement for sound management. Refusal or inability to provide the requisite information will cause the continued echoing of the prophecy of Matthew:

Everyone who hears these works of mine and does not obey them will be like a foolish man who built his house on the sand. The rain poured down, the rivers flooded over, the winds blew hard against that house, and it fell. What a terrible fall that was. (Matt. 7:26, 27)

SUMMARY

The net effect of wave and current forces impinging upon a shoreline is to change the morphology of that coastline through processes broadly grouped as sedimentation. This process of sediment transport has been active since the beginning of geologic time but becomes of concern to man when it interferes with or affects what man wants to do. Design of engineering works considers numerous factors pertinent to the processes and effects of sediment transport, including geomorphology, waves and currents, littoral materials and drift, foundation characteristics, environmental quality, and demography.

A need for engineering in the coastal zone is predicated upon socioeconomic factors. With today's knowledge and technology any engineering need can be met, but usually a tangible payout is required such that benefits exceed costs by more than unity over the projected life of the works, usually considered as 50 years. Economic analysis involves consideration of intangibles as well as life expectancy of materials; damage and loss due to failure; sediment transport; and secular changes in water level.

Water level fluctuates with varying periodicity and magnitude, depending on characteristics of one or more driving forces. On marine coasts the change in sea level over 50 years is usually not a significant factor in design, being on the order of 2 cm per decade. Conversely, levels of large lakes (inland seas) can change as much as 120 cm per decade and should be considered in design.

Changes in coastal geomorphology are the result of disequilibrium in sediment supply. Erosion occurs if the volume of sand leaving a coastal sector exceeds that entering, and accretion results if the volume of sand leaving a coastal sector is exceeded by the volume of sand entering. Coastal stability represents the equilibrium state. The volume of sediment in transit and the direction it moves along a coast are largely a consequence of the height, period, and direction of incident waves. Methods of estimating the long-term volume of sediment transport are by use of shoreline and bathymetric change maps, volume of sediment dredged from shoaling inlets and harbors, and wave characteristics.

Material shoaling coastal harbors or inlets is derived from material in transport in the littoral stream. Therefore, the process of shoaling is intimately related to shore and nearshore processes, volume transport, and seasonal changes in volume and direction of sediment transport.

Dredged channels and jetties, in addition to aiding navigation, disrupt the flow of sand, often creating coastal erosion downcoast. Mechanically moving sand trapped upcoast and placing it in the littoral zone on the downcoast side will mitigate erosion.

Selection of a protective beach as the shore protection method and periodic replenishment means long reaches of the shore may be protected at relatively low cost compared with other shore protection methods. Such a treatment remedies the basic cause of most erosion problems, a deficiency of natural sand supply. Design of a protective beach considers the geometry of the fill, littoral transport, native sand characteristics, and most importantly, the specification of sand characteristics for the beach fill material.

Enough knowledge of coastal and nearshore processes now exists to describe, with varying accuracy and precision, the processes active in this environment. Therefore it is possible to provide qualitative predictions of the effect of engineering works on those processes and subsequent consequences; in some cases it is possible to provide quantitative predictions.

ACKNOWLEDGMENTS

Permission granted by the U.S. Army Corps of Engineers' Coastal Engineering Research Center to use several illustrations from their publications is appreciated. Discussions with former colleagues, R. A. Jachowski, J. R. Weggel, and W. R. James, aided in the expression of certain stated concepts, and an earlier draft of this paper benefited by a critical and thoughtful review by W. R. James. The philosophy expressed and any errors in the text are the responsibility of the author.

REFERENCES

Abele, R. W. (in press). *Beach Morphology at Plum Island, Mass.* U.S. Army Corps of Engineers, Coastal Eng. Res. Cent., Ft. Belvoir, Va.

Ahrens, J. P. (1967). *A Model Study of the Entrance Channel, Depoe Bay, Oregon,* TM 23, U.S. Army Corps of Engineers, Coastal Eng. Res. Cent., Washington, D.C., 17 pp.

Baines, W. D. (1957). Tidal currents in constricted inlets. *Proc. 6th Conf. Coastal Eng.,* ASCE, pp. 545–561.

Bajorunas, L. (1963). Natural regulation of the Great Lakes, *Proc. 6th Conf., Great Lakes Res.* University of Michigan, Ann Arbor, pp. 183–190.

Bajorunas, L. and D. B. Duane (1967). Shifting offshore bars and harbor shoaling. *J. Geophys. Res.,* **72:** 6195–6205.

Balsillie, J. H. and R. O. Bruno (1972). *Groins: An Annotated Bibliography.* MP 1-72, U.S. Army Corps of Engineers, Coastal Eng. Res. Cent., Washington, D.C., 249 pp.

Berg, D. W. (1968). Systematic collection of beach data. *Proc. 11th Conf. Coastal Eng.,* ASCE, pp. 273–297.

Berg, D. W. and D. B. Duane (1968). Effects of particle size and distribution on stability of artificially filled beach, Presque Isle Peninsula, Pennsylvania. *Proc. 11th Conf., Great Lakes Res.* Int. Assoc. Great Lakes Res., pp. 161–178.

Berg, D. W. and G. M. Watts (1967). Variations in groin design. *ASCE J. Waterw. Harbors Div.,* **WW2:** 79–100.

Bodine, B. R. (1971). *Storm Surge on the Open Coast: Fundamentals and Simplified Prediction.* TM-35, U.S. Army Corps of Engineers, Coastal Eng. Res. Cent., Washington, D.C., 55 pp.

Boyd, M. B. et al. (1972). *Disposal of Dredge Spoil: Problem Identification and Assessment and Research Program Development.* U.S. Army Corps of Engineers, Waterways Experiment Stn., Vicksburg, Miss., 121 pp.

Bruno, R. O. and L. W. Hiipakka (1973). Littoral environment observation program in the state of Michigan. *Proc. 16th Conf., Great Lakes Res.* Int. Assoc. Great Lakes Res., pp. 492–507.

Caldwell, J. M. (1966). Coastal processes and beach erosion. *J. Boston Soc. Civil Eng.,* **53:** 142–157.

Carr, A. P. (1965). Shingle spit and river mouth: Short term dynamics. *Inst. Brit. Geogr. Trans.,* **36:** 117–129.

Courtenay, W. R., Jr., et al. (1974). *Ecological Monitoring of Beach Erosion Control Projects, Broward County, Florida, and Adjacent Areas.* TM 41, U.S. Army Corps of Engineers, Coastal Eng. Res. Cent., Ft. Belvoir, Va., 88 pp.

Das, M. M. (1972). Suspended sediment and longshore sediment transport data review. *Proc. 13th Int. Conf. Coastal Eng.,* ASCE, pp. 1027–1048.

Davis, R. A., Jr. and W. T. Fox (1972). Coastal processes and nearshore sand bars. *J. Sediment. Petrol.,* **42:** 401–412.

DeCooke, B. G. (1968). Great Lakes regulation. *Proc. 11th Conf., Great Lakes Res.* University of Michigan, Ann Arbor, pp. 627–639.

Duane, D. B. (1970). Synoptic observation of sand movement. *Proc. 12th Conf. Coastal Eng.,* ASCE, pp. 799–813.

Duane, D. B. (in preparation). *Application of Radionuclide-Tagged Sand for Coastal Engineering and Research.* U.S. Army Corps of Engineers, Coastal Eng. Res. Cent., Ft. Belvoir, Va.

Everts, C. H. (1973). Beach profile changes in western Long Island. In *Coastal Geomorphology, Proc. 3rd Ann. Geomorphol. Symp. Ser.* Binghampton, N.Y.: SUNY, pp. 279–301.

Field, M. E. and D. B. Duane (in press). Post Pleistocene History of the Inner Continental Shelf: Significance to Barrier Island Origin. *Bull. Geol. Soc. Am.*

Galvin, C. J., Jr. (1967). Longshore current velocity: A review of theory and data. *Rev. Geophys.,* **5:** 287–304.

Harris, D. L. (1972). Wave estimates for coastal regions. In D. J. P. Swift, D. B. Duane, and O. H. Pilkey, eds., *Shelf Sediment Transport: Process and Pattern.* Stroudsburg, Pa.: Dowden, Hutchinson & Ross, pp. 99–126.

Hart, E. D. (1969). *Radioactive Sediment Tracer Tests, Houston Ship Channel, Houston, Texas.* H-69-2, U.S. Army Corps of Engineers, Waterways Exp. Stn., Vicksburg, Miss., 33 pp.

Hayes, M. O. (1971). Forms of sediment accumulation in the beach zone. In *Waves on Beaches and Resulting Sediment Transport.* New York: Academic Press, pp. 111–132.

Hayes, M. O. and J. C. Boothroyd (1969). Storms as modifying agents in the coastal environment. In *Coastal Environment, NE Massachusetts and New Hampshire.* University of Massachusetts, Amherst, pp. 245–265.

Hicks, S. D. (1972). On the classification and trends of long period sea level series. *Shore and Beach*, **April**: 20–23.

Hicks, S. D. and W. Shofnos (1965). Yearly sea level variations for the United States. *ASCE, J. Hydrodyn. Div.*, 23–32.

Ingle, J. C. (1966). *The movement of beach sand*. Developments in Sedimentology, Vol. 5. Amsterdam: Elsevier, 221 pp.

Ippen, A. T. (1966). *Estuary and coastline hydrodynamics*. Engineering Societies Monographs. New York: McGraw-Hill, 744 pp.

Jagschitz, J. A. and R. S. Bell (1966). *Restoration and Retention of Coastal Dunes with Fences and Vegetation*. Bull. 382, Agric. Exp. Stn., University of Rhode Island, Kingston, 43 pp.

James, W. R. (in press). Beach fill stability and borrow material texture. *Proc. 14th Conf. Coastal Eng. Res., ASCE*.

James, W. R. (in preparation). *An Eulerian Tracer Experiment*. U.S. Army Corps of Engineers, Coastal Eng. Res. Cent., Ft. Belvoir, Va.

Keith, J. M. and Skjei, R. E. (1974). *Engineering and Ecological Evaluation of Artificial-Island Design, Rincon Island, Punta Gorda, California*. TM 43, U.S. Army Corps of Engineers, Coastal Eng. Res. Cent., Ft. Belvoir, Va., 76 pp.

Keulegan, G. H. (1967). *Tidal Flow in Entrances, Water-Level Fluctuations of Basins in Communication with Seas*. TB-14, Comm. on Tidal Hydraulics, U.S. Army Corps of Engineers, Vicksburg, Miss., 89 pp.

Kraft, J. C. (1971). *A Guide to the Geology of Delaware's Coastal Environments*. Publ. 2GL039, College of Marine Studies, University of Delaware, Newark, 220 pp.

Kraft, J. C. and R. L. Caulk (1973). *The Evolution of Lewes Harbor*. Trans. 71, Del. Acad. Sci., Newark, pp. 79–125.

Krumbein, W. C. (1957). *A method for Specification of Sand for Beach Fills*. TM 102, U.S. Army Corps of Engineers, Beach Erosion Board, Washington, D.C., 43 pp.

Krumbein, W. C. and W. R. James (1965). *A Lognormal Size Distribution Model for Estimating Stability of Beach Fill Material*. TM 16, U.S. Army Corps of Engineers, Coastal Eng. Res. Cent., Washington, D.C., 17 pp.

Krumbein, W. C. and W. R. James (1974). *Spatial and Temporal Variations in Geometric and Material Properties of a Natural Beach*. TM 44, U.S. Army Corps of Engineers, Coastal Eng. Res. Cent., Ft. Belvoir, Va., 79 pp.

Longuet-Higgins, M. S. (1970). Longshore currents generated by obliquely incident sea waves, I & II. *J. Geophys. Res.*, **15**: 6778–6801.

MacLean, W. F. (1963). Modern pseudo-upwarping around Lake Erie. *Proc. 6th Conf., Great Lakes Res.* University of Michigan, Ann Arbor, pp. 158–168.

Orme, A. R. (in press). *Effect of a Low Impermeable Groin on Shore Zone Geometry and Texture, Pt. Mugu, California*. U.S. Army Corps of Engineers, Coastal Eng. Res. Cent., Ft. Belvoir, Va.

Pararas-Carayannis, G. (1973). *Ocean Dumping in the New York Bight, an Assessment of Environmental Studies*. TM 39, U.S. Army Corps of Engineers, Coastal Eng. Res. Cent., Washington, D.C., 159 pp.

Rector, R. L. (1954). *Laboratory Study of Equilibrium Profiles of Beaches*. TM 41, U.S. Army Corps of Engineers, Beach Erosion Board, Washington, D.C., 38 pp.

Rowe, R. R. (1969). Lake Michigan–Huron stage-frequency and trend. *Proc. ASCE J. Waterw. Harbors*, 261–274.

Savage, R. P. (1958). Wave runup on roughened and permeable slopes. *J. Waterw. Harbors Div. ASCE*, **WW3**: Paper No. 1640.

Savage, R. P. and W. W. Woodhouse, Jr. (1969). Creation and stabilization of coastal barrier dunes. *Proc. 11th Conf. Coastal Eng., ASCE*, pp. 671–700.

Saylor, J. H. (1966). *Currents at Little Lake Harbor, Lake Superior*. Res. Rep. 1-1, U. S. Lake Survey, Detroit, Mich., 19 pp.

Saylor, J. H. and E. B. Hands (1970). Properties of longshore bars in the Great Lakes. *Proc. 12th Conf. Coastal Eng.*, pp. 839–853.

Scott, T. (1954). *Sand Movement by Waves*. TM 48, U.S. Army Corps of Engineers, Beach Erosion Board, Washington, D.C., 47 pp.

Seibel, E. (1972). Shore erosion at selected sites along Lakes Michigan and Huron. Unpublished. Ph.D. Thesis, University of Michigan, Ann Arbor, 175 pp.

Shepard, F. P. (1950). *Longshore Bars and Longshore Troughs*. TM 15, U.S. Army Corps of Engineers, Beach Erosion Board, Washington, D.C., 32 pp.

Sonu, C. J. (1968). Dynamic behavior of subaerial beach sediment on the Outer Banks, North Carolina. *EOS Trans. AGU*, **49**: 190.

Sonu, C. J. and J. L. VanBeek (1971). Systematic beach changes on the Outer Banks, North Carolina. *J. Geol.*, **79**: 416–425.

Stafford, D. B. (1971). *An Aerial Photographic Technique for Beach Erosion Surveys in North Carolina*. TM 36, U.S. Army Corps of Engineers, Coastal Eng. Res. Cent., Washington, D.C., 115 pp.

Szuwalski, A. (1970). *Littoral Environment Observation Program in California, Preliminary Report*. MP 2-70, U.S. Army Corps of Engineers, Coastal Eng. Res. Cent., Washington, D.C., 242 pp.

Teleki, P. G., J. W. White, and D. A. Prins (1973). A study of oceanic mixing with dyes and multispectral photogrammetry. *Proc. ASP Symp. Remote Sensing in Oceanogr., ASP*, pp. 772–787.

U.S. Army Corps of Engineers (1963). *Report on Operation 5-High, March 1962 Storm*. U.S. Army Corps of Engineers Div., North Atlantic, New York, 298 pp.

U.S. Army Corps of Engineers (1966). *Shore Protection, Planning, and Design*. Tech. Rep. 4, U.S. Army Corps of Engineers, Coastal Eng. Res. Cent., Washington, D.C., 401 pp.

U.S. Army Corps of Engineers (1971). *Report on the National Shoreline Study*. U.S. Army Corps of Engineers, Washington, D.C., 59 pp.

U.S. Army Corps of Engineers (1973). *Shore Protection Manual*, Vols. I–III, U.S. Army Corps of Engineers, Coastal Eng. Res. Cent. Washington, D.C.: U.S. Govt. Printing Office, 1160 pp.

Vallianos, L. (1970). Recent history of erosion at Carolina Beach, N.C. *Proc. 12th Conf. Coastal Eng., ASCE*, pp. 1223–1242.

Van de Kreeke, J. (1967). Water level fluctuations and flow in tidal inlets. *J. Waterw. Harbors Div., ASCE*, **93**: 97–106.

Watts, G. M. (1963). Sediment Discharge to the Coast as Related to Shore Processes. *Federal Interagency Sedimentation Conference of the Subcommittee on Sedimentation, ICWR, Jackson, Miss. 28 Jan.–1 Feb. 1963*, paper 74, 15 pp.

Watts, G. M. (1965). Trends in sand transfer systems. In *Coastal Engineering, Santa Barbara Spec. Conf., ASCE*, pp. 799–804, Chapter 34.

Weigel, R. L. (1964). *Oceanographical Engineering.* Englewood Cliffs, N.J.: Prentice-Hall, 532 pp.

Williams, L. C. (1969). *CERC Wave Gages.* TM 30, U.S. Army Corps of Engineers, Coastal Eng. Res. Cent., Washington, D.C., 117 pp.

Woodard, D. W. et al. (1971). *The Use of Grasses for Dune Stabilization Along the Gulf Coast with Initial Emphasis on the Texas Coast.* Rep. 114, Gulf Univ. Res. Corp., Galveston, Texas, 74 pp.

Woodhouse, W. W., Jr., E. D. Seneca, and S. W. Broome (1974). *Propagation of Spartina Alterniflora for Substrate Stabilization and Salt Marsh Development.* TM 46, U.S. Army Corps of Engineers, Coastal Eng. Res. Cent., 155 pp.

Yasso, W. E. (1965). Fluorescent tracer particle determination of the size velocity relation for foreshore transport, Sandy Hook, N.Y. *J Sediment. Petrol.,* **35:** 989–993.

CHAPTER 22

Sedimentation and Ocean Engineering: Structures

HAROLD D. PALMER

Westinghouse Ocean Research Laboratory, Annapolis, Maryland

> . . . Man marks the earth with ruin—
> his control
> Stops with the shore.
>
> (Lord Byron: 1788–1824)

Although Byron's verse was a severe indictment of early nineteenth century man's impotence in repulsing the forces at work upon shorelines, his observation is still relevant in many coastal regions. Headlines in newspapers dated September 23, 1973, note "Park Service Giving up Fight Against Nature on Beaches" with regard to the destruction attending storms on the outer banks of Cape Hatteras. Duane (Chapter 21) treats aspects of beaches and harbors with regard to coastal protective structures and remedial practices, and his conclusions are at least more encouraging than those of the National Park Service. In this chapter we consider the interactions of sediments and structures beyond the nearshore zone described by Duane.

This chapter attempts to relate the studies of marine geology to those of the practicing civil engineer who is called upon to assess the suitability of sites on the continental shelf for the placement of man-made structures. Collaborations between marine geologists and engineers are frequently plagued by frustrations arising from differences in methodology, nomenclature, and philosophical outlook and it is hoped that this volume can help narrow the interdisciplinary gap between closely allied activities. With this purpose in mind, we shall define the collaboration of civil engineering and marine geology as *geotechnical ocean engineering*, or *marine geotechnics*, a term adopted by Richards (1967) to describe this interdisciplinary field of activity. It is the study of the scientific and engineering aspects of soils or sediments and rocks on the ocean floor. In engineering parlance, all sediments are termed "soils" but we shall here use the two terms interchangeably. The subject includes the properties affecting the soil system of the seafloor and the response of this system to applied static and dynamic loads. These loads are created when any structure, vehicle, or other object rests upon or penetrates the seafloor. Here we will also include perturbations in the sedimentary regime created by the presence of a structure.

In this chapter, we are concerned with the continental shelf, and thus our attention is focused on sediments that are often characterized by sand-sized materials. For example, reports on shelf sediments of the Atlantic continental shelf reveal that from New Jersey to southern Florida medium-fine sands are the most abundant size class, with coarse sand the next most common size group (Milliman, 1972).

Granular materials are termed *cohesionless* by engineers, and the strength of these sediments is a function of frictional forces developed at grain contacts, the interlocking effects of collections of grains under differing packing schemes, gravity, and the nature of pressure distributions in the interstitial waters. Fine-grained sediments have different physical properties, primarily due to aspects of *cohesion* developed by pore fluids, cation exchange capacity, and particle packing. These materials, the silts and clays of the marine geologist, are termed *cohesive* soils. For purposes of introduction, we may define *strength* as those characteristics of the sediment which support a load or contain a force without the soil experiencing excessive deformation (a *failure*). These topics are examined in greater detail in a later section.

In all, fine-grained sediments account for less than 1% of the Atlantic shelf soils (Milliman et al., 1972). Thus, it is surprising to note the disproportionally small number of references in marine geotechnics which treat saturated cohesionless granular soils—the materials that constitute the foundations for most shelf structures. But as offshore activities move out of the regions where most structures rest, or in fine-grained sediments (Lake Maracaibo, the Gulf of Mexico), and into sand-floored shelves (North Sea, the U.S. and Canadian Atlantic shelves), the importance of determining the engineering properties of sands becomes more critical.

Knowledge of the distribution and properties of continental shelf sediments has been gained through the efforts of marine geologists concerned with regional relationships, and from engineers whose studies require the evaluation of a local site for a specific purpose. In the latter case, attention is usually focused on the three-dimensional aspects of soil distribution, and borings are required to provide materials for foundation design. The geologist's approach to three-dimensional study usually employs seismic reflection profiling, and in areas where borings are available, collaboration between the two disciplines can yield a fairly accurate picture of spatial relationships. Even under these ideal conditions, one is often confronted with a complex and confused stratigraphy inherited from repeated Pleistocene and Holocene emergence and submergence of the shelf. For example, Stahl et al. (1974) have shown through profiles of the upper 50 m of shelf sediments off New Jersey that the shoreface and an associated shoreface-connected ridge consist of a mosaic of lagoonal and barrier deposits lying between the Holocene transgressive sand sheet and the deeper Pleistocene sediments. As pointed out by McClelland (1975), the mechanical properties of shelf strata can vary markedly, depending on the degree of their exposure (and hence desiccation and shrinkage) during marine regressions, the amount of fine materials

incorporated within each unit, and the effectiveness of ocean currents in reworking shelf sediments.

PREVIOUS WORK

The literature of marine geotechnics is voluminous, spanning scores of books and journals and thousands of consulting engineer's reports, both open file and proprietary. This latter aspect of geotechnical documentation is necessary, but it is also unfortunate since many important findings will appear belatedly if at all. However, excellent general reviews of the field can be found in papers such as those by Noorany and Gizienski (1970), Noorany (1972), and Fukuoka and Nakase (1973). Three symposia volumes (Richards, 1967; Proc. Int. Symp., 1971; Inderbitzen, 1974) cover a variety of geotechnical subjects, but many are concerned with fine-grained sediments at depths beyond the continental shelf. Perhaps the best brief introduction to the entire field of geotechnical endeavor can be found in R. F. Legget's 1967 presidential address to the Geological Society of America (Legget, 1967).

Standard texts which provide an excellent background to all phases of marine geotechnics include the classic volume by Terzaghi and Peck (1967), a more recent assessment by Scott and Schoustra (1968), the widely used standard text of Lambe and Whitman (1969), and a lucid presentation by Schofield and Wroth (1968) which treats the mechanical behavior of saturated remolded soils—essentially the situation in all marine sediments recovered for analyses.

The increased interest in cohesionless materials is evident in many recent papers, and perhaps the most significant of these is that of the late L. Bjerrum (1973) which treats geotechnical problems of foundations in the North Sea. Lade and Duncan (1973) examine testing of cohesionless soils, and Tejchman (1973) studies through models the effects of pile groups in sands. An earlier paper, with relevance to submarine slides, is presented by Anderson and Bjerrum (1967), and Shepard and Dill (1966) devote a chapter to the physical properties of submarine canyon sediments. In a recent report Singh (1974) discusses the effects of waves on ocean sediments, a treatment that includes discussion of sand-sized materials. Problems associated with scour and liquefaction of sandy soils supporting and containing a large-diameter pipeline are addressed by Christian, et al. (1974). A recent summary by Richards et al. (1975) examines current technology from the standpoint of sampling and the shortcomings inherent in present practice, and includes nearly 200 references to work with cohesionless soils.

STRUCTURES ON THE SHELF

Man's activities on the shelf beyond the nearshore zone are increasing in both number and diversity. There are several reasons for the accelerated pace of operations: (1) coastal population density has created a demand for real estate which cannot be met and offshore sites for industry (such as floating nuclear power plants, Fig. 1) appear practical from an economic standpoint; (2) the demand for petroleum products and the uncertainty of maintaining import levels adequate to meet demands have led to a resurgence in the development of shelf fields. Such activity requires diverse structures, from the exploration drilling rig through production installations that may include storage and pumping facilities. Table 1 lists some of the structures that accompany a variety of endeavors on the shelf.

Selection of a foundation design for a given structure depends on a variety of factors, such as design life, size, dynamic and static loads, wave, current, and tide regime, and substrate conditions. This latter category is pertinent to this chapter, and we can identify four main types of structure-sediment interactions on the basis of foundation type:

1. Embedded: friction or end-bearing piles, generally in clusters. Grouted pipes. Tunnels, both cut-and-cover and excavated. Support is gained from structural members which penetrate the seafloor.

2. Surface supported ("gravity structures"): structural members with relatively large surface area resting on the seafloor; includes spread, mat, ring, and strip footings.

3. Floating: either moored or tethered to other structures of type 1 or 2; heave on mooring lines important.

4. Moored: anchored facilities partially submerged. Anchored to bottom; breakout forces for anchors critical.

The last two categories are important to a number of structures, but are peripheral to the subject of this chapter. We may note in passing that breakout forces are directly related to cross-sectional area of the object embedded in the direction of the force acting to extract the object, the properties of the soil, and to some degree the duration of embedment. Readers interested in this subject are urged to consult a series of papers which appeared in the Proceedings volume from the American Society of Civil Engineers Symposium on "Civil Engi-

Two floating nuclear stations moored within the breakwater are shown in an artist's view as they appear looking towards the shoreline. From the shore they will appear much like a ship passing on the horizon.

FIGURE 1. *Artist's conception of the current plans for twin 1150 MW floating nuclear generating plants off New Jersey. Each unit is mounted on a square barge, 400 ft x 400 ft, behind a breakwater capped by 40 ton cast concrete dolos.*

TABLE 1. Offshore Structures, Systems, and Operations

Anchoring and Mooring systems
  Cables
  Floating structures
  Large ship moorings
  Monobuoys
  Pipelines

Bottom-sitting (gravity) installations
  Habitats
  Mariculture structures
  Scientific instrumentation
  Petroleum storage

Cables
  Communications
  Military
  Power transmission

Dredging
  Borrow uses
  Channel maintenance

Islands
  Airports
  Deep-water terminals

Mining
  Aggregate uses
  Freshwater aquifers
  Marine placer

Pipelines
  Effluent outfall lines
  Freshwater aqueducts
  Oil and gas distribution lines
  Pumping installations

Platforms
  Drilling and production rigs
  Storage systems
  Tanker terminals
  Navigational aids

Transportation
  Anchored buoyant tunnels
  Cut-and-cover tunnels
  Bridges

Source. From Richards et al. (1975).

neering in the Oceans II" (Liu, 1969; Bemben and Kalajian, 1969; Vesic, 1969), and those of Bemben et al. (1973), Lee (1973), and Muga (1968).

For the remainder of this chapter, our attention will be focused on structures that are embedded within or supported by shelf soils. Selected examples of such structures are presented in Table 2, and in considering any of the installations cited therein, we may apply two new criteria with regard to these structures. A man-made object such as a drilling platform, floating power plant, pipeline, etc., is an intrusion into the marine environment. Such intrusions cause perturbations from the standpoint of a *hazard* and an *impact*. Here, with our concern being directed toward sediment-structure interactions, we may define a *hazard* as an adverse effect that the environment imposes *upon a structure*, and an *impact* as an effect (perhaps beneficial) of the structure *upon the environment*. Examples of both categories are presented in Tables 3 and 4. To the civil engineer and marine geologist, concern is divided between both categories. Hazards are significant because the structural integrity of the installation is predicated on the geological and engineering analyses of the site. Impacts carry equal weight, since they may affect structural stability as well as alter the ecological regime in the vicinity of the in-

TABLE 2. Selected Examples of Offshore Structures

Structure	Water Depth, Maximum, m (ft)	Soil-Structure Interaction	Selected References
Oil drilling platforms, Gulf of Mexico, off California, and elsewhere	114 (373)	Long piles	McClelland et al. (1969), Sullivan and Ehlers (1973)
Offshore tanker terminal Bantray Bay, Ireland	30 (100)	Long piles	Fox (1970)
Oil storage tank, Khazzan Dubai, Persian Gulf	47 (154)	Tank rests on soil; anchored by long steel pipes grouted into drilled holes	Chamberlain (1970), Wees and Chamberlain (1971)
Oil storage tank, Ekofisk I, North Sea	70 (230)	Concrete tank-island rests on soil without anchoring	Gerwick and Hognestad (1973), Bjerrum (1973)
Sewage outfalls, Hyperion Plant, Los Angeles	91 (300) 61 (200)	Effluent outfall; sludge outfall; trenched, gravel mat placed beneath and on sides of pipe	Narver and Graham (1958)
Nuclear power generating plant, designed for location off New Jersey	12 (40)	Semienclosed breakwater resting on bottom; anchored floating plant	Fischer et al. (1973), Kehnemuyi and Johansen (1972), Kehnemuyi and Harlow (1974)
Chesapeake Bay Bridge and tunnel	30 (100) (max. dredged depth)	Trestle, islands, and cut-and-cover tunnels	Peraino (1963), Sverdrup (1963)

Source. After Johnson et al. (1972).

TABLE 3. Environmental Hazards to Offshore Structures

Structure Instability Due to	Cause	Information Required for Design (See Text for Definitions)	Selected References
Soil flow slides	Liquefaction and other soil movements of coarse or fine-grained soils resulting from cyclic loading: earthquakes, storm waves, tsunamis, wind vibration of above-water structure, machinery loads	Elastic soil properties, cohesion, friction angles, density, ground accelerations, hydrodynamic loads	Bea and Arnold (1973), Demars and Anderson (1971), Seed and Izzat (1971), Spriggs (1971), Christian et al. (1974)
Soil slumps and slides	Slope steepening from erosion and deposition, effects of internal waves, scour	Density, cohesion, slope geometry, friction angle, elastic soil properties, texture, currents	Demars and Anderson (1971), Spriggs (1971)
Undermining	Erosion (scour); burrowing animals (fish and invertebrates)	Texture, density; currents; kinds and distribution of burrowing animals	Demars and Anderson (1971), Muraoka (1970), Posey (1971)
Insufficient soil-bearing capacity	Inadequate shear strength (cohesion and/or friction)	Friction angle, cohesion, density	Kretschmer and Lee (1970), McClelland et al. (1969), Terzaghi and Peck (1967)
Excessive total or differential settlement	Compressible soil strata, areal variation of compressibilities under structure, earthquake loading	Density, compression index, coefficient of consolidation, elastic soil properties	Schmertmann (1970), Terzaghi and Peck (1967)
Lateral loading	Excessive horizontal loads on structure due to winds, waves, currents; slope instability	Density, elastic soil properties, friction angle, cohesion; wires, waves, and currents	Wiegel (1964), Yegian and Wright (1973)
Breakout	Low breakout resistance, low shear strength	Cohesion, friction angle, density	Bemben et al. (1973), Lee (1973), Muga (1968), Vesic (1969)

Source. From Richards et al. (1975).

TABLE 4. Environmental Impacts of Offshore Structures, Construction, etc.

Problem	Cause	Information Required for Analysis	Selected References
Alteration of littoral drift patterns	Placement of structure hinders littoral transport processes	Texture, location and geometry of structure, bathymetry, currents, wave and breaker patterns	Jachowski (1966), Johnson (1956)
Removal of borrow or placement of fill material	Dredging and mining operations	Texture, density, volume of material, bathymetry, currents, wave patterns	Johnson et al. (1972)
Scour or deposition	Change in waste and/or cooling-water outfall velocities	Texture, density, bathymetry, currents, wave patterns, effluent velocities	Baumgartner and Trent (1970), Kohler and Moore (1972), Torum et al. (1974)

stallation. In some cases, compliance with regulations for the safe operation of offshore facilities requires the installation of supplemental equipment (such as BOPs—blowout preventers) in contact with the seafloor. These devices must also be evaluated from the aspect of sediment-structure interactions, since they are an integral part of the offshore structure.

ENGINEERING PROPERTIES OF COHESIONLESS SHELF SEDIMENTS

When any structure, vehicle, or object is placed on the seafloor without impact velocity the following will occur immediately (Richards, 1961): Either the soil will have insufficient strength and fail in shear or the soil will have sufficient strength to support the object, which will then remain on the surface. After either of these events, a relatively slow and gradual settlement of the structure may occur because of consolidation in the soil. This reduction in volume generally accompanies the expulsion of pore water from the interstices—a reduction in porosity—and in some cases actual deformation of the grains themselves. The applied load causing these events is associated with the placement of a structure on or within the seafloor materials, and the sediment response, depending on the nature of the materials, may occur over a period of minutes or months. Here we are generally dealing with a two-phase system: a liquid, essentially incompressible, and a semielastic solid comprised of granular materials. In local areas gas, either as interstitial vapor or as bubbles, may constitute a minor but significant component of subaqueous soils. We will not consider it here, but interested readers may consult Anderson and Hampton (1974) and Kaplan (1974). The engineer is concerned with the "strength" of this material, that is, its ability to support a load without excessive deformation. This load may be either resting upon the surface (gravity structures) or embedded in the sediment (pile-supported structures, pipelines, etc.). Both cases are considered here, but first it is important to review the concept of "strength" in cohesionless seafloor soils.

The internal structure of a sand body may be considered to consist of a porous skeleton formed by grains in contact with one another which are surrounded by a network of channels formed by the interparticle voids. For our purposes, these voids are considered to be flooded with seawater. When a load is applied to the surface of this grain-water mixture, the soil resists deformation through friction acting at the grain contacts and to some degree by interlocking of adjacent grains. We may portray the situation by noting the distribution

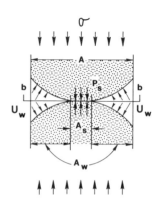

FIGURE 2. *Forces acting on saturated granular materials. See (1) and text. From Scott and Schoustra (1968).*

of forces acting in the soil (Fig. 2). Here, the compressional force σ, or external *stress* acting on the small system under scrutiny, is countered by resistance to deformation (vertical arrows pointing up), a stress of similar magnitude developed internally within the soil.* For hydrostatic pressure directed perpendicular to the surface of the grain, we assign a value, termed *pore water pressure* u_w, which is that pressure, in excess of 1 atm, created by the external stress σ. According to Scott and Schoustra (1968, p. 63), considering the balance of forces acting across the cross section bb, we note that

$$\sigma A = P_s A_s + u_w A_w \tag{1}$$

where σA is the external stress acting over area A, P_s is the contact pressure, A_s is the contact area, u_w is the pore water pressure, and A_w is the area of the grain surface ($A_w = A - A_s$).

The force u_w acts perpendicular to the surfaces of the grains, but if we calculate the component of u_w in the σ direction, and sum these for the perimeter of the grain, the result is equal to the product $u_w A_w$. Dividing (1) by A, and recalling that $A_w = (A - A_s)$,

$$\sigma = P_s \frac{A_s}{A} + u_w \left(1 - \frac{A_s}{A}\right) \tag{2}$$

Although the ratio A_s/A is small, P_s can be very high, up to the yield point where grains begin to crush (greater

---

* Symbols and units in this chapter generally conform to accepted metric and ASTM notation. Recent efforts to establish a universal standard notation have resulted in the widespread acceptance and usage of the *SI system* (Système International d'Unités) which is now required by the American Society of Civil Engineers (ASCE) and most other professional engineering organizations. Inasmuch as the symbols used in this chapter are simple and clearly defined, it was deemed expedient to refrain from the use of some SI units [for example, Pascals (Pa) for pressure] which would be alien to many readers. Richards (1974) provides the best introduction to this subject and cites the appropriate international references defining the SI system.

than 7 kg/cm², or about 100 psi). Thus the first factor to the right in (2) is not zero. In the second term, u_w is a stress on the same order of magnitude as σ (since water is incompressible), and in this case the term in parentheses approaches unity, and we may rewrite (2) as

$$\sigma = \sigma' + u_w \qquad (3)$$

with

$$\sigma' = P_s \frac{A_s}{A} \qquad (4)$$

σ' is termed the *effective stress* (where applied vertically the *normal effective stress*) and it is a major factor in discussing sediment strength. It represents the load borne by the soil skeleton.

Again from Scott and Schoustra (1968, pp. 64–65) we note that in the absence of an external stress, the condition in a soil at depth z may be considered as a function of the total pressure due to the column of soil and water $\gamma_t z$. The hydrostatic water pressure at z is equal to the unit weight of water times the height, or $\gamma_w z$ (assuming no flow is occurring and no excess u_w exists). Thus, applying these factors to (3),

$$\sigma' = (\gamma_t - \gamma_w)z = \gamma_b z \qquad (5)$$

where γ_b is the buoyant unit weight of the sediment, z is the depth in unstressed situations, γ_t is the saturated unit weight of soil, and γ_w is the unit weight of water.

With the application of an external stress (a "load") we transfer stress (pressure) through a volume of soil. If an applied load is excessive and overcomes the strength of the sediment, the result is a *failure* of the soil, often with the loss of the structure being supported. Failure in cohesionless soils generally occurs along planes of slippage between grains, termed failure planes, where shear stresses at the points of grain contact overcome the frictional and interlocking forces present within the sediment. The forces tending to maintain rigidity are jointly termed *shear strength* (τ) up to the point of failure (τ_f). Results of many laboratory tests indicate that the

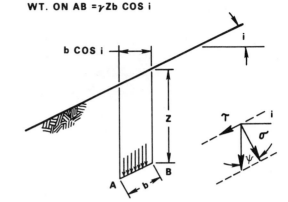

WT. ON AB =γZb COS i

FIGURE 4. *Forces acting on an area within a saturated granular deposit on a slope i. b is the segment width at depth Z in the deposit, AB is the "failure plane," P_v is vertical pressure, ψ is the resultant of τ (the shear stress acting downslope) and σ (the direct stress normal to the shear plane), and γ is the submerged density of the sediment (see text). From Shepard and Dill (1966, Fig. 137) after Moore (1961).*

shear strength in sands is nearly proportional to the effective stress (σ') on the plane of failure and, as one might imagine, the porosity of the sediment, expressed as the *void ratio* used by engineers (volume voids/volume solids), affects the shearing strength of a sediment. This relationship is shown in Fig. 3.

As pointed out by Terzaghi (1956), the relationship between shear strength and effective stress may be summarized as

$$s = c + (\sigma - u_w)\tan \phi \qquad (6)$$

an adaptation from the equation developed by Coulomb in 1776. Here, c is cohesion, or cohesive force (strength), of soil, $\sigma - u_w$ is the effective stress σ', u_w is the excess hydrostatic pressure (excess pore pressure), and $\tan \phi$ is the tangent of the angle of internal friction. In cohesionless sediments, the value of c is zero, since the plastic resistance to shear afforded by fine particles is absent. The angle of internal friction ϕ becomes the equivalent of the coefficient of friction. These relationships are shown in Fig. 4, taken from Shepard and Dill (1966, Fig. 137, modified from Moore, 1961). For cohesionless soils, we may now indicate strength by the general equation

$$s = \sigma' \tan \phi \qquad (7)$$

Recalling from (3) that the external stress σ is countered by an opposing force consisting of the combined pore water pressure u_w and the effective stress σ', we may examine the stress vectors present in a simple grain pair (Fig. 5). The vector pair formed by σ and the tangential (shear) stress τ may be reduced to a resultant, labeled *stress* in Fig. 5A. Angle ϕ is the angle of internal

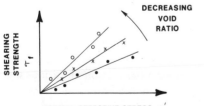

FIGURE 3. *General plot of shearing strength at failure (τ_f) compared to normal effective stress (σ'). Data points shown by similar symbols are for various void ratio values, decreasing counterclockwise. From Scott and Schoustra (1968).*

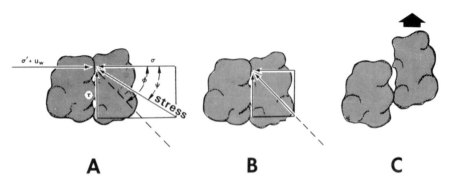

FIGURE 5. *The external stress σ and tangential (shear) stress τ can be presented as a resultant, here simply labeled "stress." It acts at an angle of obliquity ψ at the grain contacts. As this angle approaches φ, the material becomes more unstable and when ψ = φ (B) failure occurs (C). In (A), the effective stress σ' and excess pore water pressure u_w are portrayed as the force (reaction) that opposes σ (see text). Modified from Krynine (1957).*

friction; ψ is the angle of obliquity; both are tangent functions of τ and σ. As ψ approaches φ, the strength of a cohesionless material diminishes, as evidenced by the effective shortening of σ in Fig. 5B. In this example φ is given as 45° and τ has been kept constant. The diminution of σ, which might result from removal of overburden through excavation, scour, slump, etc., has resulted in equality between shear and external stress, and failure through excessive shear stress occurs (Fig. 5C). A similar situation will occur if σ were held constant and τ increased. This is a common occurrence under seismic shock, and the result is the same (Fig. 5C)— failure under excessive shear stress.

If we express (3) as the difference between external stress and pore water pressure, i.e.,

$$\sigma' = \sigma - u_w \tag{8}$$

the importance of excess pore water pressures is evident. Considering the vector $\sigma' + u_w$ in Fig. 5A, it can be seen that with an increase in u_w, a balance of forces requires a decrease in σ' if continuity is to be maintained. Inasmuch as the resistance to failure is proportional to σ', less and less stress is borne by the soil skeleton as u_w approaches σ. When u_w equals σ, σ' reaches 0, and by the definition in (7), strength vanishes. This is the case of *liquefaction*, a situation to be discussed later.

The scrutiny of forces and reactions operative on the intergranular scale reveals much regarding *why* and *how* cohesionless soils behave under stress, but it does not complete the study of sediment response with regard to structures. The behavior of a volume of material must be investigated, and one of the most useful properties derived by engineers is that of *relative density*. The calculation of this property, as defined in an excellent summary by Tavenas and LaRochelle (1972), yields a relative expression of the state of "compactness" of soils which compares the actual porosity with the possible extreme, or limiting, values (maximum and minimum compactness achieved under laboratory conditions). This relationship is often expressed as

$$D_r = \frac{e_m - e}{e_m - e_M} \tag{9}$$

where *e* represents the compactness or void ratio, i.e., (volume of voids)/(volume of solids), e_m is minimum compactness (greatest porosity), e_M is maximum compactness (least porosity), *e* is compactness as measured in the field (in situ), and *r* is relative density, expressed as percent.

The relative density of cohesionless soils has become a basic descriptive parameter in marine geotechnics. For example, the angle of internal friction has been shown to be approximately proportional to relative density up to values of 75 to 80% (see Richards et al. 1975). Other empirical proportionalities are known to exist for shear strength and a variety of deformational properties under static and dynamic loads (Schmertmann, 1970).

Although relative density is a meaningful concept, it is prone to a number of errors that diminish its usefulness unless special care is exercised in all phases of measurement. The greatest problem lies in obtaining in situ density. This is usually done either with a piston sampler employing very thin walls (to diminish disturbance) or with radioactive techniques which measure gamma radiation attentuation at various levels in a bore hole. Variations in laboratory techniques add a complicating factor to the determination of maximum and minimum compactness, and together with the uncertainties of in situ measurements, cumulative errors in determination of the relative density of materials at a given site may reach 50% (see Tavenas and LaRochelle, 1972).

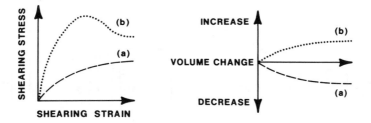

FIGURE 6. *Failure modes (left) and volume changes (right) in sediments undergoing strain with increasing shearing stress. Left plot indicates reaction (a, dashed) of "stable," or loose to medium-dense sands and (b, dotted) "unstable" or dense sands to shearing stresses imposed by loads. Under increased strain, unstable dense soils (b, dotted) increase in volume (dilate) while looser "stable" sediments decrease in volume (compact). From Scott and Schoustra (1968).*

SOIL BEHAVIOR AT SHEAR FAILURE

A soil that fails under a critical shear stress (τ_f) will generally behave according to its relative density. As the shearing stress increases, the deformation, or *strain* due to shear stresses, increases. In general, strain may be considered as the *ratio* of dimensional change to the original (unstrained) dimension. As shown in Fig. 6 a limiting shearing stress is reached at which point failure occurs. In "stable" materials (dashed line *a*, Fig. 6) consisting of loose to medium-dense sands, we note a proportional increase in shearing strain up to failure, but in "unstable" materials (dotted line *b*, Fig. 6) such as dense sands the peak shear stress at failure, the "peak strength," is followed by a diminishing shear stress as strain continues to increase. This "residual strength" results from reorientation of the particles after abrupt failure.

Unstable materials tend to increase in volume (*specific volume*, or the total space occupied by particles and water) as shear increases, whereas stable materials decrease in volume (Fig. 6, right). As would be suspected, the change in volume is accompanied by changes in the pore water pressure. In the analogy drawn by Schofield and Wroth (1968), if a "stable" soil were remolded in the hands, the structure would give way and generate a positive pore pressure. In this case, the hands would become wet. Conversely, if the soil were dense, remolding would tend to dry the hands as the soil structure expanded. This latter case can be observed as one steps on saturated beach sands in the upper swash zone. The sand tends to whiten as the structure dilates under pressure, and it momentarily appears to "dry."

This concept leads us to a final discussion of stress-induced failure, that of spontaneous liquefaction. As described by Terzaghi (1956) and summarized by Shepard and Dill (1966, p. 301) if a loosely packed sand is subjected to a suddenly applied shearing stress, the structure may collapse. Should drainage of the pore water be restricted, the pore water pressure will increase as the volume of the voids decreases through realignment of the particles into a more dense packing. The dissipation of this excess pressure will be hindered if the permeability is low or the deposit is large, and support of the soil mass may be thrown from the integranular support by the excess pore water pressure. This abrupt transfer causes a sudden loss of shear strength and permits the sand to flow as an emulsion of water and grains until the excess pore water pressure is relieved through consolidation. In experiments reported by Anderson and Bjerrum (1967), liquefaction may be expected when porosities in sands reach 44%, when very low strains (fractions of a percent) were noted at failure.

SEDIMENT-STRUCTURE INTERACTIONS

As stated by Scott and Schoustra (1968) much of soil mechanics is concerned with the response of soils to stresses developed in them by external forces (structures). Two factors are paramount: (1) the amount of deformation, or *strain*, that will occur, and its results upon structural integrity, and (2) determination of the maximum stress that can be imposed before failure occurs. Both cases may be considered as part of a sequence of loading. In Fig. 7, the abscissa shows the stress (force) applied by a structure resting upon or within a sediment, and the ordinate gives the settlement, or displacement, of the structure. As shown by line *A*, we may anticipate a certain amount of displacement, or movement of the structure, as the load (force) increases until we reach a certain point where unlimited movement occurs. This condition is termed *yield*, or *failure*, and if we are considering a surface-supported ("gravity") structure the value of force along the vertical line F_u becomes the *bearing capacity* for curve *A* which, as it becomes asymp-

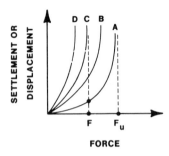

FIGURE 7. *The relationship between force (stress) applied by a structure and displacement is in part a function of time. In this diagram, the time for application increases from A to D for a given load. Failure occurs when the settlement curve becomes asymptotic to a given force, F*$_u$ *for curve A, and F for curve C. Here the force* F$_u$ *for load A can be termed the "bearing capacity" for that soil, load, and period of stress. If the period of loading is brief, failure may not occur, the case at the intersection of curve A and F. For this example, if the load applied is smaller than an asymptotic force line to curve D, failure will never occur if other conditions remain the same. From Scott and Schoustra (1968).*

totic to F_u, approaches infinite displacement. Thus, in Fig. 7, curves B, C, and D may reflect displacement behavior of the same sediment at different rates of the application of force. In this figure, curves A to D represent rates of application progressively slower by orders of magnitude. Conversely, a constant load (F in Fig. 7) applied to the surface will not produce a constant rate of settlement, but one that may gradually increase with time. This situation can account for circumstances where structures, which appeared sound for months after installation, began to settle at increasing rates.

Bearing capacity calculations depend on the dimensions of the structure, the vectors of stress application, and soil properties such as internal friction (ϕ) and cohesion. In dense sands, the value of ϕ decreases when the normal stress over a large foundation area, or "footing," is increased (Bjerrum, 1973). Modes of failure under footings are shown in Fig. 8 (McClelland, 1975). If the structure is exposed to wave forces, terms for eccentric loading must also be introduced, as well as those for horizontal stresses imposed by waves and currents. In the latter case, the structure must exert a vertical force upon the soil (the structure's weight in water and air) which will preclude lateral displacement (Fig. 8A). Eccentric loads that apply uneven stress to the seafloor may cause failure through a local loss of bearing capacity, as shown in Fig. 8B. Many bearing capacity calculations involve the assumption that the soils are free-draining, that is, pore pressures induced by applied loads (structure plus waves, current, etc.) will dissipate as quickly as they are set up. In granular soils, this is often the case, but later when we consider wave

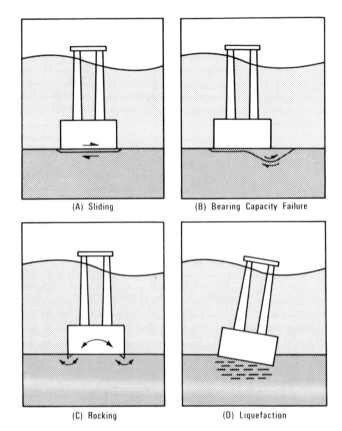

FIGURE 8. *Modes of foundation failure for gravity structures. From McClelland (1975).*

effects on seafloor sediments we will examine the "undrained case," where pore water pressure is considered.

For pile-supported structures, we are again dealing with sediment strength but in a different sense from bearing capacity. Here the support may be gained through a combination of friction along the length of the pile (enhanced where cohesion is high) as well as "end bearing" support if the pile tip rests against a firm surface (rock or dense sediments). A contour map of stress distribution about a single pile and cluster is shown in Fig. 9, where the envelope of stress is commonly referred to as the "pressure bulb." Note that for a cluster,

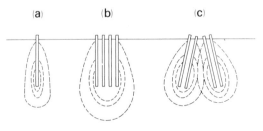

FIGURE 9. *Pressure bulbs for piles in sands. Dashed lines indicate distribution of stress in soil, increasing in value with proximity to piles (see text). Modified from Scott and Schoustra (1968).*

the pressure bulb is defined by interactions of the single-pile bulbs, so that in some volumes the stresses are additive. This situation can be partially avoided, and a stronger foundation may be established, if "batter" piles (piles driven at an angle to the vertical) are employed (Fig. 9*C*). Here the pressure bulbs are displaced, and the structure can support a greater load than one of similar surface area placed atop a vertical cluster. An excellent review of model studies with pile clusters in sand is provided by Tejchman (1973), and extensive studies of recent load tests of circular piles in sand are reported by Reese et al. (1974) and Cox et al. (1974).

In considering soil strength, we must realize that determinations of the physical properties of samples leading to calculations for foundation criteria are based on approximations of behavior in the field extrapolated from carefully controlled laboratory tests. The experimental results depart from the field situation in two major ways. First, most theory in soil mechanics treats the substrate as an isotropic medium, or in some cases as layers of isotropic media. Soils are not homogeneous, especially on the continental shelf where the Holocene transgression has created a patchwork of unconformable and disconformable relationships that are difficult to interpret. Second, sampling techniques do not recover "undisturbed" samples, since the barrel of the sampler must often be forceably driven into the soil by repeated blows of the drill string. Various vibrating coring devices now employed in shelf studies likewise seriously alter the

mechanical properties of soil samples. Although the laboratory analyses provide essential information, we often lack some significant data such as effects which may be caused by stresses applied from other sites in the vicinity of the structure in question. Another desirable aspect in analytical studies in the laboratory is an acceptable technique for applying three-dimensional stresses with an arbitrary variation in stress components during the test. Progress is being made in the latter case (see Lade and Duncan, 1973) but devices to permit such tests are at present only costly prototypes.

As Scott and Schoustra (1968 pp. 115–116) point out:

. . . it has not so far been possible in soil mechanics to introduce any theoretical concepts which enable us to predict the strength of a soil for any given change in the soil stress state from soil properties such as unit weight, water content, and grain-size distribution. It is always necessary to test the soil in the laboratory by bringing it to its "undisturbed" stress state and then subjecting it to a change in stress state as close as possible to that which it would undergo in the field under the applied loads.

WAVE EFFECTS ON SANDS

Wave effects on the floors of continental shelves are not restricted to sediment transport, although this manifestation of wave energy is most commonly considered by marine geologists. We may consider two cases where the interactions between structures and sediments become a function of the wave regime at a site. The first, liquefaction, has been discussed previously, but will be reviewed here with examples. The second effect, that of scour, will then be addressed from the standpoint of sediment transport on the shelf.

Figure 10 has been taken from Henkel (1970, as reproduced by Singh, 1974) to indicate the nature of wave-induced pressure variations with diminishing depth. The test wave in this case is a deep-water storm wave 9 m (30 ft) high with a period of 10.9 seconds and a wavelength of 183 m (600 ft). The abrupt break in pressure and height corresponds to the breaking of the wave. Considering a soil profile, we may estimate that for a sediment subject to liquefaction, a slump may occur as shown in Fig. 11 when a critical state of excess pore water pressure u_w is reached. The number of cycles (waves) required for such failure depends on the state of u_w prior to wave passage.

A severe case of potential liquefaction exists in the North Sea where large concrete gravity structures such as the Ekofisk oil tank are placed in an extreme wave regime. Here, according to Bjerrum (1973), the 100 year

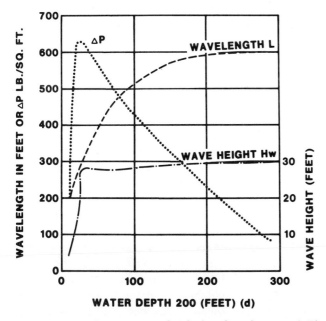

FIGURE 10. *Changes in wave height, length, and pressure* (ΔP) *for a 30 ft (9 m) wave with period of 10.9 seconds. From Singh (1974, Fig. 2) after Henkel (1970).*

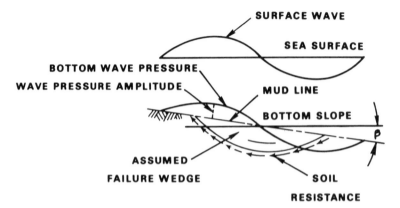

FIGURE 11. *Schematic of a submerged slope failure related to cyclic wave pressures. Soil resistance is overcome by slope angle β and pressure amplitude. From Doyle (1973).*

wave has a period of 14 to 18 seconds, a wavelength of about 600 m, and a height of 25 to 30 m. Besides the horizontal force of 78,600 tons on the structure, this implies an enormous increase in pore water pressure in the fine sand substrate. Permeability is such that we have an undrained situation, and with the size of the structure the stresses will alternate on either side through a series of oscillations in effective stress as the buoyant weight of the structure changes (see Fig. 8C). However, the fine sand at this site is fairly dense and liquefaction may not be a problem, although it is a consideration. The problem rests in the cyclic loadings imposed by a severe storm where the stress applications will act as a sine wave, first "positive" (under a crest) and then negative (under a trough). According to Bjerrum (1973, p. 347), some soils, and especially uniform fine sands, have a tendency to consolidate further when subjected to shear reversals which promote a buildup of pore water pressure much the same as cyclic loads under earthquake conditions. In the case of waves, accelerations can approach 0.4 g, but under storm conditions these forces may last for hours rather than the tens of seconds common in a seismic event. Tests of undrained samples under cyclic loads have shown that a small but constant net rise in pore water pressure accompanies a series of hundreds of shear stress reversals. This situation may ultimately lead to spontaneous liquefaction, as described by Lee and Seed (1967). Furthermore, the lifetime of gravity structures such as the Ekofisk tank ensures that the structure will undergo a number of storm events, and in the interim calm periods the excess pore pressures will dissipate. Such "preshearing" is retained as a "history" or "memory" of cyclic loadings, such that the liquefaction potential is reduced (the generation of pore pressure has been diminished for a given number of cycles) by a factor ranging from 20 to 50 (Bjerrum, 1973, p. 350).

An opposite situation to settlement under liquefaction is described by Christian et al. (1974) with regard to a steel pipeline in Lake Ontario. In this case, a steel pipe 3 m in diameter in 9 m of water failed several times during storms. The nature of the failure was a buoyant rise of the pipe through a 2 m backfill followed by subsequent lateral translation and rupture. Here, it is believed that the backfill liquefied under excessive wave-induced stress in the pore water. Calculations based on a 3 m diameter steel pipe with 2 cm wall thickness filled with fresh water yield specific gravities of 1.14 for the cross section of the pipe. The specific gravity of liquefied sand and seawater is estimated at 1.75 to 1.80, indicating that the steel pipe would float in a liquefied soil. Indeed this appears to have been the case in the Lake Ontario situation.

Scour effects in the vicinity of installations on the shelf can cause severe erosion adjacent to structural members which are in contact with the seafloor. Platform legs, the toes of breakwaters, and various footings all form obstructions to the flow of water generated by waves and currents. Sediment particles in the vicinity of these obstructions are subject to displacement by vorticity and turbulence in flow as it accelerates to pass the object. The instantaneous flow field for a cylindrical object is shown in Fig. 12, where it will be noted that the oncoming flow, either a wave-induced surge or a unidirectional current, is broken into several flow regimes. Vortices develop within the scoured pit (separation vortex) and along the sides of the cylinder where the flow creates a low-pressure field (the "wake plume") in the zone of reversed flow behind the obstruction (Masch and Moore, 1960). The most vigorous flow occurs at the base of the leading edge of the obstruction, where a jet of fluid descends to meet the seafloor. Upon contact with the sediment surface, particles are thrown into suspension

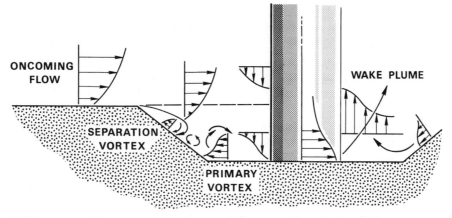

FIGURE 12. *Diagrammatic presentation of flow around a cylindrical obstruction. Increased length of arrows implies higher flow rate. This situation prevails under unidirectional flow and during the interval of one-half the wave period for oscillatory flow at the seafloor. Flow reversal under wave-induced surge maintains turbulent flow in the vicinity of the cylinder and perpetuates a scouring situation.*

and immediately entrained in the turbulent flow passing the sides of the obstacle. In the case of wave-induced surge, the situation is reversed at a rate equal to one-half the wave period and extreme turbulence is perpetuated in the vicinity of the scour pit (Palmer, 1969, 1970). Under such conditions, scour pits usually assume a symmetrical shape about the obstruction.

Scour pits developed in rivers and other unidirectional flows are generally asymmetric, but pits developed by wave-induced surge coupled with strong currents may assume an elongate form parallel to the constant reversal of wave-induced surge and net current. This is evident at the Diamond Light Tower off Cape Hatteras where scour in the medium to coarse sands has prompted an investigation of the rate and extent of local erosion (Robert Au, personal communication, 1974). At this site, the overall scour beneath the platform has reached a depth of 3.7 m (from a depth of 16.5 m to as much as 20.2 m). The major increase in water depth results from a general loss of material from beneath the entire structure. However, individual pits surrounding the base of each tower leg have reached more than 1 m, indicating that localized scour is also effective at this site. Model studies have shown that clusters of obstructions will produce scour effects similar to those developed around a single large obstruction having a diameter equal to that of the cluster (Palmer, 1970). Similar studies were performed by Posey (1971) in his analyses of an extensive scour depression which developed under and around a drilling platform off Padre Island, Texas. Elongation in the trend of the major and local depressions at Diamond Light Tower suggests that impingement of the swift Gulf Stream, as well as waves, is causing regional erosion in the vicinity of this structure.

An example of extreme wave-induced scour is provided by Wilson and Abel (1973) who describe conditions which occurred under the Sedco-H semisubmersible drilling rig on the Nova Scotia shelf. Scour due to currents was anticipated at this site, and protective nylon mesh mats were placed adjacent to the pontoons. A major storm struck the site in October, creating waves up to 13 m in height. The rig commenced rolling and pitching and began to settle at a rate of 2.5 cm/hr. By morning, settlement had reached 0.7 m. A diver's inspection of the seafloor showed extensive scour under the pontoons, and the site was abandoned.

Scour is a major concern in the installation of pipelines, since undermining of the pipe removes the supporting soils and creates bending stresses in the pipe. The failure of at least one sewer outfall in California was caused by such activity, and numerous cable and pipeline breaks have been attributed to a loss of support due to scour. Much of the erosion takes place during storms, when turbulence at the bottom is maximum and currents are accelerated.

Over a period of time, scoured depressions will become stabilized and retain their shape and volume. Infilling may occur, but accretion is temporary and subject to removal at the next severe storm. Large objects on the seafloor, such as the wreck of the Lusitania in 92 m of water south of Ireland (Haines, 1973), display large permanent scour pits which must be attributed to strong current flow at the seafloor. Measures to retard scour include placement of heavy stone and concrete forms on the seafloor, or structural mats which protect the sediments from the direct action of currents. In the case of the floating nuclear power plant (Fig. 1), the toe of the breakwater containing the barges will be embedded some 6.1 m below the seafloor to ensure stability of the foundation should scour occur along the sediment-structure contact (Kehnemuyi and Harlow, 1974).

SUMMARY

With the increasing pace of offshore activity, studies of the sediments atop the continental shelves of the world reach beyond geological exercises in sedimentology to include the geotechnical evaluation of these materials for their ability to support structures that are placed upon or penetrate this surface. In most shelf areas the sediments are cohesionless, and consist of sands, shell debris, and gravels whose behavior under long-term and cyclic stress is still imperfectly understood. The intergranular forces that tend to maintain rigidity of a deposit are collectively referred to as the shear strength, and when these forces are overcome by external stresses (a load), failure of the supporting sediment occurs. External stresses may be developed through structural loads and/or natural processes such as wave pressures, currents, or seismic shock.

If flow of the interstitial fluid is retarded, the pore water pressure within the intergranular spaces may be elevated above the ambient hydrostatic pressure with the result that part of the weight of the sediment column is borne by the pore water. If this excess pore water pressure accumulates, the sediment may reach a point where the effective stress (that weight supported by the sediment structure—the "soil skeleton") diminishes to zero. At this point, all strength is lost and a state of liquefaction exists under which loads can no longer be supported.

Scour effects are also of concern to the engineer designing structures which rest upon or are embedded in the cohesionless sediments of the shelf. The acceleration of water around obstructions to flow creates localized erosional regimes adjacent to the structures and the subsequent loss of supporting materials can adversely affect the integrity of the structure. The prevention of such erosion around and beneath pipelines has become a major concern in the development of offshore petroleum deposits.

A knowledge of the stratigraphic relationships within the upper tens of meters of shelf material is critical to proper design of platforms, pipelines, breakwaters, and other offshore structures. Inasmuch as most shelves are drowned extensions of coastal provinces, they may contain fine-grained marsh or lagoonal deposits whose strength properties are decidedly different from the "relict" sands and gravels typical of many shelf surfaces (see Chapter 15). The complex three-dimensional array of discontinuous deposits renders each shelf site unique in its distribution of engineering properties, especially in the in-place density of the materials. Adequate design parameters require both seismic reflection profiling and confirmatory borings for a complete understanding of site conditions.

Until recently, the civil engineer and marine geologist have often pursued separate or mutually evasive courses in the area of offshore activities, primarily because of ignorance of the other's field and a concern to contain one's activities in one's own discipline. However, the growing emphasis on shelf exploitation has required a closer collaboration with the result that an exchange of information and techniques, as well as an appreciation for each other's approach to data collection and analyses, has led to cooperative efforts between the two disciplines.

ACKNOWLEDGMENTS

Robert Au of the Atlanta Office of Dames and Moore kindly provided information regarding the Diamond Light Tower. Comments by Drs. Adrian F. Richards of Lehigh University and Ole S. Madsen, Massachusetts Institute of Technology, were most helpful during the preparation of initial manuscripts. Preparation of this chapter was supported by the Ocean Engineering Program, Westinghouse Ocean Research Laboratory.

SYMBOLS

A area

A_s contact area

A_w area of grain surface ($A_w = A - A_s$)

e compactness as measured in field

e_M maximum compactness

e_m minimum compactness

P_s contact pressure

r relative density

u_w pore water pressure

γ_b buoyant weight of sediment

γ_t saturated unit weight of soil

γ_w unit weight of water

ϕ angle of internal friction

σ external stress

σ' effective stress

τ_f shear stress at failure

REFERENCES

Anderson, A. and L. Bjerrum (1967). Slides in subaqueous slopes in loose sand and silt. In A. F. Richards, ed., *Marine Geotechnique*. Urbana: University of Illinois Press, pp. 211–229.

Anderson, A. L. and L. D. Hampton (1974). In situ measurement of sediment acoustic properties during coring. In *Deep Sea Sediments: Physical and Mechanical Properties*. New York: Plenum Press, pp. 327–345.

Baumgartner, D. J. and D. S. Trent (1970). *Ocean Outfall Design, Part I, Literature Review and Theoretical Development:* U.S. Dept. of Interior, Federal Water Quality Adm., Northwest Region, Corvallis, Oregon, (N.T.I.S. PB 203749) 129 pp.

Bea, R. G. and P. Arnold (1973). Movements and forces developed by wave induced slides in soft clays. *Offshore Technol. Conf. Prepr.*, **2:** 731–742.

Bemben, S. M. and E. H. Kalajian (1969). Vertical holding capacity of marine anchors in sand. In *Proc., Civil Eng. in the Oceans—II*. Miami Beach, Fla.: ASCE, pp. 105–116.

Bemben, S. M., E. H. Kalajian, and M. Kupferman (1973). The vertical holding capacity of marine anchors in sand and clay subjected to static and cyclic loading. *Offshore Technol. Conf. Prepr.*, **2:** 871–880.

Bjerrum, L. (1973). Geotechnical problems involved in foundations of structures in the North Sea. *Geotechnique*, **23:** 319–358.

Chamberlin, R. S. (1970). Khazzan Dubai No. 1: Design, construction, and installation. *Offshore Technol. Conf. Prepr.*, **1:** 439–454.

Christian, J. J., P. K. Taylor, J. D. C. Yen, and D. R. Erali (1974). Large diameter underwater pipeline for nuclear power plant designed against soil liquefaction. *Offshore Technol. Conf. Prepr.*, **II:** 597–606.

Coulomb, C. A. (1776). Essai sur une application des règles de maximis et minimis à quelques problèmes de statique, rélatifs à l'architecture. In *Mémoires de Mathématique et de Physique*. Paris: Académie Royale des Sci., pp. 343–382.

Cox, W. R., L. C. Reese, and B. R. Grubbs (1974). Field testing of laterally loaded piles in sand. *Offshore Technol. Conf.*, **II:** 459–472.

Demars, K. R. and D. G. Anderson (1971). *Environmental Factors Affecting the Emplacement of Seafloor Installations*. U.S. Naval Civil Eng. Lab. Tech. Rep. R-744, 83 pp.

Doyle, E. H. (1973). Soil-wave tank studies of marine soil instability. *Proc., Offshore Technol. Conf.*, **II:** 753–766.

Fischer, J. A., M. Kehnemuyi, and H. Singh (1973). Geotechnical considerations of site selection for an offshore nuclear power plant. *Offshore Technol. Conf. Prepr.*, **1:** 673–675.

Fox, V. S. (1970). Deep water construction problems in offshore terminal. *Civil Eng.*, **40:** 63–65.

Fukuoka, M. and A. Nakase (1973). Problems of soil mechanics of the ocean floor. *8th Int. Conf. Soil Mech. Found. Eng., Moscow*, Preprint, 18 pp.

Gerwick, B. C., Jr. and E. Hognestad (1973). Concrete oil storage tank placed on North Sea floor. *Civil Eng.*, **43:** 81–85.

Haines, G. (1973). Instrumentation gains wider application. *Meerestechnik*, No. 5, Oct.

Henkel, D. J. (1970). The role of waves in causing submarine landslides. *Geotechnique*, **20:** 75–80.

Inderbitzen, A. L., ed. (1974). *Deep-Sea Sediments: Physical and Mechanical Properties*. New York: Plenum Press, 497 pp.

Jachowski, R. A., tech. ed. (1966). *Shore Protection, Planning and Design*, 3rd ed. U.S. Army Coastal Eng. Res. Cent. Tech. Rep. 4, 401 pp.

Johnson, J. W. (1956). Dynamics of nearshore sediment movement. *Am. Assoc. Pet. Geol. Bull.*, **40:** 2211–2232.

Johnson, S. J., J. R. Compton, and S. C. Ling (1972). Control for underwater construction. In *Underwater Soil Sampling, Testing, and Construction Control*. Am. Soc. Test. Mater. Spec. Tech. Publ. 501, pp. 122–180.

Kaplan, I. R., ed. (1974). *Natural Gases in Marine Sediments*. New York: Plenum Press, 324 pp.

Kehnemuyi, M. and E. H. Harlow (1974). Breakwater for Atlantic generating station. *Offshore Technol. Conf. Prepr.*, **II:** 921–932.

Kehnemuyi, M. and B. A. Johansen (1972). Site considerations associated with offshore generating stations. In *Ocean 72, IEEE Int. Conf. on Eng. in the Ocean Environment*. New York: Institute of Electrical and Electronic Engineers, pp. 412–415.

Kohler, M. L. and J. R. Moore (1972). Sedimentation and scour off nuclear power plants. *ASCE Ann. Natl. Environ. Eng. Meet., Houston*, Reprint 1861, 21 pp.

Kretschmer, T. R. and H. J. Lee (1970). In situ determination of seafloor bearing capacity. In *Proc., Civil Eng. in the Oceans—II*. New York: ASCE, pp. 679–702.

Krynine, D. P. (1957). *Principles of Engineering Geology and Geotechnics*. New York: McGraw-Hill, 730 pp.

Lade, P. V. and J. M. Duncan (1973). Cubical triaxial tests on cohesionless soil. *J. Soil Mech. Found. Div., ASCE*, **99:** 793–812.

Lambe, T. M. and R. V. Whitman (1969). *Soil Mechanics*. New York: Wiley, 553 pp.

Lee, H. J. (1973). Breakout of partially embedded objects from cohesive seafloor soils. *Offshore Technol. Conf. Prepr.*, **2:** 789–802.

Lee, K. L. and H. B. Seed (1967). Cyclic stress conditions causing liquefication of sand. *J. Soil Mech. Found. Div., ASCE*, **93:** 47–70.

Legget, R. F. (1967). Soil: Its geology and use. *Geol. Soc. Am. Bull.*, **78:** 1433–1460.

Liu, C. L. (1969). Ocean sediment holding strength against breakout of partially embedded objects. In *Proc., Civil Eng. in the Oceans—II*. Miami Beach, Fla.: ASCE, pp. 105–116.

McClelland, B. (1975). Geologic engineering properties related to construction of offshore facilities on the Mid-Atlantic continental shelf. In *Proc., Estuarine Research Federation, Outer Continental Shelf Conference and Workshop on Marine Environmental Implication of Offshore Oil and Gas Development in the Baltimore Canyon Region of the Mid-Atlantic Coast*, pp. 217-242.

McClelland, B., J. A. Focht, Jr., and W. J. Emrich (1969). Problems in design and installation of offshore piles. *J. Soil Mech. Found. Div., Proc. ASCE*, **95**(SM6): 1491–1514.

Masch, F. D., Jr. and W. L. Moore (1960). Drag forces in velocity gradient flow. *J. Hydraul. Div., ASCE*, **86:** 1–11.

Milliman, J. D. (1972). Atlantic continental shelf and slope of the United States—Petrology of the sand fraction of sediments, northern New Jersey to southern Florida. *U.S. Geol. Surv. Prof. Pap. 529–J*, 40 pp.

Milliman, J. D., O. H. Pilkey, and D. A. Ross (1972). Sediments of the continental margin off the eastern United States. *Geol. Soc. Am. Bull.* **83:** 1315–1334.

Moore, D. G. (1961). Submarine slumps. *J. Sediment. Petrol.*, **31:** 343–357.

Muga, B. J. (1968). *Ocean Bottom Breakout Forces, Including Field Test Data and the Development of an Analytical Method*. U.S. Naval Civil Eng. Lab. Tech. Rep. R-591, 140 pp.

Muraoka, J. S. (1970). *Undermining of Naval Seafloor Installations*. U.S. Naval Civil Eng. Lab. Tech. Note N-1124, 18 pp.

Narver, D. L., Jr. and E. H. Graham, Jr. (1958). Los Angeles enlarges its sewage facilities, Part III. *Civil Eng.*, **28**: 38–43.

Noorany, I. (1972). Underwater soil sampling and testing, a state-of-the-art review. In *Symp. Underwater Sampling, Testing, and Construction Control.* Am. Soc. Test. Mater. Spec. Tech. Publ. 501, Philadelphia, pp. 3–41.

Noorany, I. and S. F. Gizienski (1970). Engineering properties of submarine soils: State-of-the-art review. *J. Soil Mech. Found. Div., Proc. ASCE*, **96**(SM5): 1735–1762.

Palmer, H. D. (1969). Wave-induced scour on the seafloor. In *Proc., Civil Eng. in the Oceans—II.* Miami Beach, Fla.: ASCE, pp. 703–716.

Palmer, H. D. (1970). Wave-induced scour around natural and artificial objects. Doctoral Dissertation, Dept. Geol. Sciences, University of Southern California, 172 pp.

Peraino, J. (1963). Chesapeake Bay Bridge-Tunnel, a series: Two tunnels and four islands. *Civil Eng.*, **33**: 47–50.

Posey, C. J. (1971). Protection of offshore structures against under-scour. *J. Hydraul. Div., Proc. ASCE*, **97**(HY7): 1011–1016.

Proceedings (1971). *The International Symposium on the Engineering Properties of Seafloor Soils and their Geophysical Identification, 1971, Seattle, Washington, July 25, 1971* (sponsored by UNESCO, National Science Foundation, University of Washington), 374 pp.

Reese, L. C., W. R. Cox, and F. D. Koop (1974). Analysis of laterally loaded piles in sand. *Offshore Technol. Conf. Prepr.*, **II**: 473–483.

Richards, A. F. (1961). *Investigation of Deep-Sea Sediment Cores. 1. Shear Strength, Bearing Capacity, and Consolidation.* U.S. Navy Hydrographic Office Tech. Rep. TR-63, 70 pp.

Richards, A. F., ed. (1967). *Marine Geotechnique.* Urbana: University of Illinois Press, 327 pp.

Richards, A. F. (1974). Standardization of marine geotechnics symbols, definitions, units and test procedures. In A. L. Inderbitzen, ed., *Deep Sea Sediments: Physical and Mechanical Properties.* New York: Plenum Press, 497 pp., pp. 271–292.

Richards, A. F., H. D. Palmer, and M. Perlow, Jr. (1975). Review of continental shelf geotechnics: Distribution of soils, measurement of properties, and geological hazards. *Mar. Geotechnol.*, **1**(1): 33–67.

Schmertmann, J. H. (1970). Static cone to compute static settlement over sand. *J. Soil Mech. Found. Div., Proc. ASCE*, **96**(SM3): 1011–1043.

Schofield, A. and P. Wroth (1968). *Critical State Soil Mechanics.* London: McGraw-Hill, 310 pp.

Scott, R. F. and J. J. Schoustra (1968). *Soil Mechanics and Engineering.* New York: McGraw-Hill, 314 pp.

Seed, H. B. and M. I. Izzat (1971). Simplified procedure for evaluating soil liquefaction potential. *J. Soil Mech. Found. Div., Proc. ASCE*, **97**(SM9): 1249–3616.

Shepard, F. P. and R. F. Dill (1966). *Submarine Canyons and Other Sea Valleys.* Chicago: Rand McNally, pp. 295–309, 381 pp.

Singh, H. (1974). The effect of waves on ocean sediments. In *Dames & Moore Engineering Bulletin 44*, Los Angeles, pp. 11–21.

Spriggs, T. F. (1971). Conceptual design of a man-made island for a seawater desalting and power plant. *Offshore Technol. Conf. Prepr.*, **1**: 435–450.

Stahl, L., J. Koczan, and D. J. P. Swift (1974). Anatomy of a shore-face-connected sand ridge on the New Jersey shelf: Implications for the genesis of the shelf surficial sand sheet. *Geology*, **2**: 117–120.

Sullivan, R. A. and C. J. Ehlers (1973). Planning for driving offshore pipe piles. *J. Constr. Div., Proc. ASCE*, **99**(C01): 59–79.

Sverdrup, L. J. (1963). Chesapeake Bay Bridge-Tunnel project, a series: Engineering design. *Civil Eng.*, **33**: 44–46.

Tavenas, F. and P. LaRochelle (1972). Accuracy of relative density measurements. *Géotechnique*, **22**: 549–562.

Tejchman, A. F. (1973). Model investigations of pile groups in sand. *J. Soil Mech. Found. Div., ASCE*, **99**: 199–217.

Terzaghi, K. (1956). Varieties of submarine slope failures. *8th Texas Conf. on Soil Mechanics and Foundation Engineering.* Spec. Publ. 29, Bur. Eng. Res., University of Texas, Austin, 41 pp.

Terzaghi, K. and R. B. Peck (1967). *Soil Mechanics in Engineering Practice*, 2nd ed. New York: Wiley, 729 pp.

Torum, A., P. K. Larsen, and P. S. Hafskjold (1974). Offshore concrete structures—Hydraulic aspects. *Offshore Technol. Conf. Prepr.*, **I**: 131–142.

Vesic, A. S. (1969). *Breakout Resistance of Objects Embedded in Ocean Bottom.* U.S. Naval Civil Eng. Lab. Rep. CR 69.031, 44 pp. [Also in *J. Soil Mech. Found. Div., Proc. ASCE*, **96**(SM9): 1183–1205, 1971.]

Wees, J. A. and R. S. Chamberlain (1971). Khazzan Dubai No. 1: Pile design and installation. *J. Soil Mech. Found. Div., Proc. ASCE*, **97**(SM10): 1415–1429.

Wiegel, R. L. (1964). *Oceanographical Engineering.* Englewood Cliffs, N.J.: Prentice-Hall, 532 pp.

Wilson, N. D. and W. Abel (1973). Seafloor scour protection for a semi-submersible drilling rig on the Nova Scotian shelf. *Offshore Technol. Conf. Prepr.*, **2**: 631–646.

Yegian, M. and S. G. Wright (1973). Lateral soil resistance-displacement relationships for pile foundations in soft clays. *Offshore Technol. Conf. Prepr.*, **2**: 663–676.

CHAPTER 23

Sedimentation and Ocean Engineering: Placer Mineral Resources

DAVID B. DUANE*

Coastal Engineering Research Center, Fort Belvoir, Virginia

Most concepts and aspects pertaining to exploration and exploitation of a particular marine placer mineral deposit are common to marine placers in general, differing in degree depending on the geology and sedimentology of the particular mineral commodity in question. From the standpoint of engineering in the coastal zone or the continental shelf, interest in open marine placers is confined to those materials used in construction, namely sand and gravel. The discussion that follows is therefore weighted to sand and gravel "placers."

Exploitation of some kinds of minerals on the continental shelves of the world has been taking place for many tens of years using land-based techniques whereby workings of deposits on land have been extended out under the seafloor. Such has been the case for coal, iron, and petroleum. In recent decades exploration and exploitation have been initiated at sea in attempts to increase reserves and production, mostly of hydrocarbons. Technology has developed to the point where production of oil, gas, and sulfur is relatively commonplace from a variety of marine platforms resulting in operations for prospecting, drilling, and producing from such diverse areas as the U.S. Gulf coast, Pacific border areas, Middle East, North Sea, and the Arctic and near-Arctic. Such deposits are not placers, and while exploi-

tation of placer deposits on land preceded in many instances the winning of ores from bedrock deposits, exploration for and the exploitation of *open marine* placers have developed only relatively recently.

Mining law in the United States recognizes placers as those deposits other than veins in place. The process involved in the development of a placer deposit is the mechanical concentration by moving water of one or more minerals into a concentration. Historically placer minerals were considered to be those of high specific gravity, chemical stability, and physical durability (Bateman, 1955). At present, particularly in reference to the marine environment, the term placer generally is applied to any unconsolidated accumulation of minerals or rocks on or near the seabed. Such a definition of placers would apply to such diverse mineral deposits as gold, magnetite, ilmenite, diamonds, or sand and gravel on the continental shelves and to manganese nodules (formed in place by chemical means) on the deep ocean floor. The broader definition is used in this discussion.

MARINE LAW APPLICABLE TO PLACER EXPLOITATION

The fact that exploration for and exploitation of open marine placers by the mining community have lagged behind the operations of the petroleum industry can be

* Present address: National Sea Grant Program, National Oceanic and Atmospheric Administration, Rockville, Maryland 20852.

attributed to a number of factors: high cost of offshore exploration and exploitation, few major discoveries, and the legal unknowns facing the mining entrepreneur. Even if problems attendant to operational costs and poor discovery statistics are not well known to a geologist or engineer as a result of personal experience, the effect on an exploration or research and development program can be readily visualized. Aspects of marine law applicable to placer exploitation are not as obvious to the uninformed, and a synopsis follows.

International Law

Christy (1968) has noted that there are no universally recognized jurisdictional limits concerning the rights of coastal states to sea bottom resources. In general, the farther seaward, the less authority a coastal state has. The basic framework of present international law is based on four conventions adopted by a 1958 United Nations Law of the Sea Conference. Based on those four conventions, five terms are important: (1) internal waters; (2) territorial sea; (3) contiguous zone of the high seas; (4) high seas; and (5) continental shelf.

Internal waters are rivers and lakes within the land area together with marine waters on the landward side of the baseline of the territorial sea. The normal baseline for determining width of the territorial sea is the low water line. Where the coast is irregular, or fringed with islands or shoals, or prominent capes or archipelagos, "straight" baselines may be drawn. Such a system in some locales will of course convert broad areas from high seas to internal waters.

The Convention on the Territorial Sea and Contiguous Zone does not specify the width of the territorial sea and claims vary from 3 to approximately 200 nautical miles. The convention recognizes that the sovereignty of a coastal state extends to the belt of sea adjacent to its coast (i.e., the territorial sea) and to the air above, as well as the bed and subseabed surface.

The contiguous zone is that portion of the high seas contiguous with the territorial sea and within which the coastal state can extend and protect its customs, sanitary, fiscal, and immigration regulations. While the convention goes on to specify that the contiguous zone may not extend beyond 12 miles from the baseline, not all nations are signators.

High seas are those waters seaward of the territorial waters. The Convention on the High Seas provides for certain freedoms of use such as navigation, fishing, and overflights. These freedoms extend to waters above those parts of the continental shelves seaward of territorial seas even though a coastal state may exert sovereignty over the seafloor. Freedom to explore and exploit min-

eral resources of the seabed and subsoil is expressly lacking. Such freedoms, beyond the limits expressed by the Convention on the Continental Shelf, are governed by general principles of international law (Commission on Marine Science, Engineering, and Resources, 1969).

As used in the 1958 Convention, the continental shelf is not strictly a geologic term, and refers to "... the seabed and subsoil of the submarine areas adjacent to the coast but outside the territorial sea to a depth of 200 meters or beyond that limit to where the depth of superjacent waters admits of the exploitation of the natural resources of the said areas; (or) the seabed and subsoil of similar submarine areas adjacent to the coasts of islands" (Commission on Marine Science, Engineering, and Resources, 1969). All nations therefore have control of their continental shelves for as far seaward as technology permits exploitation, and as technology improves the limit would continuously expand. Such a definition is a prime example of situation ethics. The importance of this factor in terms of national area and consequent potential resource is illustrated by Table 1, which refers only to the U.S. shelf region.

TABLE 1.   United States Continental Shelf Areas*

Coast	Limiting Boundary		
	3 Nautical Miles	100 Fathom	1000 Fathom
Atlantic	6	140	240
Gulf	5	135	210
Pacific	4	25	60
Alaska	20	250	755
Hawaii	2	10	30
Puerto Rico and Virgin Islands	2	2	7
Totals	39	862	1302

Source. From the Commission on Marine Science, Engineering, and Resources (1969).

* Thousands of square statute miles.

The 1974 Law of the Sea Conference recently held at Caracas, Venezuela did not act to modify any of the conventions adopted at the 1958 Geneva Conference with respect to mineral exploitation. However, several concepts received widespread agreement which indicates some changes might be agreed upon and adopted at the next conference. There was widespread agreement to extend the territorial sea (now 3 miles) to 12 miles. As that boundary would coincide with the seaward limit of the present contiguous zone, some states believed the contiguous zone would be superfluous.

A majority of states supported an economic zone extending a maximum of 200 nautical miles from the baseline. Acceptance of such a zone would most probably give the coastal state exclusive rights for the purpose of exploration and exploitation of living and nonliving resources. However, acceptance of such a limit would also certainly have to provide for freedom of navigation and overflight, as well as such other third-nation rights as laying and maintaining submarine pipelines and cables.

Domestic Law

The Submerged Lands Act of 1953 set the seaward boundaries of the states as 3 nautical miles from the baseline, or as the boundaries that existed when the state entered the Union. Texas and Florida seaward boundaries were later established as 3 leagues (approximately 10 miles). Several New England states are currently contesting the 3 mile limit on the basis of the seaward boundaries granted by England to those colonies prior to the War of Independence. There are 18 boundaries common to adjacent states in the conterminous United States. Of these only one (Florida–Alabama) is completely delimited; three (New Hampshire–Massachusetts, California–Oregon, and Oregon–Washington) are substantially delimited (Griffin, 1969). In any case, ownership of the seabed and its resources landward of the seaward edge of the territorial sea was vested to the coastal states by the Submerged Lands Act; the coastal states are therefore empowered to regulate the use of the seabed and its resources.

Various offshore mining laws of the coastal states have been collated by Goodier (1972) and a much encapsulated tabulation of regulations and management of hard mineral mining is presented by Grant (1973).

The Outer Continental Shelf Act, passed by Congress in 1953, states that the subsoil and seabed (seaward of the territorial sea) were subject to the jurisdiction of the federal government. Furthermore, the act empowered the Secretary of the Interior to lease these lands for the exploration and production of minerals. Within the Department of Interior, the Geological Survey and Bureau of Land Management administer the leasing program and resulting operations. Leasing of submerged land for oil and gas production is done on a competitive basis.

Other federal laws pertain to regulation of operations that would lead to extraction of placer minerals. The Rivers and Harbors Act of 1899 empowers the Secretary of the Army to regulate the performance of activities that take place in navigable waters. The permit program is administered by the Corps of Engineers which acts in cooperation with state agencies, and by virtue of the Fish and Wildlife Coordination Act, is coordinated with elements of the National Oceanic and Atmospheric Administration and Environmental Protection Agency. A permit is issued (or denied) on the basis of the effect the proposed activity will have on navigation, fish and wildlife, water quality, impact on ecosystems, economics, and other factors. In revised procedural directives the U.S. Army Corps of Engineers will deny a permit where state approval has been previously refused.

EXPLORATION

Minerals, such as oil and gas, are present in the earth in finite quantities. Thus shortages can be expected in time as use continues, unless of course substitute materials are developed. Advances in technology may increase reserves of certain ores; rising prices will add to reserves as once unprofitable concentrations become profitable; and new discoveries may occur in old or new mineral districts. Continental shelves of the world are thought to contain large quantities of minable materials, as summarized in Table 2. These minerals (commodities), for which projected demand exceeds reserves, are those that may provide an economic or strategic incentive for the development of an exploration program on selected areas of the continental shelf. In fact, during the past decade dwindling land supplies (because of diminution of deposits or constraints against extractions) have caused users and resource developers to turn to the shelves seeking to find and profitably develop some mineral deposits. Methods and technologies required for mining vein minerals possibly extant on continental shelves below 20 to 20,000 ft of water dictate that initial mining efforts be directed toward unconsolidated surficial deposits, which might be recovered using marine plant or systems requiring no or little modification. Such constraints lead to programs of exploring for and exploiting placer deposits using conventional dredging techniques. It was just such a set of circumstances (potential high-volume requirement; unavailable land sources) that led the U.S. Army Corps of Engineers to initiate a program to locate and assess sand deposits on the U.S. inner continental shelf suitable for use in beach nourishment and protection projects (Duane, 1968). Other reasons for turning to the seabed are geopolitical in nature. For example, imports to the United States supply 75% or more of requirements for 20 different mineral commodities, including aluminum, chromite, manganese, nickel, and platinum ores (Pings and Paist, 1970). Insofar as an accessible and adequate supply of these materials plays

TABLE 2. U.S. Land Reserves and Resources of Minerals of the Continental Shelves

Commodity	Minable Reserves* United States	Minable Reserves* World (Including U.S.)	Resources† United States	Resources† World (Including U.S.)	Projected Cumulative Demand, 1966–2000 (rounded from Bureau of Mines estimates) United States	Projected Cumulative Demand, 1966–2000 (rounded from Bureau of Mines estimates) World (Including U.S.)
Aluminum (bauxite, millions, long tons)	45	5,800	300	9,600	440§	840
Barite (millions, short tons)	60	130	100	‡	80	190
Beryllium (short tons, equiv. beryl)	‡	‡	1,000,000	1,650,000	360,000	540,000
Bromine (million pounds)	Vast	Vast	Vast	Vast	22,000	30,000
Borates (millions, short tons, B_2O_3)	95	110	‡	‡	4	14
Chromium (millions, long tons, chromite)	0	2,000	8	Several billion	22	74
Copper (millions, short tons)	86	210	65	‡	140§	400
Cobalt (thousands, short tons)	50	2,200	‡	‡	600	1,300
Gold (millions, troy ounces)	50	1,000	400	‡	670	2,370
Helium (billions, cubic feet)	154	‡	42	‡	60	‡
Industrial diamonds (million carats)	0	‡	0	‡	1,700	3,600
Iron ore (millions, long tons)	8,000	250,000	100,000	250,000	6,400	35,000
Lead (millions, short tons)	35	83	15	‡	57	180
Manganese ore (millions, long tons)	0	3,800	1,000	15,000	50	450
Mercury (thousands of flasks)	200	7,000	500	10,000	3,600§	11,400
Nickel (thousands, short tons)	250	60,000	1,400	‡	14,500§	31,700
Niobium (thousands, short tons, Nb_2O_5)	125	9,800	165	8,600	190	370
Phosphate (millions, long tons)	12,000	48,000	48,000	‡	500	2,000
Potash, (millions, short tons, K_2O)	1,400	72,000	5,000	Very large	300	1,100
Platinum group (millions, troy ounces)	3	280	‡	‡	144§	250
Rare earths (millions, short tons, Re_2O_3)	5	‡	‡	5	0.5	1
Salt (trillions, short tons)	60	Vast	‡	‡	0.003	0.009
Silver (millions, troy ounces)	1,400	5,500	500	‡	14,000§	31,000
Sulfur (millions, short tons)	‡	‡	500	2,000	690	2,600
Tantalum (short tons, Ta_2O_5)	2,100	170,000	‡	95,000	54,000	104,000
Tin (thousands, long tons)	9	5,600	43	11,400	3,200§	9,200
Titanium (millions, short tons, TiO_2)	100	500	‡	Vast	50	100
Thorium (thousands, short tons, Tho_2)	0	82	200	1,000	5	15
Tungsten (thousands, short tons)	70	1,500	200	‡	620	2,040
Uranium (thousands, short tons, U_3O_8)	210	742	675	2,700	1,500	3,750
Vanadium (thousands, short tons)	200	3,500	1,300	20,000	650	1,000
Zinc (millions, short tons)	29	100	60	‡	90	280
Zircon (millions, short tons)	6	30	‡	‡	3.0	10

Source. Compiled by Geological Survey commodity geologists. From Commission on Marine Science, Engineering, and Resources (1969).

* Minable reserves are materials that may or may not be completely explored but that may be quantitatively estimated and are considered to be economically exploitable at the time of the estimate.

† Resources are materials other than reserves that are prospectively usable and include undiscovered recoverable resources as well as those whose exploitation requires more favorable economic or technologic conditions.

‡ Unknown.

§ Demand figures include a significant quantity of recycled metal.

TABLE 3. Possible Supply Requirements for Various Minerals

World Supply Problems May Appear During the Remainder of This Century	World Supply Problems Are Likely to Appear Early Next Century	World Resources Are Vast and No Early Supply Problems Are Envisaged
Base metals (tin, copper, lead, zinc)	Nickel	Iron ore
Silver	Titanium (ilmenite and rutile)	Bauxite and other aluminum source rocks
Cadmium	Zircon	Manganese ore
Tungsten	Uranium (possibly)	Chromite
Gold		Phosphate
Fluorspar		

Source. From Dunn (1964).

a critical role in our economy or defense posture, pressures will develop to explore and exploit minerals from the continental shelves. Even though the continental shelves are still largely unexplored, deficits in the world supply of some minerals over the next 50 years have been predicted by Dunn (1964); see Table 3.

Exploration for minerals involves utilization of standard geologic and geophysical techniques adapted for use in the marine environment. However, like their land-based analogs, the basic purpose is to locate and evaluate, which essentially means mapping and sampling. Because of the high costs involved in working at sea, it is essential that field operations begin only after thorough and careful planning has taken place. This involves a library study of factors affecting the formation of the commodity and its geologic occurrence; regional aspects of the continental shelf in question; factors of the geology and bathymetry of the inferred or target location of the exploration program; field sampling plan; and types and characteristics of exploration, sampling, and positioning equipment.

Processes are most suited for concentrating metallic placer minerals on beaches, both foreshore and backshore. As a result of field and flume studies, Everts (1972) concluded that in the marine environment several types of placers can be recognized but that eroding beaches are optimum sites for heavy mineral preservation.

Drowned beaches are potential repositories of large volumes of sand (and gravel). Known drowned beaches on the continental shelf are not too common. However, other morphologic features resulting from marine processes are much more widespread and are comprised of sand and gravel in vast quantities. Linear planoconvex shoals (Fig. 1) occurring on the Atlantic continental shelf to depths of 40 m are interpreted as originating in the nearshore environment (Duane et al., 1972). These are known to contain large volumes of sand, and similar appearing features out to the edge of the Atlantic shelf presumably are also composed of sand. Shoals associated with capes and estuary entrances can also be traced across the Atlantic continental shelf of the United States (Chapter 15) and their nearshore sectors contain large volumes of sand (Field and Duane, 1974). Other deposits of sand and gravel occur as sheetlike deposits, "outcrops" of semilithified Quaternary or Tertiary strata, and channel fill, but the shoals of various types created by marine depositional processes contain the largest volumes of sand.

In high latitudes, glacial processes have produced morphologic elements on the seafloor which might resemble, in some cases, shoals of submarine origin. They are not as likely to be repositories of large volumes of "washed" sand and gravel, but may contain heavy metals and in the Bering Sea have been explored as possible sources of gold (Nelson, 1971).

Offshore exploration programs seeking metal placers should be directed toward locating and delimiting trends of old beaches as well as drowned terrestrial environments, particularly fluvial. On the other hand, if sand and gravel are the objective of an exploration program, positive topographic features 10 m or more in relief with length or width on the order of several hundreds of meters should be sought.

Mapping

Mapping requires knowing position (X, Y) in a Cartesian coordinate system or distance $(R_1, R_2,$ or $R_3)$ to two or more fixed geodetic points and the third dimension (Z). Results are analogous to mapping with an alidade or transit, but except when nearshore, distances normally require some form of electronic positioning (navigation) system. Modern electronic navigation systems are highly automated. Numerous systems are available from a variety of companies as a glance through trade magazines

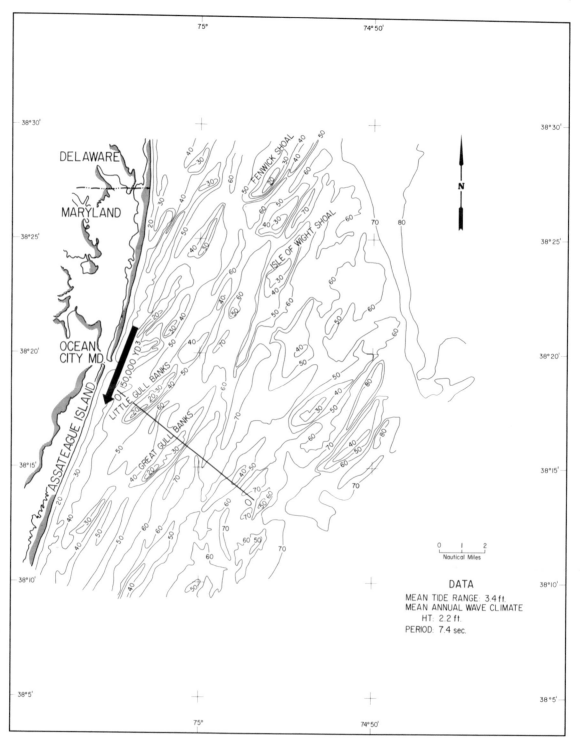

FIGURE 1. *Bathymetry of a portion of the Maryland inner continental shelf depicting shoals that are accumulations of sand. Preliminary studies indicate an excess of* *1 billion yd³ of sand occur in the shoals illustrated. Contour interval is in feet. From Duane et al. (1972).*

540

such as *Oceanology International* or *Sea Technology* will attest. A user has a wide range of operating characteristics and accuracies from which to select. The third dimension (Z), related to some parameter such as depth, density, gravity, or magnetic field strength, can then be located at some point in the map plane through the positioning system selected.

One of the first maps required is bathymetry, the topography of the seafloor. Bathymetry is now determined by techniques of echo sounding, whereby a sound source is generated at the sea surface and its echo, returned from the seafloor, is sensed at the surface and recorded on a strip chart. The sensed water depth is a function of the two-way travel time and the speed of sound in water, which varies with salinity and temperature. Maps developed by such techniques are useful in locating geomorphic features which are features of economic potential, such as shoals which contain large volumes of sand (Fig. 1). Under certain circumstances the character of the return echo pulse is indicative of surface sediment characteristics such as gravel, sand, or mud.

Sidescan sonar differs from standard bathymetric techniques only in the fact that the pulsed beam is directed at an angle to the seafloor (and normally from a transducer towed near the seafloor) rather than vertically downward. Sidescan sonar is particularly sensitive to surface roughness (i.e., degree of reflectivity) and therefore to surface sediment textures and bed forms. Also it "sees" a wider portion of the seafloor and has been used to map distribution of facies on portions of the continental shelf (Newton et al., 1973).

Geophysical surveys carried out on land have also been adapted to marine use. Marine gravimeters can be ship-mounted or encapsulated for deployment on the seafloor. Like their land counterpart, gravimeters provide information on regional subsurface structure and hence are best applicable for reconnaissance surveys; they are probably of limited value for placer deposit exploration. Marine magnetometers may be operated while under way, and as gravimeters, they provide information of structural features on a regional scale. However, since ferromagnetic minerals of various species form heavy mineral placer accumulations, magnetometer surveys are applicable to exploration for placer mineral deposits. Nonferromagnetic heavy minerals such as gold are often associated with the ferromagnetic species; the magnetometer may therefore find application in a search for nonmagnetic minerals.

Continuous seismic reflection profiling (CSRP) is a technique in wide use for delineating subbottom structures and bedding planes in seafloor sediments and rocks. As with sonic bathymetric surveys, continuous reflections

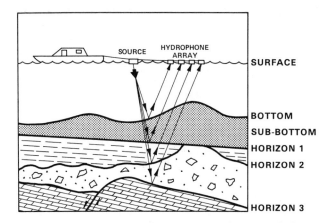

FIGURE 2. *Schematic diagram showing the essential elements used in continuous seismic reflection profiling techniques. Sound generated by a single pulse is illustrated being reflected from different acoustic interfaces and recorded at some increment Δt after the first return.*

are obtained by generating repetitive high-energy, but lower frequency, sound pulses near the water surface and recording "echoes" reflected from the bottom water interface and subbottom interfaces between acoustically dissimilar materials (Fig. 2). In general, the compositional and physical properties which commonly differentiate sediments and rocks also produce acoustic impedance contrasts. Thus, an acoustic profile is roughly comparable to a geologic cross section. Seismic reflection surveys of marine areas are made by towing sound-generating sources and receiving instruments behind a survey vessel which follows predetermined survey tracklines. For continuous profiling, the sound source is fired at a rapid rate, and returning signals from bottom and subbottom interfaces are received by one or more hydrophones. Returning signals are filtered, amplified, and fed to a recorder which graphically plots the two-way signal travel time. Assuming a constant but different velocity for sound in water and shelf sediments, a vertical depth scale can be constructed to the chart paper. Horizontal location is obtained by frequent navigational fixes keyed to the chart record by an event marker and by interpolation between fixes.

Depth of sound penetration decreases with decreasing energy of the pulse as well as with increasing frequency. Because resolution of acoustically contrasting strata improves with higher frequency pulses, a compromise must be made between penetration and resolution. For exploration of placer mineral deposits, CSRP systems emitting sound pulses in the range of 20 Hz to 12 kHz are quite satisfactory and have been used to locate marine sand and gravel placer deposits on the U.S. inner continental shelf (Fig. 3). They have also been used for stratigraphic extrapolation from land to the seafloor

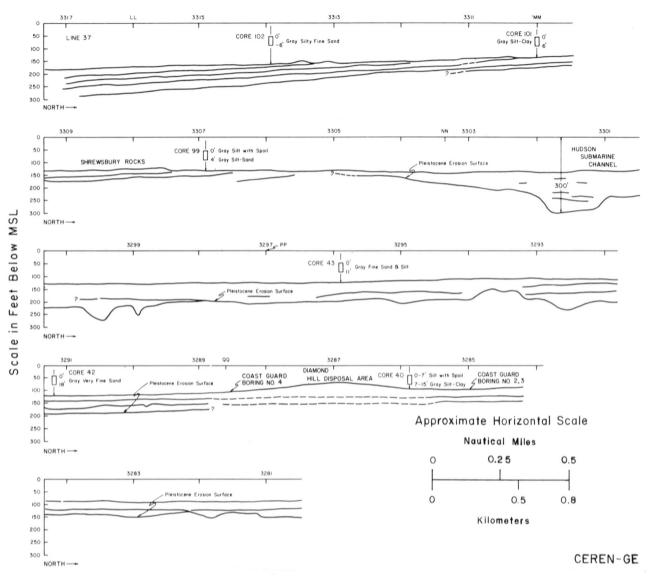

FIGURE 3. *Reduced continuous seismic reflection profile showing coastal plain stratigraphy, shoals, and buried channels in addition to bathymetry. Line runs from approximately 16 km offshore Long Beach, New Jersey north to approximately 1 km offshore Rockaway Beach, New York. Water depth is that from horizontal datum (sea surface) to first (seafloor) reflector drawn. From Williams and Duane (1974).*

(e.g., Field and Duane, 1974), as well as for delineating buried channels (e.g., McMaster and Ashraf, 1973) that might be the repository for heavy mineral deposits. A more detailed discussion of seismic profiling techniques can be found in a number of technical publications, e.g., Miller et al. (1967) and Moore and Palmer (1968). A very good single reference for more information on the geophysical techniques mentioned in the foregoing paragraphs is either the first or second editions of *Introduction to Geophysical Prospecting* by Milton Dobrin (1960).

Other techniques pertinent to exploration and assessment are on site observations. Deep-towed instrument packages or remote-controlled submersible vehicles provide a means to obtain photographs or television views of the seafloor. Other direct observations are made possible by one or more diving techniques and by the use of manned submersibles.

Sampling

Confirmation of the material comprising the seafloor and shallow subbottom is made by sampling the material. Normally this process involves obtaining a sample, raising it to the surface, and analyzing it in a shipboard laboratory or returning it to a shore-based laboratory.

Recent developments in instrumentation and nuclear technology are making it possible to use former laboratory techniques in situ for elemental analysis. For example, x-ray fluorescent analysis techniques can be developed for in situ analysis of elements comprising the material on the seafloor which was in the path of the energizing beam (Pickles, personal communication). Analogously, use of neutron activation analysis of in situ seafloor sediments has been reported by Noakes et al. (1974) and Senftle et al. (1969). Noakes et al. (1974) also report on a system for using measurements of natural radioactivity emitted by thorium (occurring in the heavy minerals monazite, epidote, sphene, and zircon) and uranium (associated with apatite comprising phosphorite deposits). These techniques, like dredge haul samplers, are restricted to assessment of the minerals comprising sediments on the surface of the seafloor.

Cores, which can be obtained by a variety of methods, are still necessary for obtaining samples of material from below the seabed surface and hence the third dimension. Gravity, or free-fall cores, provide limited penetrability of granular materials. Vibratory-type corers, those that use electrically, pneumatically, or hydraulically operated vibrating heads, have been used very successfully in obtaining cores of coarse granular material in excess of 40 ft long. Basically, the apparatus consists of a standard core barrel, liner, shoe, and core catcher with the driver element fastened to the upper end of the barrel. These are enclosed in a self-supporting frame which allows the assembly to rest on the bottom during coring, thus permitting limited motion of the support vessel in response to waves. Power is supplied to the vibrator from the deck by means of a flexible conductor. After the core is driven and returned, the liner containing the cored material is removed and capped. Information on this and numerous other seafloor sediment sampling techniques is provided in a survey by Ling (1972).

EXPLOITATION

It is generally assumed that exploitation of marine placer mineral deposits will utilize some form of a dredge defined as ". . . an earth moving machine specialized to remove bottom material from under water to . . . gain the bottom material" (Mohr, 1974). Two fundamental types of dredges (mechanical and hydraulic) have been hybridized to numerous specialized designs as operations evolved from dredging of onshore placers (and other unconsolidated mineral deposits) to protected marine

waters and subsequently to deeper and open marine waters.

A dredging plant that operates from the sea surface (in contrast to a submerged or a seabed crawler type) is subject to wave, wind, and current action leading to six forms of motion relative to the seabed (Donkers and deGroot, 1973). The six forms—heave, pitch, roll, sway, yaw, and surge—may take place while the dredge head is in contact with the sea bottom. Such motion and consequent stresses placed on the equipment create severe limitations to the sea and weather conditions under which operations can take place. It is these potential limitations, and consequent economic impact, that has led to the design of submerged systems.

Technologically exploitable deposits may not be economically exploitable. Using 1972 prices and a U.S. Geological Survey assay for gold, magnetite, platinum, ilmenite, and chromite, Wilcox and Mead (1972) calculated the cost versus revenues involved in mining a 400 km^2 deposit 5 m thick off the California–Oregon coast. Assuming a 54 ft^3 bucket on a bucket ladder dredge and 291 working days a year, operating costs were greater than revenues each year of a 20 year operation!

Dredging operations will induce some turbidity, the aerial extent, concentration, and duration of which will vary with numerous ambient factors. Sediment placed in suspension will be comprised of organic and inorganic materials. Increased levels of organic matter may benefit mobile species who are attracted to feed; sessile species may be adversely affected by decrease in oxygen, induced by increased chemical oxygen demand. Turbidity will be created at the dredge head, and in many operational techniques will occur at the surface surrounding the floating plant as well. Regardless of the particular size distribution of the material lost overboard in handling or processing, it will in time settle to the seafloor where many sessile forms may be affected by burial from the fallout. The National Oceanic and Atmospheric Administration initiated some research to assess the ecological effects of marine mining. The project, called NOMES (New England Offshore Mining Environmental Study), was terminated before all research programs were completed; a final report on work accomplished is in preparation and will be published (Padan, NOAA, personal communication). In the nearshore zone, where natural processes create periods of high turbidity, the effect of dredging operations on the biota may not be noticed (Thompson, 1973; Courtenay et al., 1974).

Some dredging operations will skim only a thin layer of sediment off the ocean bottom, whereas other operations will create large long furrows or deep holes. The latter operations will increase the surface area of the bottom, increase roughness, and consequently improve the habitat for some forms and associations of marine life (Yancey, CERC, personal communication). If "skimming" operations removed all of a particular type of bottom material and exposed bottom of a different texture or composition, they could very well cause profound changes in the sessile or mobile benthic fauna.

Reefs and reef communities are particularly sensitive to turbidity. In some regions, such as Miami, Florida (Duane and Meisburger, 1969), the Hawaiian Islands (Campbell et al., 1971), and the Caribbean, interreef areas are potential sources of sand for beach nourishment and construction purposes. In an attempt to develop a dredging system for use in and near the reefs of Hawaii

which would not create significant turbidity, Casciano (1973) has developed and tested a device that employs a suction head buried by hydrojet action in the sand; sediment is removed as the cone of sand above the head collapses toward the pump. Use of the system requires that clear water be delivered to the buried dredge head.

Conventional Dredging Techniques

Dredging techniques can be lumped into two major categories: mechanical and hydraulic. Because of differences in sea-bottom characteristics, operating environment, and economics, the basic dredge types have evolved into a variety of forms (Fig. 4, Tables 4 and 5). No one dredge type is likely to ever be superior to others.

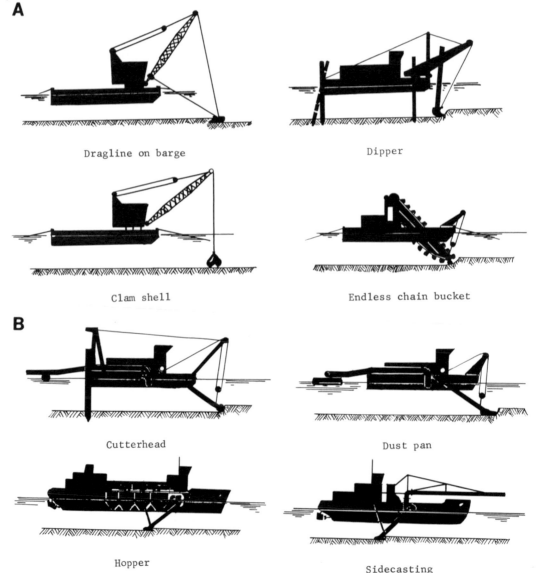

FIGURE 4. *Silhouettes of dredges in common use throughout the world.* (A) *Mechanical dredges;* (B) *hydraulic dredges. Courtesy of W. R. Murdin, Office of the Chief of Engineers.*

TABLE 4. Characteristics and Types of Mechanical Dredges

	Dredge Type			
	Dragline on Barge	Dipper Dredge	Clam Shell or Orange Peel Bucket Dredge	Endless Chain Bucket Dredge
Dredging principle	Scrapes off material by pulling single bucket over it toward stationary crane. Lifts bucket and deposits dredged material in a conveyance or on a bank	Breaks off material by forcing cutting edge of single shovel into it while dredge is stationary. Lifts shovel and deposits dredged material in a conveyance or on a bank	Removes material by forcing opposing bucket edges into it while dredge is stationary. Lifts bucket and deposits dredged material in a conveyance or on a bank	Removes material by forcing single cutting edge of successive buckets into material while dredge is slowly moved between anchors. Lifts buckets and deposits dredged material in a barge or own hopper
Horizontal working force on dredge	Medium intermittent force toward bucket	High very intermittent force away from bucket	No forces	Medium constant force away from bucket
Anchoring while working	Dragline crane can be on shore or on barge. If on barge, latter can be secured with spuds or anchors	Several heavy spuds	Several spuds or anchors	Several anchors
Effect of swells and waves	Can work up to moderate swells and waves	Very sensitive to swells and waves	Can work up to moderate swells and waves	Very sensitive to swells and waves
Material transport	Transport occurs in barges, trucks, or cars. Crane does not transport material. Material disposal occurs in many ways	Transport occurs in barges, trucks, or cars; dredge does not transport material. Material disposal occurs in many ways	Transport occurs in barges, trucks, or cars; dredge does not transport material. Material disposal occurs in many ways	Transport normally occurs in barges. Dredges equipped with hoppers are limited to material disposal by bottom dumping
Dredged material density	Approaches in-place density in mud and silt. Approaches dry density in coarser material	Approaches in-place density in mud and silt. Approaches dry density in coarser material	Approaches in-place density in mud and silt. Approaches dry density in coarser material	Approaches in-place density in mud and silt. Approaches dry density in coarser material
Comments	The term "dredge" is questionable for this machine, since it is not exclusively built for underwater excavation and is frequently used for material removal above water. It is suitable for all but the hardest material and has a low production for its size	Special hard material dredge of simple principle. Rudimentary machine can be assembled for temporary service by placing power shovel on spud barge. Low production for size of plant and investment	This machine is simple in principle. It can be assembled in rudimentary form for temporary service by placing a crane on a barge. It is suitable for all but the hardest materials and has a low production for its size	Highly developed machine. Not used in United States (other than as part of mining plant), but used extensively in other countries. It is suitable for all but the hardest materials and has a high production for its size

Source. From Mohr (1974).

545

TABLE 5. Characteristics and Types of Hydraulic Dredges

	Dredge Type			
	Cutterhead Dredge	Dustpan Dredge	Hopper Dredge	Sidecasting Dredge
Dredging principle	Material is removed with a rotary cutter (or plain suction inlet in light material) picked up with dilution water by the suction pipe, and transported through the pump and the discharge line. While working, dredge swings around spud toward an anchor	Material is removed with water jets, picked up by a wide but shallow suction opening, and transported through the pump and the discharge line. While working, dredge is slowly pulled toward two anchored spuds or anchors	Material is removed and picked up together with dilution water by draghead sliding over bottom (or stationary) and flows through suction piping, pump, and discharge piping into hoppers of vessel	Material is removed and picked up together with dilution water by draghead sliding over bottom and flows through suction piping, pump, and discharge arm over side of vessel back into the water
Horizontal working force on dredge	Medium intermittent force opposing swing to side	Medium constant force opposing forward movement	Slight constant force opposing forward movement	Slight constant force opposing forward movement
Anchoring while working	Two spuds and two swing anchors (one working spud and one walking spud)	Two spuds or anchors secured upstream while working	Dredge moves under own power to dig a channel or is anchored to dig a hole	Dredge moves under own power to dig a channel
Effect of swells and waves	Very sensitive to swells and waves	Very sensitive to swells and waves	Little affected by swells and waves	Little affected by swells and waves
Material transport	Transport occurs in pipeline. Length of discharge line depends on available power, but can be extended with booster pump units to a total length of several miles	Transport occurs in pontoon-supported pipeline to side of dredge. Spoil discharges into water. Booster pump units are not used with this plant	After material is in hoppers, transport is over any suitable waterway. Material can be bottom dumped or pumped out (if so equipped). Pump-out is similar to pipeline dredge operation	Transport occurs in pipeline on discharge boom over side of dredge. Material discharges into adjacent water
Dredged material density	Diluted to an average of 1200 g/l	Diluted to an average of 1200 g/l	Diluted to an average of 1200 g/l	Diluted to an average of 1200 g/l
Comments	Highly developed machine with intricate horizontal moving procedure used throughout the world. Suitable for all but very hard materials. High production for size of plant	Special sand dredge used only in United States in Mississippi River. Floating line is positioned with rudder in discharge stream. High production for size of plant	Highly developed machine used throughout the world. Suitable for all but very hard materials. Production depends on traveling time to dump and mode of discharge	Special sand dredge. Sand transport is limited to length of discharge boom. Used in coastal inlets or where material discharge into water is not objectionable. High production for size of plant

Source. From Mohr (1974).

Instead a dredge will be selected which is best suited for a particular application. Of existing conventional dredges, the hopper dredge seems to be more nearly a "universal" type suited for work in open marine waters as well as protected waters. The need to operate in open waters was the prime reason why in the United States, the Corps of Engineers (which has Congressionally delegated responsibility for maintaining navigable waters) was the prime developer of this type of dredge and still has 16 in operation (Mohr, 1974). The marine sand and gravel dredging fleet of the United Kingdom consists of 75 vessels representing 32 companies and is comprised almost exclusively of hopper dredges (Hess, 1971).

Operating depths of mechanical dredges are virtually unlimited. Hydraulic dredges are capable of operating to about 200 ft water depths although most present dredges are designed for work in 90 ft water depths or less.

Water and air-jet methods might be considered a modified hydraulic system, and they greatly increase operating depth capabilities. The basic principle is to create a density difference in the collecting pipe by introducing air, water, or a less dense liquid near the cutter head. The density gradient creates flow up the pipe which induces a powerful suction at the cutter head. Depth of operation of this type of dredge is approximately 600 m (Herbich, 1971).

Specialized Dredging Techniques

Submersible tractor-type dredges have been developed for use by firms in at least two countries, Japan and the United States. The U.S. device (Ocean Science & Engineering, Inc.) was designed primarily for use in borrowing sand for beach nourishment projects. It was used in the Fort Pierce, Florida area several years ago with apparently marginal results.

Two schemes for dredging the deep ocean for manganese nodules have been described: endless bucket chain and hydraulic dredge with submerged pumps. The endless bucket chain concept is being developed by a group of Japanese firms in concert with other non-Japanese companies. In a demonstration test the system operated in 12,000 ft of water (Donkers and deGroot, 1973). The hydraulic dredge with submerged pumps has been tested in 2700 ft water depth by Deepsea Ventures, Inc. Another system whose details are not known is under development by the Hughes Corporation (with Lockheed). These few attempts being undertaken by corporations with large resources (capital and skilled manpower pool) point up the very high costs involved—reported at $5 to $10 million for Deepsea Ventures, Inc. (Herbich, 1971)—in attempts to create capabilities for deep ocean mining.

RESOURCES

Prospecting for marine placer deposits has historically taken place by extending onshore "placer" deposit workings into the offshore. Similarly, raised beach accumulations have caused a search for submerged beaches offshore. Location of onshore metallogenic provinces is usually the first clue to the possibilities of offshore placers. Except for nonmetals, no mining of placers of any significance is being done in the U.S. continental shelf although considerable exploration effort has been expended (e.g., Jenkins and Lense, 1967; Clifton, 1968; Moore and Silver, 1968; Padan, 1971). In contrast several "mining" operations for heavy minerals placers were under way on continental shelves of Europe, Japan, and eastern Asia.

A visualization of the generalized world wide distribution of potential marine placer mineral deposits as known in the late 1960's is provided by the series of maps, at a slightly varying scale of 1:39 million, published by the USGS (McKevey and Wang, 1969). The potential aspect of the placers on the McKelvey and Wang maps is clearly corroborated in the activity summary of Table 6 and in the following section.

Metallic Placers

Onshore marine placers are worked much more extensively than submerged accumulations. Australian beaches, for example, provide 95% of the world's rutile but probable offshore deposits are not being worked (Padan, 1971); see Table 6. Other onshore "marine" placers of heavy minerals which have been worked for many years are those of Florida and India.

Manganese nodules have been known to be on the deep ocean floor since the Challenger Expedition of the 1870s. Present knowledge, summarized by Horn et al. (1972), indicates the nodules are widespread but in varying concentrations and sizes. In addition to manganese, manganese nodules also contain iron, cobalt, nickel, and copper. High-grade nodules assay at 27 to 30% Mn, 0.2 to 0.4% Co, 1.1 to 1.4% Ni, and 1.0 to 1.3% Cu (Hammond, 1974). Nodules represent therefore a considerable resource for those other materials as well as manganese and are in reality the minerals sought in the nodules.

A marine placer off southeast Japan, containing titaniferous magnetite, was worked from the mid-1950s to the early 1970s (Anderson, 1972). The deposit covers an area of 1 km², and runs 3 to 5% titaniferous magnetite. It is estimated to contain approximately 15 million tons of ore assaying up to nearly 50% ferric oxide and 12% titanium oxide (Oceanology Int., 1971). Extraction

TABLE 6. Recent Marine Mining Activities

Location	Activity*	Water Depth (ft)†	Interest
Africa			
Red Sea	Exploration	6,000±	Sulfide muds
Territory of Southwest Africa	Dredging	100−	Diamonds
Republic of South Africa	Exploration	‡	Phosphate
Asia			
Borneo	Exploration	600−	Tin
Indian Ocean	Exploration	‡	Chrome, iron
Indonesia	Dredging	150−	Tin
Japan	Dredging	30−	Iron sands
Malaysia	Exploration	600−	Tin
Papua and New Guinea	Exploration	600−	Iron sands
Philippines	Exploration	600−	Iron, gold, titanium
Thailand	Dredging	150−	Tin
Europe			
Iceland	Dredging	150±	Shell sands
Great Britain	Dredging	100−	Tin, sand
North America			
Bahamas	Dredging	‡	Aragonite
Canada (British Columbia)	Exploration	12,000±	Manganese nodules
Caribbean Sea	Exploration	200−	‡
Mexico	Exploration	600−	Phosphate sands
Pacific Ocean	Exploration	12,000+	Manganese nodules
United States			
Alaska	Exploration	200−	Gold
Blake Plateau	Exploration	600–2,400	Manganese, phosphate
California	Dredging	30±	Shells
California	Exploration	600±	Phosphate
North Carolina	Exploration	600−	Phosphate sands
Oceania			
Australia	Exploration	600−	Phosphate, heavy metals, gold
Fiji	Exploration	600−	Gold
New Zealand	Exploration	600−	Heavy metals, gold, phosphate
Solomon Islands	Exploration	600−	Tin
Tasmania	Exploration	600−	Heavy metals

Source. From Padan (1971).

* Dredging operations generally include exploration activity, does not include mines originating on land and drifted out under the seafloor.

† Less than is represented by −; more than is represented by +; approximately is represented by ±.

‡ Unknown.

operations were conducted with a hopper-type hydraulic dredge in water variously reported at 10 to 40 m depth.

Placer tin in the nearshore of southeast Thailand has been worked since 1907 where continued exploration has discovered additional deposits off southwest Thailand which are also now being worked. The ore, consisting of cassiterite, is reported to be alluvial in origin. Endless bucket chains have been the primary dredge type but a hopper-type hydraulic dredge is now also operational (Oceanology Int., 1971). Dredging is conducted in water depths ranging from approximately 20 to 30 m with a 3 m tide range.

The northwest portions of the Malaysian Archipelago are also sources of tin from offshore placer deposits. Their history and operations are similar to the Thailand tin resources. Some tin had been dredged off the Cornwall coast of Great Britain but operations have apparently ceased (Anderson, 1972).

Nonmetallic Placers

Dredging of diamondiferous gravels off the west coast of the Republic of South Africa during the 1960s gave great publicity to the marine mineral business. The diamonds originated in the Kimberlite district and were transported to the coast by the Orange River, where they were first worked from raised marine beaches. Later exploration offshore indicated accumulations of diamond-bearing gravels were associated with gullies and potholes in bedrock or as blankets just above bedrock. Recovery operations began just seaward of the breaker zone and extended in some cases 3 nautical miles seaward of the coast using a modified air lift technique. Mining operations ceased in 1971 (Oceanology Int., 1971).

Bahamian oolitic aragonite is mined and marketed by an American firm licensed by the Bahamian government. Ocean Cay, 20 miles south of Bimini (50 miles east of Miami), is a 6 acre island comprised of aragonite dredged for the specific purpose of siting the dredging, handling, and shipping facilities needed for the oolite operation. A hydraulic dredge is used in working the leases which are reported to contain an 850 year supply (Oceanology Int., 1971). The oolite is now used primarily in industrial or agricultural applications. It has been proposed as a source of "sand" for beach nourishment/restoration purposes.

Sand and gravel account for the largest marine mining operations extant today. European countries, notably Great Britain and the Netherlands, have the largest commercial operations. Dredging of navigation channels (both domestically and abroad) to maintain navigable depths oftentimes produces materials used for one purpose or another, but such materials are not considered here as a placer resource.

In 1970 the marine sand and gravel industry of the United Kingdom produced and marketed 14 million tons of aggregate representing 13% of total production (Hess, 1971). As noted by Hess, the bulk of the U.K. material is gravel, used for construction purposes. Origin of the placers vary, depending on the location of the activity, but run the gamut of alluvial, glacial, and marine genesis. Six major resource areas are recognized within which are 50 licensed tracts and 80 dredging sites with perhaps 100 landing points in the United Kingdom and western Europe (Hess, 1971).

Oceanographic conditions within which the U.K. sand and gravel dredges work vary but are often extreme. Nearshore tides of 10 m or more in height influence schedules and operating techniques as do tide-induced currents, which can exceed 2 knots. Open water operations too are affected by tides, particularly currents. Dredges have worked in waves of 3 m height (Hess, USGS, personal communication). Most dredging grounds are in water 30 m or less deep, although the hopper-type dredges are capable of operating in depths exceeding 40 m.

The marine sand and gravel industry of the United States is not yet so active as that of the United Kingdom or the Netherlands although marketing potential is beginning to attract considerable interest and causing legitimate programs and systems to be proposed by commercial operators (Zeigler, IHC-Holland, personal communication). A requirement for construction aggregate in several portions of the northeast megalopolis (particularly New York, Boston, Washington, D.C., and Norfolk, Virginia) indicates those regions may see the first true marine sand and gravel mining ventures (Bureau of Land Management, 1974) although Lake Erie has supported a sand and gravel industry for some years (Hartley, 1960). The potential requirements for sand and gravel, which can only be met from a marine source, were the reason for the joint NOAA–State of Massachusetts-funded New England Offshore Mining Environmental (NOMES) Project.

The sand and gravel industry in the United States is surprisingly large in terms of dollar value. Production in 1972 was valued at \$1.2 billion, exceeded only by three nonfuel commodities (Pajalich, 1972). Volume produced in 1972 was slightly in excess of 900 million short tons (approximately 670 million yd^3) of which 96% was used for construction purposes. Over the past five years production has been fairly constant although dollar value increased nearly 10% (Pajalich, 1972). Representative 1972 prices per ton of sand for three major east coast

TABLE 7. Areas Surveyed and Estimated Beach Sand Resources of the United States

Geographic Area	Seismic Miles	Cores	Area Surveyed (mile^2)	Sand Volume ($\times$ 10^6 cubic yards)
New England				
Maine			10	123
Massachusetts (Boston)[1]			175	57
Rhode Island			25	141
Connecticut (Long Island Sound)			50	130
Area Totals	1900	280	260	531
Southshore Long Island				
Gardiners–Napeague Bays			100	162
Montauk to Moriches Inlet			160	1912
Moriches to Fire Island Inlet			350	2404
Fire Island to E. Rockaway Inlet			125	1359
Rockaway			50	1031
Area Totals	955	122	785	6868
New Jersey				
Sandy Hook	255	10	50	1000
Manasquan	86	11	25	60
Barnegat	200	32	75	448
Little Egg	389	38	120	180
Cape May	760	107	340	1880
Area Totals	1660	198	610	3568
Virginia				
Norfolk	260	57	180	20
Delmarva	435	78	310	225
North Carolina	734	112	950	218
Florida				
Northern				
Fernandina—Cape Canaveral	1328	197	1650	295
Southern				
Cape Canaveral	356	91	350	2000
Cape Canaveral—Palm Beach	611	72	450	92
Palm Beach—Miami	176	31	141	581
Area Totals	2471	391	2591	2673
California				
Newport–Pt. Dume	360	69	140	491
Pt. Dume–Santa Barbara	145	34	90	90
Area Totals	505	103	230	599
Hawaii[2]	Unknown	Unknown	Unknown	Unknown
Great Lakes				
Erie[3]	Unknown	Unknown	Unknown	14
Grand totals	8920	1341	7266	15,011

Data sources. Published and unpublished reports of U.S. Army Corps of Engineers Coastal Engineering Research Center, except as follows: (1) includes results of NOMES project; (2) Campbell et al. (1971); (3) State of Ohio, 1956.

cities are Boston, \$2.00; New York, \$0.75; and Washington, D.C., \$2.75 (Manheim, 1972 in Pajalich, 1972) while nationwide a short ton averaged \$1.31 (Pajalich, 1972). Growth of the sand and gravel industry is predicted at about 4% (36 million short tons) per year (Padan, NOAA, personal communication) and the Bureau of Land Management (1974) estimates that production from the outer continental shelf could reach 40 million short tons 10 years after production began.

In the United States, a potential user of large volumes of sand is the Corps of Engineers who, together with other federal and local governmental agencies, uses large quantities of sand to restore or nourish eroding coastlines. For example, the several beach erosion control studies made on the east coast of Florida indicate approximately 180 million yd (240 million short tons) of sand would be required for initial and 50 years' maintenance nourishment. Anticipating this future need to be met by offshore sources, the Corps in 1964 initiated a survey of the inner continental shelf of the United States to seek sand deposits suitable for use on beaches (Duane, 1968).

In a program still under way, the CERC has to date surveyed approximately 7000 miles^2 of the inner continental shelf, locating in the order of 15 billion yd^3 of sand (Table 7). In the Great Lakes, Ohio (1956) has identified several sand and gravel deposits in Lake Erie, estimating 14 million yd^3 off Fairport. Several billion yards of sediment (potential sand) has been located in the Hawaiian Islands (Campbell et al., 1971). These several cited programs have located in excess of 16 billion yd^3 of sand in water depths generally less than 40 m. Surveyed areas represent approximately 0.1% of the U.S. continental shelf enclosed by the 100 fathom contour. Deposits so far located have in greatest measure resulted from nearshore marine processes (Duane et al., 1972) although deposits due to glacial processes (Hartley, 1960; Williams and Duane, 1974; Meisburger, in press) as well as fluvial processes (Williams and Duane, 1974; Meisburger and Field, 1975) also exist.

In contrast to the European sand and gravel industry, exploitation of open water placers in the United States does not yet occur in any sustained manner even though capabilities exist. Extraction of large volumes of sand for special requirements does take place, however, and is discussed in the following section.

Engineering Use of Placers

Sand and gravel used as aggregate in construction are most commonly size sorted prior to use, so that in cement, concrete, or asphalt mixes, it no longer resembles the placer material in bulk physical properties. There are, however, other engineering uses of placers in which the sand and gravel are essentially unmodified and retain their natural in-place characteristics. These uses are the basis for the following discussion.

With the passage of time, more and more planning studies relating to location of deep-water ports, nuclear-powered electrical generating plants, and desalinization and waste treatment plants, point to the inner continental shelf as the site for such facilities (Bos Kalis, Westminster Group, 1973; Kelly, 1973; Nutant, 1973; Stigter, 1973). Any structures placed on the shelf would utilize sand and gravel as fill, the actual volume involved depending on the design employed and the size and dimensions of the facility (see Chapter 21). Sectional views of two typical breakwater designs are illustrated in Fig. 5.

Such structures could accept fill directly after recovery from the seafloor. Large volumes would be required and the sites would generally be on the open shelf as, for example, in the North Sea (Bos Kalis, Westminster Group, 1973) or the New York Bight (Kelly, 1973; Williams and Duane, 1975). Consequently, sea-going hopper dredges with pump-out capability would be the single system most adaptable to extraction and placement. However, once the outer breakwater was completed it might be possible to employ other rehandling and distribution techniques.

Offshore island construction and associated marine sand and gravel uses are still in the future, awaiting real requirements. However, in other engineering uses such as beach erosion control, there are existing programs to meet present needs and requirements. As noted in Chapter 21, large sectors of the U.S. coastline are, on an economic basis, in a serious state of erosion and consequently in need of shore protection. Marine sand and gravel (especially sand) deposits figure predominantly in mitigating measures for reasons stated in Chapter 21.

Factors considered in the design of beach protection were discussed in Chapter 21 and in detail by the U.S. Army Corps of Engineers (1973). Volumes of sand required are large, since both initial fill to project dimensions and subsequent periodic nourishment are involved. If a land source is utilized, the protective beach operation is conducted by truck haul from borrow site to beach with subsequent primary distribution to project dimensions by bulldozer; waves, longshore currents, and nearshore transport processes act to distribute sand and modify the submarine portion of the offshore profile. Where an offshore source is used, a dredge is required to pick up material from the offshore bottom and transport it to the beach, either directly as in the case of a pipeline dredge, or to a point of transfer in the case of a hopper-type dredge.

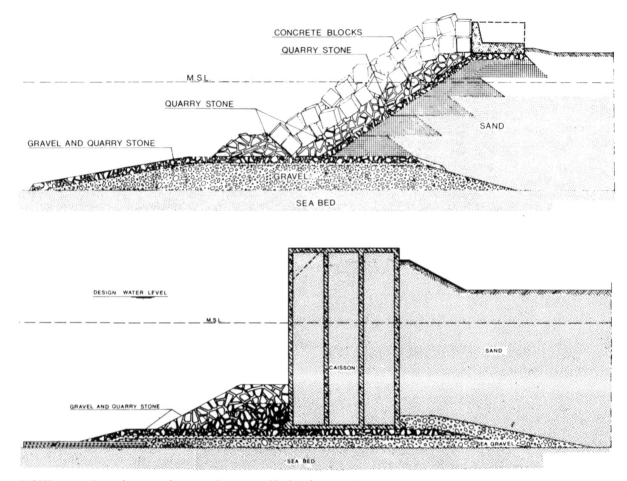

FIGURE 5. *Partial sectional views of two possible breakwater designs for an island sited on the continental shelf. From Bos Kalis, Westminster Group (1973).*

The feasibility of using a hopper dredge with pump-out capability as a means of transferring sand to the beach from an offshore placer source was confirmed as a result of a beach nourishment experiment at Sea Girt, New Jersey in 1966 (Mauriello, 1967). This experiment utilized the Corps dredge Goethals with a bin capacity of approximately 5500 yd³. The Goethals dredged sand in approximately 45 ft of water 1.5 miles from the pipeline terminus offshore of the beach to be nourished, and discharged 250,000 yd³ to the beach through a 2000 ft long submerged pipeline (Fig. 6). Sand dredging and pump-out operations were conducted as long as swells did not exceed 6 ft in height. Considering all factors in the operation, it was estimated that sand could have been placed at a cost of approximately \$1.50 yd³, a competitive price at that time and place.

The first project in the United States to successfully employ a hydraulic pipeline dredge in the open ocean took place off Redondo Beach (Palos Verdes), California in 1967 (Fisher, 1969). Operating in 30 to 40 ft water depths approximately 1200 ft offshore, a commercial dredging company pumped 1.4 million yd³ to restore the beach to berm elevation of +12 ft MLLW and approximately 200 ft width. The dredge had a cutter head modified with water jets that created a sand slurry pumped ashore through a 16 in. diameter pipe that was a combination of floating and submerged line. The operator (Shellmaker Corp., Long Beach, California) was low bidder on the project, delivering the sand at a cost of \$1.07 yd⁻¹ considerably less costly than alternative land sources (Fisher, 1969). Following the success of these projects, other open ocean sources of sand have been utilized in conjunction with coastal engineering projects at Pompono Beach and Treasure Island, Florida.

Means other than those reviewed above for transferring sand to the beach have been discussed, proposed, or attempted. One such system suggests placement of a grid of pipes placed on the seabed, containing upward-directed openings paired with high-pressure eductor

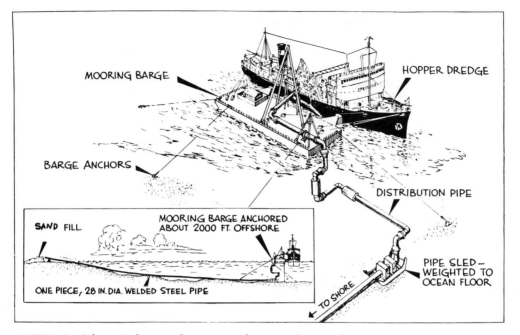

FIGURE 6. *Schematic drawing depicting a technique for employing a hopper dredge with pump-out capabilities for beach nourishment. System illus-* *trated was used in 1966 at Long Beach, New Jersey in a beach nourishment operation. From Mauriello (1967).*

nozzles. In operation, a hopper dredge or barge would float over the pipe grid and release its load directly onto the pipe grid. Water would be forced out of the nozzles under high pressure, resulting in liquefication of the descending sand. The resulting slurry would be forced by a venturi effect into the pipe-grid system and then ashore through the delivery pipe (Govatos and Zandi, 1969).

Another possible means of transferring material to the beach might be natural processes. Shoreward movement can be predicted in theory (Ippen, 1966), and has been directly observed in the laboratory (Rector, 1954) and on a limited scale in nature (Fig. 7). In 1948 the Corps of Engineers placed approximately 600,000 yd³ of sand in nearly 40 ft of water 0.5 mile offshore of Long Branch, New Jersey (Harris, 1954). The beach had a history of recession and the experiment was conducted to determine the feasibility of natural nourishment if sand were placed offshore. Four hydrographic surveys extending from 2 to 50 months after placement ceased were compared to interpret bottom changes. Harris (1954) concluded there was no evidence the material moved ashore from the stockpile and suggested that in order to benefit a beach, material should be placed in water less than 20 ft deep.

Offshore of Virginia Beach, Virginia, the Corps of Engineers (Norfolk, Virginia, District and Coastal Engineering Research Center) is currently stockpiling (and

monitoring the stockpile) sand suited for beach nourishment. Preliminary results based on initial surveys indicate this stockpile, in approximately 40 ft water depths, is stable. The Corps (Wilmington, North Carolina) is currently planning a program to test and evaluate beach nourishment using a self-propelled split hull barge. During nourishment operations the barge, loaded with sand from a remote area, would approach the beach much like a landing craft and deposit the sand landward of the breaker zone.

CONCLUSIONS

Placer mineral resources, metallic and nonmetallic, are today largely potential resources. Uncertainties in the realm of law (international, national, and intranational) create risks which render tenuous any determination of rate of return of investments required for economical extraction. Dredging technology has not experienced any major improvements or innovations in decades although "glamorous" operations (such as diamond and manganese nodule mining) have induced some noble attempts to develop the technology necessary for working, at a profit, in the open marine environment.

Need for sand and gravel, once considered ubiquitous, continues to grow. Because it can be used essentially as found with little or no need for beneficiation, an economic

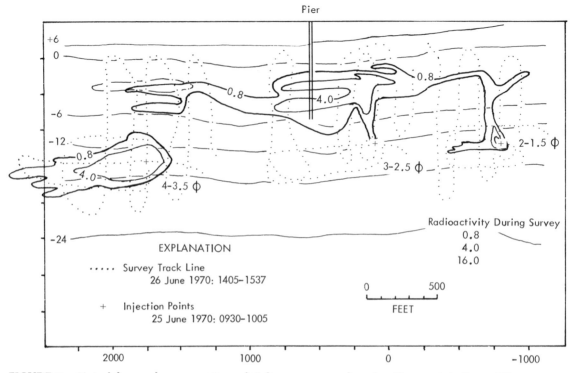

FIGURE 7. *Map of the nearshore area at Oxnard, California (Point Mugu Naval Air Station), depicting transport of radionuclide-tagged sand grains. At time of injection three different size range of particles were placed, nearly synchronously at different locations (1950 ft; 100 ft; −800 ft) along the −15 ft MLLW contour. Subsequent surveys, such as that illustrated, indicate different transport patterns and behavior among the particles tagged. The radioisotope sand tracer (RIST) system used is explained in Duane (1970) and references therein. Contours are counts per second (× 10²) corrected for background and decay. From unpublished data, CERC.*

analysis is more straightforward, and hence it is more easily exploited and marketed. Indeed, marine sand and gravel placers are a currently exploitable resource and will see increasing use for a multitude of engineering purposes in the near future. Despite continued depletion of nonrenewable resources, a move to the sea and extraction of placer deposits will not occur in any large scale until risks to men, machines, investments, and the environment can be better justified.

If marine mining activities to date can be considered to represent the result of managers "seeing through a glass darkly," then perhaps the light to see better is provided by the remarks of the Assistant Secretary of the Interior, John Kyl, who noted that the United States is heading for a minerals and materials crisis that will make the fuel crisis look like a Sunday school picnic (*Washington Post*, Sept. 30, 1974).

SUMMARY

Most concepts and aspects pertaining to exploration and exploitation of a particular marine placer mineral deposit are common to marine placers in general. Lagging

exploration for and exploitation of open marine placers can be attributed to high operational costs, few major discoveries, and the legal unknowns facing the entrepreneur.

Potential new mineral districts, the continental shelves of the world, are thought to contain large quantities of minable materials. Dwindling land supplies are creating an interest in exploiting mineral resources on the continental shelf, initially most likely unconsolidated surficial deposits, particularly those that could be recovered using present marine dredging equipment and technology.

Exploration involves utilization of standard geologic and geophysical techniques adapted for marine use. Deposits located, which may be technologically feasible to exploit, may not be economically exploitable because of low or otherwise unsatisfactory rate of return of investment or environmental concerns (dredging operations will create turbidity). In the nearshore, where natural processes create periods of high turbidity, the effect of dredging operations on the biota may not be noticed.

From the standpoint of engineering in the coastal zone or the continental shelf, interest in open marine placers is confined to those materials used in construction, namely sand and gravel. Potential repositories of large volumes of sand (and gravel) on the continental shelves are drowned terrestrial geomorphic features such as those created by fluvial, eolian, or glacial processes. Morphologic features resulting from marine (including nearshore) processes are much more widespread, occurring across the shelf from the shore to the slope. Such features as shoals associated with linear segments of barrier islands and beaches, or shoals associated with capes and estuary entrances contain large volumes of coarse-grained sediments.

The two major types of dredges are mechanical and hydraulic. Of existing conventional dredges the hopper dredge, in wide use in the United States, United Kingdom, and the Netherlands, seems to be a more nearly universal type suited for work in open (rough) marine waters as well as protected waters.

Marine placers of metallic ores have been, or continue to be, worked in protected waters. No known metallic placers in the open marine environment are currently exploited. The converse is true for marine placers of nonmetallic ores, i.e., sand and gravel. In the United Kingdom 13% of total sand and gravel production in 1970 came from marine sources. In the United States, marine sources of sand and gravel are still largely potential. Less than 0.1% of the U.S. continental shelf has been subjected to detailed surveys. However, these surveys indicate on the order of 15 billion yd^3 of sand (including a small amount of gravel) exist, or enough to supply the entire U.S. demand for approximately 15 years.

Present use of major quantities of U.S. sand is for construction purposes. Additional planned uses for large volumes of sand are for beach protection and nourishment programs. Other potential uses for large volumes of sand and gravel as fill are for structures planned to be built on the continental shelf for siting deep-water ports, nuclear-powered electric generating plants, or desalinization and waste treatment plants.

REFERENCES

Anderson, R. J. (1972). Recent developments in offshore mining. *Fourth Ann. Offshore Conf.*, Paper OTC (preprint), I-703-708.

Bateman, A. M. (1955). *Economic Mineral Deposits*. New York: Wiley, 916 pp.

Bos Kalis, Westminster Group (1973). Building a multipurpose island in the sea. *Ocean Ind.*, **4:** 187–194.

Bureau of Land Management (1974). *Draft Environmental Impact Statement, Proposed Outer Continental Shelf Hard Mineral Mining Operating and Leasing Regulations*. DES 74-4, U.S. Department of Interior, 326 pp.

Campbell, J. F., B. R. Rosendahl, W. T. Coulbourn, and R. Moberly, Jr. (1971). *Reconnaissance Sand Inventory: Off Leeward Molokai and Maui*. University of Hawaii, Institute of Geophysics, Rep. HIG 71-17, 8 pp.

Casciano, F. M. (1973). *Development of a Submarine Sand Recovery System for Hawaii*. UNIHI-Sea Grant-AR 73-04, University of Hawaii, 14 pp.

Christy, F. T., Jr. (1968). The legal aspect of the exploitation of offshore mineral deposits. *Mining Eng.*, **20:** 149–153.

Clifton, H. E. (1968). *Gold Distribution in Surface Sediments on the Continental Shelf off Southern Oregon: A Preliminary Report*. USGS Circular 587, 6 pp.

Commission on Marine Science, Engineering, and Resources (1969). Panel reports. Vol. 1, *Science and Environment*, 340 pp., Vol. 2, *Industry and Technology. Marine Resources and Legal-Political Arrangements for Their Development*. 308 pp. Washington, D.C.: Govt. Printing Office.

Courtenay, W. R., D. J. Herrema, M. J. Thompson, W. P. Azzinaro, and J. van Montfrans (1974). *Ecological Monitoring of Beach Erosion Central Projects, Broward County, Florida, and Adjacent Areas*. T.M. 4, U.S. Army Corps of Engineers, Coastal Eng. Res. Cent., 88 pp.

Dobrin, M. (1960). *Introduction to Geophysical Prospecting*, 2nd ed. New York: McGraw-Hill, 446 pp.

Donkers, J. M. and R. deGroot (1973). Dredging at sea. *Fifth Ann. Offshore Technol. Conf.*, Paper OTC 1762 (preprint), I-379-390.

Duane, D. B. (1968). Sand deposits on the continental shelf: A presently exploitable resource. In *Transactions of the National Symposium on Ocean Sciences and Engineering of the Atlantic Shelf*. Marine Technology Society, pp. 289–297.

Duane, D. B. (1970). Synoptic observation of sand movement. In *Proc. 12th Coastal Eng. Conf.*, ASCE, pp. 799–813.

Duane, D. B., M. F. Field, E. P. Meisburger, D. J. P. Swift, and S. J. Williams (1972). Linear shoals on the Atlantic inner continental shelf, Florida to Long Island. In D. J. P. Swift, D. B. Duane, and O. H. Pilkey, eds., *Shelf Sediment Transport: Process and Pattern*. Stroudsburg, Pa.: Dowden, Hutchinson & Ross, pp. 477–498.

Duane, D. B. and E. P. Meisburger (1969). *Geomorphology and Sediments of the Nearshore Continental Shelf, Miami to Palm Beach, Florida*. U.S. Army Corps of Engineers, Coastal Eng. Res. Cent., T.M. 29, 48 pp.

Dunn, J. A. (1964). Fundamental metal and mineral problems. *South Africa Mining Eng. J.*, **75:** 970–977.

Everts, C. H. (1972). *Exploration for High Energy Marine Placer Sites: Part I—Field and Flume Tests, North Carolina Coast*. Report WIS-SG-72-210, The University of Wisconsin Sea Grant Program, University of Wisconsin, Madison, 179 pp.

Field, M. E. and D. B. Duane (1974). *Geomorphology and Sediments of the Inner Continental Shelf, Cape Canaveral, Florida*. T.M. 42, U.S. Army Corps of Engineers, Coastal Eng. Res. Cent., 87 pp.

Fisher, C. H. (1969). Mining the ocean for beach sand. *Proc. Conf. on Civil Eng. in the Oceans*, II, ASCE, pp. 717–723.

Goodier, J. L. (1972). *U.S. Federal and Seacoast State Offshore Mining Laws*. Washington, D.C.: Nautilus Press, 221 pp.

Govatos, S. and I. Zandi (1969). Beach nourishment from offshore sources. *Shore and Beach*, **37**(2): 40–49.

Grant, M. J. (1973). *Rhode Island's Ocean Sands: Management Guidelines for Sand and Gravel Extraction in State Waters*. Marine Technol. Rep. 10, Coastal Resour. Cent., University of Rhode Island, Kingston, 51 pp.

Griffin, W. L. (1969). Delimitation of ocean space boundaries between adjacent coastal states of the United States. In *International Rules and Organization for the Sea, Proc. 3rd Annual Conference of the Law of the Sea Institute*, University of Rhode Island, Kingston, pp. 142–155.

Hammond, A. L. (1974). Manganese nodules (II): Prospects for deep sea mining. *Science*, **183**: 644–646.

Harris, R. L. (1954). *Restudy of Test Shore Nourishment by Offshore Deposition of Sand, Long Branch, New Jersey*. U.S. Army Corps of Engineers, Beach Erosion Board, T.M. 62, 18 pp.

Hartley, R. P. (1960). *Sand Dredging in Lake Erie, Ohio*. Division of Shore Erosion Technol., Rep. 5, 79 pp.

Herbich, J. B. (1971). Dredging methods for deep-ocean mineral recovery. *J. Waterw., Harbors, and Coastal Eng. Div., ASCE*, WW2, 385–398.

Hess, H. D. (1971). *Marine Sand and Gravel Mining Industry of the United Kingdom*. TR ERL 213-MMTC 1, National Oceanic and Atmospheric Administration, 176 pp.

Horn, D. R., M. Ewing, B. M. Horn, and M. N. Delach (1972). World wide distribution of manganese nodules. *Ocean Ind.*, **Jan.:** 26–29.

Ippen, A. T. (1966). Estuary and coastline hydrodynamics. *Engineering Societies Monographs*. New York: McGraw-Hill, 744 pp.

Jenkins, R. L. and A. H. Lense (1967). *Marine Heavy Metals Project Offshore Nome, Alaska*. U.S. Bureau of Mines, Heavy Metals Program, Dept. Interior, 11 pp.

Kelly, J. L. (1973). Offshore terminals and the national perspective. *Proc. 9th Ann. Conf. Mar. Technol. Soc.*, pp. 9–18.

Ling, S. C. (1972). *State of the Art of Marine Soil Mechanics and Foundation Engineering*. Tech. Rep. S-72-11, U.S. Army Engineer Waterways Experiment Station, Vicksburg, Miss., 153 pp.

McKelvey, V. E. and F. H. Wang (1969). World subsea mineral resources, miscellaneous geologic investigations. *U.S. Department of Interior, Geological Survey*, Map I-632.

McMaster, R. L. and A. Ashraf (1973). Subbottom basement drainage system of inner continental shelf off southern New England. *Bull. Geol. Soc. Am.*, **84:** 187–190.

Mauriello, L. J. (1967). Experimental use of a self unloading hopper dredge for rehabilitation of an ocean beach. In *Proc. WODCON 67, First World Dredging Conf.*, pp. 369–395.

Meisburger, E. P. (in press). *Geomorphology and Sediments of Western Massachusetts Bay*. U.S. Army Corps of Engineers Coastal Eng. Res. Cent., T.M. (number not yet assigned).

Meisburger, E. P. and M. E. Field (1975). *Geomorphology and Sediments of the Inner Continental Shelf, Cape Canaveral to Florida–Georgia Border*. U.S. Army Corps of Engineers Coastal Eng. Res. Cent., T.M. 54, 119 pp.

Miller, H. J., G. B. Tirey, and G. Mecarini (1967). Mechanics of mineral exploration. In *Underwater Technol. Conf. Am. Soc. Mech. Eng.*

Mohr, A. W. (1974). Development and future of dredging. *J. Waterw., Harbors, and Coastal Eng. Div., ASCE*, WW2, 69–83.

Moore, D. G. and H. P. Palmer (1968). Offshore seismic reflection surveys. In *Civil Eng. in the Oceans, ASCE*, pp. 780–806.

Moore, G. W. and E. A. Silver (1968). *Gold Distribution on the Sea Floor off the Klowath Mountains, California*. U.S. Geol. Surv. Circ. 605, 9 pp.

Nelson, C. H. (1971). Northern Bering Sea, a model for depositional history of Arctic shelf placers; ecological impact of placer development. *Proc. 1st Int. Conf. on Port and Ocean Eng. under Arctic Conditions*, **1:** 246–254.

Newton, R. S., E. Seibold, and F. Werner (1973). Facies distribution patterns on the Spanish Sahara continental shelf mapped with side scan sonar. *Meteor Forsch.-Ergeb.*, R.C., S55–77.

Noakes, J. E., J. L. Harding, and J. D. Spaulding (1974). Locating offshore mineral deposits by natural radioactive measurements. *J. Mar. Technol. Soc.*, **8:** 36–39.

Nutant, J. A. (1973). An environmental assessment of floating nuclear plants. *Proc. 9th Ann. Conf. Mar. Technol. Soc.*, 367–374.

Oceanology International (1971). Ocean mining comes of age. *Parts I and II, Oceanology International*. November, pp. 34–41, December, pp. 33–38.

Ohio, State of (1956). *Fairport Sand Pumping Grounds Study*. Dept. of Natural Resources, Div. Shore Erosion, 19 pp.

Padan, J. W. (1971). Marine mining and the environment. In D. Hood, ed., *Impingement of Man on the Oceans*. New York: Wiley, pp. 553–561.

Pajalich, W. (1972). Sand and gravel. In *Minerals Yearbook, 1972*, Vol. I, *Metals, Minerals, and Fuels*. U.S. Department of Interior, Bureau of Mines, pp. 1103–1121.

Pings, W. B. and D. A. Paist (1970). Minerals from the ocean. *Minerals Industry Bulletin*, Parts I and II, Vol. 13. Colorado School of Mines, No. 2, pp. 1–18, No. 3, pp. 1–29.

Rector, R. L. (1954). *Laboratory Study of Equilibrium Profiles of Beaches*. T.M. 41, U.S. Army Corps of Engineers, Beach Erosion Board, Washington, D.C., 38 pp.

Senftle, F. E., D. Duffy, and P. F. Wiggins (1969). Mineral exploration of the ocean floor by in situ neutron absorption using a californium-252 source. *J. Mar. Technol. Soc.*, **3:** 9–16.

Stigter, G. (1973). The building of islands in the open sea offers possibilities for the industrial development in the near future. *Proc. 9th Ann. Conf. Mar. Technol. Soc.* (preprint, 7 pp.).

Thompson, J. R. (1973). *Ecological Effects of Offshore Dredging and Beach Nourishment: A Review*. U.S. Army Corps of Engineers, Coastal Eng. Res. Cent., M.P. 1-73, 39 pp.

U.S. Army Corps of Engineers (1973). *Shore Protection Manual*, Vols. I-III. U.S. Army Corps of Engineers, Coastal Eng. Res. Cent. Washington: U.S. Government Printing Office, 1160 pp.

Wilcox, S. M. and W. J. Mead (1972). Economic potential of marine placer mining. *Ocean Ind.*, 27–28.

Williams, S. J. and D. B. Duane (1974). *Geomorphology and Sediments of the Inner New York Bight Continental Shelf*. U.S. Army Corps of Engineers, Coastal Eng. Res. Cent., T.M. 45, 81 pp.

Williams, S. J. and D. B. Duane (1975). Construction in the coastal zone—A potential use of waste materials. *Mar. Geol.*, **18:** 1–15.

CHAPTER 24

Sedimentation and Ocean Engineering: Ocean Dumping

CARL G. HARD

Project Management Branch, New England Division, Corps of Engineers, Waltham, Massachusetts

HAROLD D. PALMER

Westinghouse Ocean Research Laboratory, Annapolis, Maryland

One of the remarkable conclusions to Mitchell's (1971) intriguing study of the public's opinions regarding causes of coastal erosion along the Atlantic shoreline was that the majority of coastal resource managers perceived the ocean to be an unknown quantity that is largely beyond human control or understanding. Unfortunately, this impression prevails throughout much of the effort now termed "coastal zone management." Within such activities ocean disposal, or ocean dumping, is currently the most politically and socially sensitive issue.

Most of the world's population is concentrated in narrow belts along the shores of the continents. In the United States, about 35% of the population occupies only 8% of the land area—a coastal belt some 50 miles wide. With such a concentration of urban and industrial activity, it follows that the production of waste will be proportionally higher in the coastal zones, and the adjacent continental shelf is considered by many to be the ideal repository for these materials. Several aspects of ocean dumping make such procedures attractive. Wastes discharged into the sea are rather quickly assimilated (Fye, 1974), and they generally vanish within a period of hours. Costs are minimal in that transportation and maintenance of equipment are the only direct expenses incurred by a disposal contractor.

According to Webster's Third New International Dictionary, the first definition of *disposal* is ". . . an orderly or systematic placement, distribution or arrangement," while the fourth meaning is stated as ". . . a discarding or throwing away" (Grove, 1971, p. 654). We cannot consider ocean dumping as "throwing away," since the waste materials remain, in some form, within the sea, its inhabitants, or its sediments. Throwing away implies no return, and this is only possible if the material is accelerated to exceed the force of gravity and directed toward space. Materials dumped into the sea will be recycled, and we cannot consider them to have been discarded.

Returning to the first definition, waste disposal at sea may be politically systematic, since a designated time and place for the release of the materials may be specified. However, ocean dumping is not orderly, for once released these fluids or solids are distributed by oceanographic agents, many of which are at present poorly understood. It is at this point that the marine geologist may, or *should*, participate in making decisions regarding ocean dumping policy. Because of their experience in the study of natural agents and processes, and because they generally deal with long-term phenomena, geologists are in a position to provide some balance to the controversial subject of waste disposal in the ocean. Other chapters within this volume demonstrate the broad base of knowledge that can be applied to predictions of the fate and effects of materials dumped at sea.

The geologist, working closely with physical and biological oceanographers, can contribute to the establishment of reasonable criteria for regulating the quality and quantity of wastes and assist in selecting sites for their release.

TYPES OF WASTE MATERIALS

Materials released to the sea may be considered under two major categories, *solids* and *liquids*. Quite often the waste substance released is a mixture of both. For example, municipal sewage sludge discharged in the midshelf region of the Middle Atlantic Bight is 95% liquid, and similar materials in the New York region contain 3 to 6% solids on a dry weight basis (Gross, 1972). Other industrial wastes may have a higher solid content, and in some cases (cellar dirt, demolition debris, etc.) the materials may consist largely of particles of rock, concrete, ash, brick, and other building materials and manufacturing by-products.

Solids

The annual production of solid wastes from five major sources has been tabulated (Office of Science and Technology, 1969) and is reproduced here as Table 1. Some of these materials are recycled, reclaimed, or otherwise reused, but significant portions of such wastes are disposed of in either land fill operations or ocean dumping. A third but minor disposal technique involves incineration, but restrictions imposed by air quality regulations generally discourage burning.

For the purposes of this chapter, solid wastes will imply a composition of material that yields a specific gravity significantly greater than seawater. It thus excludes a variety of plastics which are becoming common as flotsam recovered in oceanographic work (see p. 454). The most recent summary of plastic residues (Colton et al., 1974) lists six types of materials, the most common being sheets and pieces of plastic packaging materials. They conclude that overboard disposal of trash and garbage by ocean-going vessels is mainly responsible for the observed concentrations.

Gross (1972) has examined the geological aspects of waste solids and liquids for the New York metropolitan region, and has concluded that man's activities provide a major source of sediment to the adjacent continental shelf. As he points out (p. 3163):

Waste disposal operations are the largest sediment transport and depositional process now active in the mid-Atlantic region. Production of waste solids in the metropolitan region exceeds

TABLE 1.   Yearly Generation of Solid Wastes—Five Sources

Source	Solid Waste Generated	
	lb/cap./day	10^6 ton/year
Urban		
Domestic	3.5	128
Municipal	1.2	44
Commercial	2.3	84
Subtotal	7.0	256
Industrial	3.0	110
Agriculture		
Vegetation	15.0	552
Animal	43.0	1563
Subtotal	58.0	2115
Mineral	30.8	1126
Federal	1.2	43
Total	100.0	3650

Source. From the Office of Science and Technology (1969).

sediment yields per unit area of any other major drainage basin in the New England–Middle Atlantic area.

Although such statements appear startling, they may portend the coming norm for continental shelf areas off coasts which support a population and industrial complex on a scale similar to that in the Baltimore–New York corridor. Clearly, man now has the ability to impose significant modifications to natural sedimentary processes active on continental shelves. For example, because of sundry regional geologic and oceanographic factors, relatively little suspended sediment reaches the Atlantic continental shelf of the United States (see discussion by Meade, 1972a and b). If one compares reports on annual tonnage of solid wastes (about 84% dredge spoil) dumped on the Atlantic shelf—31.65 million tons (Smith and Brown, 1971)—with the tonnage of suspended sediment contributed by tributaries to this region—14.20 million tons (Curtis et al., 1973)—one must conclude that at least in local areas we have a significant and locally dominant new agent (man) in marine sedimentology and a unique, generally fine-grained deposit (dredge spoil). According to Williams and Duane (1974), one area off the New Jersey coast has been used as a disposal site for over a century, and local accumulations of dumped materials exceed 50 ft in thickness. Considering the volume of spoil and other solid material that is routinely dumped on the shelves of the United States and other "developed" countries, we might term these deposits the *homogenic facies* (homo:

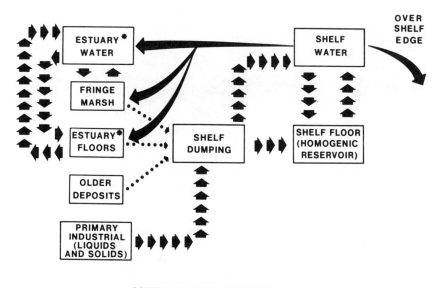

DOTTED LINE PATHS ARE DREDGE ACTIVITIES
＊ ESTUARIES INCLUDE HARBORS.

FIGURE 1.   *Recycling of the homogenic facies. Dotted lines indicate barge activities to a shelf dumpsite.*

man; -genic: produced by, formed from). It would seem an appropriate companion term to *biogenic*, identifying the waste materials as to agent and process and thus distinguishing them from sediments of natural origin.

Acknowledgment of such an input raises an interesting point with regard to the future. If Meade's (1969) concept of prevailing landward transport of bottom materials into coastal estuaries is correct, by dumping on the shelf we are beginning a process that will eventually recycle dredge spoil and other fine particulate matter. Although this may appear to be a slow and unidirectional process from the political and/or regulatory perspective, it will ultimately evolve into a closed cycle from estuaries and harbors, to the shelf, and return (Fig. 1). Man's activities can now create significant changes in regional shelf sedimentation over a period measured in decades, and in so doing alter patterns that required the millenia of the Holocene epoch to reach the current state of quasi equilibrium.

Liquids

Liquid wastes are of lesser concern to the geologist than are solid or particulate materials, since these fluids are generally mixed with and disperse in the surface waters above a pycnocline. Thus we will briefly review their relevance to geological studies and then focus on solid wastes.

In certain instances, such as dumpsites where several metallic ion concentrations are high, waste materials may form particulate matter or precipitates which can settle to the seafloor. For example, acid wastes dumped off Maryland have an iron content of about 4.5% (dry weight) and direct observations, sampling, and photography of the seafloor and water mass in the dump area suggest that a concentration of reddish particulate matter both at the thermocline and in ripple troughs and other depressions on the seafloor may be the products of chemical reactions between seawater and the liquid wastes (Golden and Champ, 1974). An identical situation has been observed in the New York Bight area (D. Drake, personal communication, 1974) where orange and red particles have been recovered through filtration and bottom sampling in acid dump areas. We hesitate to employ the term flocculation, since the electrochemical implication associated with the term suggests reactions which at this time cannot be proved and are indeed open to question (see discussion, p. 137). Studies of acid dumpsites in both the New York Bight and the area off Maryland (Folger et al., personal communication, 1974; Lear and Palmer, 1974) suggest that biochemical processes active in the surface waters are responsible for the rapid formation of iron-rich "agglomerates" which are observed as suspended threadlike matter in the water column. Orange to red collodial to sand-sized ($>62\ \mu$ diameter) bottom materials attributed to acid dumping are also recognized from both areas. In the New York Bight, these iron oxide particles are noted in such abundance that they are used as a tracer for circulation patterns, and the amorphous nature (as revealed by x-ray

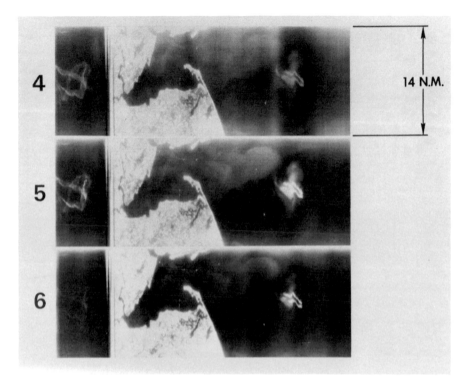

FIGURE 2.   *Multispectral image from ten-channel color scanner flown by NASA's high-altitude research aircraft. This image shows 3 of the 10 channels recorded for a large acid waste dump (left frame segment) off Maryland, while the longer strip (right segments) shows a more recent acid dump in the apex area of the New York Bight (headland is Sandy Hook, New Jersey). Width of segments (vertical dimension) is about 14 nautical miles (26 km). Aircraft was at 65,000 ft; image recorded approximately 1100 hours, July 22, 1974. Spectral bandwidth is 26 nm, with channel 4 centered on 554 nm (green), channel 5 on 582 nm (yellow), and channel 6 on 622 nm (red). Note natural sediment plume issuing from coastal bay. Courtesy of Warren A. Hovis, NASA.*

diffractometry) of these particles points to a biological assimilation of iron by the planktonic community. Apparently these "precipitates" may reach the seafloor within the dumpsites when a pycnocline is weak or absent, since an orange discoloration of the seafloor has been observed in some acid dumpsites. However, the diffusion of acid wastes may be confined to surface waters (Fig. 2) or to the water volume above a thermocline during the summer season. For liquid wastes whose specific gravity approaches that of seawater, mechanisms of dispersion in the four dimensions are dominated by regional/temporal thermohaline conditions in the dumpsite area. Figure 3 shows in schematic form the two thermal regimes which dominate the water column in the Middle Atlantic Bight.

Besides the usual 3 to 4 hour acidification shock to local receiving waters, turbidity induced by the formation of particulate water may essentially block downward light transmission through the water mass. Transmissometer data yielding a measure of turbidity (Goodenow and Herberlein, 1973) and observations from a submersible by Folger (1974) indicate that in certain instances light transmission (and visibility) can be reduced to zero as a result of suspended particulate matter in an acid dumpsite. The implication to primary productivity by phytoplankton is clear, since this process depends on incident sunlight for maintenance of the basic lowest link in most marine food chains. This material must certainly settle and accumulate as a bottom deposit at *some* site, but by its very nature it is susceptible to resuspension and transport. As fine suspended material moving as a "nepheloid layer" near the bottom (see p. 442), it may conform to Meade's (1969) model of onshore transport into estuary and bay mouths, and thus return to the coastal zone. Inasmuch as toxic metals, chlorinated hydrocarbons, and/or other undesirable or hazardous matter are often borne upon such particulate matter, there is a justifiable concern regard-

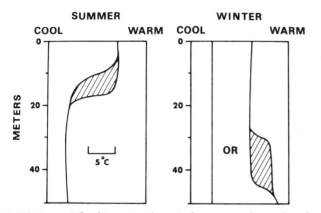

FIGURE 3. *The dispersion of particulate matter dumped at the surface is to a large degree a function of the density structure of the water mass. Thermal conditions in the Middle Atlantic Bight are shown here in schematic style, with the cross-hatched zone denoting the general range of thermocline depth. In winter, iso-thermal conditions may permit dumped materials to sink directly to the seafloor, while intrusions of warmer slope water may form temporary inversions, as shown on the right.*

ing their short- and long-term fate, and geologists are in a position to aid in the determinations of trajectories, sources, and sinks.

The biological impacts of liquid waste discharge are probably of greater concern than sedimentological effects. Large volumes of acid liquors having a pH of 0.01 are routinely dumped at sea, and other materials may give rise to an oxygen depletion of the upper layers of the water mass as oxidation of certain waste compounds takes place.

In some areas the generation of power through both nuclear and fossil fuel power plants creates a large volume of heated water which is discharged into the sea. This "thermal pollution" forms another aspect of liquid waste disposal which may require the passage of water through cooling towers or canals prior to its release to the sea.

Yet another area of concern surrounds the discharge of liquid sludge, treated sewage, or raw municipal wastes through pipelines discharging to the sea. The return of such materials to a coastline may pose a serious health hazard, and closure of beaches to recreational use involving water contact activities is a familiar and growing problem in coastal communities throughout the world.

OCEAN DUMPING STUDY PROGRAMS

The scientific basis for today's ocean dumping research consists of a collection of facts and opinions that are contained in a relatively few works from various fields

and from the disciplines of the earth sciences, especially sedimentary petrography and physical oceanography [see the comprehensive report by Pararas-Carayannis (1973) treating the New York Bight]. Study methodology is either (a) dynamic or (b) synoptic-statistical. The bulk of the work is of the latter type which, because of budget constraints, is still generally weak in statistical adequacy. The dynamic type of study is difficult to undertake because of local or regional facets of the problem, the number of complex variables, and the scarcity of inter-disciplinary scientists who can work in the relevant areas. Studies, then, generally fall under four categories: (a) site selection involving some sort of criteria; (b) monitoring; (c) specific tasks related to a broad, organized program; and (d) special problems.

Site selection usually consists of nautical chart inspection to select likely areas for reconnaissance. Selection criteria are apt to vary but are likely to consist of, or at least resemble, some that were suggested in the First Ocean Disposal Conference at Woods Hole, Massachusetts (Andreliunas and Hard, 1972). Examples of such studies are those performed by the Maine Department of Sea and Shore Fisheries in Penobscot Bay, Maine (Kyte, 1974) and New England Aquarium (Gilbert et al., 1973) in Buzzards Bay, Massachusetts. Political considerations such as state boundaries and federal jurisdictions such as the Regional Division System of the Environmental Protection Agency (EPA) are also factors that determine the location of an ocean dumpsite.

Monitoring activities include pre-, during-, and post-dumping observations, usually for a set of fixed parameters such as those set forth by the Environmental Protection Agency in the *Federal Register* of 1973. No reasonable basis for their interpretation appears to have been published. There were numerical limits set on maximum values for certain presumed toxicants as set forth in draft EPA regulations, which subsequently were liberalized somewhat to include latitude of judgment on a case-by-case basis at the discretion of regional authorities. In the case of dredge spoil, principal criteria to this point were arbitrary and focused on sediment analyses. They were superseded by the present criteria which center on the "shake" or "elutriation" test which involves shaking one volume of sediment from the dredge site with four parts seawater from the dumpsite, testing the elutriate, and comparing it with analysis of the dumpsite water. An excellent critique of all of the criteria, which essentially discounts every criterion except the shaker test and credits the latter only in the event that it is improved upon, has been published by the U.S. Army Corps of Engineers Waterways Experiment Station (Lee and Plumb, 1974).

For most wastes discharged into the sea, emphasis is directed toward toxicology involving minor constituents (termed by EPA as "major constituents": *Federal Register*, 1973) and nutrients, BOD loading, and physical burial as they relate to biological productivity relevant to fisheries resources. On occasion, the toxic effects can be regarded as local situations tied to specific events that may be identified by monitoring. On the other hand, diminution of fish resources may be attributed to over-fishing and cyclical changes in population densities and distributions rather than to waste disposal (e.g., see Sissenwine and Saila, 1973).

Special problems are likely to be related to specific situations that develop relative to individual construction projects or to the nature of the materials being discharged. Examples are studies made in connection with public concern over the effects of the dumping at sea of materials believed to contain chicken entrails from processing plants (Dean and Schnitker, 1971) or studies made to quell stories implicating dredging with an outburst of red tide (Martin and Yentsch, 1974).

Most studies thus far have been of the monitoring type using synoptic-statistical methodology; however, laboratories of NOAA and EPA have been carrying on methodical studies of the long-term effects, including genetic effects of sublethal concentrations of heavy metals on marine organisms as well as their short-term effects on larvae and plankton. Such studies are in progress at the Sandy Hook and Milford Laboratories of NOAA and the Narragansett Laboratory of EPA. Significant advances by European workers in this field will be found in Ruivo's (1972) indispensable volume on marine pollution [see Portmann (1972) and other contributors]. Sulfide fixation of heavy metals appears to have been considered only to the point of having a report prepared on the state of knowledge of the phenomenon (Anonymous, 1973). A study has been made, however, of the relationship between various mineralogic types of clays spiked with mercury and concentrations of the metal in elutriates after agitation (Feick et al., 1974). Similar studies with chlorinated hydrocarbon uptake rates on clays are reported by Huang and Liao (1970). Fine suspended matter generally carries a surface charge often described in terms of its cation exchange capacity (see Carter and Wilde, 1972) and these materials act as "scrubbers" of heavy metals and organochloride substances in the marine environment. The complexity of saltwater chemistry makes interpretation of this kind of study extremely difficult.

One of the more promising studies now under way is a series of bioassays conducted under conditions of turbidity using kaolinite mixes of varying concentrations (Figs. 4 and 5) run concurrently with a separate series in aquariums where various controlled concentrations of nutrients are maintained. After sufficient baseline data have been established, another series will combine the nutrients with the turbidity to obtain the synergistic effects (McLeod and Riser, in progress). The latter studies may perhaps be described as more dynamic than those that have been carried out with baled solid waste (Pratt et al., 1973; Loder et al., 1973) which fall in the monitoring class.

MODELING

While monitoring will have to continue, and perhaps increase in importance and extent, there appears to be a danger of monopolizing the time of the people who are needed to perform the dynamic studies that are essential to understanding phenomena which previously have been observed, albeit not well understood. The essence of what remains to be done in dumping research appears to lie in two areas: (a) modeling of an event, and (b) biogeochemical modeling of biological productivity. Two types of models to be considered are (a) mathematical models that are verified by field investigation and (b) those that are constructs made after field studies and then further verified. The first type appears to have constraints similar to the type of study referred to earlier which was described as resembling typical planning methodology, while the latter type enables real world conditions to determine the structure of our construct and thus provide a parallel to classical research. In the former type one structures things in advance because one anticipates a specific end product, i.e., one is goal oriented. Such a pragmatic approach is appropriate to applied science in a well-defined field that is not complicated with too many variables. It is most appropriate, perhaps, as a procedural frame of reference in planning field or laboratory studies.

Mathematical dispersion models indeed have been considered in the planning of studies of dumping events (Johnson, 1974). Edge and Dysart (1972) provide a simulation of a negatively buoyant jet discharged into a stratified fluid which follows the materials through dispersion and settling to the bottom. Descriptions of hydrodynamical numerical model experiments in the Strait of Gibraltar (Wolff et al., 1972) and dispersion of fluid wastes by ship's propellers (Abraham and van Dam 1972) can be found in Ruivo (1972). Such studies can serve to guide one in planning, but for solid waste disposal it is questionable whether they can be expected to be very useful after observation of the first well-designed field experiments primarily because little is known about

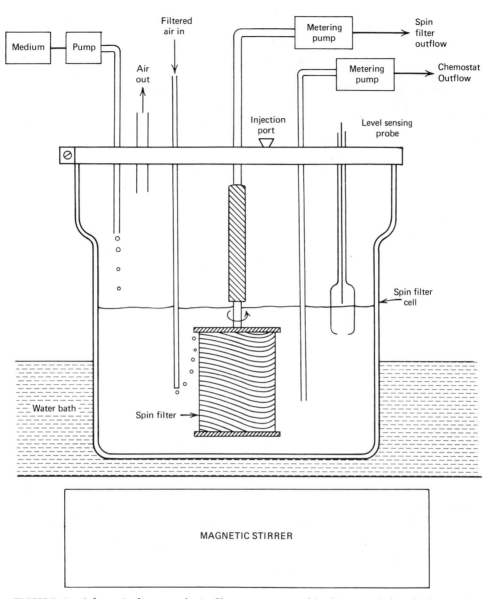

FIGURE 4. *Schematic diagram of spin filter apparatus used for bioassay of phytoplankton under controlled turbidity conditions. Courtesy of G. McLeod, New England Aquarium.*

the application of standard sediment laboratory techniques to dumping activities in a full-scale hydrographic regime. To assume otherwise would be akin to assuming simply that Stokes' law or Waddell's formula could be applied to interpret a dumping event in the ocean. Ample evidence exists to demonstrate that this is far from true.

URBAN WASTE DISPOSAL

At present, the total volume of materials being dumped in the ocean is increasing five times more rapidly than the general per capita production of wastes (McFarlane and Kelly, 1974). By far the greatest volume, at least 80%, consists of dredge spoil from harbors and navigable waterways. The advent of deep draft shipping has required deepening of many ports and approaches, and we must assume that dredge spoil will remain the dominant material dumped on the continental shelves. The relative amounts and costs of ocean disposal are summarized in Table 2, taken from Smith and Brown (1971, p. 21). A more recent summary of dumping impacts (NOAA, 1974) has resulted from requirements of the Marine Protection, Research, and Sanctuaries

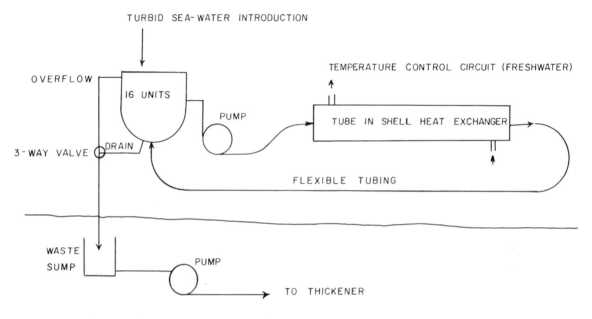

FIGURE 5. *Schematic diagram of tank apparatus used for bioassay of benthonic animals under controlled turbidity conditions. Courtesy of R. Stone, Northeastern University.*

Act (P.L. 92-532) to provide research on ocean dumping and other man-induced changes to ocean ecosystems.

A coastal zone management scheme should not address dredging, ocean mining, and urban waste disposal separately. Imagine a dredged borrow pit, backfilled with bales of refuse or sewer sludge and covered over to original ground level with dredged material. Much of the nearshore continental shelf is flat and surfaced with marketable granular materials. If we take into account what is being learned about the reestablishment rates of benthic organisms on dumping grounds, it appears feasible to exploit the ocean bottom systematically for construction materials, follow that up with the dumping of municipal refuse in the cavities, and ultimately permit nature to restore her biological resources. Meanwhile, all but relatively small areas of the shelf can be reserved for commercial fishing activities. Such a scheme would require plans involving prior coordination between coastal states, internal agencies within each state, federal agencies, and commercial fishing organizations. No such coordination exists and it would be very difficult to establish such a liaison in view of the confusion that exists because of overlapping jurisdictions, confusing if not contradictory legislation at all levels, and the independent nature of the United States fisherman. The pressure to establish new sources of construction materials in the Northeast Corridor inevitably will become irre-

sistible despite the environmentalists' apprehensive opposition to such activities. The marine scientist has, therefore, a responsibility to see that this conflict is resolved in the most beneficial manner, which means that he has the responsibility to make decisions involving the conquest of some of his own apprehension because his expertise is in the embryonic state.

Baled Waste

A start has been made in studying the effects of placing baled wastes in the sea (Pratt et al., 1973) and shipboard incineration and sea disposal of residues have also been considered (First, 1972). There is not much interest in financial support of the former, for apparent reasons, but the results of the latter studies are reassuring from the standpoint of indicating no major environmental hazard. The composition of municipal wastes throughout the United States has been surveyed (Anonymous, 1969; Smith and Brown, 1971) and the pilot studies have attempted to incorporate representative bales. Although compressed, ballasting has been required. A start has been made at Woods Hole Oceanographic Institution on observing bales placed off the continental shelf edge with the intention of making further observations with the deep submersible Alvin. The approach has been to determine the practicability of the simple placement

TABLE 2. Summary of Type, Amount and Estimated Costs of Wastes Disposed of in Pacific, Atlantic, and Gulf Coast Waters for the Year 1968

Waste type	Pacific Coast		Atlantic Coast		Gulf Coast		Total		Total	
	Annual Tonnage	Estimated Cost ($)	Annual Tonnage	Estimated Cost ($)	Annual Tonnage	Estimated Cost ($)	Annual Tonnage	Estimated Cost ($)‖	% Tonnage	% Cost
Dredging spoils	8,320,000	3,608,000	30,880,000*	16,810,000	13,000,000	3,228,000	52,200,000	23,646,000	84	63.5
Industrial wastes										
Bulk	981,000	991,000	3,011,000	5,406,000	690,000	1,592,000	4,682,000	7,989,000	8	21.7
Containerized	300	16,000	2,200	17,000	6,000	171,000	8,500	204,000	<1	<1
Refuse, garbage†	26,000	392,000					26,000	392,000	<1	1
Sewage sludge‡			4,477,000	4,433,000			4,477,000	4,433,000	7	12
Miscellaneous	200	3,000					200	3,000	<1	<1
Construction and demolition debris			574,000	430,000			574,000	430,000	1	1
Explosives			15,200	235,000			15,200	235,000	<1	<1
Total, all wastes§	9,327,500	5,010,000	38,959,400	27,331,000	13,696,000	4,991,000	61,982,900	37,332,000	100	100

Source. From Smith and Brown (1971, p. 21). (Table revised and updated by James L. Verber, FDA.)

* Includes 200,000 tons of fly ash.

† At San Diego 4700 tons vessel garbage at $280,000 per year were discontinued in November 1968.

‡ Tonnage on wet basis. Assuming average 4.5% dry solids, this amounts to approximately 200,000 tons dry solids per year being barged to sea.

§ Radioactive wastes omitted. There were no dumps during 1968. Average annual disposal in 1969–1970 was 4.2 tons.

‖ Estimated costs were increased proportionately for each area from the original tonnage per cost data.

operation and to confirm that the bales will hold together when dropped. Apprehension has been expressed, however, that the extremely low rate of biodegradation at such extremes of refrigeration and pressure militates against the idea of accumulating a "time bomb" that might "explode" at some indeterminate future date, not unlike the accumulation of radioactive industrial waste products (Fye, 1974). The other studies include seawater flume tests and observations and sampling of submerged bales placed near the coast. Accumulations of methane and hydrogen sulfide are found to be more acute in plastic-wrapped bales than in unwrapped, simply strapped bales, leading to the inference that unwrapped bales probably provide less of a hazard. The bales soon developed a bacterial coating of unidentified species, but apparently healthy organisms attached themselves to the bales shortly after their placement.

Geological State of the Art

A number of geological techniques are available which may be applied to the prediction of transport rates and magnitudes for solids dumped on the shelf. For many years marine geologists have employed "tracers," either natural materials or "tagged" particulate matter, in studies of sediment transport. As early as 1962, marine geologists and engineers in Los Angeles were using tomato seeds as tracers for sewage sludge discharged at the head of Santa Monica submarine canyon. Similar studies are now in progress in England (Shelton, 1973) which trace coarse fraction transport from the London sludge dump in the Thames River estuary.

Fluorescent dyes have been employed on beaches (Ingle, 1965) and in the nearshore area by divers (Vernon, 1966). More recently, radioactive tracers (see Duane and Judge, 1966) have been used to monitor long-term sediment dispersal patterns on the inner continental shelf. Swift et al. (1974) have applied these techniques to the measurement of sand transport at dumpsites in the New York Bight. One aspect of such studies which requires further analysis is that of applying radioactive tracers such as gold and ruthenium to fine materials.

Geological techniques may also be employed to evaluate the vertical mixing of trace metals into shelf substrates. Gross (1972), in his tabulation of major and minor cation components in man-made materials (Figs. 6 and 7), provides a basis for comparing sludge and harbor mud metal concentrations with natural materials. Cores of surficial sediments will generally reveal decreasing concentrations of metals such as lead, copper, mercury,

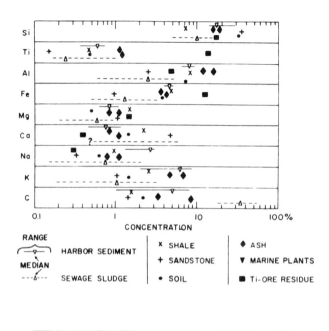

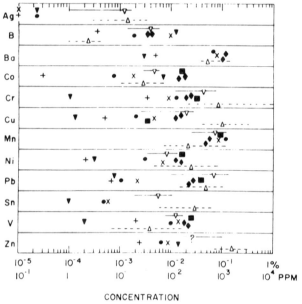

FIGURE 6. *Chemical composition of waste solids, sediments, and soils from the New York Bight. Upper plot shows major element composition; lower, the minor elements. Concentrations are expressed as dry weight. From Gross (1972, p. 3168).*

etc. (SCCWRP, 1973, p. 160; Bascom, 1974) which are directly related to marine activity in dumping practices.

A theoretical avenue of approach that might be devised to predict the configuration of dumped granular materials could be adapted from a few relationships that have been observed in beach behavior studies. It would involve the construction of computed composite grain-

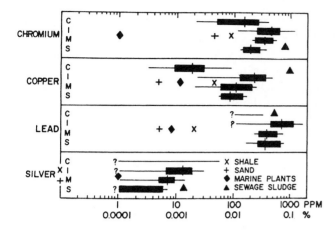

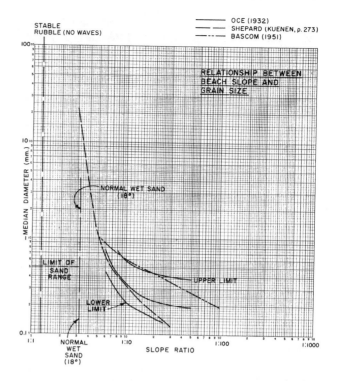

FIGURE 7. *Concentrations of four minor elements in sediments and waste deposits of the New York metropolitan region. The line represents the range, the vertical line marks the median concentration, and the heavy bar indicates the concentration limits for 70% of the samples analyzed. C, continental shelf, outside the area of waste disposal operations; I, inner harbor; M, deposits in dredged waste disposal area; S, deposits in the sewage sludge disposal area. Note that lead concentrations are up to 50 times that of natural materials. Question marks indicate approximate detection limits (optical emission spectroscopy) for various elements. From Gross (1972, p. 3171).*

FIGURE 8. *Selected observed data relating grain size to angle of repose of beach sands. Bascom's (1951) observations are in mid-tide zone; hence probably reflect somewhat coarser lag material. The range between limits is presumed to reflect the degree of site exposure.*

size distributions using, for example, Krumbein's method (Krumbein, 1957), basing them on an extensive pre-dredge sampling program (Krumbein and Slack, 1956) that would yield a composite median that could be applied to various beach slope observations of Shepard, Bascom, and the Corps of Engineers (see Fig. 8). These values, while appearing to conform reasonably to the author's observations in nature and on construction projects, have not been verified by controlled experimentation but they do appear to have value for approximation. Observational data in Holland (Kuenen, 1950, p. 256) and data such as those presented by Komar and Miller (1973) on wave-induced current velocities required to lift various grain sizes then could be coupled with the beach slope data (see Table 3) from which one might postulate an apparent relationship $V \propto M_d, \phi$ where V is velocity (meters per second) 15 cm from the bottom at the dumpsite; M_d is median diameter (millimeters) based on computed composite of the dredged material as predetermined in situ; and ϕ is anticipated slope of repose of the spoil. We then might conclude that $V < 1$ m/sec will deposit M_d 0.5 material on $\phi = 1:5$–$1:7$ (probably influenced within that range by the relative presence of vertical turbulence as affected by depth and extent of open fetch); then $V = 7.5$ m/sec would deposit M_d 0.3 on $\phi = 1:10$–$1:35$; and $V < 0.50$ m/sec would

deposit M_d 0.25 on $\phi = 1:15$–$1:60$ or less, depending on the degree of open exposure of the dumpsite. This type of rationale obviously would apply only to granular materials. Bulk materials, such as uncovered baled waste, would have to be related somehow to experimentation with studies of the type of Irribaren and Hudson (Anonymous, 1966) as they might apply to submerged pell-mell rubble mounds of low specific gravity. Fine-grained materials are still being studied for erodibility characteristics (Nacci et al., 1974) but an understanding of their behavior is complicated by rapid bioturbation processes in the upper few centimeters.

When considering the propriety of applying beach slope criteria to angle of spoil pile repose, it should be understood that we are not assuming that shore wave conditions are applicable to submerged disposal sites. Rather, we are going on the assumption that given granular materials subjected to sufficient disturbance will fall within a characteristic range of slope of repose. Fine-grained materials, however, provide quite a different set of circumstances, particularly when the particles involved are small enough for electrical forces

TABLE 3.   Estimated Relationship Between Beach Slope Data with Respect to Grain Size and Current Regime in a Dumping Ground

Lift Threshold Velocity* (m/sec)	Median Diameter (mm)	Slope of Repose		
		Shepard	Bascomb†	Corps of Engineers
1.0	.50	1:7	1:15	1:7
0.75	.33	1:10	1:35	1:10
0.50	.25	1:40	1:60	1:15

* Kuenen (1950, p. 256).

† Flatter slopes for corresponding grain size are attributed to sampling based on midtide grain size rather than composite. Values taken off curves in Fig. 4.

to influence their behavior even under strong gravitation. When the materials are cohesive (electrically attracted) they cohere and behave to a greater or lesser extent as agglomerates. This has been observed to the extent that after dumping, they have retained shapes and teeth-marks acquired in the buckets that dredged them (Pratt, verbal communication). The variables that affect the degree to which they achieve this state are mineralogic clay content, in situ density (inferred from dry unit weight/core tube volume parameter), and organic content, all reflected in the Atterberg limits (Anonymous, 1953) which, by themselves, are not sufficiently diagnostic to make reliable predictions regarding behavior.

Much time has been wasted by trying to apply settling velocity data to dumped sediment behavior, information that would be useful only under conditions of complete dispersion, where the materials were noncohesive, in a medium free of electrolytes, and where particle crowding was not a factor. Such conditions obviously do not apply to the dumping of fine-grain sediments in the ocean. These circumstances also seriously hamper any attempt to either formulate predictions on the basis of dispersion models or draw conclusions from the observation of dye tracers.

Perhaps the most valuable aid to predicting the effects of dumping is simply periodic monitoring of the bathymetry of different kinds of spoil piles deposited at various sites, and assembling enough of such experiences to be able to forecast conditions on the basis of hindcasts. This is why it is desirable to develop data for the in situ soil mechanics properties at both the dredge and dumpsites prior to operations, in order to correlate them with subsequent bathymetrric infomtaion gathered over a

period of time at the dumpsites. Preliminary information (Pratt et al., 1973; Kyte, 1974) suggests that distinct piles tend to develop (Figs. 9–11) and remain fairly well in place under a variety of exposures, at least in depths below fair-weather wave base, barring strong currents at the site. Quantities in the dumps appear to correspond with quantities determined by postdredging surveys at the dredge sites, but there is little information to establish that in-place densities at both dredge and dumpsites are the same, or would be expected to correspond to one another. If a compact material were to be excavated, then it should experience an increase in volume in the dumpsite. The determination of such underwater bulking factors will require considerable experimentation. Experience with beach placement of granular materials, however, suggests that a one-to-one relationship exists in sands and may support the application of beach slope data to submerged spoil piles, as we have suggested. Field observations to date indicate that only a small amount of organic silt is placed in suspension by the operation of a clamshell dredge, and the amount received in the spoil piles, as determined by bathymetric surveys, shows that there is not a significant contribution to the suspended sediment load of the surrounding water masses, differences in density notwithstanding. An exception may exist in the instance of highly organic fines which have a very high water content. If we can assume that the material does descend directly to form a pile, then the pollutants, which for the most part are minor percentages to begin with, can expose only a small fraction of their total bulk to the overlying ocean water. Such a limited exposure permits only a slow release (which is primarily what is to be expected after the initial drop) of these pollutants. Assuming that the pollutants are in some available form, the release is reduced by a large factor with the passage of time.

It has been demonstrated by Rhoads (1974) (Fig. 12) that shallow reworking of a sediment surface by benthic organisms rapidly results in a thin, more stable, new fabric that is considerably more resistant to erosion and substantially helps to stabilize the deposit. The reworked fabric consists of worm tubes and fecal pellets, the latter protected by a film of mucus, and the total mass assembled into a structure containing roughly 15 to 20% voids. Any contaminants very near the surface then are available principally through ingestion by these fauna which are mostly worms and amphipods, leading to the thought that the analysis of bottom feeders in a dumping ground vicinity provides an effective form of monitoring for pollution from the spoil.

Typical analyses of organic silts involved in harbor dredging of New England estuarine harbors are shown

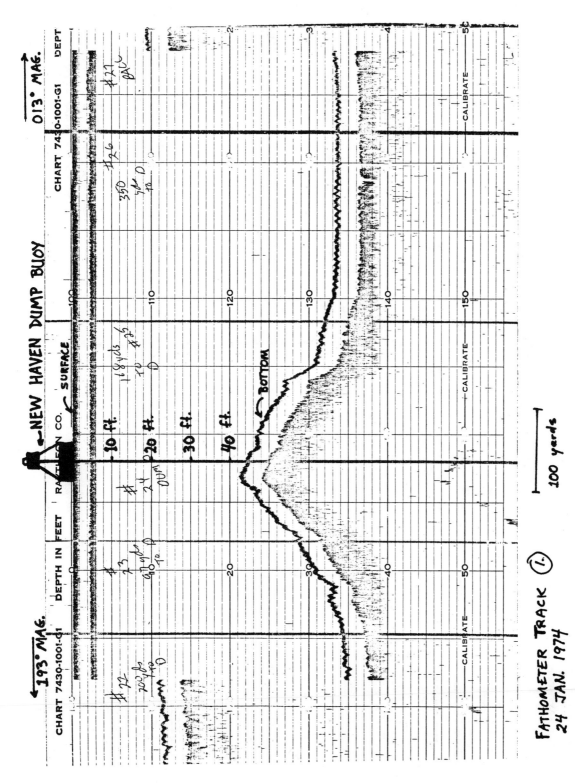

FIGURE 9. *Fathogram of dumpsite after roughly one-half million cubic meters of largely organic silt were dropped.*

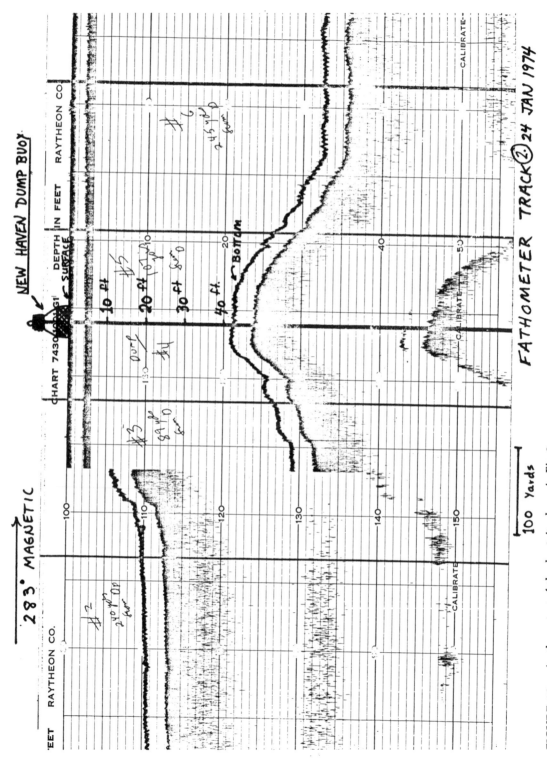

FIGURE 10. *Another aspect of the dumpsite shown in Fig. 9.*

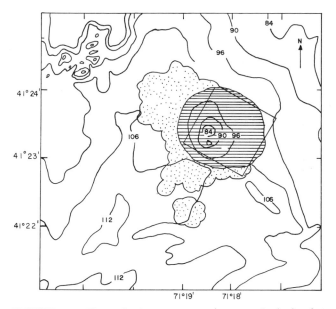

FIGURE 11. *Dumpsite in open ocean (contours in feet) after dumping roughly three-fourths million cubic meters of largely organic silty fine sands. Shaded area represents spoil over 1 ft deep. Dotted area is locus containing patches of spoil (University of Rhode Island).*

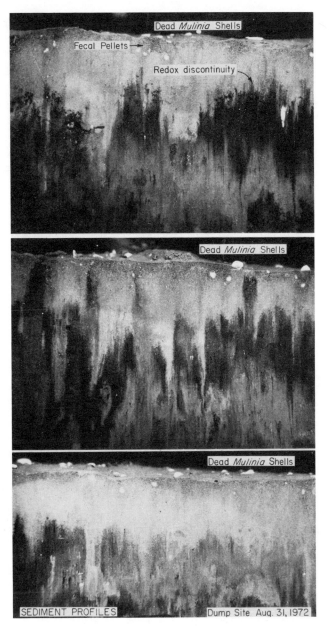

FIGURE 12. *In situ interface camera photographs of upper surface of organic material deposited in dumping ground, showing reworked fabric, partially oxidized by benthic organisms. Scale is one-half natural size. Courtesy of D. C. Rhoads, Yale University.*

in Table 4. Organic silts are illustrated because, unlike granular materials, they constitute the principal matrix for pollutants. Boston Harbor lies in the Gulf of Maine tidal system and has a tide range of roughly 10 ft, while Fall River Harbor lies well up the Taunton River estuary and is in the North Atlantic tidal system, flowing first into Mount Hope and Narrangansett bays, and then into Rhode Island Sound, where the tidal range is under 4 ft. Bridgeport Harbor, also in the North Atlantic tidal system, is on Long Island Sound and has a tide range of about 9 ft. All three are important industrial harbors, and hence tend to represent a higher type of pollution potential for sediments, particularly with respect to metallic wastes and sewage. Similar analyses have been completed for approximately 100 New England harbors and while they are in the process of statistical analysis, the obstacles to meaningful analyses and correlations are formidable in terms of:

1. incompatibility of atomic absorption spectrophotometric results by different analysts,
2. cost of adequate statistical sampling coverage,
3. problem of statistical weighting of samples with respect to different recovery depths, methods of batching the samples, field layouts, etc.,
4. lack of stratigraphic information required for geological interpretation of conditions for each site,
5. lack of knowledge of the significance of individual parameter values with respect to mineral composition and ambient background contents of metals and other substances, and
6. contamination of sampling gear.

These obstacles militate, to a great extent, against all proposals for eclectic data banks. Eight Fall River organic silt core samples obtained from fairly widespread locations within a single geologic stratum, including the materials analyzed and shown in Tables 4 and 5, were

TABLE 4. **Analysis of Bottom Sediments from Three Representative Industrial Harbors in New England**

SAMPLE LOCATION	BOSTON HARBOR	FALL RIVER HARBOR	BRIDGEPORT HARBOR
SOIL CLASS	BLACK ORGANIC SILT	BLACK ORGANIC SILT	BLACK ORGANIC SILT
MEDIAN PARTICLE DIA.	.018 mm.	.008	.008
Q_1 - 75%ile size	.032	.023	.016
Q_3 - 25%ile size	.005	.002	.001
% <200 mesh (U.S.)	92.0	94.0	95.7
Liquid Limit	72	89	134
Plastic Limit	33	40	55
Plastic index	39	49	79
Spec. Grav. Solids	2.58	2.57	2.40±
Wet Unit Wt. (#/c.f.)	77.04	72.17	70.60±
Dry Unit Wt.	32.59	34.27	18.60±
% Solids	27.84	46.89	20.00±
pH	7.04	7.64	6.80
Redox Potential (M/N)			-4.85
Volatile Solids % - EPA	9.87	15.12	11.57
Volatile Solids % - NED	8.81	7.64	9.32
Total Volatile Sol.% - EPA	16.32		16.60±
Chem. Oxygen Dem.	15.31	14.08	15.50±
Tot. Kjeldahl Nitrogen %	0.506	0.270±	0.656
Hexane Soluble %	0.940	0.459	0.429
Hg ppm	0.71	3.58	0.85
Pb	174.3	426.3	2140
Zn	377.6	468.9	5630
As	26.7		90.0
Cu	225.1	149.2	6560
Ni	36.3		
V	79.9		1070

subjected to C_{14} analysis and yielded an average age of 4700 years BP. Specific dating for the material shown in these tables was 5444 ± 255 years BP, which appears to provide somewhat persuasive evidence that the materials are nonanthropogenous in origin, and contradicts the widespread belief that these materials originated largely as sewage sludge. Furthermore, plating methods showed no evidence of fecal coliforms in dilutions of 1:100 and 1:1000. Thus, there were less than 100 fecal coliforms per gram of sample. Four day bioassays in sediment concentrations of 15, 1 and 0.1% of the same material yielded no fish mortalities. The mercury in the sediment could be either old and of natural origin or of contemporary origin. A dyestuff manufacturing firm situated a few miles up the major tributary is known to have discharged mercury and since it is traceable in increasing concentrations from the harbor to this probable source, a contemporary origin is preferred. The fact that mercury concentrations at depths in excess of 1 ft in the stratum are comparable to surface concentrations suggests that the pollutant entered old sediments via interstitial water rather than through diagenetic processes or other means. The sediments comprising the deposit were not shoaling materials but deposits lying in a horizon involved in new dredging work. Evidence that sufficient interstices exist for interstitial water movement occurs in a low percentage of solids (<50%). At worst, the material could have been adsorbed on the mineral particles (specific gravity 2.6) which presumably, in the light of local geology, are fairly stable silicates. More probably the mercury was in the interstitial water brought up inside of the core tube.

Biological State of the Art

The biological approach to the study of the environmental effects of ocean dumping has been to make

TABLE 5. Analysis of Sediments and Organisms from Fall River Harbor, Massachusetts (see also Table 4)

	SEDIMENT MATRIX (WORMS ONLY)	WORMS IN SEDIMENT (WET WT. BASIS)	WORMS IN SEDIMENT (DRY WT. BASIS)	DREDGED CLAMS (WET WT. BASIS)	DREDGED CLAMS (DRY WT. BASIS)
Hg (ppm)	0.25	4.21	0.52	4.67	0.51
Pb	68.7	< 1.62	< 0.02	< 183.2	< 20.0
Zn	149.8	2.19	0.27	595.2	60.5
Cu	87.6	0.89	0.01	100.7	10.1
Cd	1.10	< 0.16	< 0.02	< 18.3	< 2.0
Cr	61.5	< 0.65	< 0.08	< 73.3	< 8.0
As		< 0.01	< 0.01	2.27	0.25
Ni		< 0.65	< 0.08	< 73.3	< 8.0
V		< 3.24	< 0.40	< 366.3	< 40.0

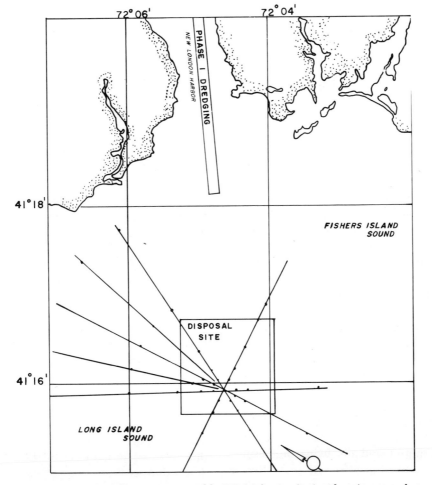

FIGURE 13. *Sampling transects used by NOAA for monitoring dumping ground at New London, Connecticut, 1974.*

predumping, concurrent, and postdumping assessments of benthic populations. Initial activity consists of taking grab samples along transects that extend sufficiently beyond the limits of the dumping ground to assess the contextural populations of the site (Fig. 13). The information thus gained is considered in the light of long-term fisheries data which must be obtained by an experienced fisheries biologist who is familiar with the local fisheries and their seasonal and cyclical fluctuations. Agency statistics are inadequate. The benthos are cataloged in terms of species diversity, frequently employing standard diversity indices, population densities, and total biomass. During dumping peripheral organisms are surveyed and examined for distress and mortality. Monitoring of bacteria, chlorophyll A, and phytoplankton accompanies these activities. The conditions are compared with those in control areas which have been selected previously for their resemblance to the dumpsite, in terms of all of the parameters being observed. Related laboratory studies involve observations of the effects of turbidity and long-term effects including the genetic influence of heavy metals uptake in representative organisms of all trouble levels.

Perhaps the weakest area of current research, and one that could prove most useful from a management point of view, is that of relating biomass to productivity via nutrient cycles, particularly those of nitrogen, phosphorus, carbon, and sulfur, all complex and enormously difficult to relate synergistically. For management purposes it might be possible to establish a simplistic but reasonable index of "charged neutral exhausted" or "oxidized neutral reduced" states, aligning parameters such as low pH, negative redox, ADP (adenosine diphosphate), luciferin, ammonia, and sulfides, and polarizing them against high pH, positive redox, ATP, luciferase, nitrates, sulfates, etc. Such comparisons might establish index values that could be used to evaluate the effect on productivity in a specific environment of materials to be dredged. For example, dead fish have a characteristic pH and E_h, and luciferin and luciferase (firefly extract) are used successfully to evaluate biomass in the drug industry. Another significant parameter is the nitrogen-to-carbon ratio which should be relatively low for healthy productivity (generally C/N < 18). The biological recovery of a dumping ground is explicable on the basis that "Carbon forms the link in the interaction between the inorganic environment and the living organisms" (Stumm and Morgan, 1970, p. 343), and that "Redox processes are slow, only partial and essentially reversible at each stage. . . . The implicit assumptions are . . . that there is a metastable steady-state that approximates the partial equilibrium state for the system under consideration" (Stumm and Morgan, 1970, p. 239).

Bacteria are the redox catalysts. The reversibility is why areas can recover readily. Organic "pollution," then, should be considered in this light and there is thus considerable reason to reject the idea that any area should be considered "lost" for future generations since the whole stability of life is dependent on a state of flux confined within a steady flow of antientropic energy which still obeys the second law of thermodynamics.

CONCLUSIONS

The EPA has designated 118 ocean dumping sites off the coasts of the United States. Most are for dredge spoil and chemical wastes although disposal of toxic wastes is permitted in several deep-water sites off the shelf. The discharge of solid and liquid wastes should not necessarily be considered adverse to marine life, for in some instances it may even be beneficial. In any case, dumping produces a perturbation in a natural ecosystem. If the magnitude of the disturbance does not exceed the natural capacity of the local area to assimilate the wastes, then harmful effects can be avoided. As Shelton (1973, p. 435) points out: ". . . Provided the dumping is properly monitored and controlled, . . . waste disposal, like controlled fishing operations, . . . has a legitimate place among man's uses of the marine environment."

In a book review several years ago, Peter Fenner (1973) pointed out that "pollution is a resource out of place." This concept is the heart of the recycling, or reclamation, philosophy. At present, approximately 45% of the nation's annual copper demand is satisfied through recycling of scrap containing that metal. Perhaps subgranular dredge spoil will one day be considered an economic deposit rather than a nuisance, and the rate of increase of ocean dumping will decline as new uses for such materials are identified.

At present, fine dredge spoil is being considered as a raw material in ceramic processes and as a component in construction aggregates. Combining organic-rich harbor muds with waste hydrogen peroxide from electrochemical processing plants produces heat, and this reaction might be assessed as an alternative to natural fuels in heating metropolitan areas. Granular dredge spoil is often employed as hydraulic or barged fill in the construction of artificial islands and piers. Possibly, all dredged materials ultimately can find use on land, and thus become a natural resource too valuable to be dumped at sea.

SUMMARY

Materials dumped in the ocean are not irretrievably lost, but are retained, in some form, in the water masses, organisms, or sediments. Solid wastes in some instances are recycled in the slow diagenetic, long-term sense familiar to the geologist, whereas organic and soluble materials are altered rapidly and readily distributed by oceanographic agents. These latter materials commonly enter biological cycles in the form of nutrients or catalysts. In the case of dumpsites on continental shelves, the inorganic, insoluble, or slightly soluble particulate matter may be returned to estuaries by local onshore bottom currents. Thus, sediments comprising a dumpsite deposit are locally significant in shelf and coastal environments, and we term these deposits the *homogenic facies*.

The release of waste materials into the marine environment results in significant local perturbations. These include changes in pH, oxygen, and turbidity levels, enrichment of nutrients, introduction of toxicants, importation of biota, burial of benthic fauna, and alteration of bottom physical characteristics. The lowering, for instance, of the relatively constant oceanic pH by the introduction of acidic materials induces local redox reactions, whereas an increase in the concentration of nutrients can produce plankton blooms. One can predict that changes of this type plus the introduction of catalysts incorporated from soluble metals may affect the progression of organic cycles. Oxygen sag, the introduction of toxicants, and concentrated turbidity, if acute, produce genetic or direct sublethal or lethal effects on biota. The transplantation of dredged material resulting in the importation of biota and alteration of bottom characteristics affects habitat and can modify fishing grounds and the potential value of fish nursery areas.

The significance of perturbations is dependent on their severity or acuteness, duration, and geographic extent, or their synergistic combination. Evidence is increasing that the effects are of limited duration. Monitoring suggests that dumping grounds usually become recolonized, sometimes at a reasonably predictable rate; although species diversity tends to be limited for some time on dumping grounds, such areas often become favorable for fishing (including commercial lobster exploitation). The concentration of toxic metals in commercial species is a real concern; but fortunately the life of many of these organisms is short enough to preclude assimilation of hazardous quantities of metals.

Active dumpsites generally are coming under closer examination by fisheries and health authorities. It appears that perturbations resulting from dumping activities tend to be limited in geographic extent, particularly in open shelf areas: liquid wastes are diluted into the surrounding water masses, whereas solid wastes (baled trash or dumped sludge) are localized either through mechanical containment or cohesion of the fine-grained material. The impact of dumping should be estimated beforehand and subsequently can be evaluated by monitoring, yet problems in interpreting or predicting the effects of various perturbations, such as turbidity on organisms, are bound to arise. Admittedly, laboratory biota sampling constitutes an insignificant fraction of a regional shelf population, and flume tests designed to determine the erodibility of sediments do not necessarily yield results that accurately reflect phenomena on the ocean bottom at a disposal site. Additional problems lie in the areas of testing accuracy, statistical analysis based on eclectic data banks, and theoretical modeling without adequate site information. As emphasized in this and other chapters in this volume, sedimentologists can play a major role in developing rational criteria for use of the shelf through collaboration with physical oceanographers, chemists, engineers, and other environmental scientists in the application of theoretical and laboratory results to actual situations.

ACKNOWLEDGMENTS

We thank D. J. P. Swift and D. E. Drake of the Atlantic Oceanographic and Meteorological Laboratories of NOAA for providing unpublished information regarding investigations in the New York Bight. D. Folger of Middlebury College and the U.S. Geological Survey kindly furnished accounts of observations made by a party of diving scientists in the dumpsites off Maryland. S. Pratt contributed information regarding the appearance of dredged and dumped cohesive sediments, and D. C. Rhoads of Yale University provided photographic records of dumping grounds. Dr. Warren A. Hovis of NASA kindly provided the high-altitude imagery of dumping activity.

REFERENCES

Abraham, G. and G. C. van Dam (1972). On the predictability of waste concentrations. In M. Ruivo, ed., *Marine Pollution and Sea Life*. Surrey, England: Fishing News (Books) Ltd., pp. 135–140.

Andreliunas, V. L. and C. G. Hard (1972). Dredging disposal: Real or imaginary dilemma? *Water Spectrum*, **4:** 16–21.

Anonymous (1933). U.S. Corps of Engineers data to accompany *Report on Shore Protection*. Dated April 15, 1933, Office of Chief of Engineers, Washington, D.C.

Anonymous (1953). *The Unified Soil Classification*. Tech. Memo. No. 3-357. U.S. Army Corps of Engineers Waterways Exp. Stn., Vicksburg, Miss., 30 pp.

Anonymous (1966). *Shore Protection Manual*, Vol. 2, 3rd ed. U.S. Army Corps of Engineers, Coastal Eng. Cent., Ft. Belvoir, Va., pp. 7–169, 525 pp.

Anonymous (1969). *Pollution Control and Management*. A report prepared for the New England Regional Commission, Charles River Associates, Cambridge, Section II: 56–135.

Anonymous (1973). *Interaction of Heavy Metals with Sulfur Compounds in Aquatic Sediments and in Dredged Material.* U.S. Army Corps of Engineers, New England Division, Waltham, Mass., (unpublished), 14 pp.

Bascom, W. N. (1951). The relationship between sand size and beach-face slope. *Am. Geophys. Union Trans.,* **32**(6): 866–874.

Bascom, W. (1974). The disposal of waste in the ocean. *Sci. Am.,* **231**: 16–25.

Carter, R. C. and R. Wilde (1972). Cation exchange capacity of suspended material from coastal sea water off central California. *Mar. Geol.,* **13**: 107–122.

Colton, J. B., Jr., F. D. Knapp, and B. R. Burns (1974). Plastic particles in surface waters of the northwestern Atlantic. *Science,* **185**: 491–497.

Curtis, W. F., J. K. Culbertson, and E. B. Chase (1973). *Fluvial-Sediment Discharge to the Oceans from the Conterminous United States.* U.S. Geol. Surv. Circ. 670, 17 pp.

Dean, D. and D. Schnitker (1971). *Report of the Studies Related to Dredging in Belfast Harbor, Maine, and Deposition of Spoil South of Isle AuHaut.* U.S. Army Corps of Engineers, New England Division. Contract No. DACW33–71–C-0013, Waltham, Mass. (unpublished), 9 pp.

Duane, D. B. and C. W. Judge (1966). *Radioisotopic Sand Tracer Study.* Point Conception, Calif.: U.S. Army Corps of Engineers, Coastal Eng. Res. Cent., Misc. Pap. 2-69, 194 pp.

Edge, B. L. and B. C. Dysart, III (1972). *Transport Mechanisms Governing Sludges and Other Materials Barged to Sea.* Report, Civil Eng. and Environmental Systems Eng., Clemson University, Clemson, S.C., 26 pp.

Feick, G., E. Johanson, and D. S. Yeaple (1974). U.S. Environmental Protection Agency, Contract No. 68-01-0060, Washington, D.C., 156 pp.

Fenner, P. (1973). The earth and human affairs. *Book Review, Geotimes,* **18**(5): 34.

First, M. W., ed. (1972). Summary. In *Municipal Waste Disposal by Shipborne Incineration and Sea Disposal of Residues.* Harvard University School of Public Health, Boston, 546 pp.

Folger, D. W. (1974). Memorandum—Submersible Operation of 1974. Unpublished memorandum, 22 pp.

Fye, P. M. (1974). To use the oceans wisely. In *Research in the Sea.* Woods Hole Oceanographic Institution, Woods Hole, Mass., pp. 4–5.

Gilbert, T., A. Clay, and A. Barker (1973). *Site Selection and Study of Ecological Effects of Disposal of Dredged Materials in Buzzards Bay, Massachusetts.* Contract No. CACW33-73-C-0024, U.S. Army Corps of Engineers, New England Division, Waltham, Mass. (unpublished), 70 pp.

Golden, P. C. and M. A. Champ (1974). Monitoring ocean disposal sites. In *Proc., 10th Ann. Conf. Mar. Technol. Soc.,* pp. 107–113.

Goodenow, J. and R. Herberlein (1973). The penetration of light and the depth of the euphotic zone. In M. A. Champ, ed., *Operation SAMS, Sludge Acid Monitoring Survey.* American University, Washington, D.C., pp. 27–38.

Gross, M. G. (1972). Geologic aspects of waste solids and marine waste deposits, New York metropolitan region. *Geol. Soc. Am. Bull.,* **83**: 3163–3176.

Grove, P. B. (1971). *Webster's 3rd New International Dictionary.* Springfield, Mass.: G & C Merriam, 2662 pp.

Huang, J. C. and C. S. Liao (1970). Adsorption of pesticides by clay minerals. *Proc. ASCE, J. Sanit. Eng. Div.,* **96**: 1057–1078.

Ingle, J. C., Jr. (1965). The movement of beach sand; an analysis using fluorescent grains. *Developments in Sedimentology,* Vol. 5. Amsterdam: Elsevier, 221 pp.

Johnson, B. H. (1974). *Investigation of Mathematical Models for the Physical Fate Prediction of Dredged Material.* Tech. Rep. D-74-1, U.S. Army Corps of Engineers, Waterways Exp. Stn., Vicksburg, Miss., 54 pp.

Komar, P. D. and M. C. Miller (1973). The threshold of sediment movement under oscillatory water waves. *J. Sediment. Petrol.,* **43**: 1101–1101.

Krumbein, W. C. (1957). *A Method for Specification of Sand for Beach Fills.* Tech. Memo. No. 102, U.S. Army Corps of Engineers, Beach Erosion Board, Office of the Chief of Engineers, Washington, D.C., 43 pp.

Krumbein, W. C. and H. A. Slack (1956). *Relative Efficiency of Beach Sampling Methods.* Tech. Memo. No. 90, U.S. Army Corps of Engineers, Beach Erosion Board, Office of the Chief of Engineers, Washington, D.C., 43 pp.

Kuenen, Ph. H. (1950). *Marine Geology.* New York: Wiley, 568 pp.

Kyte, M. (1974). *Site Selection and Study of Ecological Effect of Disposal of Dredge Material in Penobscot Bay, Maine.* Contract No. DACW33-73-C-0020, U.S. Army Corps of Engineers, New England Division, Waltham, Mass., 45 pp. (unpublished).

Lear, D. W. and H. D. Palmer (1974). Ocean disposal: Middle Atlantic Bight. In *Proc., 10th Ann. Conf. Mar. Technol. Soc.,* pp. 115–126.

Lee, G. F. and R. H. Plumb (1974). *Literature Review on Research Study for the Development of Dredged Material Disposal Criteria.* Tech. Rep. H-72-8, U.S. Army Corps of Engineers, Waterways Exp. Stn., Contract No. DACW39-74-C-0024, Vicksburg, Miss., 145 pp.

Loder, C., F. E. Anderson, and T. C. Shevenell (1973). *Sea Monitoring of Emplaced Baled Solid Waste.* Rep. UNH: SG-118, University of New Hampshire, Durham, 107 pp.

McFarlane, C. F. and S. T. Kelly (1974). Research methods for the study of ocean waste disposal. In *Proc., 10th Ann. Conf. Mar. Technol. Soc.,* pp. 127–137.

McLeod, G. C. and N. Riser (study in progress). *A Study of the Effects of Turbid Mixtures on Biological Materials.* U.S. Army Corps of Engineers, New England Division, Contract No. DACW33-74-C-0101, Waltham, Mass.

Martin, C. and C. S. Yentsch (1974). *Evaluation of the Effects of Dredging in the Annisquam River Waterway on Nutrient Chemistry of Seawater and Sediments and on Phytoplankton Growth.* U.S. Army Corps of Engineers, New England Division, Contract No. DACW33-73-M-0944, Waltham, Mass. (unpublished), 45 pp.

Meade, R. H. (1969). Landward transport of bottom sediment in estuaries of the Atlantic coastal plain. *J. Sediment. Petrol.,* **39**: 222–234.

Meade, R. H. (1972a). Transport and deposition of sediments in estuaries. In B. W. Nelson, ed., *Environmental Framework of Coastal Plain Estuaries.* Geol. Soc. Am. Mem. 113, Boulder, Colo., pp. 91–120.

Meade, R. H. (1972b). Sources and sinks of suspended matter on continental shelves. In D. J. P. Swift, D. B. Duane, and O. H. Pilkey, eds., *Shelf Sediment Transport: Process and Pattern.* Stroudsburg, Pa.: Dowden, Hutchinson & Ross, pp. 249–262.

Mitchell, J. K. (1971). Perceptions of coastal erosion on the Atlantic shore. In *Abstracts, Second National Coastal and Shallow Water Research Conference*. Office of Naval Research, p. 159.

Nacci, V. A., W. E. Kelly, and R. C. Gularte (1974). *Critical Environmental Parameters and Their Influence on Geotechnical Properties of Estuarine Sediments as Related to Erosion, Transportation and Deposition*. Contract No. DACW33-74-M-0746, U.S. Army Corps of Engineers, New England Division, Waltham, Mass. (unpublished), 23 pp.

NOAA (1974). *Report to the Congress on Ocean Dumping and Other Man-Induced Changes to Ocean Ecosystems*. Report by National Oceanographical and Atmospheric Administration, U.S. Dept. of Commerce, 96 pp.

Office of Science and Technology (1969). *Solid Waste Management— A Comprehensive Assessment of Solid Waste Problems, Practices and Needs*. Report, Ad Hoc Group, OST, May 1969.

Pararas-Carayannis, G. (1973). *Ocean Dumping in the New York Bight: An Assessment of Environmental Studies*. U.S. Army Corps of Engineers, Coastal Eng. Res. Cent., Tech. Memo. 39, 159 pp.

Portmann, J. E. (1972). Results of acute toxicity tests with marine organisms, using a standard method. In M. Ruivo, ed., *Marine Pollution and Sea Life*. Surrey, England: Fishing News (Books) Ltd., pp. 213–217.

Pratt, S. D., S. B. Saila, A. G. Gaines, Jr., and J. E. Krout (1973). *Biological Effects of Ocean Disposal of Solid Waste*. New England Regional Commission, Boston, 146 pp.

Rhoads, D. C. (1974). Organism-sediment relations on the muddy sea floor. In *Oceanography and Marine Biology: An Annual Review 1974*, H. Barnes, ed., **12**: 263–300.

Ruivo, M. (1972). *Marine Pollution and Sea Life*. Surrey, England: Fishing News (Books) Ltd., 624 pp.

SCCWRP (1973). *The Ecology of the Southern California Bight: Implications for Water Quality Management*. So. California Coastal Water Research Program, SCCWRP-TR104, 531 pp.

Shelton, R. G. (1973). Some effects of dumped, solid wastes on marine life and fisheries. In E. D. Goldberg, ed., *North Sea Science*, Boston: Colonial Press, pp. 415–436.

Sissenwine, M. P. and S. B. Saila (1973). *Rhode Island Sound Dredge Spoil Disposal and Trends in the Floating Trap Fishery*. University of Rhode Island, Marine Exp. Stn., Kingston (unpublished), 42 pp.

Smith, D. D. and R. P. Brown (1971). *Ocean Disposal of Barge-Delivered Liquid and Solid Wastes from U.S. Coastal Cities*. Publ. SW-19C, Solid Waste Management Office, Environmental Protection Agency, 119 pp.

Stumm, W. and J. J. Morgan (1970). *Aquatic Chemistry*. New York: Wiley, 583 pp.

Swift, D., A. Cok, D. Drake, G. Freeland, W. Lavelle, T. McKinney, T. Nelson, R. Permentor, and W. Stubblefield (1974). *Sedimentation in the New York Bight Apex and Application to Problems of Waste Disposal: An Interim Assessment*. AOML Spec. Rep. No. 1, Geological Oceanogr. Section, NOAA, U.S. Dept. of Commerce, 68 pp.

Vernon, J. W. (1966). Shelf sediment transport systems. Ph.D. Thesis, Dept. Geol. Sci., University of Southern California, Los Angeles, 300 pp.

U.S. Government, *The National Archives* (1973). *Federal Register*, 38, No. 198, Title 40, Chapter 1, Subchapter H.

Williams, S. J. and D. B. Duane (1974). *Geomorphology and Sediments of the Inner New York Bight Continental Shelf*. U.S. Army Corps of Engineers, Coastal Eng. Res. Cent., Tech. Memo. 45, 81 pp.

Wolff, P. M., W. Hansen, and J. Joseph (1972). Investigation and prediction of dispersion of pollutants in the sea with hydrodynamical numerical (HN) models. In M. Ruivo, ed., *Marine Pollution and Sea Life*. Surrey, England: Fishing News (Books) Ltd., pp. 146–150.

Epilogue

CHAPTER 25

Perspectives of Shelf Sedimentology

K. O. EMERY

Woods Hole Oceanographic Institution, Woods Hole, Massachusetts

Sedimentology of continental shelves is one of the most rapidly expanding subfields of geology, and it probably has not yet reached its peak of growth. Most of our knowledge about shelf sedimentology was obtained from the shelves off the industrialized nations, but scientists of these same nations have learned much about the sediments on shelves off underdeveloped nations through cruises designed for basic research. Many of the findings on all shelves bear upon the discovery of petroleum and other mineral resources, and they also help in understanding the ecology of benthic food resources. New plateaus of knowledge may be reached by future close cooperation of sedimentologists with biological, chemical, and physical oceanographers in well-organized team efforts that undertake comprehensive investigations of large shelf areas. Many such studies are needed for shelves off underdeveloped nations, and they probably will occur unless prevented or made difficult by rulings that may be included in new international laws of the sea.

INTRODUCTION

In any sort of endeavor, perhaps especially in science, there is some value in occasional reviews of the past, summaries of the present, and predictions of the future. These overviews have different values, depending on the

Contribution No. 3445 of the Woods Hole Oceanographic Institution.

stage of development currently reached by that particular field of subfield of science. For example, when the field is young (especially when it has not yet been recognized in terms of bibliographic entries, endowed professorships, or census-listed occupations), its past has little to review, its future is dimly perceived, and few workers are involved. When the field is old its past is rich, its future may contain few surprises, but it is overstaffed by workers who do not wish (or are unable) to embark upon possible new adventures. An overview is not particularly useful for either of these extremes but has its best potential for fields near the middle stage o their development—when the future still is large enough to be influenced by possible new concepts that are beginning to become evident from past and current efforts.

Shelf sedimentology probably is in late youth, as illustrated by the faster doubling time of the number of its publications (10 years) than of publications in all geology (13 years) (Fig. 1). Marine geology appears to be in the early youth stage of growth, as the doubling time of its publications is only five years. Note also that publications in shelf sedimentology began their major increase much more recently than did those for all geology. Dips in the graphs occurred during World Wars I and II, as shown in more detail by plots of annual publication data by Emery (1951).

An attempt is made here to look at the subfield of shelf sedimentology in terms of certain questions that appear to be worth asking.

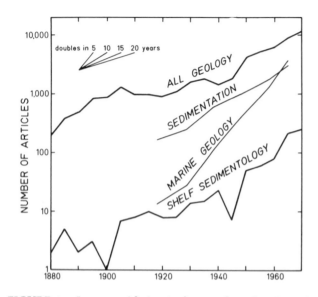

FIGURE 1. *Increase with time in the annual number of articles published in North America. The plot for* All Geology *was estimated from the* Bibliographies of North American Geology *at five year intervals by counting the entries on about 10% of the pages.* Shelf Sedimentology *was obtained by actual count of titles (in the broad sense but excluding beaches and descriptions of samplers and techniques) in the* Bibliographies of North American Geology *at five year intervals. For comparison, the curves for* Marine Geology *and* Sedimentation *(sic) were taken from a previous compilation from* Bibliographies of North American Geology *by Menard (1971, Figs. 3.7 and 3.8).*

WHAT IS IT?

The term *shelf sedimentology* (or sedimentology of the continental shelf) is the study of surface sediments (largely in order to learn their distribution patterns and the methods and dates of deposition). However, the work that is done in the name of shelf sedimentology can be much broader. For example, we know that many parts of the shelf are floored by relict sediments that are unrelated to the present environment; only a small further step is required to include those that are more deeply buried. This inclusion of older sediments is needed to gain more complete understanding of the evolution of a given shelf. Additional breadth of knowledge about provenance and transporting mechanisms and about biological and chemical processes is required to explain the presence of detrital, biogenic, and authigenic components in the sediments.

The physiographic region of shelf sedimentology is that of the continental shelf, but studies are made of the sediment suspended in the overlying water. Many sediments do not remain on the shelf but are transported into adjacent lagoons and estuaries or in the opposite direction onto the deep-sea floor, commonly via sub-

marine canyons. Before the claim is made that all marine science is included in shelf sedimentology, we must recognize that the central area of effort has many appendages that reach into whatever diverse fields can provide pertinent information. As a result, most active workers may reasonably claim fields of effort somewhat different from those of other workers, although all may be interested primarily in sediments of the continental shelf.

WHY IS IT DONE?

One simple reason why continental shelves are important subjects for investigation is their very size. They total about 23 million km², nearly as much as the land area of North America and about one-sixth the total land area of the world. In a sense, this is one of several reasons that can be grouped together as "pursuit of knowledge," or basic research. Some projects also may be termed basic research where application is not yet evident or foreseen, but there is an old truism that all good basic research contains parts that can be applied. Presumably, if there is no hope of applying any part of one's work to problems of economic interest, one should be more concerned with better uses to society of funds and time required for his research than in being proud (?) that his vocation can be termed basic research. Perhaps, also, the question of basic versus applied research is a function of timing. For example, when the sediment characteristics of particular environments have become known during a phase of basic research, there is less need for making still more of the same studies except where needed for specific applications. This means that a previously well-studied and widespread sedimentary environment (such as soils) is likely to receive less study (or at least less publicized study) at the same time that unusual or relatively inaccessible environments (such as the continental shelf) may be heavily studied. When the point of diminishing returns (or discoveries) is reached, the less adaptable investigators are likely to continue doing the same kind of research but perhaps in different regions, while the more adaptable ones look for new fields to conquer.

A direct application of shelf sedimentology in the broad sense is the search for thickly filled sedimentary basins that may constitute potential oil provinces; an indirect application is that knowledge of the blanket sands on the continental shelf may aid in the recognition and prediction of the patterns of petroleum reservoir beds in the geological column. Other direct and indirect applications include the mapping, identification, and inferred origin of deposits of sand and gravel, heavy minerals, and phosphorite. To date, however, these particular sediments have been found of economic value only near the shore where facilities already had been established for successful mining operations on the adjacent land. Similarly, most engineering installations that require knowledge about foundation strength are near the shore, but some studies have been made and more are likely to be made of foundation strength for drilling platforms, prospective oil terminals, and nuclear power plants on the outer shelf. More widespread are environmental studies to recognize distribution patterns of suitable habitats for bottom-living fish, mollusks, and crustaceans and to establish baseline populations and associations intended as guides for assessing possible damage caused by oil spills, atomic tests, or dumping of industrial and other wastes. Most ecological baseline studies are doomed to be unsuccessful, because they have only short time spans whereas populations under natural conditions undergo long-term fluctuations and movements.

Finally, considerable knowledge about the distribution patterns of continental shelf sediments was gained during the early part of World War II through the need to evaluate acoustic methods for detection of submarines. Differences in the ranges of detection were found to be controlled by differences in acoustic reflectivity and reverberation of sound over sand, mud, and rock bottoms of the shelf, and so sediment distribution maps were made for very large shelf regions off Asia, Europe, and the United States largely on the simple basis of hundreds of thousands of navigational chart notations. This interest has ended because future submarine warfare probably will occur mainly in deep waters beyond the shelf. Lest naval research be considered immoral and something to be banned, let us remember the naval funding and facilities that permitted Captain James Cook's voyages of 1768 to 1779, Charles Darwin's cruise on Beagle during 1832 to 1836, the U.S. Exploring Expedition of 1838 to 1842, the Challenger Expedition of 1873 to 1876, the Meteor Expedition throughout the Atlantic Ocean in 1925 to 1927, the Snellius Expedition to the East Indies in 1929 to 1930, and the Bikini bomb tests in 1946. These and many other naval cruises provided the expensive means to establish many parts of oceanography; in fact, nonmilitary funds for oceanography prior to about 1950 were relatively nonexistent.

WHO DOES IT?

The overwhelming majority of published articles on sediments of the continental shelves are by university professors and their students. Many of these articles are parts of doctoral dissertations that represent the only publication in the field by that person, presumably indicating a transfer of interest to other subjects after the student left the influence and help of his professor. Applied research of most kinds is done by scientists attached to large industrial companies or government laboratories; their results are much less often published than are those of university people because of disinterest or even disincentives imposed by industry and bureaucratic delays by government agencies. Occasionally, an energetic scientist in industry or government may eventually publish his results after the possible economic advantage has passed or after inertia is overcome.

Some interesting conclusions about sources of shelf sedimentology can be reached by analysis of articles published in appropriate journals. The *Journal of Sedimentary Petrology*, begun in 1931, includes articles in many fields of sedimentation, and it is the longest running journal that contains many articles in shelf sedimentology, mostly by U.S. scientists. *Sedimentology*, started in 1962, is edited and printed in England, and thus it contains articles mainly from European contributors, but largely on methods, laboratory experiments, and interpretation of ancient sedimentary strata. *Marine Geology*, dating from 1964, is published in the Netherlands but has a truly international board of editors and international contributors. *Limnology and Oceanography*, started in 1956, is heavily concentrated in American biological limnology, although it does contain some articles on shelf sedimentology. *Deep-Sea Research*, begun in 1953 and printed in England, has an international board of editors, but intentionally discriminates against continental shelves in favor of the deep ocean. Lastly, the *Journal of Marine Research*, founded in 1937, largely concentrates upon American marine biology. Of these publication media that contain some to many articles on shelf sedimentology, the *Journal of Sedimentary Petrology*, *Sedimentology*, and *Marine Geology* were deemed the most appropriate for statistical analysis.

Some results of the analysis (Fig. 2) show that the articles in the field of shelf sedimentology are greatly dominated by authors in universities (including oceanographic institutions) of the United States and western Europe. Very few articles are by authors in underdeveloped nations of Africa, Asia, and South America, and most of these few represent work done at universities in the United States and western Europe. Objection that the authors from African, Asian, and South American nations publish their articles elsewhere is inadequate. Actual experience in compiling background summaries of many regions shows that major studies have not been

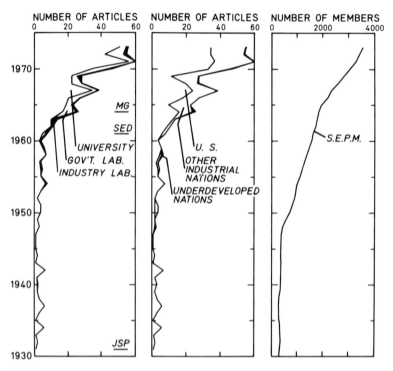

FIGURE 2. *Annual number of articles at five year intervals on shelf sedimentology (in the broad sense, but excluding beaches and descriptions of samplers and techniques) that were published in the* Journal of Sedimentary Petrology, Sedimentology, *and* Marine Geology *(starting dates are shown in the left panel). Left panel: Plot of cumulative numbers of articles by authors from universities, government geological surveys and laboratories, and industry. Middle panel: Cumulative numbers of articles by authors at organizations in the United States, in other industrialized nations (mainly western Europe), and in underdeveloped nations. Right panel: Growth in membership of the Society of Economic Paleontology and Mineralogy; a different curve (date at which present members joined) is given by Russell (1970).*

made by scientists from the underdeveloped nations of these three continents. While many findings from local studies in underdeveloped nations are published locally, even more of the findings from authors in industrialized nations are published in journals other than the three whose contents were analyzed in Fig. 2.

The potential economic benefits of shelf sedimentology as well as other kinds of science accrue to the industrial nations that support the research in one way or another, and these nations have built up a large cadre of scientists. Moreover, later research is largely built upon earlier and more descriptive research. An illustration of the effects of past experience in the field is provided by the kinds of subjects that are investigated by authors from industrialized versus underdeveloped nations. Authors from the former nations include in their writings the results of advanced research employing a variety of sophisticated equipment, but those from the latter are almost entirely restricted to rather prosaic descriptive work. This division is more emphatically borne out by the experience of editors in reviewing and often rejecting articles that are submitted for publication.

The difficulty of underdeveloped nations in "catching up" is further illustrated by Fig. 3, which shows the rate of growth of published articles in geology with time in the United States; essentially the same logarithmic growth can be seen in the increase of sedimentologists, journals, and gross national products (Menard, 1971). The curve implies that even if the rate that was experienced by a now industrialized nation could be duplicated by a now underdeveloped one, the latter must lag farther and farther behind, at least for the foreseeable future. The possibility of instant transfer of science or technology by gift from industrialized to underdeveloped nation is considered virtually impossible. Essentially, the moral is "For to him who has will more

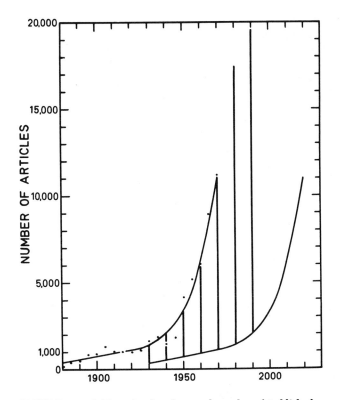

FIGURE 3. *Arithmetic plot of annual number of published articles in* All Geology *from data in the semilog plot of Fig. 1. The curve on the left is for North America (and at least 95% of the articles are from the United States). The same curve is replotted on the right. Vertical lines at 10 year intervals between the two curves show that if an underdeveloped nation could duplicate the growth of U.S. geology as exhibited by publications, it still would lag farther and farther behind the United States with the passage of time.*

both in width and length. This small-area detailed approach is typical of much university dissertation work where the objective may be more that of learning the techniques of the work than in solving important sedimentological problems. Thus, a tendency is evident for each worker to consider his small patch of shelf to be typical and the conclusions reached to be general. In effect, such studies are similar to the ones that have been made of sediments in 48 rather different estuaries and lagoons along the Atlantic and Gulf coasts of North America (Emery and Uchupi, 1972, pp. 338–341). Only a few authors had studied the sediments of more than a single estuary or lagoon, and yet many of them stated or implied generalizations based on their limited areas. The same relationship exists for studies of water circulation in the estuaries and lagoons.

Larger section of the shelf, of the order of 200 km length and the full width, have been studied by combining several small student dissertations or by making repeated cruises off given states or small nations. Most of the men who are responsible for this intermediate approach are from universities or government laboratories. Sedimentologists who are familiar with the general literature know well whom or what laboratory to contact for additional information about many sections of the continental shelves.

The largest area studies of continental shelves are those for lengths of 1000 km or more, and most of these have been made by members of oceanographic institutions or large government (national) laboratories whose facilities and funding resources are not restrictive as to region of work. During the course of these studies syntheses usually are made of previously published more local work.

Examination of Fig. 4 shows that the continental shelves that border the industrialized nations are well known from the viewpoint of topography, lithology, and structure, as well as sedimentology. All of the well-known shelves off underdeveloped nations reached their level of knowledge through the efforts of oceanographic institutions, government laboratories, and universities of industrialized nations. The distance from home ports and laboratories caused many of the cruises to be long enough and intense enough to produce broad comprehensive studies of large shelf areas that span the offshore regions of several nations. Few of the shelves off underdeveloped nations have been studied except very locally or cursorily by the scientists or agencies of these nations, owing to insufficient funds, facilities, education, or energy. A sort of exception is provided by studies made by visiting scientists from industrialized nations who spend several years in a given underdeveloped nation

be given, and he will have abundance; but from him who has not, even what he has will be taken away" (Matt. 13:13). From the viewpoint of the developing nations the much later remark of the Red Queen to Alice of the Looking Glass must seem particularly appropriate: "It takes all the running you can do, to keep in the same place. If you want to get somewhere else, you must run at least twice as fast as that!"

WHERE IS IT DONE?

By far the greatest amount of shelf sedimentology is done in nearshore waters largely because of lower costs and easier accessibility. In a general way, also, the length of the shelf investigated in a given study is proportional to the width of the study area. Thus, many studies are local

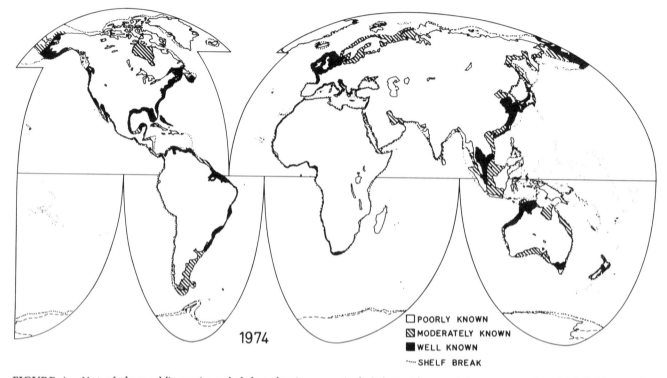

FIGURE 4. *Map of the world's continental shelves showing a subjective impression of the level of knowledge of the shelves as of early 1974 on the basis of published articles. Additional knowledge held in proprietary form by oil companies or manuscript form by government agencies is not available for scientific study and is not* *included. Best known shelves are ones for which fairly complete information is available on topography, sediments, rocks, and structure; poorly known ones generally have only some topographic information.*

as a faculty member of a local university or as an oil company employee.

Studies of shelves off some of the most underdeveloped nations during cruises by scientists of industrialized nations have been prevented by suspicion or bureaucratic delays in giving permission by the governments of these underdeveloped nations. Communist-socialist bloc nations provide many examples of denials of access to Western marine scientists and ships presumably because of political fears. As a result, few, if any, of the world's shelves have been studied by both Western and Communist-socialist bloc marine scientists, and the technology and results of the two groups are rather different. In general, the Western scientists have included at least shallow geophysical profiles, sophisticated geochemical work, and conclusions about sediment classifications and origins, whereas the Communist-socialist bloc studies have concentrated upon labor-intensive grain size and chemical analyses of cations in very large numbers of surface samples. In addition to political inhospitality, climatic inhospitality has largely prevented sedimentological or other investigations of shelves at high latitudes, particularly because of hindrance by bergs and sea ice.

In summary, logistics, politics, and climate have caused studies of sedimentology to be concentrated on perhaps 10% of the length of the world's continental shelves. Already most of the knowledge of shelf sedimentology is from these shelves, and most of the published knowledge from the remaining shelves was obtained through cruises or work by scientists from industrialized nations. As a matter of fact, this bias continues, because new methods of study are applied mainly to shelves that already have been studied intensively, and new programs are being organized by the U.S. Geological Survey for the shelves that border the United States.

HOW IS IT FUNDED?

Funding of small-area studies of shelf sedimentology may represent little problem if only small boats and inexpensive equipment are needed, especially where general laboratory facilities of universities are available and nearby. This is particularly true where small supplementary funds for students can come from research grants of their professors. Intermediate-area programs usually require at least a small ship, and ship costs generally are too high to be covered by university departmental budgets. In many instances the shipboard work can be supported by piggy-backing or by use of a regional research vessel.

For large-area shelf work long-term access to a ship generally is needed, and these ships are available almost only from oceanographic institutions, government laboratories, naval facilities, or company sources. Funding for shipboard, laboratory, and reporting phases of government, military, and economy research is provided by these sponsoring agencies, and these agencies also may control the degree of publication. More interesting, however, is the funding for research by the civilian sector, particularly the universities and oceanographic institutions. Their funds are obtained competitively from research-funding organizations, such as the National Science Foundation (including the office for the International Decade of Ocean Exploration) and the Office of Naval Research in the United States, Forschungsgemeinshaft in Western Germany, and Centre Nationale pour l'Exploitation des Océans in France. In addition, government agencies in these and other industrialized nations supply funds as subcontracts for work that may be closely related to their interests. A multitude of other private and public sources are available for funding of portions of the research and for equipment in Western industrialized nations, although presumably not in the Communist bloc nations.

The multiplicity of funding sources gives a decided advantage to the Western sedimentologist over his Communist counterpart, perhaps accounting for the greater innovation displayed in the equipment, work, and ideas of the former. Worst off by far are sedimentologists in underdeveloped nations, whose budgets may be so strained by more urgent demands on governments funds (such as for food, public schools, and economic development) that research on the sediments of continental shelves has little priority for the near term. Moreover, salaries of university professors in most underdeveloped nations are so low with respect to costs of living that the professors can scarcely afford the luxury of research and must instead obtain second jobs to support their families. This, of course, is the chief reason for the brain drain to the industrialized from the more advanced underdeveloped nations (the ones that can least afford the loss of trained personnel).

Even though shelf sedimentologists in industrialized nations are more secure in their positions and funding than are the relatively few in the underdeveloped nations, they still should not be smug and satisfied. Their problem is that shipboard work is very expensive compared with most kinds of land geology, and during a period of economic strain their funds are more likely to be cut than are those of land geologists. At such times government funding is likely to drop sharply, particularly for basic research (usually the first item to be cut). In addition, government laboratories quickly divert their remaining funds from grants or contracts for nongovernment organizations to in-house work in order to retain their staffs as intact as possible. A period of this kind of reduction in funding began about 1972 and it probably will continue throughout much of the decade; however, its effect has largely been obscured by additional funds committed to the establishment of biological baselines and to other considerations related to protection of the environment from pollution. If public disenchantment with the generally low-level science that is being done in the name of the environment occurs before the ending of the general broad dip in government funding, the expensive subfield of marine geology (including shelf sedimentology) can have some very difficult years.

HOW IS IT AFFECTED BY INTERNATIONAL LAW?

During both prehistory and history the land has been used in turn (but with overlaps) for hunting and gathering, then for herding, and eventually for farming. Mining occurred throughout the period of these changing food-securing occupations, but reached its highest development during the farming epoch, when large concentrations of miners could be fed. In the ocean, the sequence is different mainly because no one owns the fish, many of which swim freely from region to region, and no one occupies (lives on) the parts of the ocean that he may claim. Moreover, the technical difficulties of herding and farming in the ocean are so great that the ocean's hunting economy has continued for a much longer time than on land. Although the hunting stage at sea has reached a far higher level of sophistication than on land, it is being faced with diminishing returns relative to cost (the problem of the commons), and is facing conflicts with mining, which is one of the objectives of shelf sedimentology. Mining of the ocean floor, chiefly for oil and gas, has increased from a few million dollars per year to more than $30 billion during a span of only 30 years, in contrast with the present $10 billion for the fisheries that were developed during thousands of years. Inevitably, these industries (mining and fishing) will continue to expand at vastly different rates with consequent adjustments required in their conflicts, just like the conflicts on land occurred and were settled in favor of the higher level industries. Illustrations on land are well known as the conflicts between hunter and herder (fences), herder and subsistence settler (water and more fences), subsistence settler and staple-crop industry (huge areas with no fences), and now staple-crop industry and mining (the "environment"). Difficulty in settling the conflict in the ocean is due to the increasingly higher total value of petroleum produced from the ocean floor above the value of fish, in contrast with the world's need for more food and the existence of more fishermen than miners (voters). Further complication results from the use of several calories of fossil fuel to obtain and deliver one calorie of fish, just as about three calories of fossil fuel is required to raise one calorie of agricultural products on land and about 10 more calories to build containers and to process and transport the food to the consumer.

The increasing demand for petroleum, other minerals, and fish (and for the taxes on them, too) has led to seaward movement of national boundaries during the past few decades. During the seventeenth to middle of the twentieth century the boundaries of most coastal nations were 1 league, or 3 nautical miles, from the shore—"the hypothetical range of an imaginary cannon" of Hugo Grotius in 1609. This was considered the limit of defense by shore batteries, and the width of the territorial sea (for which a coastal nation has nearly all the rights that it exercises on land). By 1945 the ranges of the guns, the technology, and the demand for seafood and minerals had increased, and the "Proclamation of President Truman Claiming Jurisdiction over Resources of the Continental Shelf" off the coasts of the United States was published. A convenient compilation of documents relating to the continental shelves, including this one, was assembled by Padelford (1968). In 1953 came the "United States Submerged Lands Act, an Act to Confirm the Title of States to Lands Beneath Navigable Waters Within State Boundaries and Control over Resources of the Seabed and the Continental Shelf Seaward of State Boundaries." So many claims and counterclaims of the then 86 nations of the world resulted that an international conference was called in 1958 to try to resolve some of the questions of width of territorial seas (subject to national sovereignty) and widths of seafloor mining and other special limited rights.

The United Nations Conference on the Law of the Sea held at Geneva during 1958 approved even greater claims for signatory coastal nations, and the Conventions were ratified by the necessary number of nations by 1964. The four Conventions that resulted from the conference are one each for the high seas (beyond the territorial seas), the territorial seas and contiguous zones, the continental shelf, and fishing and conservation living resources of the high seas. The Convention on the Continental Shelf assigned the mineral rights to the adjacent nation to a depth of 200 m or to whatever depth could be exploited, but the terminology was imprecise, reflecting the lawyer's lack of knowledge about the ocean and its expanding technology. This Convention bears upon shelf sedimentology in several important ways. One is the requirement to obtain the consent of the coastal nation prior to making studies of the adjacent continental shelves. Although permission normally is to be granted if the request is submitted by a qualified organization for scientific research purposes and the results are to be published and freely available, in fact, government inertia, unfamiliarity of officials with scientific research, and fears of military or economic disadvantage have produced many denials, mainly through inaction. An estimated 35 denials have occurred since 1970; more would have occurred had not Western scientists learned that application to study shelves of Communist-socialist bloc nations is useless. Second, the rights of the coastal nation to the shelf do not include control of the water or airspace above the bottom. This had been interpreted as meaning that oceanographers can freely do geophysical work that has no contact with the bottom, but that they cannot collect cores and dredge samples. There is no restriction upon fishing on the shelves, but trawling for bottom fish often obtains rocks; must these rocks be thrown back?

Since the adoption of the Geneva Conventions of 1958 the number of nations has greatly increased, to 148 as of 1973 (Cline, 1973). Most of the coastal nations claim a 3 or 12 nautical mile territorial sea but some claims range to 200 miles; most nations also claim a 12 mile width of exclusive fishery jurisdiction (that includes the territorial sea), but some claim 200 miles; many nations also claim various widths of ocean for fisheries conservation, sanitation, customs control, criminal jurisdiction, defense, and other interests (*The Geographer*, 1974). About one-fifth (29) of the nations are landlocked with no frontage on the ocean, and 12 more have less than 100 miles of ocean frontage. Many of the latter are shelf-locked (i.e., the seaward extension of their boundaries following the rule of median lines), which permits them to claim little shelf area. Western Germany, Belgium, and Zaire are three examples of shelf-locked nations, and Western Germany filed and won a suit against its neighbors to obtain a "fairer share" of the presumed oil- and gas-rich shelf of the North Sea. Similarly, when the oil potential of the East China Sea (Emery et al., 1969) was realized, Japan, South Korea, and Taiwan immediately claimed portions of the shelf, but, curiously, each nation's interpretation of the position of median-line boundaries overlapped that of its neighbor. Later, when mainland China began to look beyond its shores, it recognized none of the previous claims.

The shallow reefs of the deep China Basin made natural drilling platforms above folded strata having possible oil potential (Emery and Ben-Avraham, 1972).

Most of the reefs are just below low tide and to date have not been legally claimable as islands, but one of them contains the Paracel Islands that had previously been claimed by mainland China, Taiwan, Philippines, South Vietnam, and France. Less than two years after the oil potential was indicated, peace-loving mainland China sank ships of South Vietnam in January 1974 to reinforce its claim, and a few months later it may have started drilling. Continuing southward, a study of the structure of the South China Sea and Gulf of Thailand (Parke et al., 1971) revealed areas of high oil potential that gave rise to a movement named the "Mothers for Peace," which asserted that the Vietnam War was being fought in the interests of the so-called military-industrial complex in order to claim the offshore oil. The assertion was laid to rest when it was pointed out in the U.S. Senate that the discovery study was made in 1969 and the war had started years earlier. As usual, the claims of adjacent nations overlapped so that a well drilled on a shelf area leased from South Vietnam was to have been destroyed by a gunboat from Khmer Republic during September 1974 until protection was provided by a larger South Vietnam ship. An indication of the economic promise (and thus of national competition) is provided by 26 discovery wells drilled in the general region of Southeast Asia during the first half of 1974 (Anonymous, 1974).

Other differences of opinion continue. For example, Chile and Peru which have narrow shelves claimed essentially that they had been cheated and that their shelves should be redefined to extend 200 nautical miles from shore. This, even though the existing narrow shelves may be very rich in oil potential and certainly are the richest of the world in phosphorite! For various reasons still other nations, particularly underdeveloped ones, wish to widen their areas of mineral rights largely in order to be in a position to collect more taxes. One way of doing this is to redefine the continental shelf to mean whatever depth or width of ocean floor is desired—4000 m, 200 nautical miles, or any other. The term continental shelf as a physiographic feature was in use during the last part of the nineteenth century, so its recent redefinition for legal purposes gives rise to the need for prefixes, such as the legal continental shelf in contrast with the geological (or perhaps the illegal!) continental shelf that is studied by sedimentologists and other geologists.

In 1968 Arvid Pardo, then ambassador of Malta to the United Nations, carried the conflict over the law of the sea to a new level by proclaiming that the resources of the ocean floor beyond the limits of national jurisdiction are *the common heritage of mankind*. This says, essentially, that the resources that can be mined only by the few nations that have the capital, knowledge,

interest, and ability to extract the minerals belong to all mankind—presumably meaning to the United Nations. The land-bound and shelf-bound nations immediately acquiesced in hopes of obtaining large tax revenues, and many plans have since been proposed for dividing the deep ocean floor. One of these was presented to the United Nations Seabed Committee in 1970 by the United States as its "United Nations Draft Convention on the International Seabed Area." It proposed jurisdiction by coastal nations to 200 m depth, an area of trusteeship administered by the coastal nation for the rest of the continental margin beyond the continental shelf, and an international jurisdiction beyond the continental margin. This was quickly tabled at the United Nations as not being generous enough, in spite of domestic protests that it was a giveaway program. The problem of subdivision is a complex one, because representatives of the numerous (about 120) underdeveloped nations know little about the ocean floor, and because an enormous volume of literature on the subject has been produced but with little substance. Moreover, the imagined revenues are far greater than those that are likely to be obtained, the mining products will compete with those now obtained on land within some of the underdeveloped nations, and the cost of patrolling, accounting, and rulemaking undoubtedly will exceed the actual revenues. Interesting as these deep-sea legal problems are, they are beyond the continental shelves and are not considered further here.

In 1974 the first formal meeting of the new United Nations Law of the Sea Conference took place at Caracas, Venezuela. Among the problems considered were questions of widths of territorial seas, jurisdictional (and tax-collecting) rights over mineral resources, widths of fishery jurisdictions, freedom of passage for military and commercial traffic through straits, and freedom of scientific research on the shelves and elsewhere in the ocean. Again much rhetoric and little action resulted, so that additional conferences are scheduled for Geneva, and Caracas during 1975. The apparent trend is toward a 12 mile territorial sea, a 200 mile economic resource zone to be controlled by the adjacent coastal nation, and a deep-sea residue under the jurisdiction of a new seabed authority (presumably an agency of the United Nations). To the shelf sedimentologist the most important problem is that of freedom of scientific research (Knauss, 1974). On the United Nations' decisions (over which scientists have very little control) will come answers to questions such as the following: To what extent will scientific research be approved and controlled by the coastal nations? Will applications for permits be so onerous and time-consuming that the marine scientist will prefer to turn to other interests? Will the coastal nation be able to greatly modify the cruise plan? What will be the requirements for reporting results? Will training of representatives from local nations be required so that they will be able to understand the results and participate in their interpretations in a significant way?

The real problems are that the governments of many underdeveloped nations know that their national scientific capability is so low that they cannot take much advantage of the knowledge that may result from sedimentological or other research on their continental shelves. How much more, then, does the acquiring of that scientific knowledge benefit the nation that funds and conducts the research than it does the underdeveloped coastal nation? Can these benefits provide substantial military and economic advantage to those that can use the knowledge? If there are no benefits, why should an industrialized nation spend considerable money and effort on such investigations? How can the coastal nation ensure that reports of the cruise results always are transmitted back to it and how soon? Before the sedimentologist ridicules this point of view, he should try to place himself in the position of an administrator in such an underdeveloped nation.

In compensation for their lack of scientific ability, the underdeveloped nations can use their voting power in the United Nations to force (up to a point) the adoption of a cartellike international control over the deep ocean floor that can establish areas to be mined, prices to be charged, and quantities to be mined. This sort of control favors the producer more than the consumer (an evident opposite view to the stated objective of using the ocean-floor resources for the welfare of all mankind). Such conditions do not favor the private enterprise that developed the means of mining and producing to their present levels and necessarily must also provide the capital to do the mining. If the international regime proves to be too uncooperative, private enterprise still can license or can sell its technology to nations that are not signatory to the final treaty. Quite clearly, the Law of the Sea Conferences are complex mixtures of social as well as economic issues important for the future of shelf sedimentology.

Even bureaus of the United States government are capable of providing obstacles to free scientific research on the shelves in their search for power. Hearings were held during July 1974 in Washington to determine whether among other controls the U.S. Geological Survey should be empowered to issue permits (and thus denials) before any individual or organization including other government agencies would be allowed to take samples or geophysical measurements on the outer continental shelf (beyond the territorial sea). In addition, a requirement was present that all data including

completely processed geophysical measurements were to be available for inspection within 60 days after their collection. Final decisions have not yet been reported, but continued restrictions or attempts at restrictions are to be expected in view of the present energy muddle (even though the muddle itself largely is due to past poor legislation) and to the fact that annual tax revenue as bonuses and royalties from oil and gas production on United States shelves averaged about $2 billion during the early 1970s and may be $5 billion by 1975. Government bureau interest appears to be more centered upon tax returns than upon solving the fossil energy shortage, and one U.S. Senator (Fannin, 1974) admits that some of his colleagues take well-published cheap shots at oil companies in order to curry favor with voters.

WHAT IS ITS FUTURE?

Most attempts to predict the future of any field of science are ineffective, as they tend to accent just more of the same kind of work that is current at the time the prediction is made. To go beyond the limits of more intensive use of existing instruments and methods is very difficult and requires an exceptionally clear crystal ball. The whole view of the future can be changed by the advent of new methods of collecting samples (as in the past by piston coring, deep-sea drilling, inexpensive precise positioning, scuba gear, and effective research submersibles), by powerful new tools for analysis (as scanning electron microscopy of grain surfaces, radiocarbon dating, stable isotopes, and electron probes), new concepts (as turbidity currents, relict sediments, and seafloor spreading), new methods of handling large quantities of data (as trend surfaces and computer acquisition, processing, and plotting), or new objectives (as paleoenvironments, normal ecological relationships, new correlations with bottom currents, relationships with suspended matter, and baselines for estimating effects of pollution and dumping of industrial and domestic wastes). None of these devices or ideas were available when the author began in the field, although they are commonplace now.

One can predict with assurance that these and even newer methods, tools, and ideas will be applied to the same continental shelves that previously were most studied by more primitive approaches. This means, of course, that we will learn more and more about the most accessible shelves—those that border the nations that are most industrially developed. Many of the studies will continue to be small and intended mainly for familiarization of students with tools and methods. However, the ability to analyze quickly and accurately large numbers of samples probably will lead to far larger areas of study than have been practicable during the past. These areas must include the shelves off underdeveloped nations, and the major question here is that of future accessibility and the role of international law. If underdeveloped coastal nations prevent study of their shelves, or if required procedures are too onerous, we can expect shelf sedimentologists to concentrate still newer kinds of studies upon the shelves that are accessible. These may include much better investigations than those of the past relative to the diagenesis of sediments (including their organic matter), to exchange of interstitial waters with the overlying ocean, and ecology and biomasses relative to sediments. Certainly we can look toward far more interdisciplinary work in shelf sedimentology than ever existed before, including coordination with biologists, chemists, physical oceanographers, archaeologists, and economists well beyond the levels heretofore reached. Thus, the trend is likely to be toward larger studies made by teams of experts in areas that are politically and climatologically accessible.

Perhaps the most fruitful sort of work that a shelf sedimentologist can embark upon now is that of synthesis of the results obtained by the far more numerous descriptive sedimentologists. For example, a huge number of sediment studies has been published by the marine scientists of the various European nations. Synthesis in Europe, however, has been little attempted perhaps primarily because of language and political barriers between the scientists of the different nations.

Unlike many subfields of geology, that of shelf sedimentology is thinly enough populated that a few energetic and capable individuals can still have a large influence upon the direction and manner of its future growth. The present tendency accepted by many young sedimentologists toward specialization rather than toward generalization is inevitable because of the growing flood of details with which they must cope; but the specialization route leads to even more minutae than at present, repetition of surveys, and confirmation of processes. However, new generalist sedimentologists familiar with rapid methods, use of computers to handle large and diverse kinds of data, and cooperation with other scientific disciplines will penetrate present ceilings to find new base levels for future important investigations that cannot now be visualized. Obviously, the most successful shelf sedimentologists must have some competence in such diverse general activities as politics, finance, technology, and science (perhaps in nearly equal parts).

At present the need for shelf sedimentologists is great, largely because of demands associated with attempts to assess the effects of pollution by oil spills and waste dumping. The large funds from government agencies for this purpose are likely to be only short-term ones; therefore, new candidates for shelf sedimentology should be informed that the field may contract within the decade. New candidates should be encouraged only if they are really interested and capable in shelf sedimentology. The danger of the present rapid expansion is that it may encourage the entrance of more new workers than the long-term growth can support. A high percentage of all present shelf sedimentologists are in universities, so that should not replicate themselves endlessly and mindlessly to produce a modern version of the sorcerer's apprentice.

SUMMARY

Sedimentology of continental shelves is one of the most rapidly expanding subfields of geology, and it probably has not yet reached its peak of growth. In the broad sense, the term sedimentology includes the distribution patterns, properties, and sources of both past and modern sediments. Fundamental interests are those of the usual "why," but practical ones abound connected with foundation engineering, acoustics, sand and other mineral resources, original environment of oil formation, and ecology of benthic and other plants and animals. Most publications have been by scientists of universities, followed distantly by those of government bureaus and industry. Most workers have concentrated upon inshore parts of the shelf (beyond the surf zone) and in small areas. Others have taken larger regions partly through the use of well-equipped ships used for long periods or through compiling of previous smaller studies. New plateaus of knowledge may be reached in coming years by close ooperation of sedimentologists with biological, chemical, and physical oceanographers in well-organized team efforts that undertake comprehensive investigations of very large shelf areas. Most of our knowledge about shelf sedimentology was obtained from the shelves off the industrialized nations, but scientists of these same nations have learned much about the sediments on shelves off underdevelopped nations through cruises designed for basic research. Many such studies are needed for the shelves off underdeveloped nations, and they probably will occur unless prevented or made difficult by rulings that may be included in new international laws of the sea. Assuming that international cooperation will continue, we can look forward to new plateaus for future studies that can be possible through the use of new instruments, methods, and frameworks of thought.

ACKNOWLEDGMENTS

Appreciation is due the International Decade of Ocean Exploration, National Science Foundation Grant GX-28193 for funding of this study.

REFERENCES

Anonymous (1974). Southeast Asia has 26 shows this year. *Offshore*, **34**(11): 65–70.

Cline, R. S. (1973). *Status of the World's Nations.* U.S. Dept. State, Bur. Intelligence and Research, Publ. 8735, 20 pp.

Emery, K. O. (1951). Trends in literature on sedimentology. *J. Sediment. Petrol.*, **21**: 105–108.

Emery, K. O. and Z. Ben-Avraham (1972). Structure and stratigraphy of China Basin. *Am. Assoc. Pet. Geol.*, **56**: 839–859.

Emery, K. O., Y. Hayashi, T. W. C. Hilde, K. Kobayashi, J. H. Koo, C. Y. Meng, H. Niino, J. H. Osterhagen, L. M. Reynolds, J. M. Wageman, C. S. Wang, and S. J. Yang (1969). *Geological Structure and Some Water Characteristics of the East China Sea and the Yellow Sea.* Econ. Commission for Asia and the Far East, Committee for Co-Ordination of Offshore Prospecting, Tech. Bull. 2, pp. 3–42. See also Wageman, J. M.,

T. W. C. Hilde, and K. O. Emery (1970). Structural framework of East China Sea and Yellow Sea. *Am. Assoc. Pet. Geol.*, **54**: 1611–1643.

Emery, K. O. and E. Uchupi (1972). *Western North Atlantic Ocean: Topography, Rocks, Structure, Water, Life, and Sediments.* Am. Assoc. Pet. Geol. Mem. 17, 532 pp.

Fannin, P. (1974). Politicians really know better! *Oil Gas J.*, **9**: 230, 232, 234, 236, 238.

The Geographer (1974). *Limits in the Seas; National Claims to Maritime Jurisdictions*, 2nd ed. U.S. Dept. State, Bur. Intelligence and Research, Office of the Geographer, Publ. 36, 141 pp.

Knauss, J. A. (1974). Marine science and the 1974 Law of the Sea Conference—Science faces a difficult future in changing Law of the Sea. *Science*, **184**: 1335–1341.

Menard, H. W. (1971). *Science: Growth and Change.* Cambridge, Mass.: Harvard University Press, 215 pp.

Padelford, N.J. (1968). *Public Policy and the Use of the Seas*, rev. ed. MIT. Dept. Naval Architecture and Marine Engineering, 361 pp.

Parke, M. L., Jr., K. O. Emery, R. Szymankiewicz, and L. M. Reynolds (1971). Structural framework of continental margin in South China Sea. *Am. Assoc. Pet. Geol.*, **55**: 723–751.

Russell, R. D. (1970). SEPM history. *J. Sediment. Petrol.*, **40**: 7–28.

Index